MW00998944

ALLE ZEIT WACH
1842

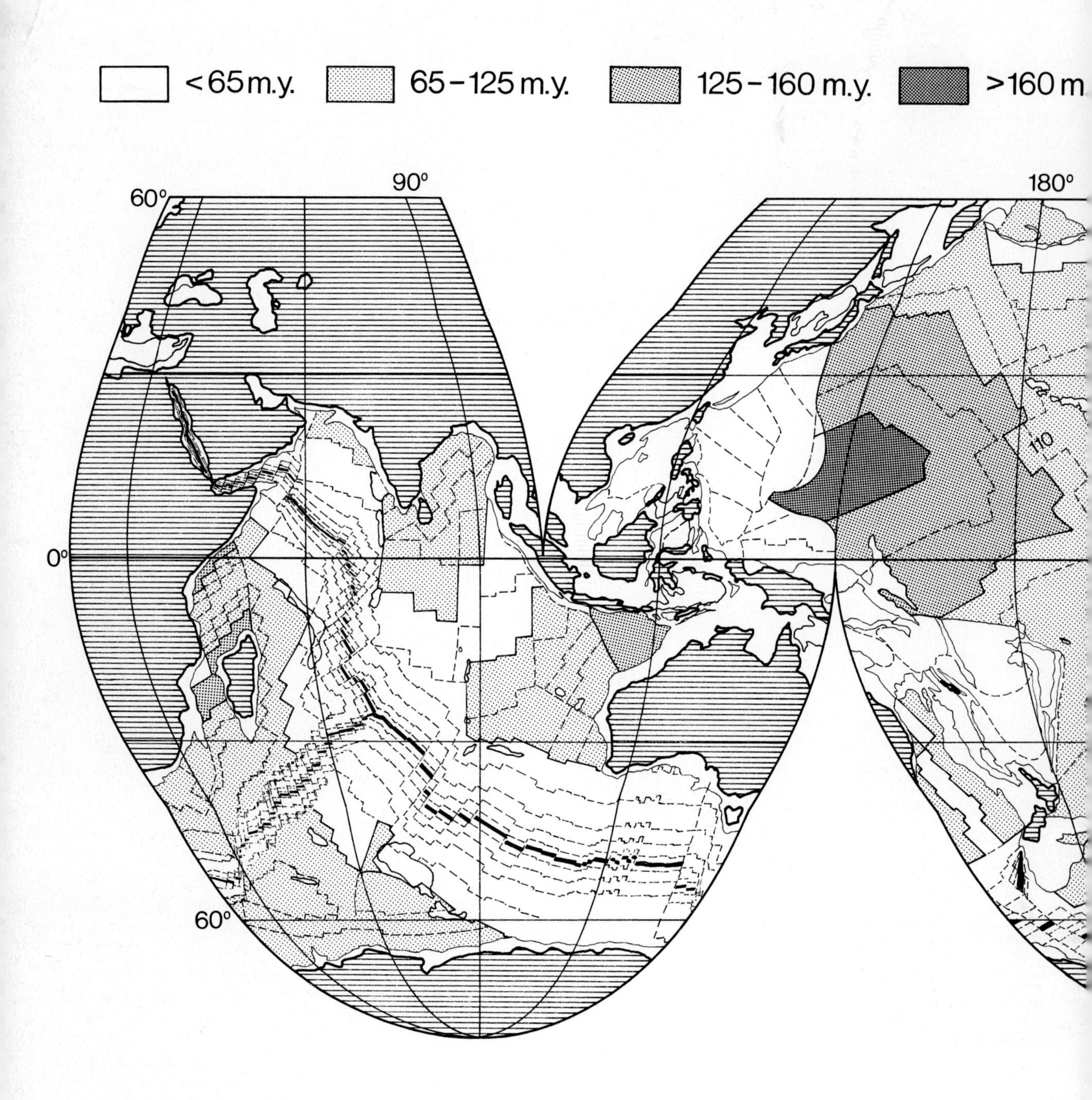

The Age of the Oceans (After Sclater et al., 1981)

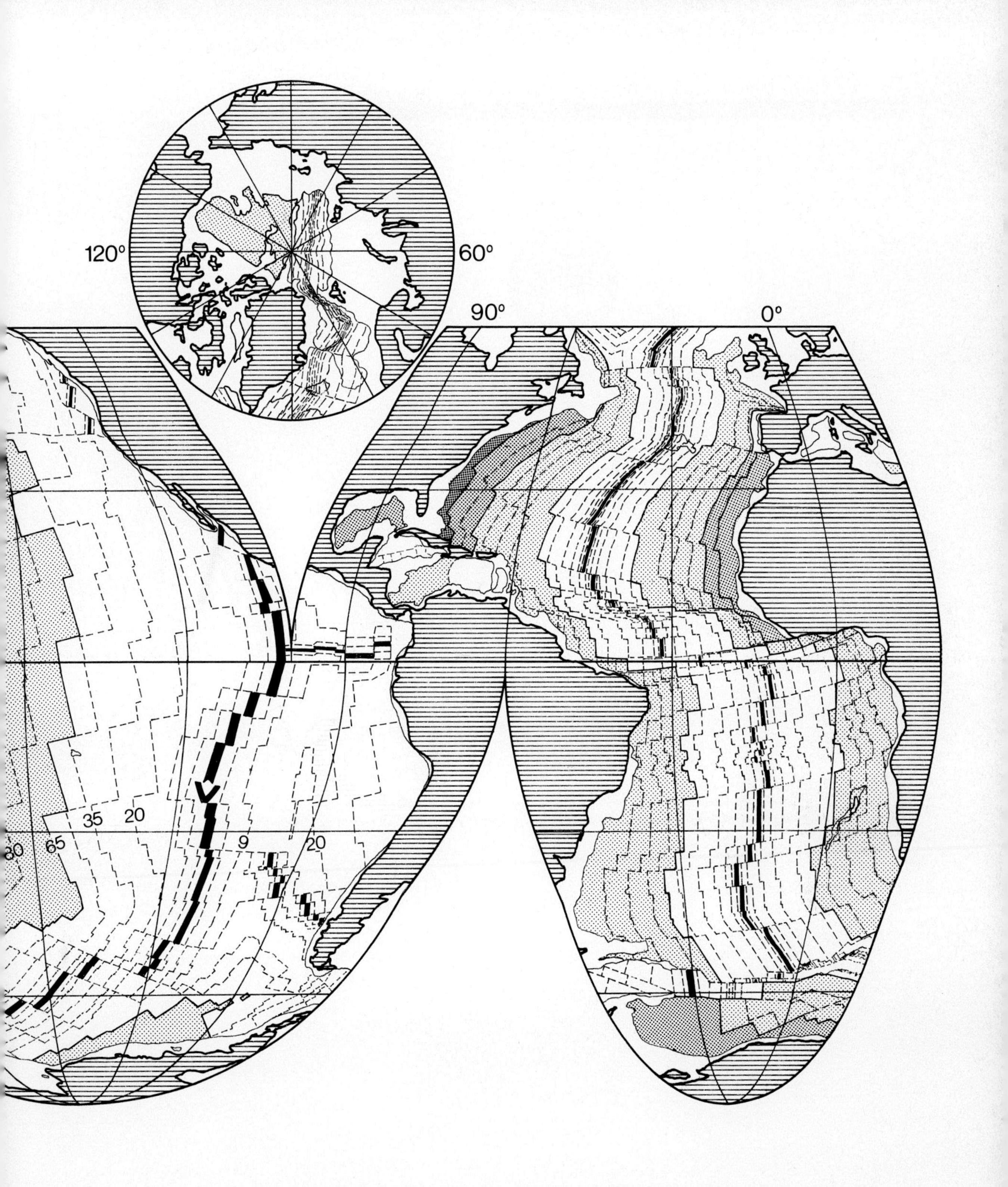
120°
60°
90°
0°
35
20
80
65
9
20

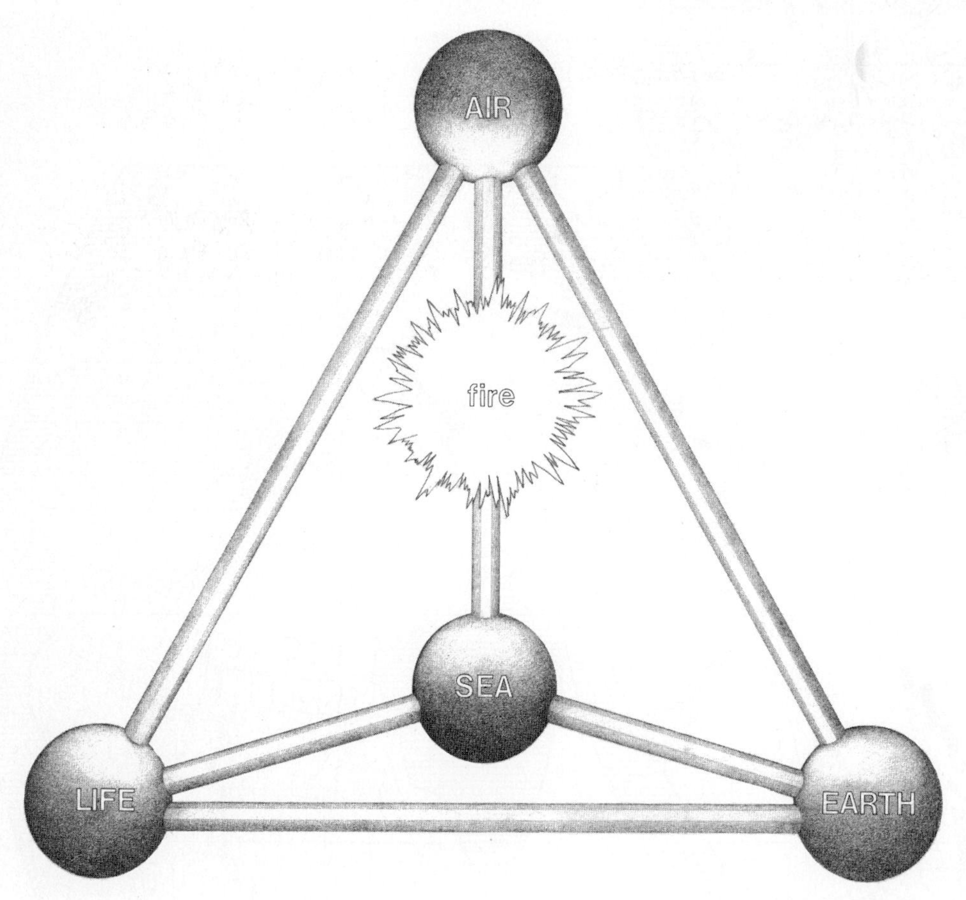

Design: H. Gudenau

Air – Sea – Earth – Life interactions are driven by the *fire* of the Sun, the *heat* derived from the decay of radiogenic elements within the Earth, and the *energy* released from phosphate bonds in the living cell. The alchemistic symbol for fire Δ can be likened to the four triangular faces of a tetrahedron with *fire* at its center, coordinating the flow of matter, be it gaseous, liquid, solid or living.

Egon T. Degens

Perspectives on Biogeochemistry

With 296 Figures

Springer-Verlag
Berlin Heidelberg New York
London Paris Tokyo

Professor Dr. Egon T. Degens †
formerly
Institute of Biogeochemistry
and Marine Chemistry at the
Center of Marine Research
and Climatology
University of Hamburg
Bundesstraße 55
2000 Hamburg 13, FRG

ISBN 3-540-50191-6 Springer-Verlag Berlin Heidelberg New York
ISBN 0-387-50191-6 Springer-Verlag New York Berlin Heidelberg

Library of Congress Cataloging-in-Publication Data. Degens, Egon T. Perspectives on biogeochemistry. Bibliography: p. Includes index. 1. Biogeochemistry. I. Title QH343.7.D44 1989 574.5'222 89-5925

Typesetting: Fotosatz & Design, Berchtesgaden; Printing and binding: Appl, Wemding
2132/3145-543210 – Printed on acid-free paper

Preface

Universal matter – according to Empedocles (483-424 B.C.) – is composed of four elements: air, fire, water and earth. At that time, and for many years to come, this classification scheme represented the best macroscopic description of matter in our universe.

Today, we use a molecular approach. The element air is described by molecular kinetics and statistical physics. The 'simple' substance fire is thermodynamically defined as heat or energy. Quantum mechanics, solid state physics and chemistry refer to matter rather than to earth. The problem child, however, is water, because so far no equation can thermodynamically describe its reaction and properties at the molecular level.

So for all practical purposes we are still referring to the same antique elements as defined by Empedocles, except that currently we look at the element fire in a more refined way, according to which heat or energy is the principal driving force for gaseous, liquid and solid matter. Inasmuch as the living cell may be considered a microcosmos, it should be treated as a separate 'element'. I propose to use as a guiding principle in the forthcoming discussion the air-water-earth-life system to describe forces and mechanisms at work on Planet Earth.

The present is a transitory state between the past and the future. It is a boundary upon which humans stand to reflect on the history and future development of Earth and Universe, using the present as a point of reference. Thus, the element time provides the structural framework in which all events should be placed in chronological order.

The book starts with an elementary discussion on the origin and evolution of matter and forces during the formative minutes and years of the cosmos. This is followed by a description of events leading to the generation of galaxies, stars and chemical elements. A chapter on the birth of our solar system, its inventory and structure brings to a close our brief survey of the cosmos at large.

Some readers may wonder why such a 'long day's journey into night' is required. The answer is simple. In the first place, chemical elements, mineral matter, and the bulk of simple organic molecules have their point of origin in outer space. Furthermore, modern science knows virtually no disciplinary bounds, and as far as geology is concerned, Earth, to paraphrase Thomas Aquinas (1226-74), is just a 'tiny drop on God's bucket'. At least a glimpse into the inside of that bucket is needed to appreciate achievements in neighboring fields and understand why Earth is as it is. This cosmological excursion is not intended for the specialist. To him or her, many sections would appear superficial, many topics slighted or even omitted. It is only a springboard to acquaint the general reader with the exciting research into space and its implication for earthbound phenomena and problems.

The next section is concerned with the coordination of atoms, ions and molecules and the way they arrange themselves structurally in gases, liquids and solids. I would like to convey the message that terrestrial matter is not only composed of chemical elements, but that these elements occur in just a few simple structural units which are the building segments of a construction-set leading to air, water, earth and life. Structure, origin, and evolution of the four antique elements, plus life as a fifth member, follow the established line of discussion.

In the last section, the information presented previously is summarized. From a wide range of prospects, a few biogeochemical case studies are chosen to bring to light the interaction and delicate balance between ocean, air, earth and life in the course of time.

The approach is didactic, so that the reader may gain sufficient information and develop an open frame of mind with which she and he can critically examine further evidence and ideas. Thus, while I had to be selective because of space limitations, I also tried to give a balanced account by including relevant information from several scientific disciplines.

In this book, I abstracted what I learned on bio-geo-chemical interactions in more than 30 years of research and teaching, and hope it will be of interest to students and colleagues in the natural sciences trying to understand phenomena existing on the periphery of their discipline.

Throughout the volume I frequently relied on major research papers, several review articles and some books I wrote in the past – alone or jointly with friends – on various aspects of general geology, low-temperature geochemistry, and the environmental sciences. Thoroughly revised, updated, and integrated, these publications became part of a broad spectrum on air-sea-earth-life interactions which I took the liberty of highlighting in this treatise. While the bias in choice of subject matter is entirely mine, I owe much to the various educational and research institutions with which I had the pleasure of associating, both in the United States and Europe, in shaping this choice.

I am grateful to Ms. Doris Lewandowski for her artwork and to Ms. Inge Jennerjahn for her secretarial assistance in the final preparation of this manuscript. For the numerous stimulating discussions among our small research team at the SCOPE/UNEP International Carbon Unit at Hamburg I express my thanks to How Kin Wong, Walter Michaelis, Stephan Kempe, Venu Ittekkot, and Alejandro Spitzy.

Hamburg, January 1989 EGON T. DEGENS

Contents

Part I

THE COSMOS AT LARGE

Galaxy in Ursa Major, M 81, about nine million light years away, gives us an idea of how our galaxy, the Milky Way, would look if seen from the outside (drawing from photograph in Arp, 1963). About a hundred billion stars are contained in this magnificent spiral and quite a few may have planets like the Sun. Our solar system would occupy a place close to the median plain of the spiral about one third the distance of the total diameter removed from the galactic center. We need roughly 250 million Earth years to circum-navigate the galactic center. Accordingly, this period of time has been defined: a cosmic year.

Chapter 1

Matter and Forces

If I Have Seen Further

Much of what we know today of the distant past is secondhand information. Elementary particles and chemical elements are nothing but the ashes of a Big Bang or of star explosions. The background radiation known from four regions of the electromagnetic spectrum, radio waves, microwaves, X-rays and gamma rays, are just remnants of special events that once took place in the history of the universe. Molecules, rocks, and plant and animal remains or imprints are only indicators of past processes, geological situations or environmental circumstances. From this set of data, the origin and evolution of forces, matter, structure and life in the course of time are inferred.

In science we are used to observing, describing and measuring natural phenomena at all levels and dimensions. As a first approach towards solving a scientific problem we define a model which we subsequently try to verify, revise or even refute. The growth of scientific knowledge over the ages is nothing but a long series of revisions and refutations of previous theories which in their time may have even been considered laws of nature. Take, for instance, the Aristotelian and Ptolemean system with the Earth as center of the universe. For more than a millennium, until Copernicus, it remained authoritative dogma. It was refuted by Galilean and Keplerian dynamics, which in turn were revised by Newton's law of gravitation. This law rests on the assumption that G, the constant of gravitation, does not change with time, place or composition of the masses involved. P.A.M. Dirac in 1937 questioned the universal validity of this law and proposed instead that gravitation is diminishing with time. Such changes – if true – would revolutionize present cosmological concepts which are based on G being an absolute constant. For the record, the majority of evidence, as far as the absolute value of G is concerned, is still in favor of Sir Isaac Newton. Nevertheless, progress in science has advanced to a point where Newton's theory is now considered to be just an approximation. Einsteinian physics shows that in strong gravitational fields there cannot be a Keplerian elliptic orbit with appreciable eccentricity but without corresponding precession of the perihelion; this indeed is observed for Mercury (for an excellent discussion on the comparability of Newton's and Einstein's theories see T.E. Hansen, 1972). Thus, it appears that physical constants and scientific theories are up for revision and falsification at any time. This is no drawback, for to quote Karl R. Popper (1978, 1979): "Every falsification should be regarded as a great success; not merely a success of the scientist who falsified the theory, but also of the scientist who created the falsified theory and who thus in the first instance suggested, if only indirectly, the falsifying experiment". However, if science consisted only of falsification, frustration among scientists would spread. Fortunately, successful predictions have fueled our curiosity and strengthened our morale. But such success should never lead to arrogance. A quote from Newton's letter to Robert Hooke (February 5, 1675) may guide us: "If I have seen further (than you and Descartes) it is by standing upon the shoulders of Giants".

In this context it is of interest that R.K. Merton (1965) in his book "On the Shoulders of Giants: A Shandyan Postscript" lists 26 sources prior to Newton where this quotation has been used. According to Merton (1965), Bernard de Chartres (ca. 1126) first coined the phrase. Bartlett's "Familiar Quotations" give an even earlier reference. Lucan (39–65 A.D.) lets Didacus Stella say, "Pigmei gigantum umeris impositi plusquam ipsi gigantes vident" ("Pigmies placed on the shoulders of giants see more than the giants themselves"). The author himself would not be surprised if a colleague were to point out even earlier sources for this quotation, biblical for instance.

Today we want to look deeper into space and time to unravel the structure of matter, forces and life, and so the saying has not lost its point.

Symmetry

Our universe has evolved over a time span of perhaps ten to twenty billions of years. In using the presently accepted physical laws of nature, we can outline a logical sequence of events leading from a state of next to nothing to the present state of the universe. However, the transition from absolutely nothing to something is still a wide field for speculation.

A discussion on the origin and evolution of the universe generally starts with the genesis of the principal building blocks of matter and the acting physical forces. Whereas previously electrons, protons and neutrons were considered the basic ingredients of matter, quarks and leptons are presently thought to be its singular constituents. And yet, there is a theoretical basis for assuming the existence of even smaller entities termed quips (= quark inner parts; also in the sense of Webster's something queer, an oddity; Shupe, 1979) or rishons (in Hebrew rishon means first, primary; Harari, 1979). The following question may reasonably be asked: Are quips or rishons truly the final endmembers of matter or are they composite products too?

Furthermore, what are their physical properties? This reasoning leads eventually to a dimensionless point for which the properties mass, electric charge, spin etc. can no longer be implied. Nor is it valid at this point to refer separately to particles and their interacting forces. To deal with this kind of system, scientists more than a 100 years ago introduced the concept of a physical field in the first place to describe the condition in space known as an electromagnetic field.

This brings us back to James Clerk Maxwell, who – in following Aristotle – suggested that all space is filled with a medium called ether which he thought to be the substrate to transmit electromagnetic vibrations. The elegant Michelson-Morley experiment put to an end the ether hypothesis by showing that the Earth's motion does not produce an ether wind which it should if such a medium existed. The revised field concept, that is one without ether as a carrier of electromagnetic waves, has in essence become a quantum field. A field through space is needed to explain how distant particles can interact. The quantum of the electromagnetic field is the photon and the range of the electromagnetic force is infinite, since photons have no mass. In general, quantum field theories are abstract phenomena aimed at describing the intricate properties of the space and time continuum.

A powerful means to reduce the complexity of matter and forces to something more manageable is to examine the properties of structural elements that come in matched pairs. It turns out that symmetry is found at various levels of complexity. On the other hand, asymmetry is also part of nature. Its sheer existence in a case where symmetry is actually called for may shed considerable light on the origin of the world at large. But first let us turn to some characteristic patterns of symmetry.

All basic building blocks of matter come in matched pairs. There is the simple pattern of positive and negative electric charges. Which of the two we call plus and which one minus has once been arbitrarily set and has remained as a standard of reference ever since. The physical significance of opposite charges rests on their neutralization when joined together. Since nature is unaware of man's nomenclature, all its laws should be invariant to the exchange of signs of charge, which indeed is the case.

The matter-antimatter symmetry is slightly more complicated. For each particle there is an antiparticle. Every time a high energy beam in a particle accelerator collides with its target, matched pairs of particles and antiparticles arise which are identical in mass but opposite in other physical properties, such as electric charge. The beginning of the concept of matter-antimatter symmetry may be traced to the prediction of the anti-electron by P.A.M. Dirac in 1928 and the experimental proof of this prediction in 1932 when Carl Anderson discovered the antiparticle of the electron which he termed positron. Again, it is a matter of definition to speak of the electron as particle and of the positron as antiparticle, and the same is true of the whole field of particle physics. What alone counts is that when matter and antimatter come in contact they annihilate each other and their mass becomes converted entirely in accordance with the well-known Einstein equation $E = mc^2$.

Antimatter is virtually non-existent outside particle accelerators. The Earth around us is composed of protons, neutrons and electrons, and antiparticles are apparently not part of the Earth. A few minor exceptions exist, for instance in the so-called CP symmetry violation. CP symmetry (i.e., charge conjugation and parity; or mirror reflection) predicts that in the decay of the antimuon a positron should emerge instead of an electron; and the positron should almost always be right-handed. In contrast, the long-lived neutral K meson, which is its own antiparticle, decays more often into a negative pion, a positron and a neutrino, than it does into a positive pion, an electron and an antineutrino. Unfortunately, this CP symmetry violation cannot account for the preponderance of matter in the universe because it is quantitatively unimportant and does not yield protons or neutrons.

Our galaxy is definitely matter, as can be inferred from cosmic rays which carry particles only. Still, there is a small discrepancy: the existence of negligible amounts of cosmic-ray antiprotons which are probably generated by high energy collisions in the interstellar medium. For the rest of the universe there is no final proof from observational data in support of matter preference. Should antimatter exist at all, it must be hidden in some obscure place out of contact with matter, simply to escape annihilation.

Information received by way of photons from distant galaxies bears no cosmic fingerprint for matter or antimatter because the photon is its own antiparticle. All light quanta look alike, whether they were emitted from matter or from antimatter. Future work on the detection of neutrinos and antineutrinos by way of perhaps a "neutrino telescope" may open up new vistas, since a star composed of matter would mainly radiate neutrinos, whereas an antimatter star should chiefly emit antineutrinos (e.g., Wilczek, 1980). In a world structured like ours it is a formidable task to separate matter and antimatter into discrete batches confined in isolated galactic clusters. For the time being this is the most convincing – even though indirect argument – for the matter-dominance of our universe. The asymmetry should not be explained by physical separation of the two antagonistic members but by primordial processes that have created tiny asymmetries which subsequently gave rise to all the matter in the universe.

Symmetry is observed in all reactions yielding primordial material (see, e.g., Schopper, 1981). Almost! During its first microseconds the primeval fireball contained an equilibrium mixture of particles, antiparticles and photons. For each pair created by collision or decay, an identical pair would be destroyed. Consider the neutrons and protons which belong to a class of particles termed baryons that participate in the strong interaction. It has been estimated that at this early stage the ratio of baryons to antibaryons must have been roughly 10^{10} plus 1 to 10^{10}, that is a single proton in excess of 10 billion antiprotons. Actually, this value is uncertain and could be lower by an order of magnitude. As the temperature fell off and the universe expanded, the rate of pair annihilation exceeded the rate of pair generation, leaving just one unmatched baryon of the original 10^{10} pairs behind (Fig. 1.1). In this view, all matter distributed in galaxies, stars, planets and spaces in between is the remnant of the-one-part-in-a-ten-billion imbalance. What also has been left from the baryon-antibaryon annihilation are the photons that gave rise to the 3 K microwave background radiation discovered in 1965 by R.R. Penzias and R.W. Wilson. Their number should be identical with the number of baryons and antibaryons present during the initial stage of the universe. The cosmic asymmetry, even though tiny, is in need of an explanation. However, before expanding on that issue, a little information on the nature of the basic building blocks of ordinary matter is needed.

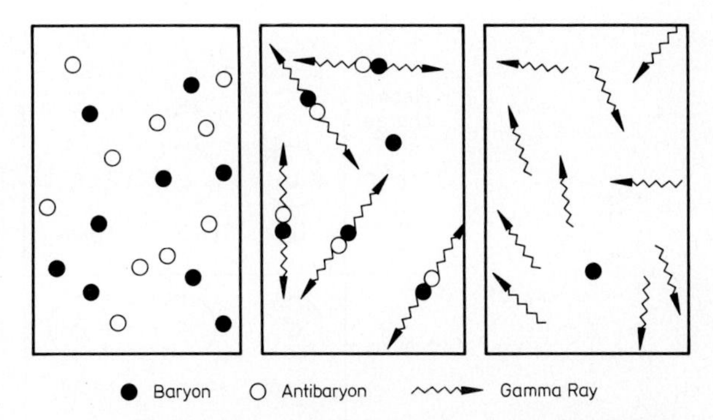

Fig. 1.1. Three-stage development of matter-predominance in our universe. Phase one (*left*) depicts thermal equilibrium between baryons and antibaryons until about 10^{-4} sec after big bang. Phase two (*center*) represents baryon-antibaryon annihilation in excess of pair generation as the universe expands. Phase three (*right*) shows left-over baryon "floating" in light quanta following pair consumption event. The one in 10^9–10^{10} surviving baryons put together account for all cosmic matter, be it planet, star, galaxy or somewhere in between (based on Waldrop, 1981b)

Tohu Va-Vohu

Matter is composed of two classes of fundamental particles: the quarks and the leptons. The quark model was introduced into physics by Murray Gell-Mann and George Zweig in 1964. Up to then, particles termed hadrons were supposed to be elemental. Now, the two familiar subclasses of hadrons, i.e., the baryons – to which the proton and neutron belong – and the mesons, are composed of quarks. Baryons have three and mesons one quark and one antiquark. No other compounds of quarks exist.

There is compelling evidence for five kinds of quarks which are referred to as flavors. The names attached to them are: *up, down, charm, strange* and *beauty* (or *bottom*); a sixth one, the *truth,* has tentatively been discovered. The masses of quarks differ but their precise values are uncertain. It has been found useful to list comparative figures representing half the mass of the lowest-lying quark-antiquark vector meson which then would yield for *u:* 0.39, *d:* 0.39, *c:* 1.52, *s:* 0.51, *t:* 15 and *b:* 4.72 GeV/c^2. The natural unit of electric charge of a quark is + 2/3 or −1/3 of the protonic charge (1.6 x 10^{-19} coulomb). In analogy to the periodic table of chemical elements, two quarks on top of one another belong to a family. This was the principal reason to look for a sixth quark, the

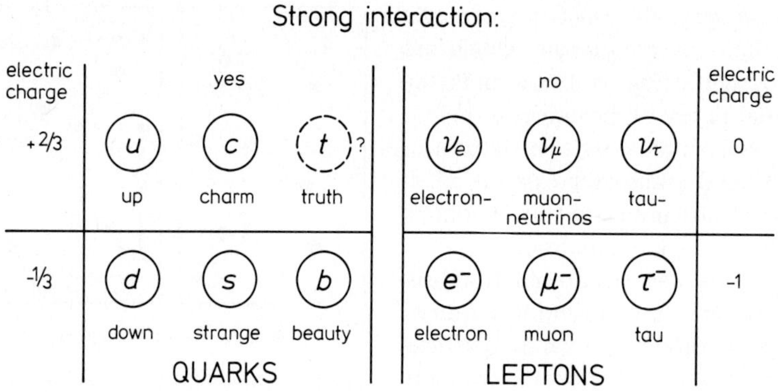

Strong interaction:

electric charge	yes			no			electric charge
+2/3	u (up)	c (charm)	t ? (truth)	ν_e (electron-)	ν_μ (muon-) neutrinos	ν_τ (tau-)	0
-1/3	d (down)	s (strange)	b (beauty)	e^- (electron)	μ^- (muon)	τ^- (tau)	-1
	QUARKS			LEPTONS			

Fig. 1.2. Classification of quarks and leptons which are presently regarded as the fundamental particles of matter. This scheme can be considered a reduction of the Periodic Table of Elements to its basic units (Schopper, 1981)

truth, above *beauty* (Fig. 1.2). The baryon number of each quark is + 1/3. The proton has a quark composition adding up to an electric charge of 2/3 plus 2/3 minus 1/3 equal 1. The neutron has one where the charges add up to zero (2/3 minus 1/3 minus 1/3 equal 0).

In addition to flavors, quarks have another property referred to as color. One should perhaps point out that flavor and color are arbitrary labels and in no way linked to taste or sight. The concept of color was introduced to reconcile a situation where the structure of observed baryons required two or even three quarks of the same flavor in the same quantum state. For spin 1/2 particles this would constitute a strict violation of the Pauli exclusion principle. Each flavor of quark comes in three colors: red, green and blue. This code or color quantum number allows up to three quarks of the same flavor to occupy a single quantum state. Inasmuch as color is the domain of hadrons and absent from leptons, the strong nuclear force is virtually an interaction between colors. Free quarks have not been isolated so far. Quarks are observed only as constituents of baryons and mesons. The properties of each antiquark are just the negative of each quark.

The leptons differ from quarks in that they do not feel the strong nuclear force. Six leptons are known and the most familiar one is the *electron.* It has a small mass, equivalent in energy units to about 5 x 10^5 eV and it has one unit of electric charge which by convention is negative (Fig. 1.2). Two other leptons, the *muon* and *tau,* seem to have properties identical to the *electron* except for mass. The *muon* is 200, and the *tau* 3,600 times as massive as the *electron.* All three particles have one partner with zero electric charge, the neutrino (*electron-, muon-,* and *tau-neutrino*).

Until recently, the neutrinos were supposed to carry no mass. However, there are two sets of experiments which seem to indicate that neutrinos have a mass that is not identical with zero. An American group (Reines et al., 1980) measured neutrinos from a nuclear reactor and masses of a few electron volts were calculated. A Russian experiment (Lubimov et al., 1980) gave ranges for neutrino masses from 14 to 46 eV. Model calculations based on closed and open Friedmann universes (Joshi and Chitre, 1981), using an age estimate of 20 x 10^9 years, resulted in neutrino masses of 18.91 and 3.38 eV respectively. Tentative proof for neutrino "oscillations" ("one type of neutrino can turn into another as it moves along") lends support to a small mass for the neutrino. However, application of more sensitive techniques than those previously applied have found no evidence for neutrino oscillations.

Neutrino weight watching is presently being vigorously pursued by a great number of research groups. This is to be expected in the light of the multivarious cosmological implications a conclusive proof for neutrino mass or no mass would have (see, e.g., R'ujula and Glashow, 1980; Waldrop, 1981a, 1982; Cence et al., 1981).

Particles currently considered to represent the principal building blocks of matter are depicted in Fig. 1.2. In a certain way this scheme replaces the periodic table of elements; furthermore, it has the advantage of being much simpler and more basic. Quarks and leptons share one property: spin 1/2 and thus – by definition – they are all fermions. Conservation laws request that their number can neither increase nor decrease except for pair generation or annihilation of a particle and its antiparticle. There is one exception to the rule and it concerns the weak nuclear force. Interactions between particles of one family can cause transformation into one another. All atoms are electrically neutral because the charge of a proton and of an electron are identical but of opposite sign; the neutron per se has zero charge. In turn, the charges of quarks match exactly those of the leptons. It is common practice to classify the leptons and the quarks in three generations each composed of a charged lepton, the related neutrino, a –1/3 quark and a +2/3 quark. Particles of the first generation are: e^-, ν_e, d and u. Inasmuch as quarks come in three colors, a generation holds eight particles.

All attempts to find additional leptons and quarks were negative. By including the tentatively identified *truth,* we are left with six quarks and six leptons. As far as we know by experiment, there is no evidence for the transformation of quarks into leptons or vice versa (e.g., Rebbi, 1983). If the proton were observed to decay, the connection would be established. So for the moment matter is composed of only 12 fundamental building blocks.

A heuristic model has been presented in which leptons and quarks are treated as composites of only two types of fundamental spin 1/2 objects with electric charges 1/3 and 0. The model has independently occurred with Shupe (1979) and Harari (1979). In a composite model of quarks and leptons, the most economical set of building blocks has to be fermionic: one with the fundamental electric charge $Q = 1/3$ and one being neutral $Q = 0$. Accepting Harari's terminology – Shupe's naming could have served equally well – the charged particle is denoted by T (for Tohu) and the neutral one by V (for Va-Vohu); and both are called rishons.

The simplest composite fermion consists of three rishons and eight combinations emerge:

- TTT ($Q = +1$); e^+
- TTV, TVT, VTT ($Q = +2/3$); three color states of u-quark
- TVV, VTV, VVT ($Q = +1/3$); three color states of $\bar{d}$-antiquark
- VVV ($Q = 0$); ν_e.

The antiparticles of the above states are e^- $(\bar{T}\bar{T}\bar{T})$, $\bar{u}$ $(\bar{T}\bar{T}\bar{V}$, $\bar{T}\bar{V}\bar{T}$, $\bar{V}\bar{T}\bar{T})$, $\bar{d}(\bar{V}\bar{V}\bar{T}$, $\bar{V}\bar{T}\bar{V}$, $\bar{T}\bar{V}\bar{V})$ and $\bar{\nu}_e$ $(\bar{V}\bar{V}\bar{V})$ (Fig. 1.3). It has been postulated that just three and not two or four colors exist because there are three rishons in a quark. The "purity" of the lepton states could explain their low mass. Baryon-number violating processes are allowed in this model and proton decay in grand unification schemes (see p. 10) may find its explanation.

New insight into the question of matter dominance in our universe will certainly be gained should rishons truly represent the fundamental building blocks of matter. At 10^{15} GeV, so the theory goes, quarks and leptons assume similar characteristics and can transmute back and forth. Finally, one should emphasize

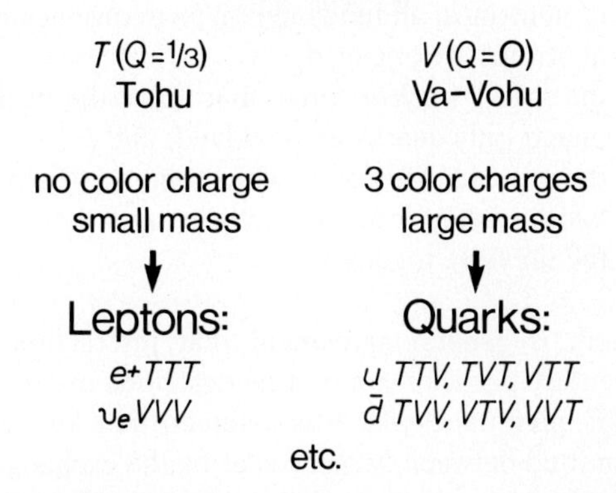

Fig. 1.3. Building blocks of quarks and leptons: rishons (Harari, 1979)

that rishons possess no individuality but should be looked upon as abstract entities.

Grand Unifications

In a routine chemical analysis one can predict the outcome of a reaction because we know that the forces operating between atoms in molecules are so weak that they will not alter the individuality of the participating chemical elements. This is simply a function of the enormous discrepancy between the values of binding energy and stationary energy of atoms in molecules. Stationary energies of atoms are several orders higher (ca. 10^{10}). Quarks are the building segments of hadrons and they themselves may consist of rishons as their fundamental particles. Because the interacting force relative to the stationary energy of the "individual" building blocks increases dramatically from atoms towards rishons, it is no longer conceivable to speak of individuality when it comes to rishons (Schopper, 1981). In the final analysis, one is dealing with point-like constituents in composite structures which are tied together in an unfamiliar way, so that the composite is also point-like (Shupe, 1979).

All this suggests that an isolated discussion of fundamental particles in matter is meaningless unless forces operating at the molecular, atomic and subnuclear levels are duly considered.

The four observable forces operating in nature individually are:

(i) the *electromagnetic force,* that governs the interactions of electrically charged particles; quarks and charged leptons (e,u,t) participate;
(ii) the *weak nuclear force,* which is responsible for the beta decay of neutrons to a proton, an elec-

tron and an antineutrino and for the interactions of neutrinos; all fundamental fermions including neutrinos are engaged;
(iii) the *strong nuclear force,* that holds the nucleus intact; only quarks are involved; and
(iv) the *gravitational force,* which operates between two attracting bodies or objects and that keeps the universe together.

There is general agreement that interactions of elementary particles can best be described in the context of quantum fields. It is assumed that a force is transmitted between two particles by the exchange of a third, intermediary particle. Whereas elementary particles are fermions (spin 1/2 fields), the intermediary particles are bosons (named for the Indian physicist S.N. Bose). A boson is a particle whose spin, when measured in fundamental units, is an integer (0, 1, or 2). For the electromagnetic and weak nuclear forces, the intermediary particle is a vector boson, whereby the term vector implies that the boson has a spin value equal to 1.

Best known is the vector boson of the electromagnetic interaction, the photon, which carries no mass. Gluons are the intermediary particles of the strong nuclear force. They resemble the photon in that they have a spin of 1 and are massless too. The corresponding force carrier in weak interactions is the intermediate vector boson. Most crucial in understanding the weak force were theories (Glashow-Salam-Weinberg) suggesting that electromagnetic and weak interactions are just different expressions of a single underlying force. The unified electro-weak theory predicts the existence of three intermediate vector bosons: W^+, W^- and the neutral Z^0. They are distinguished from all other bosons in that they have huge masses, some 80 to 90 times greater than the mass of a neutron or a proton (Cline et al., 1982). The first indication that they are real has recently been established at the European Center for Nuclear Research (CERN) in Geneva (Walgate, 1983; Close, 1983). From the few events measured in the proton-antiproton collider, the mass of the W particle was calculated to be 81 ± 5 GeV, almost exactly the value of 82 ± 2 GeV predicted by the Glashow-Salam-Weinberg model.

Following the discovery of the charged intermediate vector bosons W^+ and W^- in spring 1983, CERN scientists have recently detected *W*'s neutral partner, Z^0, with a mass closely corresponding to the predicted mass of 92 ± 2 GeV. This experiment brings to an end a 50-year Odyssey to find the force behind weak interaction. The successful vector boson hunt may have also produced data in support of the existence of the *t* quark, which should have been generated via the decay of *W*'s (Robinson, 1983). The search of particle physicists is now directed towards the identification of massive gauge field quanta, so-called Higgs bosons (named after the British physicist Peter Higgs) and which may become apparent in the area of 1 TeV. The reality of Higgs particles is essential to the Glashow-Salam-Weinberg model (see below).

Concerning gravitation, interactions are too weak to be studied in the laboratory for possible quantum effects. Should there be an intermediary particle carrying the force, a graviton, its spin would be 2.

Nature is driven by four basic interactions. At the same time there is evidence that all forces operate through the same mechanism. It all started with James Clerk Maxwell in 1864, who unified the formerly disparate electric and magnetic phenomena through the foundation of the theory of electromagnetic waves. Maxwell's equations were based on experiments taking place at classical distances in the centimeter or meter range. The second classical field theory developed by Albert Einstein for the gravitation failed to unify the two forces. By entering atomic dimensions (10^{-8} cm), Maxwell's theory developed into what we now refer to as quantum electrodynamics (QED). Unification of the electromagnetic and weak interactions was achieved by the Glashow-Salam-Weinberg model. It predicts that at distances of about 10^{-16} cm, which in terms of energy is equivalent to more than 100 GeV, the coupling strengths of both forces approximate each other. The successful generation of the W and Z^0 particles in the predicted GeV range at CERN was excellent proof of the soundness of the model.

Unification of the strong, weak, and electromagnetic forces in a single mathematical framework, the grand unified theory of particle physics (GUT's), is presently of lively concern among particle physicists (e.g., Georgi, 1981). The energy needed to unite particles and forces would amount to 10^{15} GeV, a value unlikely ever to be reached in particle accelerators. This energy level would correspond to a distance of about 10^{-29} cm.

The unification of fields will not be complete without gravity. So the quest is to unify gauge fields with supergravity. Theories on supergravity are extensions of general relativity that incorporate a fundamental symmetry between bosons and fermions (Salam, 1980; Freedman, 1981). However, the strength of gravitation is so minute that an approximation of the coupling strenghts would be reached at distances (= unification scale) of 10^{-34} cm (Fig. 1.4).

To illustrate how particle physicists derive their conclusions, a brief account on gauge-invariant field

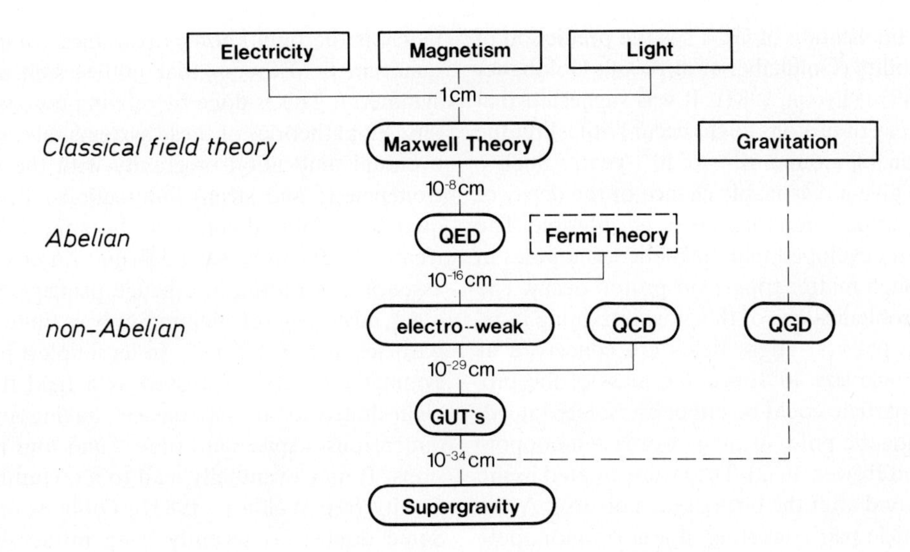

Fig. 1.4. From Maxwell to supergravity: Unification of forces. QED (quantum electrodynamics); QCD (quantum chromodynamics); QGD (quantum geodynamics); GUT's (Grand Unification Theories of particle physics). The term Abelian and non-Abelian is taken from mathematics (group theory). Abelian or commutative groups satisfy the multiplication law for every pair of elements, e.g., g and h, in the group: $gh = hg$. Non-Abelian or non-commutative groups involve rotation of an element around a certain axis in space by a certain angle (Schopper, 1981)

theories follows (see also Yang and Mills, 1954; Woolf, 1980). A gauge invariance is an internal symmetry where the symmetry operation is a rotation of the internal coordinates independent at each point of space and time. The symmetry calls for gauge fields, massless vector fields, one for each independent internal symmetry rotation. Massless bosons are the photon and the gluons. This would imply that such theories are applicable only for QED and QCD. But a phenomenon known as spontaneous symmetry breaking can occur, in which the states of the system do not have all the symmetry of the equations in motion. This in essence can explain the generation of massive gauge field quanta (Higgs mechanisms).

In gauge field theories it has been found practical to employ terms used in the mathematical theory of groups. Electromagnetism is described by a symmetry designated $U(1)$, which signifies that the electromagnetic force cannot change the identity of a particle. The strong nuclear force is characterized by a theory with an $SU(3)$ symmetry, in which the couplings of gluons to quarks can be represented by a three-by-three matrix. The 3 refers to the three colors that are transformed into one another by the gluons. The S indicates that the sum of the color charges in each $SU(3)$ family is zero. The weak force acts on doublets of particles. It is described by a theory having an $SU(2)$ symmetry, in which the two members of the doublet can be transformed into each other. It follows that the unified electro-weak theory is designated by the product of the groups it incorporates: $SU(2)$ x $U(1)$.

The $SU(2)$ x $U(1)$ theory is only a partial unification. In GUT's a superstructure has to be envisioned in which the $SU(3)$ and $SU(2)$ x $U(1)$ theories become incorporated. The symmetry of such a superstructure should place quarks and leptons close to each other. The group best fitted for that purpose is $SU(5)$ which permits all possible transformations of five distinct objects or of a five-by-five matrix. Under condition of maximum symmetry, 12 additional intermediary particles (bosons) are called for which were labeled X and are believed to mediate the interconversion of leptons and quarks. In turn, quarks and leptons should be regarded as members of a single family. Symmetries can be broken at certain energy levels or distances respectively. For the $SU(5)$ symmetry a distance of about 10^{-29} cm has been calculated for the unification scale.

X bosons should have masses roughly 10^{15} times the mass of a proton. Interactions mediated by X bosons could transform a quark into an antiquark or even a lepton. Baryon number is no longer conserved. By contrast, in the electromagnetic, the weak, and the strong interactions, baryon number remains constant. The breakdown of baryon conservation due to the interplay of X bosons and anti-X bosons could have produced the matter dominance in our universe and explain everything under the *Sun*. Other massive intermediary particles have been postulated, so-called Higgs bosons, to make the original $SU(5)$ theory more consistent with the observed asymmetries noted in the universal time-space continuum.

A further implication of GUT's is the prediction of proton instability (Goldhaber et al., 1980; Goldhaber and Nieto, 1976; Lyons, 1982). It was suggested that catalyzed decay of protons might occur with a lifetime of the proton of around 10^{31} to 10^{32} years. Such a value would give a reasonable chance of the decay of one proton per person during his or her lifetime. The idea has been developed that magnetic monopoles in passage through matter trigger off proton decay. The monopole problem is one of the most intriguing in recent particle physics. Higgs fields are conceived as having a particle size 10^{16} times the mass of the proton. Such a particle could be either an isolated north or south magnetic pole, in other words a monopole (Carrigan and Trower, 1982). They were created in the first 10^{-35} second after the birth of the universe. A particle-antiparticle pair consisting of a north monopole and a south monopole could be generated when a high-energy photon interacts with a proton. Upon collision, the mass of the monopoles is converted into photons. A certain portion of the monopoles may have survived down to the present, but their number near Earth is unknown.

Where do we go from here when the stability of matter is already put into question? Let us conclude this chapter with a quotation from Sheldon Lee Glashow (1980): "The point I wish to make is simply that it is too early to convince ourselves that we know the future of particle physics. There are too many points at which the conventional picture may be wrong or incomplete. The $SU(3) \times SU(2) \times U(1)$ gauge theory with three families is certainly a good beginning, not to accept but to attack, extend, and exploit. We are far from the end."

Beyond GUT's

The universe contains much of what cosmologists refer to as the "missing mass", which is recognized only by its gravitational effects on the galaxies. The most plausible explanation is the existence of weakly interacting elementary particles of unknown nature. However, there is some speculation that the missing mass could be related to the presence of light and electrically neutral superparticles, the gravitino and the photino, which – since long-lived – would exercise a weak gravitational effect in the intergalactic space. Such superparticles may have arisen jointly with the conventional particles, according to a fundamental idea known as supersymmetry. This idea is an extension of Einstein's global symmetry of special relativity and the local symmetry of general relativity (Freedman, 1981; Wess and Bagger, 1983). Global supersymmetry is the only known invariance compatible with quantum field theory that unifies spin and internal symmetry. This is done by relating bosons and fermions. Field theories of these systems raise hope for the eventual unification of gravity with the weak, electromagnetic and strong interactions. First found in quantum field theory, it became systematically treated by Julian Wess and Bruno Zumino in 1974. In essence it is rooted in a gauge principle but has the great advantage of yielding only a finite number of particles in a "big bang". In its simplest form, supersymmetry can be expressed as a field theory in an eight-dimensional "superspace" having four ordinary dimensions – space and time – and four new dimensions. It may eventually lead to a quantum theory of gravity (see Waldrop, 1983b; Gibbons et al., 1986). Some doubt has recently been raised that existing supergravity theories could achieve unification of all four fundamental forces. The arguments run as follows:

Grand unified theories are aimed at unifying the strong and the electro-weak interactions at an energy scale of 10^{16} GeV. They are gauge invariant point field theories which do not attempt to incorporate the gravitational force. The most favored GUT's are based on the special unitary group $SU(5)$, the special orthogenal group $SO(10)$, or the exceptional group E_6. In such schemes, quarks and leptons are unified into families.

According to Green (1985), none of the existing supergravity theories intending to unify gravity with the other forces will lead to a consistent quantum field theory. Instead of the more familiar field theories involving a structureless point particle, he proposes string-like fundamental quanta, where a single string has an infinite number of states with masses and spins which increase without limit. In the so-called superstring field theories gravity can "readily" become unified with the other fundamental forces.

Green and Schwarz (1984) have shown that the only theories anomaly-free and finite are those based on the gauge groups $SO(32)$ or $E_8 \times E_8$. The extra E_8 factor describes matter which interacts gravitationally only with the charged matter in E_6 GUT. Kolb et al. (1985) conclude from this assumption that there should actually be two forms of matter, namely ordinary matter whose interactions would be described by E_8, and "shadow matter" whose interactions would be described by E_8'. These two kinds of matter would only interact through gravitational strength interactions. The authors conclude that "the effect of shadow matter is hard to detect in everyday life; the reader could be living in the middle of a shadow mountain or at the bottom of a shadow ocean".

This would be a fitting moment to close the discussion on GUT's and superstring field theories because unification of the four fundamental forces can now be satisfactorily explained in logical orders of succession. However, a problem again appears with the universal constant of gravitation *G*, which, according to recent findings, is a function of the distance separating two interacting masses (Fischbach et al., 1986). Namely, when measuring the force of attraction in the laboratory at small distances and comparing that to the force measured by geophysical means in the field at large distances, the latter value turns out to be almost 1 percent larger. An interpretation offered is that the gravitational attraction between objects is a function of two components, i.e., normal gravitation acting upon mass only and a second relatively weak force acting upon the number of baryons. Undoubtedly, baryons (= total number of protons and neutrons) represent most of the mass of an object, but the mass of an object is only about, but not exactly, proportional to the specific baryon number. The newly discovered baryonic force seems to operate within distances up to about 200 m. A new particle, the "hyperphoton", is thought to mediate the force. Exchange of hyperphotons between nearby objects would result in a repulsion and could account for the apparent discrepancy in the value of *G* observed at short and at long distances. The massive hyperphoton has supposedly a mass of 10^{-9} eV, which is infinitesimally small when compared to the mass of an electron that corresponds to 500,000 eV. Experiments are underway to find out more about this massive hyperphoton.

Does this hypothesis alter cosmological concepts based on Newtonian gravitation, such as Einstein's theory of relativity? No, they still hold true, except that the new baryonic force has to be included. Furthermore, since hyperphotons have a small mass, they would contribute to matter in the universe and could thus profoundly influence the direction of universal evolution.

About 10 pages ago we raised the question of physical constants citing *G* the constant in the inverse square law for the gravitational force between two objects as a good example which successfully withstood objections raised ever since Sir Isaac Newton propounded it. An enlightening summary on the state of gravitation three centuries after publication of Isaac Newton's *Principia* has recently been given (Hawking and Israel, 1987). Departure from Newtonian gravitation observed at short order distances could possibly be linked to the newly discovered baryonic force, which implies that the true value for *G* is the one recognized in geophysical surveys (Holding and Tuck, 1984).

Chapter 2

Instant of Creation

Steady-State Model

The state of the universe prior to the creation of matter and forces is as yet unresolved, unless we believe that the universe was everlasting into the past as well as into the future. Taking the Bible as a base for reference, the world at large was created almost instantly by an act of God. In the end there will be a "big crunch" ("Weltuntergang") but the essence will last from eternity to eternity. In spite of all the achievements of modern science, this is still a very logical and personally satisfying conception. Admittedly, one has to believe in it. Thus in a way, when it comes to the moment of creation, religious and scientific models have many things in common. With this in mind let us expand on the problem at hand.

One idea which overcomes many of the aforementioned difficulties had been submitted by Bondi, Gold and Hoyle in 1948: the "steady-state" or the "perfect cosmological principle". According to this, the universe had the same general appearance at all times, as well as at all points in space. To explain how such a situation can arise when galaxies now appear to be moving apart, we need to generate new matter continually in between to maintain the average density of the universe at a constant value. The amount of matter generated in this fashion is about one hydrogen atom in every cubic meter of cosmos every 10^9 years. This matter will eventually give rise to new galaxies filling in the gaps created by the receding galaxies (Hoyle, 1975).

For about 20 years this was esthetically a very attractive model because it had the great advantage of making definite testable predictions. Observations of radio sources and the detection of the microwave background radiation (see p. 16), however, proved these predictions to be in error; but some still cling to their original idea of a "steady-state". In his book "Steady-State Cosmology Revisited", which appeared in 1980, Fred Hoyle questioned the validity of a young universe, which, according to various models on initial singularity has an assigned age of about 10 to 20 billion years. The primordial synthesis of enzymes he used as an argument in favor of steady-state. The logic runs as follows. The chance of a random shuffling of amino acids producing a workable set of enzymes is infinitesimally small, i.e., 10^{-40000}. This number was obtained from a calculation in which about twenty amino acids were required to be in specific sequential positions for each of two thousand enzymes. According to Hoyle (1982), it seems better to suppose that the origin of life was a deliberate intellectual act than to accept a probability as small as 1 part in 10^{40000} of life having arisen through the "blind" forces of nature. In this context "better" means less likely to be wrong. This is the strength of religion, its almost non-refutability. Nevertheless, by letting blind forces operate for eternities, any of the well-known amino acid sequences in enzymes can theoretically be fabricated. Unfortunately, Hoyle's biochemical excursion must lead into a blind alley, since the rate of synthesis of biopolymers both in the living cell and in the abiotic world is governed by epitaxis and catalysis and is simply a function of structural regulation and not of chance. In the distant geologic past the principal polymeric building blocks of life were possibly formed in almost a matter of no time along mineral templates, if this concept proposed by Bernal (1954), Matheja and Degens (1971) and Cairns-Smith (1982) is correct.

Nothing to Speak of

Steady-state is out, initial singularity apparently is in, but there might be some "strings" attached. Since conservation laws of physics forbid the creation of something from nothing, we are faced with a fundamental problem. Some light could perhaps be shed by the fact that the universe has particular values for energy, electric charge, baryon and lepton number and so on. This has led to the rather intriguing idea that the universe is a large-scale vacuum fluctuation (e.g., Tryon, 1973). Contrary to general opinion, such an event need not have violated any of the conventional laws of physics. A universe coming from the cold, from nowhere and nothing, must have certain specific properties. In particular, it should have a zero net value for all conserved quantities. For the moment this statement is correct but some adjustments have to be made later.

What is the state of true nothingness? A vacuum with energy and structure! *Gui wu*, having respect for the true nothingness is a basic principle of Daoism. From *Dao,* the way, sprung up two ancestral forces, *Yin* and *Yang,* and their interaction and permutation gave rise to the cosmos. They are in a state of flux, constantly interacting among themselves, thereby triggering processes, which are mirrored in all of nature's phenomena. *Dao, Yin* and *Yang, Wu Xing* (wood, fire, earth, metal and water) were the basic elements of Chinese cosmology until the recent past. A remarkable account on the "philosophy" of such a vacuum has been given by Trefil (1981). Here are some excerpts. Imagine a half-inch thimble filled with air. It would contain thousands of quadrillions of atoms. If placed inside a TV tube, the thimble would hold only a few billion atoms. By putting it into the best vacuum one could generate on Earth, about 500 atoms would still be left inside. On the surface of the Moon the number of atoms within the thimble would go down by a factor of ten. And finally, if the thimble were carried to interstellar space, one atom, on average, would be left behind. Except for that one isolated atom, all the remaining space in that thimble represents a perfect vacuum.

Quantum field theory predicts that any phenomenon that could happen in principle does happen from time to time, statistically. For example, QED reveals that an electron-positron pair and photon occasionally emerge spontaneously from a perfect vacuum. The particles exist for a brief interlude and then vanish. That this phenomenon is not fiction but real can be shown experimentally.

An isolated hydrogen atom inside a vacuum will be affected by electron-positron pairs appearing and disappearing in the empty space surrounding the hydrogen. That is, the electron circling the proton will feel the electrical forces exerted by the pair. The electron will be jostled, which in turn will affect the light emitted by the hydrogen atom. In short, the ghostly pairs not seen at all are brought to light. It is true that energy conservation is violated, but only for the brief particle lifetime t permitted by the uncertainty principle originally conceived in 1927 by Werner Heisenberg:

$$\Delta t \, \Delta E \geq h/2\pi,$$

where E is the net energy of the particle and h is Planck's constant (= 6.6256 x 10^{-27} erg/sec). The principle relates the uncertainty ΔE in the energy radiated to the uncertainty Δt in the time at which it radiates, i.e., the uncertainty in its lifetime.

The spontaneous, temporary emergence of particles from a perfect vacuum is called vacuum fluctuation. Thus a vacuum is not simply the absence of matter; instead, the particle-antiparticle pairs it consists of are considered to be in a transitory state. From here to the initial singularity termed big bang it is only a "small" step, provided one takes the view that the energy of the Universe with matter is lower than the energy of the Universe without, and nature always proceeds to a state of lowest energy (Fig. 2.1; see Trefil, 1981). Summarizing, going back in time and

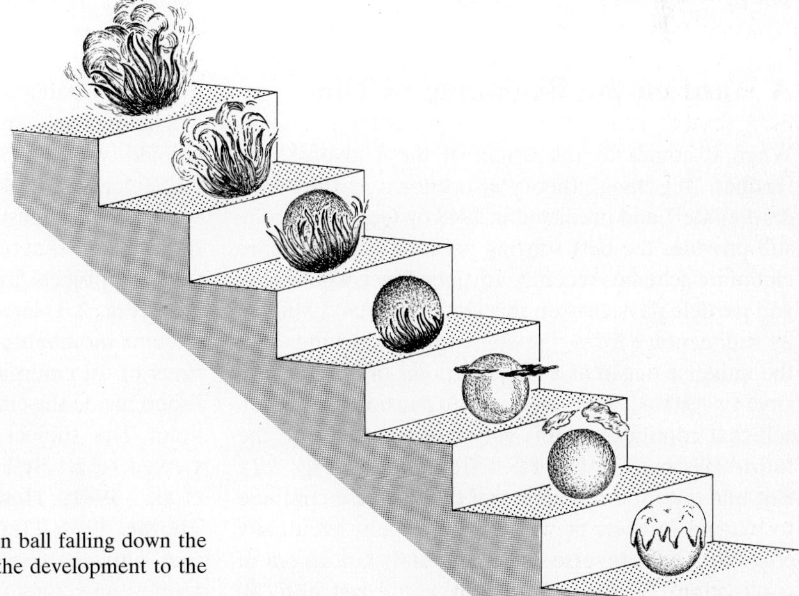

Fig. 2.1. From fire to ice. A hot cannon ball falling down the steps into an ice cellar as an analog of the development to the state of lowest energy (Trefil, 1981)

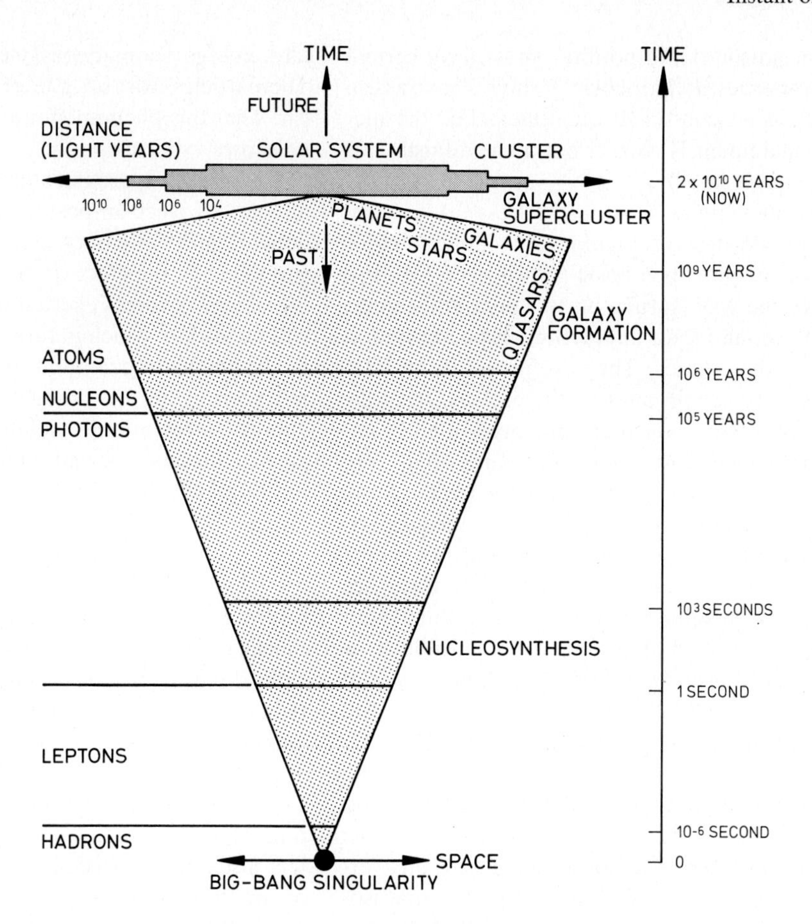

Fig. 2.2. Event cosmology in space and time. *Horizontal bar* at top of figures indicates the horizon of an "observer" at the singularity (Barrow and Silk, 1980)

space, not only forces will be unified but also dimensions. The vacuum is a one-dimensional space which upon fluctuation briefly evolves into a four-dimensional space, falling back almost instantaneously into one-dimensionality.

A Word on the Beginning of Time

When it comes to the origin of the Universe, the familiar "big bang" theory also known as the "standard model" and predicted in 1948 by George Gamov, still provides the best starting point to examine more elaborate schemes recently advanced by cosmologists and particle physicists on this issue (Gamov, 1948). In its rudimentary form, the standard model states that the universe began as a singular point of infinite density in a "titanic" explosion – an expansion of space itself that continues to this very moment and into the future (Gibbons et al., 1983; Trefil, 1983a; Fig. 2.2). But one question remains: will the universe continue to expand forever, or will the movement eventually come to a halt, reverse its course, and start an era of contraction? Such a contraction would last until all matter and radiation entered into a singularity at the origin of a universal black hole, or what John Archibald Wheeler refers to as the "big crunch" (Wheeler, 1980). Some proponents of this scenario go so far as to suggest that big bang expansion is merely the dynamical evolution of a prior phase of contraction. All objects dropped into a black hole are obliterated (Fig. 2.3) leaving behind only mass, charge and angular momentum. Consequently, the initial conditions of an emerging universe could be those established inside the curved "horizon" of a universal black hole. The subject matter has been extensively reviewed (e.g., Sullivan, 1979; Wheeler, 1980; Olive et al., 1981; Hawking, 1975, 1976, 1977, 1980; Thorne, 1974; Trefil, 1983b).

A universe characterized by infinite expansion is termed *open* and one of finite extent is referred to as

Fig. 2.3. Objects dropped into a black hole become obliterated. Only mass, charge, and angular momentum characterize the newly developed system

closed. Which choice we make for the state of our universe depends on the average density of matter and the value of cosmic deceleration we assign in model calculations. If the density falls below a critical value, gravitational effects are too small to halt cosmic expansion. In contrast, if density exceeds a critical mass, gravitation takes over and terminal collapse becomes inevitable. However, if the ratio of the actual to the critical density value equals one, which implies that the universe expands with exactly the escape velocity, the geometry of space has a zero curvature, and a *flat* or Einstein-de Sitter universe is indicated (Fig. 2.4; Box 2.1).

What kind of universe is ours? A comment of Dirac during a centennial symposium to celebrate the achievements of Albert Einstein may give a general feeling on this topic: "I want to emphasize that there is really no symmetry between the big bang and the big crunch. The big bang is very well established – I think pretty well everybody believes it. The big crunch is completely unestablished and very doubtful. I do not believe in it myself. I do not think Einstein believed in it either, because he gave a model, jointly with de Sitter in 1932, and this model is satisfactory so far as agreement with observation goes. This model definitely requires a big bang – it does not require a big crunch. It is an open universe although only just opened."

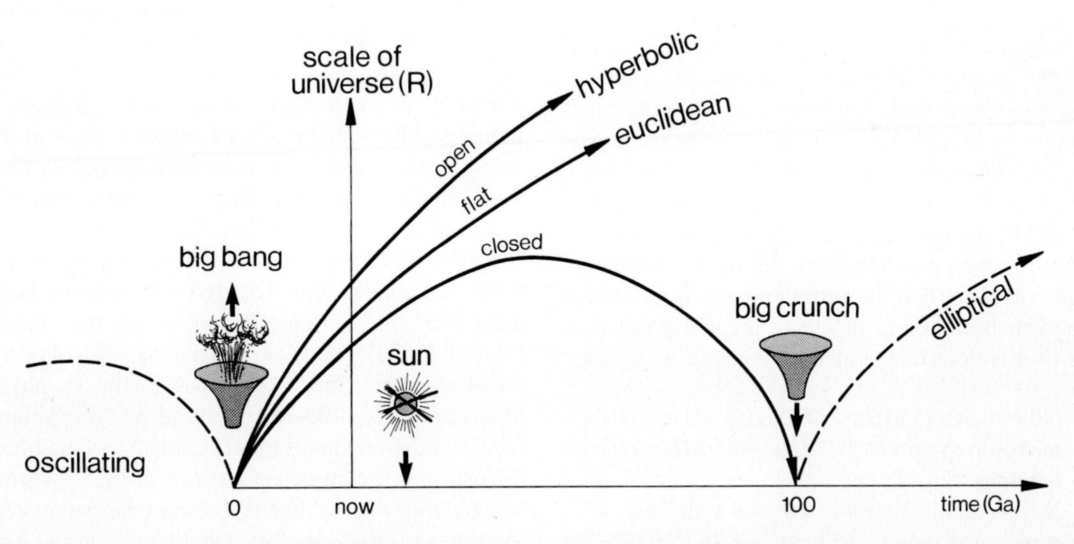

Fig. 2.4. Scale of the universe versus time: different scenarios. According to general relativity, light and matter flow unaccelerated along geodesics in space-time. The abundance of matter per unit space determines whether the universe is open and infinite or closed and finite. For an open universe, the geometry of space-time is hyperbolic; in the case of a closed universe, the curvature is elliptical. Flat space-time (Euclidean geometry) defines the curvature of the Einstein-de Sitter universe, which is said to be "just" barely open. A closed universe is expected to contract and temporarily end in a "big crunch", which could be followed by a "big bounce" leading to an expanding universe again. The expansion-contraction cycle could be repeated at intervals of 80 to 100 Ga. Alternately, the oscillating universe might gain momentum with each succeeding passage. The present value of R is fixed at 15 Ga after big bang. The Sun is expected to leave the main sequence in about 5 Ga and develop into a red giant (Dicus et al., 1983; Trefil, 1983b)

BOX 2.1

Cosmological Models

The present expansion of the universe is commonly explained by the standard model. It permits the establishment of a hierarchy of structures in which the properties of each member can be related to the structural characteristics of the underlying more fundamental submembers and so forth. At the top of this hierarchy is the universe in its entirety, and at the base stands supergravity, which is rooted in a gauge principle unifying bosons and fermions (Freedman, 1981; Ne'eman, 1981).

Much is being presently debated on the future of the universe (e.g., Trefil, 1983b; Dicus et al., 1983; Frautschi, 1982). Everything centers around the question of whether the universe will expand forever, or whether it will cease to expand at some point and start to contract. Which of the two models should be accepted, the open or the closed universe, depends on which value one assigns to the density of matter for a given volume of space. Above a critical density, amounting to roughly a mass of three protons per 1000 liters of space, the universe is said to be closed. At lower densities, it will stay open. The term flat universe (Einstein-de Sitter) describes a situation on or near the boundary of the two extremes.

In summing up all luminous matter in the cosmos, the value for the density of matter amounts to only 5 percent of the critical density, which would strongly argue for an open universe. But since there is great uncertainty regarding the amount of non-luminous matter per given volume of space, except that the value could be quite substantial, all three scenarios, open – flat – closed, must be taken seriously. All that can be said for sure at present is that the density cannot lie more than ten times above the critical density, otherwise the universe would already have contracted before reaching its present size (Dicus et al., 1983). The relationships are schematically shown in Fig. 2.4.

Cosmic Background Radiation

To a co-moving observer, that is one who "rides along" with the general expansion of the fireball, the universe would appear to be the same no matter in which direction he looked. This is in essence due to the cosmological principle which states that the universe is isotropic and homogeneous both in its rate of expansion and in the large-scale distribution of galaxies.

This observer, equipped with the right instruments, will also discover that the entire universe is flooded by a uniform background of electromagnetic radiation from four regions of the electromagnetic spectrum:

(i) radio region (1 MHz – 500 MHz; 300 m – 60 cm)
(ii) microwave region (500 MHz – 500 GHz; 60 cm – 0.6 mm)
(iii) X-ray region (2.5×10^{17} Hz; 1.3×10^{-9} m)
(iv) gamma-ray region (10^{21} Hz; 3×10^{-13} m).

Each region signals specific information on events which date back almost to the big bang (Webster, 1974). Maps of the radio sky reveal that the background radiation in this spectral region does not vary appreciably in strength between one direction and the other. The background is simply a sum-total of all universal radio emissions. In Fig. 2.5 distinct sets of peaks can be recognized from a localized segment in the sky. They represent radio galaxies or quasars (contraction of the term *quasistellar* object) towering above a background noise. Quasars are thought to be active nuclei of galaxies. Most of them exhibit large red-shifts and accordingly quasar events were more prominent in the past – about 10 to 15 billion years ago (Osmer, 1982a, b).

The discovery of the microwave background by Arno A. Penzias and Robert W. Wilson has been a milestone in the understanding of the universe (Henry, 1980; Wilson, 1979; Sciama, 1980). For a detailed report on the "accidental" discovery, physical nature, and cosmological implications, see Weinberg (1977) and Bernstein (1984). Critical for the present discussion is the observed isotropy of the background which varies by less than 0.1% when measured across the sky, and that it is a black-body radiation at a temperature close to 3 K.

The microwave background came into existence about 500,000 years after the big bang, at a time when the universe underwent a phase change from a radiation- to a matter-dominated system. Electrons were captured by the ionized gas – recombination effect –, emitting visible radiation that was subsequently

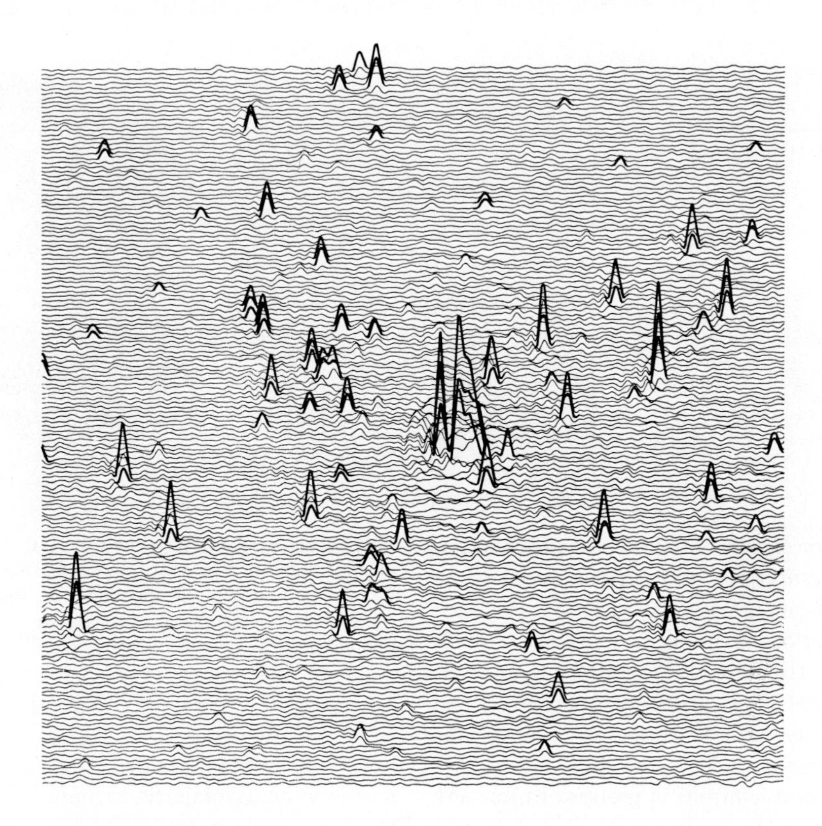

Doppler-shifted by the expansion to the microwave region of the spectrum, which smoothly cooled off from the initial 3000 K flash to the final 3 K background. In general, a shift of photon wavelength to the red occurred in the course of expansion, lowering the energy density of the universe.

Work on background radiation in the X-ray and gamma-ray region is currently of major interest to astrophysicists. Since 1962, it has been known that every part of the sky emits a uniform glow of X-rays. However, more than 20 years of research have not yet revealed the true physical mechanisms underlying the X-ray background (Pines, 1980; Margon, 1983; Margon et al., 1982). One tiny source of X-rays is the solar corona which, because of the nearness of the sun, can be readily detected. There is some tentative indication linking a large fraction, and possibly all, of the X-ray background to distant individually invisible quasars (Margon, 1983). The data bank for the gamma-ray background is even more sketchy.

To sum up, uniform radiation backgrounds are received from different regions of the electromagnetic spectrum. They underscore the cosmological principle and indicate that the universe has been isotropic at least from about 500,000 years following the big bang event to the present. The data support the standard model but leave no room for a steady-state cosmology.

Fig. 2.5. Radio galaxies (quasars) towering above background noise (map of radio sources made by the three dishes of the One Mile Telescope at the University of Cambridge; in Webster, 1974)

Cosmic Inflation

The formation of atoms and the resulting generation of the microwave background radiation roughly half a million years after the big bang is a time and phase boundary separating atomic matter from hot radiation. Judged by the uniformity of the 3 K background radiation, the universe has expanded since then in line with the cosmological principle. Under the not unlikely assumption that the universe was initially created from primordial chaos, the structural order we see manifested down to the "atomic" boundary arose within less than half a million years which is about equivalent to the age of the species *Homo sapiens*. The standard model offers no physical clues for such self-organization. There is also the problem of the cosmic horizon, where we see distant objects on opposite sides of the sky which could never have seen each other but still exhibit the same physical properties. A third puzzle concerns the present flatness of the universe. Such a geometric situation requires a

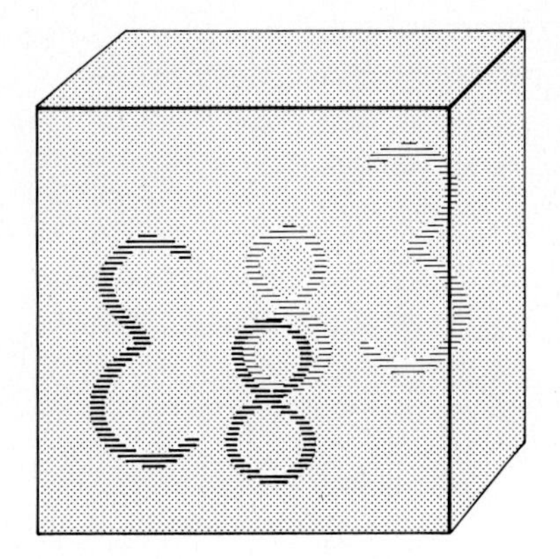

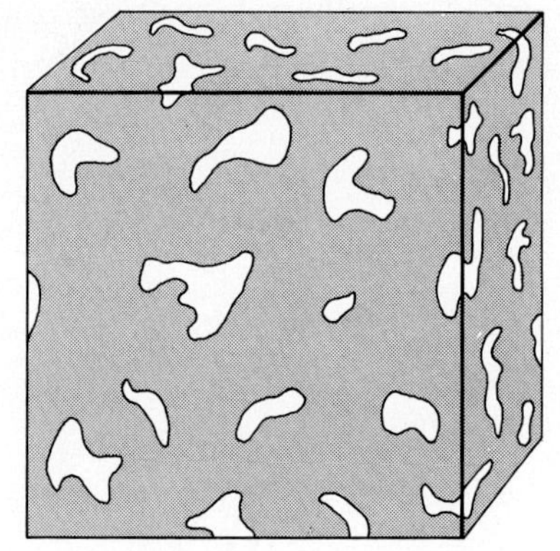

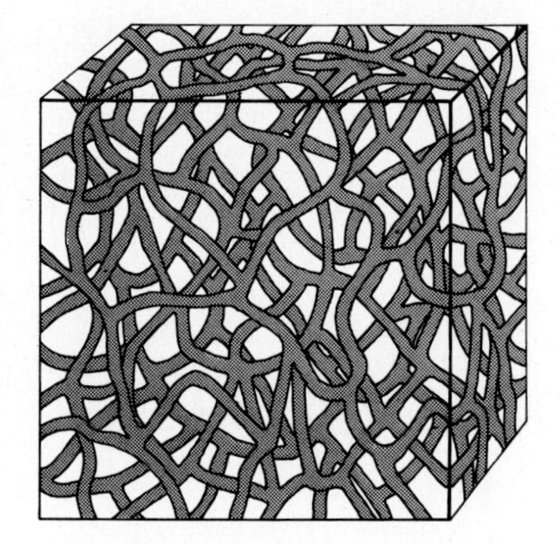

Fig. 2.6. Congealing Higgs field. Perturbations in the density of matter and energy shortly after big bang showing regions that are compressed and those that are increasingly rarefied. Large-scale fluctuations (mass 10^{16} that of our Sun) evolved into galaxy superclusters (Silk et al., 1983)

critical value in the density of matter. To arrive at this flat universe presupposes unbelievable precision in fixing the amount of matter at the onset of expansion.

Big bang scenarios also imply the existence of many exotic particles (Barrow, 1977; Barrow and Silk, 1980). Some of these are unstable and are no longer with us, whereas others, such as quarks and monopoles, survive. Quarks, as we have shown, have entered protons and neutrons in groups of three. Attempts are underway to isolate a single quark, and some tentative signals suggesting that it came from a quark in a free state were recently obtained. The same can be said for a monopole, which is an extremely heavy particle having a mass 10^{16} times that of the proton. One event has been observed (Cabrera, 1982) but the scientific community would be more at ease if a second monopole event were discovered. Monopoles ought to occur in large numbers, if the standard model is correct (Carrigan and Trower, 1982). Actually, this is the crux of the problem, because the sheer mass of predicted monopoles would have thrown the universe into the big crunch long ago.

A new theory termed the inflationary universe may resolve some of the inconsistencies inherent in the standard model. Originally conceived by Alan H. Guth in 1981, and since then many times revised, refuted, abandoned and again resurrected (see, e.g., Barrow and Turner, 1982; Lake, 1984; Linde, 1982; Waldrop, 1983a; Guth, 1984; Guth and Steinhardt, 1984), it is presently the dernier cri or the whisper of creation among cosmologists and particle physicists. According to grand unified theories (GUT's), matter had been metamorphosed many times over in the early universe, and its symmetry has thus changed. At the fusion or unification point of the strong, the weak and the electromagnetic forces – 10^{-35} seconds after the big bang – lies the phase boundary between what scientists call the *symmetric* and the *asymmetric vacuum*. Guth (1981) explained the transition from a symmetric to an asymmetric vacuum in terms of a phase transition comparable to the one observed during the freezing of water. Liquid water can briefly supercool below 0° C and will release upon freezing the latent heat – that is the energy difference between the liquid and the solid phase – spontaneously. In analogy, Guth suggested that by moving from a symmetric to an asymmetric vacuum, the system supercools, releasing its latent heat – that is the energy difference between the two vacuum states – explosively. During this phase transition, the universe underwent an exponential expansion with time, whereas expansion according to the standard model just follows the square root of time. The growth of the universe at inflated rates will lead at the end of the transition period to a sphere about 10 cm across which in essence contains everything the universe holds today.

In GUT's a quantum field known as Higgs field (see p. 9) should exist (Fig. 2.6). Such a field will slowly congeal upon freezing. As inflation continues, the Higgs field develops into a hot dense plasma, possibly generating all matter and energy during this transition, and at the same time erasing the former record. When the Higgs field finally solidifies, inflation comes to a halt and a flat, monopole-devoid universe containing a 10^{27} K plasma starts to expand along the route proposed by the standard model.

Cosmic inflation solves elegantly some inconsistencies of the standard model associated with: (i) the cosmological principle, (ii) the horizon problem, (iii) the flatness of the universe, and (iv) the scarcity of monopole events. There are some inflationary consequences with respect to size and age of the new universe. The 3 K microwave background radiation brings us back – just about half a million years short – to the big bang event which, based on the inflation model, marks the time when the latent heat of the supercool cosmos was explosively released. Everything that might have happened on the way from a vacuum to the big bang in terms of time, space and events is a wide-open field for speculation. Elucidation of events occurring during the Planck epoch, that is 10^{-43} seconds after "Time Zero", may provide the long-sought synthesis of relativity and quantum field theories.

Chapter 3

Incidents During Expansion

Que Sera Sera

Background radiation permeates the cosmos uniformly. In contrast, matter distribution on local and regional scales is patchy and irregular. Only if the universe is looked at across a larger frame will a more systematic pattern of aggregated matter and empty voids emerge (e.g., Zeldovich et al., 1982; Silk et al., 1983; Oort, 1983). What is the essential mechanism behind the heterogeneous matter dissipation in the first place? The answer is simple: the symmetry breaking incident at the dawn of time releasing gravitation as a separate force. Since then, a concept known already to Newton, gravitational instability, has been the principal driving element controlling the distribution of matter in the universe.

In accordance to a theory formulated by Sir James Jeans in the twenties, matter cannot be distributed at uniform density over an indefinitely large volume, due to universal gravitation. There is a critical mass density known as Jeans mass which will not allow matter to concentrate beyond a certain concentration per volume. As a result, matter will break up into separate entities and will lead to agglomerations of various forms and sizes that – as seen today – are spatially structured into groups, clusters and superclusters of aggregated matter.

Where is the point of departure? Actually it should be the moment when gravitation is no longer chained to the previously united strong, weak and electromagnetic forces. Small perturbations, looking like tiny wrinkles, have been proposed as seed structures. Free density fluctuations, perhaps arising from quantum and thermal fluctuations in a congealing Higgs field, might be propagated. Their perturbations can be understood as wave-like fluctuations of the density around an average value, randomly spread over all wavelengths. Thus, primeval density perturbations are nothing but fluctuations in space curvature, in which radiation and matter were jointly perturbed (adiabatic scenario), or in which only matter density was perturbed, whereas radiation remained nearly homogeneous (isothermal scenario). Although adiabatic scenarios are more en vogue, the isothermal scenario should not be lightly dismissed. A severe handicap in defining a universally acceptable model has to do with the uncertainty of the finite residual mass and the numbers of neutrinos, or the presence or non-presence of light superparticles such as the gravitino (Silk, 1982a).

The prevailing view is that following annihilation of antimatter, density perturbations in the surviving matter were propelled by adiabatic processes. All fluctuations in the mass of less than 10^{16} times that of our sun were filtered out prior to decoupling, due to photon viscosity (Silk et al., 1983). After the decoupling event, that is about half a million years after "Time Zero", matter could freely move around, and gravitational instability enhanced the growth of density. Regions of critical density expanded to maximum radius and then collapsed into what Zeldovich et al. (1982) termed "pancakes", which are compressed sheets of matter. After the initial collapse, gas will still stream into the pancakes, though at a reduced rate. This will cause them gradually to spread lengthwise in their planes, so that they eventually intersect, producing long filaments of matter. From regions that were underdense, or had a higher than average expansion, holes of virtually nothing would be generated. It is assumed that the initial fragments are clouds of masses comparable with those presently encountered in dwarf galaxies. They rapidly coagulated and evolved into galaxies, clusters and superclusters.

The inner, most dense regions of the pancakes collapse first, cool fastest, become Jeans unstable and may fragment into galaxies (Oort, 1983). It has been suggested that galaxies formed at that time have little rotation and gave principal rise to elliptical shapes. Shock wave, triggered by the gas falling into the pancakes at later stages, would tend to set up turbulence producing much angular momentum, resulting in galaxies generated in the outer regions becoming spirals (e.g., Zeldovich et al., 1982). Dekel (1982) has presented a modified picture on galaxy development. He assumes that galaxies form first from isothermal perturbations in pancake structures, with subsequent non-dissipational clustering on smaller scales. In a way it is a combination of isothermal and adiabatic scenarios, and has the advantage that it allows for an increase of clustering in relatively recent epochs.

There is a discrepancy between the existence of clusters and superclusters and the lack of corresponding fluctuations in the 3 K microwave background radiation. One way out would be to assign a mass to the neutrinos in the order of 10 eV, for example. This would allow perturbations to evolve at an earlier stage, at which the baryons decouple from radiation. The hypothesis of the heavy neutrinos as the principal source for the dark non-baryonic matter in the universe is at present the one most favored.

The cosmic pancake scenario (see Box 3.1) is consistent with the large-scale structure of the universe, which is composed of unrelaxed superclusters aligned along strings in which galaxies and clusters of galaxies are embedded (e.g., Tassie, 1986). The geometric shape of galaxies appears to be related to the position within the pancake and the time of their development. Giant voids exist between the superclusters. The pattern of huge voids and superclusters changes very slowly in the course of expansion, and implies that it has been inherited almost unaltered from the primeval structural state.

From a tangle of strings emerging from a congealing Higgs field, the strings untangle and generate sizable density fluctuations as a consequence of gravitational effects. The strings can be considered topological knots of high energy density. Just as the crystalline topology of templates determines growth pattern of minerals by providing nucleation sites for the oriented growth of minerals, the strings can be looked upon as well-defined sheets of matter from which the universe nucleates. Cellular structures come into existence in which superclusters constitute membraneous "cell walls" surrounding huge voids.

BOX 3.1

Cosmic Pancakes
(Silk et al., 1983)

Large-scale fluctuations in the density of matter and energy were able to develop once matter became decoupled from radiation a few tens of thousands of years after the big bang. From primordial fluctuations that entailed a wide range of possible scales, only those having a mass at least 10^{15} to 10^{16} times that of our Sun survived. Gravitational collapse of these structures led to what Joseph Silk, Alexander S. Szalay and Yakov B. Zeldovich have termed "flattened pancakes". Where pancakes intersected one another, long and thin filaments of matter took shape (Silk et al., 1983). Figure 3.1 illustrates the evolution of pancakes and filaments using a computer simulation model. The final outcome of this modeling is a structure that closely matches the ordering of superclusters and voids.

The pancake scenario is valid only if neutrinos have a non-zero mass (see p. 23), or if the missing 90% of cosmic matter can be provided by other, so far unknown, particles (e.g., a massive gravitino). Accord-

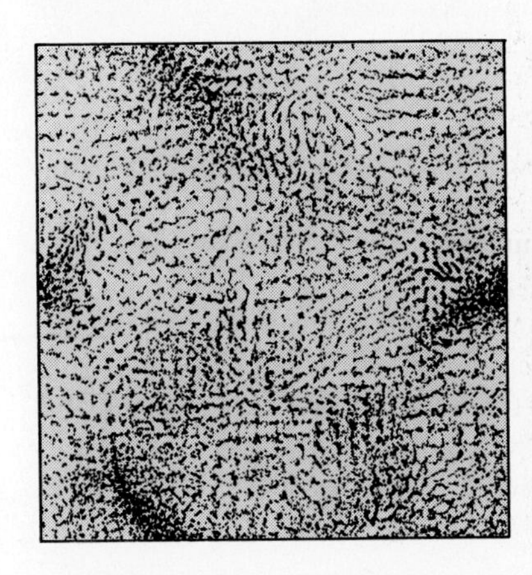
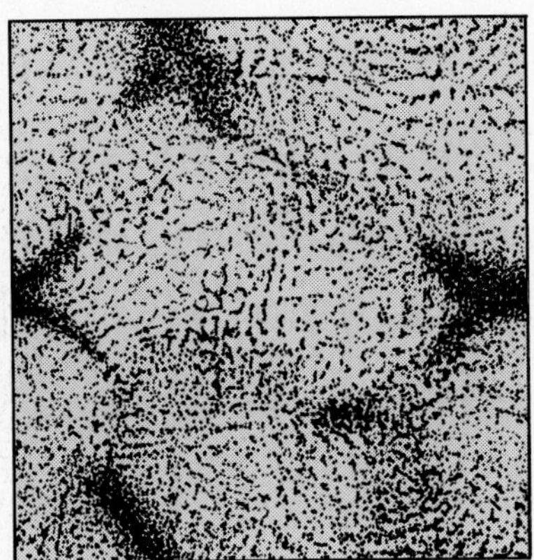
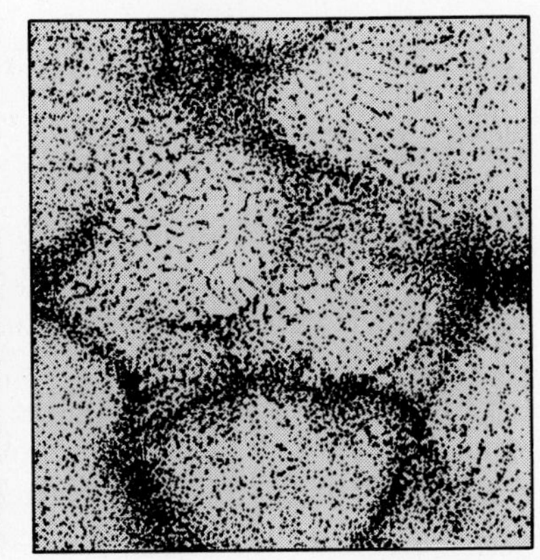

Fig. 3.1. Computer simulation showing the development of pancakes and filaments in time and space (George Efstathiou, University of Cambridge; in Silk et al., 1983)

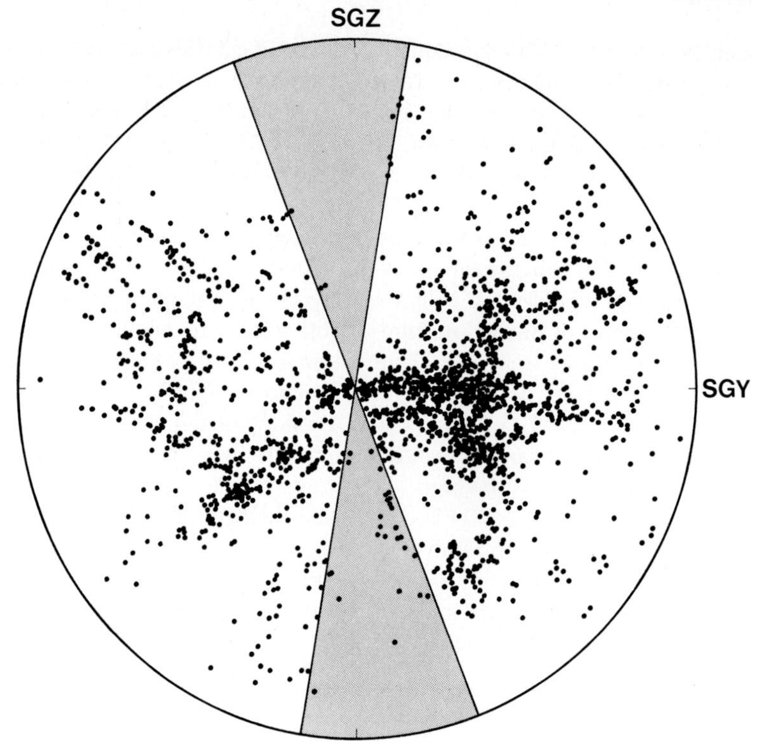

Fig. 3.2. Inferred positions of about 2,200 galaxies projected on a plane roughly coincident with the plane of the Milky Way system. The wedge-shaped regions are the exclusion zones determined by the Milky Way. The radius of the circle, inversely proportional to Hubble's constant, is 30 Mpc if the Hubble constant is 100 km s^{-1} Mpc^{-1} (Tully, 1982; Silk, 1982b)

ing to the new pancake theory, non-zero mass neutrinos may account for the dark matter in the universe, both in the region of intergalactic space and in the galactic halo.

Support for the validity of the pancake theory can be found in the work of Tully (1982), who mapped the positions of close to 2,200 galaxies from the Local Supercluster of which our Milky Way is part (Fig. 3.2). The Local Supercluster can be described as a flat disk about 2 Mpc thick and 12 Mpc in diameter (assuming a Hubble constant of 100 km s^{-1} Mpc^{-1}), which contains 60% of the luminous galaxies. The rest of it is dispersed in an inhomogeneous halo, yet concentrated in a small number of diffuse clouds. The disks' thinness and the nonrandom distribution of galaxies and voids are rather striking. Furthermore, the strings of galaxies pointing towards the Virgo cluster are suggestive of elongated filaments (see also Hart and Davies, 1982; Hogan and Rees, 1984; Reid, 1983; Strom and Strom, 1982; Gregory and Thompson, 1982).

Nature of Dark Matter

Normal matter, also referred to as baryonic matter, cannot account for the large-scale structure manifested in clusters and superclusters which are relatively stable configurations in the space-time continuum. This is because some of the individual members in clusters of galaxies are moving so fast that their gravitational attraction inferred from baryonic matter would not suffice to hold them together. There has to be some additional force to keep them going and maintain their structure.

To close the universe, critical density of matter should amount to about 5×10^{-30}g cm^{-3}, which is equal to 3 hydrogen atoms per cubic meter. However, luminous matter in the galactic space adds up to only 7.5×10^{-32}g cm^{-3} (e.g., Rubin, 1980), an amount far too small to explain gravitational dynamics, and to yield an Einstein-de Sitter universe. Thus, light is not a reliable yardstick to determine the cosmic density of mass per unit volume. There are objects which are not visible, such as dim stars of low mass, large planets or black holes. One has to accept the fact that matter in the universe can be of luminous and non-luminous types, whereby the last category is simply identified by its gravitational impact. Furthermore, even when all baryonic matter is put together, merely about one tenth of the matter present in the universe can be accounted for. The ratio of one to ten is only a mean value and varies from place to place. In general, the bulk of the matter in small galaxies appears to be baryonic. However, as the size of the galaxies increases and one finally moves to clusters and superclusters, the amount of dark and still unidentified matter becomes more and more substantial. A great deal of speculation is going on concerning its nature.

According to one school of thought, the "missing" or non-visible mass can be accounted for by non-zero mass neutrinos (see pp. 21, 23). Assuming that one assigns a mass of 10 eV to a neutrino, then shortly after "Time Zero" neutrino clouds would have decomposed into fragments of about 10^{15} solar masses, which corresponds to about the mass of the giant superclusters of galaxies. At the moment these particles become non-relativistic, perturbations can start to evolve. It should be remembered that this is an epoch during which baryons are still immersed in a "sea" of hot radiation. Adiabatic perturbations can therefore collapse quite early without inducing large fluctuations in baryon density at decoupling. Dense structures can be formed from smooth initial perturbations. Just as in the baryon-dominated case, pancakes, filaments and knots will be formed. Following the decoupling of normal matter from photons, the baryons could fall into the gravitational field formed by the relic neutrinos, and galaxies would evolve. In the presence of massive neutrinos the density fluctuations would develop much faster than they would have otherwise.

The non-zero mass neutrino scenario could: (i) account for the dark matter, (ii) explain uniformity of 3 K background radiation, (iii) lend support to the Zeldovich pancake model, and (iv) explain the fast generation of galaxies following decoupling of atomic matter from radiation. Superparticles such as the gravitino and the photino are also potential candidates for the "missing mass". For the moment, however, the massive neutrino concept is the more accepted model. For a discussion on the role of neutrinos see Silk et al. (1983), Zeldovich et al. (1982), Lubimov et al. (1980), Reines et al. (1980), Schatzmann and Maeder (1981), Schramm (1982), Hut and White (1984) and Däppen et al. (1986).

Origin of Elements

The nucleus of an element consists of two fundamental building blocks: the proton and the neutron. It is the proportions in which the two nucleons are joined in nucleonic packs termed nuclei which give rise to the various chemical elements and their isotopes. In stable nuclides with low atomic number, the neutron-to-proton ratio (N/Z) is equal or close to unity, systematically increasing to 1.5 in the direction of the heaviest stable nucleus (Fig. 3.3). Neutrons counteract electrostatic Coulomb repulsion with increase in Z. This addition is biased towards even-numbered stable nuclides yielding even-even, even-odd, odd-even and odd-odd Z/N combinations: 160, 56, 50 and 5 nuclides, respectively.

Radiogenic nuclides are those that decay by the emission of alpha, beta and gamma radiation or by electron capture into so-called daughter nuclides at half-lives that range from fractions of seconds to billions of years. To illustrate the sequence of events, the decay series of uranium-238 (^{238}U) is depicted in Table 3.1.

The periodic table contains 92 elements, of which only 90 are known from Earth, and one more, i.e., technetium, from stars (Ahrens, 1979). The missing one, i.e., promethium, named after Prometheus, who according to Greek mythology stole the element "fire" from heaven and carried it to Earth, has been synthesized. By bombarding an element with neutrons or fast-moving particles such as protons, deuterons and alpha particles, elements with atomic numbers above 92 and accordingly named transuranium elements, have been synthesized. They are members of the actinide series, from neptunium (atomic

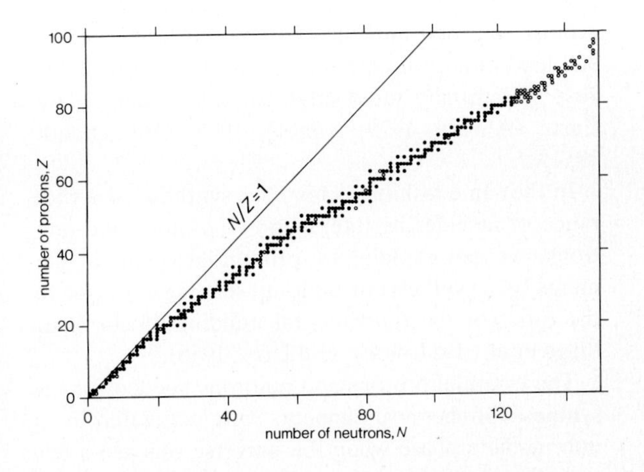

Fig. 3.3. Plot of number of protons (Z) and neutrons (N) in stable (*closed dot*) and unstable (*open circle*) nuclides

Table 3.1. Decay series of uranium-238

Element	*Symbol*	*Emission*	*Original name*	*Half-life*
Uranium	$_{92}U^{238}$	Alpha	Uranium I	4.5×10^9 years
Thorium	$_{90}Th^{234}$	Beta	Uranium X_1	24 days
Protactinium	$_{91}Pa^{234}$	Beta	Uranium X_2	1.2 min
Uranium	$_{92}U^{234}$	Alpha	Uranium II	248,000 years
Thorium	$_{90}Th^{230}$	Alpha	Ionium	80,000 years
Radium	$_{88}Ra^{226}$	Alpha	Radium	1,622 years
Radon	$_{86}Rn^{222}$	Alpha	Radon	3.8 days
Polonium	$_{84}Po^{218}$	Alpha	Radium A	3.0 min
Lead	$_{82}Pb^{214}$	Beta	Radium B	26.8 min
Bismuth	$_{83}Bi^{214}$	Beta	Radium C	19.7 min
Polonium	$_{84}Po^{214}$	Alpha	Radium C'	1.6×10^{-4} sec
Lead	$_{82}Pb^{210}$	Beta	Radium D	21 years
Bismuth	$_{83}Bi^{210}$	Beta	Radium E	5.0 days
Polonium	$_{84}Po^{210}$	Alpha	Polonium	138.4 days
Lead	$_{82}Pb^{206}$	–	Lead	Stable

number 93) through lawrencium (atomic number 103), and the transactinide elements with atomic numbers above 103. Element 106 was discovered in 1974 and work is continuing to obtain elements with larger atomic numbers. In 1981 and 1982, a team of physicists at Darmstadt headed by Peter Armbruster succeeded in synthesizing elements No. 107 and No. 109. Recently, the same group produced a few atoms of element No. 108 through fusion of ^{58}Fe and ^{208}Pb, which upon release of a neutron yielded element No. 108 with the atomic weight of 265, thereby increasing the number of artificial elements to 17. The experiment has again been verified in 1987.

In theory, for a range of elements around atomic numbers 110, 115, or 120, increased stability is predicted as a consequence of the closure of nucleon shells. In addition to new elements, artificial isotopes from virtually all common elements have been synthesized. The total number of stable nuclides is about 300, and radiogenic species amount to roughly 1,200. Just 55 naturally radioactive isotopes occur on the Earth (Ahrens, 1979; Trimble, 1975; Goldschmidt, 1954).

In the same fashion as man has synthesized a wide range of nuclides by interaction of protons and neutrons, we can explain the primordial origin of elements by a synthesis or build-up starting with one or the other of the fundamental building blocks (Burbidge et al., 1957; Suess and Urey, 1956).

The essential protons and neutrons needed for the synthesis of chemical elements were generated in an intermediate phase when the universe was just a few seconds old. At that time, temperatures around 10^{10} K allowed neutrinos to decouple and neutrons and protons to react. Among the stable nuclei generated in the course of less than 3 minutes were deuterium, helium-3, helium-4 and lithium-7, whereas the production of heavier elements could not proceed because of the absence of stable nuclei at mass numbers 5 and 8. As a result, hydrogen and helium-4 nuclei became the two most prominent members of ordinary matter in the universe in a mass ratio of about 3 to 1.

Helium generation competes primarily with the decay of free neutrons, and the faster nucleosynthesis proceeds, the more helium is obtained: the half-life of a neutron is about 10.5 minutes. It is thus crucial for any cosmological scenario to know precisely the primordial value of helium abundance. For instance, should a critical baryonic density be exceeded, the universe will close. With three neutrino species and the commonly assigned primordial helium value, the universe is at one-tenth the critical value of closure. The weight of neutrinos, the abundance of primordial helium, and the possible existence of as yet undiscovered neutrino species determine the direction into which the universe will eventually go.

Stars also contribute to the helium value and corrections must be made for this source. By examining extragalactic regions with low metallicity, and thus expected to contain less material from stellar nucleosynthesis, the mass fraction for helium in universal matter proved to be 25 percent.

The issue is further complicated by the big bang production of deuterium and lithium, whose primordial values are difficult to ascertain because of their partial stellar destruction or synthesis. Judged from the deuterium abundance in interstellar molecular clouds, space probe studies of Jupiter and Saturn, and interstellar and meteoritic helium-3 contents (^3He is generated by astration of deuterium), the loss in deuterium over the past 5 billion years must have been substantial. With respect to lithium, the cosmic mass fraction of lithium-7 — based on the abundance in meteorites — is 5×10^{-9}. Lithium content in metal-deficient old stars, having carbon and heavy metal values lower by a factor of 100 relative to the Sun, is just one-tenth the present-day lithium abundance. This value provides an upper limit to the cosmological baryon density and conforms with big bang scenarios (Spite and Spite, 1982, 1985).

In a recent study (Carr et al., 1984), doubt has been cast on the validity of the familiar big bang scenario and the principles of pregalactic element formation ingrained in it. Instead, the authors place a generation of massive stars about a hundred times the mass of our Sun at the beginning of time. Because of their gigantic weight, such stars were short-lived and allowed the production of the well-known pregalactic element spectrum. Upon termination of their star life, their remnants are assumed to have developed into black holes which, in turn, are considered a possible source for the missing mass in our Universe. Although these alternatives cannot be ruled out per se, many of the observed cosmological phenomena are more in support of the revised big bang concept of the inflationary Universe.

We have said so far that hydrogen and helium (plus trace quantities of lithium-7; Pagel, 1982) were synthesized in a fierce plasma during the first few seconds following the big bang in a mass ratio of 3 to 1. The remainder of the elements, 90 of them, have their seat of origin in certain stars and the energy levels, provided by them during their birth, life and death. The source of energy in stars has been a puzzling question for a long time. One also wondered how a star like the Sun could radiate for billions of years a rather uniform amount of energy (4×10^{26} Joule per second) without exhausting its supply. To illustrate the situation let us assume that the Sun is entirely composed of petroleum. At its present luminosity, all the potential chemical energy would be gone in less than 7000 years. Also the potential gravitational energy would last for at most 30 million years (e.g., Giese, 1981). The solution of the problem came with progress achieved in nuclear physics, i.e., the realization that a kind of "cosmic reactor" is contained in stars. Such reactors generate not only energy, but at the same time the full spectrum of chemical elements in a logical and stepwise fashion. A fascinating documentation on the origin and evolution of chemical elements is given in Burbidge et al. (1957). It seems probable that the elements all evolved from hydrogen, since

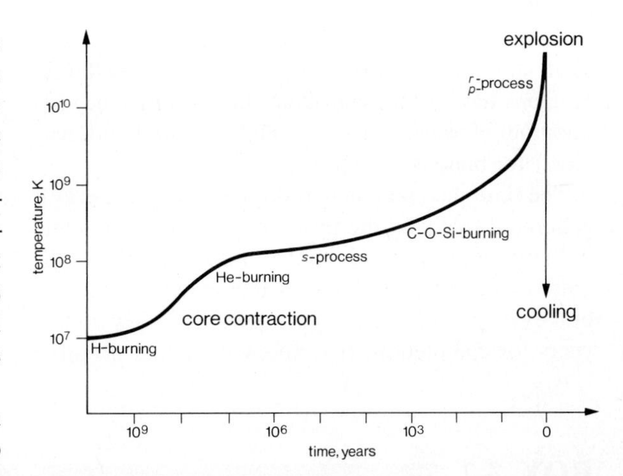

Fig. 3.4. Time scales of the various processes of element synthesis in stars. The curve gives the central temperature as a function of time for a star of about one solar mass (highly schematic) (Burbidge et al., 1957)

the proton is stable (or almost stable, see p. 10), while the neutron is not. Moreover, hydrogen is the most abundant element.

The succession of element formation can best be followed by moving up the temperature scale from a "modest" few million degrees to several billion degrees (Fig. 3.4). Reactions proceeding during stellar nucleosynthesis are depicted in Box 3.2 and Table 3.1 and the critical events in the progression of nucleosynthesis read as follows:

It all started with a process that already occurred a few seconds after "Time Zero": the fusion of hydrogen nuclei (protons) to make deuterium, helium-3 and helium-4 (*p-p* chain). During the first generation of stars, this reaction chain was fast, and yielded large amounts of energy and photons. The fact that deuterium initiated the first nuclear reactions has to do with the ease with which it "burns", a property which in the coming century might be used in fusion reactors, hopefully of an environmentally safe design.

During the first phase of star formation, almost all of the deuterium is consumed and the star stays without suitable nuclear fuel for a short period of time. Since forces of thermal expansion and forces of gravitation are counteracting, a star will shrink once the fuel valve is turned off. Upon compression of matter the less ignitable fuel – hydrogen – will take over at temperatures of about 20 million degrees. Several reaction steps promote the synthesis of helium-4. This is one of the most energy-efficient nuclear processes in which during the transformation from 1 kg hydrogen to helium about 650×10^6 megawatt seconds are liberated. By way of comparison, it would require a ton of uranium (yellow cake), or 20,000 tons of coal to release so much energy. However, in a medium-sized star, the power output per kg is just 1 milliwatt. Humans have a thousand times higher efficiency per kilogram of weight, but stars, on the other hand, may "live" for billions of years.

The time stars remain in this phase, where energy is generated solely by the fusion of hydrogen and emitted by photons, depends on their mass. The higher the mass of the star, the faster is the speed of reaction, but even medium-sized stars require a few billion years for completion. It is noteworthy that no other element except helium is produced during this phase. As soon as about 10 percent of the hydrogen sources are consumed, the fusion product helium dilutes the nuclear fuel and blocks the nuclear processes. We witness again the take-over of the gravitational forces which up to that time had established an equilibrium with the expanding energy forces, and a gravitational collapse is the result. With increasing density and temperature, the helium, i.e., the ash of the former nuclear fusion processes, starts to ignite and the star expands. This is the phase of the *Red Giants*.

BOX 3.2

Origin of Elements and Nuclides (Burbidge et al., 1957; Wagoner, 1968; Penzias, 1979; Fowler, 1984)

Stars spend most of their lifetime in *hydrogen-burning phases*. Objects about the size of our sun or smaller follow the proton-proton chain (*p-p* chain) which through fusion of four protons (^1H nuclei) yield: one helium nucleus (^4He), two neutrinos (ν) and 26.73 MeV of energy. This is a rather brief summation, inasmuch as a number of intermediates are or can be involved (^2D, ^3He, ^7Be, ^7Li, ^8Be, ^8B, e^-, e^+, γ and ν). The nucleosynthesis of lithium-7 is a consequence of this reaction. More massive stars observe the carbon-nitrogen-oxygen (CNO) cycles, which involve capture of ^1H nuclei by C, N, and O nuclei. In principle, four protons are consumed and one ^4He and two neutrinos are produced. As in the *p-p* chain, a wide variety of intermediates is generated here too. The high yield in the number of carbon, nitrogen, and fluorine nuclei in certain stars go on account of these cycles.

Helium-burning fuses three ^4He nuclei (3 α-particles) into ^{12}C and by the capture of one α-particle ^{16}O is generated: $^{12}C + {}^4He = {}^{16}O + \gamma$. A progression in helium burning along this line will lead to nuclei up to ^{24}Mg.

The *carbon-, oxygen-, and silicon-burning phases* which proceed at temperatures of about 5 x 10^8 K, 10^9 K and 3 x 10^9 K can lead to nuclei of nuclear mass number, $A = 28$, 40, and 56, respectively (Fig. 3.5). Such elevated temperatures are needed to overcome the high mutual Coulomb-potential barrier and to fuse the various starting nuclei.

The *r*- and *s*-processes (*r* stands for rapid and *s* for slow), involve neutron-capture reactions and lead to the synthesis of all elements beyond ^{56}Fe. This also includes the transuranium elements. Details on the routes of synthesis can be found, for instance, in Allen et al. (1971), and Seeger et al. (1965). Should the neutron flux per seed nucleus be large, we call the reaction an *r*-process; in the opposite case, when neutron number is limited, one speaks of an *s*-process. At proton number $Z = 94$, *r*-process synthesis comes to a halt. The resulting fission yields nuclei of intermediate mass, which may act as seed nuclei and via *r*-processes at neutron numbers, $N = 50$, 82, and 126, lead to peaks at $A = 80$, 130, and 195 (shown as *r* in Fig. 3.5).

The *p*-process proceeding at late stages in star nucleosynthesis (e.g., in red giant stars) generate proton-

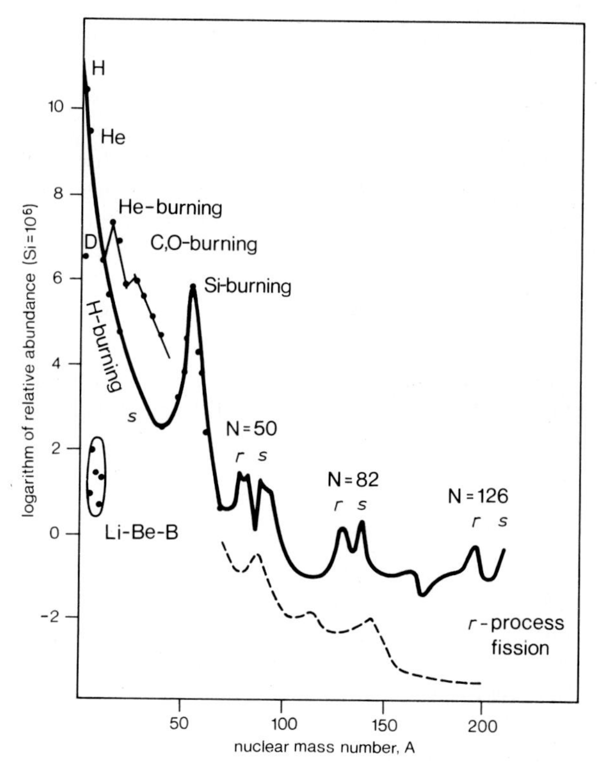

Fig. 3.5. Schematic curve of atomic abundances (relative to Si = 10^6) as a function of atomic weight (Suess and Urey, 1956)

rich heavier elements from a few of the *s*-process and *r*-process nuclei by way of (p, γ), (p, n), and (γ, n) reactions.

The *l*-process entails spallation of common nuclei, e.g., of carbon, nitrogen and oxygen by protons. They account for the presence of some rare light nuclides shown on the left lower margin of Fig. 3.5 and indicated as Li-Be-B.

The time required for the synthesis of all elements in cosmic abundance ranges from seconds to billions of years; and the temperature scales fluctuate over several orders of magnitude. The time-temperature relationships are depicted in Fig. 3.4. The various processes of element synthesis in stars are schematically labeled (see Burbidge et al., 1957; Trimble, 1975).

Helium releases per unit of mass only about 10 percent of the energy that is generated by hydrogen fusion. But the reaction results in an ash which marks an important step in the origin of chemical elements. At temperatures of ca. 10^8 K, three helium-4 nuclei (6 protons, 6 neutrons) become fused via an intermediate beryllium-8 nuclei into carbon-12 (Salpeter process). In the light of the fast decay of the ^{8}Be nucleus, the process can be looked upon as a kind of three-body reaction among helium nuclei. In a similar fashion, four helium nuclei yield oxygen (8 protons, 8 neutrons); five helium nuclei: neon, six helium nuclei: magnesium; and likewise silicon, sulfur and argon. The elements listed here are among the most abundant in the universe.

Once some catalyzing carbon is around, the so-called "CNO tricycle" (Bethe-Weizsäcker cycle) may become operative, but at lower temperatures (2 x 10^7 K), giving rise to a series of nuclei: ^{13}N, ^{14}N, ^{15}O, and ^{15}N:

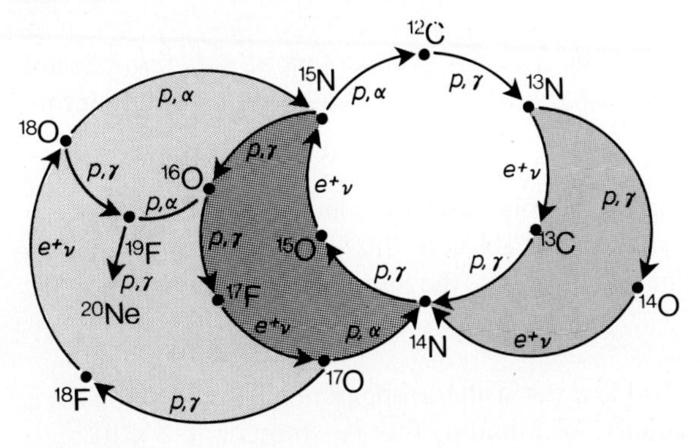

The net result is that four protons are converted to one helium-4 nucleus in addition to positrons, neutrinos and gamma rays. At temperatures around 10^9 K, the burning of two ^{12}C nuclei may yield: (i) a proton and sodium-23, (ii) a gamma quantum and magnesium-24, or (iii) helium-4 and neon-20.

When the helium reserve, the last of the valuable nuclear fuel, became exhausted, the relatively short period of the *Red Giants* came to an end. In the case of smaller-sized stars, the fusion processes after a time period of oscillations finally terminated almost like a car engine when the reserve tank runs out of gasoline. Since the gravitational collapse could not ignite new fuel, the dying star slowly cooled off over a billion years during a phase known as *White Dwarfs*. In contrast, massive stars collapsed faster, and within not more than a few seconds temperatures could rise to a few billion degrees and a shock wave ejected the outer parts of the star with sensational speeds. The energy emission of such a *Supernova* explosion had – for a short period – the same magnitude as that of a whole galaxy composed of 100 billion slowly evolving stars. In the center of such Supernovae, nuclear fusion of the previously generated chemical elements yielded new elements up to iron (26 protons, 30 neutrons), which can be considered the real ash of nuclear reactions.

Iron-56 is the most stable element and the last which can be obtained by fusion reactions that release energy. It is for this reason that it is so common in the universe, despite its cumbersome route of synthesis, for example by the burning of two silicon-28 nuclei via nickel-56 and cobalt-56 intermediates.

Matter dispersed by the first generation of Supernovae over millions of years reassembled by virtue of gravitational forces and shock waves. Again the same cycle began from protostar to star, to Red Giant to White Dwarf and eventually to Supernova. However, during a second process of this kind, the development followed a different route; for we do not start solely from hydrogen and helium, all the elements up to iron becoming involved. Supernovae of the second generation, and the neutrons released by such an explosion, produced from iron as the starting material the remaining heavy elements up to thorium and uranium. Of course, superheavy elements such as plutonium or fermium etc. could also have been synthesized, but they decayed rapidly. This contrasts with uranium or thorium which, due to their longevity, have partly survived billions of years and are found on Earth, even though the last Supernova event of this sort in our region of the galaxy dates back at least 5 billion years (see p. 39). By using uranium as fuel in conventional nuclear reactors we are liberating the energy of a star explosion. In a sense, the heavy nuclei return upon fission to the stable shelter provided around iron-56.

The chemical composition of the universe, not considering the matter in neutron stars and black holes (Bahcall, 1978), can be described as follows: of 1 million atoms there are 924,400 hydrogen, 74,000 helium, 830 oxygen, 470 carbon, 84 nitrogen, 82 neon, 35 magnesium, 33 silicon, 32 iron, 18 sulfur, 8 argon, 3 aluminum, 3 calcium, and the remaining 2 atoms for the rest of the elements.

The relative abundances of the more common elements in the solar atmosphere are depicted in Table 3.2. The remainder of the chemical elements occur only in trace quantities. If one gave an ordinary chemist a piece of the *universal matter* for analysis, he would say it was composed of hydrogen and helium, with some impurities of practically all elements.

In conclusion, the formation of elementary particles and elements follows a logical order which in turn can be used by scientists to delineate cosmic events and place time tags. Since it is easier to form light elements by fusion of hydrogen, deuterium and helium, it comes as no surprise to learn that oxygen and carbon, next to hydrogen and helium, are the two most prominent elements in our Universe.

Table 3.2. Relative abundances of the more common elements in the solar atmosphere (from Cameron, 1973, 1982)

Element	Atomic number	Relative abundances (atoms/10^6 atoms of Si)
H	1	3.18×10^{10}
He	2	2.21×10^{9}
C	6	1.18×10^{7}
N	7	3.74×10^{6}
O	8	2.15×10^{7}
Ne	10	3.44×10^{6}
Mg	12	1.06×10^{6}
Si	14	1.00×10^{6}
S	16	5.00×10^{5}
Ar	18	1.17×10^{5}
Fe	26	8.30×10^{5}

Starry Heavens

In his "Critique of Pure Reason", Immanuel Kant (1724–1804) concluded, "Two things fill the mind with ever increasing wonder and awe, the more often and the more intensely the mind of thought is drawn to them: the starry heavens above me and the moral law in me". The magic of a starlit night is more romantically revealed by Joseph Freiherr von Eichendorff (1788–1857):

Mir ist, als hätt' der Himmel
die Erde still geküßt,
daß sie im Blütenschimmer
von ihm nun träumen müßt'.

Die Luft ging durch die Felder,
die Ähren wogten sacht,
es rauschten süß die Wälder,
so sternklar war die Nacht.

Und meine Seele spannte
weit ihre Flügel aus,
flog durch die stillen Lande,
als flöge sie nach Haus.

Logic and romanticism are now followed by a few scientific details on the birth and nature of stars (e.g., Shklovskii, 1978; Wilson et al., 1981; Schwarzschild, 1970).

Stars are point sources of radiation. Their radiant flux reaching the Earth is commonly expressed as the apparent magnitude *m,* which for the faintest observable star amounts to 23.5 and for Sirius, the brightest star, –1.5. By way of comparison, the full Moon has an apparent visual magnitude of $m_\odot = -12.7$, and our Sun even one of –26.73. Knowing the difference between the *m*'s of two stars, their flux ratio F_1/F_2 can be determined:

$$F_1/F_2 = 2.512^{m_2 - m_1}.$$

For the Sun, the average flux of radiation per unit area at the Earth can be measured in absolute terms:

$$F_\odot = 1.37 \times 10^6 \text{ erg cm}^{-2}\text{s}^{-1}.$$

This is the solar constant which in turn can be used to calibrate the absolute flux density of any star having the same color as the Sun. The logarithmic ratio:

$$l_m = \log(F_m/F_\odot) = 0.4(m_\odot - \text{m}).$$

Having a star with the apparent magnitude *m* of 20, l_m equals –18.7, and its flux F_m amounts to 3×10^{-13} erg $\text{cm}^{-2}\ \text{s}^{-1}$. Where the distance *r* of a star is known, the luminosity *L* (= total radiant power) can be determined from the simple relation:

$$L = 4\pi r^2 F.$$

For example, if the star is 100 pc (parsec) away, its luminosity amounts to $L = 3 \times 10^{29}$ erg s^{-1}, which is 10,000 times lower than the total radiant power of our Sun. The term parsec is the astronomical unit for the distance of an object, whose paralax is 1 arcsecond.

Expressed in kilometers, the distance amounts to 3.0856×10^{13}, and in light years to 3.2615. One million parsecs is one Mpc (megaparsec).

Since for a star of a given luminosity the flux density of its radiation is inversely proportional to the square of its distance, a low luminosity dwarf near Earth and a high luminosity giant far away may exhibit the same apparent magnitude. Thus it is common practice to describe the luminosity of a star by its absolute magnitude M which the star would exhibit at a uniform distance from Earth, i.e., $r = 10$ pc. Absolute and apparent magnitudes are related by the simple equation:

$$M = m + 5 - 5 \log r.$$

The apparent magnitude of a star can be reliably determined. In contrast, the distance of a star from Earth is extremely difficult to measure, particularly for remote objects. This implies that the luminosity of a distant star is hard to ascertain. For this reason, astronomers determine the absolute magnitude of a star by indirect methods involving spectral characteristics.

The spectrum of a star of a given chemical composition is strongly temperature-dependent. For a comparatively cool hydrogen gas, the electron of most atoms is in the ground state $n = 1$ (n is an integer called the principal quantum number). A rise in temperature will cause hydrogen atoms to occupy higher energy levels ($n = 1, 2, 3, \ldots$). Should the temperature exceed a critical value $n = \infty$, electrons will depart from their atoms leading to a state of ionization. A hot star with temperatures close to 30,000 K will be highly ionized and will exhibit only weak absorption bands for neutral hydrogen. Stars at temperatures around 10,000 K are almost non-ionized, but a considerable quantity of electrons occupies a higher energy level, from where they can be lifted by photons to even higher levels, yielding a strong Balmer spectrum. Comparatively cool stars in the neighborhood of 3,000 K have almost all their hydrogen atoms in the ground state. Their electrons, providing sufficient energy-rich light quanta are available, can move to higher levels and the resulting spectral lines are emitted in the ultraviolet region (Lyman series). For a number of other elements, for instance helium, similar relationships exist. In general, hot stars contain in the visible range only a few lines, in particular those of hydrogen and helium, whereas cool stars have only a few hydrogen lines, but a full spectrum of lines from heavier elements (e.g., calcium and iron).

Based on these relationships, stars can be characterized by the spectra they emit, and spectral classification is based on a letter code:

O B A F G K M

("o be a fine girl (gentleman), kiss me")

in which O stars are at the hot, and M stars at the cool end of the spectrum. In addition, decimal subdivisions are used, e.g., our Sun is a type G2 star, to account for certain spectral deviations from the normal star. Also prefixes and suffixes are added, e.g., *d* for dwarf or *g* for giant, to characterize a star more fully. Almost 300,000 stars have been classified according to this scheme.

In addition to the main types, there are some particularly hot objects (Humphreys and Davidson, 1984) for which a special letter code has been adopted (e.g., WC). Furthermore, there are several side sequences for the cool stars known as Carbon and S stars, which have about the same temperature as K and M stars but differ in special spectral lines and bands.

The basic properties of stars – luminosity, surface temperature, radius, chemical composition and mass – are interdependent. In plotting luminosity L against surface temperature T of stars, characteristic patterns emerge that are related, for instance, to the stars' size, age, or evolution. These relationships were for the first time revealed in work of the renowned astronomers Ejnar Hertzsprung and Henry Norris Russell at the beginning of this century. In recognition of this contribution, the probably most important phase diagram in astronomy has been named in their honor: Hertzsprung-Russell or abbreviated H-R diagram. A wide variety of H-R diagrams exists, depending on what stellar characteristics have to be emphasized.

In plotting stars within a 5 pc range from our Sun, the majority fall within a rather narrow band termed the main sequence (Fig. 3.6). The small grouping of five stars about a hundred times less luminous than the Sun and just the size of our Earth (0.01 solar radius), is referred to as white dwarfs. By adding more distant stars with known luminosities and spectra, the main sequence will prograde and include stars of spectral type O, which are the most remote and brilliant objects within the band. This spectral classification from O to M is an expression of the changes from very hot to cool or from blue to red. The observed photoelectric color magnitude is commonly indicated by the blue minus visual color index, B-V (Fig. 3.7). About 95 percent of all stars occupy a position within the main sequence. This is a reflection of the long duration of the hydrogen-burning phase of small and medium-sized stars when compared to their subsequent phases such as helium-burning or the *r*-process (see Fig. 3.5). The remaining 5 percent of the

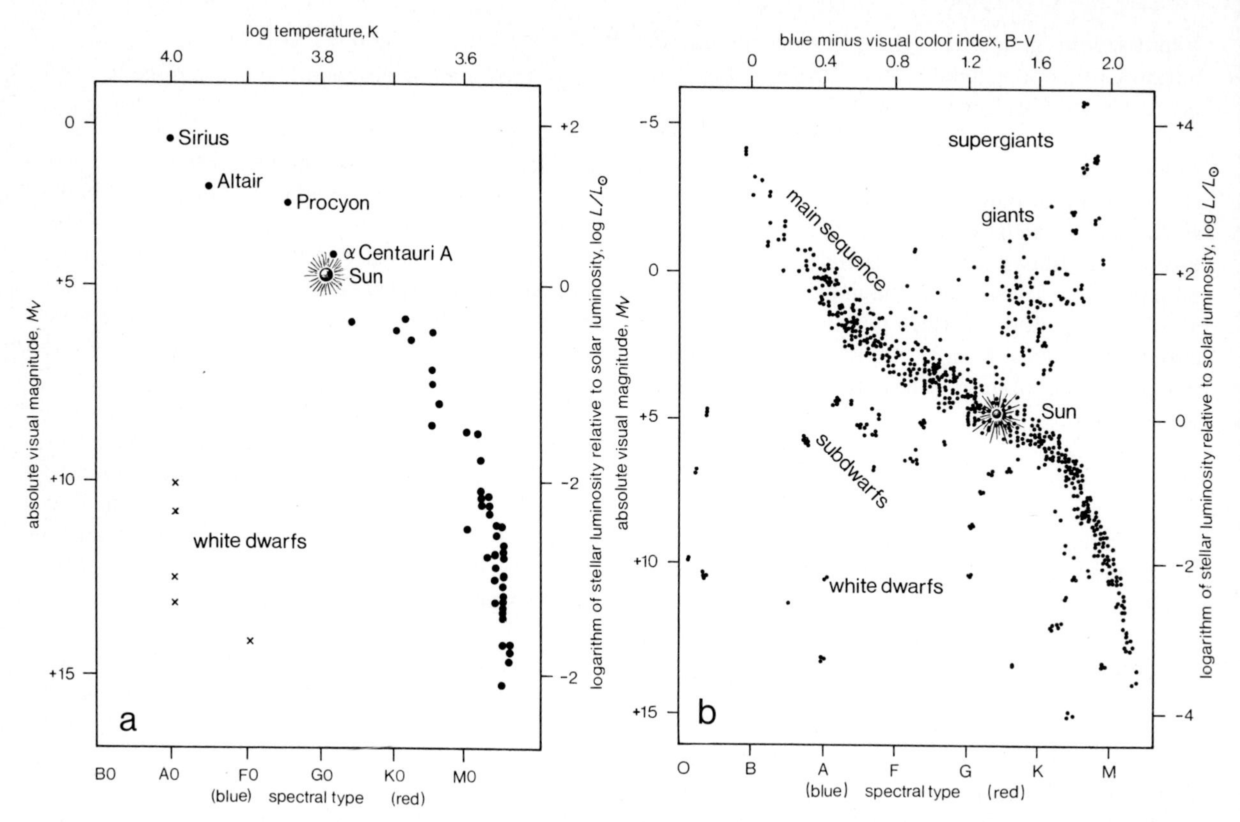

Fig. 3.6. **a** Hertzsprung-Russell diagram for stars in the vicinity of the Sun (Shklovskii, 1978). **b** Hertzsprung-Russell diagram for stars with known luminosities and spectra (Shklovskii, 1978)

stars are also not distributed randomly across the H-R diagram, but concentrated in certain regions. Distinct groups are the white dwarfs, giants and supergiants. However, it has been found useful to subdivide these groups further into subgroups to emphasize special tsrdetails and common names found in the literature include: subdwarfs, hot subdwarfs, red dwarfs, subgiants or red giants.

The residence time of stars within the band of the main sequence is principally a function of size, and ranges from a few tens of billions to a few million years. Smaller objects remain there for a longer and larger objects for a shorter period of time. In Fig. 3.8 the expected stellar residence time for variously sized bodies is indicated, using the Sun as the scale of reference for size. We can also infer from this graph the temperatures prevailing on the surface of stars in any region of the main sequence.

Following the hydrogen-burning phase, stars will leave the band of the main sequence, whereby massive stars depart at a fast and smaller-sized stars (e.g., our Sun) at a much slower rate. The track of stellar evolution for two case studies is shown in Fig. 3.9. The route track during the final developmental stages leading to either a supernova or a white dwarf is highly hypothetical.

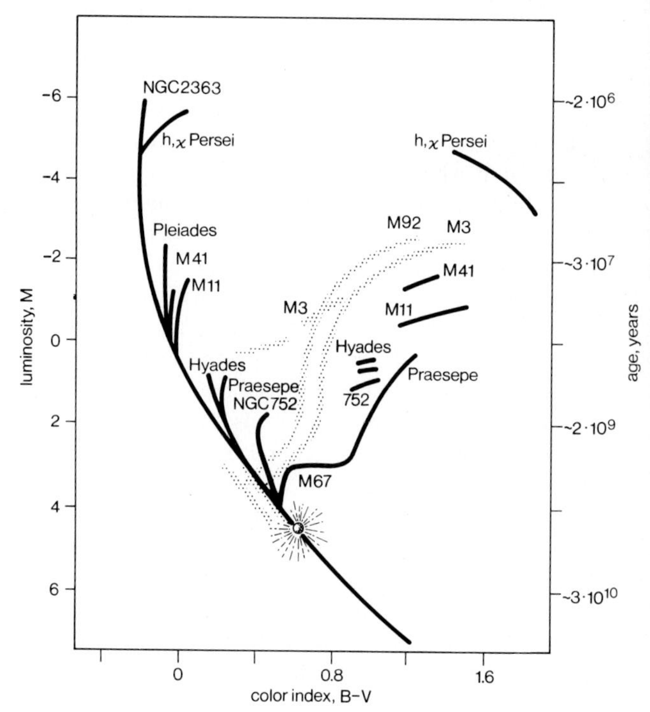

Fig. 3.7. Composite Hertzsprung-Russell diagram of a number of star clusters, both open and globular (Shklovskii, 1978; Johnson and Sandage, 1956)

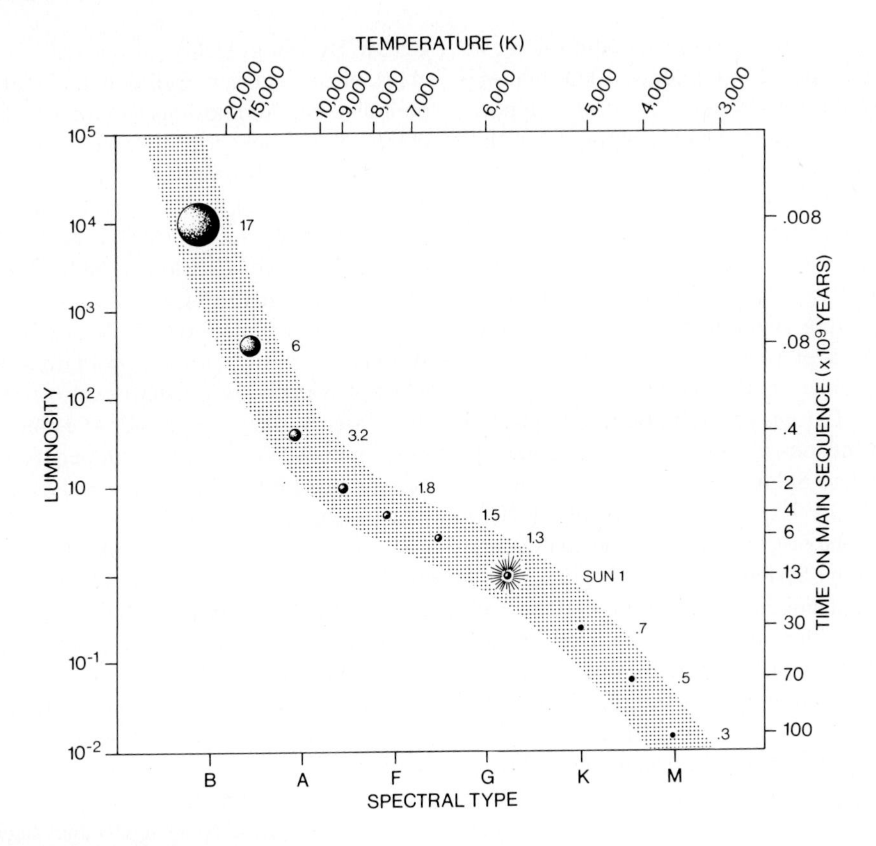

Fig. 3.8. Residence time of stars of different mass on the main sequence. Numbers next to spheres indicate the mass relative to the mass of the Sun (= 1). Surface temperatures of the various stellar objects are indicated (after Huang, 1970)

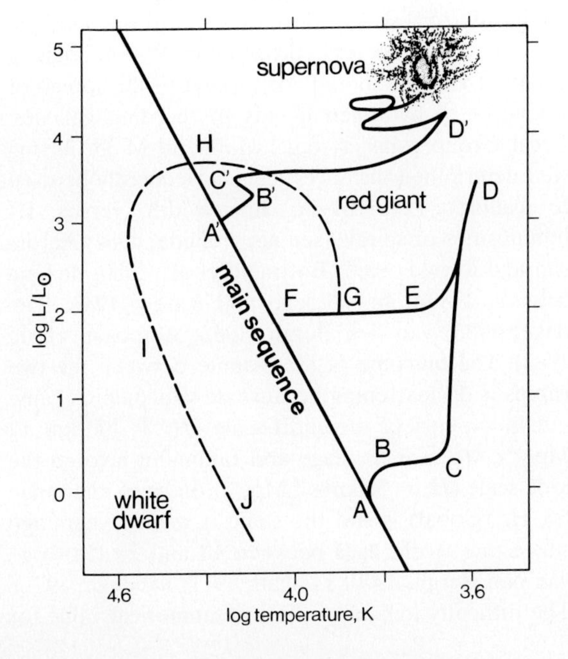

Fig. 3.9. Suggested path of evolution of stars following their departure from the main sequence (highly hypothetical). Stars having a mass about that of the sun will proceed along the route *A–J* and eventually develop into a white dwarf. More massive stars will move from A^1 to D^1 eventually to terminate as supernova (after Giese, 1981)

Age of the Universe

Astronomers and geologists alike are interested in uncovering the early history of the Universe and of Planet Earth respectively, and deducing factors and events that have led to the present state of affairs. To achieve this goal, they try to place observational data into an orderly time sequence. At first sight such an endeavor appears to be easy, because there is light as a precise yardstick of cosmic dimension and radioactive nuclei and their daughter products in rocks as an accurate clock of times gone by. But a closer look reveals that there are many obstacles to overcome "à la recherche du temps perdu".

As far as precision in time measurement is concerned, particle physics leads the way. Grand unification theories outline a series of happenings at the beginning of creation which proceed at intervals of fractions of seconds. In 3 minutes all matter is generated. In a few hundred thousand years following the big bang, the microwave background radiation is on its way to the observational cosmologists. The high degree of isotropy of this "whisper of creation" supports the cosmological principle which simply states that the world in its large-scale features is isotropic and homogeneous. Hence our position in the universe is in no way special to that of any other cosmic place. However, this knowledge is of little help in accurately determining the travel time of the 3 K radiation from its point of departure, about half a million years after the big bang to a present observer. Reported estimates vary by as much as 10 billion years. In contrast, the age of Planet Earth is known to better than within a few tenths of millions of years, and its assigned age is 4.53 Ga. For our Solar System, the beginning has been set at 4.6 Ga. Expressed differently, the Universe appears to be just two to four times older than our Earth and Solar System.

The problem of how large and how old the Universe is has been with us for quite some time. The discovery of the recession of nebulae by V.M. Slipher in 1917 gave the first hint of the non-stationary character of at least a part of the Universe. In the 10 years following this discovery, more data became available indicating that a number of galaxies were receding from the Sun. However, it was eventually Edwin Hubble, who in 1929, declared that the world at large was not static but an expanding entity. His findings opened the way for a new approach towards dating the Universe. Hubble's perceptive theory at the same time resolved an old cosmological puzzle "why is the sky dark at night", also known as Olbers' paradox (see Jaki, 1967). Under the former concept, an infinite static universe filled with stars should reveal a sky as bright as the surface of the stars.

The famous Hubble law states that the velocity V of recession of a galaxy is proportional to its distance D from us:

$$V(\text{km s}^{-1}) = H\,D\,(\text{megaparsec}).$$

In this equation, H is quasi the "cosmic stopwatch", provided the numerical value of the Hubble parameter H, which expresses the time scale on which the universe has been expanding, is linear and accurately known. V is inferred from redshifts. Light from distant galaxies is progressively redshifted so that photons are observed with energies lower than those at their point of emission. D is inferred from apparent luminosities of galaxies frequently referred to as "standard candles" or "rigid rods". But the more distant a galaxy, the dimmer the "standard candle" becomes.

The commonly applied extragalactic distance scale rests principally on the distance scale within our own galaxy. By using a class of stars, the Cepheid variables, which have a well-defined relation between their absolute luminosities and their periods, an indisputable yardstick for distance can be calculated. Cepheid variables can also be observed in nearby galaxies belonging to the Local Group – for instance the Andromeda nebula (M 31), M 33, the large and the small Magellanic Clouds, NGC 6822 or IC 1613. They too, provide primary calibrators of extragalactic distance. It appears, however, that our galaxy (Milky Way) (Bok, 1981) and the Local Group of galaxies of which it is part, are gravitationally attracted towards the center of the Virgo cluster at estimated velocities ranging between 0 and 500 km per second (e.g., van den Bergh, 1981). Thus, small-scale anisotropies and phenomena exist in the extragalactic space which make it doubtful that a locally determined Hubble parameter could be used as a reliable yardstick of time for the Universe as a whole.

To overcome difficulties associated with generalizing results from nearby galaxies, the extreme brightness of supernovae explosions in far-away galaxies was used as a measure of distance by assuming that the luminous energy, that is the apparent brightness, stems from the radioactive decay of ^{56}Ni (Arnett, 1982). The resulting H_o's suggest ages of about 13 Ga.

In recent years, the quest for H_o (global) reached new dimensions when two of the long-standing but opposing schools of thought on the Hubble flow fully set out their arguments: de Vaucouleurs (1982) and Sandage and Tammann (1984). In both articles, the Tully-Fisher (T-F) relation was one of the crucial tests to derive a numerical value for the Hubble constant (see Box 3.3). The T-F relation implies that the luminosity of disk galaxies is related to the width of their IR 21-centimeter hydrogen line. The broadening of the 21-cm line should correspond to the spread of velocities of hydrogen clouds in receding galaxies. Local Group galaxies, e.g., M 31 and M 33, having well-determined distances, can subsequently be used to calibrate the 21-cm line width versus IR luminosities of spirals (see also: Gaida, 1984; Aaronson and Mould, 1983; Bottinelli et al., 1980; de Vaucoleurs, 1983a, b; Sandage and Katen, 1983; Sandage, 1983; van den Bergh, 1982; Aaronson et al., 1982). The outcome of the dispute between the two camps is disheartening because de Vaucouleurs came out in support of the short-scale ($H_o \approx 100$ km s^{-1} Mpc^{-1}), whereas Sandage and Tammann favored the long-scale ($H_o \approx 50$ km s^{-1} Mpc^{-1}). In brief, the range for H_o (global) is still the same it was 20 years ago suggesting world ages between 10 and 20 Ga (e.g., van den Bergh, 1970; Longair, 1971; Sandage, 1971). The difficulty in finding the true numerical value for global H_o was pointedly expressed in 1982 by de Vaucouleurs himself, "The traditional approach to the extragalactic distance scale rests on a pyramid of primary, secondary and tertiary indicators of increasing range and decreasing accuracy ... fraught with the danger of cumulative errors!" In short, even the best modern techniques still fail to lift the veil hiding the correct dimensions of the Hubble flow.

A further complication arises, because the inferred Hubble time sets only an upper limit to the real age of the Universe. This is so, because expansion is decelerated by gravitation, and the deceleration parameter q_o depends primarily on the sum total of luminous and non-luminous matter distributed across the sands of time. Theoretically, q_o yields a maximum value which, when calculated in terms of years and then subtracted from Hubble time, would give a minimum age for the Universe in the range of 5 to 10 Ga, depending on whether the short or long Hubble scale is used.

At this juncture, it is appropriate to mention one further aspect of the cosmological principle: the permanence of the laws of physics. Thus, theories question the general validity of certain fundamental constants, and the most notable among the antitheses is Dirac's cosmology. He states that the similarities of dimensionless numbers of the order of 10^{40} generated by combination of fundamental constants of physics are not a coincidence but a principle of physics. For instance, if we equate the ratio of electromagnetic to gravitational forces, $e^2/Gm_e2\ 10^{40}$, the ratio of the scale of the Universe to the classical electron radius $(c/H_o)\ (e^2 m_e c^2)\ 10^{40}$, and assume e^2 constant, then G in an expanding world model is expected to vary with H, which is about proportional to $1/t$, where t is the age of the Universe. The implication of a change in G with epoch would have considerable consequences upon the evolution of Cosmos, Solar System and Earth. For details, see Longair (1971).

BOX 3.3

Tully-Fisher Relation

In 1977, the two astronomers Brent Tully and Richard Fisher presented a new technique to determine distances to the Virgo cluster and the Ursa Major cluster. In essence, they discovered a good correlation between the global neutral hydrogen line profile width (a distant-independent variable) and the absolute magnitude or diameter of galaxies. Members of the Local Group and the nearby galaxies of the M 81 and M 101 groups were used as a means of calibration (Fig. 3.10). The Tully-Fisher relation is particularly strict for members of the Local Group, of which the Milky Way is part. The data further suggest that by scanning the distances of galactic groups close to the Local Group, it appears that the term Local Group should be revised, because there is strong indication that we reside more in a local "Filament" than in a group. For instance, the spiral galaxy in Ursa Major (M 81) would belong to such a filamentous unit. Since galactic strings and filaments have recently been propounded as constituting the large-scale structural element of the universe, the findings of the T-F relation would support this idea.

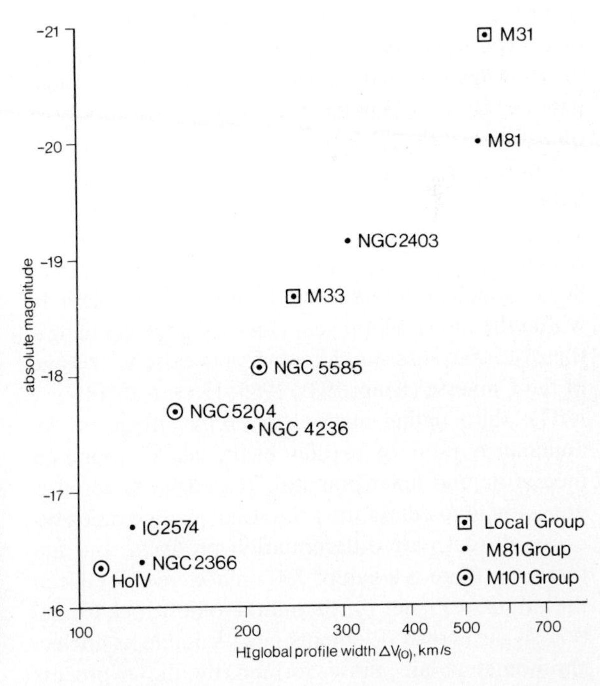

Fig. 3.10. *Tully-Fisher relation* for members of the Local Group and the M 81 and M 101 Groups. The term *M* refers to the Messier catalog (Charles Messier, 1730–1817), which is a list of nebulae and star clusters. A more comprehensive list is contained in the New General Catalog (*NGC*) and the Index Catalogs (*IC*), and the abbreviations *M, NG* and *IC* are commonly used to identify the objects by their number in these catalogs

We conclude that recession of galaxies is not a reliable distance scale for size and age of the Universe, largely because considerable uncertainty exists with regard to the numerical value of H_o and q_o.

Two other techniques are briefly mentioned which enable us to estimate the age of the Universe: (i) evolution of the stellar content of globular clusters, and (ii) isotope abundance ratios in meteorites.

Globular cluster stars are survivors from the early stage of a galaxy and they are considered some of the oldest cosmic objects known (e.g., Sandage, 1968, 1971; Johnson and Sandage, 1956; Iben, 1975; Hesser, 1980; van den Bergh, 1981; Spitzer, 1984). The ages of globular cluster stars are inferred from their mass, luminosity, composition, and surface temperature, and from a comparison of the main sequence termination point in the Hertzsprung-Russell diagram (see p. 30). Ages for the oldest globular clusters range from 10 to close to 20 Ga. The large uncertainty rests primarily on helium and heavy-element abundances adopted by the various authors (Hesser, 1980). For example, assumed high helium abundance would replace older ages for the globular clusters; supermassive stars generated during the early history of a galaxy could possibly yield high helium abundance. However, the prevalent view is that the bulk helium content has been generated within the first 3 minutes of time in an amount of 25 percent of all matter. It is true that hydrogen can be converted into helium during stellar synthesis. However, this helium immediately serves as starting material for the synthesis of heavier elements, which constitute *only* 2 percent of the universal matter. The fact that helium abundance of close to 25 percent is observed in a variety of objects can at the same time be used in support of the hot model, which permits just three stable values of helium abundance: 0, 25 and 100 percent helium.

From evolutionary tracks of stars in metal-poor globular clusters, ages of 14 to 16 Ga are inferred. Since globular clusters formed fairly rapidly – namely within the first 1 billion years following the big bang – their old star ages should come very close to the age of the Universe (King, 1981, 1985; Hesser, 1980).

The third dating method takes advantage of the abundance ratio of certain heavy nuclei found in meteoritic and lunar material. Terrestrial material is not suitable to determine primordial isotope ratios because all rocks are differentiation products, and furthermore there is a gap of 700 million years between the origin of Planet Earth and the oldest rock formation so far known. Elements most suitable as nucleochronometers are those formed by the *r* process (rapid neutron capture), which presumably operates only in supernovae (Burbidge et al., 1957). Elements of interest include ^{238}U, ^{235}U, ^{232}Th, ^{187}Re, and ^{187}Os. Their production ratios can be calculated theoretically and subsequently be compared to the abundance ratios found (e.g., Hohenberg et al., 1967; Schramm and Wasserburg, 1970; Fowler, 1984; Schramm, 1974; van den Bergh, 1981). In essence, the intercomparison provides an estimate of the time elapsed since generation of *r* process produced elements in supernovae bursts and the formation of our solar system. Some uncertainty exists concerning the rate of element production in such an event. In turn, age estimates have a certain error bar. From the available data bank, the best fit for the onset of galactic star formation is 10.7 ± 2.3 Ga (van den Bergh, 1981). Adding to this figure the time needed to produce a star and to yield a supernova, the age of the Universe determined by means of nucleochronometry emerges as about 12 ± 3/2 Ga.

In conclusion, the age of the Universe inferred by three independent techniques falls in the range of 10 to 20 Ga. Accepting only ages that conform with all three methods – which of course is a rather tentative approach – then the Universe is from 10 to 15 Ga old. In his review on the age of the elements, Schramm (1974) closed his chapter, "To be able to measure any astronomical number, particularly such an important one, to an accuracy of 33 percent is deeply satisfying."

This chapter would be incomplete, if a few notes were not added on the future of the Universe. The proposed scenarios go through extremes. All depends on whether the Universe is considered open or closed, which essentially depends on the large-scale density of the mass in the Universe (Rees, 1980; Peebles, 1984, 1986). Ice or fire are the two alternative outcomes (e.g., Gott et al., 1976; Dicus et al., 1983; Trefil, 1983b).

Up till now, the Universe has expanded, which implies that it is open and is expected to enlarge further at least for the time immediately ahead of us. Continuing the path of expansion, one can predict that our Sun will run out of its useable hydrogen fuel in roughly 5 Ga (giga anni), develop into a red giant, and thus terminate our solar system. Small stars in our galaxy have a longer lifetime but they, too, will collapse and in roughly 50 to 100 billion years the galaxies will be populated by white dwarfs, neutron stars and other faint objects. If one accepts that the missing mass exceeds luminous matter by about tenfold, the Universe has attained critical density. Expansion will decelerate, stop, and contraction will start. In this scenario, such a reversal is expected to begin in about 50 Ga. In the following period, matter will move closer together and background radiation will be shifted towards the visible wavelength, turning the

Universe into a hot fiery habitat, where eventually matter becomes dismembered into its fundamental particles and forces. The big crunch, as the final contracting state is often termed, may be followed by another expansion leading to a Universe of the type we know of today. In short, within 100 Ga, a full cycle from big bang to big crunch will come to a close and this system might last from here to eternity. There are still a number of unresolved problems, most prominent among which are the uncertainty concerning the temporal stability of fundamental physical constants, and the value of the missing mass. At present, most researchers favor the idea of an open Universe, even though some consider it only moderately open. This would mean the future of the Universe will be dark and icy, rather than bright and hot.

Actually, the first 50 Ga will look almost the same, independently of whether the Universe is open or closed. At that vantage point, however, a closed Universe will start to contract, whereas an open Universe will continue to expand, but at a progressively decreasing rate. The last stars will run out of hydrogen, their principal nuclear fuel, in 10^{14} years. Remaining planets will eventually feel the gravitational pull of a nearby passing star and be scattered into space. Such an occurrence will sooner or later afflict all planetary bodies, and in 10^{17} years from now, all planets will have vanished. The next incident in line after about 10^{18} years is the evaporation of stars from the outer regions and the collapse of galaxies, with stars from the galactic center joining into a single supermassive black hole. In the course of the next epoch, all baryons which did not enter a black hole are expected to decay, provided the proton is unstable and decays into electrons, positrons, neutrinos, and photons. Since the predicted lifetime of a proton is about 10^{31} years, one would expect the Universe in the time shortly afterwards to consist of electrons and positrons separated from one another by about 100 light years and thus without a chance of annihilating each other. In addition, there would be the photons and neutrinos from previous events and an occasional supermassive black hole. As expansion continues, all constituents become more rarefied, and little further happens till the year 10^{100}. The final event in the future of the open Universe is marked by the decay of the supermassive black holes. Black holes are considered bodies so dense that nothing from within can escape their gravitational pull. This is correct for short-term but not for long-term periods of time. Hawking (1980) showed that a black hole emits energy at rates inversely proportional to the square of its mass. Thus, small black holes vanish fast but even supermassive ones should eventually evaporate principally into photons. Evaporation time for a supermassive black hole is estimated at 10^{100} years. In case protons are stable, the evaporating black holes will leave matter behind which over eons aggregates to spheres of iron that in due course enter black holes where matter eventually radiates away.

The future of the Universe is bleak, no matter whether we adopt the open or closed scenario, or believe in a stable or unstable proton. All will end in fire or ice, and the Universe will either oscillate in a 100 Ga cycle or expand into darkness forever.

Chapter 4

Our Solar System

Denn die einen sind im Dunkeln

The presence of galaxies, stars and dark matter in all regions of the Universe is well established. The same holds true for the cosmological principle which states that the structure of the Universe is isotropic on a large scale. In contrast, planets and moons are known for sure only from our solar system. However, some tentative data exist, suggesting the existence of nearby planetary systems.

It has been known for some time that a number of stars exhibit slight perturbations in their motion. This can best be reconciled with the presence of massive objects larger than Jupiter but which are still too small to be selfluminous ("brown dwarfs"). About a third of those stars examined could have satellites of their own. Furthermore, measurements made by the new infrared astronomical satellite (IRAS) provide evidence of the existence of cold matter around the young star Vega (Lyrae), which is 8 pc distant and one of the brightest visible objects (about 40 times as bright as the Sun). Similarly, in spring 1984, a star, β Pictoris, surrounded by a disk-shaped cloud, was actually seen by astronomers of the Las Campanas Observatory in Chile.

In conclusion, three lines of evidence are all there is on planetary systems outside our own, i.e.: (i) minute perturbations in the motion of some stars, (ii) infrared sensing suggesting cold matter around Vega, and (iii) an apparent stellar disk in the constellation Pictor. This is a rather meager result, considering the wealth of published information on stars. This calls to mind Bertold Brecht's 1930 movie "The Threepenny Opera" (final song):

Denn die einen sind im Dunkeln
und die andern sind im Licht.
Und man siehet die im Lichte,
die im Dunkeln sieht man nicht.

Thus, if it comes to the elucidation of mechanisms leading to the origin of a solar system, and the description of planets, moons and asteroids, one is still confined to our Sun and its satellites. Hopefully, in the years to come, more explicit data will emerge on planets and moons from beyond our solar system (further reading: Black, 1980, 1982; van de Kamp, 1969; Gatewood, 1976; Fawley, 1979; Borucki and Summers, 1984).

Interstellar Gas and Dust

The Universe is filled with electromagnetic radiation ranging from the gamma-ray to the radio regions. It contains a wealth of elementary particles, some of which, like the neutrino, are supposed by many researchers to have a non-zero mass. And finally, looking at the end of the scale, the chemical elements, one must conclude that no place in the Universe is empty. Even in the remoteness of interstellar space there is at least one lonely atom per cubic centimeter.

Of the mass we can account for in our galaxy, about 95 percent resides in stars and the rest is interstellar gas and dust in a ratio of about 99 to 1. However, this interstellar matter is not uniformly distributed throughout the galaxy but is concentrated in regions close to the galactic plane of symmetry, where new stars are born. There are high-density regions referred to as clouds, and low-density regions known as intercloud gas.

To most of us, clouds are seen more in conjunction with weather and climate. Namely, water vapor will condense and form clouds of different sizes and shapes, which are freely moving, can become dispersed or aggregate and form a cloud cover. Some regions are more cloudy than others. In due course, they will rain out and the meteorological cycle will commence again with the evaporation of water bodies. In analogy, interstellar clouds of various sizes, from about 0.05 to 100 pc in diameter, are constantly in slow motion around the galactic center. In their perpetual wandering through galactic space, they may grow in size or collapse. During one cosmic year, which measures about 250 million earth years in the case of our solar system, an average cloud could conceivably double its size by sweeping up dispersed interstellar gas and dust. Once critical conditions are established, which are related to a number of factors such as density, size, regional position, and events in the vicinity of a cloud (e.g., supernova explosion), interstellar clouds can "rain out", i.e., collapse, yielding stars and satellites of various sorts. In the course of stellar evolution, gas and dust can become regenerated again. A dying star of medium size, before degenerating into a white dwarf, will eject a gaseous shell into its nearest environs, yielding a planetary nebula. Alternately, a massive star will go through a supernova stage and explosively release gas and dust, which then enter the pool of interstellar matter.

The gas is principally hydrogen. Should this gas be close to a star of spectral type O or B, a zone of ionized hydrogen (H^+), measuring 1 to 100 pc (Stroemgren radius) depending on the star's luminosity and the population density of hydrogen, will be generated. Such a zone of ionized hydrogen is termed the H II region. Inasmuch as H II regions can be fed from various sources of gas and dust, small regional differences in the abundance of elements do exist.

The bulk of the hydrogen in interstellar matter is not ionized. Instead, most of it, i.e., roughly 90 percent, occurs in regions where the atom is present in its ground state. There are distinct galactic regions where neutral hydrogen is highly concentrated in what is commonly referred to as giant molecular cloud complexes. These are found in a flat disk close to the galactic center at distances exceeding about 2 kpc and reaching as far as 20 kpc. Furthermore, they are prominent in the galaxies' spiral arms (Blitz, 1982; Scoville and Young, 1984).

Research on giant interstellar clouds by means of astrospectrographs and radiotelescopes was initiated in 1963 by a team of the Massachusetts Institute of Technology and the Lincoln Laboratory. The most crucial finding was the observation that cloud complexes contained, in addition to H_2, a variety of molecules. Over the past 20 years more than 50 compounds were identified. By including all isotopic species, close to 100 different types of molecules were recognized (Mann and Williams, 1980). A molecule of particular relevance is carbon monoxide. Even though its abundance in dark clouds is just one-ten-thousandth that of molecular hydrogen, its radiation properties at low temperatures make the molecule an excellent signpost for the mapping of cloud complexes (see Fig. 4.1).

The mass of the largest cloud measured exceeds that of the Sun by several orders of magnitude (2 x 10^5). The shape of the cloud is elongated and a typical structure measures 45 pc in diameter. Chemically, clouds are composed of gases and dust grains in a mass ratio of about 100:1. The most dominant molecular species is hydrogen, in a concentration of about 10^4 H_2 per cm^3. This value represents ca. 99% of the mass of a giant molecular cloud complex. The remainder of the molecules are just "impurities", but given the size of a cloud, they can pile up to a rather substantial stack of matter. For instance, trace constituents such as carbon monoxide, water, methane, formaldehyde, methyl alcohol, ethyl alcohol, hydrogen cyanide, ammonia, or the hydroxyl radical far exceed all the mass contained in our solar system.

The above listed molecules lack, however, the metal ions common to the Planet Earth: iron, silicon and magnesium. If we assume that molecular cloud complexes are the birthplace not only of stars but of planets and moons as well, these elements should hide somewhere in the complex. In short, the ions are part of fine dust particles scattered throughout the cloud complex. The actual size of the grains can be inferred from the dispersion of light rays as a function of wave length. In essence, we are looking at the reflection scattering of star light by solid particles (reflection nebulae). Inasmuch as the size of these particles is just below the wave length of visible light, a precise measuring ruler is provided: dust particles have a size of less than 0.5 μm in diameter.

Other spectral indicators (for details see, e.g., Giese, 1981; Martin, 1978; Khromov, 1977; Peimbert et al., 1984) provide a means to determine the chemistry and mineralogy of cosmic dust. In all probability graphite, carbynes, Fe-Mg silicates and native iron are the major mineral phases in dust aggregates, and grains can be covered by solid water, ammonia and methane. The presence of solid particles has numerous consequences for the synthesis and protection of organic molecules in the space environment. For example, the probability of collision between atoms and molecules is enhanced, and three-body reactions become feasible. Moreover, mineral surfaces may provide not only a convenient "resting place" for certain atoms and molecules, but by virtue of their crystalline order, catalysis and epitaxis may ensue. The generation of more complex molecules such as sugars, amino acids or the bases of the purines or pyrimidines is conceivable, but their detection awaits more sophisticated technologies.

Under the condition of identical element distribution in molecular cloud complexes and in universal matter, the observational data suggest that in the space region accessible for studies by radiotelescope all carbon is molecularly bound, that is, no atomic carbon remains. For oxygen, 30 percent has entered a complex organic molecule (Salpeter, 1974). For nitro-

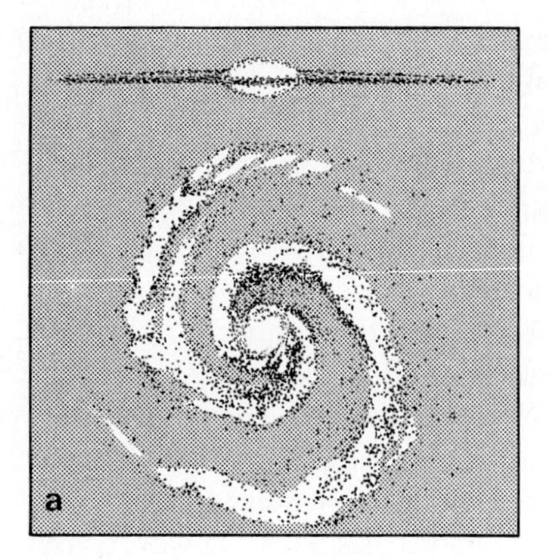
a

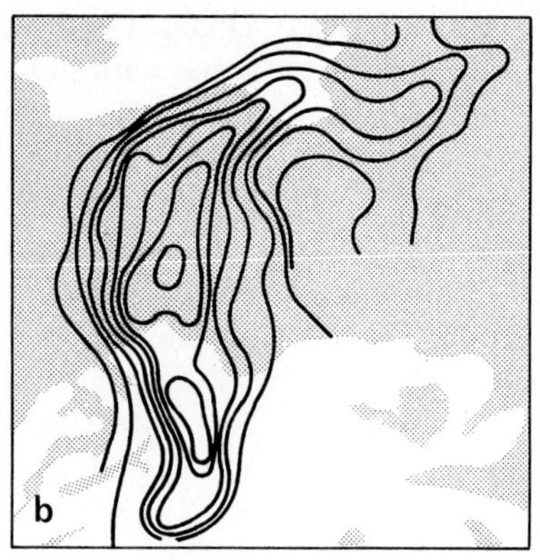
b

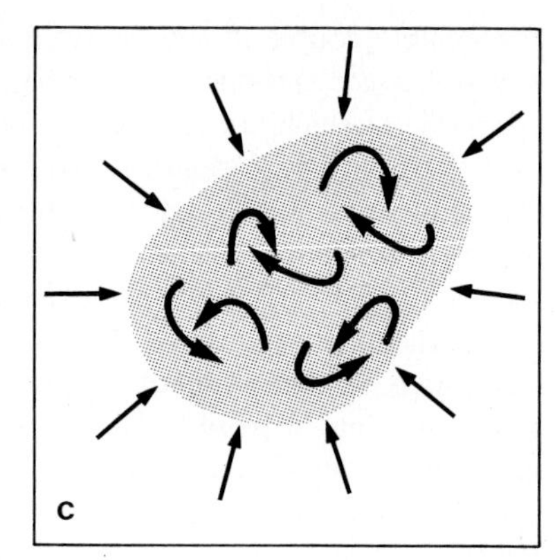
c

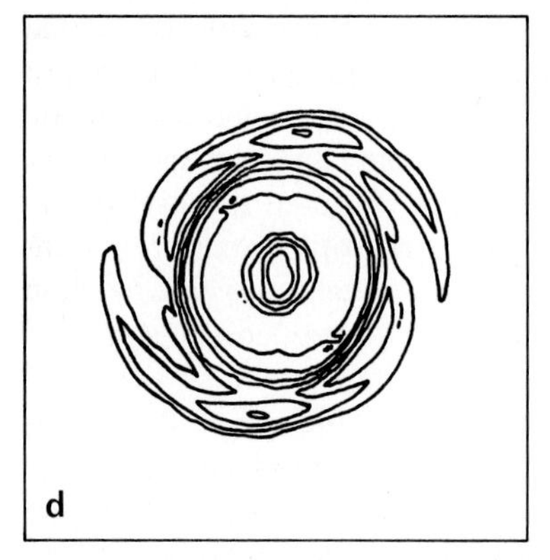
d

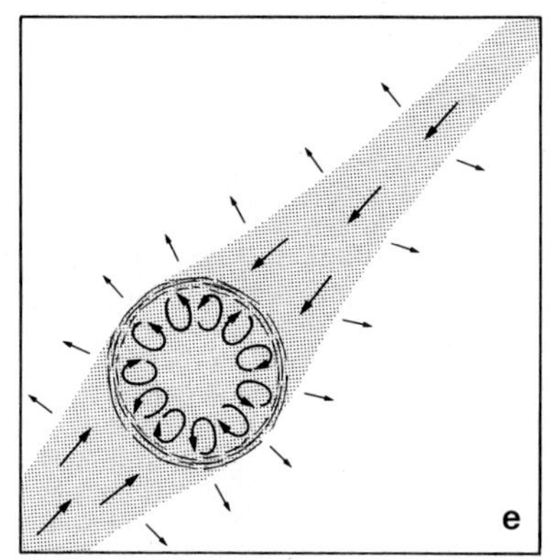
e

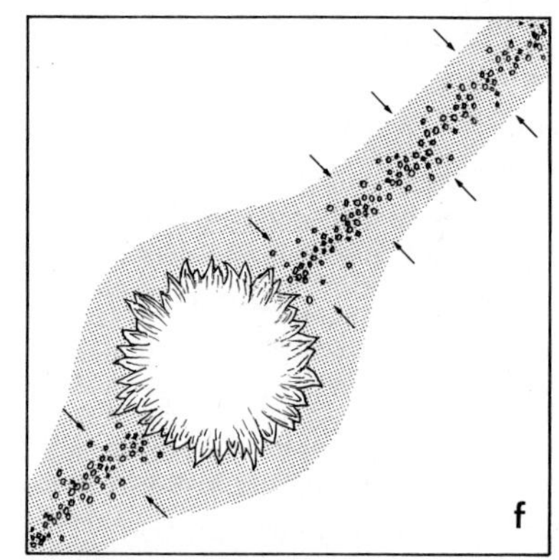
f

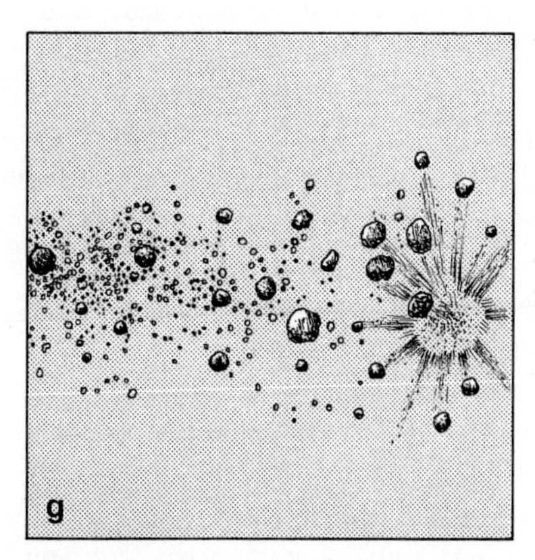
g

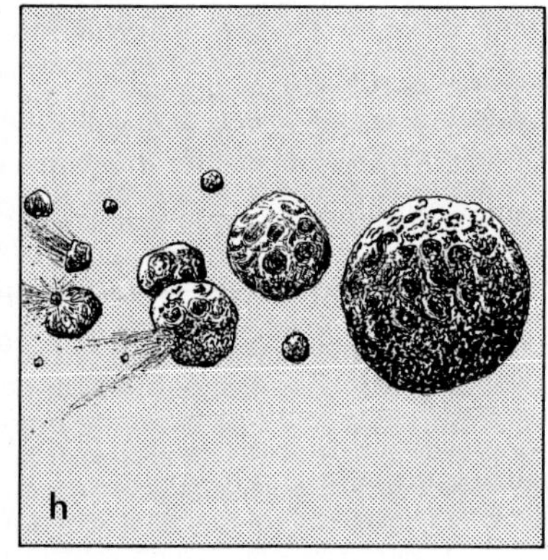
h

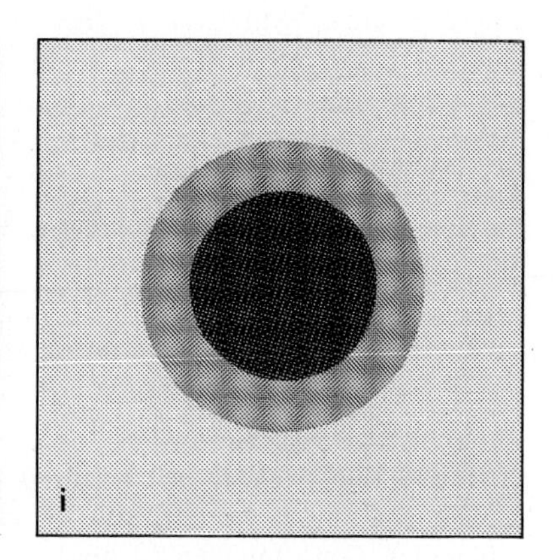
i

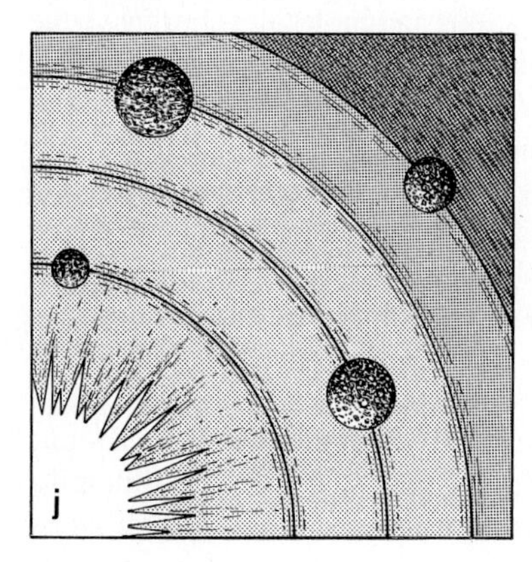
j

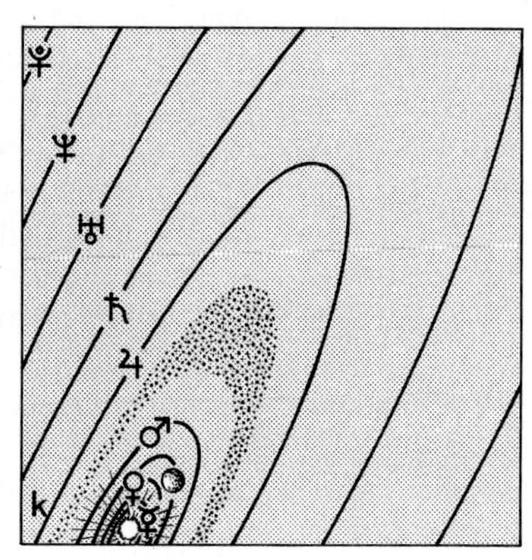
k

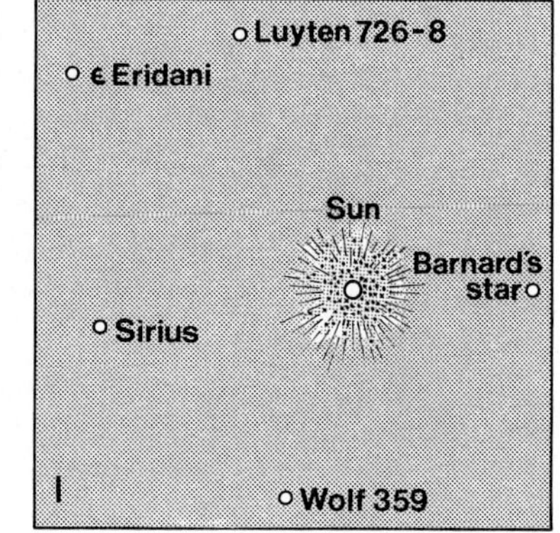
Luyten 726-8
ε Eridani
Sun
Barnard's star
Sirius
Wolf 359
l

gen and sulfur compounds present in the interstellar medium, the data bank is not sufficient to make a tentative assignment yet.

In brief, the chemistry of giant molecular cloud complexes is basically one of hydrogen, oxygen and carbon judged by the prevalence of H-O-C-containing molecules. It is noteworthy that in the presence of a mineral catalyst, simple organic molecules such as formaldehyde or hydrogen cyanide are expected to yield biochemically interesting monomers such as sugars, amino acids, purines and pyrimidines (Weiss et al., 1970; Mizutani et al., 1975).

Collapse of an Interstellar Cloud

We have previously seen that clouds circle around the galactic center, and that they will pick up gases and particles which were previously ejected from dying stars (Fig. 4.1). Once an interstellar cloud reaches a critical mass, where the two major forces operating, that is (i) *expansion,* due to pressure of the gases, and (ii) *contraction,* due to gravitational attraction, are shifting in a sense that the gravitational force becomes the dominant one, the cloud will collapse. This is a rather simplified picture, because there could be a number of additional phenomena at work. For instance, gravitational forces exerted from larger nearby objects might initiate a contraction. Magnetic fields from within or outside a cloud complex could trigger a collapse (Sofue et al., 1986). Supernova explosions in the vicinity of a cloud might generate large shock waves and cause the collapse of a cloud (Cameron and Truran, 1977). The isotope record in lunar and meteorite material (Clayton, 1977) strongly suggests that some isotope anomalies are caused by supernova events. In turn, huge shock waves derived from these sources are presently believed to be the triggering event for the collapse of the interstellar cloud that produced our solar system.

Much speculation exists on what happens following the collapse of a cloud towards the establishment of a presolar nebula (e.g., Gehrz et al., 1984). It is agreed that this depends not only on the mass and size of an interstellar cloud which may range from a few to several thousand solar masses distributed over a few to one hundred pc, but also its rotational properties. Rapidly rotating clouds collapse and fragment into a hierarchy of multiple stellar systems (Bodenheimer, 1968; Bodenheimer et al., 1980; Boss, 1985), rather than into single stars. Fragmentation will continue until sub-solar-mass-sized pieces are obtained. Consequently, this scenario does not apply to our solar system. In contrast, a slowly rotating cloud of a type that was supposedly present in the region of the galaxy where our Sun is currently located could yield a presolar nebula capable of evolving to the solar system of which we are part.

Previously we have stated that the Earth is iron-, oxygen-, silicon-, and magnesium-dominated, whereas universal matter is predominantly hydrogen and helium. We therefore have to conceive of processes by which we can differentiate universal matter to obtain terrestrial matter.

The first solid evidence of the type of mineral phases that were generated in a presolar nebula is contained in meteorites, which actually predate even the most ancient rock formations known from Earth which are "just" 3.8 Ga old. For comparison, the age of meteorites is about 4.5 Ga. Their wide range in mineralogical and chemical composition serves as a means to outline a sequence of events involved in the transformation of cosmic dust and gas into the present population of planetary bodies.

A question of considerable interest concerns the redox state of the nebula which, when hydrogen is in excess, is controlled by the carbon/oxygen ratio. The redox state of meteorites extends from the super-reduced enstatite meteorites and Si-bearing irons, via various types of reduced irons and stones, to the oxidized C 1 meteorites (J. V. Smith, 1979). This all implies that throughout the nebula an Eh gradient must have existed defining growth regions for reduced and oxidized mineral species (Fig. 4.2).

Another quest has to do with temperature and pressure regimes in the solar nebula, as well as angular momentum conservation and rotation, which determines the direction a planetary system will take. It appears that a pressure-supported central core can

Fig. 4.1 a–l. Cartoon showing major steps in the evolution of the solar system (from upper left to lower right): **a** whirlpool galaxy, similar to Milky Way, about 100,000 light years in diameter; frontal and edge-on views; **b** contours of a giant molecular cloud complex, a few light years in diameter revealed by the radiation of carbon monoxide at 2.7 mm ^{13}CO radio line; **c** collapse of molecular cloud complex possibly triggered by supernova event; **d** rotating solar nebula in statu nascendi with the proto-Sun evolving in its center; **e** formation of accretion disk; arrows indicate motion of gas; **f** aggregation of particles along mid-plane of accretion disk; **g** and **h** accretion from dust, to planetesimals, to planets; **i** retention of primordial atmosphere by a large planet (e.g., Jupiter); **j** T Tauri phase, sweeping off "excess" primordial gases from the solar system leaving atmosphere-free terrestrial planets behind; **k** spacing of the orbits of the planets (astrological symbols) and the asteroids (dotted area); **l** Oort's cloud (comet reservoir) surrounding the Sun within a radius of about 1 light year in relation to the nearest stars

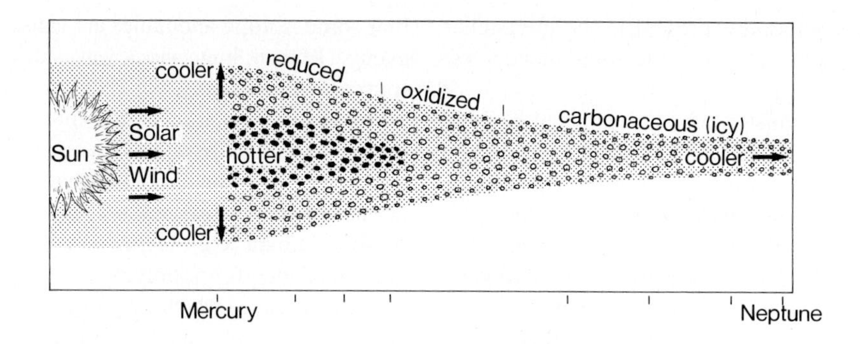

Fig. 4.2. Accretion disk across solar system from Mercury to Neptune. Variations in temperature and oxidation states in the assumed growth regions of meteorites and cometary bodies (after J.V. Smith, 1982)

form, surrounded by a strongly non-axisymmetric region of infalling and quasistatic gas. In this situation, gravitational torques between non-axisymmetric regions will lead to an outward transfer of angular momentum. The presolar nebula, now free from angular momentum interference, can contract, leading in due course to a main sequence star, our Sun (Boss, 1985).

There is a wealth of ideas on the forming years of the solar system and the conditions and mechanisms responsible for its structure at large (Sagan, 1975; Smoluchowski, 1983). In principle, two schools of thought exist on this problem area. The first one assumes a *homogeneous* condensation of matter under relatively "cool" conditions in the direction of nine proto-Planets and a proto-Sun. Due to gravitational collapse and radioactive heat generation, the protoplanets differentiated subsequently into their respective planets, i.e., Mercury, Venus, Earth, Mars, Jupiter, Saturn, Uranus, Neptune and Pluto and, of course, the Sun.

Planetary accretion is believed to have started from an intimate mixture of silicate and metal particles of the type found in chondrites. This chondritic material supposedly had formed prior to the formation of planets in a presolar nebula. With respect to the proto-Earth, this mixture accreted over an extended period of time (10^7 to 10^8 years) and allowed gravitational potential energy to radiate away. The material, composed of various metal and silicate phases, remained undifferentiated and was kept at ca. 1,000° C during this stage. Heat released by the decay of long-lived radiogenic elements subsequently caused melting first of metals and then of silicates. In due course, a differentiated Earth developed. Some initial inconsistencies of the homogeneous accretion model were "ironed-out" in a revised version (Ringwood, 1979).

The second concept operates with a *heterogeneous* accretion (J. V. Smith, 1982; Turekian and Clark, 1969). A chemically inhomogeneous solar nebula is assumed, displaying a wide range in oxidation states and temperature progressions (Fig. 4.2). Accordingly, planetesimals of various size, chemistry, and mineralogy were generated in different locations and at different times. Aggregation of such objects resulted in the present chemically diverse population of bodies in the solar system, including the parent bodies of meteorites. The period of major growth took less than 100 million years, and was accomplished in the time interval between about 4.5 to 4.6 Ga. Following completion of planet formation, the remaining planetesimals were captured in an era of heavy bombardment that lasted for about 0.5 Ga. The major impact craters seen on the Moon's surface are a vivid display of what is often referred to as "terminal lunar cataclysm" (Wasserburg et al., 1977), which peaked at the time of about 3.9 Ga. The inner planets and other moons have certainly felt this cratering event. On Mercury this record is well preserved, while on Earth it has been processed away. During the succeeding 4 Ga, clean-up was slower.

At present, the heterogeneous model is favored, since it is more consistent with the observed facts. The following, highly generalized scenario, is put forward:

A comparatively small molecular cloud positioned in one of the spiral arms of our galaxy (local arm) collapsed, possibly as a result of a nearby supernova explosion and the huge shock waves generated by this incident. Temperatures in the center of the shrinking cloud rapidly soared from 20 K at the start to 1,400 K, thereby transforming the cloud into a hot solar nebula. Similar to a magma, where differentiation begins with the crystallization of olivine, pyroxens and spinels, and ends with quartz, out of a solar nebula grains and particles of high temperature origin emerge first and aggregate to clusters and superclusters leading to condensation points of protoplanets. As the nebula cools down, the aggregates contain a different mineralogy (e.g., feldspars, magnetite). The final products that agglomerate come from even cooler regions of the nebula, that is to say, they were never heated up, but might be considered leftovers from the original solid matter in the interstellar cloud. This last layer arrives at a point where protoplanets condense, containing water, gases, organic molecules, and low-temperature minerals such as clays.

More than 95 % of all the mass retained by the solar system is collected by the proto-Sun. The rest is distributed among the protoplanets and protomoons. As soon as the proto-Sun collapsed to pre-main sequence densities, the cores' temperature was raised to the point of ignition. An enormous outburst of energy, the T Tauri wind, swept the gases, principally hydrogen and helium, from the so-called terrestrial planets (Mercury, Venus, Earth and Mars), leaving atmosphere-free terrestrial planets behind. By contrast, the two largest planets, Jupiter and Saturn, retained their original atmosphere, whereas Uranus and Neptune lost about 90 percent of their hydrogen and helium. Information on Pluto is presently too incomplete for comment on this aspect; furthermore, Pluto's orbit is quite eccentric. However, there is some speculation that Pluto is not a planet but a kind of comet or a giant dirty snowball. This property is shared by its moon Charon. The corresponding radii are 1,600 km for Pluto and 800 km for Charon; the two bodies are 20,000 km apart and circle around each other in 6.5 days, always exposing the same face (Brown and Cruikshank, 1985). They are too distant from the Sun to develop a comet's tail. So in a way one can say that our solar system is in essence bounded by the orbit of Neptune, which is 30 AU (astronomical units) away from the Sun, or 30 times more distant than Earth. The fact that Neptune's orbit is slightly eccentric was formerly reconciled with the discovery of Pluto. However, Pluto's mass – according to Tholen's calculation – is about two orders of magnitude too small to accomplish this, which in turn opens up a new quest.

In comparing Venus and Earth as the two largest of the terrestrial (= inner) planets to Jupiter and Saturn the two largest iovian (= outer) planets (= giant planets), one can say that all four of them contain about the same amount of chemical elements with atomic numbers >2 (ca. 10^{28} g); the difference in mass rests on the amount of hydrogen and helium, which gives the iovian planets a weight of about 10^{30} g. Concerning Uranus and Neptune, the mass of elements with atomic number >2 is also 10^{28} g, but they have only about 10 % of the amount of hydrogen and helium retained by Jupiter; accordingly their mass is 10^{29} g.

The mass and density of the planets do not change in accordance with their distance from the Sun. The fact that some condensation centers obtained more and others less of a certain type of material coming out of the condensing nebula can explain many of the differences in size, chemical composition, or density between the final products, that is: the Sun, the 9 planets, 54 moons, or uncountable numbers of asteroids. The series of meteorites, ranging from iron meteorites to normal chondrites to carbonaceous chondrites (reducing to oxidizing) appears to reflect in a rough way the various stages in the development of our solar system.

The above complex of information concerning composition and evolution of the solar system provides at best some tentative impressions on what really happened with cosmic dust and interstellar gas, for example, on the way to Earth. Geologists must consider this field of cosmochemistry; otherwise their understanding of processes leading to the origin and evolution of Planet Earth will be limited, to say the least.

Above all Brother Sun

The pre-main-sequence evolution of the Sun, following violent and dynamical collapse from the parent gas cloud, has been principally governed by a slow contraction commonly referred to as the Kelvin-Helmholtz contraction. In the course of this comparatively short interlude, about half of the gravitational energy set free became consumed in heating the material, whereas the other half of the energy radiated away. It is interesting to remember that at the end of the last century, energy derived from gravitational contraction was the accepted opinion for explaining even the present luminosity of the Sun. A brief calculation, however, will reveal that this cannot be the case. Assuming that the Sun's luminosity was always constant through the ages, the radius of the Sun would have to shrink about 20 m a year to account for the loss in mass. Each second, the Sun radiates energy in the amount of 3.82 x 10^{26} Joule (J). This corresponds to a loss in mass of about 4 x 10^{6} metric tons, which represents an insignificant fraction of the total solar mass. At this rate, gravitational energy could drive radiation at present value for at most 40 million years. This time period is far too short, considering the age of our solar system, which runs into billions of years. The answer as to where the solar energy stems from was given in 1939 by Hans A. Bethe in his theory on energy generation in stars by means of fusion.

Once temperature-pressure conditions moved into a regime where nuclear reactions could be triggered, the period of slow contraction came to a halt. At that point, the Sun – a star of spectral type G2 – had more or less arrived on the main sequence. This event dates back 4.7 million years, and at the same time is quasi the hour of birth of the solar system, which in the aftermath of the T Tauri incident saw suddenly and for the first time the flash of light of its exploding Sun.

The Sun will remain on the main sequence for roughly 5 billion years more. It will take as much time as this before all the available hydrogen fuel is exhausted. This implies that we are half-way through the expected residence time of the Sun on the main sequence. In about 5 billion years the Sun will leave this track, become a red giant and eventually shrink into a white dwarf (for further details see, e.g., Cameron, 1975; Shklovskii, 1978; Pepin et al., 1980).

The Sun derives energy through thermonuclear processes, and the actual fusion reactor is located in the core. In principle, two protons collide, forming a deuteron (deuterium nucleus) and emit one positron and a neutrino (proton-proton chain; Fig. 4.3). The deuterium nucleus will immediately pick up a proton, yield a helium-3 isotope and emit a gamma-ray photon. Two of these ^{3}He nuclei will combine to a helium-4 nucleus and release two protons:

Reaction	Energy released (MeV)	Reaction time[a] (log t, s)
$^{1}H + ^{1}H \rightarrow ^{2}D + e^{+} + \nu$	1.44	17.6
$^{2}D + ^{1}H \rightarrow ^{3}He + \gamma$	5.49	0.7
$^{3}He + ^{3}He \rightarrow ^{4}He + 2\,^{1}H$	12.86	13.5

[a]values log t = 0, 5, 10, 15 correspond to $t \approx 1$ s, 1 day, 300 yr, 30 million years, respectively (after Shklovskii, 1978).

There is an alternate branch of the proton-proton chain giving rise to: ^{7}Be, ^{8}Be, ^{7}Li, and ^{8}B (see Table 3.1). The proton-proton reaction chain is by no means a straightforward one. To achieve a collision, the proton must approach to within 10^{-13} cm of the nucleus and this against electrostatic repulsion, since both partners have a positive charge (the Coulomb barrier). The law of quantum mechanics, however, allows one in a hundred million protons to have energy high enough to overcome the Coulomb barrier and liberate energy equal to the observed luminosity of the Sun:

Praise to thee, my Lord, for all thy creatures
above all Brother Sun
who brings us the day and lends us his light.
(St. Francis of Assisi, 1181–1226)

The bulk of the Sun's radiation is in the visible and the bordering regions of the infrared and ultraviolet (Fig. 4.4) (Parker, 1975). A thin layer measuring ca. 350 km in height, the photosphere, is a device which controls the rate at which radiation is emitted from the Sun (Fig. 4.5). It has the great advantage of being directly accessible for observation. At its base temperatures reach 8,000 K, and at its surface they are close to 4,500 K. The luminosity of the Sun is an average between the two extremes, i.e., ca. 5,800 K. A change in luminosity by 1 percent would already raise global temperature at the Earth surface by 2° C. This explains why currently so much research is going on to find out whether the solar constant is truly a constant or whether it changes in a random or systematic way with time (e.g., Schatten et al., 1987). Observational data collected over the past decade indicate, however, that during this period the solar constant has not changed at least within a 0.1 percent confidence limit (Eddy, 1979; Eddy et al., 1982; Wilson, 1978).

The topology of the photosphere is not smooth but granular. The diameter of a single granule measures almost 1,000 kilometers. Granules can be looked upon as moving cells that come and go and have a life span of a few hundred seconds at most. This regular pattern of the photosphere can change abruptly. Solar

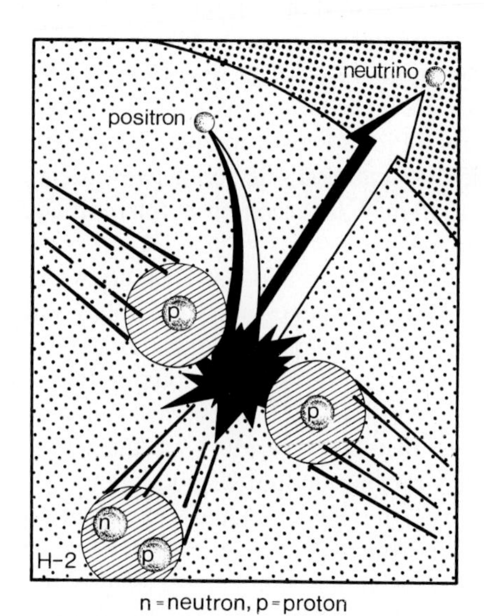

Fig. 4.3. In a reaction at the Sun's core, two protons collide, yielding a deuterium nucleus and emitting one positron and a neutrino (after Trefil, 1978)

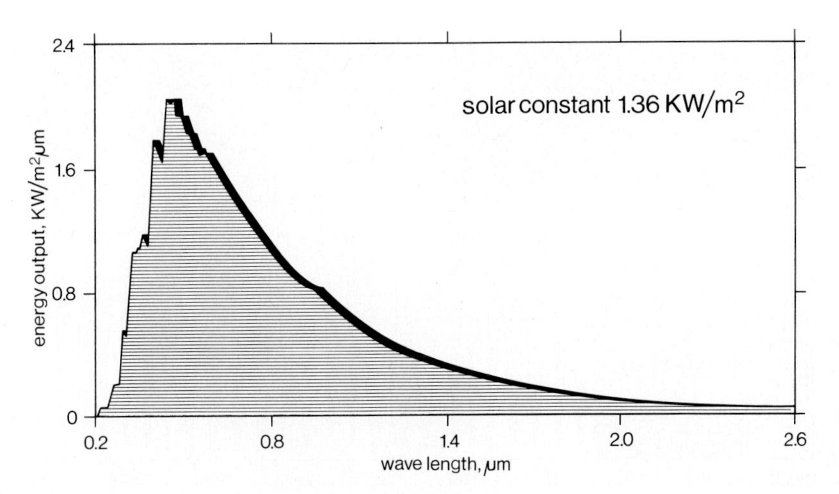

Fig. 4.4. Solar energy spectrum, measured outside of the Earth's atmosphere. Energy output in relation to wave length

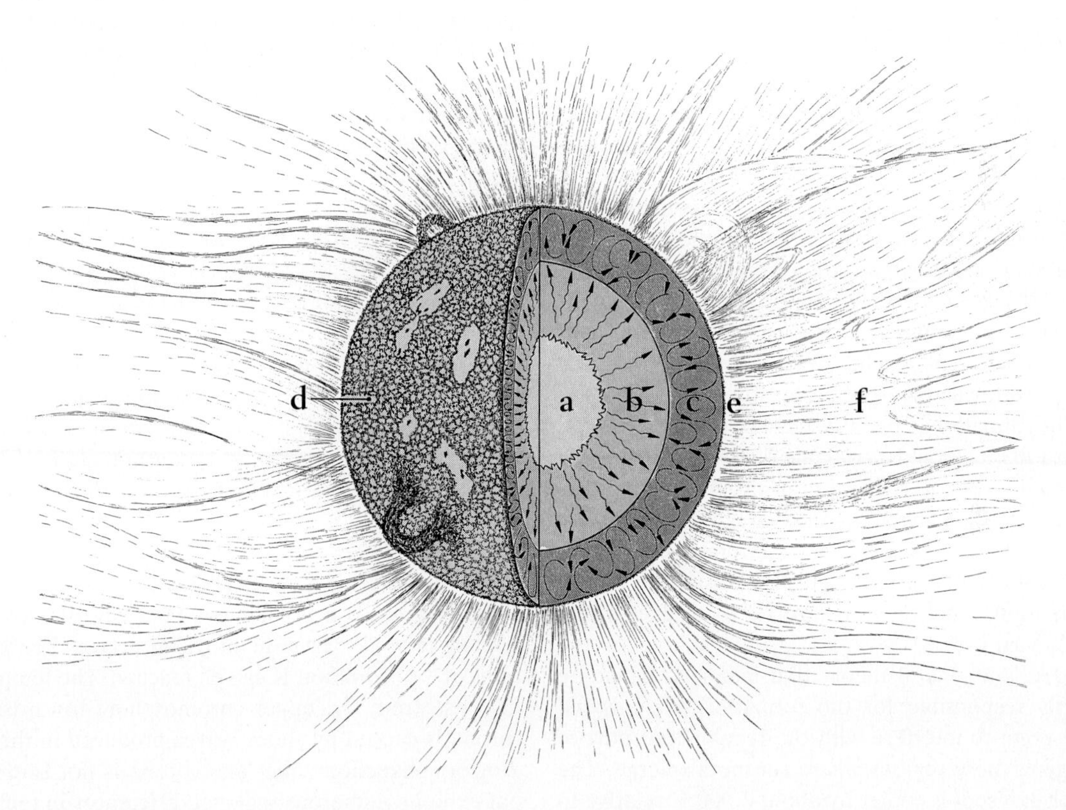

Fig. 4.5. Structure of the Sun's interior and atmosphere (from inside to outside; schematic): a nuclear reaction zone; b radiative zone; c convective zone; d photosphere showing visible granulated surface with prominences, plages and sunspots; e chromosphere only seen when disk of the Sun eclipsed, and f corona, Sun's outer atmosphere, seen during eclipse with streamers (major features), polar plumes (N and S) and loop (upper right)

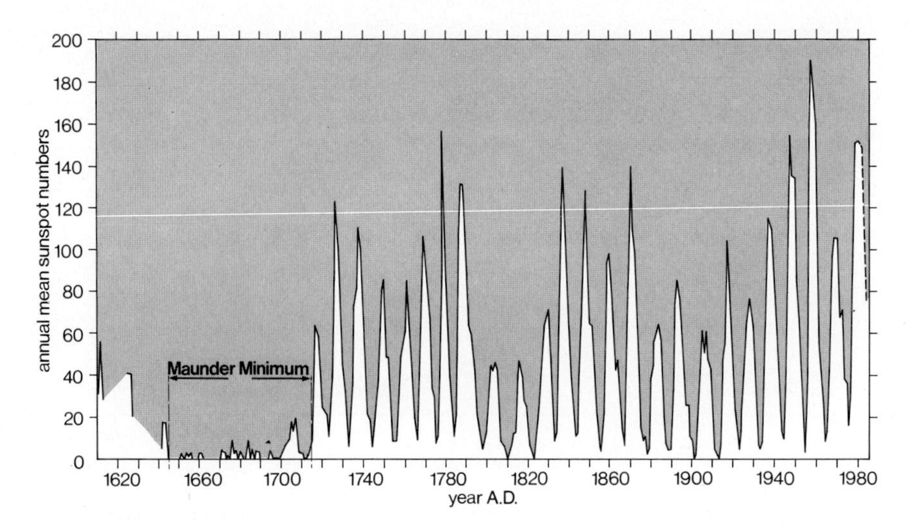

Fig. 4.6. Annual mean sunspot numbers from the days of Galileo to the present (after Waldmeier, 1961; Eddy, 1977a,b; Schove, 1983)

activities can generate sunspots and plaques which at times can cover more than 20 percent of the Sun's surface (e.g., Lawrence, 1987).

Sunspots were first seen by Galileo Galilei (1564–1642) around the year 1610. Mapping these incidents by number, areal extent, and duration revealed that from 1645 to 1715 they were almost non-existent (Maunder Minimum = little ice age). Since then, peaks of different heights were observed at roughly 11-year intervals (sunspot cycle; Fig. 4.6). Their impact on climate has been discussed by Eddy (1977a, b). In 1908, George Ellery Hale discovered that spots commonly come in pairs of opposite magnetic polarity, with the leading spots in each solar hemisphere having the same polarity. Furthermore, polarity of leading spots is reversed in successive sunspot cycles, which implies that the Sun reverses its magnetic field in a 22-year cycle.

There is a general theory that magnetic fields are directly responsible for the formation of sunspots. They seem to interfere with the regular transport of energy in those regions where sunspots emerge. The center of a spot is cooler by about 2,000 K relative to the photosphere. In brief, the convection zone organizes the solar magnetic field, and the 11-year sunspot cycle somehow mirrors these events at the surface of the photosphere (Box 4.1).

Just above the photosphere lies the chromosphere, which during a solar eclipse is seen as a rosy arc for a few seconds. The pinkish color is due to the H-line of hydrogen. The chromosphere is an irregular band a few thousand kilometers high, from which flamelike spicules shoot up to a height of occasionally 10,000 km. Their width is ca. 1,000 km, which matches that of the granulation, and suggests that they are extensional features. Above the photosphere, temperature drops to a minimum of 4,000 K at a height of about 1,500 km, but rises later to coronal temperatures (10^6 K).

The most prominent phenomena emerging from the chromosphere are the prominences which have a filamentous appearance. They are a product of magnetic fields. Jets of matter reach up several hundred thousand kilometers at speeds of up to 700 km s^{-1}, and the matter appears to be in a constant state of flux. During their stationary phase, prominences may last for months.

The Sun's outer atmosphere, the corona, can be observed only during eclipses (Pasachoff, 1977; Wolfson, 1983) (Fig. 4.5). It is seen as a faint whitish glow extending several solar radii into all directions. Temperatures of 1 to 2 million K can be reached. The temperature rise from the upper chromosphere towards the corona is caused by shock waves produced in the hydrogen convection zone. The corona is not isotropic but exhibits numerous regional differences in temperature, density, particles, degree of ionization or conductivity. The corona, and here the coronal holes, are the areas where solar winds originate that rapidly sweep matter from the Sun into the interplanetary space at velocities of several hundred kilometers per second (Conti and McCray, 1980).

BOX 4.1

Sunspot Cycle Through Time (Kempe, 1977; Williams, 1986)

Controversy exists as to the cause of the sunspot cycle and associate magnetic polarity reversals. One explanation is that internal motions of the Sun interact with the large-scale magnetic field in the Sun to generate a dynamo. Such a mechanism could readily be maintained for billions of years with little overall change. Alternately, the large-scale magnetic field is considered to be a remnant of a field the Sun acquired in statu nascendi. In such circumstances the solar activity cycle could be expected to change in the course of geologic time.

Solar signals have apparently found their imprint in the thickness of tree rings (e.g., Suess, 1968, 1970) and stable isotope oscillations observed in ice cores from Greenland (e.g., Dansgaard et al., 1971). Furthermore, the water level of Lake Van, the fourth largest closed lake on Earth, fluctuates in phase with annual mean sunspot number (Kempe, 1977). This pattern seems to be responsible for the approximately ten-year cyclicity observed in the thickness of varves, which are annual sediment deposits that generally come in sets of two, namely one dark winter and one light summer layer. In the case of Lake Van, the individual layers measure less than 1 mm in thickness. In a 10 m sediment core retrieved from a water depth of 350 m, Kempe (1977) counted more than 20,000 microlayers which add up to 10,420 years of recorded history, thus representing the entire Holocene. By means of a Fourier analysis he was also able to establish cycles of 70, 94 and about 400 years, and he concluded that such rhythms were sunspot-related.

Post-Glacial lake deposits formed about 1 million years back in time were recovered from the Black Sea deep basin. The classical varve pattern was encountered twice at a core depth of about 360 and 400 meters. On the average, four dark and four light layers per cm were counted. The layers are all about the same thickness, except for the twelfth light layer, which is two to three times wider than the average layer; this feature was explained as sunspot-related (Degens et al., 1978).

Periodic signals encoded in varved lacustrine sediments of late Precambrian time (680 million years; Elatina Formation, Australia) can best be reconciled with sunspot rhythms (Williams and Sonett, 1985; Williams, 1981, 1986; Sonett and Trebisky, 1986). Nowhere in the 19,000 years of continuous signals is there a cessation of cyclicity comparable to the alleged Maunder Minimum, the little ice age at the end of the seventeenth and the beginning of the eighteenth century (see p. 44). Principally on these grounds, Williams (1986) questions the validity of that minimum. In general, the observed sunspot cycle is not strictly periodic, but ranges from about 9 to 14 years in duration. The Elatina lamination cycle is in accord with this variation and other periods, for instance 22 to 25 years or 90 to 110 years in length; they match extremely well with the sunspot observation record. By using longer-scale periodicities, e.g., 275 to 335 years from the Elatina data, future solar activity cycles might be predicted.

Varved lake sediments, displaying solar activity signals, are apparently found throughout the geological column. However, they are restricted to times or regions exhibiting a cold and generally dry climate of the type prevailing at high latitudes or altitudes, or in lakes bordering glaciers. In such circumstances an increase in solar activity will entail a greater annual discharge of meltwater, a higher supply of nutrients to the lake, enhanced primary productivity in the lake, and consequently the deposition of thicker varves. The fact that the sunspot cycle has been with us since Precambrian times argues in favor of the dynamo principle.

Planets, Moons, Etceteras

All gaseous, liquid or solid objects encountered in the solar system from the Sun to the planets, moons, asteroids, and interplanetary particles have the same heritage: a single molecular cloud complex. Even the huge reservoir of cometary bodies, about 10^4–10^5 AU from the Sun, is supposed to have originated here. In contrast, much of the electromagnetic radiation flooding interplanetary space is from beyond and way back in time.

Over the past quarter of a century, and with the advent of space exploration, planetary science has made a great step forward. A cornucopia of new data on planets and moons is now available and its proper assessment will require years of scrutiny. The excite-

ment, looking at the colorful pictures of planets and their satellites, is shared by scientists and laymen alike. Details are absorbing but also overwhelming. Plenty of room is left for speculation, because new data accumulate at a rate which leaves scientists little time for constructive reflection.

One of the major goals in solar system studies is to find an answer as to why mass fractionated in the way it did. About 97 percent of the mass that following the T Tauri event remained in the system, is kept by the Sun, leaving a meager 3 percent for all the other planetary bodies. This 3 percent rest has accreted in the nine planets, and here principally in the four major planets – Jupiter, Saturn, Uranus, and Neptune. The terrestrial planets – Mercury, Venus, Earth, and Mars – comprise less than 0.5 percent of the mass of the major planets. Expressed differently, terrestrial planets contain only about 0.1 percent of all matter present in the solar system. All 54 moons known so far have received less than 1/3000 the mass of the planets. The etceteras: asteroids, comets and interplanetary dust are chemically speaking, "impurities", and are just a miniscule fraction of planetary matter.

Distribution of mass across the solar system and its fractionation into rocky, gaseous, and ice materials occurred quite early in the evolution of the solar nebula. In essence, planetary feeding zones allowed protoplanets to grow at the expense of nearby competitors (Fig. 4.7). Regularities in the planetary spacings appear to conform to a scheme known as the Bode or Titius-Bode Law. Actually, it is not a law but a sequence of numbers: 0, 3, 6, 12, ..., doubling each time. To the individual number one adds 4 and divides the sum by 10. The end figures, expressed in astronomical units (AU), accord remarkably well with the actual distance of the planets from the Sun. The exception is Pluto, where the calculated distance is twice the observed one. It is interesting to note that the Titius-Bode scheme presupposes a planet between Mars and Jupiter at 2.8 AU. Situated at this distance is the asteroidal belt, which is composed of small objects up to a few hundred kilometers in diameter. The total mass of all asteroids is 5 x 10^{24} g or 0.1 percent of the mass of the Earth (Fig. 4.8).

Concerning planet formation and history, there is no way to condense facts and ideas into a few pages of text. The issue is too complex. However, the subject matter has been critically reviewed by J.V. Smith (1979, 1982), who makes a strong case for the heterogeneous growth model of meteorites and planets on the basis of mineralogical data. Ringwood (1979), using geochemical evidence, pays special tribute to the Earth-Moon system and favors a revised homogeneous growth model. Bodenheimer et al. (1980) and Pollack (1984) treat the origin and evolution of the outer planets largely from a physical point of view, while Wetherill (1980, 1981) and Cox and Lewis (1980) present a careful analysis on the formative years of terrestrial planets. The topic at large has been discussed, e.g., by Murray (1975), Greenberg (1979, 1984), Greenberg et al. (1978), Harris and Ward (1982), Hartmann (1978), Matsui and Mizutani (1978), Stevenson (1982), Weidenschilling (1978), and McKinnon (1984). The body of information indicates

Fig. 4.7. Cartoon from planetesimals to planets. Feeding zones widen with time in accordance with the growth rate of the protoplanet, eventually yielding overlap of zones. Major surviving feeding zones give rise to planets and asteroids which are displayed at the square root of distance from the Sun. Planetesimals ejected from giant planets yield cometary bodies, which accumulate in the Oort cloud, from where they become episodically ejected and re-enter the solar system. Bombardment of the terrestrial planets by comets and asteroids throughout the ages has been omitted for graphical reasons (after Safronov, 1969, 1972, and J.V. Smith, 1979, 1982)

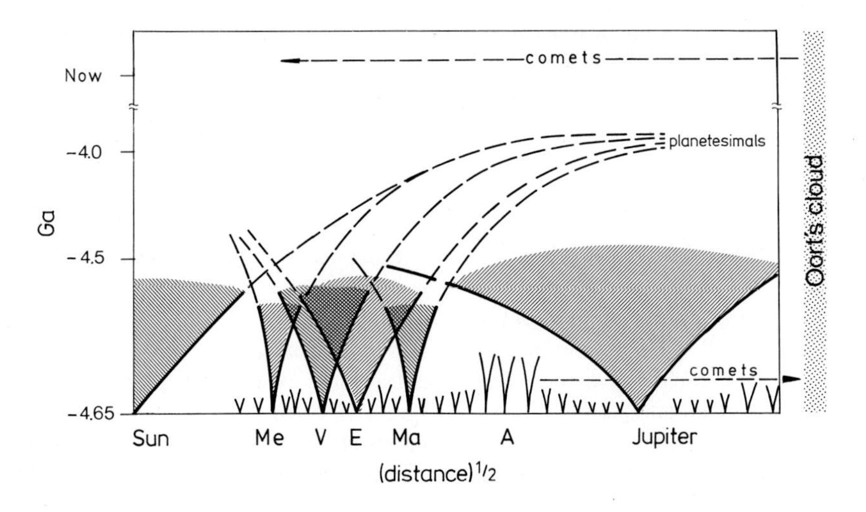

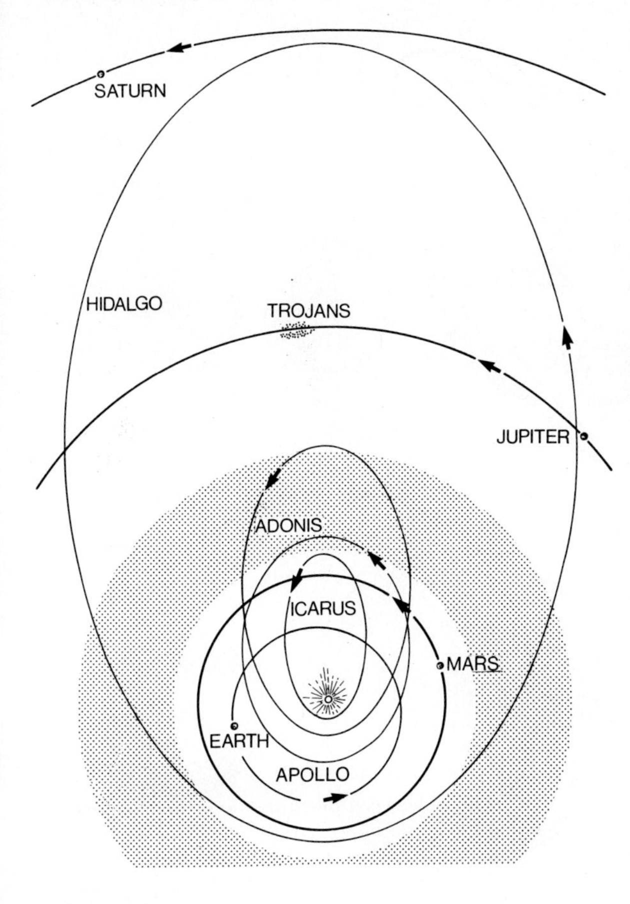

Fig. 4.8. Orbits of major asteroids in relation to the Sun, and the orbits of Earth, Mars, Jupiter and Saturn. The shaded area between Mars and Jupiter is the region of the asteroidal belt

that planet formation is not a random process but that it proceeds in logical successions of events which can account for the present composition, structure and size of an individual planetary body (Hubbard, 1981; Podolak and Reynolds, 1981). Critical factors are (i) the position of the planet in relation to the Sun at the time of aggregation and during the T Tauri event, (ii) the distance to its next planetary neighbors competing for planetesimals, (iii) the supply of planetesimals and the efficiency to retain them (growth rate), (iv) the energy released by gravitational collapse during accretion into a full-sized planet, and (v) the differentiation of matter up to the present time.

An inventory of planets and their principal characteristics is set out in Table 4.1 and their interior structures are shown in Fig. 4.9. The planets can be grouped into terrestrial and great planets (except for Pluto), or the inner and the outer planets. Based on the state of matter, one could further classify them as rocky, radiating, and icy planets. The terms rocky and icy are self-explanatory but a word is needed for the term radiating. The power Jupiter and Saturn emit in form of infrared radiation is for Jupiter 1.5 to two times the amount of solar energy it absorbs. The value for Saturn is even higher, i.e., two to three times the sunlight the planet absorbs. Thus, the heatflow from within must be considerable, but the size of the two planets is far too small to trigger nuclear fusion.

The logic behind the planetary spacings has already been briefly mentioned. Retrograde movement (indicated as a minus sign) is observed for Venus and Uranus. The mass difference between terrestrial and outer planets results principally from the mixing ratio between hydrogen + helium versus the rest of the elements. The two terrestrial planets Venus and Earth

Table 4.1. Characteristics of the Planets

	Terrestrial Planets				Great Planets				
		Rocky Planets			Giant Planets		Icy Planets		
	Mercury	Venus	Earth	Mars	Jupiter	Saturn	Uranus	Neptune	Pluto
Mean distance from Sun (millions of km)	58	108	150	228	778	1,427	2,870	4,497	5,900
Period of revolution (in years)	0.24	0.62	1.00	1.88	11.9	29.5	84.0	164.8	247.7
Period of rotation (in days)	59	–243	1.00	1.03	0.41	0.43	–0.45	0.70	6.4
Mass (Earth = 1)	0.055	0.815	1	0.108	317.9	95.2	14.6	17.2	0.003(?)
Density (g/cm^3)	5.44	5.27	5.52	3.93	1.33	0.7	1.2	1.7	2
Equatorial radius (in km)	2,439	6,052	6,378	3,393	71,400	60,000	26,000	24,500	2500 (?)
Mean temperature at visible surface (°C)	350 day –170 night	480	22	–23	–150	–180	–210	–220	–230(?)
Number of moons	–	–	1	2	16	17	15	2	1
Atmosphere (main components)	none	CO_2	N_2,O_2	CO_2,Ar	H_2,He	H_2,He	H_2,He, CH_4	H_2,He CH_4	CH_4(?)

Fig. 4.9. a, b. Interior structures of planets (tentative models). Top graph (from left to right): Jupiter, Saturn, Uranus, Neptune. Bottom graph (from left to right): Mercury, Venus, Earth, Mars. For each row scale bars are given at left-hand margins. **(a)** Top row: *Jupiter* consists of three layers: (i) iron-silicate core (radius 12,000 km), (ii) liquid metallic hydrogen (33,000 km), and (iii) liquid molecular hydrogen (25,000 km).
Saturn resembles Jupiter structurally, even though its density is just one half, i.e., 0.704 vs. 1.314. Between rocky core (radius 10,000 km) and metallic hydrogen layer (8,000 km) lies a 5,000 km thick shell composed of various kinds of frozen substances (e.g. ice). The outer layer (37,000 km) consists of molecular hydrogen.
Uranus and *Neptune,* about equal in size, have three shells: rocky core, ices, and molecular hydrogen each being close to 8,000 km thick. The core of Uranus is probably liquid.
Pluto (not shown) has a density of 2 g cm^{-3} and possibly a mass about a tenth the mass of the Earth; its structure is virtually unknown. **(b)** Bottom row: *Mercury* has an "over-sized" molten iron core, a plastic mantle, and a rigid crust. Its mean density is comparable to that of the much larger sized Earth (5.4 vs. 5.5 g cm^{-3}) which shows the difficulty for primordial volatile compounds to condense when being so close to the Sun (Weidenschilling, 1978).
Information on *Venus'* structure is scanty, but because of size, density and position in the solar system the principal internal structure could match that of the Earth (twin planets). The lack of a magnetic field is linked to the slowness of rotation.
Earth and *Mars* are principally composed of core, mantle and crust. Geophysical data and geochemical model calculations permit more detailed subdivisions for both planets (see text)

have each about the same amount of rock-forming elements as Jupiter, Saturn, Uranus or Neptune contain individually. The temperature differential (from left to right) could be used to speak of hot (Mercury, Venus), temperate (Earth, Mars), cold (but radiating Jupiter, Saturn), and icy planets (Uranus, Neptune, Pluto). On Earth and Mars, surface temperatures can rise as high as 60 and 40° C, respectively. This lies within the stability field of liquid water, and accordingly, they are often referred to as the water planets. The number of moons is highest for Jupiter and Saturn and decreases to one towards Pluto and Earth. Planetary gases will be discussed at length in the chapter: air.

Much of what we know today about atmospheres, surface environment, or the interior of planets we owe to space exploration that started in 1959 with the first mission to the Moon. Space probes have brought us to Saturn, the Lord of the Rings. Voyager 2 reached Uranus in January 1986 and will come closest to Neptune in summer 1989. Nothing is yet in sight for Pluto (c'est la vie!).

This is not the place to elaborate further on the subject of planets. David W. Hughes in a 1985 review in Nature on a book entitled "The Solar System" remarked, "Everything is mentioned, but only briefly and sketchily, and the author has thus given himself little room to point out the problems and mysteries ... and now that we are used to pages of color images of planets, the reversion to black-and-white produces a sense of disappointment". I have tried to condense some of the problems and mysteries into a few pages of text and hope that the black-and-white pictures (Figs. 4.14 and 4.15) do not leave that sense of disappointment.

To find out more on the solar system one has recourse to a wide selection of books from which to choose, not to mention the thousands of articles in this field. A highly recommendable one is that of Roman Smoluchowski (1983), "The Solar System: The Sun, Planets, and Life".

In general, planets and far-away moons are rather remote topics to the average geologist. In contrast, there is an enormous interest when it comes to the Sun, our Moon, the asteroids or the comets. There are very good reasons for this. These objects control much of what is occurring on the Earth's surface: the hydrological cycle, weather, climate, or the evolution of life, just to mention a few impacts. However, there is another aspect, a less apparent one. A meteorite, or a piece from the Moon, can be picked up, be held in one's hand and looked at. So until the moment comes that you can "catch a falling star, and put it in your pocket, keep it for a rainy day", as Perry Como sang in the late fifties, geologists can only examine extraterrestrial rocks that come from probes or that fall from the sky.

Our Moon

Our only satellite, the Moon, has not revealed its true origin yet. The same three fundamental questions that have been asked for almost 100 years are still with us: is our Moon Earth's wife, captured from elsewhere, Earth's daughter, fissioned directly from proto-Earth, or sister, accreted from the same binary system? (El-Baz, 1975). After Apollo, all three basic theories have been modified to accommodate novel findings. And still no general consensus on the Moon's origin exists. It is not unusual for reputable scientists to move from one theory to another, only to return, after many transmutations, to their original idea. The best example of a change of mind is the case of the eminent planetary scientist Harold C. Urey (1893–1981). His quest for the Moon has been kindly related by Brush (1982). Urey spearheaded the manned Apollo lunar mission in the hope of finding supporting evidence that our Moon was indeed captured by the Earth, and that to the present day it had remained a completely primitive and cold body. When the first rocks came back from the Moon, showing that it had once been hot, Harold C. Urey accepted the new facts gracefully. He was willing to change his mind and opted for the fission theory. At the same time he saw merits in other scenarios as well, and doubts about the Moon's origin were always with him.

The Moon has a radius of 1,738 km, a mass of 7.35 x 10^{25} g (ca. 1 percent of Earth), a density of 3.34 g cm^{-3} (5.52 g cm^{-3} for Earth) and a rotation period of 27.3 days. The oldest lunar rocks are 4.5 Ga, which predate the most ancient terrestrial rocks by 700 Ma (Wasserburg et al., 1977; Turner, 1977). Rock ages, combined with topographic and morphologic characteristics of the Mare and highlands, permit lunar stratigraphic mapping and the outline of major events that shaped the surface of the Moon. Shoemaker et al. (1963) started the stratigraphic work. The time-stratigraphic sequence is now well established (e.g., Mutch, 1972; El-Baz, 1977).

The idea that the Moon had been captured in toto can be dismissed. The chemical similarities between the Moon, Earth, and eucrite parent body are too striking, and exclude such a chance event (Anders, 1977). However, there is good reason to believe that the proponents of the fission theory and the binary planets theory do not differ as widely in their viewpoints as originally thought. A possible compromise is in sight (Ringwood, 1986; Binder and Lange, 1977; Taylor, 1979).

In the previous discussion it was said that the heterogeneous accretion model found wider support than the homogeneous model in explaining the growth of the terrestrial planets. In applying this concept to the Moon, J.V. Smith (1981) came up with the following scenario (Fig. 4.10):

During stage 1, the proto-Moon was melting and crystallizing as it grew by the capture of moonlets and planetesimal debris orbiting the Earth. Its volume was about half that of today, corresponding to a radius of 1,300 km. Complete melting is assumed, giving rise to a differentiated proto-Moon with FeS core, mantle and crust. Impacts were common not only on the Moon but on Earth as well. Impact-induced fission of the Earth and orbiting planetesimal debris might have supplied hot material for growth of the Moon. Critical points are that a material exchange between Moon and Earth is feasible, and that melting began before accretion was completed.

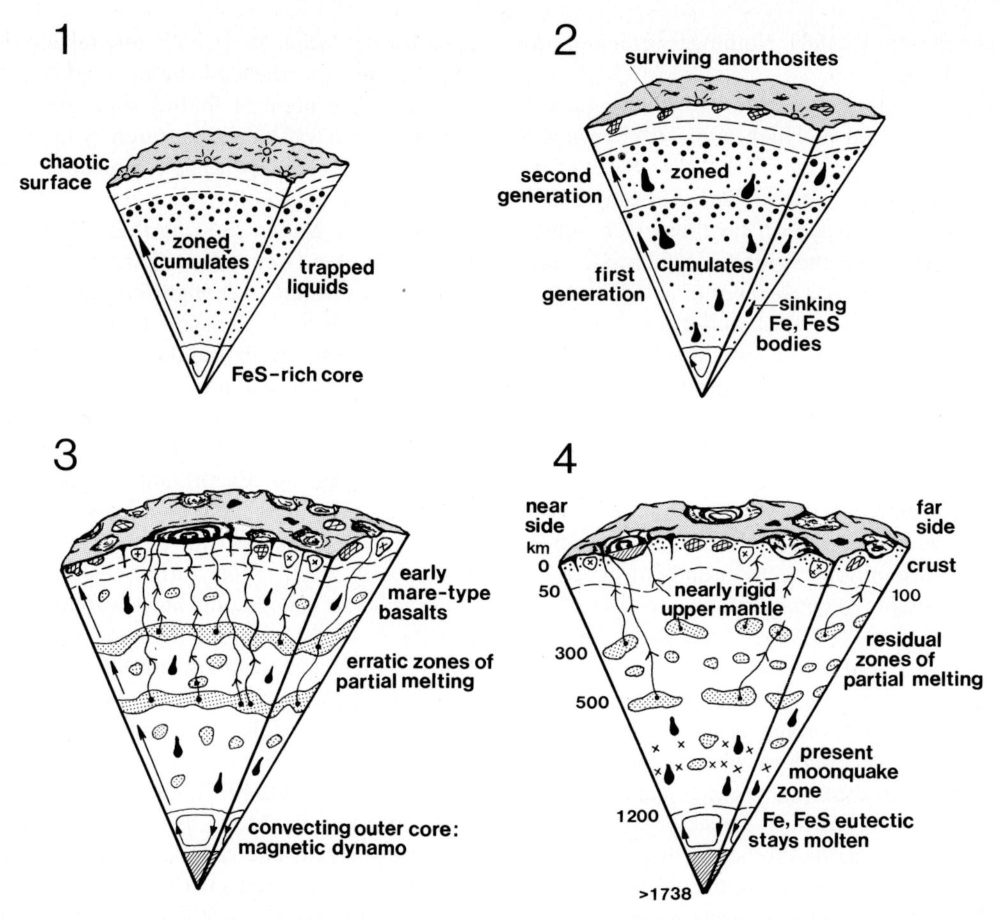

Fig. 4.10. Progressive differentiation of a growing Moon (tentative model): The initial stage marks the time the Moon had differentiated to a point where crust, mantle and a FeS-rich core became first established. Petrographically, the crust consisted of plagioclase-rich rocks and basalt, and the mantle of zoned cumulates grading from olivine-rich towards plagioclase, pyroxene, ilmenite-rich members. Arrow indicates progressive crystal-liquid differentiation. A zone of residual liquids rich in Fe and incompatible elements was intercalated between crust and mantle.
The next stage (100 Ma) is arbitrarily set at a time the volume of the Moon had grown by one-quarter. The size of the mantle had substantially increased, and the convecting core grew by "droplets" of Fe, FeS-rich material. Within the crust, surviving anorthosite bodies are indicated.
During the following stage (300 Ma), the Moon had reached full size. Early mare-type basalts were generated in zones of deep-seated cumulates. In the crust, layered plutons developed. At the age of 500 Ma (stage 4), the interior of the Moon had essentially achieved the present structural state. Because of difference in crustal thickness, only the crust nearest to the Earth received flows of mare-type magmas. ANGST (anorthositic-noritic-gabbroic-spinel-troctolitic) rocks mainly form the crustal megaregolith (after J.V. Smith, 1981)

During each of the two succeeding stages, arbitrarily ending at 100 and 300 million years, about one quarter of the volume of the Moon was supposed to have been added. These are, of course, not discrete events, but represent a continuous and dynamic growth of the Moon up to a fixed volume. The type of material received during stage 2 remains about the same as before. Fragments of anorthosites from stage 1 crust survive. Fe- and FeS-rich droplets sink towards a convecting core. The process of zone refining (Longhi, 1980) leads to a pronounced enrichment of iron and incompatible elements in the residual liquid.

During stage 3, late projectiles produce basins which become irregular because of a weak crust. Layered plutons form, and liquids trapped at depth start to migrate. Heavy FeS-rich residues sink and silicate-rich ones rise. Early mare-type basalts are generated. A crystallizing inner and a convecting outer core create a magnetic dynamo. The fact that samples of lava and breccia returned by Apollo missions possess remnant magnetism is used in support of an iron core and the dynamo principle. Since today's Moon does not exhibit a magnetic field, the Moon is thought to have lost its magnetic field, probably at about the time when volcanic activity stopped. This interpretation is at present widely challenged (see Runcorn, 1983). For instance, meteoritic and cometary impacts, electric effects in volcanic eruptions or even lightning, are considered a possible trigger mechanism to induce a magnetic field. The fundamental question is, whether such local events can generate a lunar magnetic field on a global scale.

During stage 4, heavy bombardment comes to a close, and the crust is becoming stronger, as indicated by ringed mare basins that retain their circular shape. Crustal thickness on the near side of the Moon is about 50 km, and on the far side about 100 km. This difference in size explains why mare-type magma flows into Earth-faced basins, but tends not to penetrate the far-side crust. The core differentiates into an inner core composed of iron alloys and into an outer core made up of a molten Fe-FeS eutectic. Mare-type volcanism gradually ceases, and the majority opinion is that lunar volcanism came to a halt 3 Ga ago (e.g., Wasserburg et al., 1977). A minority viewpoint is that lunar volcanism has been a relatively continuous process, perhaps starting as early as 4.3 Ga ago and gradually terminating around 1 Ga, a time synchronous with the Copernicus impact (Schultz and Spudis, 1983). In any event, the Moon is now tranquil, except for deep-seated moonquake activities.

From time to time the Moon is hit by a meteorite or comet. The rate of infalls onto the Moon as a function of time is commonly plotted in the form of temporal crater density curves for craters of intermediate size (4–10 km in diameter) (Fig. 4.11). It has also been found useful to make a comparison to Martian or terrestrial cratering events (Wetherill, 1977; Neukum et al., 1975; Neukum, 1977; Baldwin, 1971; Soderblom et al., 1974; Hartmann, 1975; Grieve and Dence, 1979). Wetherill's calculations gave a continually declining rate of infall following lunar terminal cataclysm at 3.9 Ga and a half-life increasing very substantially towards the present. In contrast, Baldwin (1985) gathered evidence consistent with a rapid decline of infalls between 4 Ga and 3 Ga ago, and a systematic rise in the number of crater-forming incidents to a modern cratering rate twice that of 3 Ga ago. The interpretation is offered that heavy pre-mare bombardment came from remnant asteroid-like bodies in eccentric orbits extending from Mercury to Saturn and perhaps beyond. Once this source became exhausted at about 3 Ga, fragmented parts of the Apollo asteroids were received by the Moon in increasing numbers.

The Moon orbits Earth at a distance of 384,400 km and recedes at 4 cm per year. This departure is a result of tidal forces. In essence, as the Moon raises ocean tides, they in turn bring the Moon to a higher orbit,

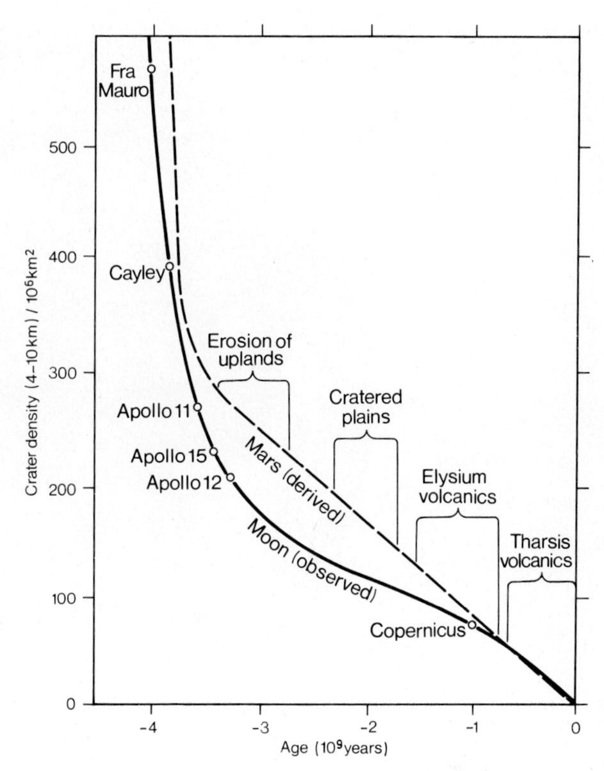

Fig. 4.11. Comparison of martian and lunar cratering events (Soderblom et al., 1974)

slow down Earth's rotation, and make the day and the month longer. Rates of energy dissipation are presently 4 x 10^{12} Watts (Lambeck, 1980), and 90 percent becomes dissipated in the ocean and here largely through bottom friction in shallow seas. Extrapolating modern dissipation rates into the geologic past should lead to a close Moon 1.5 to 2 billion years ago. The shape, alignment, and depth of modern ocean basins are optimal for the dissipation of tidal energy. Temporal changes in the distribution of continents and oceans could be a cause for a lower rate in tidal dissipation. Webb (1982), in using an "average" ocean model in relation to the evolving Earth-Moon system, concludes that tidal energy dissipation has been less in the geologic past than today. K.S. Hansen (1982) suggested that it is not so much the changing ocean basins but the changing rotation of the Earth that determines the rate of tidal dissipation. In Hansen's model the closest the Moon could come is 225,000 kilometers, which is more distant than the Roche limit, i.e., the minimum distance at which a satellite can hold itself together by its own gravitational field.

Summing up, Moon and Earth evolved side by side. Impact-induced fission might have contributed hot terrestrial material to the growing proto-Moon. Differentiation of lunar material started early, resulting

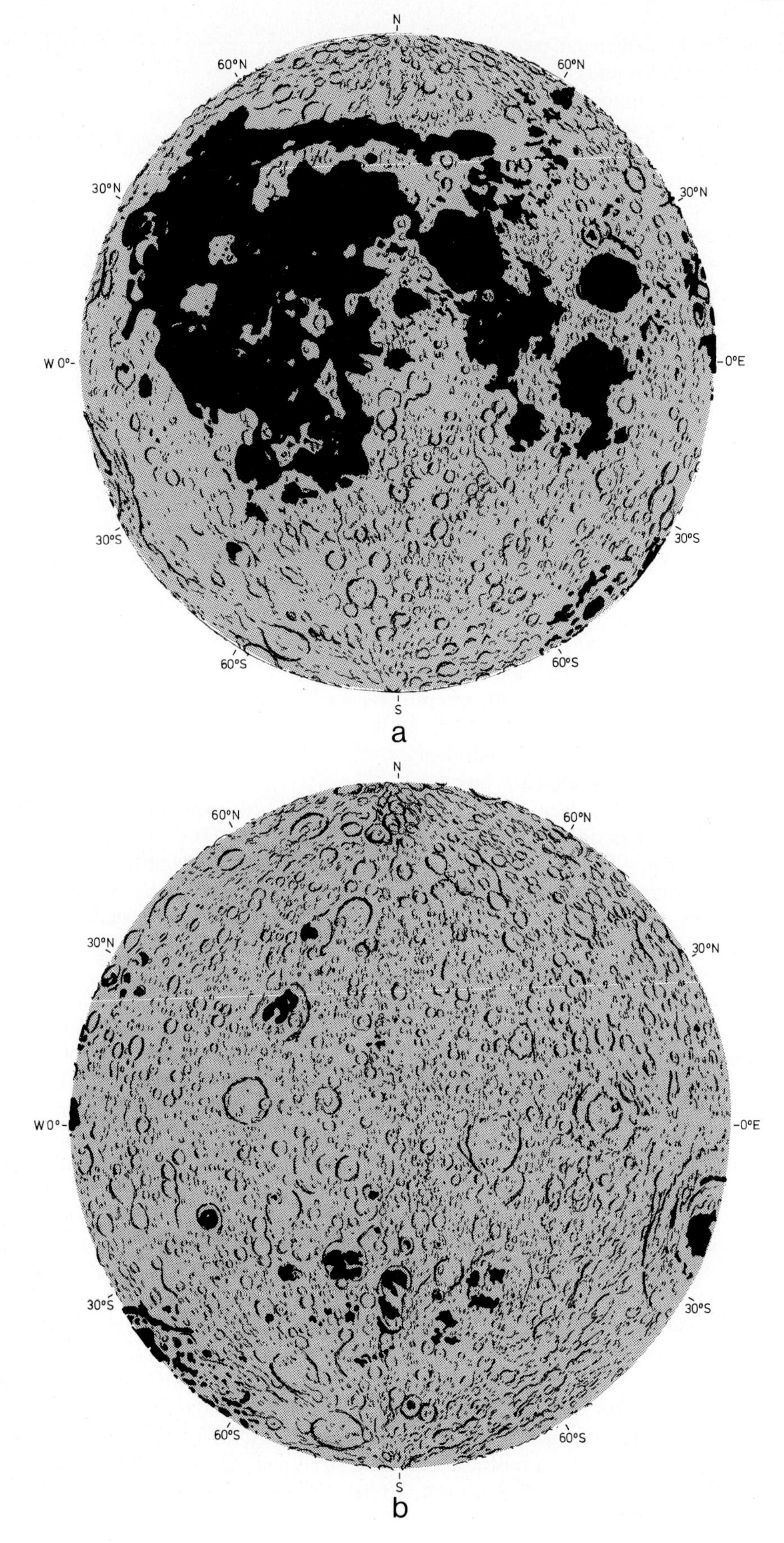
N
60°N
60°N
30°N
30°N
W 0°
0°E
30°S
30°S
60°S
60°S
S
a
N
60°N
60°N
30°N
30°N
W 0°
0°E
30°S
30°S
60°S
60°S
S
b

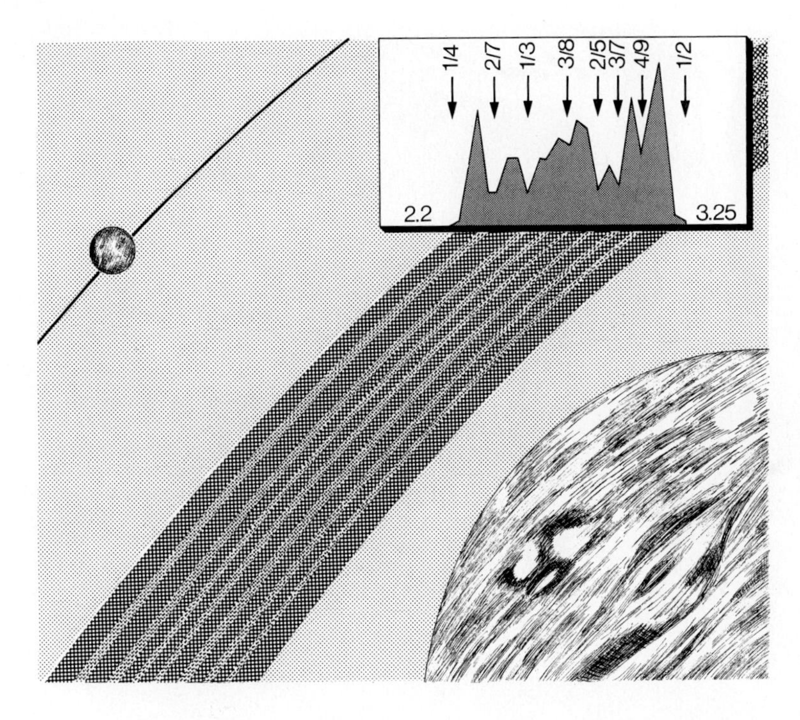

Fig. 4.13. Unevenness of distribution of bodies within the region of the asteroidal belt, which lies between Mars and Jupiter 2.2 to 3.3 AU away from the Sun. The seven valleys of lower frequency are termed Kirkwood gaps. Their position appears to be determined through orbital resonances with Jupiter

in core, mantle, and crust. The fact that the Moon's crust is thinner on the side facing the Earth is a consequence of tidal forces. The mare-type volcanism is thus an expression of this feature, as shown in Fig. 4.12 (after El-Baz, 1977). Volcanism stopped 3 Ga ago. The only activity left on the Moon is occasional moonquakes, impacts and visits from Earth. A reappraisal of the Moon's origin and history has recently been given by Hartmann et al. (1984) and Ringwood (1986).

Asteroids

The Titius-Bode scheme predicted a planet between Mars and Jupiter at a solar distance of 2.8 AU. This is about the region where the asteroidal belt is found. Asteroids are considered fragmented relics of one or many planetesimals that once accreted in the solar nebula close to the point where they presently reside. The total number of asteroids larger than 200 km in diameter is about 35. The largest, Ceres, has a diameter of 1,025 km, and a mass of 1.2 x 10^{24} g, which is roughly a quarter of all mass contained in the asteroidal belt. It has been tentatively proposed that a few of the larger-sized bodies are "failed planets" which because of their closeness to Jupiter could not aggregate beyond a certain size. Diameter and albedo of principal asteroids has been measured by a combination of radiometric and polarimetric techniques. Albedo plus visible and near-infrared reflection data permit a petrographic classification of the main asteroids. The majority are in all probability of primitive carbonaceous chondrites (75 %) and stony-iron types (15 %). It is of note that the most common meteorites that fall upon Earth, the ordinary chondrites, appear to be quite rare.

The distribution of asteroids is not random with regard to position, orbital elements, or clustering. The asteroidal belt is closer to Mars than to Jupiter, and a series of vacant lanes known as Kirkwood gaps exist (Fig. 4.13). They occur at intervals, where periods of revolution would be a simple fraction of Jupiters's period of 11.86 years (Chapman et al., 1978; Hartmann, 1975). Jupiter's gravitational field induces orbital resonance and alignment of objects. About 1,000 asteroids are clustered in a family, the Trojans, which lie in the orbit of Jupiter, with one group ahead of and one group behind the planet. Still others follow eccentric orbits, for instance the Apollos, Icarus, Adonis or Hidalgo (see Fig. 4.8). It is generally as-

Fig. 4.12 a, b. The surface of our Moon. Note the abundance of mare-type volcanism (*black*) on the side facing the Earth **(a)**, whereas the far side exhibits only a few spots of mare basalts **(b)** (after El-Baz, 1977)

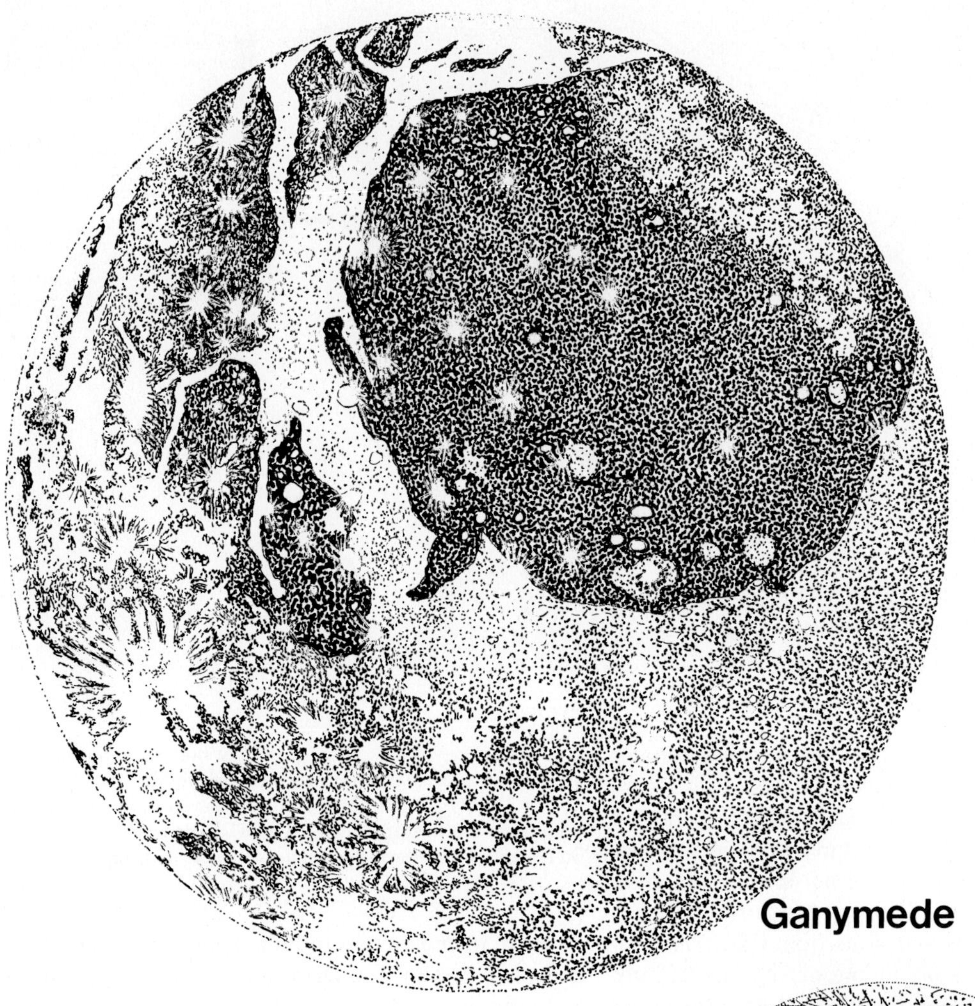

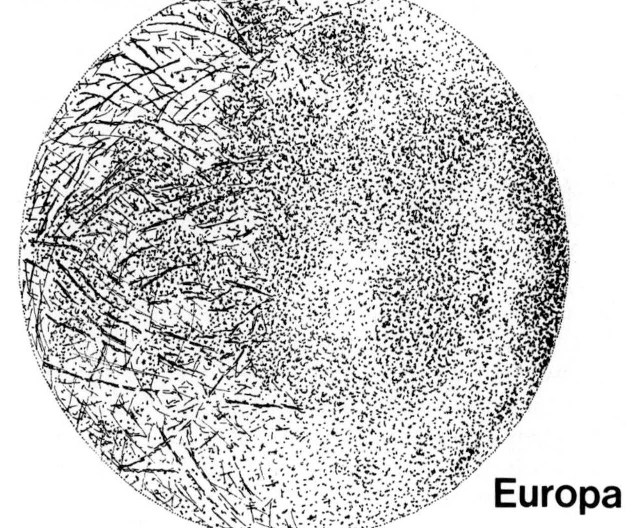

Fig. 4.14. The four large moons of Jupiter (for details and literature references see Morrison, 1982; Smith et al., 1979; Smoluchowski and McWilliam, 1984; Smoluchowski, 1983). Rocky moons: *Io* (J1, r = 1,816 km) is the most active body known in the solar system, with on-going volcanic eruptions. The internal heat source – estimated at 10^{14} W – is several orders above that expected from radioactive decay. The closeness to Jupiter will exert tidal forces upon Io with a resultant production of heat in the range of 10^{12} to 10^{14} W. The brilliant color of Io's surface represents different allotropic forms of sulfur. The highest mountain measures 9 km. *Europa* (J2, r = 1,563 km) is covered by a crust of water ice, beneath which a liquid ocean might be found. Most remarkable is the shattered surface which has possibly formed through tectonic fracturing of an icy crust and intrusion of material from below. Tidal heating, although substantial, is much smaller than that of Io.

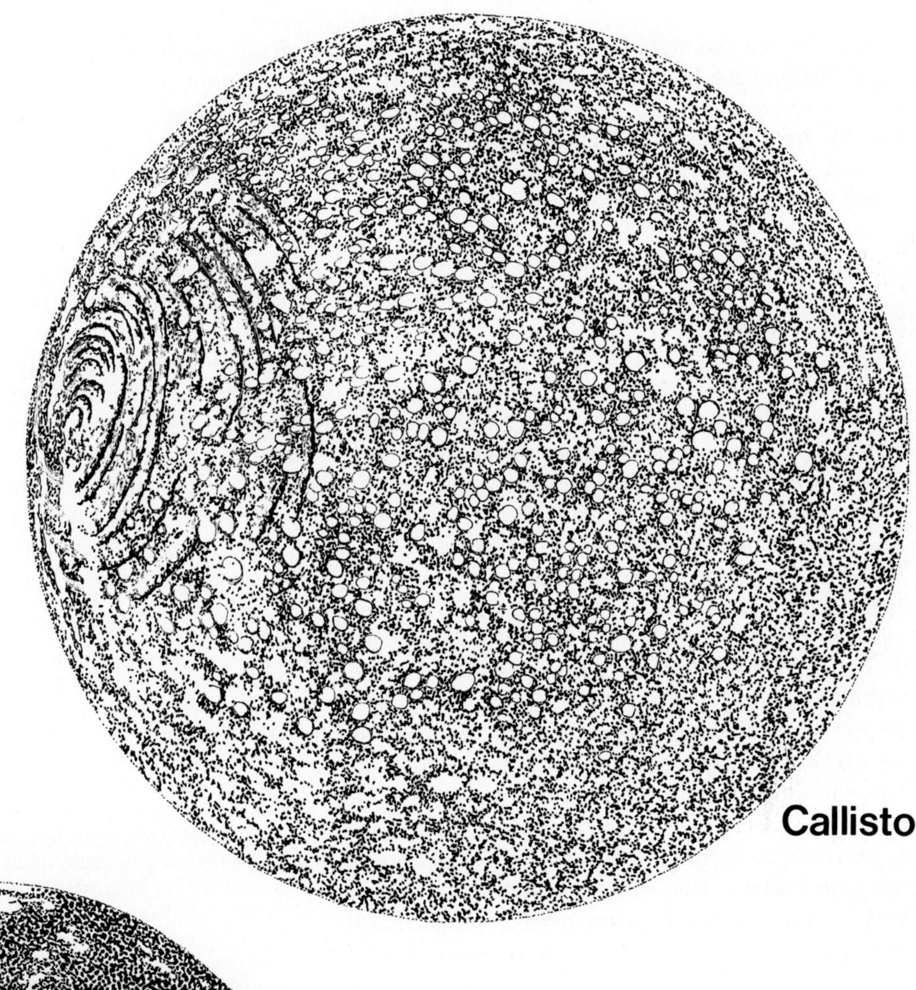

Io

Icy moons: *Ganymede* (J3, r = 2,638 km) is the largest satellite of the solar system. At first sight, it looks like our Moon with dark patches, except that here they are not the youngest features and mare-type basalts but relics of an ancient cratered crust. The spectacular bright bands of grooves represent the younger regions. It has been proposed that a kind of "plate" tectonics or faulting might have been active a few billion years ago. Since then, tectonic activity has ceased, as evidenced by bright ejecta patterns and ray systems surrounding large fresh craters. *Callisto* (J4, r = 2,410 km) is the most heavily cratered body in the Galilean system, with several impact craters measuring up to 100 km in diameter. In addition, a few large multiring basins exist with a "bull's-eye" appearance. The largest, Valhalla, has a 350 km diameter crater and concentric ridges that extend to 2,000 km from the center. Preservation of so many craters implies that Callisto has been tectonically quiet. The long duration of evaporation on the cratered surface has left a dirty crust and flattened relief on old Callisto craters, while younger craters are marked by fresh ice covers sumed that the majority of meteorites originate from asteroidal parents. Moreover, the cratering history at the surface of terrestrial planets and their moons is intimately linked with some happenings in the asteroidal belt, or asteroids in unusual orbits, such as collisions or changes in orbital parameters (e.g., Shoemaker, 1983).

Far-Away Moons

Phobos and Deimos, the two Martian moons, are considered by some investigators to have formed in the vicinity of Mars, i.e., on or close to the orbits they presently occupy. Alternately they are viewed as captured asteroids. In the latter case, Jupiter or drag in the solar nebula were suggested as possible causes to have brought the two bodies near Mars (see: Hunten, 1979). Both moons have irregular-shaped geometries, with the longest axis measuring 27 km for Phobos and 15 km for Deimos. Phobos is closest to Mars at a distance of about 9,400 km; for Deimos the value is 23,500 km. A basaltic composition was originally assumed, but in recent scannings the surface material has the appearance of carbonaceous chondrites, which again argues for an asteroidal origin (Cazenave et al., 1980; Pang et al., 1980). The groves of Phobos are seen in connection with large meteoritic impacts which caused fracturing as the rotation properties of the small body adjusted to such events (Weidenschilling, 1979).

1610 was the year when Galileo discovered the four Jupiter moons: Io, Europa, Ganymede and Callisto. The five large moons of Saturn were detected by Huygens (1655: Titan) and by Cassini (1671: Iapetus; 1672: Rhea; 1684: Tethys and Dione). Up to 1974, 15 additional Jupiter and saturn moons were found. With the advent of the Voyager mission, the number of Jupiter and Saturn moons has now reached 33 (Cruikshank and Morrison, 1976; Burns, 1980; Morrison, 1982; Soderblom and Johnson, 1982; Owen, 1982; Hartmann, 1980; B.A. Smith et al., 1979, 1981, 1982). For Uranus, ten additional moons were discovered on the January 1986 Voyager fly-by (Box 4.2); and with respect to Neptune, we have to wait until August 1989 to win a clearer picture of its number of moons, when Voyager 2 zeros-in on Neptune.

Physical and orbital data of the satellites of Jupiter and Saturn are summarized in Morrison (1982). Chapman et al. (1979) prepared a "solar system fact finder", which can be consulted for details on the satellites of Uranus (Miranda, Ariel, Umbriel, Titania and Oberon), Neptune (Triton and Nereid), and Pluto (Charon). A first assessment of the Voyager imaging science results for Jupiter and Saturn was prepared by B.A. Smith et al. (1979, 1981, 1982).

Size, shape and position of the moons of Jupiter and Saturn are graphically shown in Figs. 4.14 and 4.15. Jupiter's four large Galilean satellites, plus the four inside the orbit of Io, have a regular orbit (prograde, coplanar, small eccentricities). The eight satellites remaining exhibit irregular orbits. It appears that the moons with regular orbit formed from a proto-Iovian nebula, while the rest represent captured bodies (e.g. asteroids, cometesimals, fragmented parent bodies). For Saturn, all except Phoebe and Iapetus exhibit regular orbits. In turn, Jupiter and Saturn have each generated a kind of "miniaturized" solar system, except that the mass of the two planets was too little to create a luminous object. It is possible that the structural lay-out for the Jupiter, Saturn and Uranus system followed the same blueprint used for

Fig. 4.15. The seven large moons of Saturn (Morrison, 1982; Cruikshank, 1979; Tyler et al., 1982; Smith et al., 1981; Smoluchowski, 1983): *Mimas* (S1, r = 196 km) is a small icy satellite with high crater density and one spectacular crater about 9 km deep with a 6 km central peak. No signs of surviving geological activity. *Enceladus* (S2, r = 255 km) displays high geological activity. Crater density is low with no craters, >30 km in diameter. Some terrains are essentially free of craters down to resolution limits of the Voyager 2 images. High albedo supports the inference of a pure water ice surface with no dirt contamination as observed in other icy satellites. *Tethys* (S3, r = 530 km) and *Dione* (S4, r = 560 km) are heavily cratered and have an ice surface. Troughs, extending hundreds of kilometers in length may be explained as tensional features caused by freezing and expansion of a water-rich body. Fracturing by impact is an alternate view. Episodes of resurfacing are indicated by regions of reduced impact crater density and attest to the geological activity of Tethys and Dione. *Rhea* (S5, r = 765 km) and *Iapetus* (S8, r = 730 km) are nearly identical in size, mass, and density. For Rhea, crater density is high but the ice crust appears to be more rigid than that of, for example, Callisto and in consequence the topography resembles that of the cratered highlands of the Moon. The leading hemisphere of Iapetus is dark and its trailing face bright (remember Kubrick's movie 2001: Space Odyssey). Whereas water ice can explain the brightness, there is much speculation as to the nature of the dark material. Since the dark material is apparently localized in crater floors, an internal origin, perhaps organic matter produced by a methane- or ammonia-driven volcanism, is quite likely. *Titan* (S6, r = 2,575 km) is the only moon in the solar system having an atmosphere that is slightly denser than Earth's (1.6 bar). The upper atmosphere is opaque to visible light, due to photochemically produced haze layers extending as high as 500 km. The top of the visible reddish cloud layer lies about half-way down to the surface. Titan's crust may be covered by oceans or glaciers of methane (up to 1 km thick), upon which massive organic deposits might rest. The interior structure is composed of core, mantle and crust. The core is considered to be metallic-rocky, the lower mantle rocky with much bound water, the upper mantle a water-ammonia "magma", and the crust a mixture of water ice and either solid or frozen methane

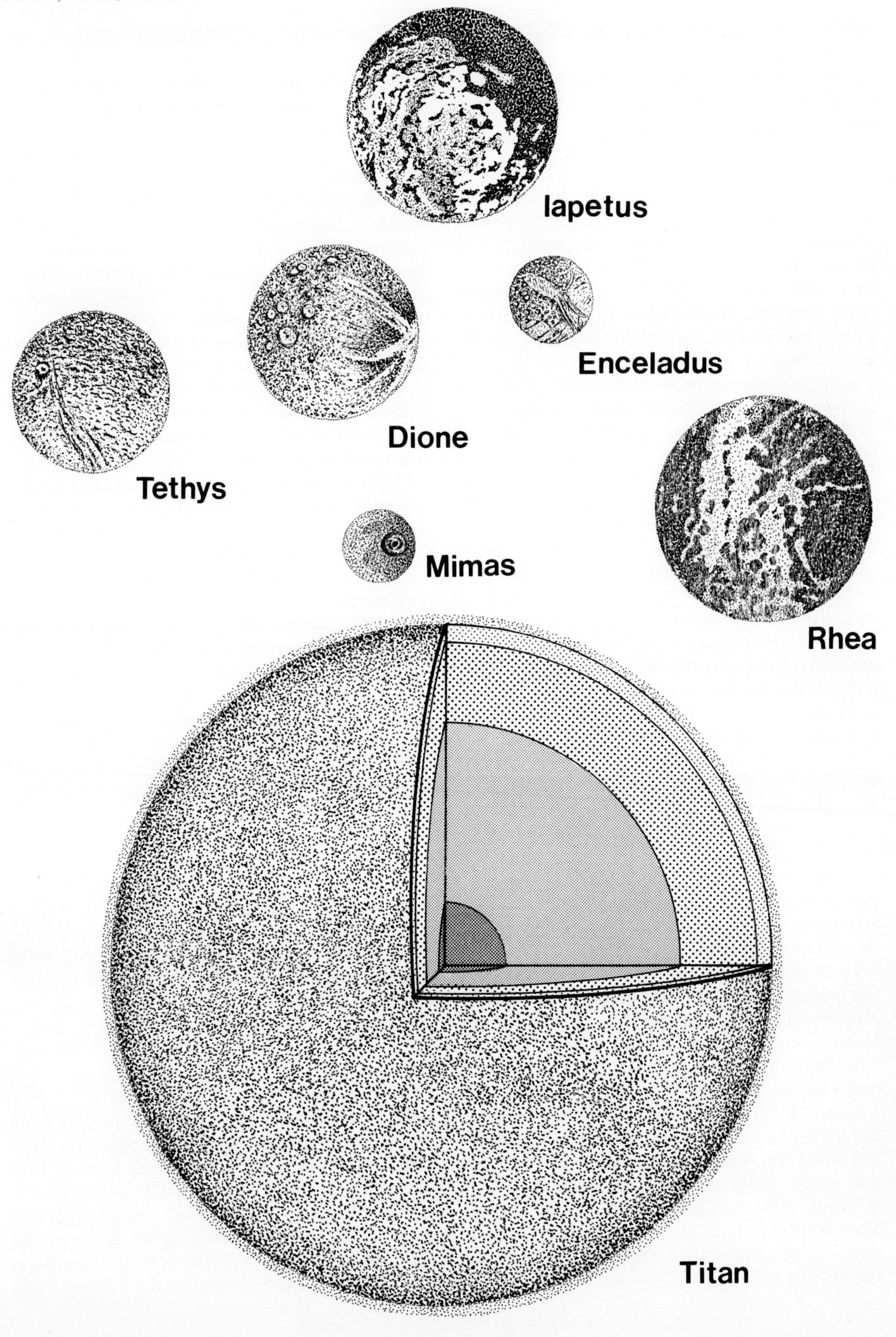
Iapetus
Enceladus
Dione
Tethys
Rhea
Mimas
Titan

BOX 4.2

Voyager 2 Encounter with the Uranian System (Stone and Miner, 1986; B.A. Smith et al., 1986; Dermott and Nicholson, 1986)

Following successful encounters with the Iovian and Saturnian system, Voyager 2 had its closest approach, at 107,000 km from the center of Uranus, on 24 January 1986. Nine new satellites were found to be orbiting between Uranus and 1985U1, the first of the recently discovered moons. All ten small satellites lie in a series of concentric, nearly circular orbits, roughly coplanar with Uranus' equatorial plane.

A major difference between the small satellites of Uranus and Saturn is that the Uranian bodies are all dark, whereas the Saturnian bodies are all bright. This may suggest significant differences in their origin, evolution, or processes operating on the respective satellites.

The composition of the dark material and its relation to dark matter found elsewhere in the solar system are major issues. A simple explanation would be that we are dealing with primitive carbonaceous chondrite material. Pure carbon is another possible candidate for this phenomenon. Furthermore, the material could be primarily radiation-darkened methane, probably in clathrate form.

As to the origin of the newly discovered moons, they may have accreted from fragments derived from collisional disruption of 1985U1 a few billion years ago. Most of them may have been broken and reaccreted once or twice since then.

With respect to the five larger Uranian moons, their surface reflectivity is much brighter than that of the minor ten satellites. Ariel, being the brightest, is therefore assumed to possess the geologically youngest surface in the Uranian satellite system. In contrast, Oberon and Umbriel have heavily cratered surfaces resembling the ancient cratered highland of Earth's moon. Titania and Ariel show crater populations different from those on Oberon and Umbriel and exhibit many extensional fault systems. They probably disrupted once and reaccreted again. Miranda is part old, cratered terrain and part younger terrain, displaying complex sets of parallel and intersecting scarps and ridges.

The nature of the flows on both Ariel and Titania is uncertain. It is conceivable that ammonia-water ice mixtures are sources of fluid in the Uranian satellites. Other volatile assemblages, such as methane or carbon monoxide clathrates, may also play a role in low-temperature resurfacing processes.

As more detailed work on the Voyager 2 data is going on, the spacecraft continues to explore the interplanetary medium and may even reach the bounds of our solar system. It should transmit data until well into the 21st century, when electric power output is expected to fade away.

---

our solar system with its one Sun and nine planets. Accordingly, when it comes to the logic behind the origin of a solar system, we can draw not only on one, but on four data banks (Sagan, 1975).

The composition of the moons is presumably a function of the mixing ratio between rocky and icy planetesimals. Io and Europa have densities like our Moon (≈ 3 g cm^{-3}), and should thus be rocky in nature. In contrast, the other moons are about "half rock – half ice" in a ratio of 4:6 with densities close to 1.2 g cm^{-3} for the smaller objects, and densities close to 1.8 g cm^{-3} for the larger ones (Ganymede, Callisto and Titan). The higher densities are due to compression and ice phase changes (Lupo and Lewis, 1979; Morrison, 1982). Io, Europa, Enceladus and possibly Titan (hidden beneath thick clouds) are geologically active. Others stopped volcanism at some point during the past (Ganymede, Dione, Tethys). For all the other large satellites there is no apparent activity, and small moons are presumed to have too little thermal energy left to drive volcanism.

The spatial relationship among all known moons of our solar system in relation to the Sun and their parent planet is depicted in Fig. 4.16.

Comets

Halley's comet, which had been studied in some detail in 1910, made its most recent appearance in April 1986. This will certainly promote research in the field of cometary science, and the first data have already appeared (Nature, 1986; Maddox, 1986). Comet Halley, returning to the inner solar system with an orbital period of 76 years, is the only comet with a short-term periodicity which can be seen by the naked eye and

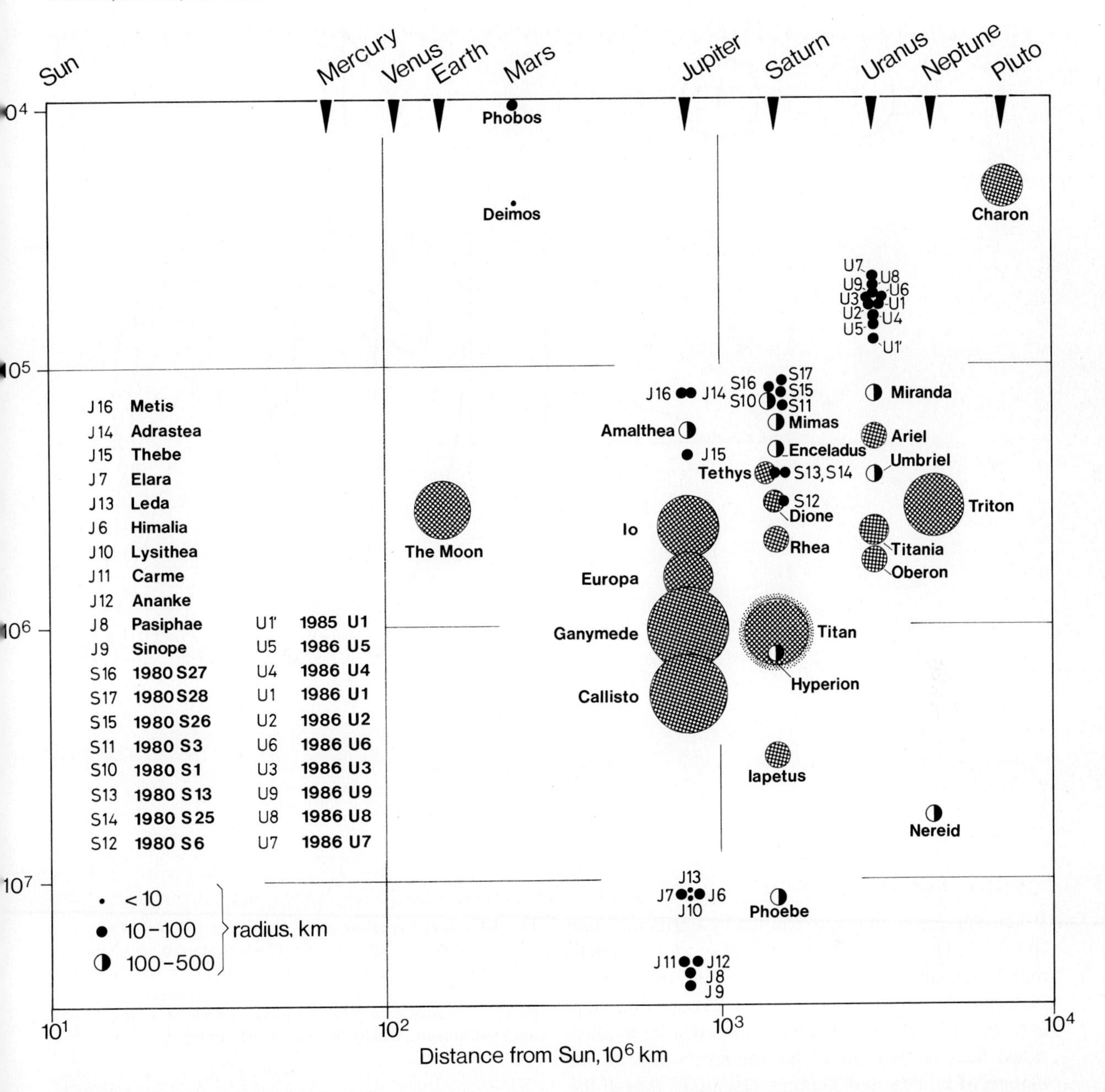

Fig. 4.16. Spacial relationship among all moons of the solar system

that still displays all features of a "new" comet (Feldman, 1983). Many questions on comets have still to be resolved, in particular those related to their origin, chemistry, structure, and possible impacts on life on Earth and maybe elsewhere.

Comets are stored in a spherical halo, the so-called Oort's cloud (named in honor of the famous Dutch astronomer J.H. Oort), which is about 5 x 10^4 AU from the Sun. This reservoir contains about 100 billion cometary bodies, which from time to time become ejected by stellar gravitational perturbations into the inner solar system. About 10 to 20 comet incidents are counted per year. An individual comet has a mass 10^{-11} that of the Earth, or figuratively all comets put together equal only the mass of the Earth.

The origin of comets is still debated. One line of thought supposes gravitational capture from interstellar clouds by the passing Sun. Another line is that comets stem from a former asteroidal planet and were released during a break-up event in the asteroidal belt (van Flandern, 1978). However, the most widely accepted viewpoint is that comets originated in the outer planetary region, and are a kind of "left over" from a time when the giant planets formed (Safronov,

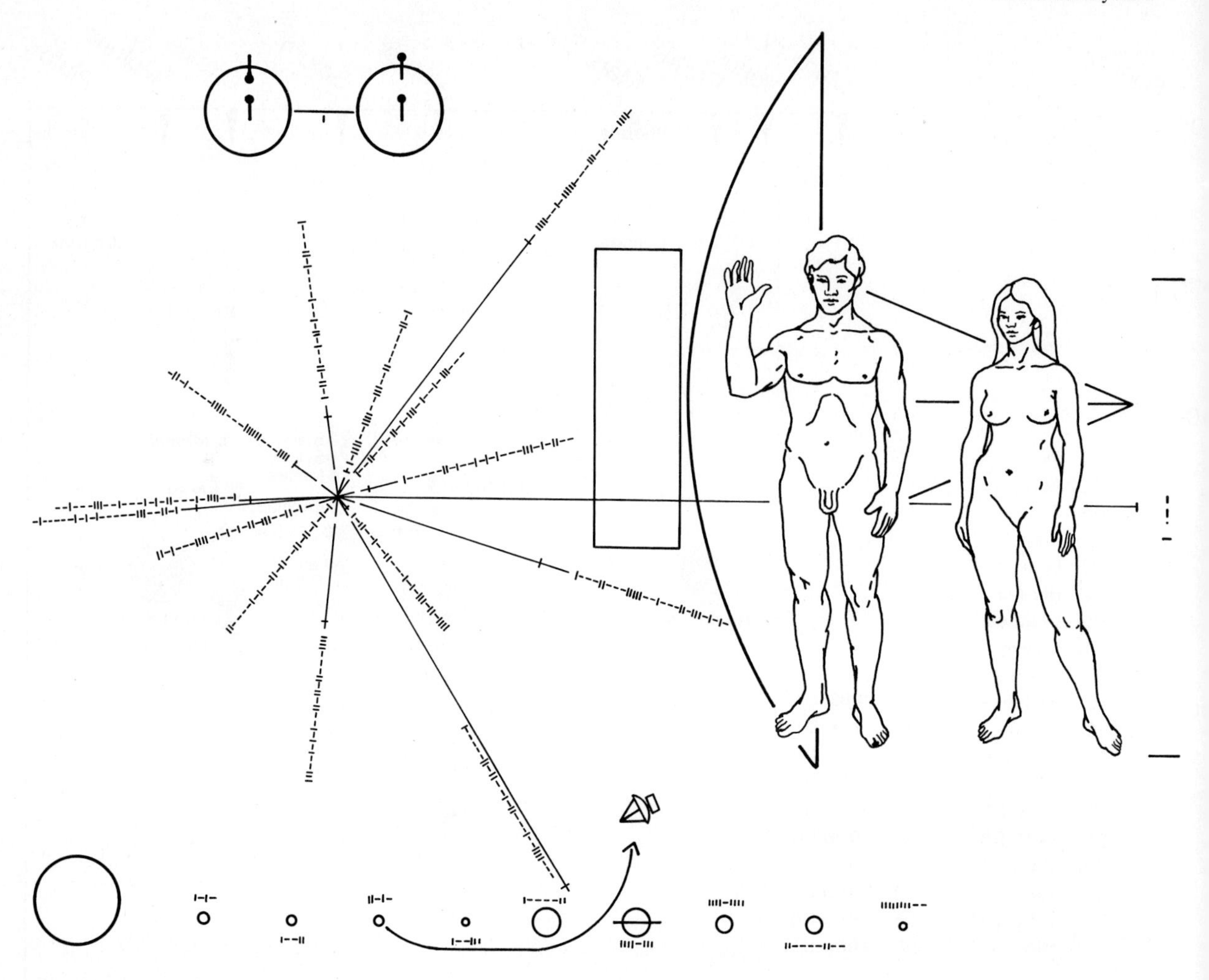

Fig. 4.17. Metal greeting cards on Pioneers 10 and 11 (Sagan, 1978). For comparison, Voyager 12 which started September 5, 1977, for a trip beyond the solar system has a gold-plated record on bord engraved with welcoming messages in 55 languages, the song of a Navajo Indian, sounds of a busy street, music of Mozart, and the cry of a newborn baby

1972). The last idea was expanded by Fernandez and Ip (1981). They suggested small bodies, or "cometesimals", composed of ice and rocks, as the starting material not only of comets but of Uranus and Neptune as well. This proposition has interesting implications, since it could mean that the mechanism of formation of Uranus and Neptune resembles that of the terrestrial planets but entirely differs from that of Saturn and Jupiter. Accretion and scattering of cometary swarms by Uranus and Neptune represent a dynamic model for the source and evolution of cometary bodies and perhaps of some of the moons of the outer planets which could represent captured cometesimals. Even some of the water in our ocean may stem from the capture of cometesimals (see p. 271).

The comet's mass resides in the nucleus, which has a diameter of a few kilometers. When passing close to the Sun, evaporation and sublimation of volatile material can generate a spectacular tail, which extends for tens of thousands of kilometers and sometimes even beyond. The nucleus is composed of condensed water and gases (e.g., H_2O, HCN, CH_3CN, CO), plus minerals, elementary carbon, and complex organics. The term "dirty snowball" was appropriately coined by Fred L. Whipple in the early fifties (Whipple, 1950).

From the data gathered during the last encounter with Halley in spring 1986, there is nothing which does not conform with the Whipple view on comets, in that they are objects composed of dust particles and ice, except that this ice appears to be a mixture of CO_2 and H_2O ices. Work in progress might perhaps reveal what many of us already believe, that comets carry the building blocks of life, since the dust particles analyzed so far are rich in the light elements H,

C, N and O, plus elements associated with common silicates (see articles in Nature, 1986). The presence of formaldehyde and hydrogen cyanide is expected to yield by self-condensation or in the presence of a mineral catalyst practically all monomeric compounds needed for the generation of more complex organic molecules.

Not only is cometary matter a possible source of biochemical molecules, atmospheric gases and water at the dawn of time, but cometary impacts have recently been suggested as a triggering event of marine mass extinctions (Hsü, 1980). The cyclicity of those extinctions during at least the past 250 million years is 30 Ma (Raup and Sepkoski, 1984, 1986). This interval corresponds closely with the time period for the solar system to oscillate vertically about the plane of the galaxy (33 ± 3 Ma; Rampino and Stothers, 1984). A close encounter with a nearby interstellar cloud could gravitationally perturb the Oort's cloud, scatter cometary bodies, increase the flux of comets to the inner solar system, and raise the chance for large-body impacts on Earth. A cometary impact had even been photographed in 1979, when the Howard-Koamer-Michels Comet crashed into the Sun. Moreover, the famous Tunguska event in Siberia on June 30, 1908, is in all probability of cometary origin (Kondratyev, pers. comm.). A fragment of Comet Encke, about 100 meters in diameter, might have been the ultimate source, but other suggestions were made to account for the Tunguska incident such as: a drop of antimatter, a carbonaceous chondrite, a meteorite rich in deuterium and tritium leading by compression to a nuclear explosion, a mini black hole; and naturally visitors from another world. For a vivid account on Tunguska see Sullivan (1979) and for the topic of comets in general Whipple (1978, 1982), Burns (1981), Weissman (1982), and Wilkening (1982).

Concluding Remarks

Our Space Odyssey has come to a close. With a kind of Ariadne's thread we have briefly commented on major cosmic events, starting from the point of singularity some 15 billion years back in time and moving towards a place closer to home: the Solar System. Man's niche in this cosmic setting is Planet Earth, where he wonders about himself and the Universe and the ultimate logic behind Microcosmos and Macrocosmos.

It would greatly help, so the argument goes, if intelligent life could be found elsewhere in the cosmos (Shklovskii and Sagan, 1966; Clarke, 1981) or contained perhaps in a bacteriophage (Yokoo and Oshima, 1979). The search for extraterrestrial intelligence (SETI) is the search, as Sagan (1978) put it, for ourselves. Radio telescopes are watching for signals from extraterrestrial civilizations (see Murray et al., 1978), and the messages attached to space probes leaving our solar system have already been sent in the form of engraved metal greeting cards (Sagan et al., 1972). These cards (Fig. 4.17), placed on Pioneers 10 and 11, locate our Sun with respect to 14 pulsars (burst pattern). The space probe is shown leaving Earth and swinging by Jupiter before abandoning the Solar System for good. A straight line behind pairs of humans, whose size is given in units of the radio emission of hydrogen, indicates the distance to the center of the Galaxy.

I have shown a picture of the metal card to my students. Carl Sagan would be amazed to hear what they had to say about the engravings. There is a good chance that in our lifetime "Contact" to ETI can only be made in a novel (Sagan, 1985). In our search for finding out more about the mysteries of this world I was led by Johann Wolfgang von Goethe:

"Willst Du ins Unendliche schreiten,
So geh' nur im Endlichen nach allen Seiten.
Willst Du Dich am Ganzen entzücken,
So mußt Du das Ganze im Kleinsten erblicken."

References

Chapter 1–4

Aaronson M, Mould J (1983) A distance scale from the infrared magnitude/velocity-width relation. IV. The morphological type dependence and scatter in the relation; the distances to nearby groups. Astrophys J 265:1–17

Aaronson M, Huchra J, Mould J, Schechter PL, Tully RB (1982) The velocity field in the local supercluster. Astrophys J 258:64–76

Ahrens LH (1979) Origin and Distribution of the Elements. Int Ser Earth Sci 34, Pergamon Press, Oxford, New York, Toronto, Sydney, Paris, Frankfurt, pp 909

Allen BJ, Gibbons JH, Macklin RL (1971) Nucleosynthesis and neutron-capture reactions. Adv Nucl Phys 4:205–259

Anders E (1977) Chemical compositions of the Moon, Earth, and eucrite parent body. Philos Trans R Soc Lond A 285:23–40

Arnett WD (1982) The cosmic distance scale: Methods for determining the distance to supernovae. Astrophys J 254:1–7

Arp HC (1963) The evolution of galaxies. Sci Am 200/1:70–84

Bahcall JN (1978) Masses of neutron stars and black holes in X-ray binaries. Annu Rev Astron Astrophys 16:241–264

Baldwin RB (1971) On the history of lunar impact cratering: the absolute time scale and the origin of planetesimals. Icarus 14:36–52

Baldwin RB (1985) Relative and absolute ages of individual craters and the rate of infalls on the Moon in the post-Imbrium period. Icarus 61:63–91

Barrow JD (1977) A chaotic cosmology. Nature 267:117–120

Barrow JD, Silk J (1980) The structure of the early universe. Sci Am 242/4:98–107

Barrow JD, Turner MS (1982) The inflationary universe – birth, death and transfiguration. Nature 298:801–804

Bernal JD (1954) The origin of life. New Biol 16:28–40

Bernstein J (1984) Three degrees above zero. The New Yorker, August 20, 1984:42–70

Binder AB, Lange M (1977) On the thermal history of a Moon of fission origin. The Moon 17, 29–45

Black DC (1980) On the detection of other planetary systems: detection of intrinsic thermal radiation. Icarus 43:293–301

Black DC (1982) A simple criterion for determining the dynamical stability of three-body systems. Astron J 87:1333–1337

Blitz L (1982) Giant molecular-cloud complexes in the galaxy. Sci Am 246/4:72–80

Bodenheimer P (1968) The evolution of protostars of 1 and 12 solar masses. Astrophys J 153:483–493

Bodenheimer P, Grossman AS, DeCampli WM, Marcy G, Pollack JB (1980) Calculations of the evolution of the giant planets. Icarus 41:293–308

Bok BJ (1981) The milky way galaxy. Sci Am 244/3:70–91

Borucki WJ, Summers AL (1984) The photometric method of detecting other planetary systems. Icarus 58:121–134

Boss AP (1985) Collapse and formation of stars. Sci Am 252/1:28–33

Bottinelli L, Gougenheim L, Paturel G, de Vaucouleurs G (1980) The 21 centimeter line width as an extragalactic distance indicator. Astrophys J Lett 242:L153–L156

Brown R, Cruikshank DP (1985) The moons of Uranus, Neptune and Pluto. Sci Am 253/1:28–37

Brush SG (1982) Nickel for your thoughts: Urey and the origin of the Moon. Science 217:891–898

Burbidge EM, Burbidge GR, Fowler WA, Hoyle F (1957) Synthesis of the elements in stars. Rev Mod Phys 29/4:547–650

Burns JA (ed) (1980) The satellites of Jupiter. Icarus 44:225–547

Burns JA (ed) (1981) Comets: a special issue of "Icarus". Icarus 47:301–522

Cabrera B (1982) First results from a superconductive detector for moving magnetic monopoles. Phys Rev Lett 4:1378–1381

Cairns-Smith AG (1982) Genetic takeover and the mineral origins of life. Cambridge Univ Press, Cambridge, London, New York, New Rochelle, Melbourne, Sydney, pp 477

Cameron AGW (1973) Abundances of the elements in the solar system. Space Sci Rev 15:121–146

Cameron AGW (1975) The origin and evolution of the solar system. Sci Am 233/3:32–41

Cameron AGW (1982) Elementary and nuclidic abundances in the solar system. In "Essays in Nuclear Astrophysics" (eds Barnes CA, Schramm DN, Clayton DD), Cambridge Univ Press, Cambridge

Cameron AGW, Truran JW (1977) The supernova trigger for formation of the solar system. Icarus 30:447–461

Carr BJ, Bond JR, Arnett WD (1984) Cosmological consequences of population III stars. Astrophys J 277:445–469

Carrigan RA jr, Trower WP (1982) Superheavy magnetic monopoles. Sci Am 246/4:91–99

Cazenave A, Dobrovolskis A, Lago B (1980) Evolution of the inclination of Phobos. Nature 284:430–431

Cence RJ, Ma E, Roberts A (eds) (1981) Neutrino 81. Proceedings of a conference, Maui, Hawaii, July 1981, 2 vols. Honolulu, 1981, pp 510, pp 512

Chapman CR, Williams JG, Hartmann WK (1978) The asteroids. Annu Rev Astron Astrophys 16:33–75

Chapman C, Matson D, Newburn R, Orton G, Scott E, Zeller B (1979) Solar system fact finder. Lunar & Planet Inst Contrib No 393:1–5

Clarke JN (1981) Extraterrestrial intelligence and galactic nuclear activity. Icarus 46:94–96

Clayton DD (1977) Solar system isotopic anomalies: supernova neighbor or presolar carriers. Icarus 32:255–269

Cline DB, Rubbia C, van der Meer S (1982) The search for intermediate vector bosons. Sci Am 246/2:38–49

Close F (1983) A golden year for the weak force. Nature 301:287–288

Conti PS, McCray R (1980) Strong stellar winds. Science 208:9–17

Cox LP, Lewis JS (1980) Numerical simulation of the final stages of terrestrial planet formation. Icarus 44:706–721

Cruikshank DP (1979) The surfaces and interiors of Saturn's satellites. Rev Geophys Space Phys 17:165–176

Cruikshank DP, Morrison D (1976) The Galilean satellites of Jupiter. Sci Am 234/5:108–116

Däppen W, Gilliland RL, Christensen-Dalsgaard J (1986) Weakly interacting massive particles, solar neutrinos, and solar oscillations. Nature 321:229–231

Dansgaard W, Johnsen SJ, Clausen HB (1971) Climatic record revealed by the Camp Century ice core. In "Late Cenozoic Glacial Ice Ages" (ed Turekian KK), Yale Univ Press, New Haven and London, 37–56

Degens ET, Stoffers P, Golubic S, Dickman MD (1978) Varve chronology: estimated rates of sedimentation in the Black Sea deep basin. In "Initial Reports of the Deep- Sea Drilling Project" (eds Ross DA, Neprochnov YP et al.), vol 42, Pt 2, US Government Printing Office, Washington, DC, 499–508

Dekel A (1982) Clustering of Lyman-alpha absorption lines in quasars: Like galaxies in proto-superclusters? Astrophys J Lett 261:L13–L17

Dermott SF, Nicholson PD (1986) Masses of the satellites of Uranus. Nature 319:115–120

de Vaucouleurs G (1982) Five crucial tests of the cosmic distance scale using the Galaxy as fundamental standard. Nature 299:303–307

de Vaucouleurs G (1983a) The galaxy as fundamental calibrator of the extragalactic distance scale. I. The basic scale factors of the galaxy and two kinematic tests of the long and short distance scales. Astrophys J 268:451–467

de Vaucouleurs G (1983b) The galaxy as fundamental calibrator of the extragalactic distance scale. II. Comparison of metric and photometric scale lengths and three further tests of the long and short distance scales. Astrophys J 268:468–475

Dicus DA, Letaw JR, Teplitz DC, Teplitz VL (1983) The future of the universe. Sci Am 248/3:74–85

Eddy JA (1977a) The case of the missing sunspots. Sci Am 236/5:80–92

Eddy JA (1977b) Climate and the changing Sun. Clim Change 1:173–190

Eddy JA (1979) The historical record of solar activity. In "The Ancient Sun" (eds Pepin RO, Eddy JA, Merrill RB), Geochim Cosmochim Acta Suppl 13:119–134, Pergamon Press, New York, Oxford, Toronto, Sydney, Frankfurt, Paris

Eddy JA, Gilliland RL, Hoyt DV (1982) Changes in the solar constant and climatic effects. Nature 300:689–692

El-Baz F (1975) The Moon after Apollo. Icarus 25/4:495–537

El-Baz F (1977) Lunar stratigraphy. Philos Trans R Soc Lond A 285:549–553

Fawley WM (1979) On the detection of jovian companions to white dwarfs. Nature 279:622

Feldman PD (1983) Ultraviolet spectroscopy and the composition of cometary ice. Science 219:347–354

Fernandez JA, Ip W-H (1981) Dynamical evolution of a cometary swarm in the outer planetary region. Icarus 47:470–479

Fischbach E, Sudarsky D, Szafer A et al. (1986) Reanalysis of the Eötvös experiment. Phys Rev Lett 56:3

Fowler WA (1984) The quest for the origin of elements. Science 226:922–935

Frautschi S (1982) Entropy in an expanding universe. Science 217:593–599

Freedman DZ (1981) Unification in supergravity. In "To Fulfill a Vision" (ed Ne'eman Y), Chapter 6, 82–97, Addison-Wesley Publ Co, Inc, Reading, Mass

Gaida M (1984) Neues von der Lokalen Gruppe. Bild Wiss 8:102–103

Gamov G (1948) The origin of elements and the separation of galaxies. Phys Rev 74:505–506

Gatewood GD (1976) On the astrometric detection of neighboring planetary systems. Icarus 27:1–12

Gehrz RD, Black DC, Solomon PM (1984) The formation of stellar systems from interstellar molecular clouds. Science 224:823–830

Georgi H (1981) A unified theory of elementary particles and forces. Sci Am 244/4:40–55

Gibbons GW, Hawking SW, Siklos STC (eds) (1983) The Very Early Universe. Cambridge Univ Press, New York, pp 480

Gibbons GW, Hawking SW, Townsend PK (eds) (1986) Supersymmetry and its Applications: Superstrings, Anomalies and Supergravity. Cambridge Univ Press, New York, pp 481

Giese R-H (1981) Einführung in die Astronomie. Wiss Buchges, Darmstadt, pp 393

Glashow SL (1980) Toward a unified theory: threads in a tapestry. Science 210:1319–1323

Goldhaber AS, Nieto MM (1976) The mass of the photon. Sci Am 234/5:86–96

Goldhaber M, Langacker P, Slansky R (1980) Is the proton stable? Science 210:851–860

Goldschmidt VM (1954) Geochemistry. Clarendon Press, Oxford, pp 730

Gott JR, III, Gunn JE, Schramm DN, Tinsley BM (1976) Will the universe expand forever? Sci Am 234/4:62–79

Green MB (1985) Unification of forces and particles in superstring theories. Nature 314:409–414

Green MB, Schwarz JH (1984) Anomaly cancellations in supersymmetric D = 10 gauge theory and superstring theory. Phys Lett 149B:117–122

Greenberg JM (1984) The structure and evolution of interstellar grains. Sci Am 250/6:96–107

Greenberg R (1979) Growth of large, late-stage planetesimals. Icarus 39:141–150

Greenberg R, Wacker JF, Hartmann WK, Chapman CR (1978) Planetesimals to planets: numerical simulation of collisional evolution. Icarus 35:1–26

Gregory SA, Thompson LA (1982) Superclusters and voids in the distribution of galaxies. Sci Am 246/3:88–96

Grieve RAF, Dence MR (1979) The terrestrial cratering record. II. The crater production rate. Icarus 38:230–242

Guth AH (1981) Inflationary universe: a possible solution to the horizon and flatness problems. Phys Rev D 23/2:347–356

Guth AH (1984) One objective for physicists. Nature 307:762–763

Guth AH, Steinhardt PJ (1984) The inflationary universe. Sci Am 250/5:90–102

Hansen KS (1982) Secular effects of oceanic tidal dissipation on the Moon's orbit and the Earth's rotation. Rev Geophys Space Phys 20:457–480

Hansen TE (1972) Confrontation and objectivity. Dan Yearb Philos 7:13–72

Harari H (1979) A schematic model of quarks and leptons. Phys Lett 86B/1:83–86

Harris AW, Ward WR (1982) Dynamical constraints on the formation and evolution of planetary bodies. Annu Rev Earth Planet Sci 10:61–108

Hart L, Davies RD (1982) Motion of the local group of galaxies and isotropy of the universe. Nature 297:191–196

Hartmann WK (1975) The smaller bodies of the solar system. Sci Am 233/9:142–159

Hartmann WK (1978) Planet formation: mechanism of early growth. Icarus 33:50–61

Hartmann WK (1980) Surface evolution of two-component stone/ice bodies in the Jupiter region. Icarus 44:441–453

Hartmann WK, Phillips RJ, Taylor GJ (eds) (1984) Origin of the Moon. Lunar & Planetary Institute, Houston, pp 781

Hawking SW (1975) Particle creation by black holes. Commun Math Phys 43/3:199–220

Hawking SW (1976) Black holes and thermodynamics. Phys Rev D 13/2:191–197

Hawking SW (1977) The quantum mechanics of black holes. Sci Am 236/1:34–40

Hawking SW (1980) Theoretical advances in general relativity. In "Some Strangeness in the Proportion" (ed Woolf H), Chapter 8, 145–152, Addison-Wesley Publ Co, Inc, Reading, Mass

Hawking SW, Israel W (eds) (1987) Three Hundred Years of Gravitation. Cambridge Univ Press, New York, pp 684

Henry PS (1980) A simple description of the 3 K cosmic microwave background. Science 207:939–942

Hesser JE (ed) (1980) Star Clusters. D Reidel Publ Co, Amsterdam

Hogan CJ, Rees MJ (1984) Gravitational interactions of cosmic strings. Nature 311:109–113

Hohenberg CM, Podosek FA, Reynolds JH (1967) Science 156:233–235

Holding SC, Tuck GJ (1984) A new mine determination of the Newtonian gravitational constant. Nature 307:714–716

Hoyle F (1975) Highlights in Astronomy. Freeman WH and Company, San Francisco, pp 179

Hoyle F (1982) The universe: past and present reflections. Annu Rev Astron Astrophys 20:1–35

Hsü KJ (1980) Terrestrial catastrophe caused by cometary impact at the end of the Cretaceous. Nature 285:201–203

Huang Su Shu (1970) Life outside the solar system. In "Frontiers in Astronomy", Readings from Scientific American, Freeman WH and Company, San Francisco, 122–130

Hubbard WB (1981) Interiors of the giant planets. Science 214:145–149

Humphreys RM, Davidson K (1984) The most luminous stars. Science 223:243–249

Hunten DM (1979) Capture of Phobos and Deimos by protoatmospheric drag. Icarus 37:113–123

Hut P, White SDM (1984) Can a neutrino-dominated Universe by rejected? Nature 310:637–640

Iben I, jr (1975) Thermal pulses, *p*-capture, *s*- process nucleosynthesis, and convective mixing in a star of intermediate mass. Astrophys J 196:525–547

Jaki SL (1967) Olbers', Halley's or whose paradox? Am J Phys 35:200–210

Johnson HL, Sandage AR (1956) Three-color photometry in the globular cluster M 3. Astrophys J 124:379–389

Joshi PS, Chitre SM (1981) Neutrinos of non-zero mass in Friedmann universes. Nature 293:679

Kempe S (1977) Hydrographie. Warven-Chronologie und organische Geochemie des Van Sees, Ost-Türkei. Mitt Geol-Paläont Inst Univ Hamburg 47:125–228

Khromov GS (1977) Planetary nebulae. In "Topics in Interstellar Matter" (ed Van Woerden H), 107–177, Reidel D Publ Co Dordrecht

King IR (1981) The dynamics of globular clusters. Quarterly J Roy. Astron Soc 22:227–243

King IR (1985) Globular clusters. Sci Am 252/6:66–73

Kolb EW, Seckel D, Turner MS (1985) The shadow world of superstring theories. Nature 314:415–419

Lake G (1984) Windows on a new cosmology. Science 224:675–681

Lambeck K (1980) The Earth's Variable Rotation: Geophysical Causes and Consequences. Cambridge Univ Press, Cambridge (England), New York, pp 449

Lawrence JK (1987) Ratio of calcium plage to sunspot areas of solar active regions. J Geophys Res 92:813–817

Linde AD (1982) A new inflationary universe scenario: a possible solution of the horizon, flatness, homogeneity, isotropy and primordial monopole problems. Phys Lett 108B/6:389–393

Longair MS (1971) Observational cosmology. Reports Progr Phys 34:1126–1249

Longhi J (1980) A model of early lunar differentiation. Geochim Cosmochim Acta, Suppl 14:289–315

Lubimov VA, Novikov EG, Nozik VZ, Tretyakov EF, Kosik VS (1980) An estimate of the V mass from the β- spectrum of tritium in the valine molecule. Phys Lett 94B:266–268

Lupo MJ, Lewis JS (1979) Mass-radius relationships in icy satellites. Icarus 40:157–170

Lyons L (1982) Proton decays and high-energy speculation. Nature 299:295–296

Maddox J (1986) First journey to a comet. Nature, Suppl 321: 366; "Encounters with Comet Halley, the first results": 259–365

Mann APC, Williams DA (1980) A list of interstellar molecules. Nature 283:721–725

Margon G (1983) The origin of the cosmic X-ray background. Sci Am 248/1:94–104

Margon B, Chanan GA, Downes RA (1982) The luminosity of serentipitous X-ray QSO5. Astrophys J Lett 253:L7–L11

Martin PG (1978) Cosmic Dust. Its Impact on Astronomy. Oxford Univ Press, Oxford, pp 266

Matheja J, Degens ET (1971) Structural Molecular Biology of Phosphates. Gustav Fischer Verlag, Stuttgart, pp 180

Matsui T, Mizutani H (1978) Gravitational N-body problem on the accretion process of terrestrial planets. Icarus 34:146–172

McKinnon WB (1984) On the origin of Triton and Pluto. Nature 311:355–358

Merton RK (1965) On the Shoulders of Giants. A Shandyan Postscript. McMillan, New York

Mizutani H, Mikuni H, Takahashi M, Noda H (1975) Study on the photochemical reaction of HCN and its polymer products relating to primary chemical evolution. Origins Life 6:513–525

Morrison D (1982) The satellites of Jupiter and Saturn. Annu Rev Astron Astrophys 20:469–495

Murray BC (1975) Mercury. Sci Am 233/3:58–68

Murray B, Gulkis S, Edelson RE (1978) Extraterrestrial intelligence: an observational approach. Science 199:485–492

Mutch TA (1972) Geology of the Moon: A Stratigraphic View (Ref ed). Princeton Univ Press, pp 391

Nature (1986) Encounters with Comet Halley, Nature 321, Suppl:259–366

Ne'eman Y (ed) (1981) To Fulfill a Vision. Addison-Wesley Publ Co Inc, Reading, Mass, pp 279

Neukum G (1977) Lunar Cratering. Philos Trans R Soc Lond A285:267–272

Neukum G, König B, Fechtig H, Storzer D (1975) Cratering in the earth-moon system: Consequences for age determination by crater counting. Proc Lunar Sci Conf 6th, 2597–2620

Olive KA, Schramm DH, Steigmann Y, Turner MS, Yang J (1981) Big-Bang nucleosynthesis as a probe of cosmology and particle physics. Astrophys J 246:557–563

Oort JH (1983) Superclusters. Annu Rev Astron Astrophys 21:373–428

Osmer PS (1982a) Evidence for a decrease in the space density of quasars at 3.5. Astrophys J 253:28–37

Osmer PS (1982b) Quasars as probes of the distant and early universe. Sci Am 246/2:96–107

Owen T (1982) Titan. Sci Am 246/2:76–85

Pagel B (1982) Discovery of pre-galactic lithium. Nature 297:456–457

Pang KD, Rhoads JW, Lane AL, Ajello JM (1980) Spectral evidence for a carbonaceous chondrite surface composition on Deimos. Nature 283:277–278

Parker EN (1975) The sun. Sci Am 233/3:42–50

Pasachoff JM (1977) The solar corona. Sci Am 229/4:69–79

Peebles PJE (1984) The origin of galaxies and clusters of galaxies. Science 224:1385–1391

Peebles PJE (1986) The mean mass density of the Universe. Nature 321:27–32

Peimbert M, Serrano A, Torres-Peimbert S (1984) Interstellar matter and chemical evolution. Science 224:345–349

Penzias AA (1979) The origin of the elements. Science 205:549–554

Pepin RO, Eddy JA, Merrill RB (eds) (1980) The Ancient Sun: Fossil Record in the Earth, Moon and Meteorites. Geochim Cosmochim Acta, Suppl. 13, pp 581

Pines D (1980) Accreting neutron stars, black holes, and degenerate dwarf stars. Science 207:597–606

Podolak M, Reynolds RT (1981) On the structure and composition of Uranus and Neptune. Icarus 46:40–50

Pollack JB (1984) Origin and history of the outer planets: theoretical models and observational constraints. Annu Rev Astron Astrophys 22:389–424

Popper KR (1978) Conjectures and Refutations. 7th ed, Routledge & Kegan Paul Ltd, London

Popper KR (1979) Truth, Rationality and the Growth of Scientific Knowledge. Klostermann, Frankfurt/Main, pp 61

Rampino MR, Stothers RB (1984) Terrestrial mass extinctions, cometary impacts and the Sun's motion perpendicular to the galactic plane. Nature 308:709–712

Raup DM, Sepkoski JJ, Jr (1984) Periodicities of extinctions in the geologic past. Natl Acad Sci Proc 81:801–805

Raup DM, Sepkoski JJ, Jr (1986) Periodic extinction of families and genera. Science 231:833–836

Rebbi C (1983) The lattice theory of quark confinement. Sci Am 248/2:36–47

Rees MJ (1980) The size and shape of the universe. In "Some Strangeness in the Proportion" (ed Woolf H) 291–301, Addison-Wesley Publ Co Inc, Reading, Mass

Reid N (1983) Galaxies in the early universe. Nature 305:97–98

Reines F, Sobel HW, Pasierb E (1980) Evidence for neutrino instability. Phys Rev Lett 45:1307–1311

Ringwood AE (1979) Origin of the Earth and Moon. Springer-Verlag Berlin Heidelberg New York, pp 295

Ringwood AE (1986) Terrestrial origin of the Moon. Nature 322:323–328

Robinson AL (1983) CERN vector boson hunt successful. Science 221:840–842

Rubin VC (1980) Stars, galaxies, cosmos: the past decade, the next decade. Science 209:64–71

Ru'jula A, de Glashow SL (1980) Neutrino weight watching. Nature 286:755–756

Runcorn SK (1983) Lunar magnetism, polar displacements and primeval satellites in the Earth-Moon system. Nature 304:589–596

Safronov VS (1969) Evolution of the Protoplanetary Cloud and Formation of the Earth and the Planets. Nauka, Moscow. Engl Transl 1972, Israel Program for Scient Transl, NASA TT F-677, pp 206

Safronov VS (1972) Ejection of bodies from the solar system in the course of the accumulation of the giant planets and the formation of the cometary cloud. In "The Motion. Evolution of Orbits, and Origin of Comets" (eds Chebotarev GA, Kazimirchak-Polonskaya EI), IAU Symposium no 45:329–334

Sagan C (1975) The solar system. Sci Am 233/3:23–31

Sagan C (1978) The quest for intelligent life in space is just beginning. Smithsonian 9/2:38–47

Sagan C (1985) Contact. Simon and Schuster, New York, pp 432

Sagan C, Salzman-Sagan L, Drake F (1972) A message from earth. Science 175:881–884

Salam A (1980) Gauge unification of fundamental forces. Science 210:723–732

Salpeter EW (1974) Organische Moleküle im interstellaren Raum. In "Analyse extraterrestrischen Materials" (eds Kiesl W, Malissa H, Jr) Springer- Verlag, Vienna New York, 203–212

Sandage A (1968) A new determination of the Hubble constant from globular clusters in M87. Astrophys J 152:L149–54

Sandage AR (1971) Semaine d'Etude, les Noyaux des Galaxies, Scripta Varia 35, Vatican: Pontificia Academia Scientiarum, 601–622

Sandage A (1983) On the distance to M 33 determined from magnitude corrections to Hubble's original Cepheid photometry. Astr J 88:1108–1125

Sandage A, Katen B (1983) On the intrinsic width and luminosity function of the M 92 main sequence. Astr J 88:1146–1158

Sandage A, Tammann GA (1984) The Hubble constant as derived from 21 cm linewidths. Nature 307:326–329

Schatten KH, Mayr HG, Omidvar K (1987) Active region influences upon the solar constant. J Geophys Res 92:818–822

Schatzmann E, Maeder A (1981) Solar neutrinos and turbulent diffusion. Nature 290:683–686

Schopper H (1981) Die jüngste Entwicklung des Bildes von der Grundstruktur der Materie. Naturwissenschaften 68:307–313

Schove DJ (ed) (1983) Sunspot Cycles. Benchmark Papers in Geology 68, Hutchinson Ross Publishing Company, Stroudsburg, Pennsylvania, pp 397

Schramm DN (1974) The age of the elements. Sci Am 230/1:69–77

Schramm DN (1982) Constraints on the density of baryons in the Universe. Philos Trans R Soc Lond A 307:43–54

Schramm DN, Wasserburg GJ (1970) Nucleochronologies and the mean age of the elements. Astrophys J 162:57–69

Schultz PH, Spudis PD (1983) Beginning and end of lunar mare volcanism. Nature 302:233–236

Schwarzschild M (1970) Stellar evolution in globular clusters. QJR astr Soc 11:12–22

Sciama DW (1980) Issues in cosmology. In "Some Strangeness in the Proportion" (ed Woolf H), Chapter 24:387–404, Addison-Wesley Publ Co, Inc, Reading, Mass

Scoville N, Young JS (1984) Molecular clouds, star formation and galactic structure. Sci Am 250/4:30–41

Seeger PA, Fowler WA, Clayton DD (1965) Nucleosynthesis of heavy elements by neutron capture. Astrophys J, Suppl 11/97:121–166

Shklovskii IS (1978) Stars, their Birth, Life, and Death. Freeman WH and Company, San Francisco, pp 442

Shklovskii IS, Sagan C (1966) Intelligent Life in the Universe. Dell Publ Co, Inc, New York, pp 509

Shoemaker EM (1983) Asteroid and comet bombardment of the earth. Annu Rev Earth Planet Sci 11:461–494

Shoemaker EM, Hackman RJ, Eggleton RE (1963) Interplanetary correlation of geologic time. Adv Astronaut Sci 8:70–89

Shupe MA (1979) A composite model of leptons and quarks. Phys Lett 86B/1:87–92

Silk J (1982a) Mapping the Local supercluster. Nature 299:577–578

Silk J (1982b) The missing mass – now it's a gravitino! Nature 297:102–103

Silk J, Szalay AS, Zel'dovich YB (1983) The large- scale structure of the universe. Sci Am 249/4:56–64

Smith BA, Soderblom LA, Johnson TV et al (1979) The Jupiter system through the eyes of Voyager 1. Science 204:951–972

Smith BA, Soderblom L, Beebe R et al. (1981) Encounter with Saturn: Voyager 1 imaging science results. Science 212:163–191

Smith BA, Soderblom L, Batson R et al. (1982) A new look at the Saturn system: the Voyager 2 images. Science 215:504–537

Smith BA, Soderblom LA et al. (1986) Voyager 2 in the Uranian System: imaging science results. Science 233:43–64
Smith JV (1979) Mineralogy of the planets: a voyage in space and time. Mineral Magazine 43:1–89
Smith JV (1981) Progressive differentiation of a growing moon. Proc Lunar Planet Sci 12B:979–990
Smith JV (1982) Heterogeneous growth of meteorites and planets, especially the earth and moon. J Geol 90:1–48
Smoluchowski R (1983) The Solar System. Scientific American Library, New York, pp 174
Smoluchowski R, McWilliam A (1984) Structure of ices on satellites. Icarus 58:282–287
Soderblom LA, Johnson TV (1982) The moons of Saturn. Sci Am 246/1:72–86
Soderblom LA, Condit CD, West RA, Herman BM, Kreidler TJ (1974) Martian planetwide crater distributions: implications for geologic history and surface processes. Icarus 22:239–263
Sofue Y, Fujimoto M, Wielebinski R (1986) Global structure of magnetic fields in spiral galaxies. Annu Rev Astron Astrophys 24:459–497
Sonett CP, Trebisky TJ (1986) Secular change in solar activity derived from ancient varves and the sunspot index. Nature 322:615–617
Spite F, Spite M (1982) Abundance of lithium in unevolved halo stars and old disk stars. Astron Astrophys 115:357–366
Spite M, Spite F (1985) The composition of field halo stars and the chemical evolution of the Halo. Ann Rev Astron Astrophys 23:225–238
Spitzer L, jr (1984) Dynamics of globular clusters. Science 225:465–472
Stevenson DJ (1982) Interiors of the giant planets. Annu Rev Earth Planet Sci 10:257–295
Stone EC, Miner ED (1986) The Voyager 2 encounter with the Uranian system. Science 233:39–43
Strom KM, Strom SE (1982) Galactic evolution: a survey of recent progress. Science 216:571–580
Suess H (1968) Climatic changes, solar activity and the cosmic-ray production rate of natural radiocarbon. Meteorol Monogr 8:146–150
Suess H (1970) The three causes of the secular carbon-14 fluctuations, their amplitudes and time constants. In "Radiocarbon Variations and Absolute Chronoloy" (ed Olsson IU). Twelfth Nobel Symp Upsala 1969, Stockholm Almquist & Wiksel, New York, Wiley
Suess HE, Urey HC (1956) Abundances of the elements. Rev Mod Phys 28:53–74
Sullivan W (1979) Black Holes. Anchor Press/Doubleday, Garden City, New York, pp 303
Tassie LJ (1986) Cosmic strings, superstrings and the evolution of the Universe. Nature 323:40–42
Taylor SR (1979) Structure and evolution of the Moon. Nature 281:105–110
Thorne KS (1974) The search for black holes. Sci Am 231/6:32–43
Trefil JS (1978) Missing particles cast doubt on our solar theories. Smithsonian 8/12:74–82
Trefil JS (1981) "Nothing" may turn out to be the key to the universe. Smithsonian 12/9:143–149
Trefil JS (1983a) Closing in on creation. Smithsonian 14/2:33–42
Trefil JS (1983b) How the universe will end. Smithsonian 14/3:73–83
Trimble V (1975) The origin and abundance of the chemical elements. Rev Mod Phys 47:877–976
Tryon EP (1973) Is the universe a vacuum fluctuation? Nature 246:396–397
Tully RB (1982) The local supercluster. Astrophys J 257:389–422
Tully RB, Fisher R (1977) A new method of determining distances to galaxies. Astron Astrophys 54:661–673
Turekian KK, Clark SP, Jr (1969) Inhomogeneous accretion of the Earth from the primitive solar nebula. Earth Planet Sci Lett 6:346–348
Turner G (1977) The early chronology of the Moon: evidence for the early collisional history of the solar system. Philos Trans R Soc Lond A 285:97–103
Tyler GL, Eshleman VR, Anderson JD, Levy GS, Lindal GF, Wood GE, Croft TA (1982) Radio science with Voyager 2 at Saturn: Atmosphere and ionosphere and the masses of Mimas, Tethys, and lapetus. Science 215:553–558
Van de Kamp P (1969) Alternate dynamical analysis of Barnard's star. Astron J 74:757–759
Van den Bergh S (1970) Extra-galactic distance scale. Nature 225:503–505
Van den Bergh S (1981) Size and age of the universe. Science 213:825–830
Van den Bergh S (1982) In search of the Hubble parameter. Nature 299:297–298
Van Flandern TC (1978) A former asteroidal planet as the origin of comets. Icarus 36:51–74
Wagoner RV (1968) Explosive nucleosynthesis. Astrophys J 151:L103–L108
Waldmeier M (1961) The Sunspot Activity in the Years 1610–1960. Zürich, Schulthess & Co, pp 171
Waldrop MM (1981a) Massive neutrinos: masters of the universe. Science 211:470–472
Waldrop MM (1981b) Matter, matter, everywhere ... Science 211:803–806
Waldrop MM (1982) Neutrinos: No oscillations? Science 216:605
Waldrop MM (1983a) The new inflationary universe. Science 219:375–377
Waldrop MM (1983b) Supersymmetry and supergravity. Science 220:491–493
Walgate R (1983) A giant leap for physics – where to? Nature 301:285
Wasserburg GJ, Papanastassiou DA, Tera F, Huneke JC (1977) Outline of a lunar chronology. Philos Trans R Soc Lond A 285:7–22
Webb DJ (1982) The effect of polar wander on the tides of a hemispherical ocean. J Geophys J R astr Soc 68:689–707
Webster A (1974) The cosmic background radiation. Sci Am 231/2:26–33
Weidenschilling SJ (1978) Iron/silicate fractionation and the origin of Mercury. Icarus 35:99–111
Weidenschilling SJ (1979) A possible origin for the grooves of Phobos. Nature 282:697–698
Weinberg S (1977) The First Three Minutes. Basic Books, Inc, Publ, New York, pp 188
Weiss AH, La Pierre RB, Shapira J (1970) Homogeneously catalyzed formaldehyde condensation to carbohydrates. J Catalysis 16:332–347
Weissman PR (1982) Dynamical history of the Oort cloud. In "Comets" (ed Wilkening L), 637–658, Univ Ariz Press, Tucson
Wess J, Bagger J (1983) Supersymmetry and Supergravity. Princeton Univ Press, pp 180
Wetherill GW (1977) Evolution of the Earth's planetesimal swarm subsequent to the formation of the Earth and Moon. Proc Lunar Sci Conf 8th, 1–16

Wetherill GW (1980) Formation of the terrestrial planets. Annu Rev Astron Astrophys 18:77–113

Wetherill GW (1981) The formation of the earth from planetesimals. Sci Am 244/6:130–140

Wheeler JA (1980) Beyond the black hole. In "Some Strangeness in the Proportion" (ed Woolf H) Chapter 22:341–375, Addison-Wesley Publ Co, Inc, Reading, Mass

Whipple FL (1950) A comet model. I. Acceleration of Comet Enke. Astrophys J 3:374–394

Whipple FL (1978) Comets. In "Cosmic Dust" (ed McDonnell JAM) 1–73, John Wiley & Sons, New York

Whipple FL (1982) The rotation of comet nuclei. In "Comets" (ed Wilkening L) Tucson, Univ Arizona Press, 227–250

Wilczek F (1980) The cosmic asymmetry between matter and antimatter. Sci Am 243/6:60–68

Wilkening L (ed) (1982) Comets. Univ Arizona Press, Tucson, pp 766

Williams GE (1981) Sunspot periods in the late Precambrian glacial climate and solar-planetary relations. Nature 291:624–628

Williams GE (1986) The solar cycle in Precambrian time. Sci Am 255/2:80–89

Williams GE, Sonett CP (1985) Solar signature in sedimentary cycles from the late Precambrian Elatina Formation, Australia. Nature 318:523–527

Wilson OC, Vaughan AH, Mihalas D (1981) The activity cycles of stars. Sci Am 244/2:82–91

Wilson RC (1978) Accurate solar "constant" determination by cavity pyrheliometers. J Geophys Res 83:4003–4007

Wilson RW (1979) The cosmic microwave background radiation. Science 205:866–874

Wolfson R (1983) The active solar corona. Sci Am 248/2:86–95

Woolf H (ed) (1980) Some Strangeness in the Proportion. Addison-Wesley Publ Co, Inc, Reading, Mass, pp 539

Yang CN, Mills RL (1954) Conservation of isotopic spin and isotopic gauge invariance. Phys Rev 96:191–195

Yokoo H, Oshima T (1979) Is bacteriophage X174 CNA a message from an extraterrestrial intelligence? Icarus 38:148–153

Zeldovich Ya B, Einasto J, Shandarin SF (1982) Giant voids in the universe. Nature 300:407–413

Part II

DOWN TO EARTH

Satellite imagery taken during the northern summer (drawing after METEOSAT-image of August 4th, 1981, kindly provided by the European Space Agency). Westerly low pressure systems are visible as cloud spirals, the subtropical high pressure belt as cloudless sky over the Sahara Desert, and the tropic front as a zonal band of convective clouds north of the equator. Here, the tropical rain forest of Central Africa shines through. It is bordered to the south and to the north by the savanna, the domicile of wild life. The longest river on Earth, the Nile, and the large lakes of the East African rift system are clearly visible. Clouds, rivers, lakes and oceans make Earth a water planet, and liquid water is the basic element of life here on Earth and possibly elsewhere in the universe.

Chapter 5

Fire

Universal Elements

According to Empedocles (483–424 B.C.) the physical universe can be subdivided into four basic elements: air, water, fire and earth. This classification takes care of the gaseous, liquid, and solid matter in a rather profound way, except that the element fire should be regarded rather as the driving force, keeping the others going. But for many centuries to come, this ancient definition was satisfactory to describe the physical state of universal matter.

What had been found useful for characterizing the elementary condition of the macrocosmos did not work for the living cell, which constituted a microcosmos of its own. Even though life has lost some of its complexity in recent years with the advent of molecular biology, we are still eons away from understanding how individual cycles, operating in the living system, became spliced together with this magnificant machine, invented at the dawn of time and called the cell.

Thus, I propose to include life as a separate entity and expand Empedocles' scheme by placing fire at the control center of the four elements. Air – Sea – Earth – Life interactions are driven by the *fire* of the Sun, the *heat* derived from the decay of radiogenic elements within the Earth, and the *energy* released from phosphate bonds in the living cell. The alchemistic symbol of fire △ can be likened to the four triangular faces of a tetrahedron with fire at its center coordinating the flow of matter, be it gaseous, liquid, solid or living.

In ancient China, we had not four, but five universal elements. But before I comment on those, I would like to give some impressions of a river boat trip down the Yangtse as told by Li Bai, 12 centuries ago:

The rivers of China have attracted the attention of poets throughout the ages. My favorite poem entitled, "Downriver to Jiangling" was written by Li Bai in the year 726 A.D., when he was 25 years old. He recollects the impressions of a long voyage down the Yangtse from a small town, Baidi, in the northeast part of Sichuan province. We somehow become co-travellers on an exciting trip through the Qutang, Wuxia and Xiling gorges and eventually reach Jiangling, the place where Li Bai was born. Today the Gezhuo Ba High Dam near Yichang, finished in 1981, would have stopped our "dream boat's" voyage (Fig. 5.1).

Fig. 5.1. Downriver to Jiangling

Downriver to Jiangling

Leaving Baidi early under many
colored clouds
I travel back to Jiangling –
a thousand li a day.
On river banks the cries
of monkeys chatter on,
while my small boat has passed
along ten thousand mountains.
Li Bai

The clouds reveal the air, the river symbolizes the everflowing stream of water from land to sea, the mountains are solid witnesses to the eternal strength and dynamics of our Earth, and the monkeys and Li Bai himself are both life and evolution of man and mind. Air – Water – Earth – Life are the universal elements that interact and keep our Earth a habitable planet. Now back to the five universal elements (*Wu Xing*).

This idea was first conceived by a disciple of Confucius (551–479 B.C.) and appeared in written form in the Shang Shu and here in the chapter on fundamental laws of government (Fig. 5.2). Each of these five elements has its own properties: water nourishes what is underneath, fire burns what lies above, wood can be made straight or bent, metal is malleable, and earth promotes agriculture. Each of these elements also has a specific taste. Thus water is salty, fire is bitter, wood is sour, metal is spicy-sharp, and earth is sweet. Not only do the five elements promote each other – wood promotes fires, fire promotes earth, earth promotes metal, metal promotes water, and water promotes wood – but they also inhibit each other: water inhibits fire, fire inhibits metal, metal inhibits wood, wood inhibits earth and earth inhibits water. The view that the universe is composed of these five interacting elements has played a very important role in the development of ancient Chinese astronomy, medicine and science in general.

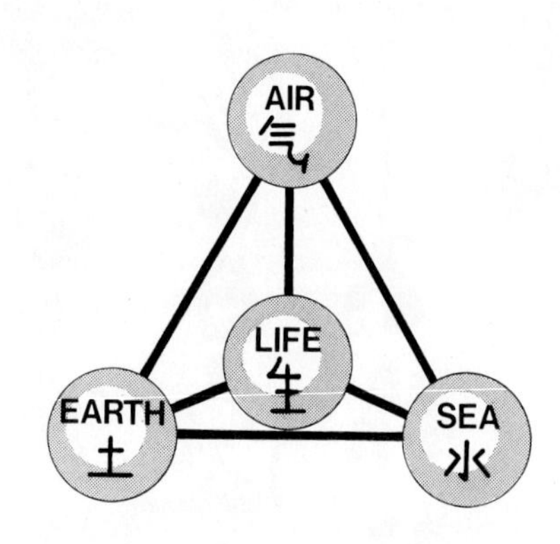

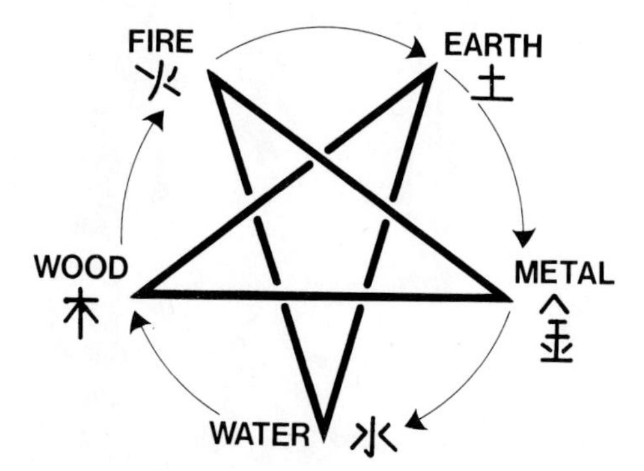

Fig. 5.2. Five universal elements

Gravitational Forces

So far we have looked at forces operating in nature from the standpoint of particle physics, and the meanings of terms such as GUT's or supersymmetry should be familiar to us. In the present chapter, we will take a different approach and examine what type of forces keep the terrestrial system going. The principal sources of energy are: gravitational forces, solar radiation, heat flow from within the Earth, and chemical bonding. The order of presentation will follow this line of thought.

Gravitational collapse of dark molecular clouds was the trigger for the formation of the solar system. In the subsequent course of accretion, Earth experienced the collapse of planetesimals of different sizes and mass over relatively short periods of time. The collapsing bodies generated kinetic energy which became a starting fundus of heat within the evolving planet. Gravitational differentiation of matter into core, mantle and crust provided additional increments in energy.

In this context, one should recall that the release of gravitational energy – past and present – accounts for the excess luminosity of Jupiter, Saturn and Neptune (Pollack, 1984). The outer planets shrank from protoplanets that – depending on the mass of the early Sun – were about two to three orders of magnitude larger than today, and contracted in a kind of stepwise fashion to their current values. For Jupiter, the temporal behavior of gravitational shrinking is best studied, and observed and predicted luminosities agree well. For Saturn, contraction models yield a factor of three less internal luminosity dissipation from gravitational energy release. It has been proposed that, in addition to gravitational energy obtained by contraction, the downward-directed segregation of helium in the planet's interior, as a consequence of helium's partial immiscibility in the outer portion of the metallic hydrogen region, is a supplemental energy source (Pollack, 1984; Pollack et al., 1977; Stevenson, 1980).

When it comes to gravitational forces operating on Earth, we no longer use the universal term gravitation, but speak of gravity instead, which literarily means having weight. Thus, gravitation is a physical process and gravity a natural phenomenon. The force of gravity is essentially the downward pull of mass over the surface of the Earth at any point. Since the distribution of mass is globally, regionally and locally not necessarily uniform, gravity anomaly maps give details of the distribution of dense or less dense bodies and, in turn, processes causing such anomalies (see examples in: Lerch et al., 1974; Gaposchkin, 1974; Wagner et al., 1977).

The Earth is roughly an oblate spheroid whose polar diameter of 12,713.5 km is almost 43 km less than the equatorial diameter. Actually, the Earth is slightly pear-shaped, as revealed in the longitude-averaged meridional profile of the geoid (King-Hele et al., 1980) (Fig. 5.3). Indeed, satellites, through changes in their orbital motion, are a powerful tool to measure the Earth's gravitational field and hence the shape of the Earth's calm sea-level surface. This, in combination with ground truth gravimetric data will provide the information on which at present maps on global geoid heights are based (e.g., Lerch et al., 1974).

At first sight, mapping of gravity anomalies appears to be a straightforward operation. However, corrections have to be made for Earth's oblateness, its rotation and topography. One can readily take care of these factors by implying that readings were "quasi" done at sea level (free-air correction), and by assuming that sea water is rocky. Appropriate subtractions in the case of overlying rocks or additions for ocean depth will give final readings for the true subsurface mass variation (Bouguer correction).

On the global scale, the residual gravity anomalies are small in comparison with those on Mars, the Moon and our twin planet Venus. The explanation is that on Earth variations in elevation of the solid surface, both on land and at sea, are compensated by mass distributions beneath in what is commonly referred to as isostatic adjustments. Thus, the principle of isostasy tells us something about the elastic properties of the lithosphere, and the buoyancy or floating of crustal blocks.

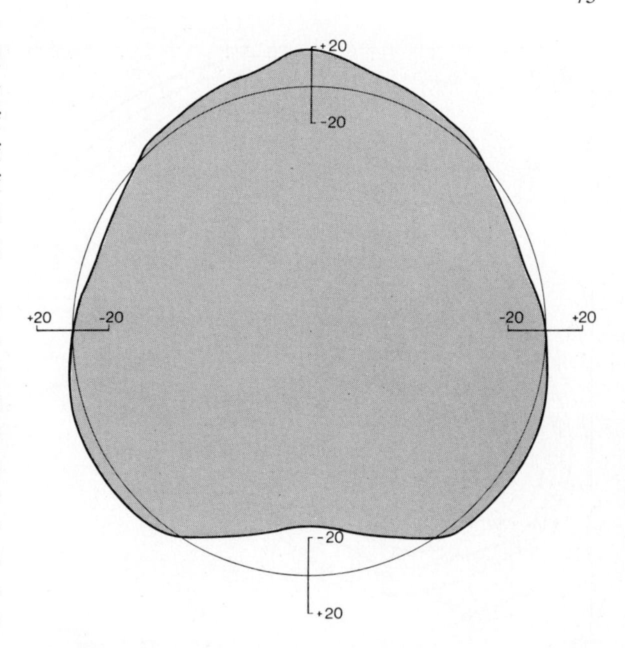

Fig. 5.3. Longitude-averaged meridional profile of the geoid (King-Hele et al., 1980)

The temporal response of continental crust to changes in weight is exemplarily shown for the crystalline shield of Fennoscandia. At the height of the last glacial, about 18,000 years ago, the ice cap at its center supposedly had a thickness of 3 km. This, in turn, could theoretically have produced a depression in the crust of max. 1 km. It is worth noting that the ice cover started to form about 60,000 years ago and gradually vanished in the time interval 14,000 to 7,000 years before the present, first rapidly (see p. 268) and then slowly. This initiated the so-called Flandrian transgression, which is still going on. Since that time, isostatic rebounding has produced about 300 m of uplift in the center of the former ice cap, located in what is now the northern part of the Gulf of Bothnia. The pattern of epirogenic uplift over the past 5,000 years is depicted in Fig. 5.4. To achieve isostatic compensation, the crust is supposed to rise another 200 m. It should be pointed out that the lowlands to the south of Fennoscandia (NW-Germany) are rapidly sinking. Furthermore, Quaternary ice ages came at intervals of about 100,000 years, and the crust rebounded in phase with interglacial periods (Yo-Yo effect). Thus, the Quaternary witnessed close to 25 ice ages, rather than only four as one commonly reads.

Isostatic equilibrium of crustal blocks implies that the mantle beneath can compensate for whatever weight we place on top. For instance, a mountain of low-density crust would root deeply in high-density mantle, whereas in the case of an ocean the crust-

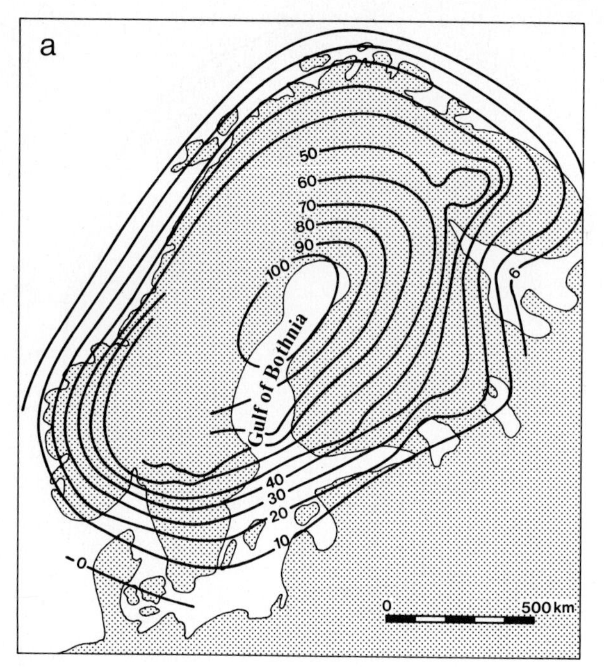

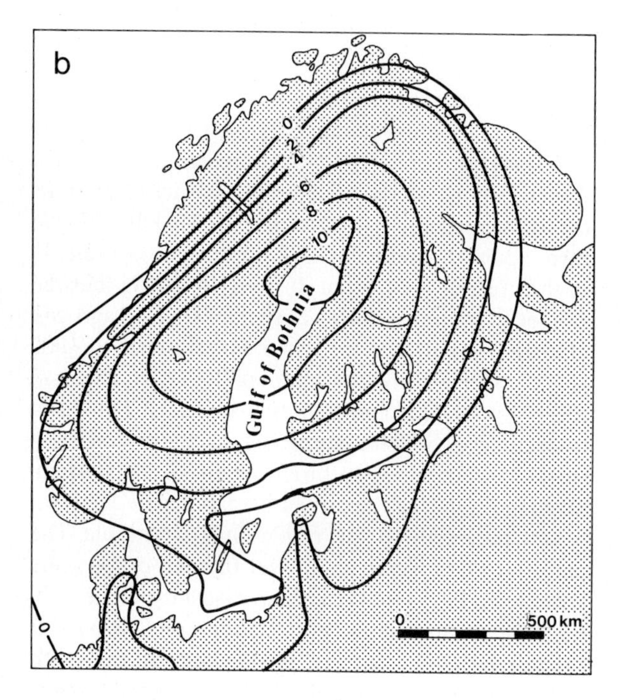

Fig. 5.4 a, b. Epirogenic uplift in Fennoscandia over the past 5,000 years in meters **(a)**; present uplift in mm a^{-1} **(b)** (after Sauramo, 1955; Gutenberg, 1941; Brinkmann, 1972)

mantle boundary below would stay high. This boundary, abbreviated Moho or M-discontinuity, has been named in honor of the geophysicist A. Mohorovičić, who discovered it for the first time in 1909 on the occasion of an earthquake near Zagreb. A Bouguer gravity anomaly curve commonly mirrors the position of the Moho by negative values over the mountain, near-zero values for the flat plain, and positive values across the deep ocean basin.

In the course of time, mountains become eroded and basins filled. Isostatic rebounding will try to compensate for the displacement of weight from one place to the other. It is elevation that ultimately provides the kinetic energy for a rock to roll down the hill, for a grain to be carried by rivers from land to sea, and for a particle in the water column to settle towards the abyss of the ocean. The downward pull of gravity maintains the water runoff and the transport of minerals. Before closing this subject, a brief reference should be made to tidal energy.

On Io, on-going volcanism is maintained by tidal forces exerted from Jupiter. However, the energy Earth derives from being so close to the Moon and the Sun is orders of magnitudes less, and just sufficient to cause a slight bulge in the solid Earth and the sea. At the same time, the Moon too is affected by tidal forces, and the difference in crustal thicknesses between the far and the near sides of the Moon is accounted for by the difference in gravitational pull received from the Earth. While Moon's impact on the solid terrestrial crust is negligible, the ocean's water is more responsive. Tides are highest when the gravitational pull from Sun and Moon come jointly (spring tide). Some regions, like the Bay of Fundy in Canada, experience enormous tidal forces with tidal ranges measuring more than 12 m. For other regions, such as the Baltic Sea, the difference between low and high water levels is only a few decimeters.

Most of the tidal energy is dissipated in the shallow sea through friction of water masses moving along the sediment floor (Brosche and Sündermann, 1982). This energy stems from the rotation of Earth and Moon, and a loss in tidal energy means a slowing down of the Earth's rotation. Days should become longer and fewer days per year should result as we move on in time. Also rotation of the Moon will decelerate, and the Earth-Moon distance will gradually widen (see p. 51).

Solar Energy

The Sun is composed of four structural units, which are from top to bottom (see Fig. 4.5):

(i) Photosphere ($\approx$350 km);
(ii) hydrogen convection zone ($\approx 10^5$ km);
(iii) radiative zone ($\approx 3.2 \times 10^5$ km); and
(iv) core ($\approx 2.8 \times 10^5$ km).

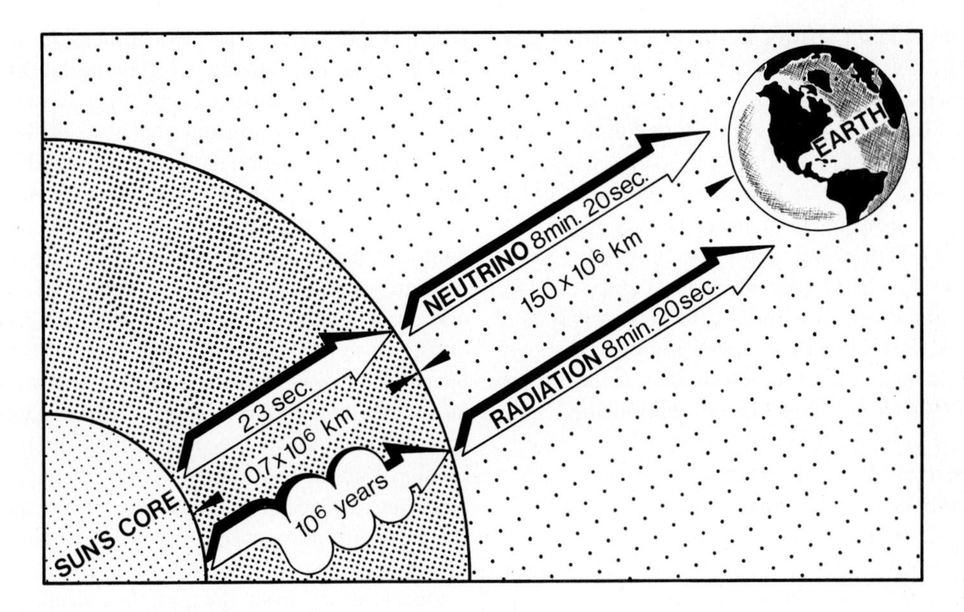

Fig. 5.5. Neutrinos and ordinary radiation (including light) take the same amount of time to travel from the Sun's surface to Earth, but the light we see today may have taken a million years to reach the photosphere (after Trefil, 1978)

Energy produced in the core is conveyed outward by photons or within the radiative zone. Inside the hydrogen convection zone hot ascending gas masses are the medium of energy transport. A switch from ionized hydrogen to neutral hydrogen proceeds as the hot gases ascend. This is accompanied by the capture of an electron and the neutralization of the hydrogen atom. The process releases additional energy and enforces the outward flow of heat. Once the photosphere is reached, transport of energy is again mediated by photons.

The time energy needs to travel from its central point of origin to the photosphere from where it radiates away will take at least a million years. It is this energy that expands the Sun to its present size and counteracts the enormous gravitational pull exercised by the sheer mass concentrated in the core. Most of this energy (95 %) is generated in the lower half of the core, where roughly 35 percent of the Sun's mass resides. Temperature, pressure, and density are here: 15.5×10^6 K, 3.4×10^{16} N m^{-2}, and 160 g cm^{-3} respectively. The corresponding values at midpoint are: 3.4×10^6 K, 4.7×10^{13} N m^{-2}, and 1 g cm^{-3}.

Some apparent anomalies exist in the Sun's interior. They have to do with the solar neutrino flux (Hartline, 1979). A small part of the energy released in nuclear reactions streams off via neutrinos. For the *p-p* chain this amounts to ca. 0.5 MeV. From what is theoretically known about thermonuclear reactions and the structure of the Sun, one can calculate the solar neutrino flux from the photon flux. However, the measured neutrino flux is just about one third of its expected value.

As opposed to photons, which require at least a million years to reach the Sun's surface, neutrinos achieve it in a few seconds because the Sun is transparent to neutrino flow (Fig. 5.5). From there to the Earth's surface, neutrinos and ordinary radiation travel at the same speed. Thus, one way to account for the low neutrino flux is to assume that today the solar fusion reactor operates only at one third of the capacity it had about one million years ago. Alternately, one could question the validity of the neutrino flux experiment, as has been recently proposed.

The solar neutrino problem may be resolved by assuming that weakly interacting massive particles, so-called WIMP's, are present in the solar center in a concentration as small as 10^{-11} by number. The appropriate scattering cross-section would reduce the predicted neutrino flux to the value observed on Earth without having to postulate significant changes in the solar structure outside the central region. Actually WIMP models provide a better fit than former standard solar models to explain the neutrino problem (e.g., Faulkner et al., 1986; Däppen et al., 1986).

However, in all probability the truth of the matter rests in the still limited knowledge about some physical properties of neutrinos, in particular neutrino mass and mass differences between low energy and higher energy neutrinos. Non-zero mass neutrinos may also provide an answer to the low neutrino flux (see p. 23).

The principal physical characteristics of today's Sun are given in Table 5.1. Standard stellar evolution models predict that 4.5 Ga ago, the Sun's luminosity was lower by about 30 percent (Newman and Rood, 1977) (Table 5.2). Accordingly, all water bodies at the surface of the Earth should have been frozen. Since liquid water has been with us at least since 3.8 Ga (Isukasia formation, Greenland), there certainly is a problem. A warming of the Earth might have been achieved via a greenhouse effect, due to increased levels in CO_2 and other climate-optimizing trace gases, or by varying the Earth's albedo (e.g., Sagan, 1977; Owen et al., 1979; Henderson-Sellers, 1983; Henderson-Sellers and Meadows, 1977; J.C.G. Walker, 1982, 1983).

There is much evidence supporting the idea that in its early history the Sun was a T Tauri star (e.g., Gaustad and Vogel, 1982; Canuto et al., 1983). T Tauri stars are the earliest optically visible state of evolution for stars about the mass of the Sun. By taking the presently observed X-ray and far ultraviolet radiation fluxes of T Tauri stars into consideration and comparing them with today's solar and earth-bound energy fluxes, one can say that a young Sun emits up to 10^4 times more UV and 10^3 times more X-rays (Table 5.2). How much of this radiation actually penetrated to the ancient Earth's surface cannot be reliably answered at present, because little is known about the composition and structure of the early atmosphere, or the extent of the ozone shield.

Table 5.1. Principal physical characteristics of the Sun

Mean distance from Earth (1 AU)	1.50×10^8 km
Radius	6.96×10^5 km
Mass	1.99×10^{33} g
Mean density	1.41 g cm^{-3}
Surface gravity	274 m s^{-2} (26 x terrestrial gravity)
Total energy output	3.83×10^{26} W
Energy flux at surface	6.34×10^3 W m^{-2}
Effective surface temperature	5780 ± 50 K
Solar constant	1.36 kW m^{-2}
Period of rotation	about 27 days; 25 days at the equator to 31 days at the poles

In addition, the early Earth witnessed more abundant meteorite and cometary impacts. This, combined with enforced volcanic activity, could have produced numerous feedbacks of various sorts, sizes and directions of the atmospheric global temperature profile. Speculation runs high on this subject matter and almost any interpretation must – unfortunately – be considered seriously, because there is a stratigraphic hiatus of about 700 million years; likewise, the first Archean rock so far gives no clue with respect to air-sea-earth-life interactions driven by $h\nu$, the radiating energy from the Sun (e.g., Pepin et al., 1980).

Observations from the past few decades indicate that the Sun's diameter had stayed constant over the past 250 years (Parkinson et al., 1980) and that the solar constant remained practically unchanged with variations amounting to less than 0.2 % (e.g., Wilson, 1978; Newkirk, 1983). This, however, does not apply to the solar UV output. According to measurements made from rocket and satellite platforms (Heath and Thekaekara, 1977), the solar UV flux varies significantly at 1,750 Å from solar minimum to solar maximum (UV_{max}/UV_{min}). These data were used by Borucki et al. (1980) and Lean (1987) to show possible implications for Earth's climate and stratospheric ozone levels. The little ice age which Earth experienced in the 17th and 18th centuries during the Maunder Minimum might have been a consequence of increased solar UV fluxes, which in turn raised stratospheric ozone levels. These two factors combined will increase the fraction of the total energy absorbed in the stratosphere and result in less energy being deposited in the troposphere. Since the total solar flux stays constant, redistribution of radiant energy towards longer wave lengths should compensate for this

Table 5.2. Energy sources averaged over entire Earth (cal cm^{-2} a^{-1}) (After Gaustad and Vogel, 1982)

Type of energy	Contemporary Earth age 4.7×10^9 a	Earth at age 5×10^8 a	Earth at age 10^7 a
Total solar radiation	265,000	170,000	132,000
Far UV (20–150 nm)	1.4	4–30	100–10,000
X-ray (0.3–6 nm)	0.2	7	70–700
Radioactivity from crust (35 km)	15.5	47	–
Heat from volcanic emission	0.15	0.15	–
Electric discharges	4	4	–

phenomenon (Pollack et al., 1979). A perturbed Sun, that is one without the 11-year solar sunspot cycle, might conceivably induce climatic change, and alter air chemistry. In turn, these data are useful, when it comes to unraveling the relationship between a "faint early Sun" and a pre-Archean life-supporting environment on Earth (Eddy, 1977).

Some information on the ancient Sun can be gained not only from Earth but from our neighboring planets, Venus and Mars. Venus was probably much cooler, perhaps having temperatures below the boiling point of water (Pollack, 1979), and Mars certainly warmer than today, as evidenced from large desiccated river channels (Toon et al., 1980a; Masursky et al., 1977). My former department chairman Bob Sharp expresses the Martian situation more cautiously, "The erosional channels seem to require that at some time, possibly in the far distant past, the surface environment of Mars has been less rigorous than it is at present. This is what the channels appear to say in spite of various models, theories, and arguments to the contrary" (Sharp and Malin, 1975).

Heat Flow

A profound gap in a person's education exists, should he or she not know the essence of the "Second Law of Thermodynamics". This is at least what one can read in C.P. Snow's "The Two Cultures". I would not go to such an extreme, but a good grasp of thermodynamic principles and processes is certainly of value when looking at the state and temporal changes of natural systems.

Our physical world is composed of a great variety of structural units, that is systems which through certain boundaries are separated from their surroundings. In general, a system which exchanges some of its properties such as mass, heat or work with the surroundings is considered to be open. In analogy, a system having no exchange with the outside is called isolated. If heat and work, but not mass, are exchanged, the system is referred to as being closed.

One can describe systems in several ways. For example, to a biologist an ecosystem is defined as a community of organisms that live together and interact among each other within the bounds of specific environmental settings. To a chemist, a system has a known volume, contains interacting chemical species of various sorts and concentration, and is controlled – among other factors – by temporal changes in temperature and pressure. To a biogeochemist, system analysis is often too complex to be of practical use. It has been found more convenient to speak of biogeochemical cycles in order to bring to light the complexity of interactions among a series of systems cross-cutting each other.

In the following, we shall consider the interior of the Earth as a heat-propagating system. Principal generators of internal heat are gravitation and radioactivity. We have already seen that gravitational collapse will yield energy which could raise temperatures within the early Earth to about, 1,000° C. This was followed by the decay of radiogenic elements such as uranium, thorium and potassium-40, which raised temperatures to a point that iron started to melt, form droplets and sink to the core. This in turn produced gravitational heat along the same line as the sinking of helium "drops" into Saturn's interior, accounting for a substantial part of its radiating energies. The sinking of iron droplets at that stage yielded energies amounting to about 2×10^{37} erg. Energy derived from all three sources, (i) gravitational collapse, (ii) radioactive decay, and (iii) sinking of iron droplets, liberated the heat needed to differentiate the Earth into core, mantle and crust.

The process of differentiation already began, if one follows the model of heterogeneous accretion of the proto-Earth, in the accreting disk. During succeeding stages of differentiation which lasted at most a few hundred million years, the structure of the Earth, as we observe it today, was generated. Everything that happened afterwards in terms of, for example, global tectonics involved adjustments in crust and asthenosphere. These took shape principally as a consequence of heat transfer from one spot to another by conduction and convection. Increments of heat liberated from the decay of radioactive elements are becoming smaller and smaller. At the onset of differentiation about 4.5 billion years ago, there were short-lived isotopes such as aluminum-26, twice as much uranium-238, and about ten times as much potassium-40. For thorium-232, having a halflife of 14.1 billion years, the loss has been comparatively small. So at present, the energy sources driving tectonisms are those saved from the early past and those derived from ongoing radioactivity.

In order to accomplish something in the crust, heat has to become transferred, and thermal conductivity of a material will control how fast heat moves along. Salt, for example, is a good conductor, and for this reason salt domes were mentioned as likely candidates for the disposal of "hot", that is radioactive waste. However, the common rock has a very small thermal conductivity. It would take about one billion years to transfer heat from the upper mantle through the lithospheric plate to emerge finally at the surface. This also explains why a granitic dome that has formed a few million years ago still emanates heat and will continue to do so for many million years to come. Radiative heat transfer as is observed in the Sun is another means of distributing energy. It will proceed whenever temperatures are so high as to cause a rock to glow.

The most effective way of transporting energy within the Earth is by convection. The process has best been studied in heated fluids, and the speed of heat transfer can readily be watched when a pot of water starts to boil. Since rocks are solid, where is the medium that can promote convective flow? The mantle beneath the rocky plates appears to a geophysicist like a rock too. Seismic waves behave as if the material were solid. Yet by observing that part of the mantle over much longer time scales through phenomena, manifested in a changing solid crust and lithosphere, one can reconcile such changes only by assuming that convection cells are transporting heat upward and cooler material downward. The idea of thermal convection in subcrustal compartments was first conceived by O. Ampferer in 1906 quasi as an antithesis to the established contraction theory of E. Suess (1885/1909). It was the first mobilistic approach to explaining the diversity of tectonic elements.

At present, mantle convection is a well-supported process, but details on height or extension of convection cells, or the time scales of their operation, whether continuous or cyclic, is still an open question. In contrast, more information exists on the amount of heat the Earth crust is dissipating through its surface.

The heat conducted at present to the surface of the Earth from below averages about 80 mW m^{-2}. Regional variation in the heat flux lies within a factor of three about the mean (Chapman and Pollack, 1975). By comparison, the Sun gives us 5,000 times as much heat. The global heat flow pattern is depicted in Fig. 5.6. Major oceanic ridge systems are represented as heat flow highs, so are the marginal basins of the West Pacific, Alpine Europe, and the American Cordillera. Here and there one notices bulges, as for instance, at the Galapagos spreading center. In contrast, continental shields, platforms and the oldest oceanic regions exhibit a low heat flow pattern. From this knowledge of regional differences of the Earth's heat flow field a better understanding of geodynamic processes can be acquired.

Studies on variation of continental and oceanic heat flow as a function of time reveal not only temporal changes related to differences in the rates of crustal heat emanation, but give clues on the cooling history of the oceanic crust, which in turn permit reconstruction of paleobathymetry, or rates of continen-

Fig. 5.6. Global heat flow pattern (Chapman and Pollack, 1975)

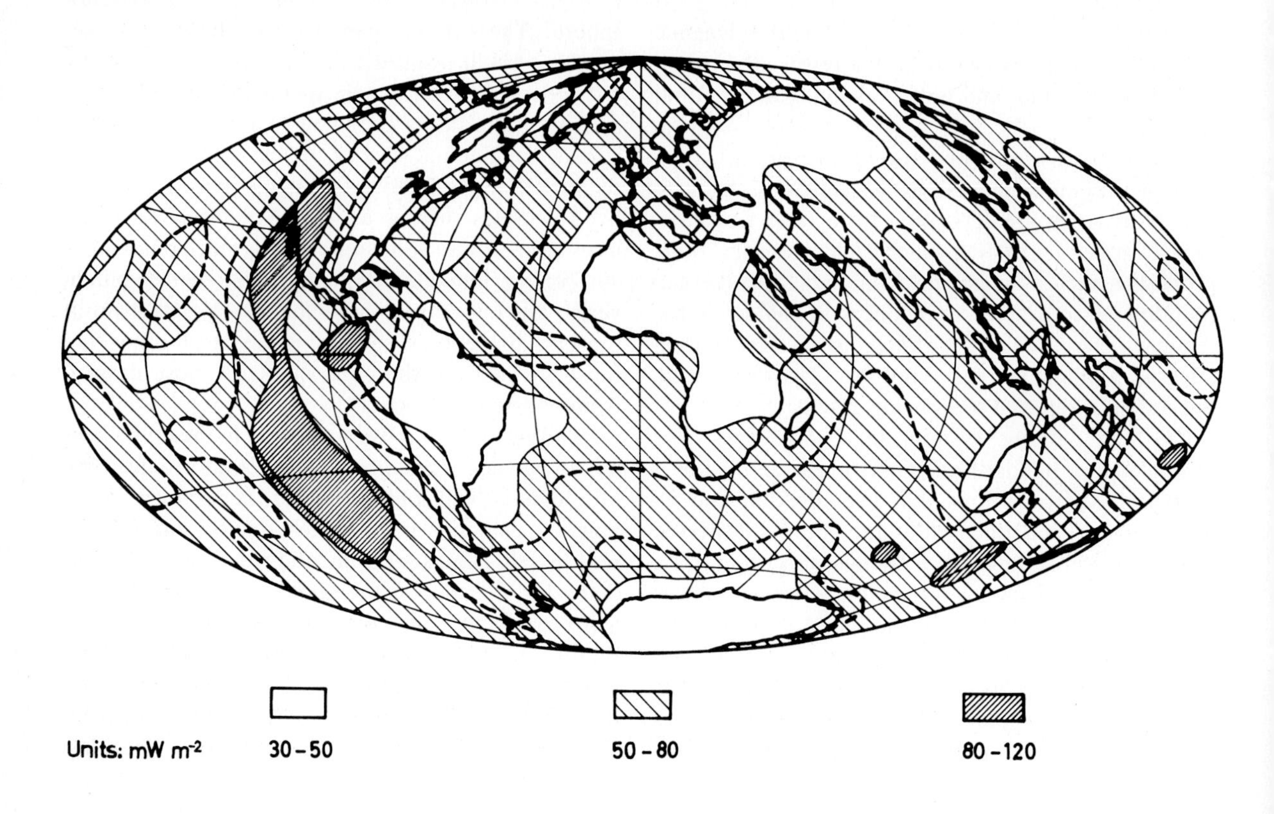

tal erosion. Following the pioneering work of Kraskovski (1961), Lee and Uyeda (1965), and Polyak and Smirnov (1968), a wealth of information has become available and is documented, e.g., in Parsons and Sclater (1977), Sclater et al. (1980), Sprague and Pollack (1980), and Vitorello and Pollack (1980).

A stratigraphic record of more than 3 Ga exists on the continents, whereas for the oceans there is no crust beyond 180 Ma of age. Old continental lithosphere is cool, thick, strong, and refractory. Moreover, crystalline shields tend to be centrally located in continental masses, and are thus less frequently affected by plate margin interactions. Modelers use this phenomenon to predict probability of tectonothermal mobilization, which is assumed to decrease exponentially with time since the previous mobilization (Sprague and Pollack, 1980). Thus, younger mountain chains have a greater chance of becoming remobilized than remote Precambrian shields.

According to Vitorello and Pollack (1980), the decrease of continental heat flow can best be reconciled with a three-component model consisting of: (i) radiogenic heat from the crust, (ii) heat derived from transient thermal perturbation associated with tectonically mobilized lithosphere, and (iii) background heat flow of possibly deep origin. Contributions derived from these three sources are illustrated in Fig. 5.7. Component (i) contributes 40 percent of the observed heat flow in terrains of all tectonic ages, whereas component (iii) stays uniform and stems in part from the thick continental conductive root (~15 mW m^{-2}), and in part from deeper zones within the Earth, possibly from the core (~12 mW m^{-2}). Component (ii) accounts for about 30 percent of the total in Cenozoic tectonic regimes, to decrease effectively to zero values in the Proterozoic.

With respect to the ocean, where no crust older than the Upper Jurassic survived plate tectonics, a different heat flow field emerges. Knowing the age distribution of the oceanic crust at a given time using established accretion and subduction rates, a modeling of temporal changes of heat flow can be made (Sprague and Pollack, 1980). Data summarized in Fig. 5.7 suggest that global heat generation was more substantial during the Cretaceous relative to times before and after, and that about three-quarters of global heat production is released to the open sea. Information beyond the 180 Ma mark are difficult to obtain, because the ancient record has been subducted to make room for a younger oceanic crust. All the same, it is safe to say that during the Archaean, Earth experienced a higher heat flow regime simply due to the much greater abundance of radioactive isotopes. As a rough approximation, heat flow figures were probably twice as high as today. Enforced spreading rates could conceivably be responsible for an effective release of the "additional" heat derived from the interior of the Earth. It would, of course, be more accurate to say that a higher marine heat flow promoted faster cycling. In contrast, Precambrian crystalline shields were certainly more resilient, as we can infer from their modern heat flow field.

In summary, global heat flow figures for the past 180 Ma are in the range of 78–101 mW m^{-2}. For most of this time, heat flow stayed uniform at about the present value of 81 mW m^{-2}. Peak values established for the late Cretaceous are linked to rapid seafloor spreading.

Chemical Bonding

"The Nature of the Chemical Bond" by Linus Pauling, published in 1948, was the classic textbook we consulted as students (3rd ed. Pauling, 1960). It has not lost its appeal, because much of what one reads nowadays on chemical bonding has its root in that treatise. Another scholar has also to be mentioned in this context: Victor Moritz Goldschmidt. In 1923, J.A. Wasastjerna presented ionic refractivity data which for the first time provided the correct size for the ionic radii of fluorine, 1.33 Å, and oxygen, 1.32 Å. These measurements were used by V.M. Goldschmidt and his co-workers as a yardstick to determine empirically the ionic radii for the majority of chemical elements (Goldschmidt, 1926). Linus Pauling followed a different route and computed ionic radii from the electronic theory of atomic structure, employing as a standard of reference the empirically derived interionic distances in crystals of the alkali halides (Pauling, 1927). Even though there were minor discrepancies in the Pauling and Goldschmidt values – see Tables 26 and 27 in Goldschmidt (1954) – their work brought a new dimension to the fields of crystal chemistry and geochemistry.

Ionic bonding is the most straightforward type of bonding among chemical elements that are joined in molecules. In principle, one or more electrons of a neutral atom are transferred to another neutral atom, resulting in ions of opposite charges. Anions and cations are mutually attracted to one another until their electron clouds are adequately accommodated, and repel one another to counterbalance the force of attraction. Resulting products are ionic crystals, and a prominent example is common salt. It consists chemically of Na^+ and Cl^- ions and has the mineral name: halite. The number of charges an ion carries is called

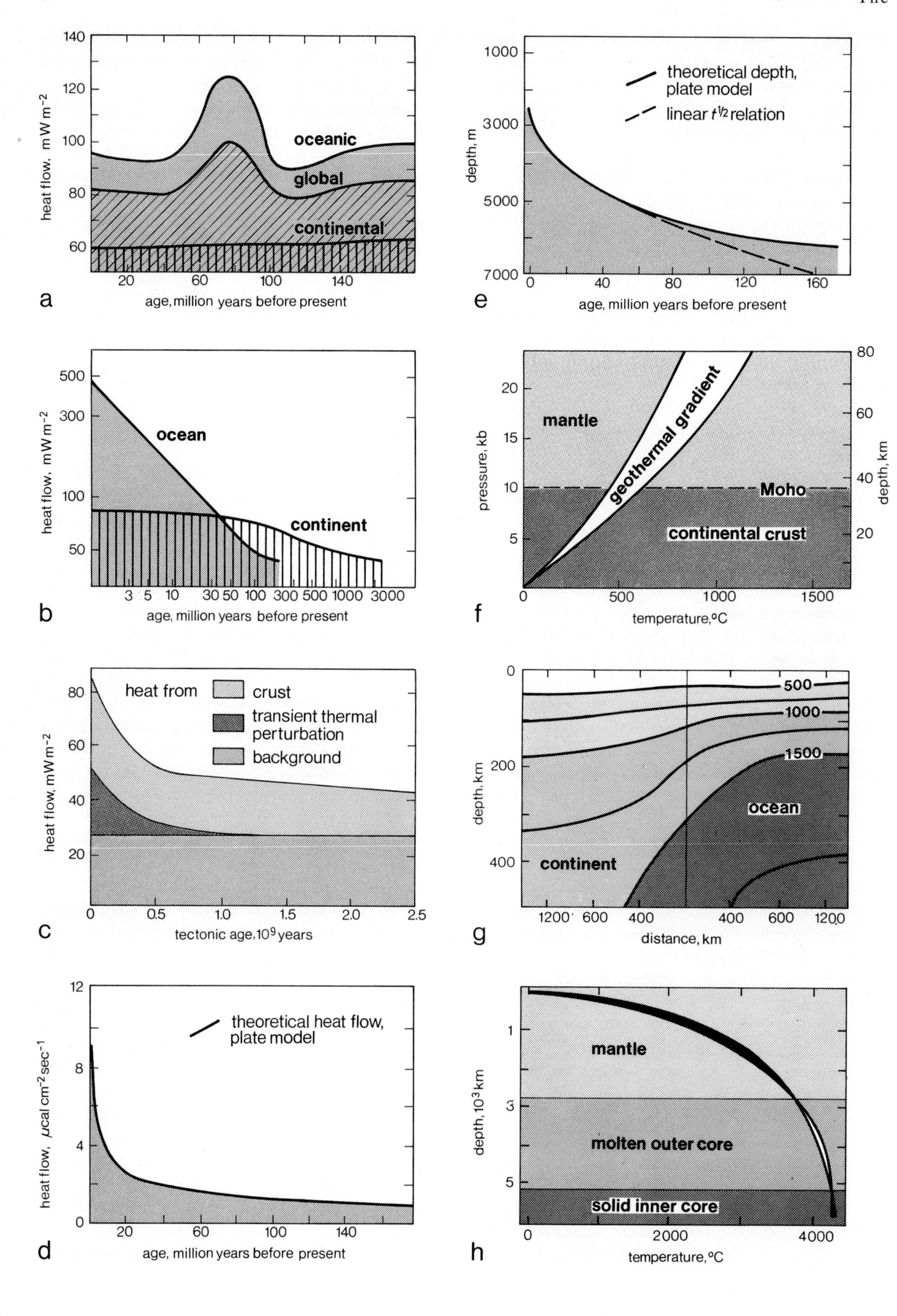

a
heat flow, mW m-2
age, million years before present
oceanic
global
continental
b
heat flow, mW m-2
age, million years before present
ocean
continent
c
heat flow, mW m-2
tectonic age, 10^9 years
heat from
crust
transient thermal perturbation
background
d
heat flow, μcal cm-2 sec-1
age, million years before present
theoretical heat flow, plate model
e
depth, m
age, million years before present
theoretical depth, plate model
linear t½ relation
f
pressure, kb
depth, km
temperature, °C
mantle
geothermal gradient
Moho
continental crust
g
depth, km
distance, km
500
1000
1500
ocean
continent
h
depth, 10³ km
temperature, °C
mantle
molten outer core
solid inner core

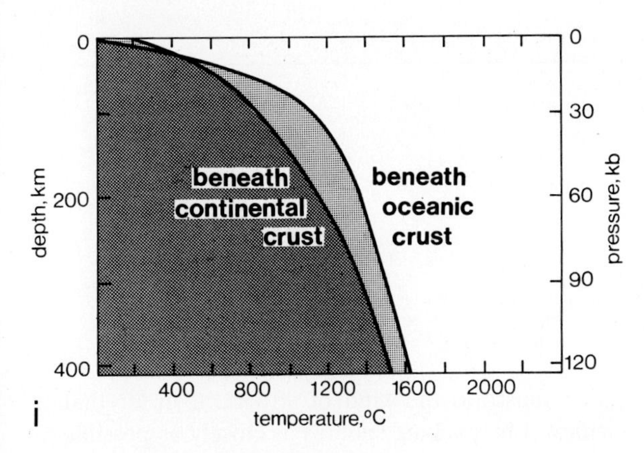

Fig. 5.7. **a** Variability of mean oceanic, continental and global heat flow over the past 180 Ma (Sprague and Pollack, 1980). **b** Heat flow versus age relationships for oceans (Parsons and Sclater, 1977) and continents (Chapman and Pollack, 1975). **c** Continental heat flow with age (highly schematic; Vitorello and Pollack, 1980). **d** Surface heat flow; plate model (Parsons and Sclater, 1977). **e** Ocean bathymetry as a function of age of ocean crust (Parsons and Sclater, 1977). **f** Generalized scheme of geothermal gradient in continental crust and underlying mantle. **g** Temperature profile beneath continents and oceans as a function of distance from shelf (MacDonald, 1964). **h** Temperature profile (geotherm) in the Earth's core and mantle (Press and Siever, 1986). **i** Temperature profile beneath continents and oceans (Clark and Ringwood, 1964)

the electrovalence of an element. In the case of sodium chloride, Na^+ has an electrovalence of +1 and Cl^- an electrovalence of –1.

There are several compounds which maintain a stable electronic configuration by the sharing of electrons, rather than by gaining or losing them. This type of bond has consequently been termed an electron pair or a covalent bond. By considering the case of carbon, the principle of covalent bonding is briefly addressed. Usually, carbon forms four covalent bonds which, in association with four other carbon atoms, are ordinarily placed at angles of about 109° to one another, giving rise to a regular tetrahedron. A simple covalent structure of this type is the diamond. Each C-atom shares an electron pair with all of its neighbors, thus generating a stable octet of electrons in the outer shell.

Covalent and ionic bonding can coexist in many compounds. For instance, in ammonium chloride the hydrogens are bound to nitrogen by electron pairs, whereas NH_4^+ and Cl^- are simple ions.

Native metals such as gold or silver are composed of cations, immersed in a "sea" of electrons which are drawn together and maintain the crystalline structure. Some of the electrons can freely move around and account for the electrical conduction of native metals.

The van der Waals force, also termed dispersion force, is a weak electrical force which is generated through the asymmetry of certain atoms and ions. An electrical dipole moment originating in one molecule will exert an electric field and act on a neighboring molecule, thereby inducing a dipole moment in that molecule. The two dipole moments will interact and establish a weak attractive force between the two molecules.

A hydrogen or a deuterium atom, covalently bonded to an electronegative atom such as nitrogen, oxygen, fluorine or chlorine, may interact with another atom via a hydrogen bond. H-bonds develop to atoms having electronegativity greater than that of hydrogen. The chemical molecule containing the hydrogen used for hydrogen bonding is strictly focused towards the nucleus of the electronegative atom accepting the hydrogen. Although the energy levels of hydrogen bonds can vary over one order of magnitude, the most common ones have energies in the range of 4–6 kcal/mol, and thus fall between the weak van der Waals forces (ca. 0.3 kcal/mol) and the covalent chemical bonds (ca. 100 kcal/mol).

We started this section with Linus Pauling and should like to close it with him. In the late twenties, in his studies on the structure of clay minerals, he became familiar with the nature of hydrogen bonds and the way they serve to stabilize the structure of clays. About 20 years later, he introduced the concept of hydrogen bonds to the field of biochemistry, when unveiling the configuration of the α-helix, and opened up new vistas for the study of the living cell.

Chapter 6

Coordination Principles

Closest Packing of Spheres

The world we live in consists of a great variety of coordination systems. A family, a nation or the human race are such entities. The units may differ in size, but all are composed of the same structural elements, i.e., individuals held together by bonds. The individual members can vary in size, status or sex, and ties between them can be weak or strong. The resulting structure determines the functional properties of the coordination complex. Social groups are programmed for change. Number, character and rank of members can alter and bonds will adjust accordingly. As a consequence, new structures and functions emerge.

Universal matter is structured according to the same principles, except that the number of different elements is small compared to the more than five billion individuals of the human race. Consequently, social systems exhibit a high degree of complexity.

In order to illustrate the basic feature of a coordination system, we will examine the relationship between element A and element B in a complex AB_n; n is the number of B's involved in the unit and is commonly referred to as *coordination number.* Should n equal four, one would write AB_4; and in the case of six neighboring B's, the complex reads AB_6. The two coordination complexes can also be drawn in the form of a *tetrahedron* or an *octahedron,* respectively (Fig. 6.1). Accordingly, the central element A may have a four-fold (tetrahedral) coordination, or a six-fold (octahedral) coordination.

Solid matter on Earth is principally crystalline in nature and chemical elements can be present as atoms or ions. As shown by high pressure experiments on solid phases, ions in the common rock-forming minerals and atoms in native metals behave as nearly rigid spheres in crystal structures. Thus, it comes as no surprise that X-ray diffraction analysis reveals a molecular architecture of crystals in the form of a geometric packing of spheres. The most prominent one is the closest packing of spheres, because such a configuration represents a minimum in free energy.

To illustrate the kind of structural order that is achieved by packing spheres as closely as possible, a series of "ping-pong" balls are stacked together in the way depicted in Fig. 6.2. We immediately find that the simplest type of packing is two-layered – the third layer repeats the first, the fourth layer the second, and so on. Such a primitive packing of spheres is geometrically described as the

A*B*A*B*A ...

stacking of layers, which is the hexagonal closest packing of spheres. The density of packing can be cal-

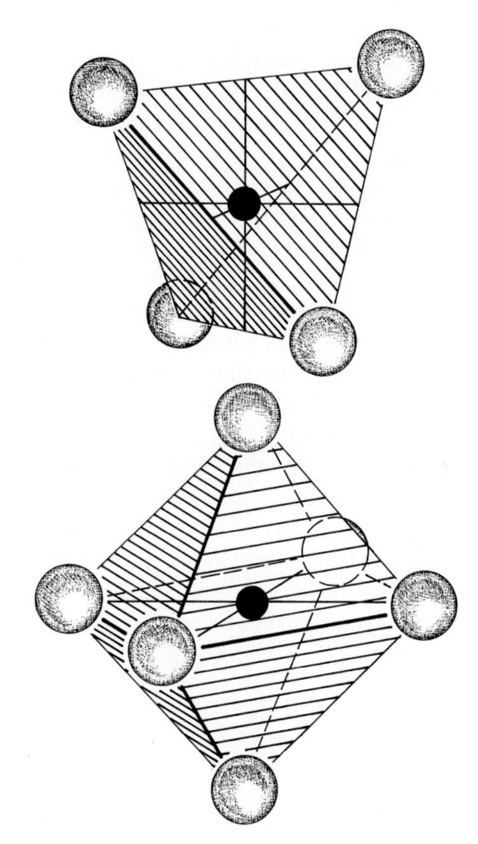

Fig. 6.1. Drawing of a regular tetrahedron (AB_4) and octahedron (AB_6). In case coordination partners B differ, for instance, in size, distorted polyhedra develop. Central element A is indicated as *black dot*

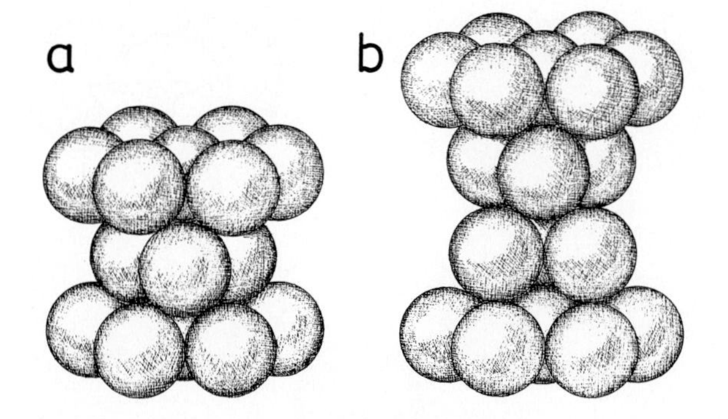

Fig. 6.2 a, b. Hexagonal **(a)** and cubic **(b)** closest packing of rigid spheres

culated to show that the spheres occupy 74 percent of the total space. Regardless of how the spheres are deposited, 74 percent is the highest density possible when packing spheres.

A three-layered stacking of spheres is the second simplest type of packing. Under these conditions, the fourth layer is the identical repetition of the first layer. This type of

A*B*C*A*B*C* ...

layering yields a face-centered cubic packing. Again, a 74 percent density similar to the 74 percent hexagonal stacking is achieved. A great number of metals crystallize according to these two packing principles. Thus, crystallographically different lattices can be constructed from identical types of layers, simply by gliding the identical layers against each other. In this way it is possible to obtain configurations such as

A*B*C*D*A,

in which the fifth layer represents a repetition of the first layer A. In extension, a sixth layer, seventh layer, n-layer arrangement is feasible; they will all create structures with different repeat distances for the unit cell.

The one property that remains unchanged is the relationship of the atoms to their nearest neighbors. In the closest packing of spheres, each atom stays in contact with 12 spheres or has a "kissing touch" with 12 neighbors. This type of a 12-fold coordination is independent of the spacing of the unit cell of the crystal. Each atom has always 12 neighbors in a three-dimensional network.

This coordination concept becomes meaningful in that it illustrates the characteristic grouping of atoms AB_n in a crystal. In contrast, a crystallographical description, in terms of structure and space group, will furnish only a mathematical point lattice, but nothing is said about the crystal-chemical properties.

Native zinc forms a hexagonal closest packing of spheres, and the coordination is described as $AB_n = ZnZn_{12}$. The coordination number n indicates the number of neighboring atoms B which surround the central atom A roughly equidistant. The coordination number indicates the number of neighboring atoms but does not consider the geometry of the atomic architecture. In the case of the closest packing of spheres, two different geometries are realized for the 12-fold coordination (Fig. 6.3).

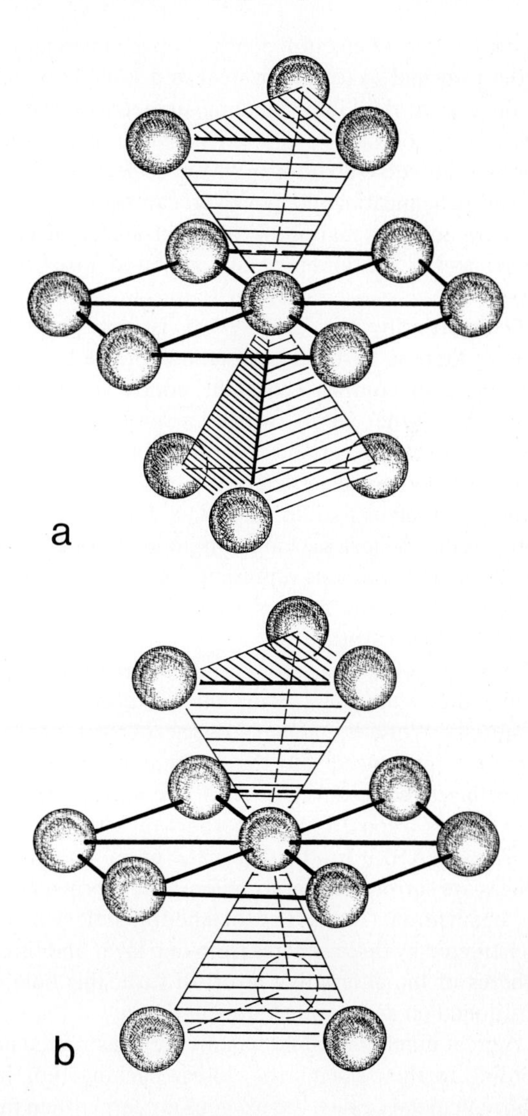

Fig. 6.3 a, b. The cubic **(a)** and the hexagonal **(b)** coordination polyhedra AB_{12}. Each atom has six neighbors within its own layer and a set of three neighbors each in the two neighboring layers. The two sets of triplets can be rotated by 60° parallel to the planar surface (n x 1/6 x 360° = 60°), because six interstitial positions are conceivable. This will result in the formation of two different coordination polyhedra. Independently of the sphere arrangement, only two twelve-fold coordination geometries can be obtained. The polyhedron **(a)** is termed cubo-octahedron

In the cubic structure, e.g., $CuCu_{12}$

$$A^*B^*C^* \ldots,$$

the third layer occupies a different position from the first layer. The centers of the spheres of the third layer do not lie above the centers of the first layer as they do in the case of the hexagonal packing.

Packing of Ionic Structures

Redistribution of electron density between two atoms is the principal factor in covalent and ionic bonding. In the case of rigid ionic bonding an electron is transferred from one atom to another, while in covalent bonding an equal sharing of valence electrons is involved. Chemical bonds, however, can also be generated by unequal sharing or partial transfer of electrons, and a continuous range of bond properties exists.

Geometrically, the ionic bond is not predetermined; there is no fixed orientation for the bonding direction. In contrast, covalent bonds control the position of the atoms within a molecule. Ionic molecules exhibit variable coordinations, whereas a covalent bond possesses a fixed coordination. A great number of metal oxides and salts form structures which reflect a close packing of rigid ionic spheres. In this connection, anions represent the closest packing of spheres, and the metal cations occupy the vacancies left in such structures.

Two types of unoccupied hole geometries are generated: the *tetrahedral* and the *octahedral* ones (Fig. 6.4). When the spheres of the nearest stacked layer are introduced into the triangular gaps of the lower layer, holes develop which are surrounded by four anions disposed at the corners of a regular tetrahedron $(A)B_4$. In the alternate case, interstitial spaces are surrounded by six anions; they form a regular octahedron $(A)B_6$. The octahedral hole will be constructed by three spheres from one layer and three spheres of the alternating layer. In turn, this hole is positioned on an intermediate plane.

A great number of ionic crystals are constructed according to the principle of closest packing. For instance in metal oxides, the oxygens are larger than the metal ions; they form the closest hexagonal or cubic packing, with the smaller-sized metal ions being placed in the unoccupied holes. Also smaller-sized cations such as Si or P will coordinate tetrahedrally, whereas larger-sized cations such as Ca or Fe commonly occupy octahedrally coordinated sites.

As an example of the general ionic type structure we present the cubic structure of sodium chloride

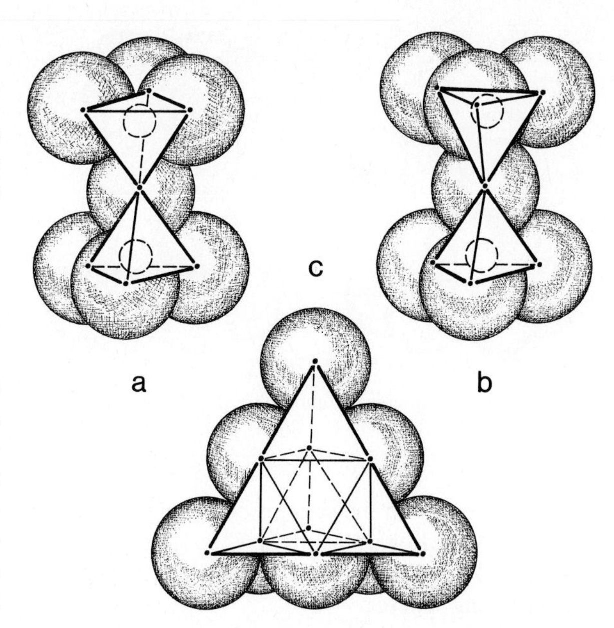

Fig. 6.4 a–c. The generation of tetrahedral and octahedral interstitials during the formation of the closest packing of spheres; by packing the spheres, the next sphere is always positioned in the triangular gaps. This will result in the formation of interstitial holes which are surrounded by four atoms; the space fabric formed this way is that of a regular tetrahedron. In the next proximity of a tetrahedron interstitial there remain three unoccupied triangular gaps which, for geometric reasons, will not be filled by spheres from the following layer. The resulting larger holes are surrounded by six neighbors. The centers of this set of spheres form a regular octahedron. The octahedral hole will be constructed by three spheres of one layer and three spheres of the alternating layer. In turn, this hole is positioned on an intermediate plane. A layer sequence in a structure may be rather complicated, yet only two types of coordination gaps are realized for a closest packing of spheres: tetrahedral (**a** and **b**), and octahedral (**c**)

(common table salt). All octahedral gaps become fully occupied by the sodium ions (Fig. 6.5). At first sight it is difficult to visualize the hexagonal layer organization of the chlorine atoms in the face-centered cubic structure, but viewed diagonally through the unit cell, the hexagonal organization of the chloride layers can be recognized (Fig. 6.6). It is noteworthy that the tetrahedral coordination gaps remain unoccupied in the NaCl structure. The drawings indicate the positions of these tetrahedra in the NaCl lattice. This principle of occupying tetrahedral and octahedral sites is followed by ions in many crystals. Cations which occupy the gaps are often larger in size than ideally provided for by the stacks of anions; as a result the anions become pushed apart. Thus, the radius ratio of cations to anions determines the stability of the ionic bonds. Furthermore, the coordination state of a metal, whether tetrahedral, octahedral or higher,

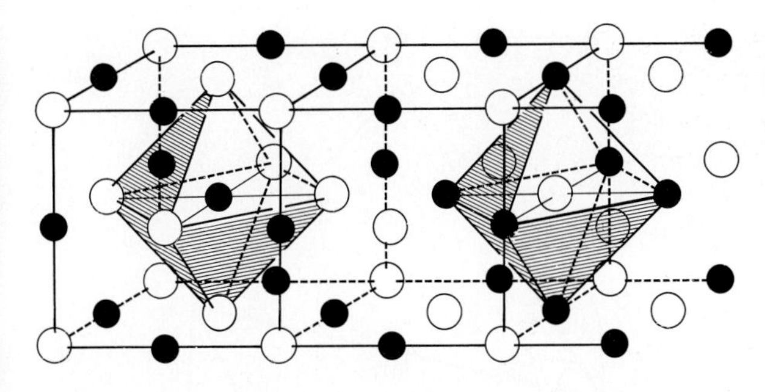

Fig. 6.5. Structure of table salt NaCl, which is an example of the general ionic type structure. A face-centered cubic symmetry is established; the corners and plane centers are occupied by anions Cl^-. The white spheres represent the chloride ions and the black dots the sodium ions.

The Na-Cl distance is 2.82 Å.

The chloride ions form a closest packing of spheres:

A * B * C * A ...

and the sodium ions occupy all octahedral interstitials. This relationship is shown in the left-hand cube, which illustrates the AB_6 coordination polyhedron for the sodium in the form of a shaded octahedron. The right-hand cube also contains an octahedron in which six sodium ions surround a chloride ion. However, as a coordination polyhedron the "Cl-octahedron" is unreal. It is shown only to demonstrate that for the closest A * B * C packing around a chloride sphere, the centers of the octahedral interstitials surround the chloride spheres in a six-fold arrangement which also yields an octahedron. The planes for the closest packing of the chlorine atoms cut the cube diagonally. Inasmuch as it is difficult to visualize these hexagonal layers inside the cube, Fig. 6.6 has been prepared

is controlled by the size of the ions. X-ray studies have revealed this simple relationship between coordination and the space requirement of ions:

$$\frac{\text{radius}_{\text{cation}}}{\text{radius}_{\text{anion}}} \rightarrow \text{coordination type.}$$

One can calculate that for the tetrahedral coordination state the minimum radius ratio of cations to anions is 0.225. Octahedral coordination becomes established when the radius ratio of cations to anions exceeds the value of 0.414. These listed ratios represent minimum values, because the anions cannot approach one another more tightly in the case of the cation being smaller than the coordination gap. Because of the electrostatic forces involved, a cation tends to approach the neighboring anions as closely as possible and this will lower the coordination state when its ionic radius is too small for the given coordination state.

The structural organization principles obeyed in crystal lattices can also be extended to ionic complexes present in aqueous solutions. As an illustration, we refer to a well-known chemical used in experimental biochemistry: ethylene diamine tetra-acetic acid (EDTA).

$$\begin{array}{lcr} HOOC-CH_2 \diagdown & & \diagup CH_2-COOH \\ & N-CH_2-CH_2-N & \\ HOOC-CH_2 \diagup & & \diagdown CH_2-COOH \end{array}$$

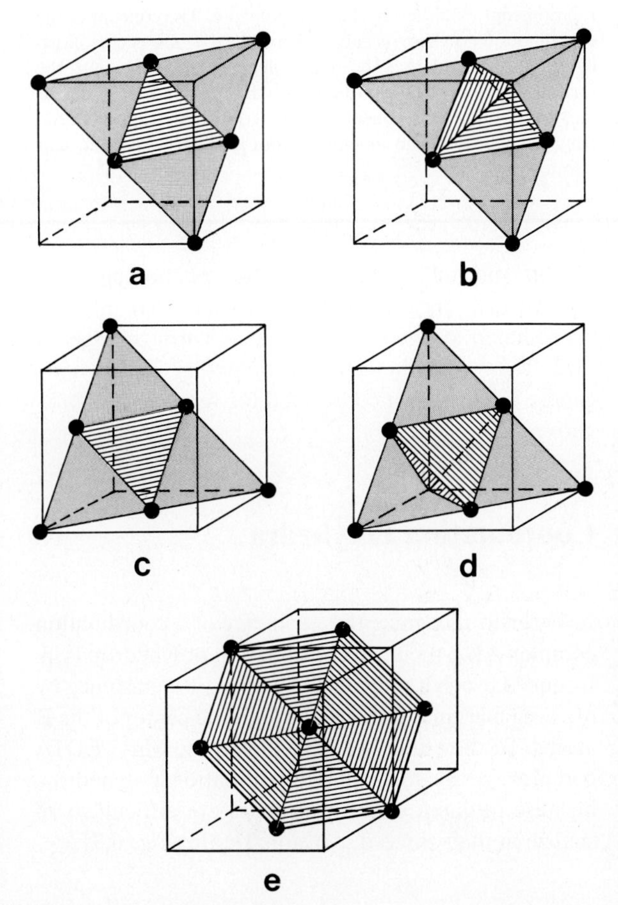

Fig. 6.6 a–e. The cubic face-centered unit cell in which the corners and centers of the planes are occupied by chlorine atoms is shown in the drawings **a** to **d.** The light gray triangular planes indicate the orientation of the hexagonal chloride layers in such a cube. The layer of closest packing of spheres thus cuts the cube diagonally, and in drawing **e**, two cubes are shown illustrating the position of this layer; the *black dots* represent the chlorides. Two successive layers are presented in drawings **a** and **c**; the smaller sized triangles (*dashed lines*) represent the marginal planes of the octahedron illustrated in Fig. 6.5 as well as the basal planes of the tetrahedral interstitials. In drawings **b** and **d**, these tetrahedra are fully shown to provide a better dimensional view of the distribution of the four-fold and eight-fold coordinated holes. In the face-centered unit cell, eight tetrahedra surround one octahedron

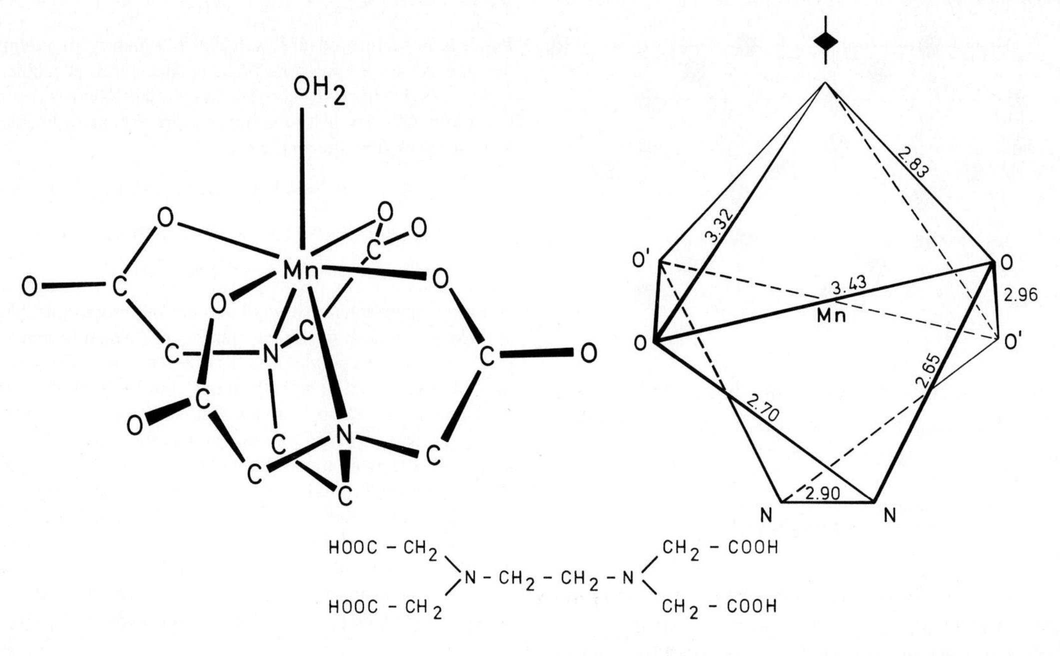

Fig. 6.7. *Left* Coordination complex of EDTA and Mn^{2+}: $Mn(EDTA)^{2-}(OH)_2$ (after Richards et al., 1964). The coordination AB_n of the manganese: $Mn(N_2O_5)$. EDTA complexes are widely used in experimental biochemistry; even lead poisoning in the human body has been treated therapeutically by EDTA with success. EDTA is able to pick up polyvalent metal cations from other oxygen coordination polyhedra and to incorporate these into its own structure. In this fashion insoluble oxygen-hydroxy metal compounds become solubilized via the apparent valence change:

$$Ca^{2+} + EDTA^{4-} \rightarrow Ca(EDTA)^{2-}.$$

The mean distances for the individual bond types are listed for comparison:

Bond type	*Mean length, Å*
C–N	1.471
C–C	1.519
C–O	1.257
Mn^{2+}–O	2.236
Mn^{2+}–N	2.377
Mn^{2+}–OH_2	2.155

Right The coordination polyhedron: AB_7 of the $(Mn\,(EDTA)^{2-})$ chelate as revealed by X-ray structure determination (Richards et al., 1964). The numbers refer to the length of the edges in Å units. This kind of idealized structure can be considered for the EDTA complex in solution. A comparison of the structure as shown here, and of the molecular complex illustrated in the above graph, shows the usefulness of the construction of a coordination polyhedron for the description of crystal chemical relationships exhibited by a given molecule. The concept of the coordination polyhedron can be considered a "molecular shorthand" which condenses the real molecular assemblies into the form of structural units which are easy to handle. As far as molecular biology is concerned, the coordination polyhedron complex represents an invaluable tool in describing molecular entities

By administering polyvalent metal ions to the solvent, EDTA will immediately incorporate the metal cation in its "clawlike" structure. The structure of the EDTA complex having manganese as a central cation is shown in Fig. 6.7. The coordination results in the formation of ionic AB_7^{2-} sphere-like complexes with a diameter of approximately 6–6.5 Å; the charge causes the solubility of such complexes in polar solvents. EDTA coordination compounds are able to form strong intramolecular hydrogen bridges. The oxygens in the carboxylated groups are virtually identical and exhibit orbital resonance. The phenomenon is noteworthy, since we have to assume that all four-coordinated oxygens possess an electric charge of –1/2. The water molecule participating in this manganese complex is incorporated for reasons of coordination.

Coordination Polyhedra

In order to recognize the geometry of a coordination complex AB_n, the term coordination polyhedron is introduced. Such a polyhedron comes into existence by drawing marginal surfaces through the center of the B atoms. To show this, we have drawn the Mn^{2+} EDTA complex in the form of a coordination polyhedron, because frequently such polyhedra are difficult to visualize in their three-dimensional state (Fig. 6.7).

Table 6.1. Minimum radius ratio for stable coordination polyhedra

Minimum radius ratio q_r	Coordination number n and orbital configuration	Coordination polyhedron AB_n	Polyhedron (Fig. 6.8)
<0.155	2 sp, dp	linear	a
<0.155	2 p^2, ds	angular	b
0.155	3 sp^2, dp^2, d^3	equilateral triangle	c
0.225	4 sp^3, d^3s, d^2sp, dp^3	tetrahedron	d
0.414	6 d^2sp^3	octahedron	e
0.528	6 d^4sp, d^5p	trigonal prism	f
		trigonal antiprism	g
0.645	8 d^5p^3	square antiprism	
0.732	8 d^4sp^3	cube	h
1.000	12	cubo-octahedron	i
		and hexagonal prism	j

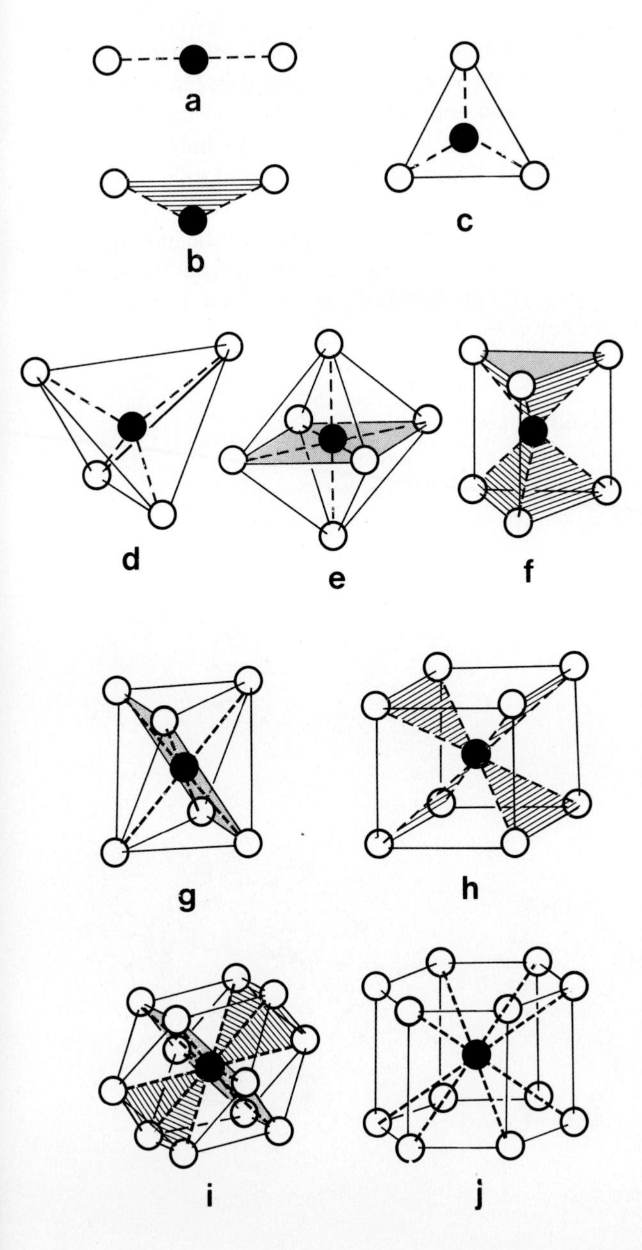

A coordination polyhedron is obtained in the following way. The individual ions are assumed to represent undistorted spheres, and their radii quotient is calculated as follows:

$$q_r = \frac{r_c\,(\text{Å})}{r_a\,(\text{Å})} \quad \text{where } r_c < r_a.$$

A comparison of these values with the geometrically inferred minimum values gives us the coordination polyhedra. In Table 6.1, the ratios of the radii are summarized for all stable coordinations. The polyhedra are plotted in Fig. 6.8.

Work on close sphere packing by Strunz (1966), Wells (1978) and Addison (1961) has shown that such a treatment represents a unifying approach for classifying and relating most inorganic structures to one another.

For biological systems where water, salt and organic compounds are fully integrated both structurally and functionally, the coordination concepts developed

Fig. 6.8 a–j. The coordination polyhedra AB_n for the bond orbital configurations presented in Table 6.1. The kind of polyhedron that will be established is a function of the geometric relationships that exist between the individual anions and cations. The coordination state is quantitatively determined by the rules of minimum radius ratios. Increase in the size of the metal cation results in an increase in coordination number, i.e., more and more neighbors can be accommodated. The majority of inorganic structures are composed of one of these coordination polyhedra:

a	linear two-fold	AB_2	f	trigonal prism	AB_6
b	angular two-fold	AB_2	g	trigonal antiprism	AB_6
c	equilateral triangle	AB_3	h	cube	AB_8
d	tetrahedron	AB_4	i	cubo-octahedron	AB_{12}
e	octahedron	Ab_6	j	hexagonal prism	AB_{12}

here are equally applicable. Gaseous particles, whether atoms or molecules, behave like point centers of mass which exert no force on one another, except for brief moments when they collide with each other. The stability of the two main gases of our atmosphere, the diatomic molecules N_2 and O_2, is caused by the equal sharing of valence electrons (covalent bonding). With respect to the structure of water, no uniform theory exists at present. It is our opinion that the structure of water also obeys the packing rules of crystalline materials without being an ice or a compressed gas structure. Since water is a key element of our discussion, its coordination state will receive special attention later.

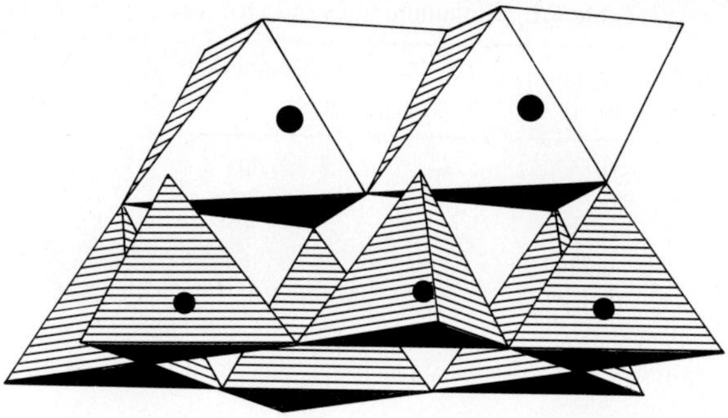

Fig. 6.9. Sharing of oxygens between SiO_4 tetrahedra and AlO_6 octahedra in kaolinite

Polyhedra Groupings in Minerals

Tetrahedra and octahedra are the two most abundant structural elements of which mineral matter is composed. Individual polyhedra can share atoms or ions and yield more complex structures.

The structural arrangement of ions in the common clay mineral kaolinite is schematically shown in Fig. 6.9 to illustrate the type of ion sharing that may proceed between different polyhedra. In essence we are dealing with silica tetrahedra and aluminum octahedra, where the structural frame is generated by layers of oxygens and hydroxyls.

The geometry of linkage of SiO_4 tetrahedra in the major silicate structures is depicted in Fig. 6.10. It is interesting to note that polyhedra other than SiO_4 are not shown. The way tetrahedra are linked to each other is used to classify this group of minerals. Five principal groupings can be recognized: (i) nesosilicates (isolated tetrahedra), (ii) sorosilicates (groups of tetrahedra), (iii) inosilicates (single or double chains), (iv) phyllosilicates (sheets), and tectosilicates (frameworks).

Fig. 6.10. Classification of silicates

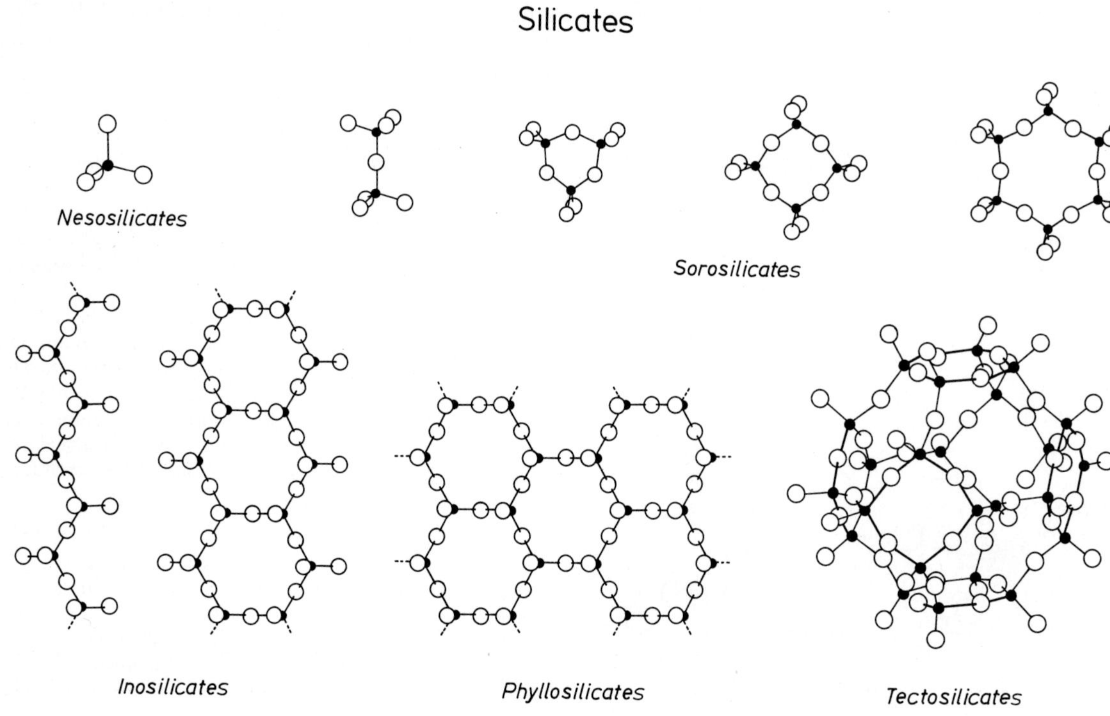

Olivine is a prominent nesosilicate. The pure magnesium olivine is forsterite, Mg_2SiO_4; the pure iron olivine (Fe^{2+}) is fayalite, Fe_2SiO_4. Magnesium and divalent iron have similar ionic radii, Mg^{2+} 0.78 Å and Fe^{2+} 0.83 Å, which permit substitution for each other, resulting in mixed compounds commonly referred to as solid solutions. However, substitution of iron for magnesium will cause an increase in the average interionic distances in a crystal lattice, and a decrease in the (negative) electrostatic energy.

Sorosilicates have finite groups of SiO_4 tetrahedra: Si_2O_7 double groups, hemimorphite; Si_3O_9 rings, wollastonite; Si_4O_{12} rings, neptunite, and Si_6O_{18} rings, turmaline.

Inosilicates have single or double chains of tetrahedra, and pyroxenes and amphiboles are familiar representatives.

Mica and clay minerals are prominent examples for the phyllosilicates.

Tectosilicates are the feldspars and zeolites.

In the crystalline Si-modification (quartz, cristobalite, tridymite), Si is always surrounded by four oxygens, whereby each oxygen is part of two tetrahedra. Framework structures are generated, but it is well to remember that quartz is not a silicate but an oxide.

Compounds that are maintained by covalent bonds, where participating atoms are sharing pairs of electrons that spin in opposite direction, yield regular polyhedra. Carbon is an outstanding example, and the coordination of carbon in diamonds is shown in Fig. 6.11. Each carbon atom in a tetrahedron shares an electron pair with four neighbors, resulting in a stable electron configuration in the outer shell.

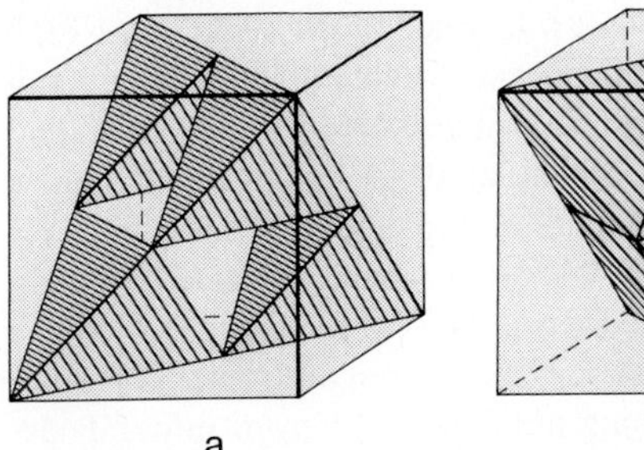

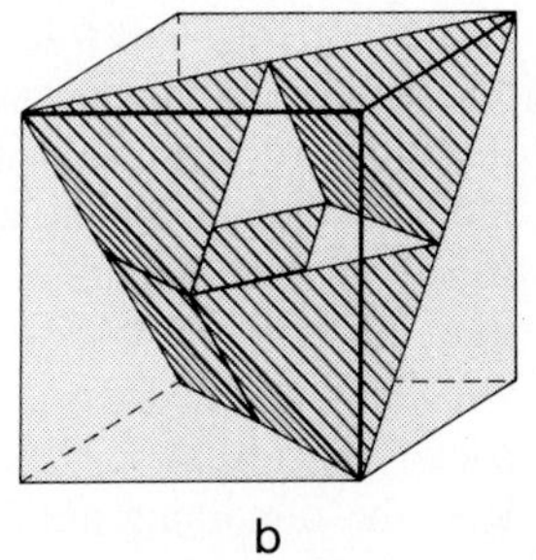

Fig. 6.11 a, b. The cubic structure of diamond is shown in **a**; it can be considered as a sphere packing array. Part of the carbon atoms form a cubic:

A * B * C * A * B * C * ...

stacking of layers; they occupy corner positions in the cube and lie in the center of the faces. The remaining carbon atoms occupy half of the tetrahedral interstitial sites, and all occupied sites are shown in this drawing. This scheme of occupation indicates that each carbon atom is bonded covalently to four neighboring carbon atoms. The C–C bond distance amounts to 1.54 Å. In drawing **b**, the remaining four tetrahedra are shown, since a cubic face-centered unit cell exhibits eight tetrahedral holes. In many crystals these sites are filled, for instance, in sphalerite (ZnS). The larger sized S^{2-} anions follow the closest packing of spheres, whereas zinc occupies the tetrahedral coordination sites. The octahedra and half the tetrahedral sites remain unoccupied

In conclusion, densely arranged configurations of spheres have greatly contributed towards understanding the behavior of solids and liquids (Rogers, 1964; Matheja and Degens, 1971; Sloane, 1984). The fact that the value of 74 % for the volume that can be occupied by the close packing of spheres is only an approximate figure, and that a precise mathematical understanding of ordinary, three-dimensional Euclidean space is far from complete, is a good reason for mathematicians to work on this problem. Furthermore, mathematical objects called *n*-dimensional spheres are of extreme significance in the branch of mathematics called group theory (Sloane, 1984).

All types of chemical bonds that hold elements together, that is to say covalent, polar covalent, coordinate covalent, ionic and metallic, produce polyhedra. Two of the most prominent ones are the tetrahedron and the octahedron. We will see later that also water hydrogen bonding will introduce a structural ordering of a similar type. Furthermore, in essential structures of life, such as the phospholipid membranes or the nucleic acids, tetrahedra and octahedra are critical building elements of the metabolic and genetic apparatus.

Rock-Forming Minerals

In 1949, Hugo Strunz published the second edition of his outstanding handbook: “Mineralogische Tabellen” (see 4th ed., 1966). He classified minerals according to their crystal chemistry, and the sheer bulk of mineral names listed on pages 238 to 306 – about 80 mostly unfamiliar names per page – impressed us students. But at the same time it scared us, because we knew that some professors would open this Pandora’s Box during examinations, leaving little hope to a student to pass in style. It is as well to remember that a mineral not only has a name, but also belongs to one of 230 space groups and 32 crystal classes; it has a chemistry which varies only within narrow bounds; it has a certain number of planes of weaknesses – cleavage directions – within the crystal structure, but may have no cleavage, as is the case for quartz; it exhibits crystal faces; it has a specific gravity and can be soft or hard; it has a color and may be magnetic or not; it can be transparent, translucent or opaque; it has a metallic or nonmetallic luster; it can be brittle, malleable, flexible or elastic; it will "react or not react" with HCl and finally may even have a taste. The fine print in the index of Strunz's handbook fortunately revealed that about 50 percent of the listed mineral names were either outdated and thus obsolete, not precisely defined "minerals", identical with others, based on wrong identification, simply a variety of, a pseudomorph, or just a commercial term. Nevertheless, the remaining number of different minerals accessible to visible examination still added up to a few thousand names.

The week before my mineralogy examination in the winter of 1950, I started to learn the book by heart. During a dress rehearsal the night before this event, my friends asked me 100 questions, and much to their surprise, I answered all of them correctly. Three months after the examination, the procedure was repeated, and I came up with only 58 right answers. A year later their number dropped to 10. I wonder what my score would be today. Luckily enough, the majority of minerals are lithospheric "impurities" and rarely encountered in the common rock. Only about 20 remain when it comes to describing the mineral content of ordinary crystalline rocks and sediments. The lesson we can draw is: it is senseless to feed students with details she or he can readily retrieve from handbooks. It is more important that they learn how to define a scientific problem, master the processes, and find an intellectually satisfactory answer. Let us now return to the rock-forming minerals:

The *feldspar group* is the most abundant mineral in the Earth's crust. Almost half of the crust is feldspathic. Two main subgroups can be distinguished: (i) potassium feldspars ($KAlSi_3O_8$) and the most prominent member is the monoclinic orthoclase, and (ii) sodium-calcium feldspars ($NaAlSi_3O_8$-$CaAl_2Si_2O_8$), also called the plagioclase feldspars which have as their sodium and calcium endmember the triclinic albite and anorthite respectively. Feldspars come in reddish, whitish and greenish colors. The main pigment is structurally incorporated iron (red) and magnesium (green).

Quartz is the single most common mineral in the continental crust but is practically missing from the basic oceanic crust. It is a major constituent in all three petrographic domains; igneous, metamorphic, and sedimentary. It is an oxide with the formula SiO_2 and occurs in two crystallographic moieties, that is the hexagonal (six-sided) high-temperature form, and the trigonal low-temperature form. Many names exist to identify a certain type, such as milky quartz, smoky quartz, rose quartz, amethyst, agate, jasper, chert and flint, to name a few common ones.

The *olivine group* shares with the plagioclase subgroup the property of complete solid solution, this time among the principal cations magnesium and iron (Mg_2SiO_4 to Fe_2SiO_4). The corresponding endmembers are forsterite and fayalite. The group is restricted to basic (gabbro, basalt) and ultrabasic (peridotite) rocks and is never found in association with quartz. Olivines are generally aggregated into clusters.

Next in line are the *pyroxene* and *amphibole groups* which are chemically complex chain silicates rich in Ca, Mg, Fe, and are quite prominent in dark-colored igneous rocks and high-grade metamorphic rocks. Monoclinic and rhombic varieties exist. Both groups can megascopically be readily distinguished from one another through their cleavage. For pyroxene, cleavage planes intersect at right angle, whereas for amphiboles the two directions intersect at 56° and 124°.

The *garnet group* is restricted to intermediate to high-grade metamorphic rocks. Three common ones are pyrope ($Mg_3Al_2(SiO_4)_3$), almandine ($Fe^{2+}_3Al_2(SiO_4)_3$) and grossulare ($Ca_3Al_2(SiO_4)_3$). They are cubic and form 12-sided crystals. They principally occur as mixed crystals, and in this combination are frequently encountered in eclogites, the high pressure form of basalt, which supposedly is upper mantle material.

Micas of some prominence in intermediate and acid magmatic rocks and low- to intermediate-grade metamorphic rocks are muscovite (white) and biotite (dark). They are monoclinic potassium-rich aluminum silicates with the respective general formulas $KAl_2(OH,F)_2AlSi_3O_{10}$ and $K(Mg,Fe,Mn)_3(OH,F)_2AlSi_3O_{10}$. They exhibit only one cleavage plane and can thus be easily split into fine, lightly tinted transparent sheets.

Clay minerals are structurally almost identical to the mica group, and are generally $< 1\ \mu$ in size. They are prominent in low-grade metamorphic rocks and sediments. Actually, the crustal sediment compartment consists to more than two thirds of clays. Because of the complexity of clay minerals and their significance in air-sea-earth-life interactions, this group will be discussed in more detail.

The classification system of the pure endmember layer lattice type clays (Fig. 6.12) is based on: (i) the structural arrangement of layers, (ii) population of the octahedral sheet, and (iii) type of chemical bonding between the layers (ionic, intermolecular, hydrogen bonding). The two-sheet, or 1:1 clays, contain one silica sheet and one sheet of either aluminum, magnesium, or iron. The three-sheet, or 2:1 clays, contain two silica sheets which enclose an aluminum, magnesium or iron sheet ("sandwich" structure). The

LAYERS	POPULATION OF OCTAHEDRAL SHEET	GROUP	SPECIES	CRYSTALLOCHEMICAL FORMULA	STRUCTURE (SCHEMATIC)
Two-Sheet (1:1)	Dioctahedral	Kaolinite	Kaolinite Dickite Nacrite	$Al_4(OH)_8[Si_4O_{10}]$	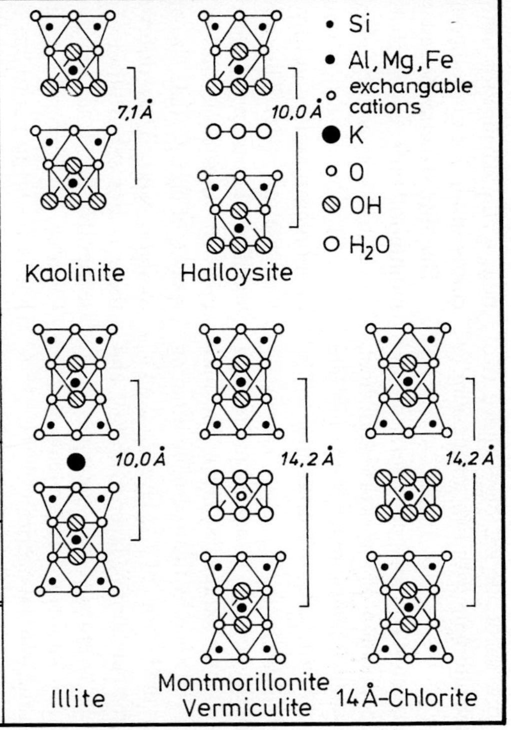 Kaolinite (7.1 Å); Halloysite (10.0 Å) • Si • Al, Mg, Fe ∘ exchangable cations ● K ∘ O ⊘ OH ○ H_2O
		Halloysite	Halloysite Metahalloysite	$Al_4(OH)_8[Si_4O_{10}] \cdot (H_2O)_4$ $Al_4(OH)_8[Si_4O_{10}]$	
Three-Sheet (2:1)	Dioctahedral	Montmorillonite (Smectite)	Montmorillonite Beidellite Nontronite	$\{(Al_{2-x}Mg_x)(OH)_2[Si_4O_{10}]\}^{-x} Na_x \cdot nH_2O$ $\{Al_2(OH)_2[(Al,Si)_4O_{10}]\}^{-x} Na_x \cdot nH_2O$ $\{(Fe'''_{2-x}Mg_x)(OH)_2[Si_4O_{10}]\}^{-x} Na_x \cdot nH_2O$	Illite (10.0 Å); Montmorillonite, Vermiculite (14.2 Å); 14 Å-Chlorite (14.2 Å)
		Illite (Hydromica)	Illite-Varieties	$(K,H_3O)\,Al_2(H_2O,OH)_2[AlSi_3O_{10}]$	
	Trioctahedral	Vermiculite	Vermiculite	$(Mg,Fe)_3(OH)_2[AlSi_3O_{10}]Mg \cdot (H_2O)_4$	
Three-Sheet + One-Sheet (2:2)	Trioctahedral	14 Å-Chlorite (Normal Chlorite)	Chlorite-Varieties	$(Al,Mg,Fe)_3(OH)_2[(Al,Si)_4O_{10}]Mg_3(OH)_6$	

Fig. 6.12. Classification of clay minerals (Degens, 1965)

three-sheet plus one sheet, or 2:2 clays, have two silica sheets, which alternate with two magnesium or iron sheets.

The system further takes into consideration the character of the population of the octahedral sheet, which may be dioctahedral according to the gibbsite pattern $Al_2(OH)_6$, or trioctahedral according to the brucite pattern $Mg_3(OH)_6$.

Clay minerals of the *kaolinite group* are made up of one octahedral and one tetrahedral layer. The individual sheets are firmly tied by the hydrogen bonding of the OH^- ions on the bottom of one layer to the O^{2-} ions at the top of the neighboring layer. Only Si^{4+} and Al^{3+} are structurally required; basal spacing is 7.1 Å. Superposition of two kaolinite layers yields dickite, and of four layers the mineral nacrite.

Structurally and chemically, the *halloysite group* resembles the kaolinite group, except that halloysite contains a layer of water between the original kaolinite layer units, causing an increase in basal spacing to 10.1 Å. Upon dehydration, metahalloysite will form.

The *montmorillonite* unit is a combination of one octahedral and two tetrahedral layers. In order to balance the structure, OH^- ions are replaced by O^{2-} ions wherever atoms belong to the tetrahedral and octahedral layer. Consequently, O^{2-} layers of each unit are faced with O^{2-} layers of the neighboring unit. Water and other polar molecules, such as organic substances, can slip between the unit layers, thereby causing expansion. Exchangeable cations exist between the structural units where they are loosely held by excess negative charges within the units. The basal spacing fluctuates strongly with the type of exchangeable cations and the thickness of the water layers. Common species are montmorillonite, beidellite, and nontronite, but because of the ease of ionic substitution, a clay having, for instance, the ideal beidellite composition, is rather uncommon in nature.

The mica structure is essentially the same as that of montmorillonite, except that the small excess of negative charge between the silicate layers is balanced by a weakly positive cation-water interlayer in the case of montmorillonite, and by neutralizing K^+ ions in the case of mica. The introduction of potassium stabilizes the structure to such an extent that the uptake of water molecules is prevented. Hydromica species of clay particle size are commonly referred to as *illitic clays*. They differ from muscovite in that there is less substitution of Al^{3+} for Si^{4+} and consequently less of an unbalanced charge deficiency. Even so, the amount of potassium ions is generally not sufficient to neutralize the small excess of negative charge between the silicate layers, with the result that other cations can be introduced.

The *vermiculite* structure consists of sheets of trioctahedral mica or talc separated by layers of water molecules occupying a definite space (4.98 Å) which is about the thickness of two water molecules. Charge deficiency between the silicate layers is satisfied by cations, largely of an exchangeable nature.

Most of the *chlorites* contain in their structure alternating mica type layers such as $Mg_3AlSi_3O_{10}(OH)_2^-$ and charged brucite type layers of the general formula $(Mg_2Al(OH)_6)^+$. But other chlorites are represented by layer structures consisting of trioctahedral 7 Å units.

In Table 6.2, eight representative chemical analyses of the common amorphous and crystalline clay minerals are presented. Of particular interest for a study of clay mineral formation are the three following features, which account for the differences in bulk chemistry of the various clay minerals listed: (i) Al:Si

Table 6.2. Chemical analyses of clay minerals (After Degens, 1965)

	(1) Allophane	(2) Kaolinite	(3) Halloysite	(4) Montmorillonite	(5) Vermiculite	(6) Illite	(7) Glauconite	(8) Chlorite
SiO_2	33.96	45.44	44.08	51.14	35.92	49.26	52.64	26.68
Al_2O_3	31.12	38.52	39.20	19.76	10.68	28.97	5.78	25.20
Fe_2O_3	Trace	0.80	0.10	0.83	10.94	2.27	17.88	
FeO					0.82	0.57	3.85	8.70
MgO		0.08	0.05	3.22	22.00	1.32	3.43	26.96
CaO	2.26	0.08		1.62	0.44	0.67	0.12	0.28
Na_2O		0.66	} 0.20	0.04		0.13	0.18	
K_2O		0.14		0.11		7.47	7.42	
$H_2O^{-105°}$	12.84	0.60	1.44	14.81	} 19.84	3.22	2.83	
$H_2O^{+105°}$	20.28	13.60	14.74	7.99		6.03	5.86	11.70
Total:	100.46	99.92	99.81	99.52	100.64	99.91	99.99	99.52

ratio, (ii) distribution of alkalies, alkaline earths and iron and (iii) total amount of water.

As for kaolinite and halloysite, the chemical compositions correspond closely with the theoretical formula (Fig. 6.12). Al^{3+} and Si^{4+} are practically the only cations to build up the structure. In montmorillonite and vermiculite, the Al:Si ratio is drastically reduced, partly as a result of the structural composition (2:1 layer), and partly because magnesium substitutes largely for aluminum in the octahedral position. Calcium and alkalies may assume exchange positions to match the deficit in electrical charge caused by substitution phenomena. The water content is the highest of all the clays under consideration.

The Al:Si ratio of the illite falls approximately between that of montmorillonite-vermiculite and that of kaolinite-halloysite. High magnesium and iron contents are related to substitution. Excess negative charge, developed between the silicate layers, is neutralized by large amounts of potassium in both illite and glauconite. Potassium acts as a structural stabilizer and prevents swelling along the *c*-axis of the clay. A low water concentration is the result.

The Al:Si ratio of chlorites may fluctuate over a wide range but generally it is of the same magnitude as in kaolinite. The high content of two-valent iron or magnesium, or sometimes of both elements, is typical of all chlorites. Since chlorite is with rare exceptions a non-swelling clay mineral, the low water concentration is in keeping with this characteristic.

The wide range in the chemical and structural composition of clay minerals indicates that there are several ways by which clays can be produced. According to Keller (1956), clay minerals may be considered fundamentally a product of parent material and the energy impressed upon it.

One may distinguish between the so-called primary and secondary parent materials. The primary category includes all igneous rocks, e.g., granites and granodiorites (K-Na rocks), basalts and gabbros (Ca-Mg rocks), and various kinds of tuffs. It is of interest that granites and granodiorites are the principal intrusive rocks and basalts the major extrusive ones (see p. 96).

Parent materials of secondary nature are those that once in their history have gone through the exogenic cycle. All sediments and some of the low-grade metamorphic rocks derived from sediments fall into this category, whereas the overwhelming portion of the metamorphic rocks have to be considered as primary parent materials. The average chemical composition of some of the parent materials is depicted in Table 6.3.

With respect to chemistry, there are rarely two natural clay specimens that are identical in composition. This chemical diversity can be attributed to either isomorphous substitution or ion exchange. Substitution may take place: (i) within the octahedral layers (six-fold coordination), (ii) within the tetrahedral layers (four-fold coordination), and (iii) between

Table 6.3. Chemical composition of igneous rocks and sediments (After Degens, 1965)

	Igneous rocks					Sediments			
	(1) Granite	(2) Granodiorite	(3) Average 1 + 2	(4) Basalt	(5) Average 3 + 4 (3/4 = 2.13)	(6) Sandstone	(7) Limestone	(8) Shale	(9) Shale (CO_2 corrected)
SiO_2	70.77	65.69	68.23	51.55	62.90	79.63	5.24	61.16	65.08
TiO_2	0.39	0.57	0.48	1.48	0.80	0.25	0.06	0.68	0.72
Al_2O_3	14.59	16.11	15.35	14.95	15.22	4.85	0.82	16.21	17.25
Fe_2O_3	1.58	1.76	1.67	2.55	1.96	1.09	} 0.55	4.23	4.50
FeO	1.79	2.68	2.23	9.10	4.43	0.31		2.58	2.74
(Fe)	2.50	3.31	2.90	8.86	4.81	0.99	0.43	4.96	5.28
MnO	0.12	0.07	0.10	0.20	0.13	Trace	Trace	Trace	Trace
MgO	0.89	1.93	1.41	6.63	3.08	1.18	7.96	2.57	2.24
CaO	2.01	4.47	3.24	10.00	5.40	5.59	42.97	3.27	0.52
Na_2O	3.52	3.74	3.63	2.35	3.22	0.46	0.05	1.37	1.46
K_2O	4.15	2.78	3.47	0.89	2.65	1.33	0.33	3.41	3.63
P_2O_5	0.19	0.20	0.20	0.30	0.23	0.08	0.04	0.18	0.19
CO_2						5.11	41.93	2.77	
Misc.						0.12	0.05	1.57	1.67
Total:	100.00	100.00	–	100.00	–	100.00	100.00	100.00	100.00

the silicate units (interlayer position). In contrast, ion exchange proceeds at the solid-liquid interface between anions and cations held in unbalanced charges at or near the surface of the solid material, and with ions present in the aqueous medium. The exchange reactions always proceed in a stochiometrical manner, but even so, there is no simple way to predict the ion exchange capacity of a given clay mineral. According to Hendricks (1945) and Carroll (1959), other controlling factors are:

Structural factors:

- Unsatisfied valences produced by "broken bonds" at surfaces and edges of clay particles;
- unbalanced charges caused by isomorphous substitution;
- dissociation of OH^- radicals, the H^+ of which may be exchanged; and
- accessibility of atoms in structural positions when brought to the exchange site as a result of a change in the environment.

Environmental factors:

- Availability of exchangeable constituents in the aqueous medium;
- pH-Eh relationship;
- general chemistry of the environment; and
- pressure and temperature conditions.

In conclusion, clay minerals are extremely variable in size, structure and chemistry. The most prominent species are: kaolinite, montmorillonite, illite, and chlorite.

Minerals of the *carbonate group* have the CO_3^{2-} ion in common. They can be conveniently classified on the basis of presence or absence of water and foreign anions such as OH^-, Cl^-, F^-, SO_4^{2-} and others. Among the about 60 different carbonates known from nature, only a few are important rock-forming minerals. It is remarkable that some carbonates are equally at home in the magmatic and metamorphic domain, e.g., in carbonatites and marbles, as in sediments, e.g., in limestones, dolomites and shell materials. Furthermore, carbonates can be a product of both organic and inorganic processes. Most prominent are the three anhydrous Ca- and Mg-carbonates: rhombohedral calcite $CaCO_3$, orthorhombic aragonite $CaCO_3$, and rhombohedral dolomite $CaMg(CO_3)_2$. Ca^{2+} in calcite is six-fold coordinated (CaO_6), and in aragonite nine-fold coordinated (CaO_9). With increasing geologic age the metastable aragonite will recrystallize into calcite, whereas dolomite can be generated from calcite through isomorphic substitution of Ca^{2+} for Mg^{2+}.

The bulk of the *sulfates* and *halides* belong to a group of sediments referred to as evaporites. The most common species are: gypsum $CaSO_4 \cdot 2H_2O$, and anhydrite $CaSO_4$. They occurred first as a major rockforming element in the late Proterozoic Amadeus Basin of Central Australia, about 700 million years ago. The same can be said for the initiation of rock salt (NaCl) deposition, which started to form close to the Precambrian/Cambrian boundary, where they are presently exposed as spectacular salt "glaciers" in the southern deserts of Iran.

In summary, about 20 minerals compose the major rock types and represent more than 98 percent of crust and upper mantle. The remainder are either accessories in common rocks, restricted to less common rock types, or associated with ore deposits. In a way, the situation is comparable to the distribution of chemical elements in the universe, where 98 percent are hydrogen and helium, leaving only a trifle for the remaining 90 elements of the periodic table.

(Suggested supplementary reading: Brindley, 1980; Grim, 1968; Turner, 1968; Strunz, 1966; Deer et al., 1972; Winkler, 1979; Suk, 1983; J.V. Smith, 1974a, b; Degens, 1965; Klein and Hurlbut, 1985).

Chapter 7

The Earth from Within

Rocks

Close to 3,000 minerals are known from rock formations, and their number increases by about 100 newly discovered minerals per year. Yet only 50 or so comprise 99 percent of the Earth's crust; the rest are – globally speaking – trace constituents. By contrast, a small group, the plagioclases and K-feldspars, less than 10 in number, account for about 50 percent of all mineral matter.

Minerals form aggregates which we call rocks, and only a handful of different ones occur in the ordinary stone you may pick up in the mountains or on the beach. Ratio of minerals and texture are the two major criteria for giving a rock a name. It is thus the kind, amount, size, shape and boundary arrangement of the various rock-forming minerals, which are important in a petrological classification. A given mineral has a chemical composition which varies only within rather narrow bounds. Furthermore, its structure signifies well-defined physical-chemical conditions at the time of crystallization. In turn, minerals of known age are excellent indicators for chemical composition and physical state of mantle, crust, or surface environments in the course of geologic time.

The list of rock names matches that of minerals in length. In spite of this diversity, all major rock formations share the predominance of oxygen which by volume constitutes ca. 94 percent of crustal matter. Another 5 percent can be added for silicon, aluminum, iron, calcium, sodium, potassium and magnesium, the main elements associated with oxygen principally in the form of silicates, oxides and carbonates. To a certain extent we are dealing with an oxysphere when it comes to crust and upper mantle, as we refer to the ironsphere when considering the Earth's core.

It has been found useful to speak of igneous, sedimentary and metamorphic rocks. A brief overall view is given on nomenclature and origin of the major rock families.

Igneous rocks: Igneous rocks solidify from hot molten material, the magma. Should crystallization proceed at a depth greater than a few kilometers, crystal seeds have ample time to grow, and may reach sizes a few millimeters to a few centimeters in diameter, where all grains can be seen by the naked eye and phaneritic texture develops. The more viscous the magma, the higher the trend towards small crystals, whereas fluid magmas permit crystals to grow larger in size. The fluidity of a magma in this respect is controlled by its general chemistry. For instance, a silica-rich melt is commonly viscous, but the addition of water will render it more fluid. The final outcome of solidification at depth are plutonic (= intrusive) rocks.

A melt that rises to the surface cools rapidly, leaving little time for a crystal to grow. The resulting volcanic (= extrusive) rock consequently appears massive and structureless and an aphanitic texture having no megascopically recognizable minerals is indicated. Where the lava flows into a lake or the sea, the magma melt would be quenched and the resulting rock represents a supercooled solution or natural glass. A situation frequently develops where a magma starts slowly to solidify, giving rise to large and idiomorphic crystals, so-called phenocrysts, and cools off more rapidly afterwards. The final product has porphyritic texture where the phenocrysts are emplaced in a matrix of either aphanitic or phaneritic appearance.

Bowen's reaction principle (Fig. 7.1) of high-temperature petrogenesis gives igneous petrologists a convenient tool to present data on geochemistry in a physically-chemically organized and methodical form. The sequence of fractional crystallization is in this case independent of whether a melt solidifies slowly or rapidly. The minerals are identical no matter whether they are of magmatic or volcanic origin; only rock textures differ. The order of crystallization of the common rock-forming minerals proceeds along two paths: (i) the continuous series of the light-colored (felsic) plagioclases and (ii) the discontinuous series of the dark-colored (mafic) varieties. Felsic is a mnemonic term for feldspathic minerals and mafic for magnesium- and iron-rich minerals. The melt as it

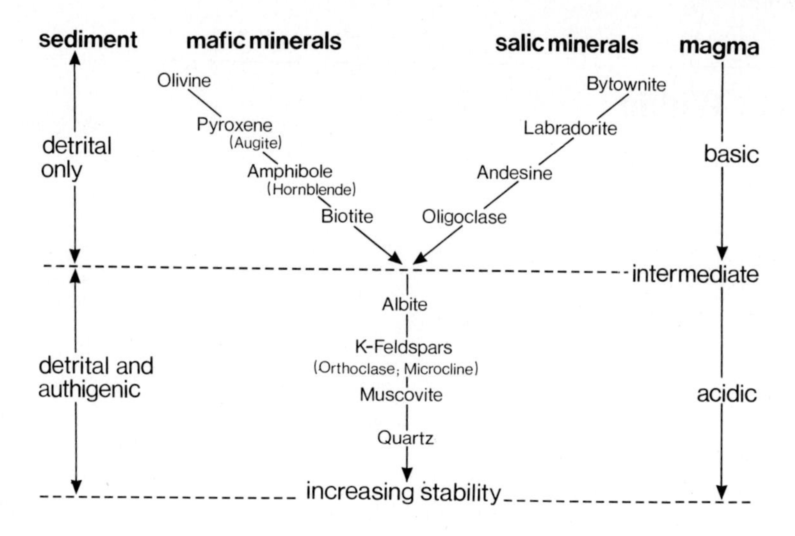

Fig. 7.1. Bowen's reaction principle

cools off and differentiates becomes more silicic, and the final products are K-feldspars, muscovite and quartz.

As illustrated in Table 7.1, the ratio of minerals is a simple means to classify plutonic and volcanic rocks. The fact that granites and basalts are the two major rock types of igneous provenance is in some way related to the difference in temperature and viscosity between acidic and basaltic melts. A basaltic magma is namely hot and fluid and has thus a better chance to rise to the Earth's surface and yield, e.g., a basalt, while the cooler silica-rich melt upon intruding a rock formation prefers to solidify right there as, e.g., a granite instead of ascending to the surface to become a rhyolite.

Sedimentary rocks: The moment a surface is exposed to air or water, a sediment forms, unless winds or currents are strong enough to sweep the infalling material away. Even common dust settling on your desk or bookshelves is a sediment. And just as that dust represents a complex mixture of all sorts of products and activities present in the local and distant environment, the ordinary sediment, too, gets its mineral grains from any preexisting source rock, its organic compounds and biominerals from a wide spectrum of plants and animals, and its various chemical precipitates from the water above or within. Furthermore, sediments contain water and some carry gases and oil (Fig. 7.2).

A sediment is also shaped by the environment of deposition (Pettijohn et al., 1972; Pettijohn and Potter, 1964). This will lead to primary structures such as ripple marks on a beach, cross-bedding in a sand dune or worm burrows in a mud. Following burial, a loosely packed sand, a watery mud or a calcareous ooze will be subjected to diagenesis, which is a collective term for processes and changes that transform a freshly deposited sediment into a sandstone, a shale or a limestone. Physical compaction and chemical generation

Table 7.1. Classification of igneous rocks (After Hamblin and Howard, 1986)

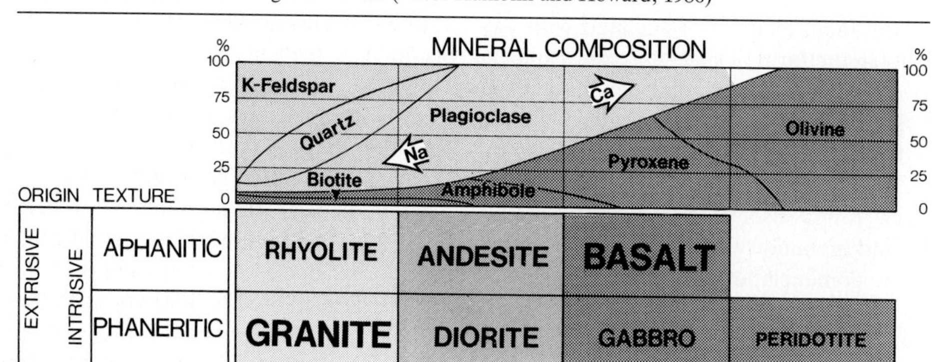

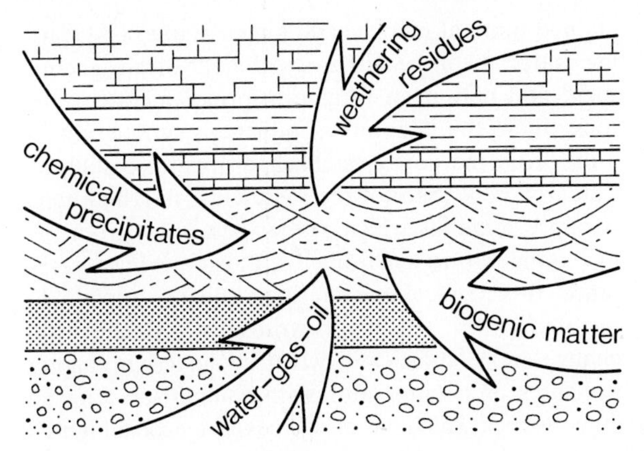

Fig. 7.2. Principal endmembers in sedimentary rocks

of a cementing matrix, for instance quartz, clay or carbonate, are the principal means to solidify the deposited material.

In the light of all the complexity involved in forming a sediment, it comes as a surprise to learn that the classification of sedimentary rocks is quite simple. In the first place this has to do with the fact that the resistance of primary minerals towards chemical weathering more or less follows Bowen's sequence of fractional crystallization, in that olivine and anorthite are the least, and muscovite and quartz the most stable minerals in the outer environment. This relationship, established by Goldich (1938), however, should not imply that olivine weathers to pyroxene and pyroxene to amphibole etc.; but it may indicate that the equilibrium conditions for olivine formation deviate more strongly from the environmental conditions established at or near the Earth surface than they do for any other high temperature minerals. This suggestion receives support from the observation that the four most stable minerals of Goldich's stability series can also occur as authigenic sedimentary mineral species (Fig. 7.1).

With respect to abrasion, which is the mechanical resistance of minerals towards erosion and transportation, it is the hardness, cleavage and tenacity of a mineral that counts. Thus, the overall chemical and mechanical stability determines the fate of mineral grains in the exogenic cycle. It is obvious that a mineral with the lowest abrasion effect and the highest chemical stability will best survive weathering and denudation.

Differential sorting during transportation and sedimentation, due to differences in size, shape and specific gravity of the individual grains, will further enhance the fractionation effect. Should chemical and physical processes continue long enough, only the most resistant endmembers will be left over. Since sedimentary matter can go through a series of rock cycles, that is: weathering – transportation – deposition – lithification – tectonics – denudation – weathering – transportation – etc., the final outcome will be a pure quartz sand, a clay, or a placer deposit enriched in resistant heavy minerals, such as zircon, rutile, diamond, magnetite and monazite.

So far we have spoken only of the detrital components that is those which are swept as suspended particles into a basin of deposition. But there is a wide range of authigenic minerals generated in situ which are thus excellent indicators of physical and chemical circumstances that prevailed in the environment at the time of deposition (Degens, 1965). The most common ones are calcite ($CaCO_3$), dolomite ($CaMg(CO_3)_2$), gypsum ($CaSO_4$ x $2H_2O$), anhydrite ($CaSO_4$), Fe/Mn/Al-oxides and -hydroxides, pyrite (FeS_2), microcrystalline quartz (SiO_2), and halite (NaCl). Nevertheless, when it comes to the relative abundance of rocks, limestones and dolomites outrank by far the rest of the authigenic rock formations. It is of note, however, that many of the world's leading ore deposits are concentrates of authigenic sedimentary minerals.

Sediments were laid down in chronological order and are thus the history book of planet Earth. The fossil record is no doubt their most valuable treasure, because our concepts on the origin and evolution of life are largely based on it. On the other hand, it was and is life which controls the environment by modifying air, water and sediment towards its special needs. Therefore, it is safe to say that an abiotic stratigraphic column would bear little resemblance to ours. Admittedly, a quartz sand may look the same everywhere, but not the muds and calcareous oozes, bearing in mind that microbial, planktonic and benthic communities digest and recycle the settling or deposited material by various biogeochemical means. Furthermore, biological activities are instrumental in the deposition of fine detritus in the open water column. What has been mentioned as regards to authigenic minerals, in that they are often of economic value, equally applies to the organic residue of life: oil, coal and gas. Without fossil fuel and sedimentary mineral deposits, civilization as we know it could not have developed.

In summary, sediments are the products of physical, chemical and biological processes operating in surface environments. Their ingredients are of either clastic, chemical, or biological origin. Of the many kinds of sediments known, only a few are common. Shale, sandstone and limestone comprise jointly more than 99 percent of all the sediment (Pettijohn, 1975).

Using geochemical balance calculations, the proportions of shale, sandstone, and limestone become about 80, 10, and 10, respectively. If estimates are based on actual measurements of many stratigraphic sections exposed on land, the ratio is more in the order of 50 to 30 to 20. The large difference between the two methods prevails on account of lack of information on pre-Cretaceous/Upper Jurassic deep-sea sediments, which have been processed away by global tectonic forces beneath the sea.

Classification of sediments is based on: (i) mode of origin, that is clastic or chemical/biochemical, (ii) grain size and shape, and (iii) whether they are consolidated or unconsolidated (Table 7.2).

Metamorphic rocks: Metamorphic rocks can be generated from any pre-existing rock formation the moment heat, pressure and the chemistry of fluids and gases have advanced to a point where the in-grained mineral matter is no longer stable. The starting material can be either a sediment or an igneous rock, and may even include a former metamorphic rock where the mineral assemblage is sufficiently remote from the newly established mineral equilibria. The argument used for the sediments, that high temperature minerals which are farthest apart from the low temperature environment have the least resistance to chemical weathering, equally applies to metamorphism. In other words, a sediment will change more drastically when subjected to higher pressures and temperatures, in comparison with an igneous rock which has already experienced a high PT regime.

Distinction is commonly made between contact and regional metamorphism. The first type has only local importance, and occurs near the contact of cooling magma generating a thermal halo a few millimeters to

Table 7.2.a. Classification of clastic sediments

grain size (log scale)		unconsolidated	consolidated
	boulder	gravel	conglomerate (rounded) breccia (angular)
256 mm	cobble		
8 mm	pebble		
4 mm	granule		
2 mm	very coarse sand	sand	sandstone arkose, (>25% feldspar) graywacke (rock fragments and clay
1 mm	coarse sand		
1/2 mm	medium sand		
1/4 mm	fine sand		
1/8 mm	very fine sand		
1/16 mm	silt	silt	siltstone (mudstone)
1/256 mm	clay	clay	claystone shale (fissility) (mudstone)

at most a few kilometers in width. By contrast, regional metamorphism is associated with global tectonics involving large terrains and thick rock sections.

Heat and pressure are the main propelling agents of regional metamorphism. Since heat passes slowly through rock masses, and rock formations sink only gradually in the direction of higher PT regimes, metamorphic changes proceed in a series from low- to high-grade metamorphism or vice versa. One should emphasize that besides overburden pressure, directed stresses are also involved. The combined action of heat, overall pressure, and directed stress will generate in time from any pre-existing rock either new minerals, overgrowths on old crystals, or distinct textures and structures. Minerals cannot grow freely as they would do in a liquid magma. In turn, rotations, distortions, irregular outlines etc. are the rule rather than the exception.

Classification of metamorphic rocks is principally based on foliation and dominance of certain minerals or sets of minerals. Foliation in this connection refers to any planar element, for instance, cleavage or banding of groups of minerals. Terms such as grade, rank, zone or mineral facies are often employed to specify to what PT or stress regime a given metamorphic rock belongs.

Typical foliated products of regional metamorphism are crystalline schists, gneisses and migmatites. Most prominent among non-foliated rocks are metaconglomerates, quartzites and marbles (Table 7.3).

Geophysical data suggest that upper mantle material consists of eclogite. This metamorphic rock, which is principally composed of omphacite, a Na-containing Mg-Ca-Al pyroxene, and of garnet, that is a Mg-Fe-Al variety, can form over a wide range of

Table 7.2.b. Classification of chemical sediments

texture	rock name	composition
coarse grained (>2 mm)	crystalline limestone	calcite, $CaCO_3$ (aragonite, $CaCO_3$)
medium grained ($^1/_{16}$ mm–2 mm)		
microcrystalline	micrite, chalk	
oozy	calcareous ooze	
rounded aggregates ($^1/_{16}$ mm–5 mm)	oolitic limestone	
abundant fossils	coquina (loosely cemented) fossiliferous limestone	
banded	travertine	
microcrystalline to coarse grained	dolomite, dolostone	dolomite, $CaMg(CO_3)_2$
cryptocrystalline	chert	chalcedony, SiO_2
oozy	siliceous ooze	opal, $SiO_2 \times nH_2O$
fine to coarse crystalline	evaporite	gypsum, $CaSO_4 \times 2H_2O$ anhydrite, $CaSO_4$ halite, NaCl other salts
	coal. oil, gas	carbon compounds

Table 7.3. Classification of metamorphic rocks

<table>
<tr><th colspan="3">Texture</th><th colspan="6">Composition</th><th>Rock name</th></tr>
<tr><td rowspan="4">Foliated</td><td rowspan="3">Non-layered</td><td>Very fine grained</td><td rowspan="2">Chlorite</td><td rowspan="3">Mica</td><td rowspan="4">Quartz</td><td rowspan="1"></td><td rowspan="1"></td><td rowspan="2"></td><td>Slate</td></tr>
<tr><td>Fine grained</td><td rowspan="3">Feldspar</td><td rowspan="3">Amphibole</td><td>Phyllite</td></tr>
<tr><td>Coarse grained</td><td rowspan="2"></td><td rowspan="2">Pyroxene</td><td>Schist</td></tr>
<tr><td>Layered</td><td>Coarse grained</td><td></td><td>Gneiss</td></tr>
<tr><td rowspan="3">Non-foliated</td><td colspan="2">Coarse grained</td><td colspan="6">Deformed grains of any rock type</td><td>Metaconglomerate</td></tr>
<tr><td colspan="2" rowspan="2">Fine to coarse grained</td><td colspan="6">Quartz</td><td>Quartzite</td></tr>
<tr><td colspan="6">Calcite or dolomite</td><td>Marble</td></tr>
</table>

pressure and temperature. Inasmuch as it can be synthesized from ordinary oceanic basalt, it is commonly assumed that upon subduction oceanic crust will be converted to eclogite, once the rock slab has reached the proper PT regime for the generation of the eclogite facies.

Suggested readings for *igneous rocks:* Daly (1933); Basaltic Volcanism Study Project (1981); Fisher and Schmincke (1984); Yoder (1976, 1979); Cox et al. (1979); Williams et al. (1982); *sedimentary rocks:* Pettijohn (1975); Degens (1965); v. Engelhardt (1977); Selley (1978); Leeder (1982); Füchtbauer and Müller (1977): *metamorphic rocks:* Turner (1968); Turner and Verhoogen (1960); Winkler (1979); Miyashiro (1978); Borradaile et al. (1982); Barth and Newton (1982).

The Formative Years

The solar system developed from a rotating discoidal nebula (stellisk). Temperature and pressure in the inner region amounted to about 1,400 K and 10^{-4} atm (10 Pa) respectively. Condensation started with Ca, Al, Ti oxides and silicates, platinum metals, W, Mo, Ta, Zr, REE, followed by Fe-Ni metal, and olivine (1,360 K); at 1,200 K the remaining SiO_2 had condensed in the form of $MgSiO_3$ (Anders, 1977; J.V. Smith, 1982). The bulk of the proto-Earth accreted rapidly during the early condensation phase of the solar nebula, resulting in gross stratification of the proto-Earth with an iron core, surrounded by a mantle of magnesium silicates. The later stages of accretion occurred after the nebula had cooled to relatively low temperatures (400 K), so that the equilibrium condensate contained oxidized iron, iron sulfide, hydrated magnesium silicates, volatiles, organics, i.e., material similar to that found in type 1 carbonaceous chondrites. Volatile-rich materials of this nature, mixed with earlier high-temperature condensates that had failed to accrete previously, are believed to have accreted upon Earth over a longer time-scale (10^5 to 10^7 years) to produce the upper mantle-crust system (Fig. 7.3).

An important feature of the *heterogeneous* accumulation model is that the upper mantle has never been in contact with the core. It is a direct consequence of the short time-scale of the main phase of accretion, which enables the core to settle to the center and be surrounded by the lower mantle before the outer layers are added. In this way the heterogeneous accretion hypothesis provides an acceptable explanation of the high abundance of siderophile elements in the upper mantle, and its oxidation state and content of volatiles. Even the revised *homogeneous* accretion hypothesis (Ringwood, 1979) appears to be unable to account for these features. To be fair, the heteroge-

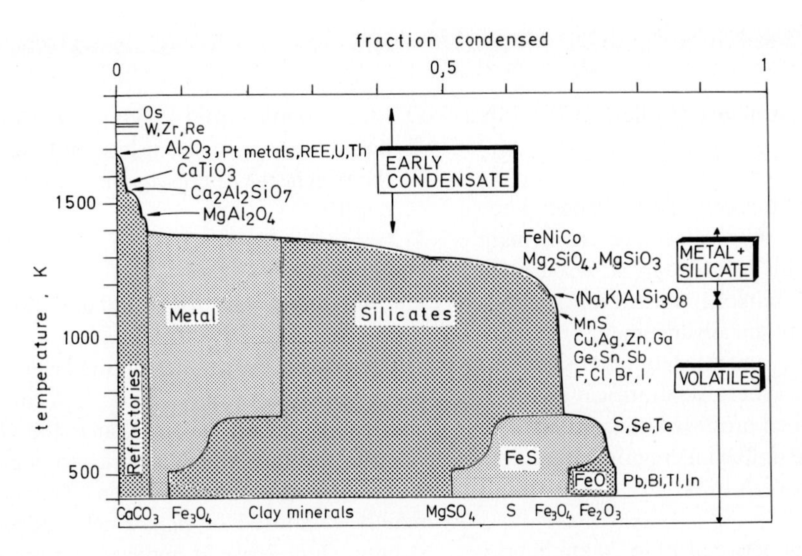

Fig. 7.3. Condensation of solar gas at 10^{-4} atm (10 Pa). Three types of dust condense from a solar gas: refractories, metallic nickel-iron, and magnesium silicates. On cooling, iron reacts with H_2S and H_2O to give FeS and FeO. Further major changes take place below 400 K (Anders, 1977)

neous accretion model has its problems too. This results particularly from the unknown mass of the nebula, which in standard models is assumed to have at most two solar masses. If we involve a less-massive nebula, we accordingly obtain lower temperatures. If, on the other hand, we assume higher masses for the nebula, then difficulties exist in sweeping the solar system free of the "excess" matter by T Tauri wind, once condensation has been achieved.

After much deliberation, I gradually warmed to the heterogeneous scenario so convincingly advocated by J.V. Smith (1982). Based on an enormous data set from the fields of planetary sciences and petrology, he condensed simple ideas into a comprehensive and consistent scheme on the inhomogeneous growth of the planets and parent bodies of meteorites. This model implies that the Earth continuously differentiated into a core, mantle and chaotic crust as hot planetesimals were captured. The early material aggregating is supposed to be "dry" and reduced, and the later material to be "wet", progressively more oxidized and carrying volatiles such as CO_2 and H_2O.

What the heterogeneous model suggests is that the layered structure of the Earth interior, namely crust, mantle and Fe/Ni core, is largely a result of differential condensation in a planetary nebula and is in part caused by a subsequent high temperature differentiation. Some critical information on structure and composition of the present Earth is given in Box 7.1.

Doubt about the validity of the heterogeneous growth of planet size objects was recently expressed by David C. Tozer (1985). When carefully reading his article "Heat transfer and planetary evolution", two sentences especially amazed me: "There are no good reasons for believing terrestrial planet accumulation was directly capable of introducing the kind of large-scale heterogeneity of metallic and silicate phases referred to as core/mantle structure", and "... the early stages of accumulation of all planets was chemically very similar". Tozer's article raises fundamental questions not only on the thermal history of planets, but also on the dependence of solid state creep on temperature, and the availability of water or hydrated mineral phases in crust and upper mantle. To end this controversial issue I would like to suggest that – as is frequently the case in a scientific controversy – the truth might perhaps be found in the middle of the argument. For a detailed account the reader is referred to, e.g., Turekian and Clark (1969), J.V. Smith (1979, 1982), Anders (1977), Ringwood (1979), Smoluchowski (1983), and Tozer (1985).

We previously said that Earth and Moon formed side by side, and that much of what we infer from the early history of the Earth, namely before 3.8 Ga, stems from first-hand knowledge of the morphology, petrology and general geology and geophysics of the Moon. By analogy, the Earth's original ocean basins were possibly mare-type basins produced by the flux of asteroid-sized objects. The greater capture cross-section and impact velocity of the Earth yielded more and conceivably deeper mare basins. It has been estimated that 4 Ga ago at least 50% of the ephemeral global crust had been converted to basin topography (Frey, 1977). Here ends the analogy, because unlike the Moon, where the basins became flooded by basal-

BOX 7.1

The Earth's Layered Structure (Bullen, 1975; Gutenberg, 1959)

Since the beginning of this century it has been known that our globe can be subdivided into crust, mantle and core. Much of this knowledge is based on the work of Wiechert, Gutenberg and Mohorovičić (see Gutenberg, 1959). With the advancement of geophysical techniques and subsequent data processing, a refined model on the spherical stratification of the Earth's interior has been proposed. According to Bullen (1963, 1975), the following regions were established using a letter code A to G:

A Crust (base varying between 10 to 70 km depth)
B upper mantle (from base of crust, i.e., the Mohorovičić Discontinuity, abbr. Moho, to 400 km depth)
C transition zone (from 400 to 1,000 km depth)
D lower mantle (from 1,000 to 2,900 km depth)
E liquid outer core (from base of lower mantle, i.e., the Wiechert-Gutenberg Discontinuity to 4,600 km depth)
F transition region(s) (from 4,600 to 5,150 km of depth)
G solid inner core (from 5,150 to 6,371 km depth).

The crust is the thinnest of the layers, having an average thickness of only 17 km. It is composed of two contrasting silicate layers, that is the aluminum-rich *sial* plus the *sima* which is enriched in mafic elements such as magnesium and iron. The continents are mainly composed of *sial,* which lies on a *sima* substratum (Fig. 7.4a). It appears that the sial layer was

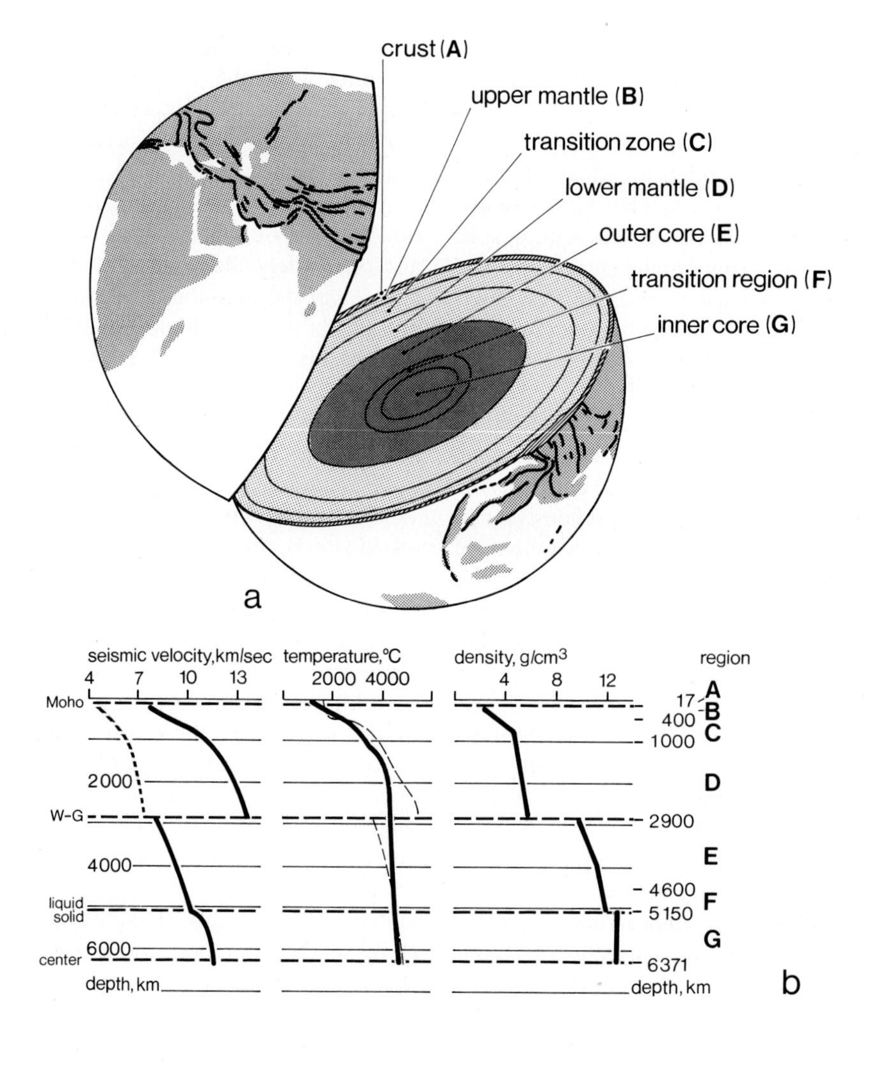

Fig. 7.4 a, b. Structural features of the Earth and related physical properties

formed quite early in view of the fact that granites as old as 3.8 Ga are known; furthermore, the oldest basic magmatic sequences, the komatiites, rest on *sial* (see p. 123).

The top upper mantle contains cherzolites (10% clinopyroxene, 25% orthopyroxene, 65% olivine), harzburgites (bronzite = orthopyroxene peridotite) and eclogites, while the middle upper mantle is believed to be composed of garnet-rich peridotites. For the bottom upper mantle spinel, stishovite and garnet have been suggested as the principal mineral phases (e.g., McElhinny, 1979a, b; J.V. Smith, 1982; Ringwood, 1982).

Those parts of crust and upper mantle that are rigid are commonly referred to as the lithosphere, which is "roughly" synonymous with the term plate. The layer below the lithosphere (80–100 km) is the asthenosphere, which in geophysical terminology is the low velocity layer and assumed to be partially molten and representing the lubricated phase boundary upon which plates move.

The simplest assumption about the lower mantle is that it also consists of peridotitic material. Alternatively, it has been proposed that its main component is perovskite, (Mg,Fe)SiO_3 (e.g. Anderson, 1977) having a predominance of magnesium.

The core is believed to be composed of iron and nickel and some lighter elements such as sulfur and silicon (Anderson, 1986). The outer core is liquid and electrical currents moving in this conducting medium are assumed to be the source of the Earth's magnetic field.

In Fig. 7.4b, main structural features of the Earth from within are schematically shown, including tectonic structures revealed at the surface of the crust (Cloos, 1936).

tic liquids, the Earth experienced at that time massive degassing, leading to a proto-atmosphere and -hydrosphere. Furthermore, Earth underwent a different thermal evolution, resulting in convective flow regimes in crust and mantle.

Large cratering events on a water-bearing planet produce an impact melt sheet overlying fractured basement rocks (Newsom, 1980). He submitted that a major fraction of the soil on Mars may consist of the erosion products of altered impact melt sheets, including hydrothermally altered clays, ferric hydroxides and large concentrations of salts. The ability of vapor-dominated hydrothermal systems to concentrate sulfate relative to chloride may explain the high sulfate to chloride ratio found in the Martian soil by the Viking landers (Clark and Baird, 1979). Brecciated crustal material, and hydrothermally produced iron-rich clays and hydroxides constituted possibly the first detrital sediments accumulating on the ephemeral

Fig. 7.5. Magmatism and cooling in the early Earth. Depth scale would be expanded for a partly grown Earth (J.V. Smith, 1982)

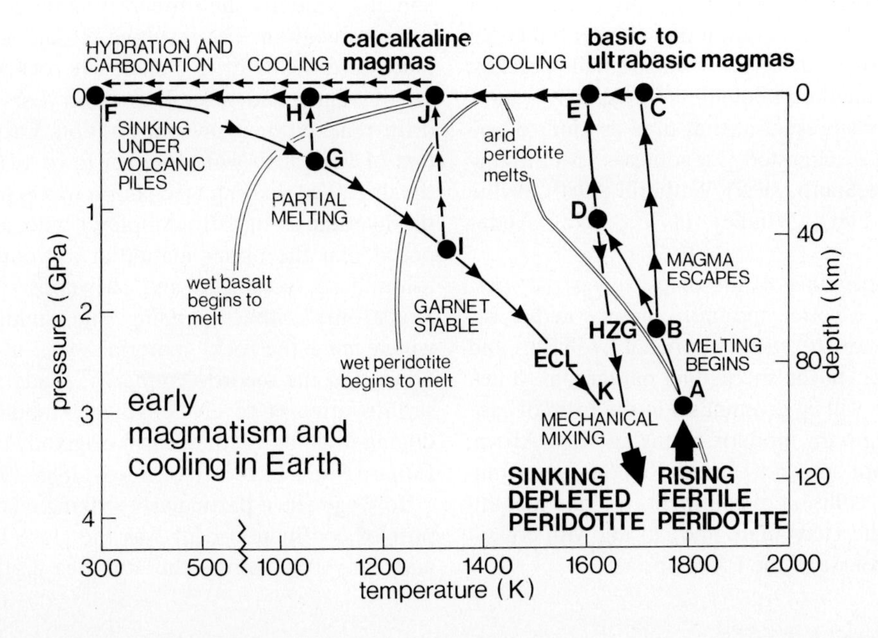

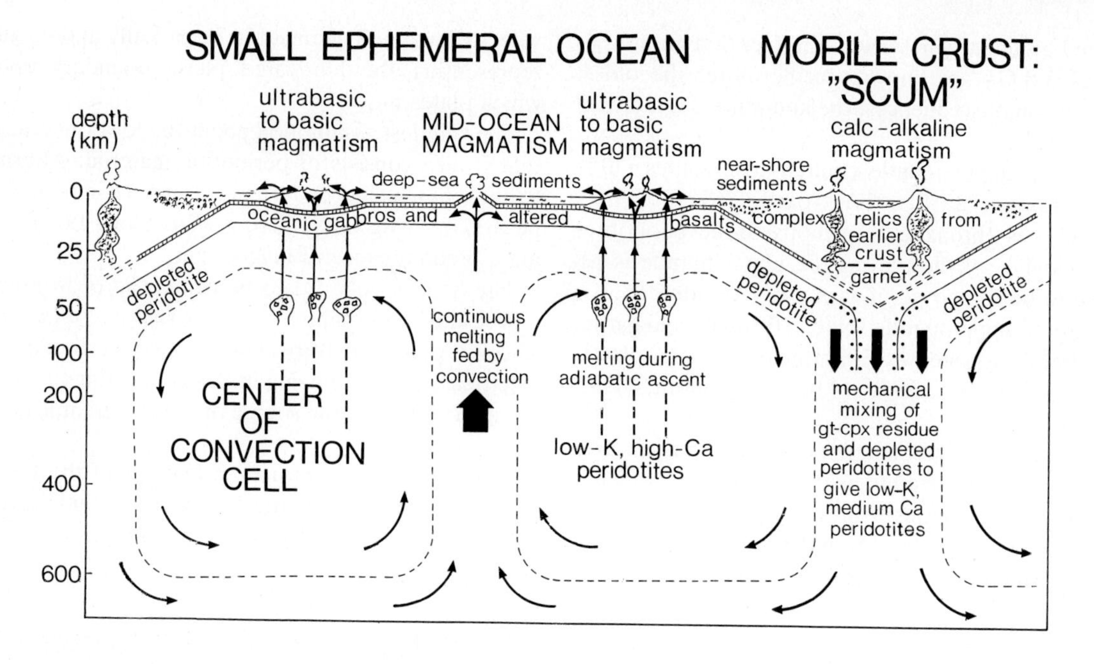

Fig. 7.6. Polygonal tectonics in the early Earth (highly schematic) (J.V. Smith, 1982)

oceanic crust. This crust of alkali-poor basalts and gabbros was generated from magma chambers fed by the melting of rising fertile peridotite along the lines of thought advanced by Green et al. (1979). The sequence of magmatism and cooling in the early Earth is schematically graphed in Fig. 7.5. Magma of different composition can be generated, and yield a suite of basic to ultrabasic rocks at the Earth's surface. Intense volcanism, vigorous "stirring" by means of polygonal convection cells (Fig. 7.6), and huge impact events inhibited the generation of a substantial continental-type crust prior to 4 Ga. Small patches, perhaps a few hundred kilometers across, might have been all that was present at that time as kinds of microcontinents; suggested readings: McElhinny (1979a, b), J.V. Smith (1982), Wetherill (1980), Wyllie (1971, 1976, 1981), Windley (1975, 1977), Yoder (1976).

With the appearance of the oldest terrestrial rock, 3.8 Ga in age, a better and more direct way has become feasible for the study of crustal evolution and global tectonics. In the succeeding paragraphs, a few relevant points will be given quasi in the form of case studies starting with modern rifting in East Africa, onset of seafloor spreading in the Red Sea a few million years ago, collision of continents and microcontinents during the Hercynian, and closing with crustal events in the course of the Precambrian.

Rifting

"Hebung – Spaltung – Vulkanismus" (uplift – faulting – volcanism) is the title of an article by Hans Cloos in 1939, which in just three words lists the critical mechanisms he thought to be responsible for the rifting of continents. Actually, the script was a belated response to the famous 1912 paper by Alfred Wegener, who suggested that our modern continents were once combined in a huge land mass known as Pangaea. This "Urkontinent" – according to Wegener – developed deep-running scars at the end of the Paleozoic. They were principally aligned north-south and the modern shelf-break roughly contours their former lineation. The resulting mosaic of continental pieces eventually drifted apart like rocks floating in a sea of magma and, after 250 million years of westward drift, reached their present position. Orbital parameters of the Earth were thought to be responsible for the drift. The floating mechanism was in part based on the hypothesis of Otto Ampferer who, in 1906, proposed that the upper mantle at its contact with the lithosphere convects and develops "Steig- und Sinkströme", that is rising and sinking currents, which cause the rocky material above to move up or down. For the record, Ampferer's model was the first serious attempt to challenge the standard textbook dogma of a contracting Earth originally conceived by Eduard Suess in 1855 (see Suess, 1885–1909).

To Wegener, a particularly striking example in support of continental drift was the close fit of Arabia and NE-Africa across the Red Sea and the Gulf of

Aden. Cloos respected Wegener highly, but in line with many other contemporary scholars he opposed continental drift. It was not so much for reasons that Wegener failed to come up with a driving force strong enough to propel the horizontal movement of continental blocks; instead it seemed to be more related to the fact that Cloos favored vertical displacements. In his studies on the Oslo-Rhine-Rhone Graben or the Red Sea-Gulf of Aden-East African Rift, Cloos interpreted the morphological features to be a result of graben tectonics and rejected continental drift as the structural mechanism behind the split of Arabia and NE-Africa.

Fifty years after Wegener's 1912-paper, and about 20 years following the publication of "Hebung – Spaltung – Vulkanismus", Harry Hess, during a colloquium at Princeton in 1961, presented a unifying concept that revolutionized our geological thinking: seafloor spreading (Hess, 1962). In a way, his basic philosophy contained a good dose of Ampferer's upper mantle convection, Wegener's continental drift, and Cloos' uplift, block faulting and volcanism. The Hess lecture spread rapidly across the geoscience community and triggered a flood of articles not only in geotectonics but in all branches of the earth sciences. Seafloor spreading advanced in 1968 to a broader term: global plate tectonics (Cox, 1973; Le Pichon et al., 1973). In mentioning the pioneering work of Dietz (1961), Vine and Matthews (1963), Wilson (1965), D.P. McKenzie (1967), Sykes (1967), Le Pichon (1968), Morgan (1968), and Sclater and Francheteau (1970), credit is given to a new generation of geophysicists and geologists, who in the light of plate tectonics trace the horizontal movement of continents and seafloors, determine the rate of accretion and subduction of oceanic crust, or offer conceptual models on the origin and evolution of oceans and continents back to the Archean. Even evolution of plate tectonics on terrestrial planets per se is a lively concern (e.g., Meissner, 1983; Head and Solomon, 1981). Since that memorable talk of Harry Hess published in 1962, the field of geosciences took fresh momentum and critical voices such as that of Beloussov (1970) are heard less and less.

Rifting is a process generally believed to precede seafloor spreading. However, the process of crustal formation at oceanic ridges is still not well understood. The most widely held opinion is that beneath the ridge crest a mixture of mantle material and basaltic melt is being injected into the crust. The basaltic melt, being lighter, migrates upwards and accumulates near the top of the injection diapir as magma chambers (Ballard and Van Andel, 1977; Bryan and Moore, 1977; Lewis, 1983). These magma chambers are very limited in areal extent – at most a few kilometers – and it is unlikely that they represent a steady state phenomenon, certainly not in the Atlantic, where the mechanisms have been studied in some detail. The extrusion of pillow basalts and lava flows from such magma chambers is episodic onto the seafloor, both in space and time. The relationships are depicted in Figs. 7.7 and 7.8.

Under the ridge axis, plate-driving forces produce an extensional strain which facilitates hydrothermal circulation, since crack propagation as a result of convective cooling is enhanced. Eventually, hydrothermal alteration of the crust and infilling of cracks by precipitation take place. Whether the Mohorovičić Discontinuity represents a petrologic boundary or whether it is determined by the depth to which hydrothermal penetration extends, is still not resolved. A schematic cross-section of the oceanic crust displayed

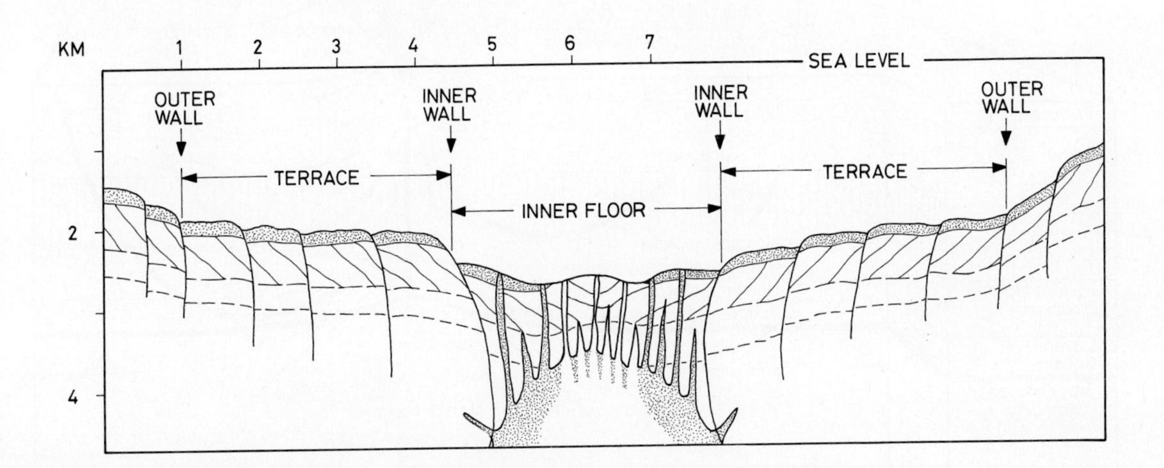

Fig. 7.7. Morphological structure of Mid-Atlantic Ridge rift valley (after Bryan and Moore, 1977)

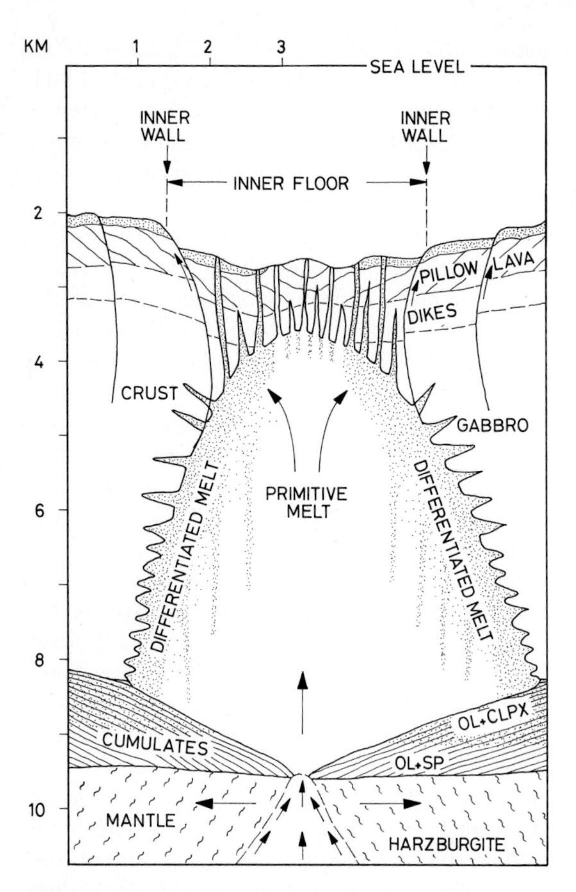

Fig. 7.8. Model of inferred magma chamber beneath median valley, in true scale. *Stipple* indicates differentiated melt, as well as lavas and cumulates generated from it. Note asymmetry of chamber (Bryan and Moore, 1977)

in Fig. 7.9 gives the main petrographic successions, and indicates suggested pathways of hydrothermal circulation through crust and central rift. Actually, it is hydrothermal ventilation which promotes the transfer of metal ions, anions and gases from the oceanic crust to air and sea. During ascent through the rock formation, ore minerals can be released (e.g., Norton, 1984).

The majority of rifts is distant from land and located in mid-oceanic ridges. They mark the zone where the seafloor spreads and has disrupted the positions of the continents (e.g., Ramberg and Neumann, 1978). For instance, the Atlantic and the Indian Ocean were born about 150 to 200 million years ago, when Pangaea started to disaggregate.

Fig. 7.9. Cycling of sea water through the oceanic crust. A simplified cross-section exhibiting the main petrographical features of crust and mantle is shown on the left side of the graph. The Moho (named after its discoverer Mohorovičić) represents the crustal base about 8–11 km below sea surface. The schematic section is principally derived from ophiolite sequences exposed on land; ophiolites are thought to be former oceanic crust. The serpentinized ultramafics are in most observed cases products of hydrothermal alteration after emplacement on land. For details see Lewis (1983). The *arrows* indicate the general direction of water movement through the "porous" basaltic crust. Zones of mineral precipitations, metal leachings, ion extractions, redox and pH changes, or biological activities both in crust and sea water are roughly sketched. The information is principally drawn from Edmond and Von Damm (1983), Edmond et al. (1982), East Pacific Rise Study Group (1981), Coleman (1977)

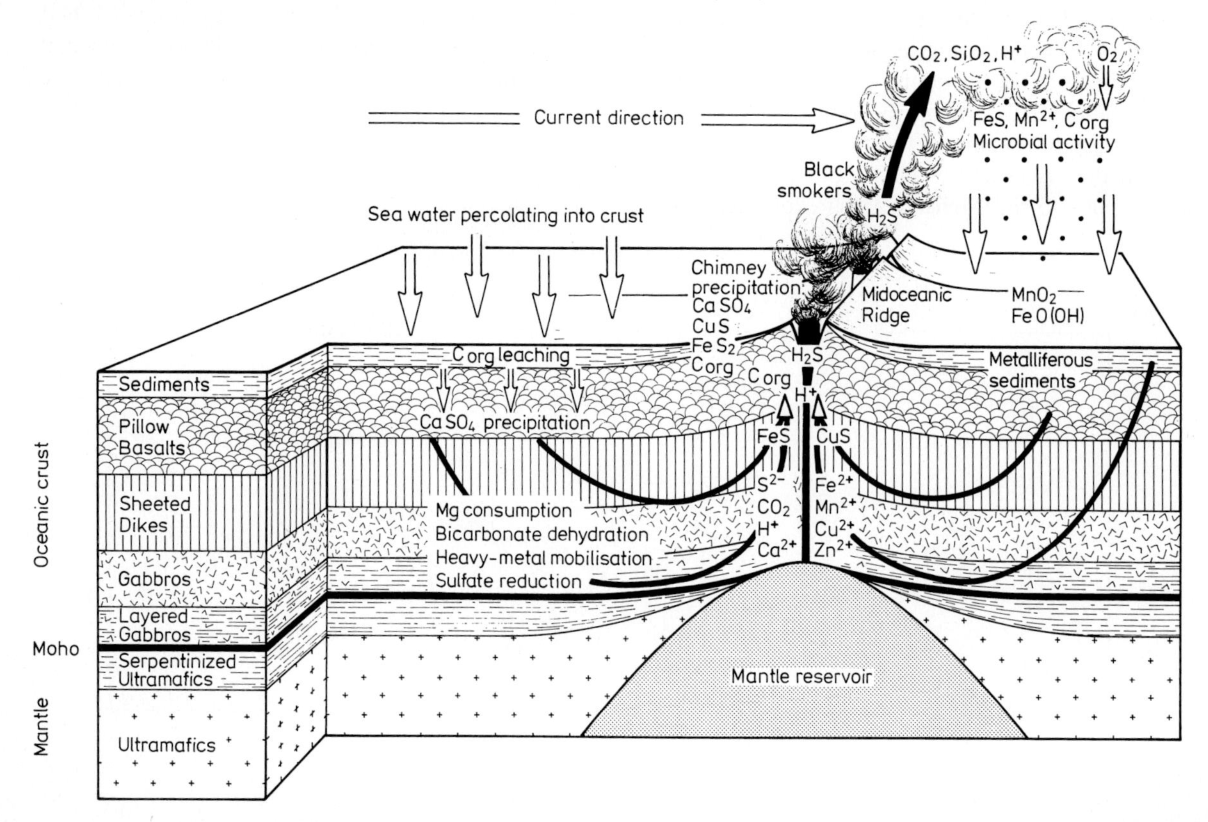

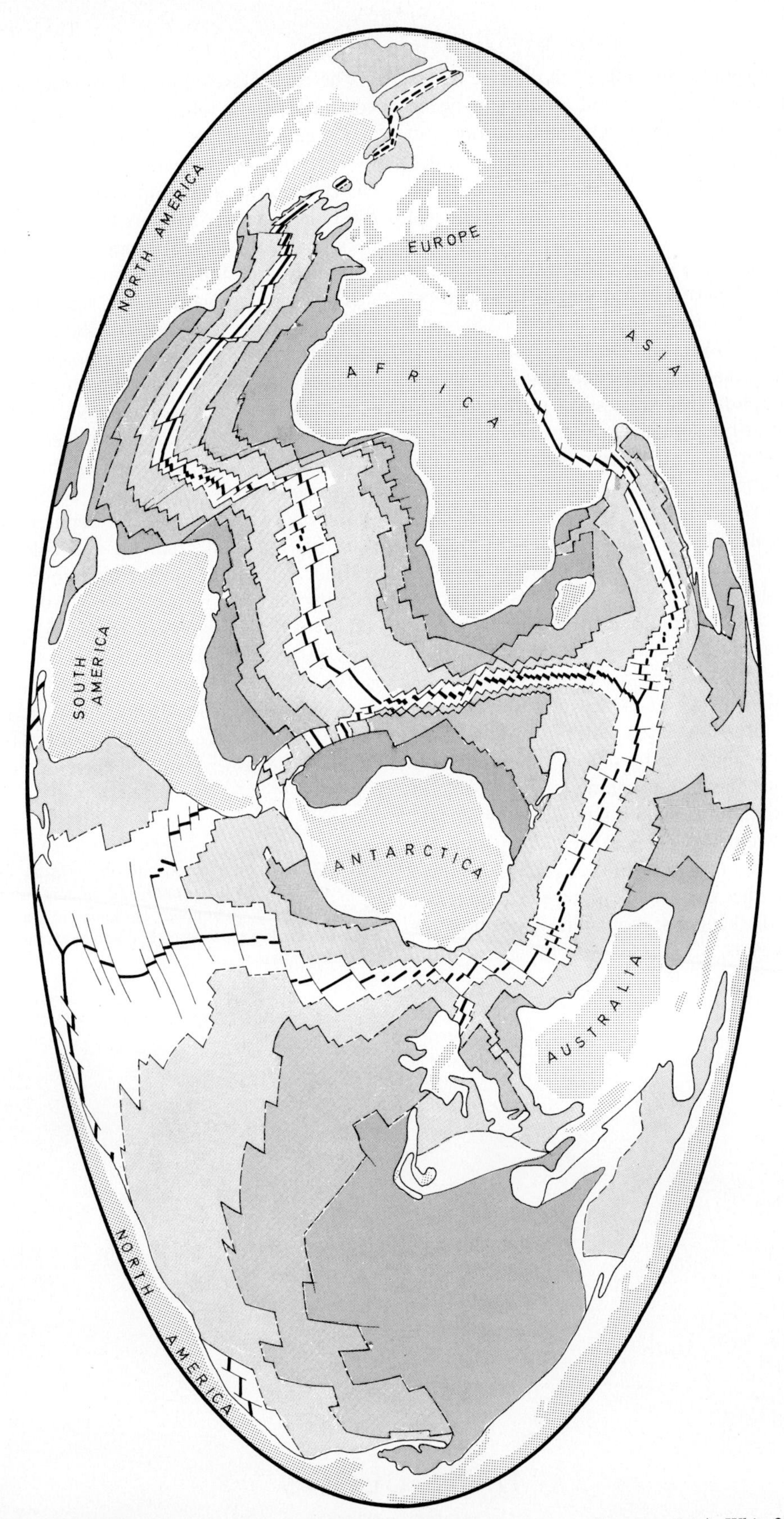

Fig. 7.10. World rift systems (after Degens et al., 1971, and updated using more recent literature data). *White* 20 Ma; *light gray* 20–65 Ma; *medium gray* 65–100 Ma; *dark gray* >100 Ma

Both oceans expanded gradually or in discrete pulses at the expense of the Pacific and an ancient ocean named Tethys which formerly separated Europe and parts of Asia from North and East Africa and India. The modern Mediterranean Sea may be viewed as a tiny remnant of what was formerly the wide open Tethys.

Other parts of world ocean started to open much later (Fig. 7.10). A particularly young rift, on-and-off active for about 25 million years, is the Red Sea and the Gulf of Aden (Wong and Degens, 1984). There is wide agreement that this rift continues into Africa, finding its morphological expression in the form of the East African rift system. So there are spreading rifts in the modern sea which have been in operation for about 150 to 200 million years, and others have been active for only a few million years. A third category is represented by the East African rift system, that is perhaps on the brink of moving apart. As a result, ocean basins are in a constant state of restructuring. They can become reduced in size or they may grow areally and expand at about the speed of a fingernail's growth. Like the continents, they too can move across the surface of the Earth and may even vanish completely. Associated with this evolution are changes in water depth and a redistribution of water masses. Should the world ocean no longer be able to hold its water within basinal bounds, continents will be flooded. In contrast, a general deepening of ocean basins will cause a retreat of the sea (see p. 280). We can almost say that not a single square centimeter of ocean floor will keep its geographical position for more than a year.

The early classical models of continental rifting invoking tensional forces were consistent with observational data available at that time. The refined version of Cloos (1939), showing the intimate association of rift valleys with gently updomed areas of the Earth's crust, lent further support to this hypothesis. Ramberg (1963, 1971, 1978) proposed lateral spreading of buoyantly injected subcrustal bodies as a trigger mechanism and gave lesser significance to crustal doming for rifting. At present, the stretching model of McKenzie (1978), involving lithospheric extension or thinning and passive upwelling of hot asthenosphere, is the most widely accepted view. Originally conceived to explain the formation of sedimentary basins in continental regions, the model has been extended by numerous investigators to account for: (i) the evolutionary stages in the development of continental rifts, (ii) the final break-up and commencement of seafloor spreading, (iii) the formation of passive continental margins, or (iv) the origin of suites of arcuate reflectors dipping seaward at passive continental margins. Selected reading: Pollack and Chapman (1977), Burke (1977), Le Pichon and Francheteau (1978), Parsons and McKenzie (1978), Le Pichon and Sibuet (1981), Sclater et al. (1981), Vink (1982), Martin (1984), Bédard (1985), Browne et al. (1985),

Fig. 7.11. Invitation card sent on occasion of Hans Cloos' 100th birthday to his friends and students showing a sketch made from his 1936 famous "Tonkuchen Experiment" in which a plastic clay layer was stretched and bent, thereby producing a tectonic graben

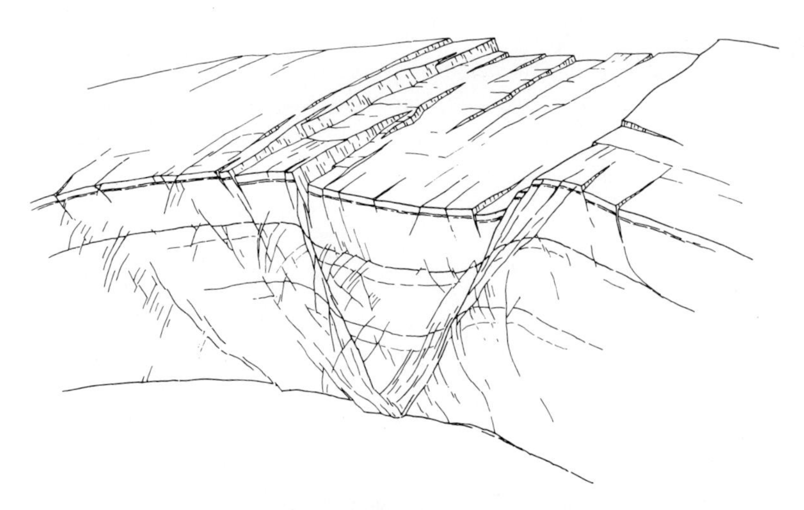

HANS CLOOS
1885 - 1951

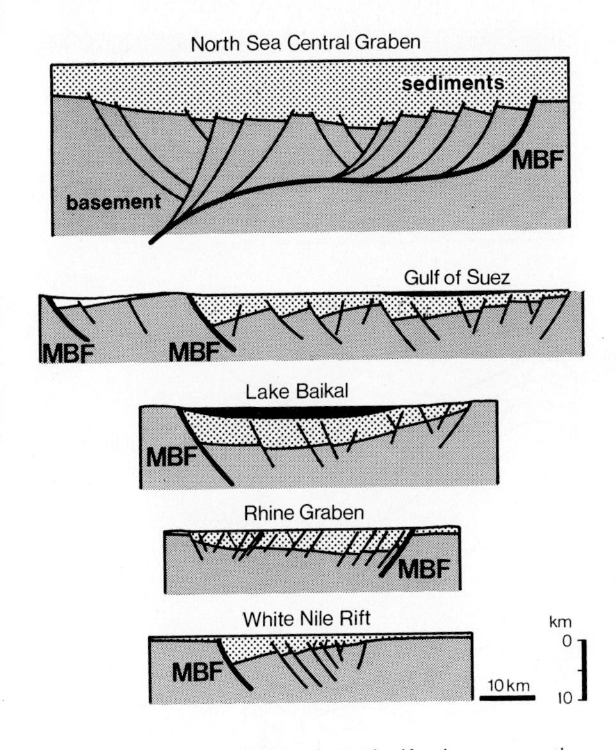

Fig. 7.12. Asymmetry of continental rifts in cross-section. Rifts commonly show half-graben-like forms in cross-sections taken normal to their long axes, with most basin relief generated by a single rift bounding fault (main bounding fault, MBF) (Bosworth, 1985)

Mulugeta (1985), Steckler (1985), Bosworth (1985), Mutter (1985), Sclater and Christie (1980), Steckler and Watts (1980).

Continental rifts normally do not exhibit a symmetric pattern of the type shown in Fig. 7.11. Instead, they are strongly asymmetrical in cross-sections normal to their long axis, as can be seen in a few representative examples depicted in Fig. 7.12. Half-graben-like forms are prominent, and displacements proceed on low-angle normal faults and along curved fans known as listric fault systems (Box 7.2). Listric faults curve only towards the hanging wall block, which limits their horizontal extension to a few tens of kilometers at most. Young rifts of the type generated in East Africa can be looked upon as large-scale half-graben which flip more or less regularly along the axis of the rift (Bosworth, 1985; Fig. 7.13). His model for the propagation of continental rifts starts with an upper crustal extension in the sense of McKenzie (1978). This finds its structural expression as a broad zone of diffuse faulting which quickly evolves into a system of listric faults which at depth bottom out at a major crustal discontinuity such as the brittle-ductile transition (see Fig. 7.13). When curving of the active listric system departs enough from the overall rift trend, a new detachment system of opposite geometric polarity may be generated within an overlapping region known as accommodation zone. It is of note that cross-faults in rifts are kinematically similar to accommodation zones, although on a smaller scale.

So far, stretching and necking – that is crustal erosion from below – plus subsidence dominate early rifting. At what point does uplift enter the scene? According to McKenzie (1978), uplift is a consequence of heating the lithosphere. Conflict exists only because calculated rates of subsidence or uplift according to this concept are much too small to account for the commonly observed rift flank uplifts of 1 km. To reconcile this discrepancy, an active asthenosphere heat source is often invoked to produce the uplift and act as a driving force for the rifting (e.g., Neugebauer, 1983; Mareschal, 1983). In short, in an active rift

Fig. 7.13 a, b. Following upper crustal extension, diffuse faulting is initiated, evolving to a few main listric faults above two oppositely directed detachments (**a**). Eventually the curving, active listric system departs enough from the overall rift trend to favor a new detachment system, which links to the old via an accommodation zone (**b**) (Bosworth, 1985). Similar curvature is seen in the bounding faults of the Basin and Range Province (Moore, 1960)

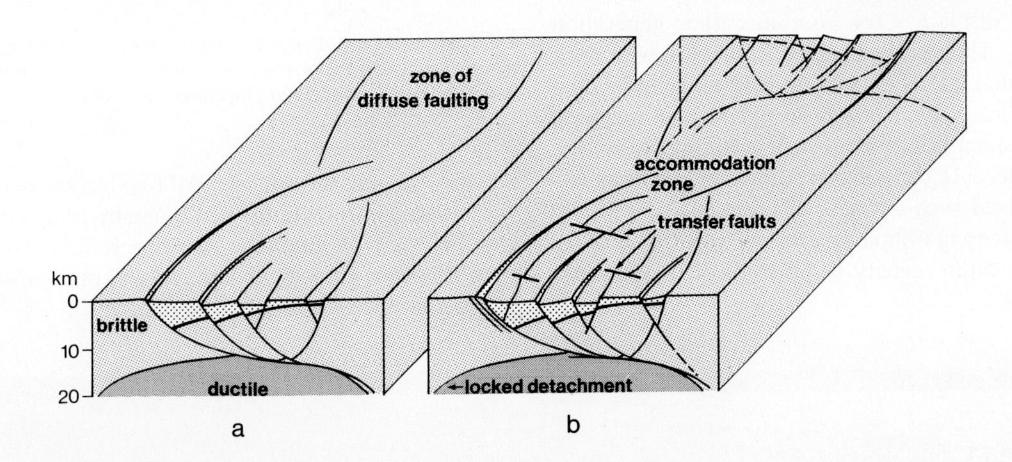

BOX 7.2

Listic Normal Faults (Bally et al.,1981)

At the beginning of this century, Eduard Suess introduced the concept of listric faults in his famous book "Antlitz der Erde". The term listric is derived from the Greek word "listricon" meaning shovel. However, for many years to come the concept gained little acceptance outside of Central Europe. It was only with the advent of extensive reflection seismic profiling for oil exploration, the execution of long and deep geophysical traverses across fold belts, and continental platforms or through seismic site survey studies as part of the GLOMAR CHALLENGER Deep Sea Drilling Project, that the phenomenon of listric normal faults became increasingly more obvious. Experimental work further illustrated the growth of listric normal faults in connection with the simulation of graben-like structures using so-called "Tonkuchen", literally meaning clay cakes, as a substrate (Cloos, 1936). However, a few decades had to pass before mapping geologists warmed to curved fault systems that flatten with depth, are shovel-like and therefore represent listric normal faults.

Listric faults are abundantly developed in rift systems or along passive continental margins (Fig. 7.13). In addition, they are generated during syn- and post-orogenetic faulting associated with the stretching and shearing of orogenetic systems and parts of their foreland. Blocks underlain by listric surfaces can be mobilized via gravity sliding and even rotate. In contrast, during compressive stages of tectonism, the movement is reversed. Curved fault planes become reactivated or form de novo and serve as thrust faults with horizontal and/or vertical displacements measuring as much as several kilometers in width. It is even conceivable that the flat part of a listric surface provides a starting point of tectonic nappes.

Listric faults generated during an orogeny come ordinarily in sets, whereby one generation of listric surfaces can displace a set from an earlier generation (Fig. 7.14). Their curved nature, combined with their often limited lateral extent – at least as an individual fault surface – frequently makes geological mapping in fold belts a rather cumbersome affair. This is the more so because there may be other complicating factors involved such as facies changes, strike and slip faults, a steep morphology, erosion, vegetation cover and quite often scarcity of outcrops (Degens et al., 1980).

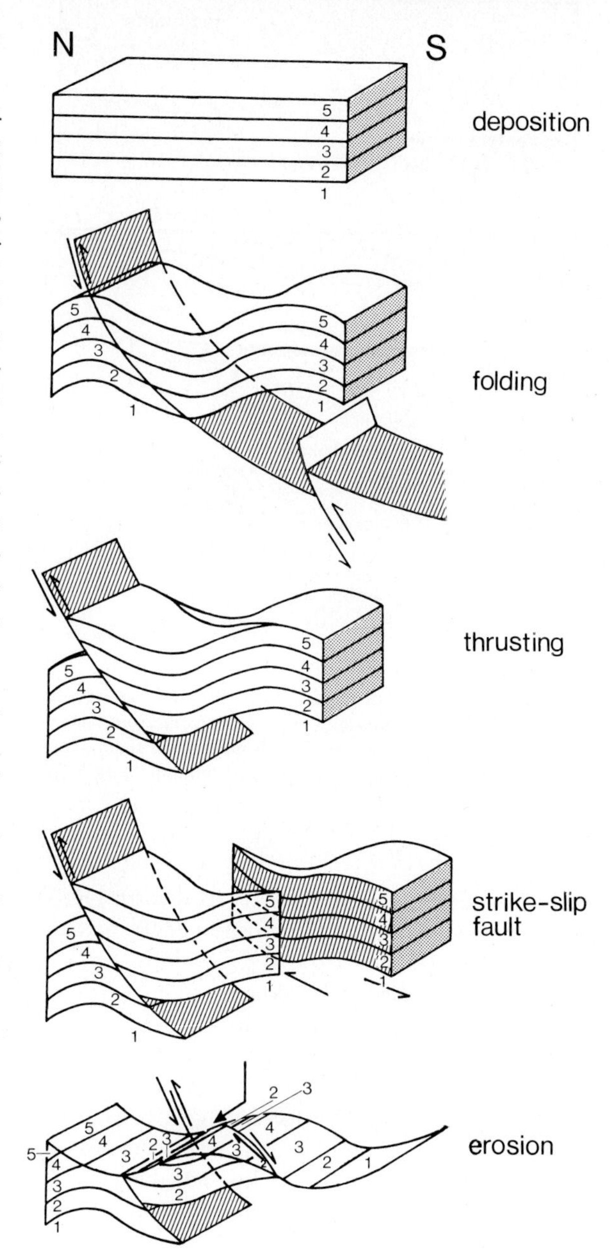

Fig. 7.14. Proposed model for the propagation of listric fault planes, strike-slip faults and erosional surfaces during orogeny and subsequent uplift (after Degens et al., 1980)

An extremely informative data collection pertaining to listric normal faults is the one by Bally (1983). On the basis of hundreds of deep seismic profiles from different geological settings, the omnipresence of listric surfaces can be ascertained.

scenario driven by a deep-seated heat source (i.e., a diapir) uplift precedes rifting.

Using the Gulf of Suez as a case study to delineate the evolution of a continental rift, Steckler (1985) was able to demonstrate that the main phase of uplift is a late event and commenced about 8 to 10 million years after the initiation of rifting. The extra heat needed to propel massive uplift at that time stems from thermal and velocity perturbations introduced by uniform rifting. These induced secondary small-scale mantle convection beneath the rift, whereby the rift flanks producing the broad uplift are the regions that become primarily heated.

The apparent lack of a pre-rift doming event has interesting geological implications. For instance, if rifting starts close to sea level, rates of sedimentation will at most be in the order of a few centimeters per thousand years because subsidence is entirely controlled by crustal thinning. In contrast, uplift caused by lithospheric heating will dramatically change the depositional pattern and result in substantially higher rates of sedimentation.

In summing up, continental rifting and break-up were formerly thought to be a result of local domal uplifts produced by upper mantle plumes. The modern scenario involves tensional forces which cause the brittle upper crust to fracture along normal and listric fault systems. Long narrow lineaments of pre-existing crustal weakness direct the pattern of rifting. In phase with crustal thinning and necking are subsidence and sedimentation in the newly formed rift basins. Passive upwarping of the asthenosphere will in time result in thermal doming, leading to volcanism first of continental plateau basalts and eventually of oceanic tholeiites. Rift-shoulder uplifts will accompany the upper mantle heating event. Gravity sliding of plates to either side of the thermally expanded upper mantle and partial melting of crustal blocks within the rift axis proper will eventually cut and separate the adjoining lithospheric plates and initiate seafloor spreading.

The sequential stages of rifting are: tensional stress – lithospheric necking – subsidence and sedimentation – passive upwarping of asthenosphere – "continental" volcanism – thermal doming – major uplift – gravity sliding – "oceanic" volcanism – seafloor spreading. These stages taken either as a time sequence or as a spatial sequence are gradational or they may overlap so that adjacent stages may co-exist (Fig. 7.15).

It has been suggested that in the early stages of continental rupture a three-arm system often develops in a more or less symmetric pattern. In the normal course of events, two of these arms develop while the

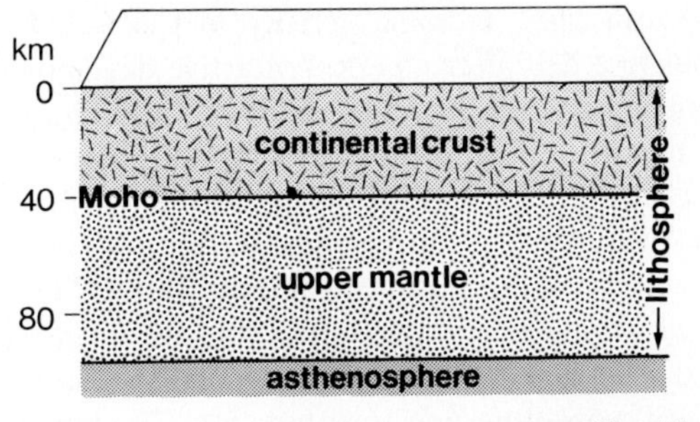

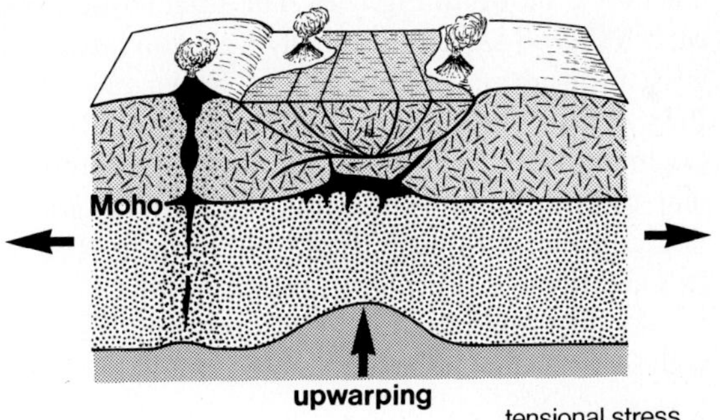

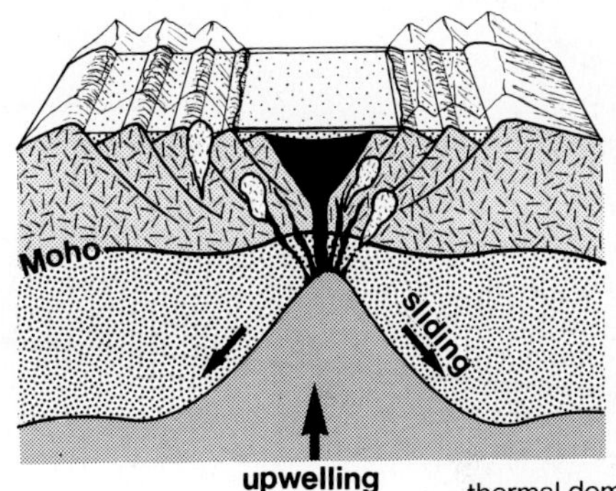

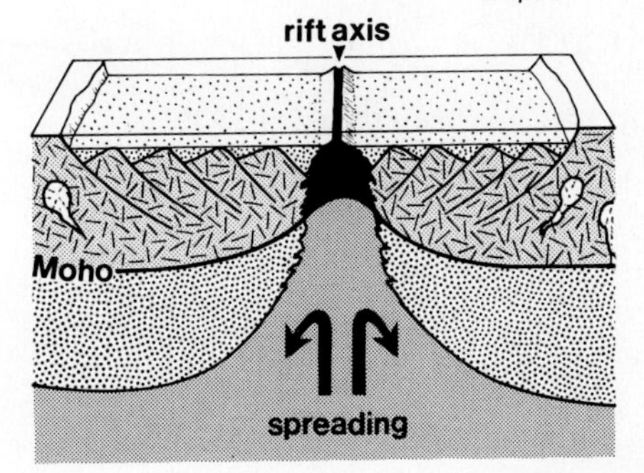

Fig. 7.15. Block diagrams showing the sequential stages in the development of rifting, leading to seafloor spreading

third arm fails. Tectonic activity in the failed arm comes to a halt after a period of active development. It could have reached the seafloor spreading stage and then the spreading axis dies out, as is the case in the Benui Trough in Nigeria. Alternately, it may not even have reached the spreading stage, as has been postulated for the East African Rift System. In East Africa a number of triple junctions have been recognized. The Red Sea, Gulf of Aden and the East African Rift constitute one such junction. The kinematic pattern developed in the early stages of that triple junction is shown in Fig. 7.16. The total motion since early Miocene agrees reasonably well with rates calculated from magnetic anomalies (Girdler and Styles, 1978; Brown and Girdler, 1982; Girdler et al., 1980; Cochran, 1981; Noy, 1978). Of particular interest is that the extension rates across continental grabens (e.g., in the Ethiopian Rift), the western Red Sea Rift and in the Gulf of Suez are of the order of 0.3 cm a^{-1}. At this slow rate, features commonly associated with seafloor spreading are not observed.

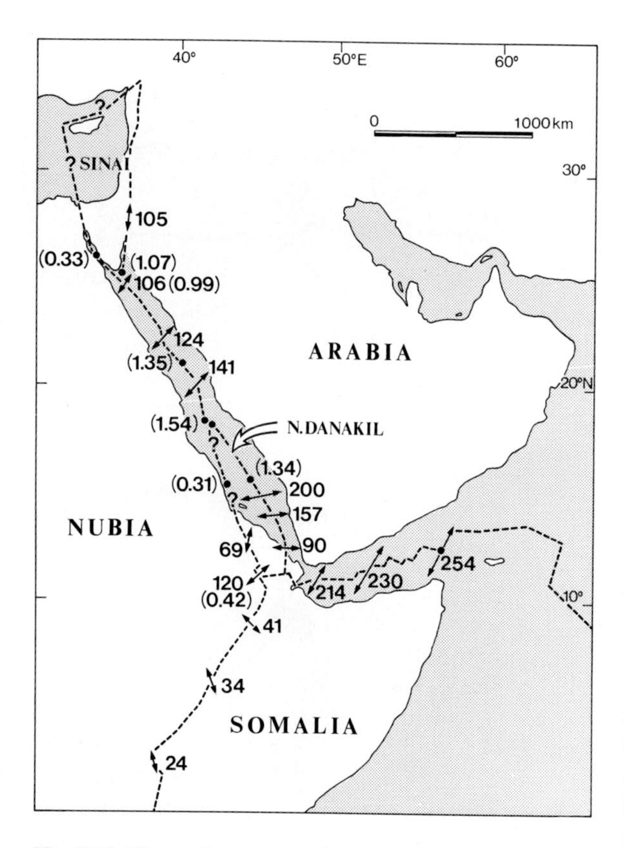

Fig. 7.16. Kinematic pattern in the Red Sea, Gulf of Aden and East African Rift area. *Dashed lines* represent plate boundaries. *Arrows* give the total motion (in km) since the early Miocene. Numbers in brackets are total opening rates in cm a^{-1} (after LePichon and Francheteau, 1978)

Collision

The Rhenish Massif, positioned in the center of Europe, is part of the Hercynian orogenic belt that stretches from Eastern North America to Northwest Africa and Europe. The European segment is commonly called the Variscan belt and the Rhenish Massif is synonymous with the expression Rhenish Slate Mountains, or to use a term more familiar to me, "Das Rheinische Schiefergebirge". This gentle hilly region of today, on the geological map of Europe looking almost like a small butterfly, is considered to be the result of the collision of the ancient continent Baltica and the microcontinents of Southern Europe, after subduction destroyed the intervening Rheic Ocean during the early Paleozoic.

When mentioning the early Paleozoic we should know that part of the Rhenish Massif contains patches of rock formations that were formed in an earlier epoch – the Caledonian – and which became subsequently drawn into the Variscan folding process. These regions are found on the left bank of the Rhine, for instance in the Brabant and Stavelot Massifs. It is of extreme importance that in one area on the right side of the Rhine, just a few tens of a square kilometer in size, early Ordovician and Silurian rocks are exposed which have not been subjected to the Caledonian Orogeny, suggesting a hybrid origin of the pre-Devonian crust.

In order to elucidate the earlier history of the Rhenish Massif, we will first make an excursion into the Precambrian geology of Central and NW Europe. Next, we will reconstruct the "wanderings" and collisions of Baltica and its bordering geosynclines through Scylla and Charybdis or to paraphrase Shakespeare (The Merchant of Venice, III, v. 17), "Thus when I shun Scylla, your father, I fall into Charybdis, your mother." And finally, we shall examine rock fragments that were broken off from the deep crust and brought to the surface by ascending volcanoes. Such pieces are found as foreign elements, so-called xenoliths, in a number of Quaternary volcanic deposits.

The crystalline basement of Europe consists of a mosaic of crustal elements which were consolidated stepwise during various Precambrian and Paleozoic orogenic cycles (P.A. Ziegler, 1977, 1982, 1984). A brief look at a geologic map of Europe reveals that the two largest regions of exposed Precambrian rock formations are the Fennoscandian Shield and its extension beneath the Baltic depression and Moscow platform and, on the other side, the Laurentian-Greenland Shield, of which the Rockall-Faeroe

Plateau and the Hebridean Craton form a part. The age of crustal consolidation ranges between 3,000 and 700 Ma, whereby the major basement province is of pre-Grenvillian age (1,100 Ma). The welding of continental crust onto the pre-existing continents Fennoscandia-Baltica and Laurentia-Greenland yielded, in the aftermath of the Grenvillian-Dalslandian orogenic cycle, a megacontinent named Proto-Laurasia (Patchett and Bylund, 1977; Fig. 7.17).

Again, there are many smaller and isolated Precambrian outcrops south of Fennoscandia, such as the Irish Sea Horst (2,400–550 Ma), the London Platform (700–530 Ma), the Armorican Craton (2,600–550 Ma), the Bohemian Craton (950–550 Ma), or the East Silesian Massif (1,410–520 Ma). They are commonly referred to as Allochthonous Terrains, because they did not grow in place but were transported as microcontinents to their present position in the course of the break-up of a larger land mass which began to crumble at about the Precambrian-Cambrian boundary.

Caledonian and Variscan elements are part of the Rhenish Massif. One might, therefore, ask when did they become united and what were the overall paleogeographic and tectonic circumstances that led to Caledonian and Hercynian crustal consolidations in general? The Caledonian orogenic cycle is composed of three main phases: (i) Grampian/Finmarkian (Late Cambrian to Early Ordovician), (ii) Taconic (Mid-to-Late Ordovician), and (iii) Main Scandinavian (Mid-Silurian to Early Gedinnium) (P.A. Ziegler, 1982; Gee and Sturt, 1985). Orogenies mobilize huge basins of sedimentation, so-called geosynclines, that are stretched in a band a few 100 km wide along coasts and continental margins, and furthermore contain thick sections of infill (10 km). The subsiding geosynclinal stage commonly lasts at least several tens of a million years before orogenic events take over. Thus, for the Rhenish Massif, we should follow the Odyssey of Baltica and its bordering geosynclines from the Early Cambrian to the Kazanian (Permian), spanning a time interval close to 300 million years. This segment about covers the beginning of the first "Caledonian" geosyncline and the end of the Variscan epoch, which timewise corresponds to the final consolidation of the supercontinent Pangaea.

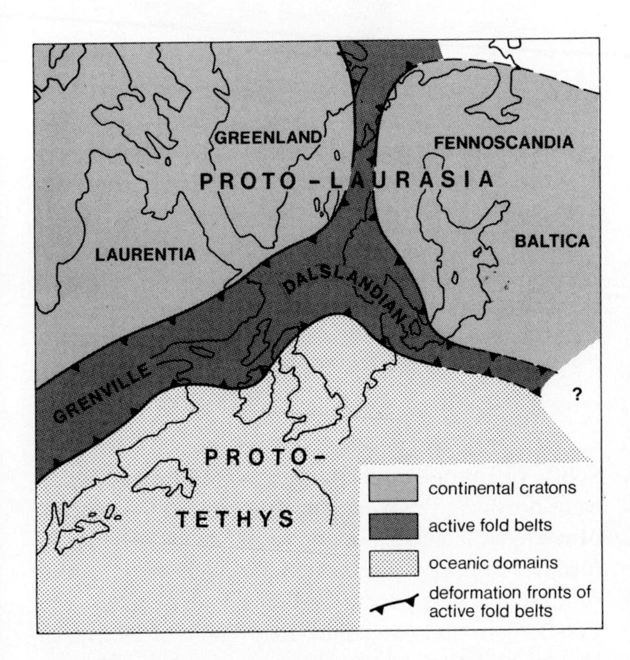

Fig. 7.17. Tentative Grenvillian tectonic framework of the Arctic-North Atlantic domain (after P.A. Ziegler, 1982, 1984; Patchett and Bylund, 1977)

As mentioned, Pangaea prevailed for only a short time and broke up into a number of continental fragments which, after about 200 million years of spreading, collision and subduction, gave rise to the land-sea distribution of today. In contrast, by retracing the Paleozoic collision history which eventually led to Pangaea, events causing the formation of the Caledonian, and Hercynian fold belts come to light.

The Paleozoic maps of Scotese et al. (1979) and A.M. Ziegler et al. (1979), which are principally based on paleomagnetic, tectonic, faunal, floral, sedimentological and related criteria, reveal three continents of special relevance in our context: Laurentia, Baltica and Gondwana. Laurentia consists of North America, Greenland, Scotland, Ireland north of the Caledonides, Spitsbergen and parts of Eastern Siberia; Florida and Avalonia (Nova Scotia and Newfoundland) are excluded. Baltica is that part of Northern Europe bounded by the Hercynian and the Uralian Sutures, with England excepted. Gondwana, the largest Paleozoic continent, was originally composed of South America, Africa, India, Arabia, Madagascar, Australia and Antarctica, as well as southern Europe, Turkey, Iran, Afghanistan, Tibet, New Zealand and Florida. Rifting began along its northern margin in the late Paleozoic, yielding the Zagros and Indus structures.

According to this paleogeography, most of the paleocontinents occupied low-latitude positions during the Cambrian (Fig. 7.18). A notable exception is Baltica, which lay between 30 and 60° S. It was separated from Laurentia by an ocean, the Proto-Atlantic called Iapetus, which in Greek Mythology is the father of the Atlantic. Southern Europe, consisting of Czechoslovakia, France, Iberia, and Italy, which un-

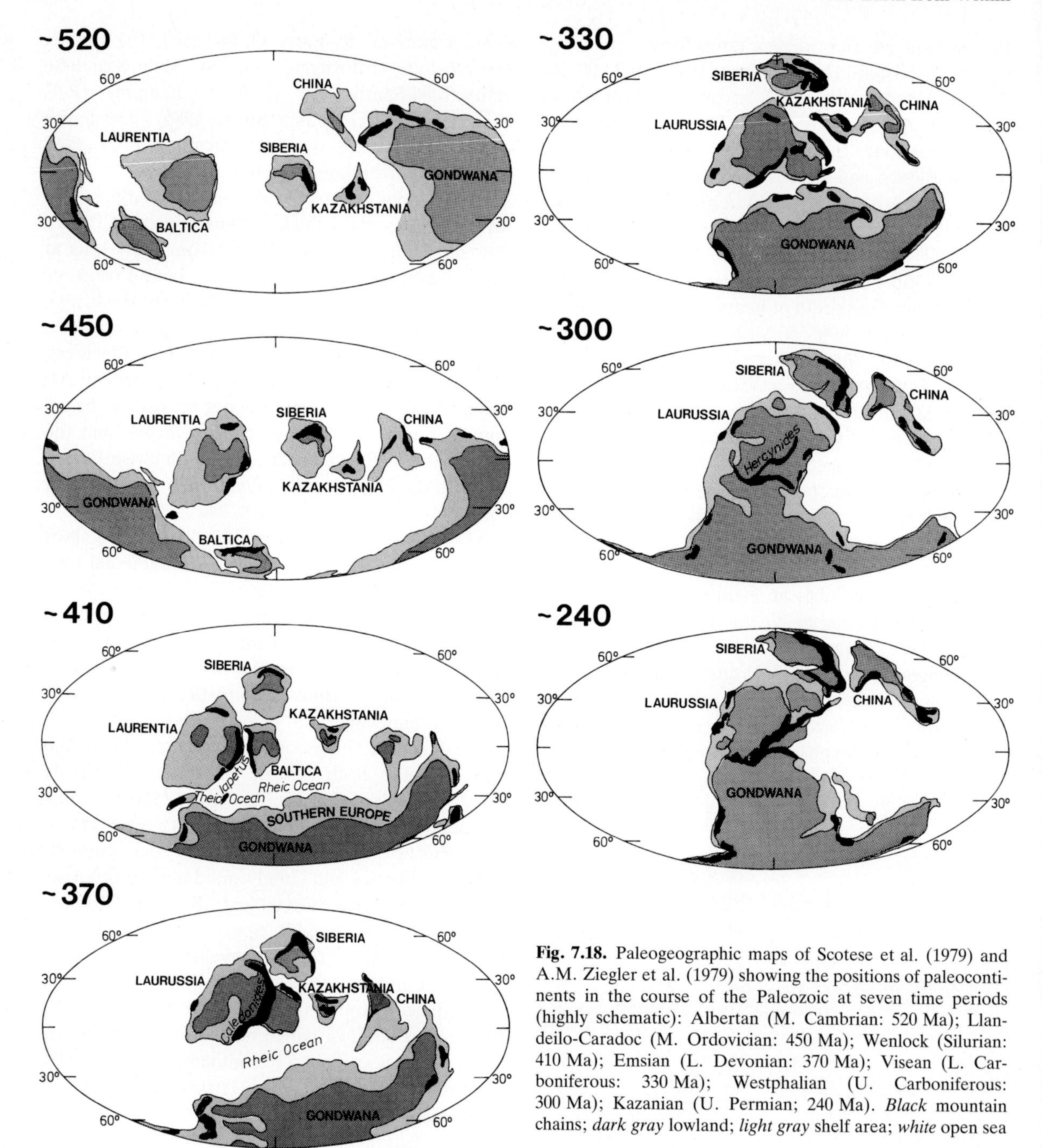

Fig. 7.18. Paleogeographic maps of Scotese et al. (1979) and A.M. Ziegler et al. (1979) showing the positions of paleocontinents in the course of the Paleozoic at seven time periods (highly schematic): Albertan (M. Cambrian: 520 Ma); Llandeilo-Caradoc (M. Ordovician: 450 Ma); Wenlock (Silurian: 410 Ma); Emsian (L. Devonian: 370 Ma); Visean (L. Carboniferous: 330 Ma); Westphalian (U. Carboniferous: 300 Ma); Kazanian (U. Permian; 240 Ma). *Black* mountain chains; *dark gray* lowland; *light gray* shelf area; *white* open sea

derwent small relative motions, was a part of Gondwana and likewise occupied a position in the high southern latitudes (45–60° S).

From the Upper Cambrian through Middle Ordovician, Laurentia and Gondwana rotated counterclockwise, while Baltica migrated slightly southwards. By Middle Ordovician, subduction of Iapetus along both of its margins was well underway. Laurentia varied little in its position, while Baltica and Southern Europe (still attached to Gondwana) underwent northward movements. This caused the Iapetus ocean to close and Laurentia to collide with Baltica to form a composite continent: Laurussia, the Old Red Continent. This event produced essentially the Caledonian and Arcadian fold belts of Northern Europe and Western North America, and probably occurred with Baltica south of its position in the Permo-Triassic Pangaea. A wide span of ocean – the Rheic

Ocean – existed between Laurussia and Gondwana, and andesitic volcanism began in the Greater Caucasus along the south-western margin of Baltica.

By Lower Carboniferous, clockwise rotation of Gondwana had closed much of the Rheic Ocean. Active volcanism occurred in Morocco, Central Europe, the Greater Caucasus, and in a belt extending from Turkey to Tibet. Southern Europe rifted away from Gondwana and accelerated its northward drift towards Laurussia. By Upper Carboniferous, collision occurred between Laurussia and Southern Europe and between Laurussia and Gondwana. The Rheic Ocean was completely subducted, and the Hercynian, Mauritanide, Appalachian and Ouachita foldbelts were formed. Readjustment also took place along the former Caledonian Suture in the form of sinistral shear as Europe was driven northwards relative to North America. The St. Lawrence and North Sea grabens opened.

The Permian paleogeography is very similar to that of the Upper Carboniferous, although the paleoclimate was rather different. Because the newly formed mountain belts blocked the moist equatorial easterlies, very dry conditions prevailed in Europe and North America. This left its imprint on the Permian deposits of these continents. This situation resembles that established during the late Tertiary where, due to the rise of the alpine foldbelt, large areas such as the Tibetan Plateau became covered by glaciers. This changed world albedo and led to a global cooling and desertification (see p. 223).

A central feature of the paleogeography discussed above is the existence of the Rheic Ocean. Supporting evidence may be found firstly in paleomagnetic data. The apparent polar wander path for Southern Europe differs significantly from that for Baltica until Carboniferous time, implying that these continental fragments were on separate "plates" that underwent relative movements until they became welded together in a Carboniferous collision (Burke et al., 1976). Secondly, biogeographic data suggest a decreasing separation in the floral and faunal species of Baltica and Southern Europe in the course of the Paleozoic. Starting with a cold water fauna in Southern Europe and one of intermediate character in Baltica in the Middle Ordovician (A.M. Ziegler et al., 1979), they evolved into faunas that belonged to the same biogeographic province by Lower Carboniferous. Thirdly, there is a marked tectonic difference north and south of the Hercynian suture. The northern zone is largely undeformed, except for a narrow paratectonic belt folded during the Sudetic phase in the Visean-Namurian. By contrast, the southern zone was deformed during the Asturic phase in the Westphalian (Dewey and Burke, 1973), but shows no sign of a Caledonian orogenic imprint (e.g., Schoenenberg, 1979). Moreover, being developed on a Precambrian (Cadomian) basement, it is characterized by low-pressure metamorphism, granites, migmatites and mantled gneiss domes as a result of local remobilization (e.g., Massif Central, Bohemian Massif). Fourthly, a striking stratigraphic contrast exists between the Upper Carboniferous paralic and limnic facies across the Hercynian suture. During the Dinantian, extensive shallow-water sedimentation with limestones and reefs prevailed to the north and to the south. With the regression in the Upper Carboniferous, isolated coral basins of a paralic (marine) facies developed in the North-bordering foreland, while clastic and coal basins of a limnic (freshwater) facies (with a complete lack of Old Red Sandstones) were laid down under intermontane conditions in the south. This lithofacies contrast suggests that the areas on either side of the Hercynian suture represent shelves on opposite sides of the Rheic Ocean in the Dinantian (Lower Carboniferous). With subduction of this ocean, the different Upper Carboniferous tectonic development to the north and south led to different sedimentary histories. Finally, the occurrence of late Ordovician glacial deposits in France and Spain as well as in North Africa (Fairbridge, 1971) suggests the proximity of these areas as well as a high paleolatitude. Hence, southern Europe probably lay closer to Gondwana than to Baltica in the Ordovician, so that the Rheic Ocean may be inferred.

While the paleogeography used here predicts an early (Carboniferous) opening of the Tethys, many authors argue from paleomagnetic (Irving, 1977, 1979; Westphal, 1977), lithofacies and tectonic considerations (Rau and Tongiorgi, 1981) that the Tethys came into existence only in the Permo-Triassic. This implies that during the Hercynian orogeny, Africa collided with Europe and not with Laurentia and that the southern margin of the Hercynides was not characterized by subduction, but was ensialic. Regardless of whichever view is accepted, the resulting geotectonic development of the European Hercynides is not substantially affected, except to the extent that the latter view requires a continent-continent collision after subduction along only one (northward-dipping) zone.

Based on the paleogeography outlined above, a plate tectonic model for the Paleozoic geotectonic development of the European Hercynides has been proposed (Wong and Degens, 1983, 1981). We envisage that this development envelopes four stages because the same four stages exist in younger mountain belts, though perhaps in a slightly different form because of a different plate geometry.

The first stage is one of lithospheric thinning, subsidence, and basinal and geosynclinal development on the opposing shelves of the Rheic Ocean in southern Baltica and northern Gondwana. It starts with a stretching of the continental lithosphere which caused crustal thinning, passive upwelling of hot asthenospheric material, faulting and rapid subsidence. This pre-rift and syn-rift subsidence may be accentuated by the presence of a high thermal gradient as a result of the previous Caledonian tectonism, or by deep-seated processes in the mantle which produce active heating of the lithosphere at the plate margin (Steckler and Watts, 1980). As the lithosphere cools, slow subsidence unaccompanied by faulting but controlled by thermal relaxation takes place (McKenzie, 1978). In response to lithospheric stretching, listric faults develop along contacts of clayey and sandy, quartzitic material. The decrease in dip of these fault surfaces with depth causes back-tilting of the subsided blocks, so that stretching is accommodated. In the course of tectonic development with the uplift of the land masses, these basins become centers of very rapid clastic sedimentation and a geosyncline develops.

The second geotectonic stage overlaps the first and begins in the earliest Devonian. Subduction of the separate Rheic plate (Rheica) (assuming that it exists as such) takes place probably at both of its opposing margins. This subduction is intermittent, particularly starting with the Late Devonian when the dextral shear component becomes dominant. Interrupting this subduction are episodes of fracturing, volcanic eruptions, subsidence of small basins, basement reactivation and local uplift.

Stage three is characterized by folding with a northward vergence and thrusting and overthrusting along listric surfaces. It consists of several phases because of the collision geometry. This is the true orogenic stage.

Massive granite intrusions accompanied by hydrothermal and volcanic events mark the last (post-orogenic) stage. Permian volcanism, which exhibits both basic and acidic properties (Jung and Vinx, 1973), may be a result of crustal thickening following continental collision and remobilization of the lower crust.

The assumed paleogeography dictates that the collision between Baltica and Rheica and between Rheica and the microcontinents of southern Europe be oblique. What tectonic processes actually occur during such a collision is dependent largely on four factors: the angle of approach, the approach velocity, the intrinsic geological characteristics of the interacting plates, and the geometry of the plate margins. We suggest that subduction takes place where the approach is head-on, the approach velocity high, and the subducted oceanic crust free of irregularities such as seamount chains. Obduction and ophiolite emplacement are assumed to be favored when during subduction light, buoyant bodies in the form of oceanic plateaus, continental fragments, island arcs or old hot spot traces are encountered (Ben-Avraham et al., 1981; Ben-Avraham and Nur, 1982; Coleman, 1977; Dewey, 1976; Nur and Ben-Avraham, 1982). They become incorporated into the continent as allochthonous terrains. When the angle of approach is low, i.e., when a large shear component is present – as is the case for the Hercynides – subduction is assumed to become intermittent, being interspersed with episodes of intense shear and readjustment of the interacting fragments to narrow down the tectonically active zone. During these episodes, volcanics may erupt, fault-bounded basins may subside, become sedimented in this complex environment and may be deformed by further subduction. The leading edges of these fragments would become sites of reactivation, uplift and erosion.

Subduction of the Rheic Ocean eventually led to Cordilleran-type orogeny (ocean-continent collision; see Box 7.3 and Fig. 7.19) or to Himalayan-type orogeny (continent-continent collision) in addition to dextral shear. This occurred at first at sites of promontories, later extending to broader fronts (Dewey, 1977). Crustal shortening is assumed to be taken up largely by thrusting along the listric surfaces which had been laid down during the early geosynclinal stage and which now became reactivated. In a sense, therefore, the pre-lithospheric-stretching positions of the tilted subsided blocks were partially restored.

Such a thrust mechanism generating listric thrust faults has been reported from many regions of the world (e.g., Bally, 1983). The Rhenish Massif, too, bears witness to thrusting during the last stages of the Variscan orogeny (Meyer and Stets, 1975; Degens et al., 1980). Meissner et al. (1981) discovered a strong seismic reflector in the form of a low velocity listric surface at 3–4 km depth dipping to the SEE, which they interpreted to be a well lubricated thrust fault along which large nappe displacements took place (Fig. 7.20). The whole aspects of reverse faults, imbricate structures, and overthrust tectonics in the Rhenish Massif have been thoroughly treated by Murawski et al. (1983). A particularly conspicuous example of a large overthrust zone is the "Faille du Midi" in the northwestern part of the Rhenish Massif (Fig. 7.21). The Caledonian deformed Brabant Massif serves quasi as a buttress against the folding "waves" emanating from the Variscan orogeny in the Ardennes.

BOX 7.3

Subduction Models (Uyeda, 1982)

A system that constantly generates oceanic crust along active rifts must invoke mechanisms to compensate for the resulting gain in surface area. Thus, to maintain a kind of steady state, older oceanic crust has to become either subducted or obducted along convergence zones where moving plates collide. The most plausible candidates for the sites of plate consumption are the deep-sea trenches. It is noteworthy that in the process of plate collision small segments of the subducted slab can be broken off and they glide upon (= obduct) the overriding plate; the ophiolites we already know of are products of obduction.

Uyeda and Kanamori (1979) classified subduction zones according to the nature of the back-arc regions. The two principal categories are:

- continental arc without back-arc basin, and
- island arc having a back-arc basin.

Island arcs can further be subdivided into those having inactive back-arc basins and others where active back-arc basins are developed. However, it has been found useful to consider the continental arcs and the island arcs with active back-arc basins as the two principal endmembers. All other subgroups may be looked upon as hybrids of the two. Commonly one refers to the continental arcs as to the Chilean- or Andean- or Cordilleran-type, whereas island arcs are more familiarly known as the Mariana-type subduction zones.

The two types of subduction zones represent two basically different modes of subduction processes (Fig. 7.19). The term mode, here, refers to the strength of mechanical coupling between the subducting slab and the upper landward plate. In the Chilean-type mode, two plates are closely coupled and for the Mariana-type they are virtually decoupled. The subject matter has been reviewed in Uyeda (1981).

Fig. 7.19. Subduction models (not to scale) (strongly modified after Uyeda, 1982, and Bally and Snelson, 1980)

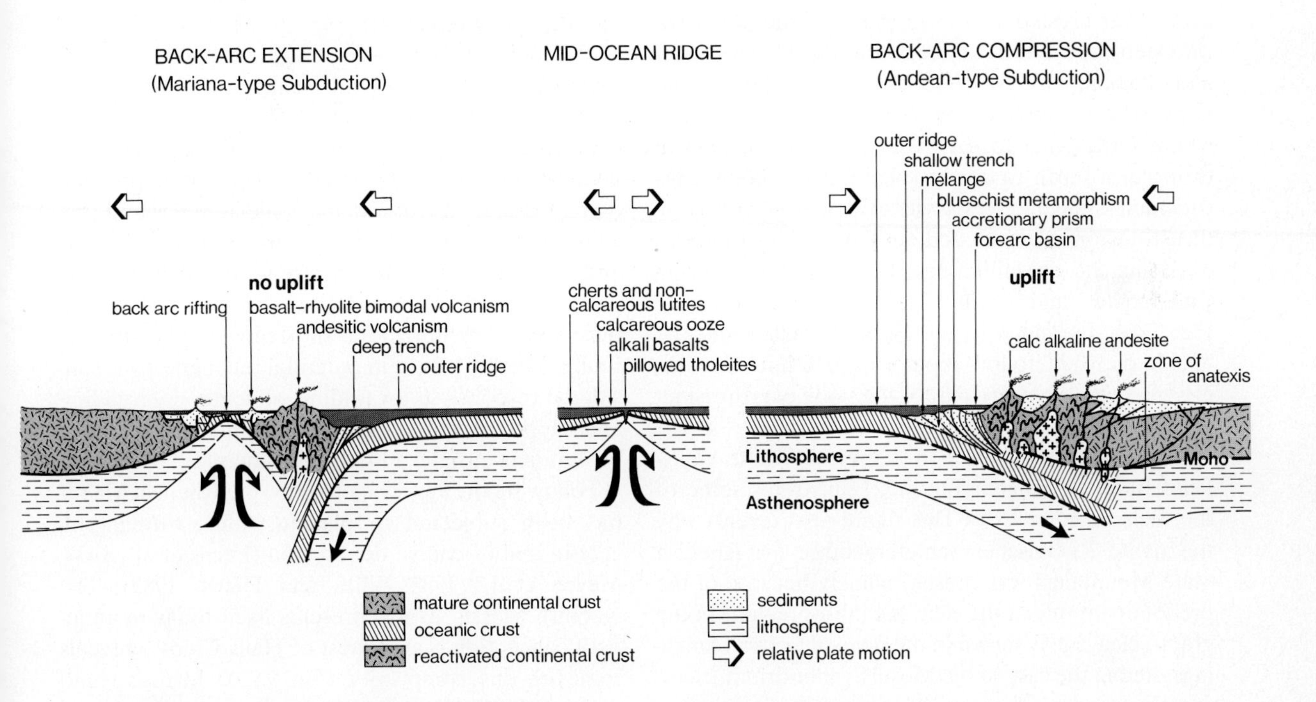

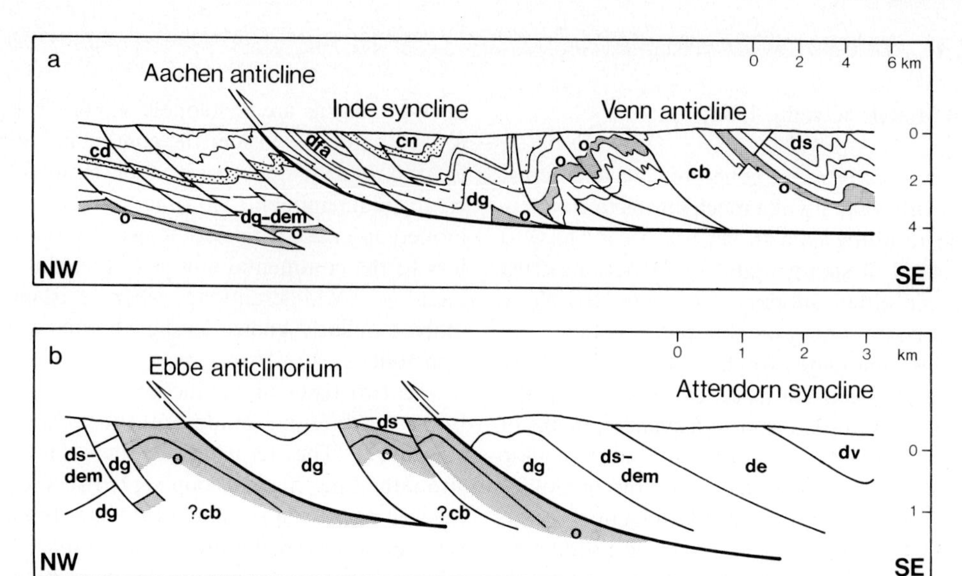

Fig. 7.20 a, b. Geological cross-section near Aachen (NW Rhenish Massif), showing the main thrust fault and accompanying listric faults as deduced from seismic reflection data (**a**) (graph after Meissner et al., 1981). Listric fault system in the Ebbe Anticlinorium (NE Rhenish Massif) (**b**) (Degens et al., 1980). *cb* Cambrian; *o* Ordovician; *dg* Devonian (Gedinnian); *ds* Devonian (Siegenian); *dem* Devonian (Emsian); *de* Devonian (Eifelian); *dv* Devonian (Givetian); *dfa* Devonian (Famennian); *cd* Carboniferous (Dinantian); *cn* Carboniferous (Namurian)

So in a way, wide regions of the Rhenish Massif became overthrusted and even parts of the Northern foredeep were overrun by such nappes. The erosional and tectonic window of the Ebbe Anticline in the northeastern section of the Rhenish Slate Mountains, where Ordovician to early Devonian sediments are thrust in a north to northwest direction along a listric fault system, gives vivid evidence of the overthrust tectonics developed in the rock formations overlying the crystalline basement generated during Caledonian and earlier orogenic events (see Fig. 7.20). Inasmuch as the Ebbe Anticline and the "Faille du Midi" follow about the same line of strike, it is likely that they are part of the same overthrust regime (Fig. 7.21).

A geological cross-section along the Rhine Valley from Bonn to Bingen (Fig. 7.22) illustrates some of the tectonic complexity. This figure also reveals why the name Rheinisches Schiefergebirge (= Rhenish Slate Mountains) was chosen, namely because of the predominant role of the cleavage in this region. In the graph, cleavage is shown in the form of vertical hatchings, and in the case of the Mosel Synclinorium, cleavage gives rise to a kind of pile structure.

Our model takes the mobilistic view. It is developed for the entire Hercynian foldbelt extending from England to Upper Silesia. Other mobilistic scenarios exist which, however, differ in regional details and tectonic conceptions; see references in: Wong and Degens (1983), Martin and Eder (1983), Fuchs et al. (1983), and P.A. Ziegler (1984).

The mobilistic school of thought is opposed by fixistic models (e.g., Krebs and Wachendorf, 1974; Krebs, 1976, 1978; Böger, 1983 a, b). According to these models, gravitative differentiation in the upper mantle gives rise to ascending mantle diapirs that produce dome structures with metamorphic aureoles and preorogenic plutonism. Synchronous with domal uplift and erosion is the subsidence of sedimentary troughs peripheral to the domes. These vertical movements induce secondary horizontal shortening via gravity tectonics. The difference in potential relief energy is considered responsible for folding, allochthonous gliding and nappe transport. Thus, geosynclinal and orogenic evolution is ensialic and intracontinental.

Following the Variscan orogeny, the Rhenish Massif has been subjected to plateau uplift, rifting, volcanism and of course denudation (Fuchs et al., 1983; Meyer et al., 1983; Illies and Fuchs, 1983). The Rhenish Massif, as it represents itself today in an artist's conception – a bequest of Hans Cloos – reveals some of this complexity (Fig. 7.23). Intrusive and high grade metamorphic rocks are virtually absent. The only sign of magmatic activity revealed at the surface are volcanoes of Tertiary and Quaternary age, and the last eruption dates back only 1,000 years. Since volcanoes "punch" holes through the crust and may have their point of origin as deep as the upper

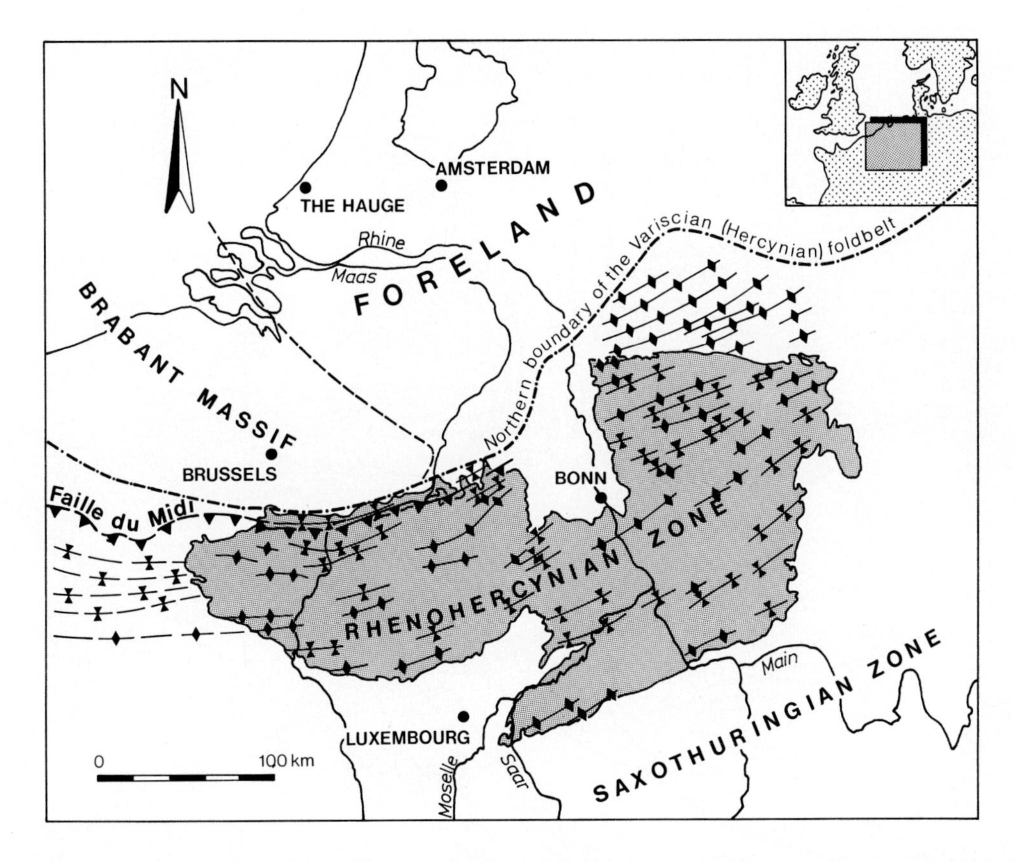

Fig. 7.21. Map of the Rhenish Massif showing the Variscan fold tectonics (after Murawski et al., 1983). The *Faille du Midi* overthrust regime finds – along strike direction – its probable extension in the north-eastern part of the Rhenish Massif in a series of listric thrust faults (e.g., Ebbe Anticlinorium). This would suggest that the Rhenish Massif to a large extent became overthrust during the Variscan Orogeny

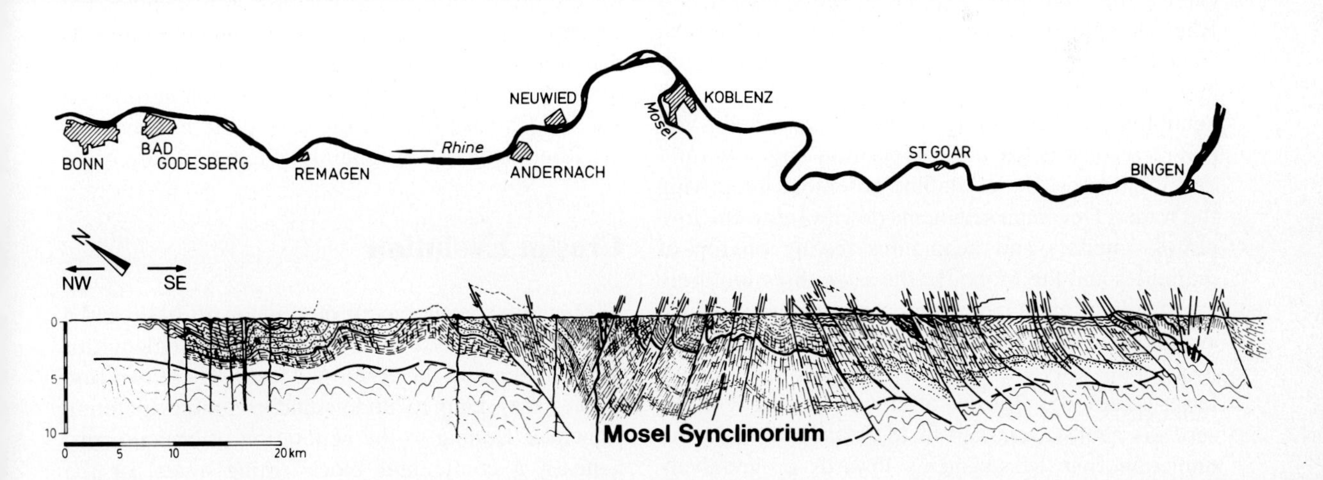

Fig. 7.22. Geological cross-section along the Rhine Valley from Bonn to Bingen (after Meyer and Stets, 1975). Sediments of the lower and middle Devonian, predominantly shales, sandstones and quartzites are thrust upon the metamorphic basement. Tertiary and Quaternary volcanism is sporadically encountered in the region between Bad Godesberg and Koblenz. The Nahe Trough at the southern periphery of the Rhenish Massif (south of Bingen) belongs to a younger tectonic regime, and coarse-grained sediments of the lower Permian (Rotliegendes) represent the basal stratigraphic unit

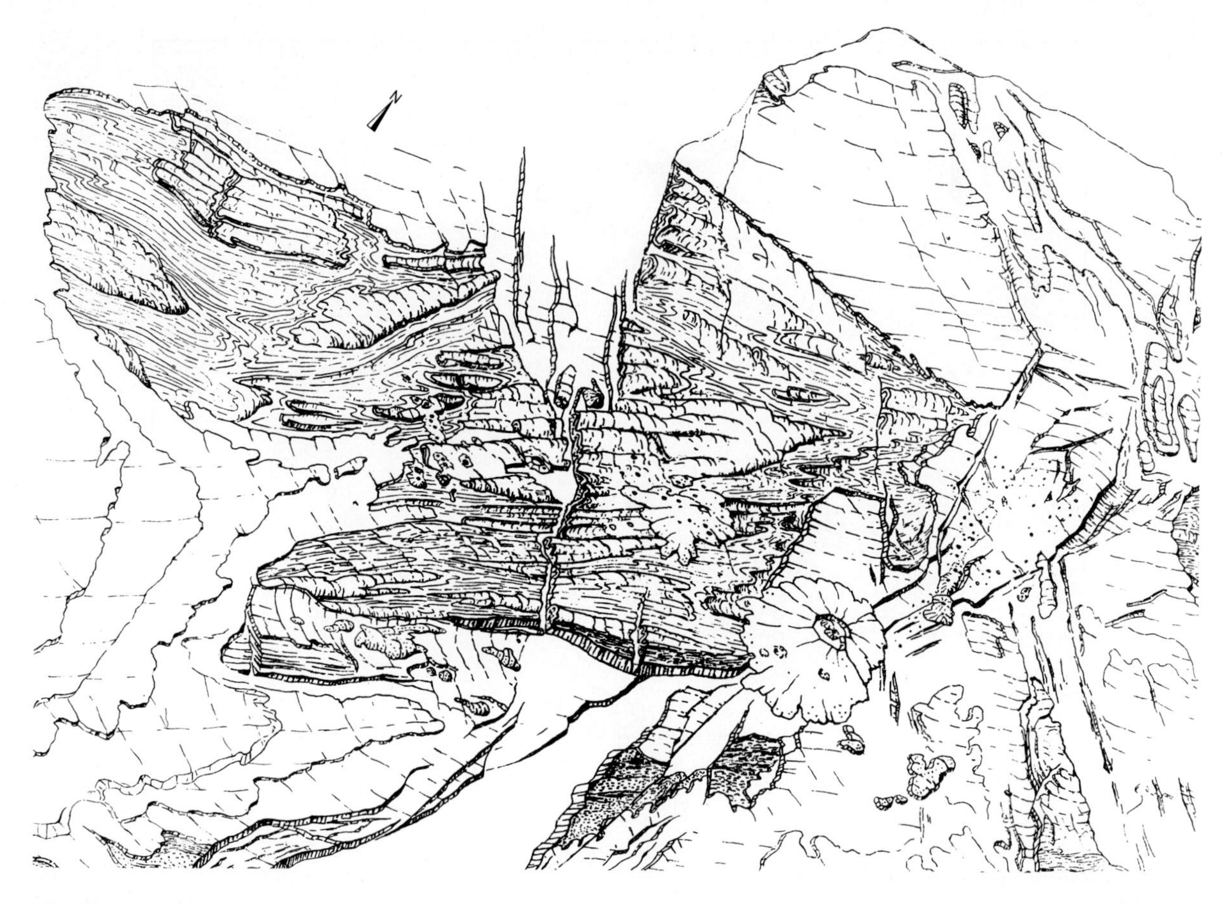

Fig. 7.23. The Rhenish Massif and bordering regions (Cloos, 1955)

mantle, they may carry rock fragments from strata through which the magma passes to the surface. Volcanoes from the Eifel District in the center of the Rhenish Slate Mountains have been searched for such fragments (Voll, 1983; Woerner et al., 1982; Stosch and Lugmair, 1984). On the basis of about 1,000 xenoliths – as these foreign elements are called – Voll (1983) reconstructed the composition and structural development of the crystalline basement underlying the folded Devonian sediments downward to the lowermost gneisses and migmatites resting on top of granulites and the Moho. In the same direction there is a steady increase in regional metamorphism, which according to Voll (1983) suggests that the whole crust is a straightforward Variscan metamorphic Barrow facies series. G. Barrow, after whom such a facies series is named, reported in 1893 that in going from unmetamorphosed sediments towards progressively higher metamorphosed rocks, systematic changes in the type of mineral assemblages and rock types do occur. Fig. 7.24 depicts the reconstructed metamorphic facies pattern beneath the Rhenish Slate Mountains.

The former sediment sequence that became regionally metamorphosed during the Variscan should be Precambrian at its base. The Moho stands at 30 km, which implies that during the Variscan orogeny the crust had a thickness of 30 km plus 4 km to account for the eroded top of the Rhenish Massif. Since the eastern part of the Rhenish Massif contains sediments that were folded during the Caledonian orogeny and furthermore overthrust tectonics is common throughout the Rhenish Massif, the crystalline basement of the Rhenish Massif is certainly of a hybrid nature.

Crustal Evolution

The tectonic development of a major rift basin and a mountain chain has been outlined in some detail for the Red Sea and the Rhenish Massif. These two case studies may help to understand common operation principles leading to the generation of oceanic crust beneath a continental block rifting apart, or the aggregation of geographically widely scattered macro- and microcontinents towards one large land mass. Tectonic scenarios from other parts of the globe and across different time frames may have served equally well. So in a way, the continental and oceanic crust is

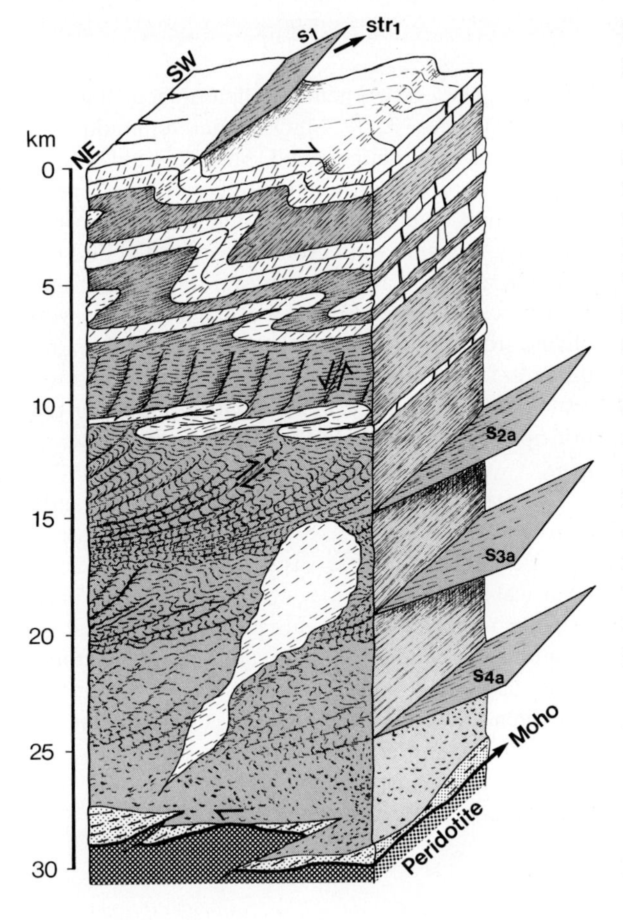

Fig. 7.24. Petrographic and structural development within the crust of the central part of the Rhenish Massif reconstructed on the basis of crustal xenoliths (Voll, 1983). Upper 4.5 km: Rhenish Devonian; 4.5 to 15 km: greenschist facies; 15–20 km: lower amphibolite facies; below 20 km: medium and high grade amphibolite facies and migmatites; 0–1.5 km above Moho: granulite facies; below Moho: peridotites, and light area between 15 and 25 km: pre-Variscan metadiorites

a product of a longer series of tectonic events spread across more than 4 Ga of Earth history. Activities presently displayed along mid-ocean ridges, deep-sea trenches, island arcs or young alpine foldbelts bear witness to ongoing crustal evolution.

The crust underlying the modern sea has been generated over a time span of not more than 200 Ma (see Fig. 7.10). A first reference section over 1 km through layer 2 of the oceanic crust has been drilled (Anderson et al., 1982). Older crust has principally vanished along subduction zones. However, a small number of basaltic "Rosetta Stones" of former oceanic crust are found on land here and there, and help to decipher the nature and the evolution of the ocean's crust and basic magmatism. For the continental crust, the data base is more substantial, since Precambrian shields or Paleozoic carbonate platforms are abundantly exposed on land (see maps on inside of book cover).

The ancient crust has been subjected to mechanical destruction, magmatic resorption or weathering and erosion from the beginning of time. During the formative stages, lasting for about 2.5 Ga, the gain in crustal material was more substantial than the losses. Later on, a kind of steady state conceivably developed where generation and destruction of crust matched each other. At the same time, other scenarios were proposed as well. These processes led to a continued differentiation of matter both regionally and globally, and an Archean crust should thus differ in many ways from a crust that has been generated more recently. It is well to remember that 4 billion years ago the Earth was a more energetic patron, due to higher heat flow; also crust and upper mantle material were more primitive; in addition, air and water looked chemically different from today, where molecular oxygen is plentiful and where the sea is enriched in sodium chloride; and finally, life had not yet evolved to a point that it could spread across the land. Air, sea, and life are active elements shaping the evolution of the crust. In return, the crust exercises a strong chemical influence upon the three environmental elements. Thus, the use of the term *evolution* in the context of the air-sea-earth-life system emphasizes the irreversibility of interactions leading to a progressive differentiation of matter and life.

We stated before that the primitive crust of the Earth was most likely of basic to anorthositic composition. Massive meteorite bombardment and associated crustal resorption prior to 3.9 Ga must have erased the former record. Therefore, the oldest crustal remnants so far discovered are the komatiitic-tholeiitic series of the greenstone belts which rest on a sialic basement (e.g., Anhaeusser, 1971; Goodwin, 1981; Werner, 1984). Greenstone belts are deformed, elongate, metavolcanic and metasedimentary rock formations. They include massive granitic intrusions and substantial gneissic elements.

At least three cycles can be distinguished. They were laid down between 4 and 2.5 Ga. The oldest terrestrial rocks from the 3.8 Ga Amîtsoq gneisses of West Greenland contain metabasaltic and metasedimentary enclaves having an age of close to 4 Ga (McGregor and Mason, 1977). These enclaves were interpreted by Glikson (1976) as relics of the early crust. Whatever viewpoint we accept, the first 500 to 700 million years of Earth history have apparently left no stratigraphic record.

Komatiites are chiefly volcanites of effusive or explosive character that formed subaquatically at water depths probably not exceeding 2 kilometers. They

BOX 7.4

Magma Ocean Model (Nisbet and Walker, 1982)

Green (1975) suggested that local diapiric episodes penetrated the Archean mantle, left the geotherm at great depth (ca. 200 km), rose to the surface and started a komatiite eruption. The "magma ocean" model instead assumes a subterranean pool of komatiitic melt. The boundary layer between the completely molten magma ocean and its buoyant refractory cap was most likely positioned in the 30–50 kbar range (ca. 100 to 150 km depth; Stolper et al., 1981). In the graph it is placed slightly above the 40 kbar level (Fig. 7.25).

Komatiite accordingly erupts from a stagnant, differentiating, magmatic drainage pool rather than from crystalline residue, from which it has freshly segregated upon melting. Eruption of tholeiite can proceed by high-level fractionation of komatiite or convection within the solid upper mantle overlying the magma ocean. Komatiite and tholeiite should be considered as end-members in a continuum between direct piping of the effluent of the magma ocean to the surface, and the ascent of diapirs from the boundary layer. The collection of diapirs at the base of the continental lithosphere could produce a heat pulse sufficient to provoke tonalitic plutonism associated with greenstone terrains.

Indicated on the left of Fig. 7.25 are mantle solidus and liquidus as well as the 32 % MgO komatiite liquidus. The experimentally determined liquidus temperature for 32 % MgO komatiite liquid (ca. 1,650° C) is a minimum value. Should liquid ascend adiabatically, eruption temperature could be higher. As the melt layer cools and differentiates, buoyant crystallized olivine will become incorporated into the cap, whereas the garnet-rich crystal fraction will sink to the ocean's base. Filling the magma ocean with olivine from the top will make it progressively more difficult for the liquid to rise. Komatiite eruption wanes and is replaced by tholeiite eruption as the principal means by which the Earth dissipates excess heat.

The magma ocean model precludes whole-mantle convection during the Archean. If the model is correct, the Archean mantle was highly differentiated on a vertical scale, but homogeneous laterally. The magmaphile radioactive elements produced the heat to maintain the fluidity of the magma ocean. Tectonic consequences are considerable, because the uppermost mantle and crust floated on a molten zone, a situation comparable to the motion of ice floes on water. For the early Moon, a similar scenario has been proposed (Warren, 1985).

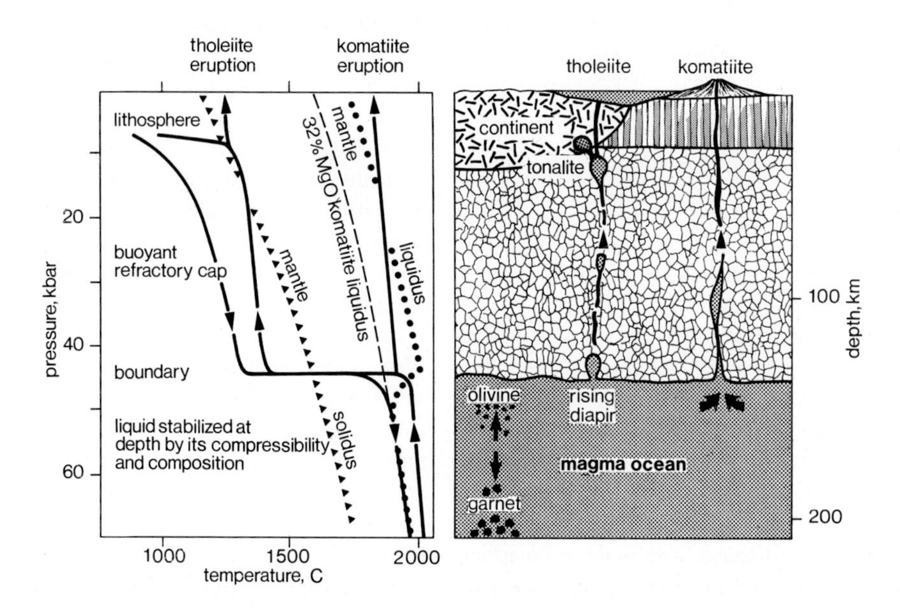

Fig. 7.25. Magma ocean model (Nisbet and Walker, 1982)

piled up on a sialic basement in flow after flow, individually measuring a few decimeters to several tens of meters to yield massive sections 10 to 15 kilometers thick. Based on mineral composition and chemistry, they are peridotitic-basaltic and are interlayered by tholeiites, andesites and various sediments. Green (1975) and Green et al. (1975) suggested high extrusion temperatures for the ultrabasic komatiites of about 1,650° C. Partial melting of a pyrolitic mantle at a depth of about 200 km was considered to be the starting material. According to an alternate model advanced by Nisbet and Walker (1982), a subcrustal "magma ocean" seemed to be the point of origin for the komatiitic lava (see Box 7.4).

In this context the term "magma ocean" has recently also been coined to describe another crustal scenario (Matsui and Abe, 1986). Under this, an impact-induced H_2O atmosphere increases the surface temperature of the accreting Earth to a point where a magma ocean is possible, involving a melting of the top crustal layer once the radius of the Earth exceeds 0.4 R_o, where R_o is the final radius (= 6,987 km) and M_o the final mass of the Earth (6.0 x 10^{29} kg).

Fractionation of chemical elements between a primitive mantle and one that has generated a crust gives geochemists an idea of how much of the primordial matter had to become "digested" and depleted in certain elements to account for all the crustal material (e.g., Ringwood, 1982; Veizer and Jansen, 1979; Engel et al., 1974; Veizer, 1983, 1985; McLennan and Taylor, 1982; Wedepohl, 1975; see Box 7.5). The common assumption is that a primitive mantle layer several hundred kilometers thick would suffice. This would imply that in the course of time the generation of a basic magma within the mantle proceeds in those parts which are progressively depleted in a certain number of chemical elements, except where mantle plumes, subduction or other factors introduce local geochemical anomalies.

For the komatiites which dominated basic magmatism for 1.5 Ga, a number of geochemical trends can be discerned by comparing the three major cycles spaced at about 0.5 Ga intervals (e.g., Naldrett and Smith, 1981; Jahn et al., 1982; Hermann et al., 1976; Wedepohl, 1981 a, b; Werner, 1984). For instance, the light rare earth elements (LREE: La-Sm) decrease towards the younger komatiites, a feature not only observed by going from cycle 1 to 3, but by looking even within one cycle, where the older komatiites are richer in LREE when compared to their younger members (Naldrett and Smith, 1981). This behavior can be explained by a progressive depletion of the upper mantle in the LREE-affine clinopyroxenes which are the first minerals to be molten during partial anatexis (rock melting). This inference seems to be supported by similar trends in the CaO/Al_2O_3 ratios which run from 2–3 (cycle 1), to 1–2 (cycle 2), to 0.6–1.2 (cycle 3).

The generation of komatiitic-tholeiitic magma in greenstone belts took place during a tectonic stage governed by global microplate tectonics (Goodwin, 1981). Convection cells principally driven by the high thermal regime caused a vigorous turnover of upper mantle and crustal matter, leaving behind patches of microcontinents which in time served as "rocky seeds" for the aggregation of larger proto-continents. This tectonic stage was consequently marked by progressive cratonization, i.e., a maximum of consolidation of the continental crust (Dobretsov, 1980; McLennan and Taylor, 1982).

Mention should also be made of a scenario by Newton et al. (1980) and J.V. Smith (1981, 1982). Following the symmetrical linear tectonics that led to the alignment of greenstone belts and high-grade metamorphic regions, an asymmetrical linear subduction of oceanic crust developed along continental margins at the end of the Archean era (3.0 to 2.5 Ga). The shallow late Archean subduction caused melting in both the subducted slab and the overlying wedge of convecting mantle (Fig. 7.28). Global stabilization of continental crust, that is cratonization, at approximately 3.0 to 2.5 Ga, might have been the consequence of flushing out H_2O by CO_2, two molecules known to be critical elements controlling the melting process in mantle and crust (see Box 7.6).

There is much speculation as to from where the "polymerizing" CO_2 might have come. Massive oxidation of organic carbon in a weakly reduced part of the upper mantle, triggered by oxidizing material transported downwards from the crust or perhaps even upwards from the lower mantle, could have been responsible for the large quantities of CO_2 needed. An alternate source is the Archean and early Proterozoic sea itself, which is envisioned as a huge reservoir of dissolved soda – baking powder, so to speak (Kempe and Degens, 1985). That pool, assumed as having a pH around 10 and containing 100 to 1,000 times more dissolved carbonates than the modern ocean, became gradually depleted in CO_2 via subduction of interstitial solutions and extraction of organic carbon. Furthermore, limestones and dolomites that once precipitated from Ca- and Mg-rich hydrothermal waters emanating from the deep ocean floor or rivers and ground waters entering the shallow alkaline sea, could in time end up in subduction zones and orogenic belts. Since it is difficult to transport H_2O deeper than 100 km, unless the sinking slab is particularly cool (Wyllie, 1981), a separation of the two molecules can

BOX 7.5

Episodicity in Petrogenesis (Engel et al., 1974)

An article that has considerably advanced our understanding on crustal evolution and global tectonics is that of Albert E.J. Engel and his colleagues published in 1974. From a broad spectrum of field and laboratory data, they traced major diagnostic rock types present on land through Earth history, made rough estimates on relative volumes, and tried to reconstruct mode of formation and petrogenic and tectonic interrelations among principal rock assemblages.

In Fig. 7.26, some of their findings are summarized. Variations reflect the episodic occurrences of significant refractionative rock-forming and -destroying processes and products. It follows that these changes must also indicate profound variations with time in the patterns, rates, and products of global dynamics. For instance, between 3 and 2.5 Ga the main cratonization event is clearly indicated, followed by a major shift in the relative abundance of certain sediments; or take the first appearance of blueschists at about the Precambrian/Cambrian boundary, which seems to be coupled to a gross thickening, isostatic equilibration, and cooling with time in the lithosphere.

Thickness of crust appears to be related to geologic age and crustal differentiation. Changes in the distribution of large cations, the "incompatible elements" in mantle mineralogy, along the line of differentiation from basic to acidic, are shown to illustrate elemental fractionation patterns that are commonly followed in high-temperature petrogenesis (Fig. 7.27).

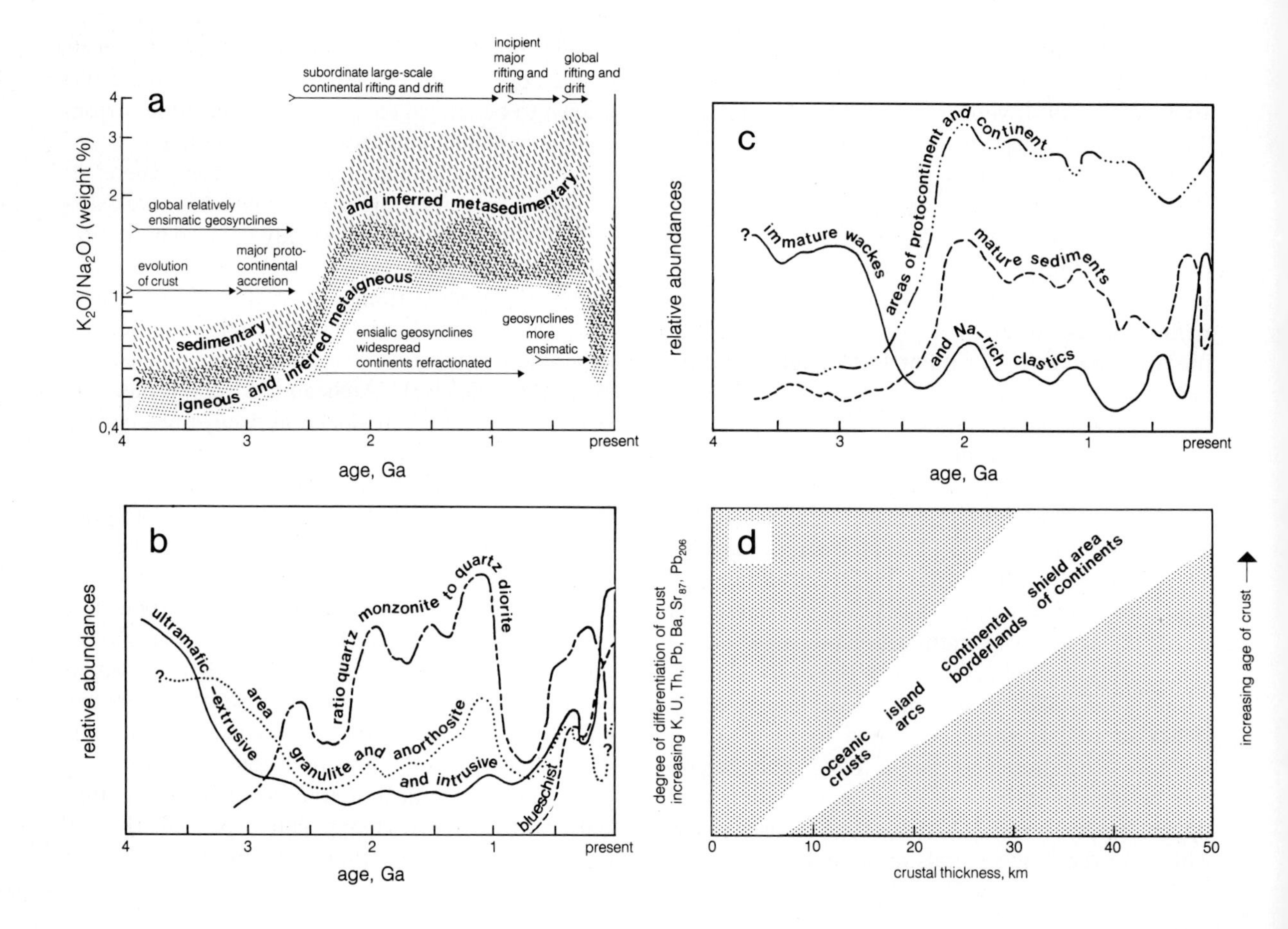

Fig. 7.26. a Variations in the petrogenic index K_2O/Na_2O of igneous, sedimentary, and metamorphic rock complexes plotted as a function of geologic time; **b** approximate relative abundances of several diagnostic rock types and facies formed and preserved in the exposed crust during known geologic time; **c** rough estimates on volumes of immature versus mature sediments and their relation to areas of proto-continent and continent; **d** interrelations in age, thickness, and composition of the Earth's crusts (after Engel et al., 1974)

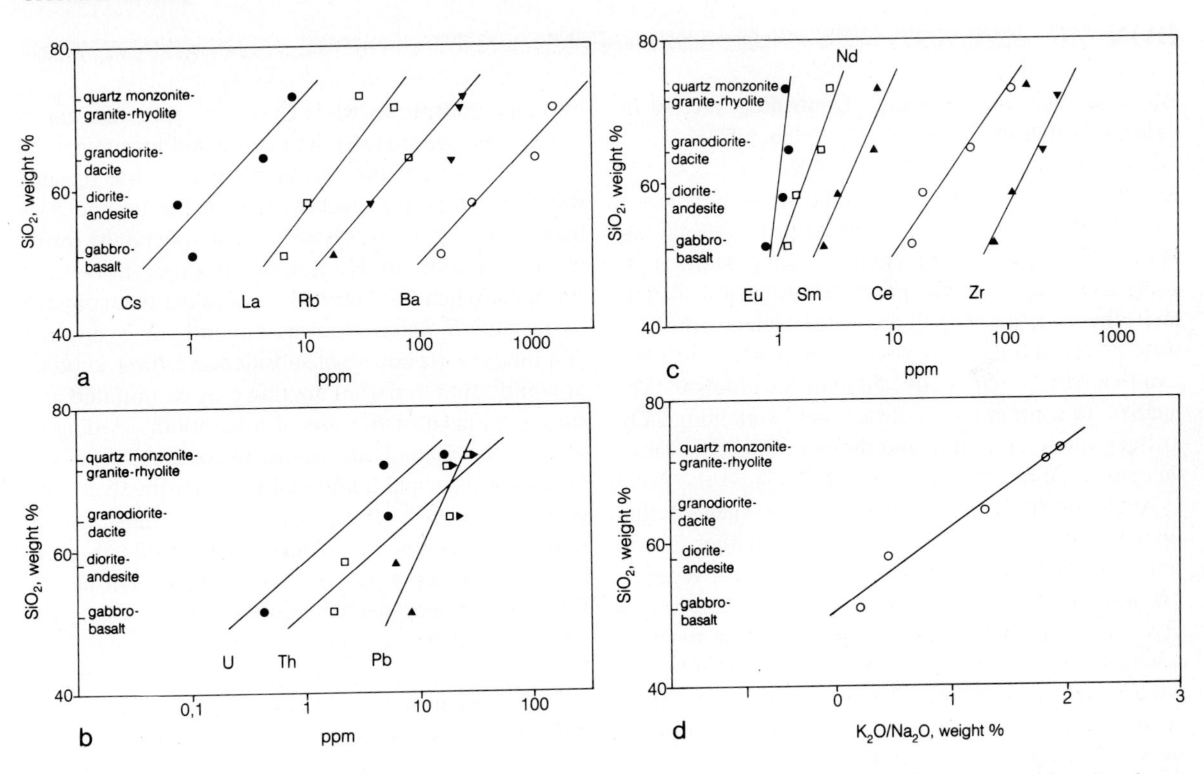

Fig. 7.27 a–d. The content of various large cations (**a–c**) and in K_2O/Na_2O ratio (**d**) in the calc-alcaline igneous rock series of continents and arcs (after Engel et al., 1974)

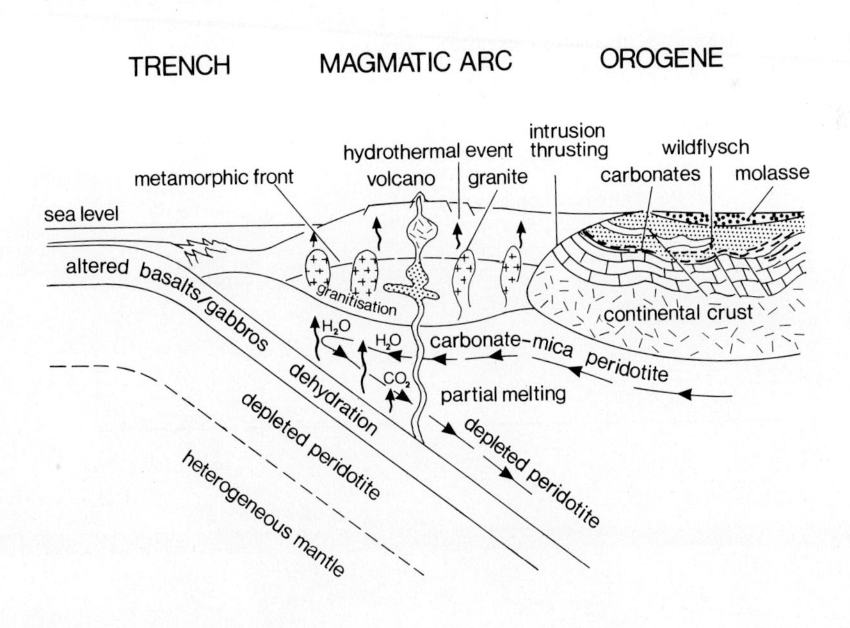

Fig. 7.28. Suggested processes at a continental margin at the end of the Archean (Newton et al., 1980)

BOX 7.6

Water and Carbon Dioxide – Controling Factors in Crustal Evolution (Mysen, 1977; Wyllie, 1977)

Crustal thickness or thickness of the lithosphere is not only a function of the geothermal processes taking place. Results in experimental petrology show that, while CO_2 intensifies the polymerization of a silicate melt, the presence of H_2O promotes its depolymerization. The resulting lower viscosity of a wet melt relative to a "dry" one, is reflected in its lower seismic velocities. In contrast, for a silicate melt containing CO_2 at the same temperature and pressure, higher seismic velocities can be expected (Mysen, 1977).

Analogously, in a peridotitic system such as the upper mantle, the process of partial melting is not only controlled by the depth and geothermal regime, but also by the presence of trace constituents (e.g., H_2O and CO_2; Wyllie, 1977). Such a system may be schematically represented by CaO-MgO-SiO_2, to which traces of H_2O and CO_2 are added (Fig. 7.29). In the presence of H_2O-CO_2, the mantle peridotite with carbonate acts as a buffer, controlling the ratio of H_2O/CO_2 as a function of temperature and pressure. In the absence of H_2O, CO_2 reacts and carbonate is produced. A polymerization of the magma along the lines of Fig. 7.29 therefore results. In the presence of H_2O and CO_2, vapor is retained in the system until melting begins, but fractionation occurs depending on the molar ratio of the two volatiles. This results in a separation and stratification of H_2O and CO_2. H_2O-rich vapor migrates upwards and may even escape into the atmosphere, while CO_2-rich vapor is retained or fixed as carbonate in the mantle and lower crust.

It follows from the arguments above that the upward or downward displacement of the liquidus/solidus of the peridotitic system may simply be the result of the addition of H_2O, CO_2 or other trace constituents. When H_2O is introduced in the form of pore or formation water, hydrous minerals or hydrothermal fluids in the course of subsidence, rifting, subduction and folding, partial melting may be initiated and start volcanism or the rise of a batholith, leading to the familiar granitic intrusions. In contrast, if carbonate is the principal additional material involved in a remelting process, the probability of granite formation would be greatly reduced and lithification enhanced. For example, a subduction of let us say Phanerozoic deep-sea clays poor in carbonates would favor depolymerization at depth, whereas late Mesozoic/Cenozoic calcareous oozes and limestones accumulated in the ocean's abyss would – upon subduction – tend to raise the solidus/liquidus of the upper mantle to higher temperatures and polymerize its upper layer (Wong and Degens, 1983; Böttcher et al., 1982).

Fig. 7.29. Mantle cross-section for the system: CaO-MgO-SiO_2 plus small quantities of H_2O and CO_2. Mantle cross-section with mixed volatiles (case a and b) retain excess vapor until melting begins, whereas if no H_2O is present (case c), all CO_2 reacts to produce calcic dolomite in the peridotite (after Wyllie, 1977)

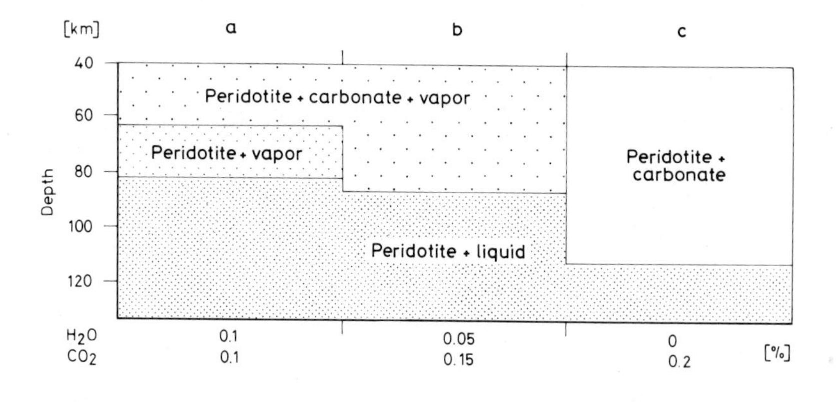

be achieved in this way. The addition of H_2O could lead to melting and the emplacement of granitic plutons, whereas that of CO_2 promoted the growth of cratonic blocks.

The cessation of komatiitic-tholeiitic magmatism at the end of the Archean – 2.5 Ga – might have been related to a thermal exhaustion and the considerable depletion in certain minerals, most notably the clinopyroxenes in the upper mantle (Werner, 1984). On the other hand, basic magmatism continued – although on a much smaller scale – in the form of continental tholeiitic lava and intrusiva, which indicates that in the course of intra-continental rifting and the break-up of land masses partial anatexis of sub-continental mantle material took place. This was the main stage of intraplate tectonics (2.5–1.3 Ga). Huge intracontinental basins of sedimentation, so-called geosynclines, eventually gave rise to ensialic orogens, that is to mountain chains born out of a continental crust. Furthermore, first indications of Benioff-type subduction are revealed. At the end of this stage, the major continental blocks we know of today were in essence created.

The activity of continental tholeiites did not cease when the main stage of intraplate tectonics came to a halt. Two additional maxima are registered, i.e., one during the Proterozoic (ca. 1.1 Ga), and one during the Meso-Cenozoic (250 to 10 Ma). Both events correspond timewise with the onset of continental rifting and drifting that eventually led to the creation of Iapetus and the Atlantic-Indian Oceans, respectively. The plateau basalts, mobilized at temperatures between ca. 1,200–1,300° C at 10–15 kbar in the presence of H_2O, are prominent members of the third cycle: Karroo province (165–200 Ma), Paraná basin (120–150 Ma), Deccan trap (50–65 Ma), and Columbia River province (10–15 Ma). The production of magma reached individually 10^5 to 10^6 km^3.

A small but distinct group are the continental alkali basalts which are tectonically associated with rifts and deep-seated fracture zones. Partial anatexis of mantle material low in H_2O or rich in CO_2 gives rise to a wide spectrum of rocks, for instance alkali-olivine-basalt, kimberlite, carbonatite, and phonolite. Sporadically found in the Archean, and more continuously since the middle of the Proterozoic, they reach their climax during the Cretaceous and Tertiary (Werner, 1984; J.C. Green, 1981).

There is much speculation as to since when plate tectonics has taken over the global tectonic regime. From a petrologic point of view the first occurrence of ophiolites would be a good time marker because ophiolites are thought to represent a segment of obducted oceanic crust of the type presently encountered beneath the modern sea (Lewis, 1983). A schematic cross-section is depicted in Fig. 7.9. If we follow this line of thought, then the oldest complete ophiolite sequence ranging from tholeiitic pillow basalts to gabbros occurs in the Baikal-Muja-Zone, and has an age of 1.8 to 2.0 Ga (Dobretsov and Kepezhinscas, 1981). On the other hand, ophiolites became prominent only since the late Proterozoic (1.3 Ga) and especially since Cambrian time (e.g., Werner, 1981 a, b; Basaltic Volcanism Study Project, 1981). With the crumbling of Pangaea and the gradual emergence of the present sea, the activity of oceanic tholeiites reached its peak. Magma generation along mid-ocean ridges proceeds at a maximum depth of 30 km, 10 kbar and at ca. 1,300° C. It is well to remember that rates of seafloor spreading and subduction are not constant but of episodic nature.

Within oceanic plates, but removed from active plate margins, basic igneous rocks occur which differ from oceanic tholeiites and exhibit instead a certain similarity to intracontinental volcanics. They are bound to local "hot spots" in the mantle. Just as a blow torch can easily burn a hole through an iron plate, a hot spot, too, could cut a small passage way through the lithosphere above, allowing magma to rise to the surface (Anderson, 1981). The outcome are volcanic islands and seamounts. Since plates are moving across hot spots, magma "burns" through the lithosphere, creating chains of volcanoes that crop up from the ocean floor, the older ones long extinct, while those immediately above a hot spot are still alive. A particularly attractive chain of volcanoes developed in the NW Pacific, where the Emperor Seamounts stretch south from Kamchatka Peninsula towards the Hawaiian Ridge. The youngest member of this system is the Hawaiian volcanic group, which started to erupt 5 million years ago.

Spreading will lead to subduction along converging plate margins. On its way down a Benioff zone, the lithospheric slab will interact with the rock formation above. The rate of melting – aside from temperature and pressure – will depend on the type of rocks involved and whether water or CO_2 is present or not. Characteristic magmatite series develop as a function of mode of collision, be it ocean-ocean, ocean-continent, or continent-continent. For example, island-arc subduction will lead to up to four rock series with increasing depth: tholeiitic (IATH), calcalkaline (IACA), H-K (IAHK) and shoshonitic (IASHO). Each series contains basalt (49–51 % SiO_2), basaltic andesite (53–55 % SiO_2) and andesite (58–60 % SiO_2); even more acidic members such as dacite, rhyodacite or rhyolite may be present. Major chemical differences are among others the K_2O contents which increase in basalts from: 0.5 % (IATH), to 1.4 (IACA), to 2.2 % (IAHK), to 3–4 % (IASHO).

Recently, attention has been given to a special rock type found in island-arc settings and commonly associated with ophiolites: the boninites (Cameron et al., 1979; Hickey and Frey, 1982). First described in Cenozoic lavas from the Bonin islands (Western Pacific) by Y. Kikuchi in 1890, their chemical composition is close to that of magnesian andesites. However, mineralogy and texture, as well as high chromium and nickel contents, set this rock type so far apart from regular andesites, that a separate term, namely boninite, is justified. The conditions under which boninite formed were experimentally determined by Green (1975). A water-saturated partial melting of pyrolite (ca. 35 % melting) produced a liquid at 1,200° C and 10 kbar, which chemically conformed with boninite: 54 % SiO_2, 14 % MgO, a Mg/(Mg+Fe) ratio of 0.73, and a CaO/Al_2O_3 ratio of 0.88.

Boninites bear remarkable resemblance to basaltic komatiites, and their similarities in texture are particularly striking (Cameron et al., 1979). It has been suggested that melting of a mantle segment that has been depleted in incompatible elements by previous melting events would require very high temperatures to produce such magnesian liquids (Green, 1975). To overcome this problem a "wet" process is postulated for the boninites, whereas melting of a less depleted Archean mantle under dry conditions and fostered by higher geothermal gradients is thought to yield a Mg-rich komatiitic liquid. Since basaltic komatiites were probably generated jointly with peridotitic komatiites, it is tempting to speculate that perhaps traces of water were still responsible for the formation of basaltic komatiites.

Summing up, the global development of basic magmatism reveals that the individual rock types spread over 4 Ga involved different tectonic settings, geothermal regimes, and mantle properties. Thus, intercomparison with respect to cyclicity, chemical trends, or volume and rates of magma emplacement and crustal growth are at best tentative. In Fig. 7.30 some historic trends are depicted which are based on the work of Dobretsov (1980). For a critical review on the global development of basic magmatism see Werner (1984).

There is considerable uncertainty as to whether the rate of magma emplacement into continental crust matches the rate at which continental crust grows, or whether continental erosion is tuned to magma emplacement. This would – if true – represent a kind of steady state between endogenic and exogenic forces. Recent data has shed considerable light on this problem area by showing that in one representative Archean greenstone – granite terrain (Abitibi-Wawa belt, Superior Province, Canada), 1,500 km in length and 225 km in width – the rate of magma emplace-

Fig. 7.30. Development of basic magmatism in the course of Earth history and its suggested relative intensity (after Dobretsov, 1980, and Werner, 1984). Associated tectonic stages (after Goodwin, 1981) are indicated on the *left margin* (highly schematic and tentative)

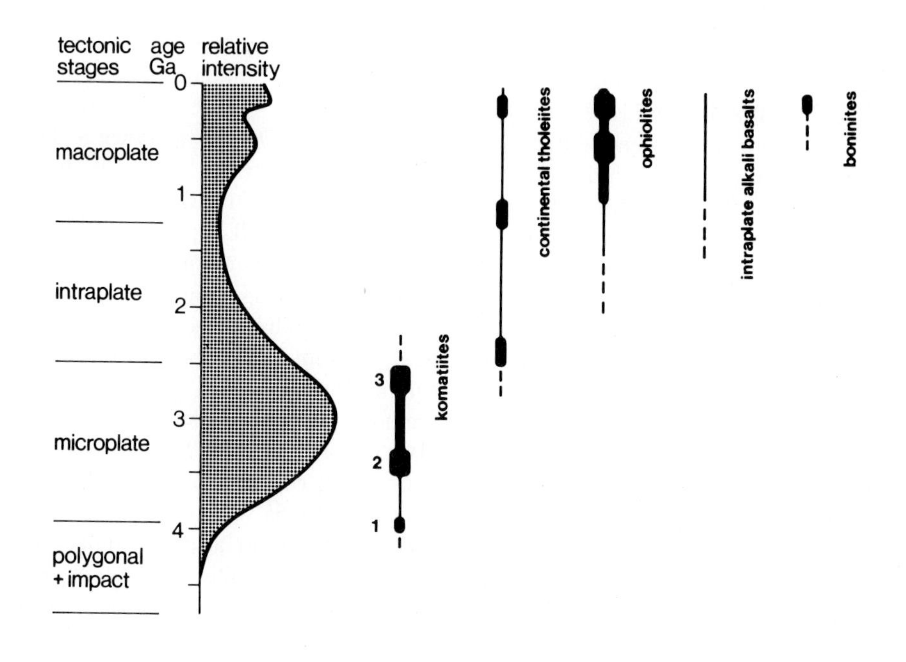

ment is similar to that of Mesozoic-Cenozoic island-arcs (Dimroth, 1985). Both rates exceed estimated growth rate of the continental crust by at least a factor of 2, which argues for rapid recycling of land masses.

For the Mesozoic and Cenozoic, magma emplacement and volcanic output were calculated by Nakamura (1974), Fujii (1975) and Crisp (1984). Crisp's estimates have the most substantial data base and are thus accepted as reliable by many volcanologists (e.g., Fisher and Schmincke, 1984). Accordingly, magma emplacement into the continental crust falls between 3.03 and 10.2 km^3 a^{-1}, and into accreting plate margins between 2.9 and 8.6 km^3 a^{-1}. The length of present-day accreting plate margins is 37,000 km (Reymer and Schubert, 1984). In using Crisp's (1984) normalized rates of magma emplacement for the Mesozoic-Cenozoic, i.e., (8–28) x 10^{-5} km^3 a^{-1} per km strike length or (8–28) x 10^{-5} km^2 per year, there appears to be no substantial difference to the rate of Archean magma emplacement into the Abitibi-Wawa belt at 21 x 10^{-5} km^2 a^{-1} (Dimroth, 1985). At modern rates of magma emplacement, the total continental crust (7.76 x 10^9 km^3, Reymer and Schubert, 1984) could have formed within an interval of 0.76–2.6 Ga.

The disproportion between the rate of magma emplacement and the rate of crustal growth could be related to the episodicity of magmatism. Dimroth et al. (1984), however, assume that globally magmatism is a continuous process, and that a return flow of crustal material into the mantle is responsible for the above disproportion. They suggest that at present rates the equivalent of the mass of the continental crust is recycled through the mantle in roughly 1 Ga, whereby mainly young rocks are involved.

In conclusion, a wealth of data exists on chemical trends in crust and mantle through time. There is no room to spread out fully the arguments on this rather controversial issue. Further selected reading: e.g., Arndt (1977), Arndt et al. (1977), Allègre et al. (1977), Clark and Ringwood (1964), Ronov and Yaroshevsky (1969), Garrels and Mackenzie (1971), Ronov (1972), Veizer and Compston (1976), Windley (1977), Veizer and Jansen (1979), Veizer (1985), Fyfe (1976, 1981), De Paolo (1980), Goodwin (1981), McLennan and Taylor (1982), Ringwood (1982), Reymer and Schubert (1984), Kröner (1985), Kröner et al. (1984), Head and Solomon (1981), Head et al. (1977), Taylor and McLennan (1985).

Chapter 8

From Land to Sea

Boundary Phenomena

Borders separate political blocks, nations, states, communities and also your house from that of your next-door neighbor. Some boundaries are easy to cross, others more difficult to pass. But no matter how liberal or restricted neighboring policies are, all boundaries slow down or may even stop communications. At the other hand, the "lines of defense" displayed, for instance, along a national border will throw considerable light on state philosophies behind it without having first-hand knowledge about the two governments in question. Or take a vicious dog gnashing his teeth on the other side of the fence. He certainly does not encourage a friendly neighboring visit, but will reflect something about his master's state of mind.

A heuristic theorem implies that the physical-chemical phenomena established at the boundary between two different states characterize these two states. For instance, refraction and reflection of a light beam proceeds between two different states. Both states are characterized by boundary surface phenomena which are indicated as circles in Fig. 8.1a. Or the example of the parallelogram of forces that is established between the boundary points of individual states (vectors); it will describe the physical system of a pulley (Fig. 8.1c).

The individual spheres depicted in Fig. 8.2 are well separated from one another. It is a fact that the most critical reactions will not proceed within a sphere, but at or close to its boundaries to other spheres. At the contact air-solid earth we observe weathering. Where water touches ground there is growth of minerals, erosion on land and sediment deposition in the open sea. Vegetation and the human race too are spread out across a thin veneer on land and sea. However, life in its ecological complexity stands on rather shaky ground because of the ongoing destruction of the Earth's surface. Air-sea interactions are responsible for the exchange of water, gases, heat, to name just a few. Biological membranes which can be considered phase boundaries between the organic cell and the outside environment, be it air or water, are chemically and structurally involved in channeling photosynthesis, osmosis, charge transfer (= electron donor-acceptor reactions) and hydration and dehydration processes. Moreover, catalysis including enzyme regulation is a

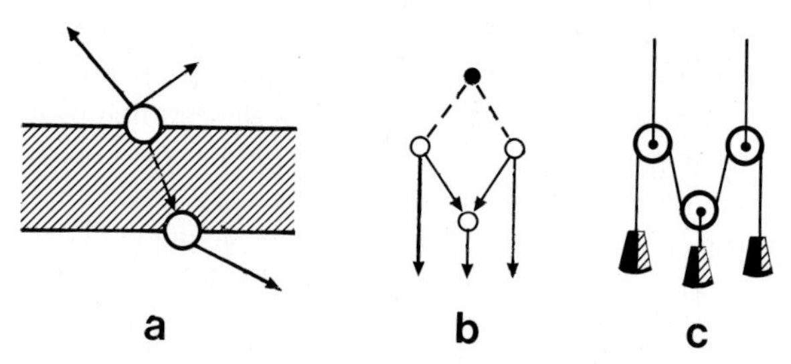

Fig. 8.1. **a** Refraction and reflection of a light beam proceeds between two different states. Both states are characterized by boundary surface phenomena which are schematically shown as open circles; **b** the parallelogram of forces established between the boundary points of individual states (vectors) describes **c** the physical system of a pulley

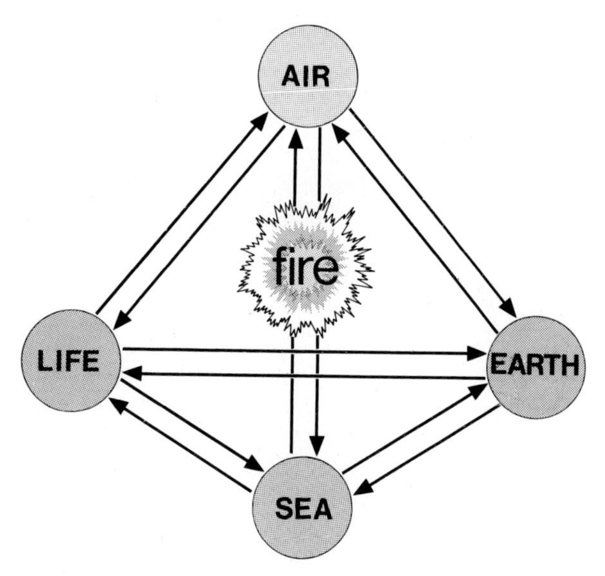

Fig. 8.2. Atmo-, hydro-, litho- and biospheres are separate entities. Without interaction along the peripheries of these spheres driven principally by solar radiation, heat flow or chemical bonds, the air-sea-earth-life system would cease to evolve and eventually "freeze"

boundary phenomenon. It represents a process by which a solid state surface "tries" to establish a thermodynamically favorable phase transition structure with the molecule to be catalyzed. In short, boundary phenomena keep the world alive and going. This is the case for the sciences, too, where the most interesting problems may no longer be found in the centers of physics, chemistry, biology or geology, but on the peripheries of these sciences (e.g., biogeochemistry). Thus, we have to examine closely the phase boundaries between air, sea, earth and life upon which the interactions are written as a kind of menetekel.

We can go even further and inspect the individual sphere. A wide range of layers and boundaries show up. Take, for example, the atmosphere, which is grossly stratified into the familiar: troposphere (0–12 km), stratosphere (12–50 km), mesosphere (50–100 km), thermosphere (100–300 km), and the exosphere. And yet an even closer view will reveal many more layers with distinct physical and chemical characteristics. For instance, the ionosphere at about the boundary of upper stratosphere-mesosphere, is a region of electrification, where molecules become ionized by infalling X-rays and UV-radiation, giving rise to multiple layering. This is a result of different ionization processes and changes in the density of electrons; or take the ozone layer in the stratosphere or the cloud layers or a weather front and so forth. With respect to the solid earth, a multitude of layers and boundaries exist as we have already seen. Concerning the living cell, it is probably no overstatement to say that it consists of a series of individual reaction boxes separated from one another by boundaries (= membranes) that are screening and regulating the flow of matter and energy within the microcosm. And finally the sea, which is physically and chemically stratified, and where the individual layers, are in constant motion, like the sea itself.

Each major boundary certainly deserves attention in a discussion on air-sea-earth-life interactions but obviously not all of them can be dealt with in our text. Since the action of water on weathering, erosion and sedimentation is a central theme in this chapter, major boundaries in and around water will serve as case studies to illustrate the general structural and functional pattern of boundaries. The focus will be on the ocean, and we start with the boundary between air and water.

Air-Sea Exchange

To cross a gas-liquid interface, a molecule needs energy to overcome surface tension. This energy is derived from the kinetic energy of either (turbulent) eddy diffusion or molecular diffusion. Capillary waves were mentioned as a reservoir of kinetic energy (Hasse and Liss, 1980). Experiments reveal that at the phase boundary eddy kinetic energy contributes little – if anything at all – to gas transfer relative to molecular diffusion. As a consequence, the air-sea interface is seen as a strictly laminar layer where only molecular diffusion transports a gas molecule. Air-water interface is a composite of a liquid and a gaseous laminar layer or "film". The air above and the water below such films are well mixed by turbulent eddy motion; molecular diffusion is negligible. The transition from laminar to turbulent is certainly continuous. Nevertheless, it has been found useful to work with a simplified concept where a sharp discontinuity is applied between a purely molecular diffusive film and a purely turbulent reservoir above and below this film (Whitmann, 1923; Danckwerts, 1970; Bolin, 1980).

The gas-flux F through a laminar film of thickness dz is, according to Fick's first law of diffusion, proportional to the concentration gradient dC/dz across the film:

$$F = D(dC/dz), \tag{1}$$

where D is the coefficient of diffusion of gas in the film. Since D has dimensions L^2T^{-1}, a "transfer velocity" $k = D/dz$ (dimensions LT^{-1}) can be defined. For steady state transport of gas across the two layers of the interface we may then write (Liss and Slater, 1974):

$$F = k_g(C_g - C_{sg}) = k_l(C_{sl} - C_l), \tag{2}$$

where k_g and k_l, are the transfer velocities for the gas and liquid phases respectively. C_l is the concentration of the gas in the well mixed liquid phase below the liquid film, C_{sg} and C_{sl} are the concentrations of the gas at the boundary of the two films in the gas and the liquid phases, repectively.

If the exchanging gas obeys Henry's law, then:

$$C_{sg} = HC_{sl}^{max}, \tag{3}$$

where H = equilibrium concentration in gas phase (g per cm^3 of air)/equilibrium concentration of non-ionized dissolved gas in liquid phase (g per cm^3 of water). The inverse of k is a measure of the "resistance" to gas transfer, and has dimensions (TL^{-1}).

Eliminating C_{sg} and C_{sl} between Eqs. (2) and (3), one arrives at an expression for the overall resistance R to gas transfer through the air-sea interface which may be written as the sum of the apparent series resistances (Liss and Slater, 1974):

$$R = R_L + R_G, \tag{4}$$

where R_L and R_G are the liquid and gas film resistance, respectively.

They may be expressed as:

$$R_L = l/k_l = dz_l/D_l \quad (5)$$

$$R_G = l/k_g = dz_g/HD_g, \quad (6)$$

with dz_l and dz_g as thicknesses of the liquid and gaseous laminar film and D_l and D_g the molecular diffusivities of the gas in the respective films.

To illustrate gas exchange between air and sea, the case of CO_2 is taken, where transfer is liquid resistance controlled, i.e., $R_L \geq R_G$. (Examples for the opposite are H_2O, SO_2, or NH_3). For gases, R_L depends on their solubility and chemical reactivity in the aqueous phase. Liquid resistance control on CO_2 exchange is supported by its long atmospheric residence time of several years, whereas that of H_2O is just several days, although their molecular diffusivities in the gas phase are similar (Bolin, 1980).

Because CO_2 transfer is liquid film-controlled, the film's thickness apart from concentration difference between air and sea determines exchange velocities. Film thickness depends on wind speed (e.g., Kanwisher, 1963; Merlivat and Memery, 1983). A roughly quadratic increase of gas transfer with wind speed was observed (Kanwisher, 1963; Liss, 1973) as well as a distinctly linear one (Broecker et al., 1978). At the onset of rough waves there is a sudden increase in gas transfer rates (Jähne et al., 1979; Flothmann et al., 1979). Dampening of waves by a monolayer of oleyl alcohol led to a 80 % reduction in gas tranfer. For wind speeds greater than 9 m s^{-1} the presence of bubbles created by breaking waves is responsible for a jump of gas transfer velocities up to factors of 3 (Merlivat and Memery, 1983).

Evidently the laminar film concept of the gas-liquid interface is an oversimplification of nature, as was nicely expressed by Hasse and Liss (1980): "It is safe to say that the physics of gas exchange processes remains largely unresolved."

Mid-Water Stratification

The ocean water column exhibits a series of mid-water boundaries of physical, chemical or biological nature. They separate water masses and control the flow of matter from surface to sediment. The best-documented examples of layer formation are those introduced by temperature and salinity gradients. For instance, density boundaries, so-called pycnoclines produced by temperature and salinity gradients, cut the water column into an upper and a lower layer. Where such an interface is caused by temperature, we speak of a *thermocline;* should salinity be responsible for it, the term *halocline* is used.

Systematic temperature-depth measurements have been made in the sea for more than 100 years (e.g.,

Fig. 8.3. Temperature profile across South Atlantic in °C (16° S: Brazil to Angola; Fuglister, 1960). A thermocline is established close to a water depth of 200 m. The sharp incision close to 13° W is the central rift valley in the Mid-Atlantic Ridge

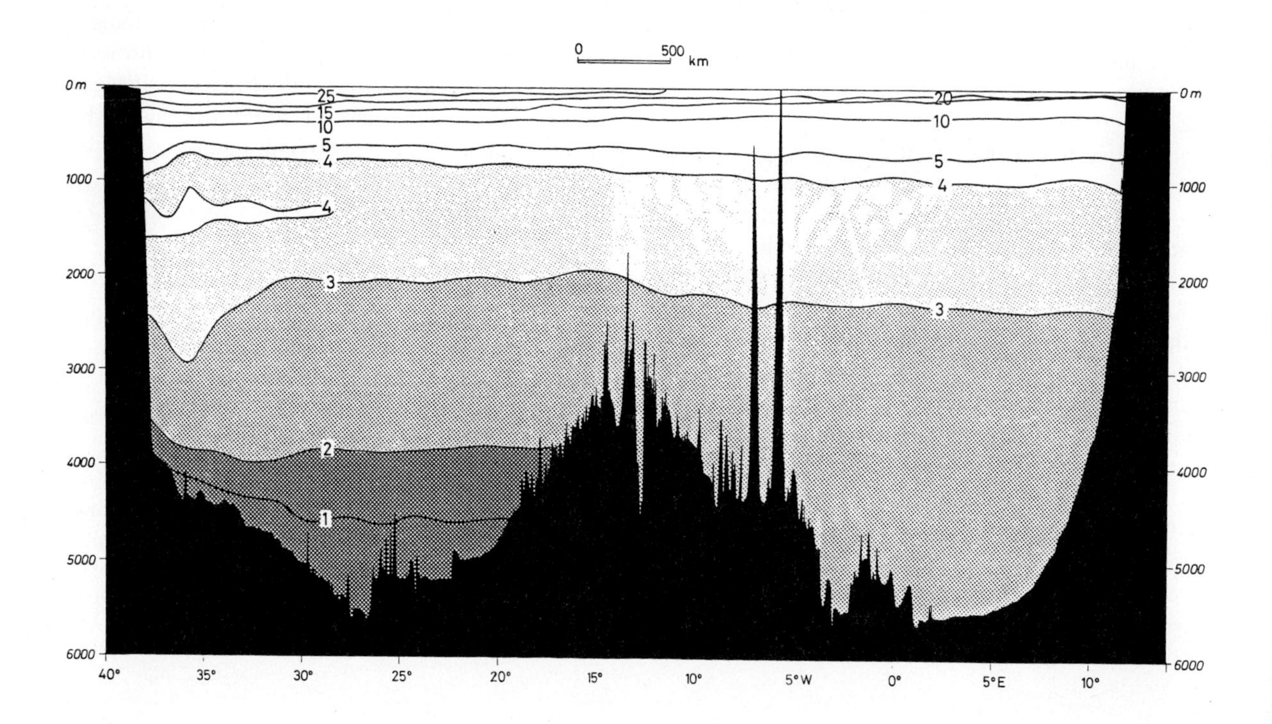

Worthington and Wright, 1970; Fuglister, 1960). A common feature that emerges from that record is that the sea's vertical temperature profile shows a steep decline at a water depth of around 200 m. This is the level of the *thermocline* which decouples the warm upper layer from the cold lower layer. The term euphotic zone is frequently used for that part of the upper layer of the ocean – as well as of a lake – having sufficient light for photosynthesis. As an example of the graduated behavior along the main thermocline in the sea, a temperature-stratified east-westerly section across the South Atlantic is shown (Fig. 8.3).

Thermoclines can prevail for years and centuries without ever breaking up. However, there are others that episodically come and go and by doing so set the stage for the onset of a plankton bloom in the euphotic zone, or the development of anoxia in restricted basins such as the Black Sea or the Cariaco Trench. To illustrate the intimate relationship between the development of a seasonal thermocline, biota and mineral nutrients, the Fladen Ground Experiment, FLEX 76, is briefly discussed (Meeresforschung in Hamburg, 1977).

FLEX 76

The experiment took place within a 100-km-side square almost in the center of the North Sea, where horizontal gradients of physical conditions are small and, therefore, the vertical exchange and transport processes can be properly assessed. The depth in the Fladen Ground area is sufficient (150 m) to permit the development of a dynamically and biologically decoupled upper layer during the time of the experiment (Steele, 1956, 1957). Some critical data from a central station continuously collected over a period of three months are depicted in Fig. 8.4 to show the intricate relationships between water temperature, mineral nutrients, phytoplankton, zooplankton and bacteria. Samples were taken at intervals of 2 to 6 hours; gaps in the profile are related to difficult ship logistics.

In general, phytoplankton growth depends on light intensity, availability of mineral nutrients, turbulent mixing, remineralization, secondary production and adaptation processes (Steele, 1974). Optimal conditions may, for example, arise during the spring season. The principal factor to start the spring bloom in our experiment is the development of a thermocline during the second half of April (Fig. 8.4). The interface permits planktonic cells to remain in the eupho-

Fig. 8.4. Depth and time profiles of water temperature, phosphate, chlorophyll, *Calanus finmarchicus,* and colony-forming bacteria (CFU) at a central station of the Fladen Ground Experiment (FLEX 76) positioned in the center of the North Sea (58°55'N, 0°32'E) (after "Meeresforschung in Hamburg". Illustrated documentation of the "Sonderforschungsbereich 94" at the University of Hamburg, FRG, 1977)

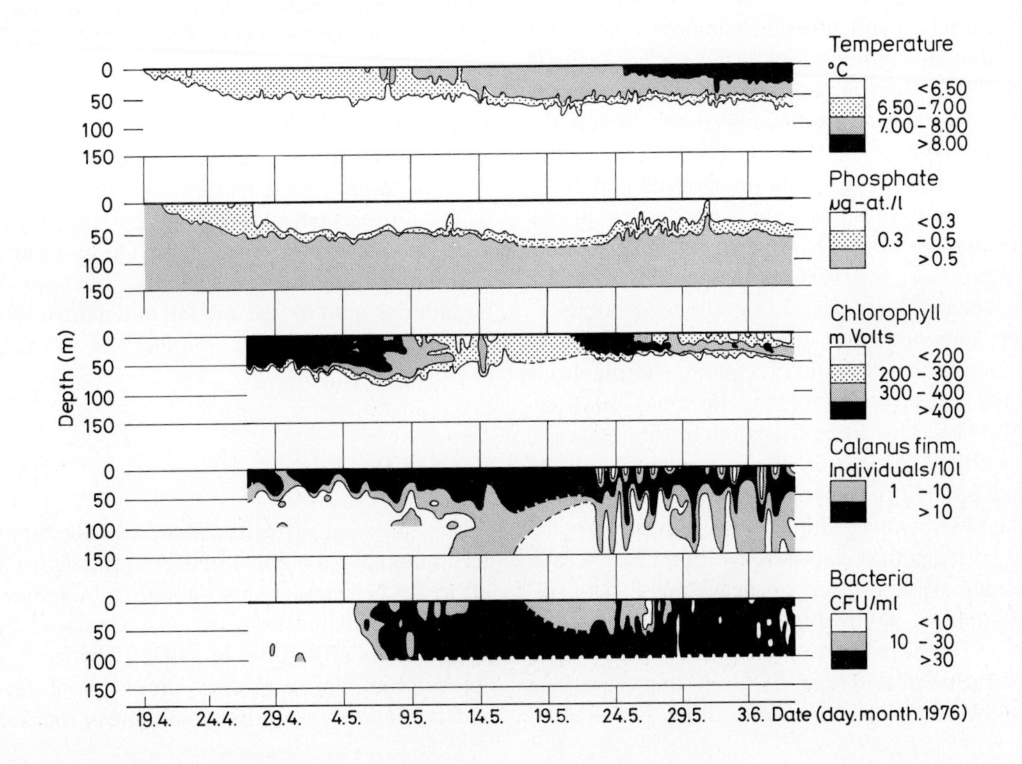

tic layer for an extended period of time and to generate first a micro- and then a macroenvironment for the orderly growth of a plankton community. In the initial phase, massive excretion of dissolved organic matter by the cells seems to be needed to remove toxic metals from the aquatic environment through metal-organic complexation. Following the first phytoplankton maximum (29 April to 2 May), grazers and bacteria appear in large numbers. Whereas changes in mineral nutrients – here exemplified by phosphates – and variations in phytoplankton density – shown as chlorophyll values – are restricted to the upper layer, zooplankton and bacteria living on organic matter traverse freely through the thermocline.

In late summer, with more stormy weather conditions taking over, the whole water column will turn over, and planktonic cells have to await less turbulent conditions and the development of a new thermocline around April the following year to prosper again.

FLEX 76 clearly demonstrates that phase boundaries in mid-water have a pronounced effect on the pattern of mineral nutrients, which in turn control the temporal establishment of a balanced marine ecosystem.

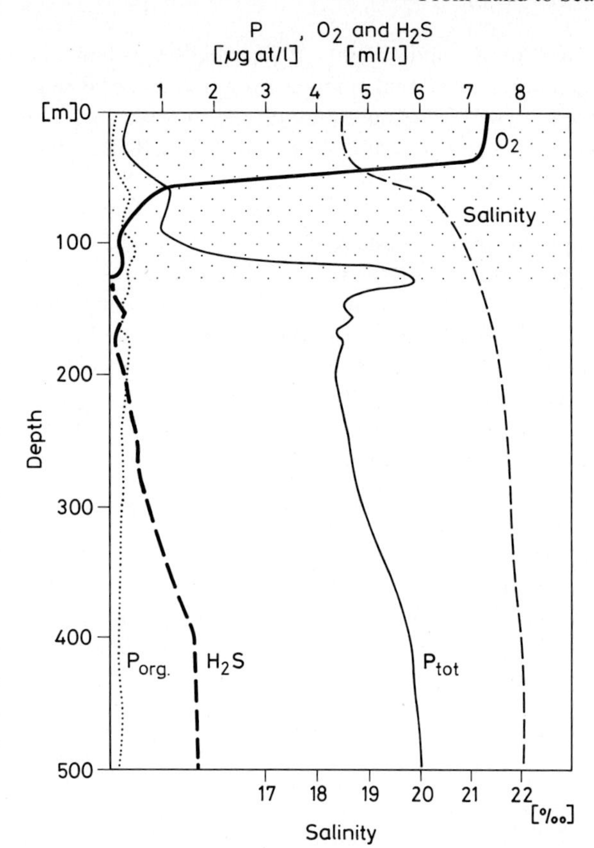

Fig. 8.5. Phosphorus, salinity, oxygen and hydrogen sulfide distribution at station 1466 of 1969 ATLANTIS II Black Sea cruise (Fonselius, 1974). In the Cariaco Trench, the second largest anoxic marine basin known, an identical pattern can be observed for sulfur, oxygen and phosphorus, except that molecular oxygen disappears at a water depth of 375 m (Richards, 1975)

Black Sea

Mechanisms and rate of molecular exchange across a well-developed pycnocline have been studied thoroughly (Craig, 1969; Spencer and Brewer, 1971). Vertical advection and diffusion will move deep water through the density boundary to the surface layer. A particularly well-developed pycnocline exists in the Black Sea which, like a submarine dome, covers the entire sea at a water depth of ca. 120 m in the center and up to 250 m close to shore (Degens and Ross, 1974). Since salinity is the main factor separating the upper from the lower layer, we are principally dealing with a halocline, but vertical temperature changes provide an additional stabilizing element; some researchers thus prefer the term thermo-halocline. As can be seen in Fig. 8.5, the O_2 content sharply drops off at the phase boundary and becomes quasi replaced by H_2S. The dramatic switch in the redox potential within a few meters has a considerable influence on a number of mineral equilibria and the biota. For example, dissolved iron and manganese form hydroxide particles that can scavenge other heavy metals forming a thin layer of heavy-metal-rich suspended matter along the oxicline. Brewer and Spencer (1974) have determined a vertical eddy-diffusion coefficient of 0.014 $cm^2 s^{-1}$. From this value, one can readily calculate the net upward flux of dissolved species such as hydrogen sulfide. Conversely, there is a downward diffusion gradient and transfer of, e.g., oxygen. Models using a constant eddy-diffusion coefficient for the vertical transfer of passive properties in a continuously layered medium (Riley et al., 1949) have only limited applicability, because density boundaries tend to rise and fall and many other physical perturbations exist (Kraus and Turner, 1967; Denman, 1973).

Red Sea Hot Brines

Along the axial rift of the Red Sea, hot brines have accumulated in a number of deeps ranging from about 19° to 24° N. These deeps – about 20 in number – are located in isolated axial troughs, separated by intertrough zones (Bäcker et al., 1975). The most remarkable feature in many of those sites is a well-developed mid-water density stratification. One of the deeps, Atlantis II, at about the height of Mecca, is particularly well-suited to present the general features of such hot brines (Fig. 8.6). (The rather involved story behind its discovery is told in Box 8.1).

At a depth of about 2,000 m, normal Red Sea deep water with a temperature of 22° C, a salinity of 40.6 per mil, and a dissolved oxygen content of 2 ml per liter is decoupled from the underlying brine having temperatures ranging slightly above ambient to over 65° C, salinities of 100 to several hundred parts per mil, and an oxygen content of about zero. A series of mid-water phase boundaries is revealed by sharp acoustic echoes between water layers of different densities (Fig. 8.7). The existence of such layers is characteristic of liquids which are stabilized with salt but are made unstable by heating from below. According to J.S. Turner and Stommel (1964) and J.S. Turner (1969), a stratified system can be maintained in this state because the much larger molecular diffusivity for heat compared with salt allows heat to escape from a layer to cause a convective stirring above, while most of the salt is left behind to maintain a net stable density difference across the interface.

The main interface is located approximately at a depth where Reflector S outcrops (Fig. 8.7). It has also been noted that heat flow along the coast may yield temperatures in the order of 100° C at a depth of less than 2 km. These two observations combined suggest that the discharge of salty water occurs principally along the flanks of the brine pool rather than in the central region of the axial trough. Hydrostatic pressure at times of sea level rise may induce a pulsating flow of formation waters entrapped in the sediments overlying Reflector S, whereby rock salt (NaCl) may be leached from the underlying evaporite sequence. In contrast, heavy metals are principally derived from the hydrothermal leaching of basaltic rock material or were supplied during active spreading phases from deeper regions. Heavy metal fluids and gases are emanating from the sediment/water interface and generate small black smokers which have been dredged from the sea floor.

To summarize, the Red Sea brines can be interpreted as paleowaters (or interstitial waters) which have circulated in the upper lithosphere as hydrother-

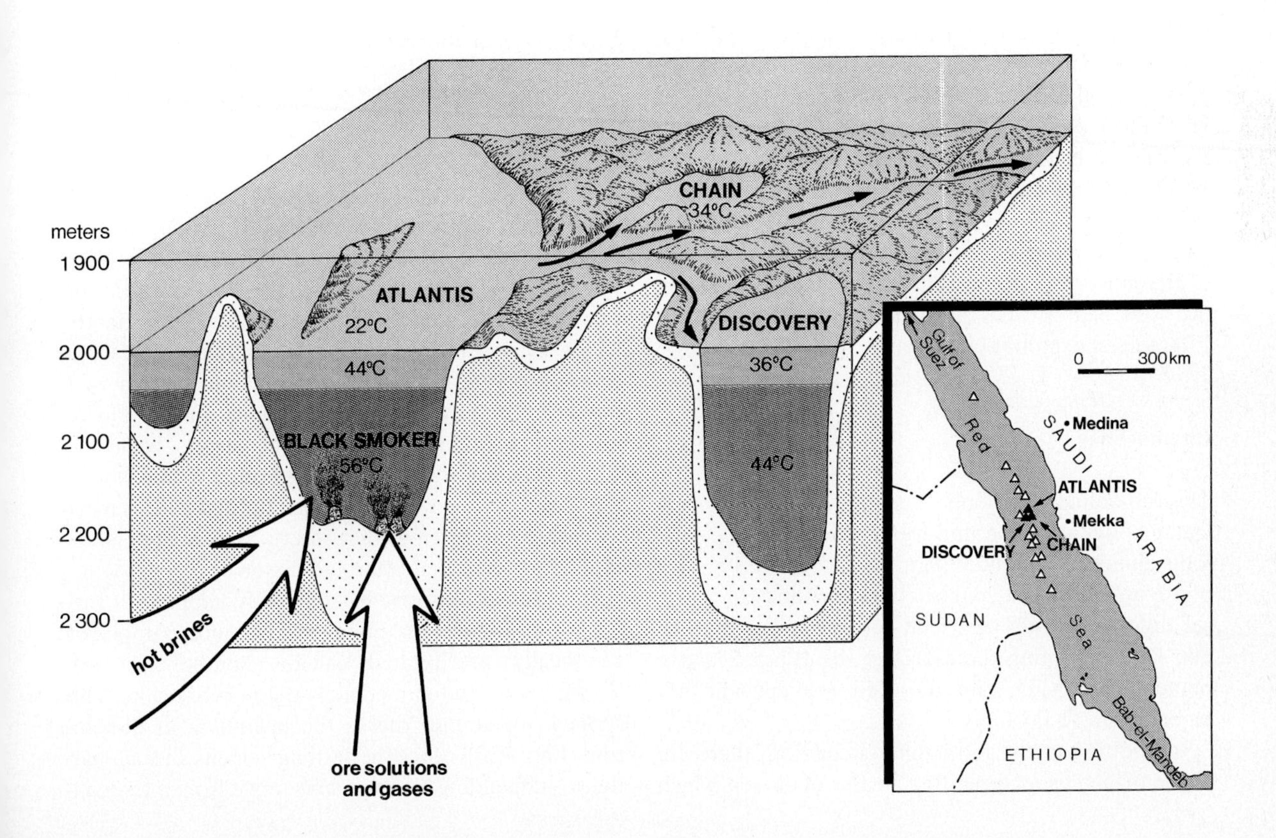

Fig. 8.6. Thermo-haloclines in the central Red Sea axial trough region (after Degens and Ross, 1970). An overflow system is established from the hydrothermally active Atlantis Deep to deeps nearby (*black arrows*). Small-sized "brine geysirs" have been dredged from active Red Sea deeps (Bäcker, pers. comm.) and are schematically drawn in graph (not to scale). Map on the right shows position of known deeps (after Bäcker et al., 1975)

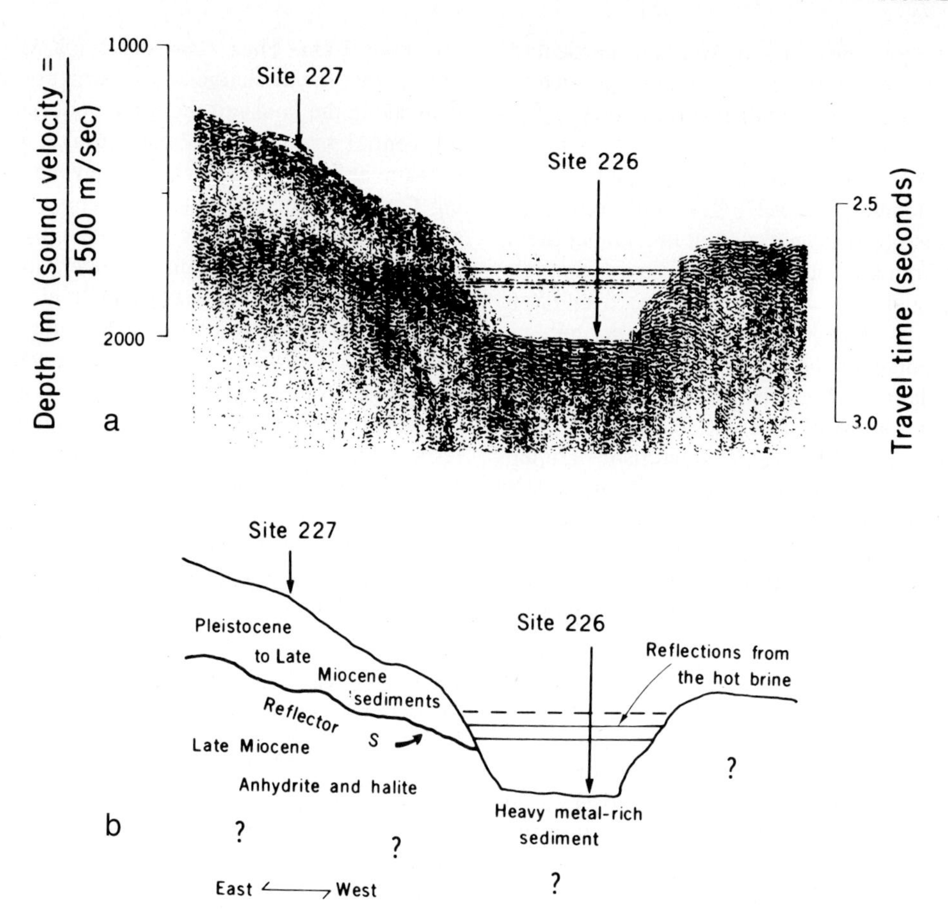

Fig. 8.7 a, b. Echo-sounding record showing acoustic echo between hot brine water and Red Sea bottom water (after Ross et al., 1973; Degens and Ross, 1976)

mal fluids. Their rate of discharge is in part controlled by climatic factors and in part associated with events of active plate accretion (Degens and Ross, 1970, 1976; Ross et al., 1973). Enrichment in helium with the mantle helium isotopic signature (Lupton et al. 1977), and depletion in ^{18}O in some of the deep brines indicative of freshwater contributions (Schoell and Faber, 1978) lend further support for the dual origin of the hot brines and recent heavy metal deposits in the Red Sea. Efforts have been expanded towards commercial exploitation (Bäcker, 1976, 1979).

Chemoclines

Thermo-haline gradients can physically stratify a water body. Should a mid-water stratification persist long enough, the water regimes above and below a phase boundary will acquire characteristics which also set them chemically apart, as was shown for the Fladen Ground spring thermocline, the Black Sea permanent pycnocline and the Red Sea episodic hot brine thermo-haloclines.

In addition to thermo-haline circulation, there are other mechanisms operating in the open sea which profoundly alter water chemistry and introduce a mid-water chemocline. An especially prominent one is the lysocline (Heezen and McGregor, 1973; W.S. Broecker, 1983). In the ocean, the concentration of the carbonate ion (CO_3^{2-}) decreases with depth, which causes an increase in the solubility of calcite (Fig. 8.9). In a very simplified way one can say that the intersect between calcite solubility and carbonate ion content represents the level of the lysocline, where calcite starts to dissolve. In other words, above the lysocline, sea water is supersaturated with respect to calcite, and below that interface it is undersaturated and thus calcite starts to dissolve.

In order to illustrate what actually happens at this chemical boundary, scanning electron micrographs of chemically precipitated calcite rhombohedra are shown above and immediately below a lysocline. The spongy appearance marks the beginning of dissolution (Fig. 8.10). Since the settling velocity of such particles – depending on size – may range from a few cen-

BOX 8.1

Events Leading to Discovery of Atlantis II Deep (cum grano salis)

Oceanographic research can be pursued in a number of ways. A conventional approach is to make thousands of kilometers of traverses across the sea, stop at certain intervals (e.g., 10–100 km), collect a series of water samples from surface to bottom, and measure, for example, temperature, salinity, molecular oxygen content and pH. Every tenth station, a small sediment core may be taken and brought home for inspection and chemical analysis. An alternative approach is to go straight ahead, and often for long distances, to a particular spot, where a scientist has the feeling that some critical events are happening which may throw new light on a general oceanographic phenomenon. The discovery of the Red Sea hot brines and the events that followed are a consequence of a combination of both approaches.

"The trip through the Red Sea was rather uneventful." This statement can be read in an account of the 1948 Swedish ALBATROSS expedition. However, when reading carefully through the list of routine water profiles taken in the Red Sea at that time by ALBATROSS on her way home from the Indian Ocean, one will find the first hint at warm bottom waters in a central deep of the Red Sea at a water depth close to 2,000 m (Bruneau et al., 1953). And yet, it took 10 more years following the ALBATROSS cruise before this finding was appreciated and considered factual after the old ATLANTIS of the Woods Hole Oceanographic Institution revisited the place and confirmed the anomaly (Neumann and Densmore, 1959). In 1963, the new R/V ATLANTIS occupied the same station and again established the presence of unusually warm saline water in one of the isolated axial troughs (Miller, 1964). Since the deep water was just a few degrees warmer and only a few permil saltier than the normal Red Sea water commonly found at such a depth, the interpretation was offered that the salty waters came from coastal evaporating pans and were transferred to the isolated "salty pools" in the form of high-density fluids moving down the slope.

Interpretations changed following the 1964 visit of the British R.R.S. DISCOVERY (Charnock, 1964; Swallow and Crease, 1965). She made the startling discovery of hot, 44° C brines with salinities ten times that of sea water. A few months later, the R/V ATLANTIS II on her way to join the International Indian Ocean Expedition had to cross the newly discovered deep hot brine pool. Even though 4 days behind schedule, the chief scientist Rocky Miller could not resist the temptation to make a hydrographic station and sample the anomalous water. During the searching operation for the already familiar Discovery Deep, the vessel drifted by a few miles and was put above what is now known as the Atlantis II Deep, where at a water depth of 2,160 m, the temperature rose to 56° C and the salinity to 317 per mil (Miller et al., 1966; Miller, 1969). During this operation the two geologists aboard had the excitement of getting their first sediment, quasi as a by-product, because a short corer was luckily used in place of the hydrographic lead weight. Retrieved was a hot core full of a black greasy-looking material. Rumor spread that oil had been struck; for its appearance was more like sludge from an automobile repair shop rather than the deep-sea cores the crew were accustomed to.

The R/V ATLANTIS II was well equipped with all sorts of gear used in physical oceanography. But since no chemical work was envisioned before leaving home, there was nothing aboard, except a bottle each of sulfuric, nitric, and hydrochloric acids as well as one of ammonia. The ingenuity of the two geologists was put to the test to provide analytical chemicals from such materials as one could find in galley provisions and other ships' stores. A vivid account of the hectic days following the retrieval of the hot brine sediments aboard ATLANTIS II and during a stop at Aden has been given by Densmore (1965). In a wire to the director of the Woods Hole Oceanographic Institution one can read the essence of this highly significant finding, "On February 18, 1965, we discovered hydrothermal anoxic brines – 56° C and 300‰ salinity – emanating from the sea floor at a depth of about 2,000 m. The brines are enriched in heavy metals, most notably iron, manganese, copper and zinc, which upon precipitation produce a metal-rich sediment composed of various sulfides and oxides. The sediment appears to be a recent example of some major stratiform ore deposits found in the geological past."

In October the following year, the R/V CHAIN embarked on a 6-week research cruise to conduct a detailed geophysical, geological, geochemical and microbiological study of the Red Sea hot brine area (Hunt et al., 1967; Emery et al., 1969; Degens and Ross, 1969).

Since then, the brine pools have been revisited over and over again, and have become a kind of Mecca to all those interested in hydrothermal ore deposits. No research vessel will pass that spot without retrieving a "bottle" of that unusual hot salty water. In the mean-

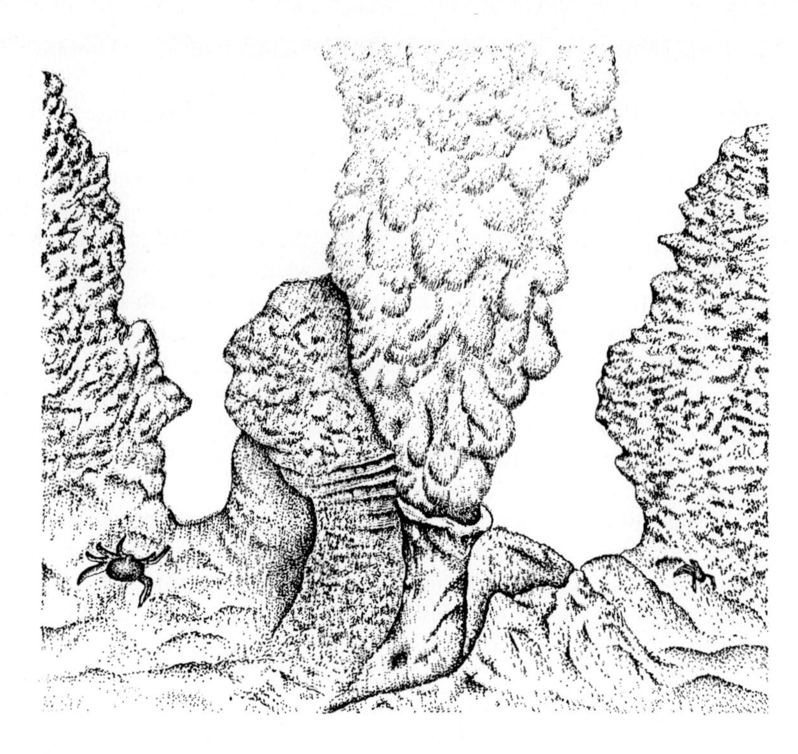

Fig. 8.8. Hydrothermal vent on top of East Pacific Rise at a water depth close to 2,500 m discharging heavy metal solutions from so-called "black smokers" at temperatures of a few hundred °C. Microbial and benthic activity in the vicinity of such vents is high (sketch drawn after photograph in Ballard, 1979)

time, many more brine pools have been discovered in the Red Sea, but none has aroused more scientific curiosity and excitement than Atlantis II Deep. Recent hydrothermal events in other parts of the deep ocean, notably the Galapagos area, have further enhanced our knowledge on the intricate interplay between the sea, the crust, and the deep-sea biota (Fig. 8.8; Ballard, 1979; Jannasch and Mottl, 1985). To be on the fair side, the heavy metal deposits in the central valley of the Red Sea were first discovered by Natterer aboard the Austrian H.M.S. POLA in 1897 (Natterer, 1901). At that time, however, nobody appreciated that find.

timeters to a few hundred meters per day, there is a broad zone in the water where $CaCO_3$ dissolution is going on. The level at which carbonate input from the euphotic zone is balanced by dissolution is termed the calcium carbonate compensation depth, or in short, CCD. Actually, we are dealing with kinetic boundaries which adjust to the amount of infalling calcareous detritus, degree of upwelling or volcanic emanations. CCD occupies positions several thousand meters below sea level (commonly 3–4 km). Since aragonite ($CaCO_3$) is more soluble than calcite ($CaCO_3$), its dissolution starts much earlier. It is of interest to note that CCD can be as much as 5 km below the ocean surface, for instance in the highly productive equatorial Pacific. The consequences to the type of bottom sediments that accumulate are depicted in Fig. 8.11. CCD tends to rise towards the continents, which is most likely related to higher primary productivities stimulating the activity of benthic life and causing high respiration rates (Berger and Winterer, 1974).

In the past, the ocean has experienced many rises and falls of pycnoclines, lysoclines, CCD's and oxiclines (W.S. Broecker, 1983; Van Andel, 1975, 1983; Van Andel et al., 1977; Degens et al., 1984a). These changes have profoundly affected the global carbon budgets in many ways. Their presence and level of position appear to be one of the most determining environmental factors controlling the CO_2 system – including life – in the course of time.

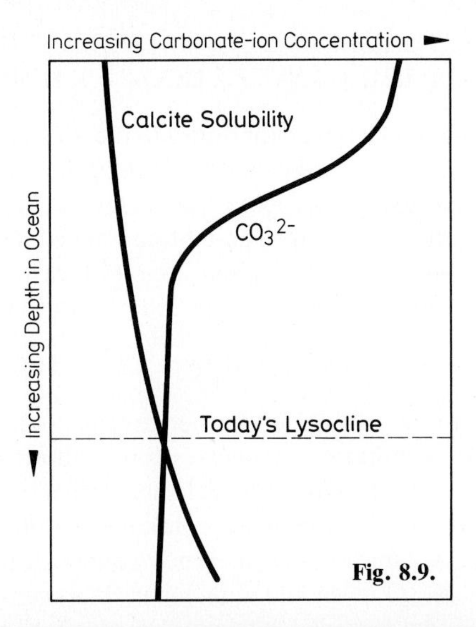

Fig. 8.9.

Fig. 8.9. Calcite solubility in world ocean in relation to CO_3^{-2} content. Today's lysocline is positioned at about 3 km in the Central Atlantic (Broecker, 1983)

Fig. 8.10. Scanning electron micrographs of chemically precipitated calcites (*right* deposited above Black Sea pycnocline; *left* their appearance below pycnocline); note onset of dissolution indicated by spongy appearance. Size of individual crystals about 5 to 10 μm (Degens and Stoffers, 1977)

Fig. 8.11. Dynamic model (schematic) showing the effect of the carbonate compensation depth (CCD) on the kinds of bottom sediments that accumulate. Note the movement of new crust which is indicated by a vertical striation pattern reflecting magnetic reversals. Seafloor spreading will move areas into and out of the influence sphere of the CCD. In the event of euxinic conditions developing, due to density stratification, the lysocline and CCD will gradually move upward towards the pycnocline and cause dissolution of infalling $CaCO_3$ detritus (after Heezen and McGregor, 1973)

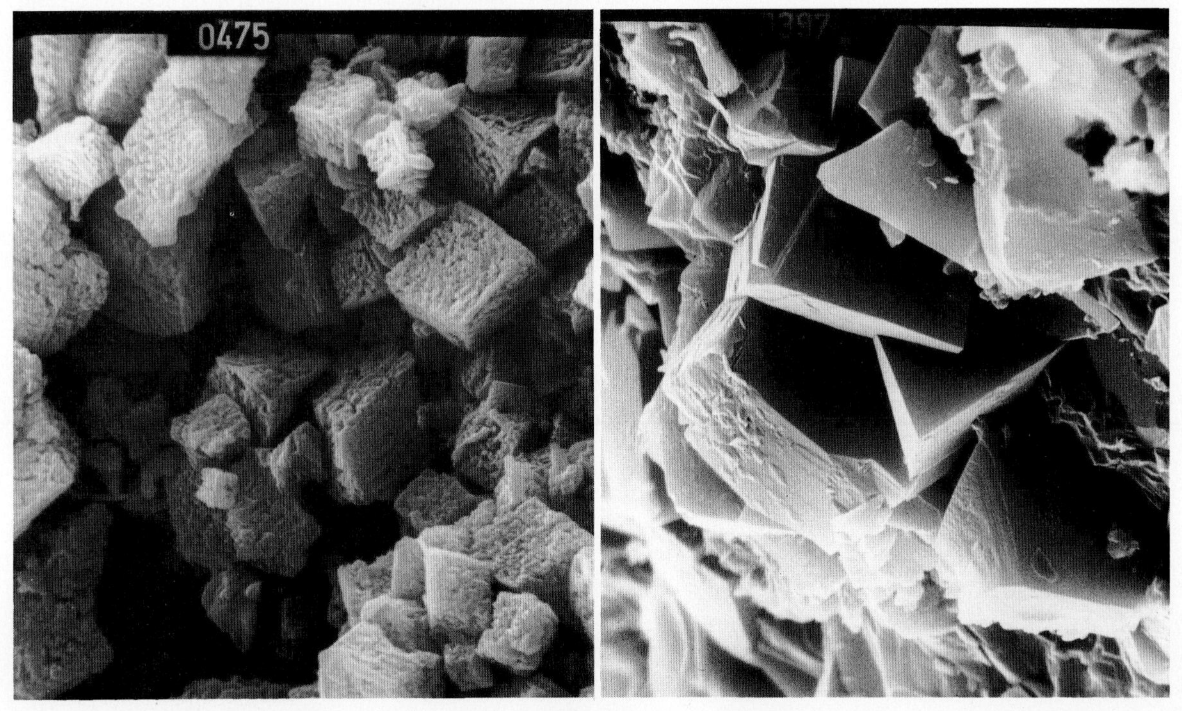

Fig. 8.10.

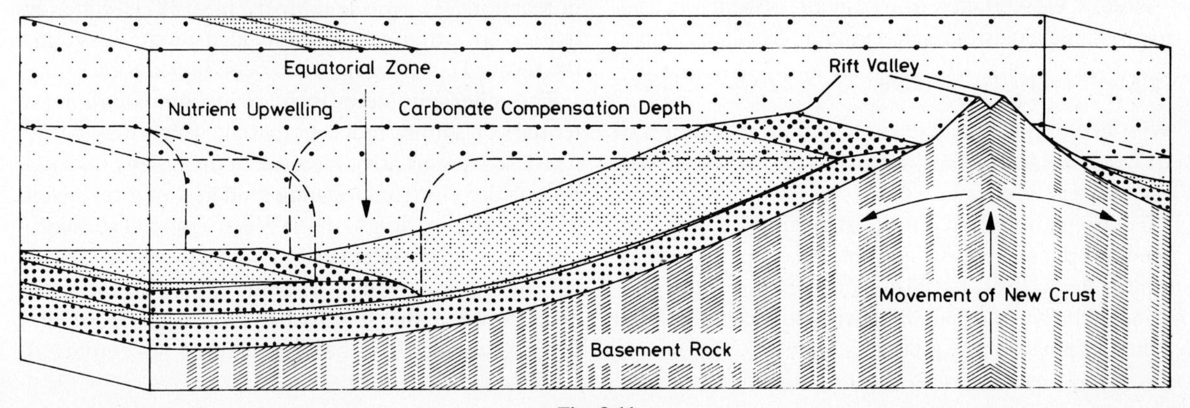

Fig. 8.11.

Weathering and Erosion

The land surface, a coast line, or the seafloor represent boundaries to water and/or air. And yet, life is omnipresent in those lines and layers too, even in remote deserts, on ice caps, or in mountain "high" and abyss "low". So in a way, the Earth's skin is an intimate contact place for air, water, minerals and life to meet. Consequently, many things happen along and within this thin layer measuring a few decimeters or meters at most. The formation of soils and sediments is a result of this interaction, and weathering, erosion, and sedimentation are the processes involved in the transformation of a solid rock into soil or a sediment.

Weathering

Textbooks usually distinguish between mechanical and chemical weathering. The physical break-up of rocks and minerals frequently follows lines of weakness such as joints, fissures, or mineral cleavages. Furthermore, freezing and thawing or heating and cooling may cause fragmentation of solid rocks. However, chemical weathering commonly sets the stage for the mechanical crumbling of rock masses and the development of a soil.

The most important chemical ingredients involved in the weathering of a rock are: water and CO_2. The first is supplied by rain, the second stems principally from root respiration and microbial degradation of organic matter, which varies the CO_2 partial pressure in sediment and soil air to a level several orders of magnitude above that found in our atmosphere. Let us now take a few common primary minerals and follow their fate during weathering. In the presence of carbonic acid the Mg-olivine forsterite disintegrates into:

$$Mg_2SiO_4 + 4\,CO_2 + 4\,H_2O \rightarrow 2\,Mg^{2+} + 4\,HCO_3^- + H_4SiO_4$$

4 moles CO_2 are taken from soil air, of which two return when magnesium is precipitated as magnesite:

$$2\,Mg^{2+} + 4\,HCO_3^- \rightarrow 2\,MgCO_3 + 2\,CO_2 + 2\,H_2O.$$

With every mole of forsterite weathered, 2 moles of CO_2 become fixed from the gaseous pool in the form of carbonate.

Feldspars, the most abundant group of minerals in the lithosphere, are principally members of the ternary system: $NaAlSi_3O_8$-$KAlSi_3O_8$-$CaAl_2Si_2O_8$. The weathering of the Na-feldspar albite will produce an alkaline solution, plus the clay mineral kaolinite:

$$2\,NaAlSi_3O_8 + 2\,CO_2 + 11\,H_2O \rightarrow Al_2Si_2O_5(OH)_4 + 2\,Na^+ + 2\,HCO_3^- + 4\,H_4SiO_4.$$

The consumed CO_2 will not return to the atmosphere. For the K-feldspar orthoclase the weathering pattern is the same, except that potassium is set free. In contrast, weathering of the Ca-feldspar anorthite will yield kaolinite, 1 mole Ca and 2 moles bicarbonate, but no dissolved silica. Upon precipitation of $CaCO_3$:

$$Ca^{2+} + 2\,HCO_3^- \rightarrow CaCO_3 + CO_2 + H_2O$$

1 mole of CO_2 is returned to the gaseous pool. Consequently, weathering of silicates forms an effective sink for CO_2. It has been calculated that silicate weathering on the continents consumes within 7,000 years one volume of CO_2 presently contained in the atmosphere. One should remember, however, that this CO_2 is made available to a soil via respiration and organic decay and does not come directly from the atmosphere.

Kaolinite is not the only clay mineral generated during weathering. Actually the wide range in the structural and chemical composition of clay minerals indicates that there are several ways by which clays can be produced. According to Keller (1956), clay minerals may be considered fundamentally a product of the parent material and the energy impressed upon it.

The average chemical composition of some of the common parent materials is presented in Table 6.3. The data are largely taken from a review paper by Wickman (1954), from where also the calculation procedures for the numbers in the columns of Table 6.3 can be obtained.

Most dominant as a factor of energy in aqueous environments are the activities of electrons and protons; they will ultimately determine the direction of the alteration processes. As the activity of hydrogen ions is involved in, or can with little effort become involved in reactions with dissolved species, it has been found convenient to use the a_{H^+} as a characterizing variable to provide a common reference activity for a variety of reactions (Garrels and MacKenzie, 1971). Since activities are often expressed as logarithmic functions, the term pH = (-log a_{H^+}) has been introduced.

The other parameter is the reduction-oxidation potential, or simply redox potential, which may be defined as a quantitative measure of the energy of oxidation or the electron-escaping tendency of a reversible oxidation-reduction system. When referred to hydrogen, the redox potential is commonly expressed as Eh, in terms of volts – E being the potential difference between the standard hydrogen electrode and the system which is being measured. According to

Baas Becking et al. (1960), four types of reaction are conceivable:

(1) neither electrons nor protons are involved:
$Fe_2O_3 + H_2O \rightarrow 2\,FeO \times OH$

(2) protons only are involved:
$H_2CO_3 \rightarrow H^+ + HCO_3^-$

(3) electrons only are involved:
$Fe^{2+} \rightarrow Fe^{3+} + e^-$

(4) both electrons and protons are involved:
$FeSO_4 + 2\,H_2O \rightarrow SO_4^{2-} + FeO \times OH + 3\,H^+ + e^-$.

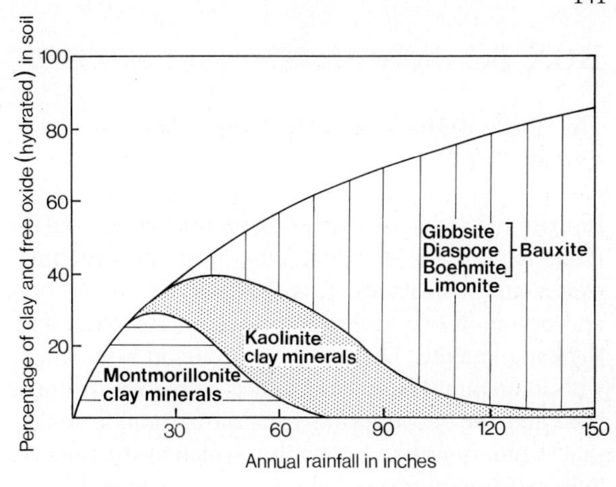

Fig. 8.12. Distribution of clay minerals, and bauxite and laterite minerals in soil samples from Hawaii as a function of rainfall (Sherman, 1952)

Most reactions in natural environments are of type 4 and therefore depend on both the pH and Eh of the system in which they take place, and water seems to be a prerequisite.

Concerning the mechanism of clay mineral formation from a suitable parent material, the most pronounced chemical changes are the following: hydration and hydrolysis of mineral matter, such as feldspars, is the first step towards altering parent materials in the direction of clay minerals. By substitution or ion exchange hydrogen ions will replace some of the ionic constituents present in the parent material, thereby initiating a gradual breakdown of the original silicate structure. This process is accompanied by equilibrium reactions along the intergranular film, which will preferentially remove some of the original ions present in the silicate structure but leave others unaffected. It all depends on the activity of the dissolved species which, in an ideal solution, would be proportional to its concentration in terms of mole fraction.

In conclusion, it can be said that the efficiency of the various chemical reactions leading to the hydrous decomposition of parent materials and the subsequent formation of clay minerals from the decomposition products is a function of: (i) parent material, (ii) general environment in terms of chemical composition, pH, Eh, and temperature, and (iii) water circulation, i.e., removal of soluble materials and supply of "fresh" rain water to the system activated by biogenic CO_2. Some of the parameters under (ii) and (iii), i.e., temperature, rainfall and humidity, are more popularly known by the term climate. (Supplemental reading: Keller, 1957; Degens, 1965; Selley, 1978; Leeder, 1982).

To assist in the evaluation of the factor climate in the generation of clay minerals, data collected by Sherman (1952) of Hawaiian soil samples are taken, where the parent material, a basalt of the same petrography and age, had been subjected to a wide spectrum of weathering conditions ranging from a desert to a moist tropical environment (Fig. 8.12). At no or only moderate precipitation, montmorillonite is the principal clay obtained. A further increase in annual rainfall causes the formation of kaolinite at the expense of both montmorillonite and basaltic rock fragments. In the final stage, montmorillonites disappear completely from the soil, kaolinites decrease sharply in abundance and a progressive development of bauxites and laterites takes place. That is, basalt and clay minerals are gradually eliminated and replaced by Al, Fe and Ti oxides and hydroxides. Bauxite and laterite are the terra rossa of tropical and subtropical regions (Box 8.2).

Erosion

What drives this rather powerful force called erosion that wears down mountains? In the first place it is high relief which provides the kinetic energy to let rocks roll down the hill, keep rivers flowing, start glaciers moving or provide surfaces for the wind and the surges to act upon. Thus the higher the mountain, the more forceful should erosion be – unless water is a limiting factor.

The high mountain ranges of the world, the Rockies, the Andes, the Alps or the Himalayas, are geologically young features compared to the more hilly Appalachians or the Rhenish Slate Mountains, that formed a few hundred million years ago. Actually, we are still witnessing the elevation of mountains in the alpine fold belts, but at rates of at most a few millimeters a year. The reason why mountains grow only

BOX 8.2

The Early Tertiary Bauxite Event (Valeton, 1983; Prasad, 1985)

Bauxite is a residual weathering product composed of more than 50 % of aluminum-, iron- and titanum oxides and hydroxides. It is the main aluminum ore and occurs in two varieties: (i) lateritic bauxite, and (ii) karst bauxite. The first type occurs in association with aluminum-silicate rocks such as basalts or granites, and the second kind rests on carbonate rocks, that is limestones and dolomites which at the time of bauxite formation stood close to the ocean.

Lateritic bauxites are globally by far the most predominant ones and cover wide areas of South America, Africa, Australia and India (Frakes, 1979; Fig. 8.13). In contrast, karst bauxites are less abundant but common in Europe (Bardossy, 1981).

The chemical steps involved in the generation of bauxites from Al-silicates are schematically shown in Fig. 8.14. The origin of karst bauxite is less certain. One line of thought is that bauxite minerals precipitated from CO_2 charged weathering solutions when these passed through carbonate rocks. Dissolution of carbonates due to high PCO_2, resulting changes in pH

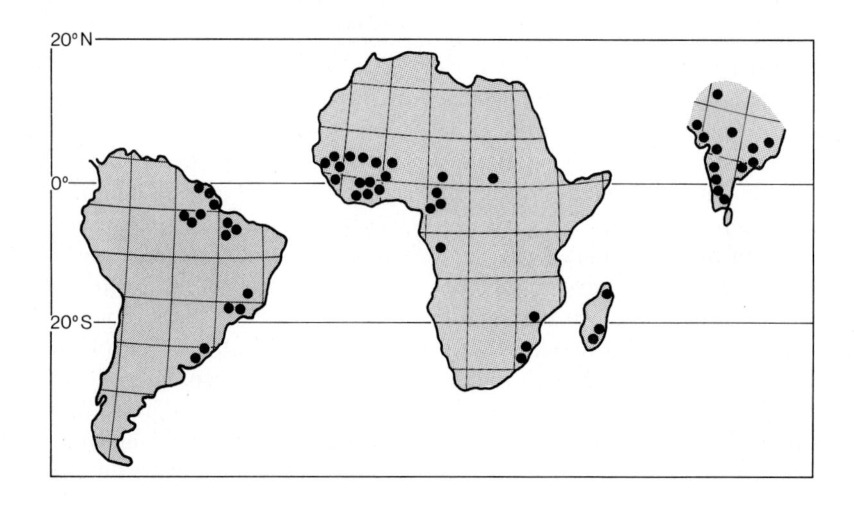

Δ
Fig. 8.13. Paleogeographic position of continents during the Eocene and the distribution of bauxites in South America, Africa and India (Prasad, 1985)

Fig. 8.14. Weathering of primary aluminum-silicates such as basalts or granites under tropical and subtropical conditions (after Frakes, 1979)
∇

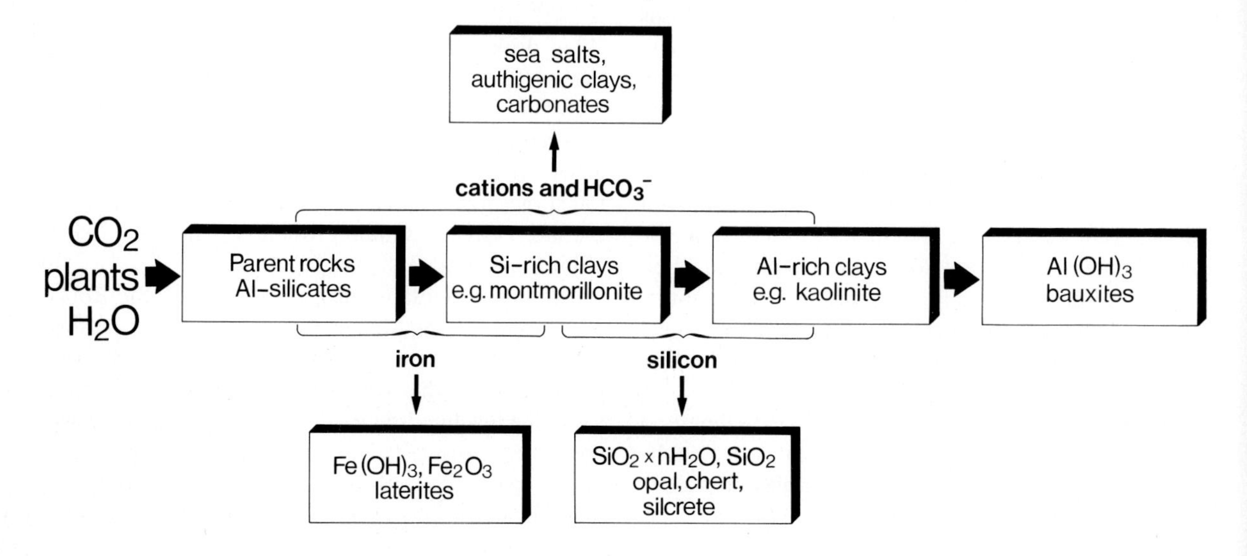

and redox potential plus salt effects appear to be chiefly involved in the generation of karst bauxites.

Only once in Earth history, that is during the lower and middle Eocene – about 50 million years ago – was bauxite formation widespread. This early Tertiary bauxite event must be seen in conjunction with coinciding factors: (i) the favorable geographical position of South America, Africa and India with respect to the tropical climatic zone (Fig. 8.13), (ii) wide-spread rain forests far beyond their present areal range, (iii) global temperatures about 5 to 10° C above those of today, and (iv) relative stability of continental platforms allowing the ancient erosion surface to be preserved.

It is of note that much of the silica removed from the continents during the bauxite event ended up in marine cherts and the Eocene chert event may largely go on its account. Inasmuch as silica-depositing organisms such as diatoms or radiolaria monitor dissolved SiO_2 in the sea, bauxite formation on land must certainly have had an impact on marine planktonic life.

on continents and not in the open sea is related to the difference in crustal thickness and densities of rocks. Whereas the oceanic crust is thin and dense, continents are thick and light and so they float high (see discussion on isostatic equilibrium on p. 73).

The marine counterpart of mountains is the trenches, so that the maximum elevation of the Earth's solid surface measures 20 km (Fig. 8.15). The hypsometric diagram of the globe also reveals that most of the area is occupied by the continental and deep sea platforms. This distribution pattern is a consequence of the ongoing endogenic activity which counteracts erosion. If one could stop orogeny, epeirogeny, seafloor spreading, subduction, in one word, all tectonic forces acting in crust and mantle, the continents would become eroded to sea level in about 10 million years, the trenches would eventually be filled up with sediment and thus disappear, and the hypsometric curve would look more gentle and smooth.

Whereas high relief provides the kinetic energy for erosion, it is solar radiation which maintains the dynamics of land denudation. The Sun causes oceans to circulate, water to evaporate, or winds and clouds to move. At any one time, the air carries 0.13×10^{14} tons of water vapor, a volume which turns over in less than a month. Most of that water stems from the equatorial sea which evaporates at a rate which would lower the world ocean by about 120 cm a year. However, a large fraction of that moisture becomes refluxed to the sea before reaching land. The rest rains out over the continents, where two thirds of that amount return quickly to the air by straightforward evaporation or by evapo-transpiration of land plants. Rivers, lakes and groundwater aquifers carry the remaining one third of continental precipitation back to the sea and thus close the hydrological cycle.

At the same time, continental runoff consists not only of water but of suspended and dissolved material liberated during weathering of rocks and destruction and decomposition of the vegetation cover. Garrels and MacKenzie (1971) have estimated that at present

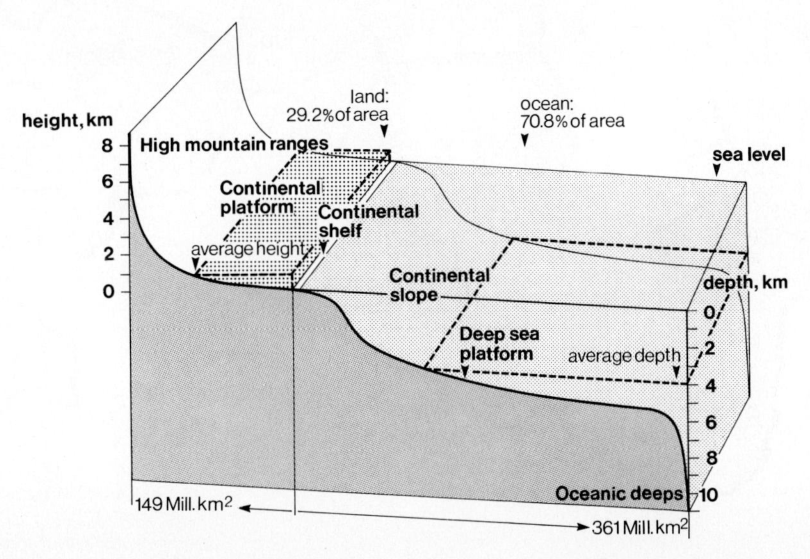

Fig. 8.15. Hypsometric diagram of the Earth's solid surface (Kossinna, 1921)

streams provide roughly 90 percent of the global material transport, which is followed by glacial flux amounting to 7 percent of the total; this contribution is principally derived from the movement of the antarctic ice shield. Groundwater transport and wave erosion along crustal shorelines add 1 to 2 percent, and desert wind and submarine discharge account for less than 1 percent of the global erosional flux.

Cosmic dust falls on the Earth's surface in an amount of 3 to 4 million tons a year (Wassen et al., 1967). At first sight, this contribution from "cosmic erosion" may not seem substantial. But it would take a few thousand train loads of 100-car trains or, figuratively, a continuous line of railroad freight cars extending from Copenhagen to Rome to transport such an amount of extraterrestrial debris.

Since rivers are the principal erosional force shaping the landscape, we will now turn to the riverine system in some more detail.

Global Riverine Transport

Annually, about 40,000 km^3 of water run off the Earth's continents. At this rate, one volume of ocean water becomes recycled via rivers in roughly 40,000 years. In addition to water, rivers carry salts, suspended materials and dissolved organic compounds to the sea, but their individual quantities are difficult to assess. Thus, estimates of data on global riverine transport are at best tentative (e.g., Clarke and Washington, 1924; Livingstone, 1963; Holeman, 1968; Garrels and Mackenzie, 1971; Martin et al., 1980; Meybeck, 1982; Milliman and Meade, 1983; Degens et al., 1984b).

Many of the inferences drawn in these studies rest principally on data collected randomly, that is, at different points in time and space. With the exception of a few rivers in the temperate regions, none has been studied in detail for flux variations throughout a year, not to mention for interannual changes or episodic events. Moreover, it is increasingly being recognized that the rivers of Asia and Oceania carry huge quantities of suspended matter in their loads, which totally amount to about 80 percent of the global average (Fig. 8.16). However, direct measurements of their fluxes are rare so that the 80 percent figure represents only a rough approximation. An additional problem stems from the use of different analytical techniques to break down the quantities of matter, both dissolved and suspended, carried by different rivers, which makes a global assessment of their measured transport rates quite problematic. As

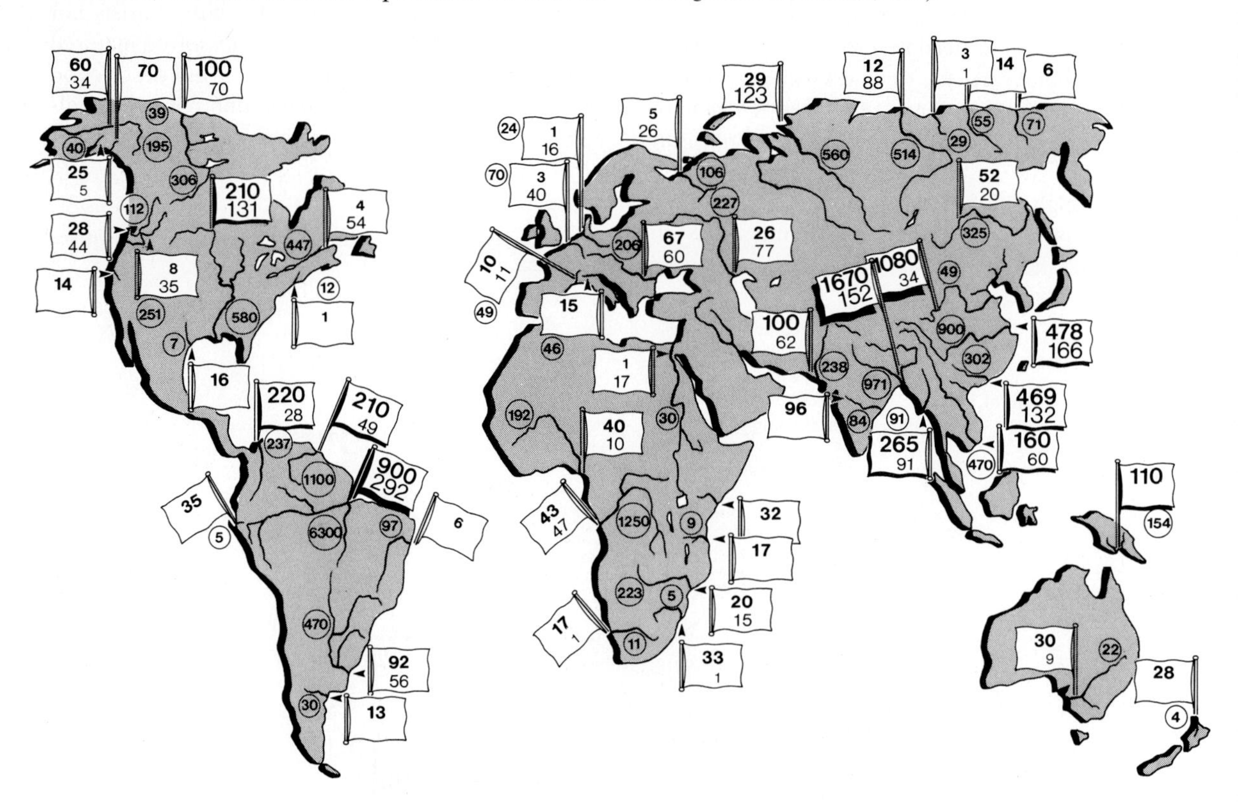

Fig. 8.16. Discharge of detritus (*boldface*) and salt (*lightface*) into world ocean. Figures are in million tons per year (various literature sources). River run-off in km^3 per year (*circle*) (after Baumgartner and Reichel, 1975)

a consequence, the global figures given in Fig. 8.16 are considered just a first approximation.

The following major trends can be discerned. The Arabian Sea and the Bay of Bengal receive the erosion products from the south side of the Himalayas. Indus and Ganges sediment cones are a vivid expression of the enormous quantities of detritus that for millions of years were carried down the southern slopes of that mountain chain. Shape and lateral extent of a deep sea sediment cone or fan generated this way is schematically depicted in Fig. 8.17. In contrast, rock material liberated from the northern and eastern part of the Tibetan plateau becomes principally discharged into the East and South China Sea. So in a way, the highest mountain belt on Earth delivers the bulk of the global detritus due to its height and its strategic position within the monsoonal influence sphere. It is fair to say that the Monsoons and the Tibetan plateau are two of the most influential factors controling world climate (see p. 223). They, in turn, are responsible for the global bulk of erosional runoff to the sea.

Whereas large streams are the principal means for the removal of detritus and dissolved matter from the Tibetan plateau, many smaller rivers are responsible for the high erosion rates of Oceania. Since in this part of the world massive deforestation is presently going on, and furthermore forest fires are frequent and widespread, soil erosion is here at its peak. An accurate assessment, however, is difficult to achieve because of seasonal and interannual variations, as well as the episodicity of major erosional events and the paucity of data.

The only other region where riverine transport is substantial is South America, and the Amazon is the chief contributor of dissolved and particulate matter. Australia, Africa, Europe and the Arctic jointly contribute only 5 percent of the global river load. In this context, it may come as a surprise that the longest river on Earth, the Nile, carries virtually no detritus to the Mediterranean. The Assuan Dam is the culprit, restricting the flow of detritus from south to north. Moreover, the discharge of nutrient minerals is at a minimum because they are already utilized in large measure for the eutrophication of Lake Nasser. This artificial lake behind the Assuan Dam is evaporating at a high rate, so that much of the Nile's water is lost there, raising the salt level and alkalinity. In short, this dam is by no means only beneficial to society (Kempe, 1983). On the other hand, that lake can store water to be used at times when rainfall in the source area of the Nile is minimal, as is the case today.

Fig. 8.17. Block diagram (highly schematic) illustrating the path of erosion and sediment deposition from high mountains to ocean depths

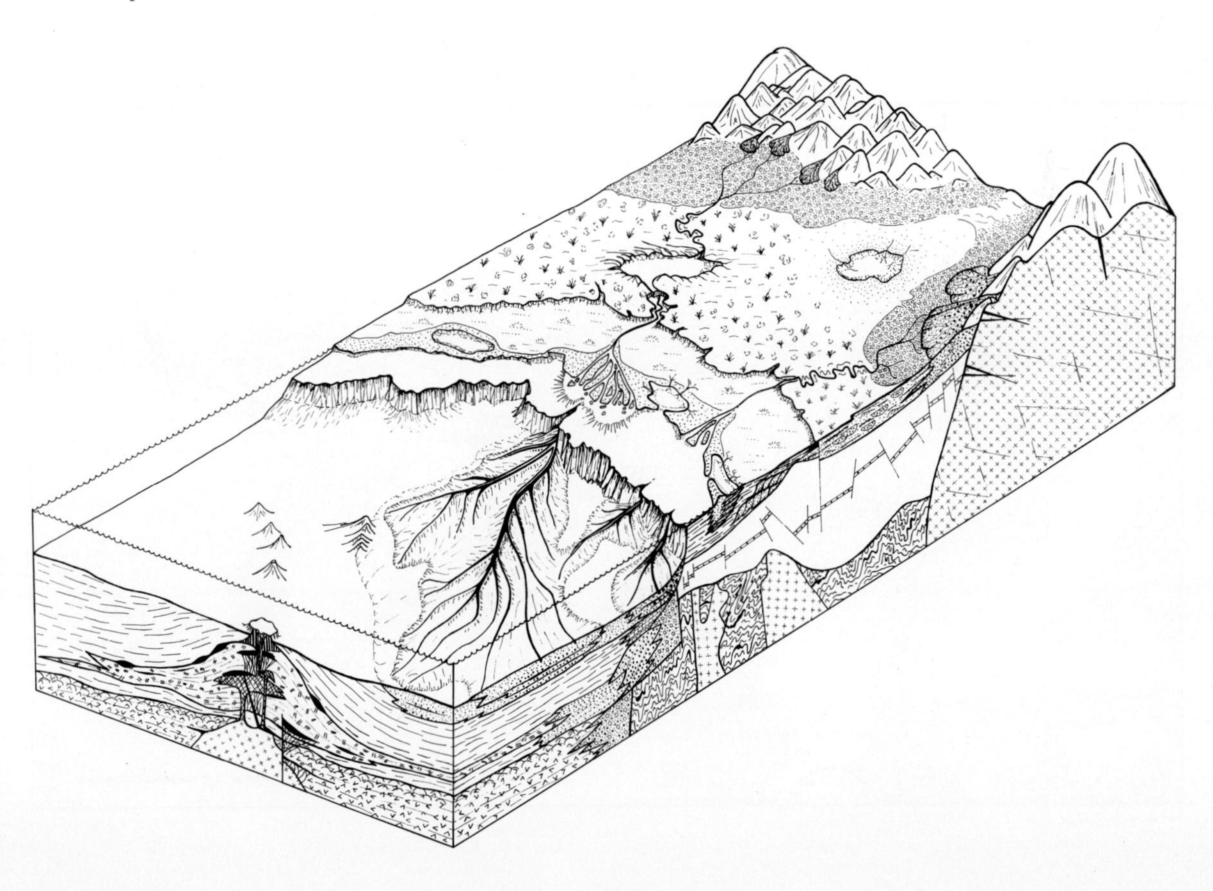

The ocean is the principal catch basin for river discharge. Over geologic time, it has received all products of weathering and soil erosion. In her book "The Sea around Us", Rachel L. Carson tells the epic of oceans and rivers in a poetic way:

"For the sea lies all about us. The commerce of all lands must cross it. The very winds that move over the lands have been cradled on its broad expanse and seek ever to return to it. The continents themselves dissolve and pass to the sea, in grain after grain of eroded land. So the rains that rose from it return again in rivers. In its mysterious past it encompasses all the dim origins of life and receives in the end, after, it may be, many transmutations, the dead husks of the same life. For all at last return to the sea – to Oceanus, the ocean river, like the ever-flowing stream of time, the beginning and the end."

(Oxford University Press, New York, 1951)

While at present the regional diversity, combined with the lack of representative time series experiments, give global assessments of riverine transport an enormous error bar, the situation is entirely different for a small number of enclosed seas such as the Mediterranean or the Black Sea, where the data base on river runoff to those seas is optimal. In the following, the Black Sea region will serve as a small-scale denudation model, where input and output over time can be readily controlled.

Black Sea Catch Basin

At present, half of Europe and part of Asia drain into the Black Sea (Fig. 8.18). The source area is well defined and can be subdivided into several orographic provinces which are distinguished by geography, extension, climate, and elevation. The subdivisions shown in Table 8.1 seem to be appropriate. These figures are accurate within a few percent: they represent only the planar projection of the individual terrain and not the actual surface area exposed to denudation.

Fig. 8.18. Orographic relationships in the Black Sea region. Arrows indicate annual flux of sediment (*boldface*) and salt (*lightface*) (after Shimkus and Trimonis, 1974)

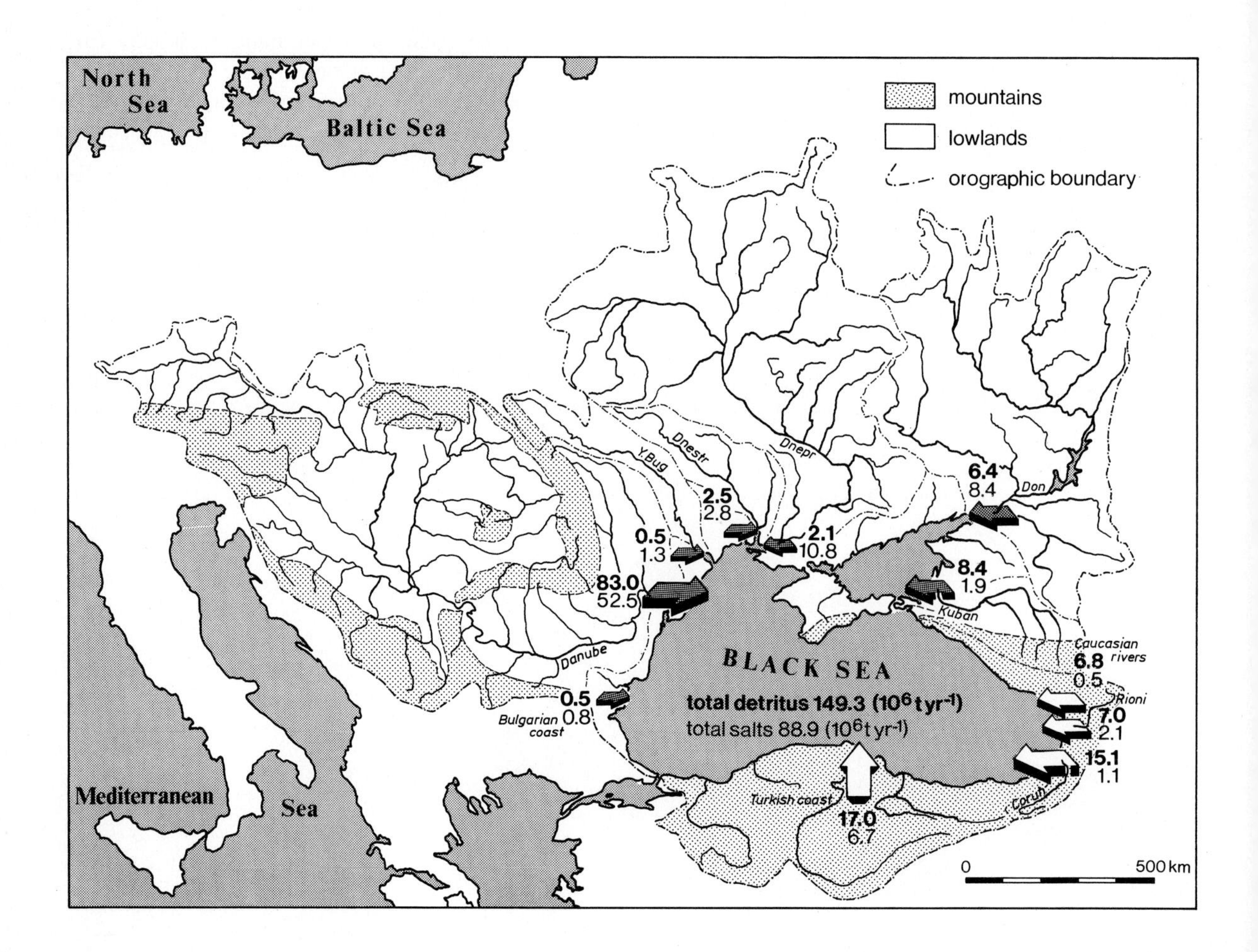

Table 8.1. Rivers entering the Black Sea and their drainage areas

River	Size of drainage area (km^2)
1 Danube	836,000
2 Dnestr	61,900
3 Y. Bug	34,000
4 Dnepr	558,000
5 Don	446,500
6 Kuban	63,500
7 Caucasian rivers	24,100
8 Rioni	15,800
9 Coruh	16,700
10 Turkish rivers	213,500
11 Bulgarian rivers	22,200
Total	2,290,200

The region under study is characterized by its diversity. We encounter all transitions between arid and humid climates and between lowlands and mountainous areas. The issue is further complicated, because the Pripjet swamps, the Panonic flatland, and the lowlands along the Black Sea coast are regions of sedimentation rather than denudation.

Shimkus and Trimonis (1974) present data on the sediment and salt load which is annually carried by the major rivers into the Black Sea basin. The total load divided by the size of the individual drainage area gives the total amount of yearly denudation in tons per square kilometer (Table 8.2). From these figures, Degens et al. (1976) extracted the mean volume of eroded material, applying some "cosmetic" corrections to account for the areas of sedimentation within the Black Sea source region. The average denudation rate for the entire source is 0.063 mm a^{-1} or about 100 t $km^{-2}a^{-1}$.

From a sedimentological point of view, it is of interest to know at what points most of the detritus enters the Black Sea. Fig. 8.19 illustrates that: (1) the Danube carries about 60 % of the detrital load; (2) the Don and Kuban jointly discharge close to 10 % of the total into the Sea of Asov where the material comes to rest; (3) the Caucasian rivers, Rioni and

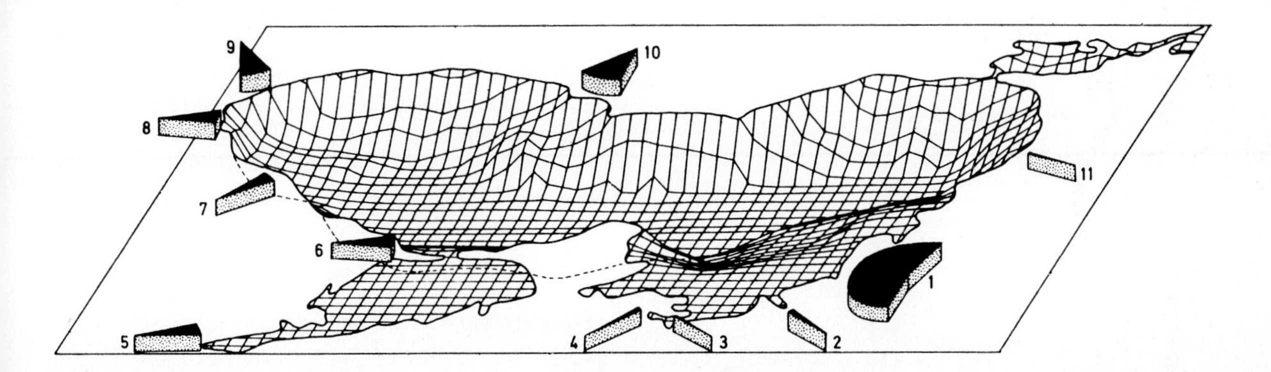

Fig. 8.19. Points of entrance of detritus into Black Sea Basin. Size of individual section represents weighted percentage of total discharge (numbers correspond to rivers as listed in Table 8.1)

Table 8.2. Denudation in source area of Black Sea basin (Shimkus and Trimonis, 1974)

River	Detritus (10^6t a^{-1})	Salts (10^6t a^{-1})	Total load (10^6t a^{-1})	Size of area (km^2)	Amount of weight (t km^{-2} a^{-1})	Denudation volume (m^3 km^{-2} a^{-1})	Denudation rate (mm a^{-1})
Danube	83.00	52.51	135.51	681,000*	199.0	124.4	0.125
Dnestr	2.50	2.79	5.25	61,900	85.5	53.5	0.054
Y. Bug	0.53	1.35	1.88	34,000	55.4	34.6	0.035
Dnepr	2.12	10.79	12.91	383,500*	24.0	15.0	0.015
Don	6.40	8.43	14.83	446,500	33.2	20.8	0.021
Kuban	8.40	1.95	10.35	63,500	163.0	102.0	0.102
Caucasian rivers	6.79		7.3	24,100	303.0	189.5	0.190
Rioni	7.08	2.16	7.6	15,800	481.0	301.0	0.301
Coruh	15.13		16.2	16,700	971.0	607.0	0.607
Turkish coast	17.00	6.70	23.70	231,500	102.4	64.0	0.064
Bulgarian coast	0.50	0.80	1.30	22,200	58.5	36.6	0.037

*reduced area

Coruh, contribute 20 % of the total detritus; (4) Turkish rivers supply about 10 %; and (5) Dnestr, Y. Bug, Dnepr and the Bulgarian rivers yield very little material.

Rivers in general have a rather short memory, since in a matter of weeks or months they lose all their water, ions, molecules, and suspended particles to the sea. Here the detritus will settle to the seafloor assisted by the action of organisms (see p. 153). The riverine input is thus kept on file in the form of sedimentary layers.

The Black Sea basin with a maximum depth of 2,200 m, an area of close to 500,000 km^2 and a volume of about 500,000 km^3 is physiographically a kind of micro-ocean with: shelf, basin slope, basin apron and abyssal plain (Ross et al., 1974). Continental slopes are steep, and a bathtub shape is indicated (Fig. 8.19). In the deep-water area, isolated cyclonic currents are developed having distinct seasonal signals (Zenkevich, 1963). These two factors combined, i.e., physiography and current pattern, are responsible for the distribution of detritus across the Black Sea. Since, with the exception of a small outlet, the Bosporus, the Black Sea is enclosed by land, all

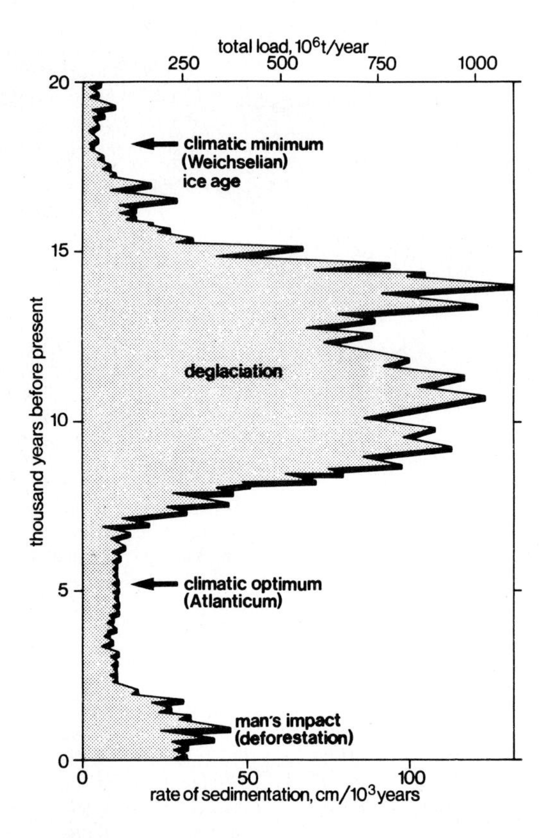

Fig. 8.20. Riverine discharge into the Black Sea for the past 20,000 years

riverine detritus has to become processed within the basin proper.

Close examination of the basinal record reveals that annual layering (varves) can be traced as far back as the last climatic optimum (Atlanticum), and that thousands of microlayers often measuring less than 1 mm can be traced across the entire abyssal plain. The historical sediment record was able to "survive" intact because the Black Sea below a depth of 100 to 200 m was anoxic for the last 5,000 years, thus inhibiting the invasion of deep-water benthic communities such as burrowing worms, which would have destroyed the varve pattern. With the aid of radiocarbon techniques and narrowly spaced samples from sediment cores, rates of sedimentation can be precisely measured for the past 20,000 years; the results are depicted in Fig. 8.20. The data is quite informative:

- detritus carried by the Black Sea rivers is homogeneously spread – almost like a blanket – across the deep basin floor;
- modern sedimentation rates are 30 cm/10^3 years, not considering planktonic remains (e.g., coccoliths, diatoms); annual variations are considerable;
- 2,000 years ago rates were three times lower with no significant fluctuations for the time period 5,000 to 2,000 years B.P. (before present);
- rapid rises in rates of sedimentation during the time envelope 15,000 to 7,000 years B.P. are related to deglaciation; and
- low rates of sedimentation (<10 cm/10^3 years) around 18,000 years B.P. are a consequence of wide areal coverage of ice sheets and permafrost in a reduced drainage area.

Typically, varves in Black Sea cores consist of dark and light layers and reflect a seasonal sedimentation pattern related to the primary productivity in the water column. Since one light and one dark layer each are formed per year, a simple count will reveal the age of the sediment, and the thickness of the individual laminae will give the sedimentation rate which in this case is a function of river load and primary productivity. The three-fold increase in rate of sedimentation over the past 2,000 years is chiefly a consequence of deforestation in the regions bordering the Black Sea; episodic events of high riverine discharge are discernible (Fig. 8.21). This is solid evidence that man's agricultural activities have substantially increased soil erosion.

Today's total suspended load in the Black Sea is in the order of 200 x 10^6 t a^{-1}. Since that load will produce annually a blanket 0.3 mm thick, all the observed changes and rates of sedimentation recorded in the sediments deposited over the past 20,000 years simply reflect changes in the amount of river runoff and the land surface characteristics in terms of vegetation, soil and climate. All this suggests that we have to be rather cautious when using a globally monitored one-year river record as a steady state figure. In the case of the Black Sea, we have observed 20-fold changes in sediment transport during an interval of only 20,000 years.

The Black Sea has existed as a catch basin since the Cretaceous (Brinkmann, 1974; Degens et al., 1986; Degens and Ross, 1974). In taking the whole sedimentary column on top of the basement as a measure of all material that has been deposited since then, a mean sedimentation rate (with respect to the compacted sediments) of 10 cm per 1,000 years can be calculated. The 1 km sediment section obtained during the GLOMAR CHALLENGER drilling operations in the Black Sea substantiates this inference (Ross and Neprochnov et al., 1978).

So in a way, the rates of subsidence and rates of sedimentation are, in the long run, balanced to each other. Yet, in shorter terms (10^6 years) climatic and tectonic pulses will slightly put out of phase the general steady state. Man's input upon this system will in all probability be of shorter duration. Roughly 2 percent of the global land surface is in the influence sphere of the Black Sea catch basin. The remaining 98 percent will be more cumbersome to tackle with respect to the interplay of exogenic and endogenic forces, since the real ocean is at stake.

Sedimentation

The preceding paragraph may have left the impression that river suspension will eventually settle in the ocean. This is certainly correct on geological time scales but it is wrong to assume that a suspended particle we see today in a river's load will reach the ocean in a few days or weeks, where it sinks to the sea floor and jointly with other particles forms a sediment layer. Meade (1982) has clearly formulated the problem of modeling riverine sediment transport, "We can probably model the local movement of sediment in rivers at time scales on the order of seconds to hours because of the large number of experiments that have demonstrated the physics of particle movement under the influence of fluid forces. We might even be able to model sediment movement in rivers on a geological time scale by allowing enough time to reach whatever sort of steady state we chose to assume." However, "Time scales on the order of years to centuries are too long for us to apply the predictive power of Newtonian physics and too short for us to make the comforting assumption of a steady state."

At present, estuaries, coastal marshlands and the continental shelves retain temporarily most of the detritus discharged by rivers, except in those cases where the shelf break is close to a river's mouth or tidal and longshore currents are sufficiently strong to sweep suspended particles into canyons, from where they may be episodically carried in the form of turbidity currents towards the abyss (Box 8.3; Fig. 8.22).

Fig. 8.21. Rates of sedimentation in the Black Sea from 800 B.C. to present, deduced from thickness of annual varve deposit along continental rise, off Bulgaria (Degens et al., 1976)

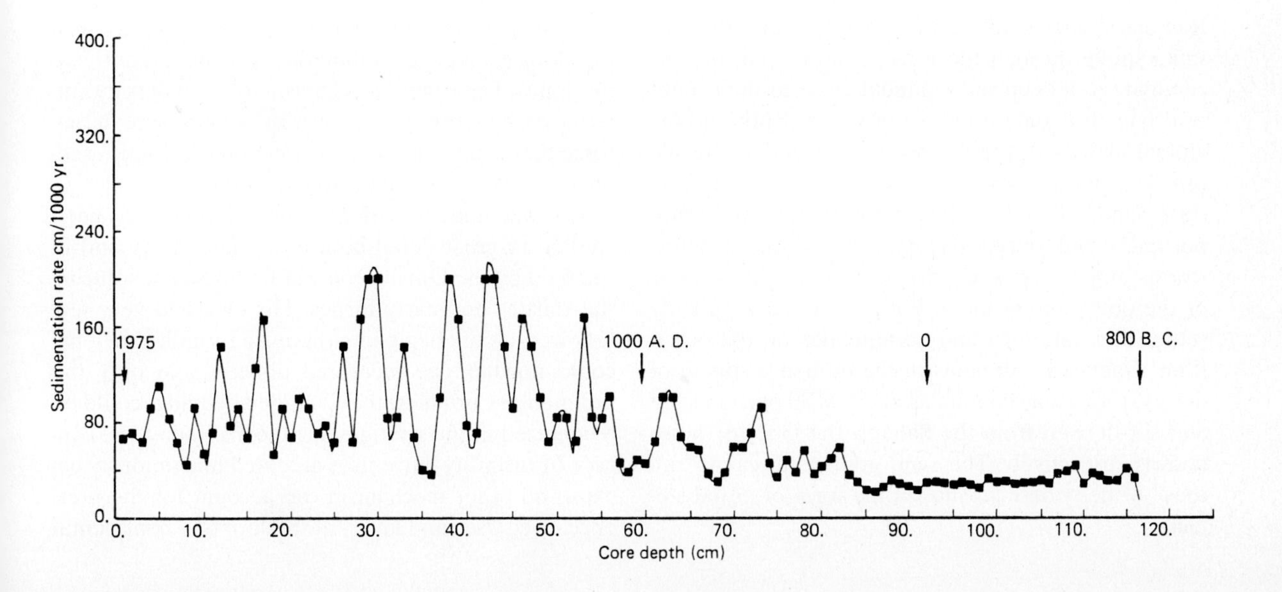

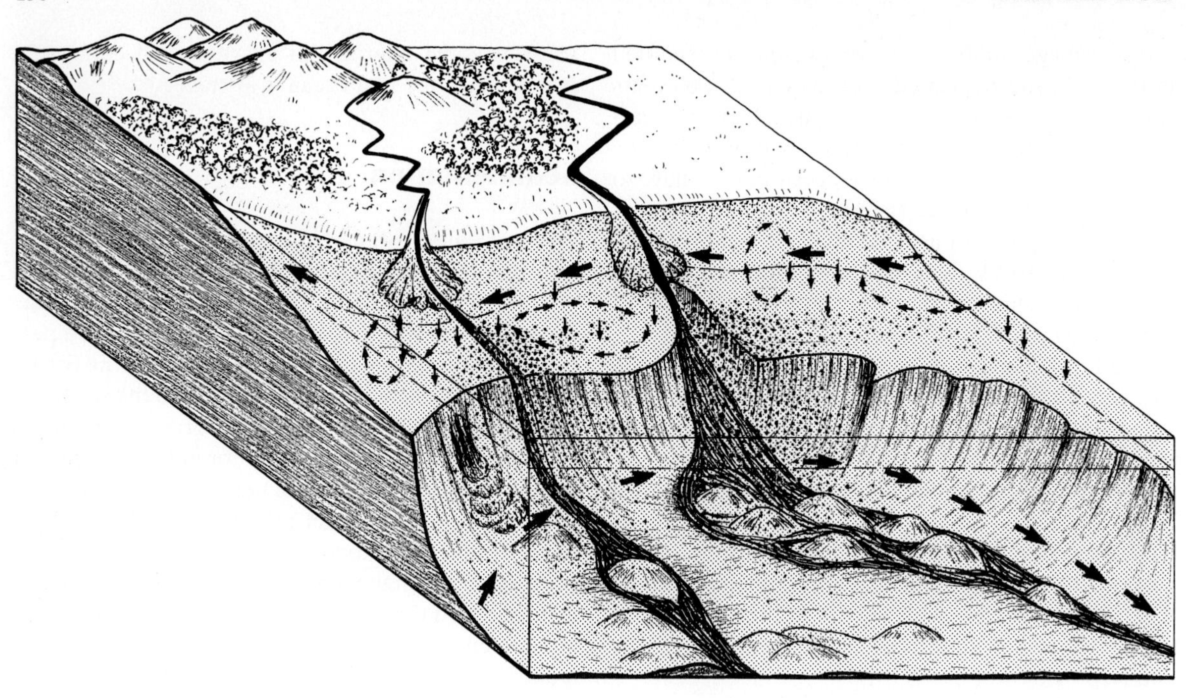

Fig. 8.22. Block diagram (highly schematic) showing transport mechanism responsible for the removal of sediment particles from the shallow to the deep sea. *Arrows* indicate longshore currents (*from right to left*) and direction of water movement along continental slope (*from left to right*). Canyons act as temporary sink to sediment; episodic turbidity currents and mud slides along slope will eventually sweep material into the ocean's abyss. In estuaries and coastal environments, fluffy material rich in organic matter is kept in suspension through wave action, currents, tidal forces and storm events, or will settle and enter the benthic layer under quieter conditions. Those organic particles surviving chemical oxidation and biological consumption are resuspended over and over again. In time, they may (a) cross the shelf break or (b) become trapped in canyons or (c) be monitored by planktonic organisms. All three mechanisms result in the transfer of particles in transit from the euphotic zone to the deep sea. At times of low sea-level stand, for instance during glacial times, routes of transport from land to open sea are considerably shortened. Counter-currents are indicated (*arrows from left to right*)

BOX 8.3

Down the Continental Slope

Radiolaria and planktonic foraminifera and diatoms sink rapidly through the water column and may accumulate as a deep-sea sediment close to their point of origin. It is only a matter of water depth and microbial activity in the top sediment whether the siliceous and calcareous ooze becomes fossilized. In contrast, sand, silt and mud have a more involved transportation and burial history. For instance, desert storms may sweep sand particles across wide regions of the ocean where the airborne material eventually settles as a fine dust and accumulates on the ocean floor. Deep-sea sediments in the influence sphere of desert storms may contain as much as 30 percent sand and silt derived from the Sahara, the Gobi or other deserts respectively. The sand grains still carry "tattoos" of hefty desert winds in the form of pitted surfaces.

Sand and silt may also reach the abyssal plain by means of turbidity currents which are propagated by the resuspension of sediment material formerly resting along shelf edges, submarine canyons or continental slopes. The trigger mechanism for such density currents can be earthquakes, overburden pressure or any force that can release the friction brake holding a sediment to its resting site (Holler, 1986).

The phenomenon of turbidity currents has been studied in great detail both in the laboratory and in the field by the Dutch geologist Philip Kuenen during the thirties and early forties. His idea had very few followers at the beginning, because turbidity currents could nowhere be measured directly, and only the sedimentary product, that is graded bedding, could be recognized in the stratigraphic record. Today, the concept of turbidity currents is accepted unanimously because no other mechanism can account for the presence of coarse sand layers at the foot of a continental

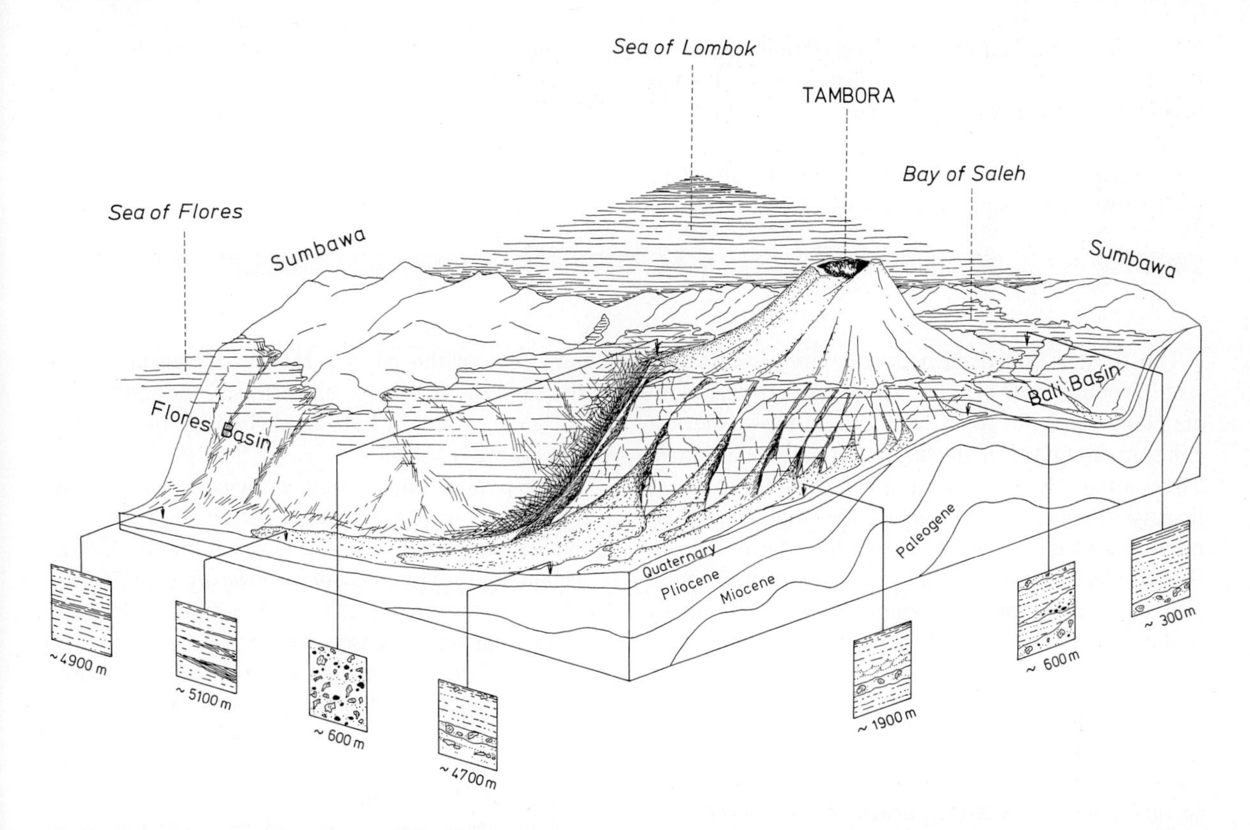

Fig. 8.23. Generations of turbidites and mass flow following the eruption of Tambora Volcano in 1815. The deposits from sediment tongues hundreds of kilometers long extending eastwards into the deep Sea of Flores are indicated. Box cores showing the top 50 cm of sediment reveal incidents of turbidity flows

slope, or the sand waves spread here and there across an abyssal plain.

Submarine fans and cones of the type shown in Fig. 8.17 are essentially composed of turbidite sequences. An especially attractive example is the Tambora Volcano on Sumbawa Island, Indonesia, which erupted in 1815 on April 10–11. It has long been regarded as the greatest eruption in historic time (Self et al., 1984). About 175 km^3 of pyroclastic material (equivalent to about 50 km^3 of dense rock) was ejected in 24 hours, and the ash fall 1 cm thick covered 500,000 km^2 of the Java Sea and surrounding islands. During the SNELLIUS II expedition in fall 1984, turbidites composed of Tambora ash and unconsolidated sediment fragments eroded by turbidity currents from the sediment strata exposed in the canyons were sampled. They have been carried for hundreds of kilometers downwards and eastwards along the continental slope and rise of the Flores Basin (Fig. 8.23), attesting to the enormous mass transport in the deep ocean. It is of note that this carpet of sand, silt and mud resulted in a Waterloo for deep-sea life, which for 100 years became extinct following the 1815 catastrophe.

Whereas sand and silt in the deep sea is of aeolian or turbiditic origin, clay-sized particles use different vehicles for vertical and horizontal transport, namely an organic substrate. Filter-feeding plankton sequesters clay in the form of fecal pellets, while phytoplankton and bacterial communities excrete dissolved organic matter which glues the finely suspended particles together, resulting in macroflocs. Estuaries and shelf regions are generally fertile environments, and consequently macroflocs are constantly produced by the trillions. Furthermore, the 1815 explosion wiped the bordering shelf area clean of its Quaternary sediment cover. Since this event only a few centimeters of ooze have come to rest on the Neogene substratum. However, due to tidal effects, longshore currents, waves and other physical factors introducing turbulence into the shallow marine environment, the light flocculate material is constantly removed from the sandy-silty fraction and carried horizontally for considerable distances. This reminds me of the situation in a desert where strong winds enable loosely structured grass balls to tumble across sand dunes. Should macroflocs be swept into a canyon or cross the shelf edge, they may be trapped on the canyon floor or settle as fine marine snow to the abyss (Honjo, 1976). In Fig. 8.22, two scenarios are depicted, one of today and the other at glacial times, where sea level was almost 100 m lower.

In conclusion, transport of particles into the deep sea follows a number of avenues which are related to provenance, size, composition, and specific density of the individual particles. However, the regional hydrographic, geologic and environmental conditions determine their final burial place (Global Ocean Flux Study, 1984; Degens et al., 1987).

The flow of fluids can be either laminar or turbulent. Groundwaters observe laminar flow, whereas rivers, lakes and the top ocean layer (above thermocline) maintain turbulent flow practically at all times. Thus, a particle has to have a certain size, shape or density to overcome convection and the eddy movement of a water body in order to sink down.

If you believe textbooks, the settling of small particles less than 0.1 mm in diameter follows principally Stoke's law:

$$v = \frac{2}{g} \frac{d_1 - d_2}{n} gr^2,$$

whereby v is the settling velocity in centimeters per second, r the radius of the particle in centimeter, g the acceleration of gravity, n is the viscosity of the liquid in poises, and d_1 and d_2 are the densities of the particle and fluid, respectively (Krumbein and Sloss, 1963; Garrels and Mackenzie, 1971). For particles greater 1 mm, the settling velocity is proportional to the square root of the particle radius (impact law). For particles in the size range 0.1 to 1 mm, the settling velocity is controlled by both fluid viscosity and particle size (Fig. 8.24).

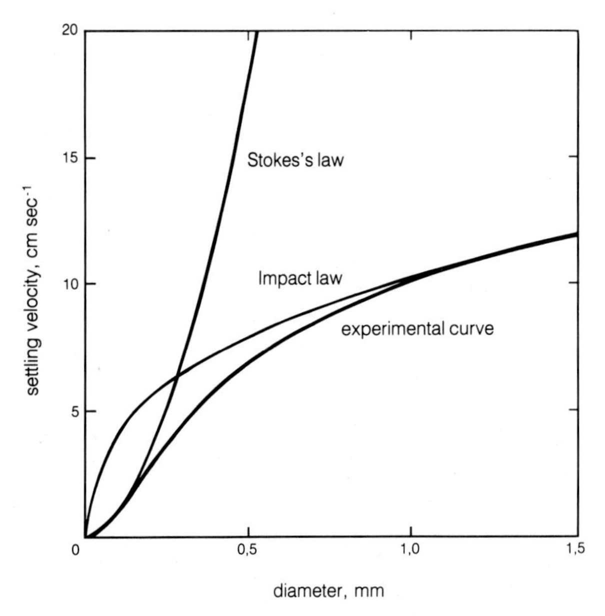

Fig. 8.24. Settling rates of particles of different diameters in the water column (after Krumbein and Sloss, 1963)

These relationships should principally allow us to predict how much time is needed for a particle of a given size, density and shape to reach the seafloor at a certain depth. This concept is, however, complicated by the fact that the ocean's water column is not stagnant but that water masses move around and are driven by winds and currents. As a very rough approximation, one can say that the upper water layer, about 200 m thick, is recycled mole by mole in a matter of 50 years, whereas the lower water column of world ocean is mixed in roughly 500 years.

A sand grain, about 1 mm in size, will fall some 3.5 m per second and reaches a water depth of 100 m in half a minute. Should the grain have a diameter of 0.1 mm, 5 hours are needed for it to sink through 100 m of water. At a current velocity of 10 cm per second the 1-mm diameter sand grain would have traveled 3 m before getting to a water depth of 100 m, whereas the 0.1-mm diameter sand grain would be found almost 2 km past the point of entrance for the same traverse. For clay minerals with diameters of 1 micron and less, and which constitute the bulk of the global detrital flux, settling velocities are in the order of a few centimeters per day at most (Fig. 8.25).

We will now make an experiment and place a clay particle into the Equatorial Current of the Pacific near the Galapagos Islands off Ecuador. It takes roughly one year for that current to cross the whole Pacific. Neglecting convection and only assuming that our clay settles according to Stoke's law, the particle has sunk during that voyage to a water depth of 5 to 10 m at the most. In short, other mechanisms have to be invoked to remove the fine detritus from the sea.

We have known for a few years that there is not a constant flow of material from top to bottom but that there are seasonal or episodic events during which the particles become scavenged from the water column. Such events last a few weeks, interrupted by a few months' interval where little material reaches the seafloor. The consensus is that biological processes are involved in the vertical transport of mineral and organic matter.

About 10 years ago it was recognized that zooplankton was actively involved in this transfer mechanism (e.g. Honjo, 1976; Deuser et al., 1983; Scheidegger and Krissek, 1982). For instance, tiny shrimps (copepods) filter clay-sized particles and organic detritus from the water column and subsequently release the indigestible parts in the form of fecal pellets having a size of about 100 micron. The velocity of sinking of such fecal pellets can be as high as 500 to 600 m per day, as shown in Fig. 8.25. A typical fecal pellet is depicted in Fig. 8.26.

However, simple calculations have revealed that the fecal pellet mechanism cannot account for all the fine suspensions found deposited in the form of recent marine muds. Actually 5 to 10 percent at the most can be attributed to this process. The rest is principally monitored by phytoplankton communities which excrete specific dissolved organic matter as a defense mechanism against various environmental hazards such as heavy metal pollution, high water temperatures, life-interfering organic compounds in the ambience, or high abundance of particles in the euphotic zone limiting light penetration. These excretion products are biochemical molecules with a

Fig. 8.25. Settling rates of particles common in the marine habitat in cm per day according to Stoke's Law. Note exponential scale. Speed of convection in the euphotic zone is indicated

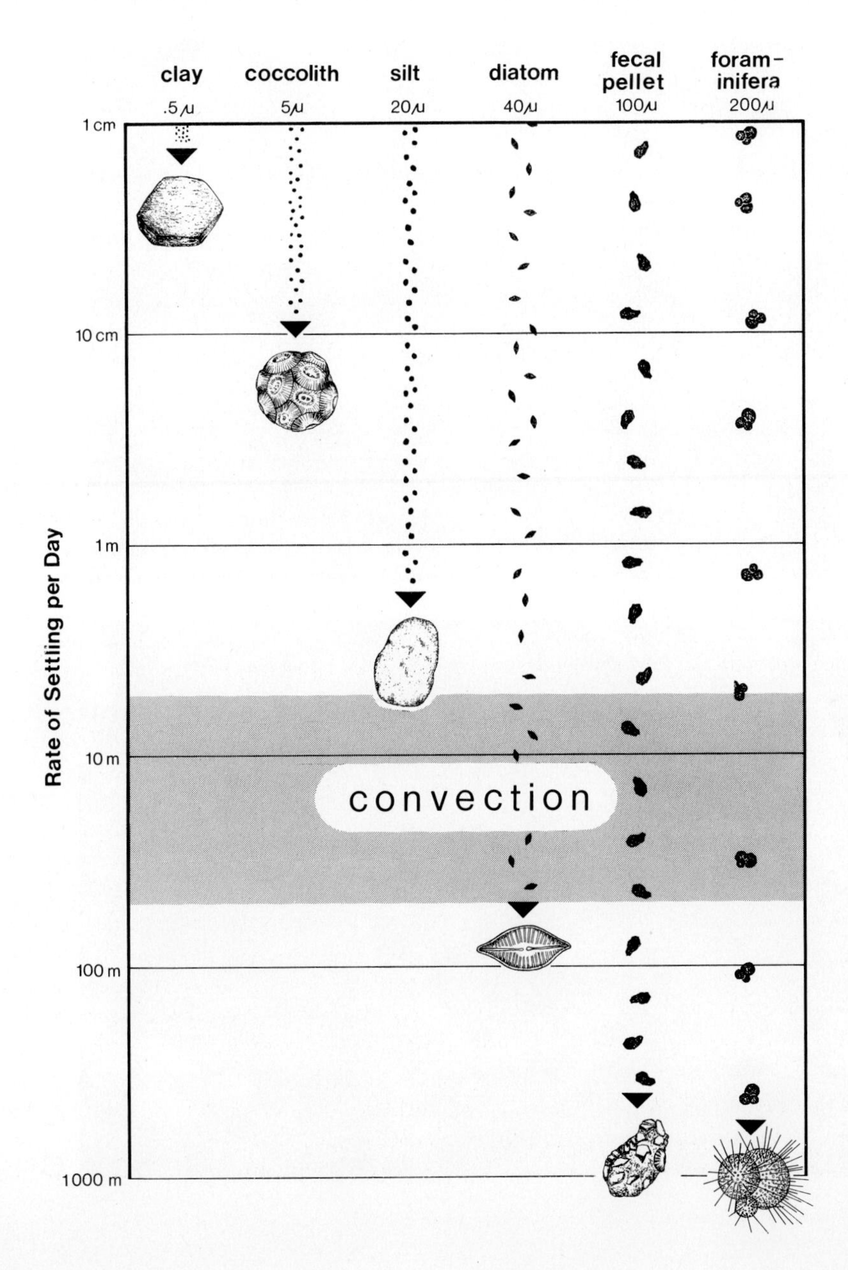

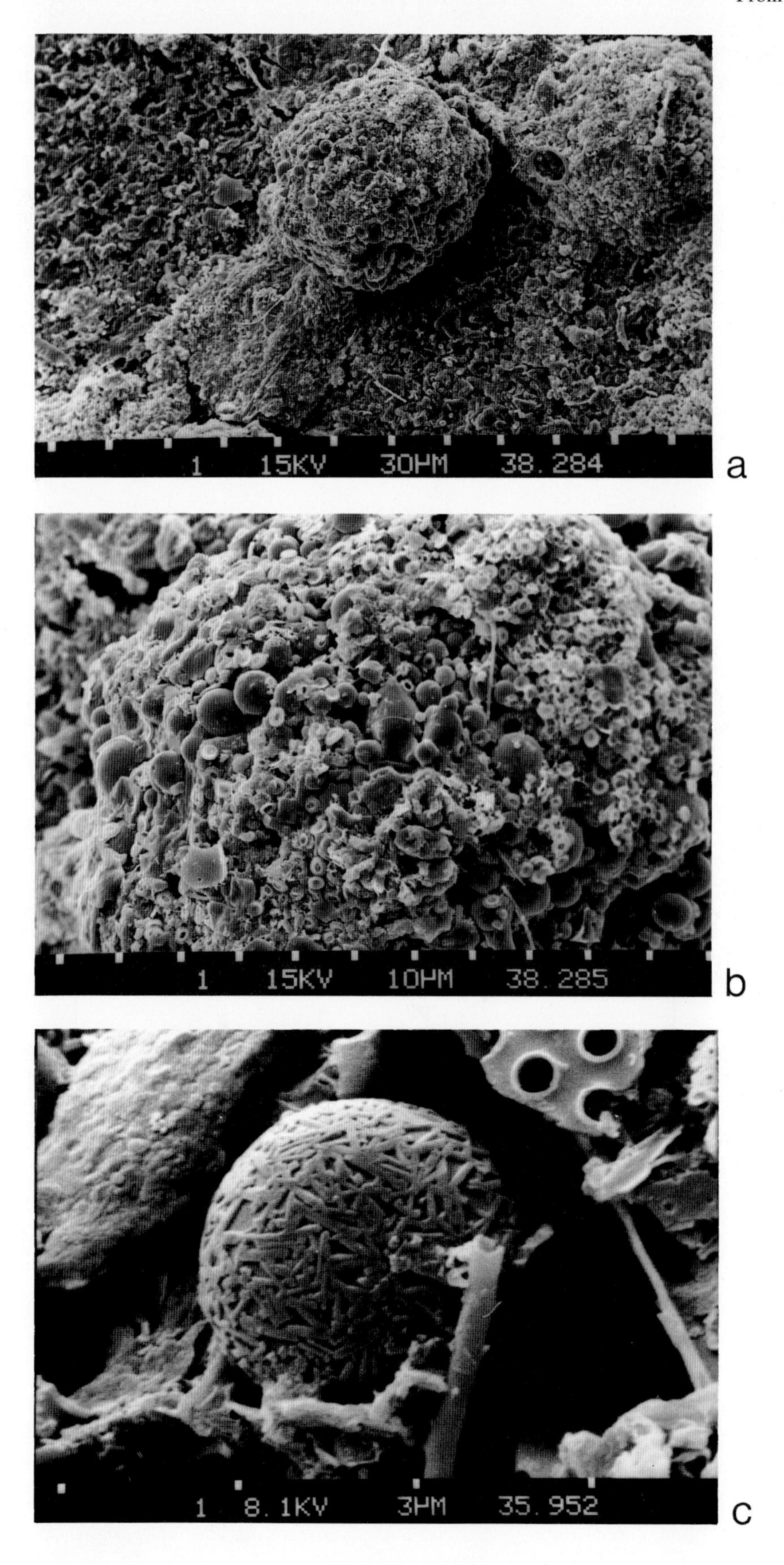
1 15KV 30ΜM 38.284
a
1 15KV 10ΜM 38.285
b
1 8.1KV 3ΜM 35.952
c

molecular weight of about 500 to 2,000 dalton which are principally composed of certain sugars and amino acids. They can complex metal ions and agglomerate dispersed particles to so-called macroflocs (Kranck, 1980; Eisma et al., 1985; Honjo et al., 1982; Degens and Ittekkot, 1986; Cadée and Laane, 1983; Degens et al., 1987; Anonymous, 1984).

Macroflocs are aggregates about 50–200 micron in diameter, in which mucopolysaccharides and small peptides represent the glue for the collection of mineral and organic matter.

In conclusion, active participation of planktonic organisms in aquatic sedimentation processes has far-reaching implications. Firstly, it ensures that even clay-sized particles will reach the deep seafloor within a few days, which would have otherwise remained suspended for years. Vehicles of transport are metabolic products of planktonic organisms like fecal pellets or mucopolysaccharides, which are produced in the surface layers of the ocean. This transport mechanism affects not only the vertical transport; even horizontally advected clay particles are translated into a vertical flux by such mechanisms.

These processes affect the sedimentation of riverine particulates too. Mud deposition on a river's flood plain or in intertidal zones is linked to organic-inorganic interactions driven by the aquatic biota. Sediments can be temporarily stored on flood plains or in small river basins before being released years later in an episodic event and placed into the estuaries and shelf environments, where they may remain for hundreds or even thousands of years (Meade, 1982). Eventually, turbidity currents may mobilize the accumulated shelf and slope sediment again and carry it to its final resting place, a marine geosyncline or the abyssal plain.

Mountain-building events may in time resurrect geosynclinal sediments in the form of foldbelts. This will initiate a new global cycle of weathering, erosion and sedimentation. In contrast, abyssal sediments are expected to inch their way to a subduction zone, where they become swallowed, metamorphized, and granitized, thus erasing their sedimentary heritage.

Fig. 8.26 a–c. Scanning electron micrograph of fecal pellet collected in mid-water in the Black Sea; particle size is about 100 μ in diameter **a**. Spherical bodies in **b** are fly ash particles and the small oval bodies, ca. 2–3 μ, are calcareous plates of coccolithophorid algae. Lower picture **c** shows single fly ash particle with typical mullite fabric

Facies

A person's face can change in many ways, and happiness or sorrow may find different reflections in people's features. Therefore it requires a certain amount of intuition, understanding, and sentiment to fathom somebody's feelings from a facial expression.

The Earth has a face, too, as was so nicely put by Eduard Suess more than a century ago, when naming his book "Antlitz der Erde". However, it was the Swiss geologist A. Gressly, 20 years before the "Face of the Earth" was published, who in 1838, introduced the term facies to describe the multifarious patterns of Upper Jurassic sediment sequences he mapped in the Jura Mountains.

In following Walther (1894), Teichert (1958), Middleton (1973), Murawski (1964) and Hallam (1981), the term facies is presently used as the sum total of lithological and faunal characteristics of a given stratigraphical unit. It should also include the chemical aspects of a sediment, because in the final analysis facies is the environmental impact of physical, chemical and biological processes that turned to stone.

Definition

Two hundred years ago, James Hutton coined the phrase "The Present is the Key to the Past". This concept of uniformitarianism became adopted by Charles Lyell in his classical book "Principles of Geology" that was published in 1830. Ever since, it has influenced the minds of generations of geologists and paleontologists.

In our treatise we have continuously stressed the point that the "System Air-Sea-Earth-Life" is evolving and not kept in a steady state. So many features of today have no counterpart in the past. And yet weathering, erosion, and sedimentation have always involved the same elements, namely water, wind, currents, and organisms. Oceans and continents, or mountains and valleys were there almost from the start. And finally, the genetic and metabolic apparatus operates along the same lines from the most ancestral to the most highly evolved organism. In a word, the basic principles of uniformitarianism are still valid in spite of evolution.

The physical environment shaping a modern sediment involves currents, water depth, or wind, to name just a few important factors. They will control the distribution of particles and yield distinct external and internal structures once the material has come to rest (Fig. 8.27). Should we find the identical pattern in an ancient sediment, the same environmental situ-

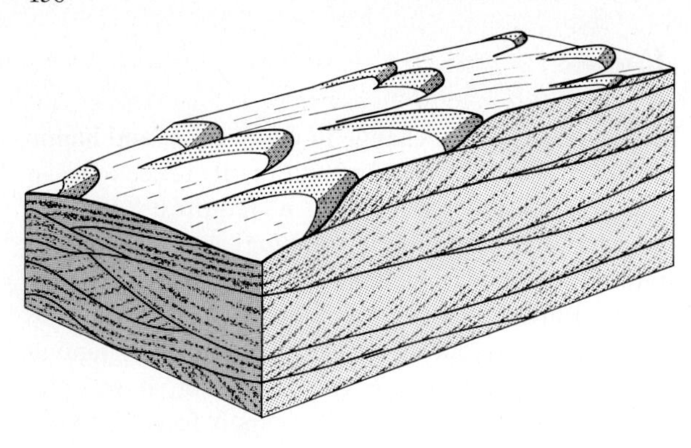

Fig. 8.27. Development of crossbedding in a field of traversing sand dunes (highly schematic after Wurster, 1958)

ation as that of today is indicated. It may allow us to distinguish between aeolian, river, lacustrine or marine environments and even to reconstruct the direction of transport (e.g., Pettijohn, 1975; Potter and Pettijohn, 1977; Füchtbauer and Müller, 1977; Blatt et al., 1972; Folk, 1974; Scholle, 1979; Wurster, 1958).

The chemical environment covers a wide range of situations. Variations in salinity, element distribution, redox potential or pH are critical variables. This, together with changes in water temperature and pressure, will profoundly affect chemical imprints upon a sediment.

The biological environment, the biotope, is characterized by a distinct fauna and flora, as well as ecosystem(s). Since organisms populate only an ambience best suited to their metabolic demands, fossils are excellent facies indicators. Even a lack of fossils in a sediment which normally should be rich in them may reflect upon facies.

Salinity is a particularly sensitive regulator of aquatic life. A freshwater, brackish, marine or hypersaline facies can readily be identified by the type of fossils left behind in the strata. On the other hand, life processes can influence environmental parameters. Take organisms that produce or utilize molecular oxygen, and consume or release carbon dioxide; they will change the redox potential and pH of a system. This has in turn a decisive impact on mineral equilibria and will find its precipitation in the chemofacies of a rock.

Modern environments can be described quite accurately in terms of physical, chemical or biological parameters. Recent sediments originating in such habitats accordingly acquire certain shapes, compositions and biological communities, all of which are useful fingerprints of facies (Reading, 1981; Reineck and Singh, 1980). Unfortunately, however, the stratigraphic record is full of sedimentation gaps and estimates on times of non-deposition run as high as 90 percent for all of Earth history, although figures of one third to a half are probably more realistic (Badgley et al., 1986).

A hiatus between two layers often provides more insight into the state of a depositional environment than the stratum itself. And yet, it is common practice to pay less attention to a gap covering a time span of perhaps a millennium than to a layer that formed in an instant. This situation reminds me of a history book where a one-day battle is glorified and reported on in great detail, whereas a century of peace is just a filler between two wars, even though that long-lasting peace time probably holds the secret as to why a day's battle arose in the first place. Take – but with a grain of salt – the case of the "Tirpitz Plan" proposed by Admiral Alfred von Tirpitz (1849–1930) at the turn of the century to bring the Imperial German Navy up to par with His Majesty's Fleet, the British Royal Navy. A few years later a tax was imposed on salt partly to finance this ambitious plan, which in spring 1916 led to the battle of the Skagerrak, where many a good sailor and ship was lost in the salty sea. All that is left today of the "Tirpitz Plan" is a massive iron facies at the bottom of the Skagerrak and the tax levied on table salt.

Two Case Studies

An illustration of what can happen on the way to a new sediment layer, formation and evolution of a black shale facies is presented in Fig. 8.28. Here I outline a scenario where a fully oxygenated environment (a) changes to a stratified one (b); with a lowered pycnocline, massive carbonate extraction will take place (c). An upward progression of the interface (d) will extend anoxic conditions to shallower parts of the basin, with established benthic communities becoming extinct and carbonates starting to dissolve because the level of carbonate compensation in time becomes identical to the O_2-H_2S interface. This process may continue repeatedly. The cyclicity of sedimentation is consequently tuned to the speed and frequency at which a pycnocline rises or falls. It can be a seasonal effect or 100 years' event. In the last instance, it is the time of the hiatus during which the critical environmental changes leading to a new sediment layer proceed, and it is here where major gaps in our knowledge exist.

Let us now turn from the local to a more global scene, that is the regional distribution of modern pelagic deposits (= open sea sediments) in world ocean (Fig. 8.29). About half of the deep sea is covered by calcareous ooze, and one third by the non-calcareous red clay. The rest is siliceous ooze, glacial de-

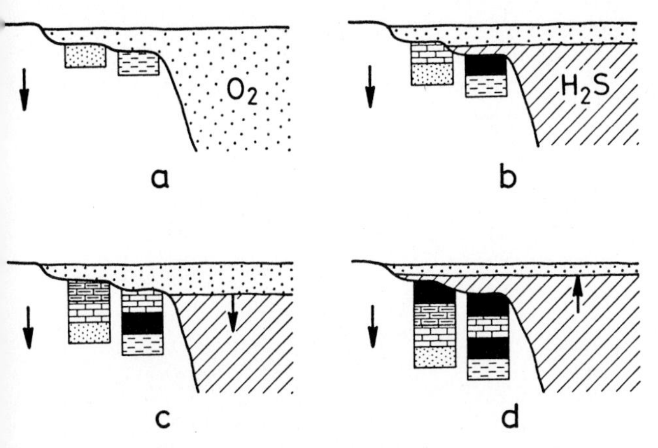

Fig. 8.28 a–d. Generation and development of stratified waters (see text; Degens and Stoffers, 1977)

posits, and continental-margin sediments, plus some minor varieties such as phosphorites and manganese nodules. In the light of pronounced ocean circulation in surface waters which would readily carry, for instance, sea water near Galapagos to Australia in a year's time (Fig. 8.30) the neat arrangement of the pelagic sediment carpet comes unexpected.

An explanation for this distribution pattern has been presented by Heezen and Hollister (1971), Davies and Gorsline (1976), Jenkyns (1978), and Hallam (1981). From their discussion, it appears that primary productivity, ocean currents, and position of carbonate compensation depth are the three governing factors. All others such as closeness to river outlets or shelf edge, climatic zonation, volcanism, desert winds, earthquakes, hydrothermal activity, ocean-floor ecology, or submarine weathering (cf. halmyrolysis) may be viewed as feedbacks or are of a modulating nature only. In short, the living cell is considered to be the main force weaving the facies carpet of the sea. Carbonate ooze is chiefly composed of coccolithophorids and foraminifera (single cell zooplankton). Siliceous ooze consists principally of diatoms in high, and radiolaria (unicellular zooplankton) in low latitudes. Red clays are in part altered volcanic ash and aeolian dust, but still carry clays monitored by organisms in the form of marine snow. All this suggests that pelagic sediments are a mirror image of the kind and intensity of aquatic life at any one time in the geologic past.

Fig. 8.29. Distribution of the principal types of sediment on the floors of the oceans (Davies and Gorsline, 1976)

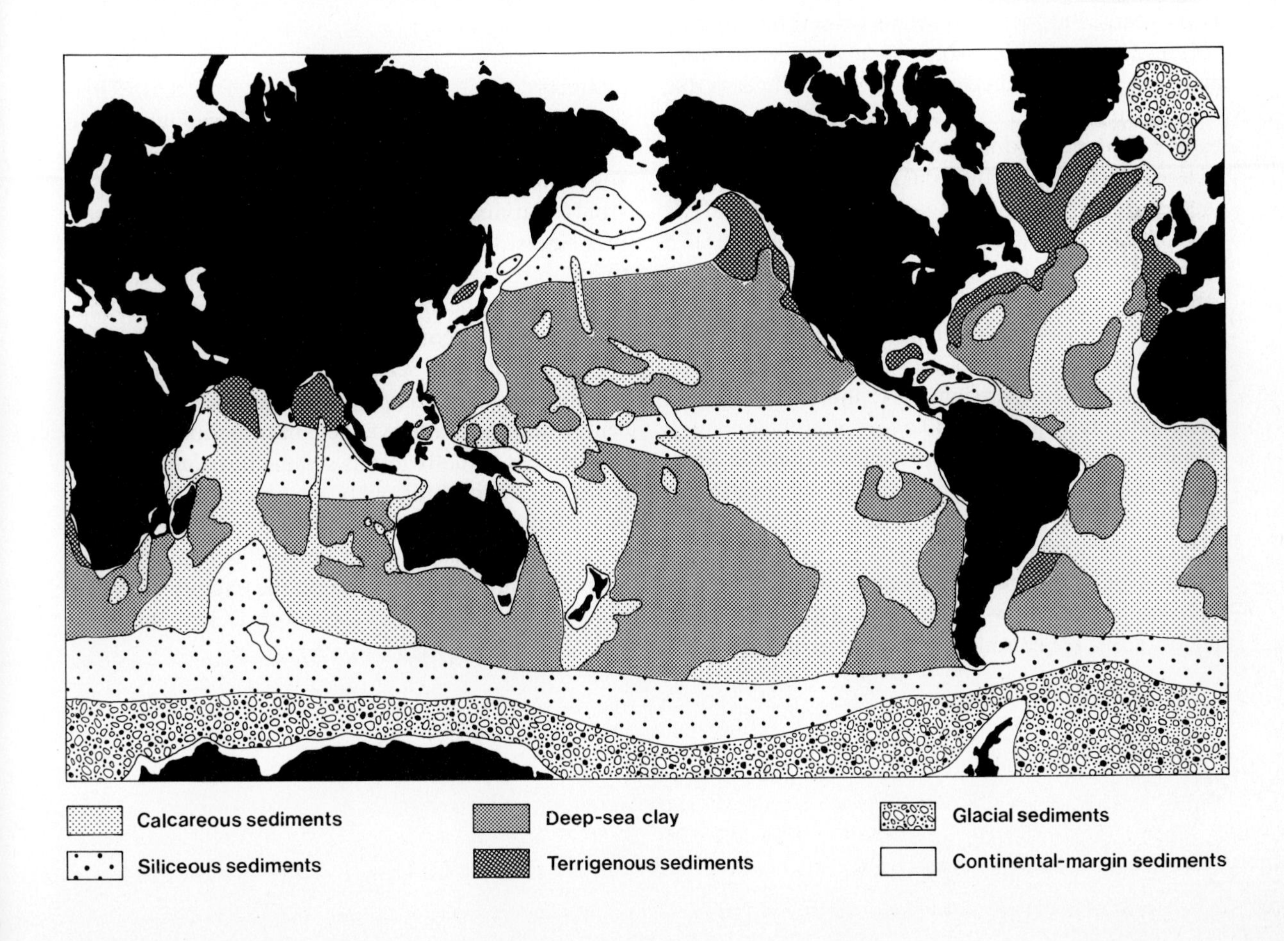

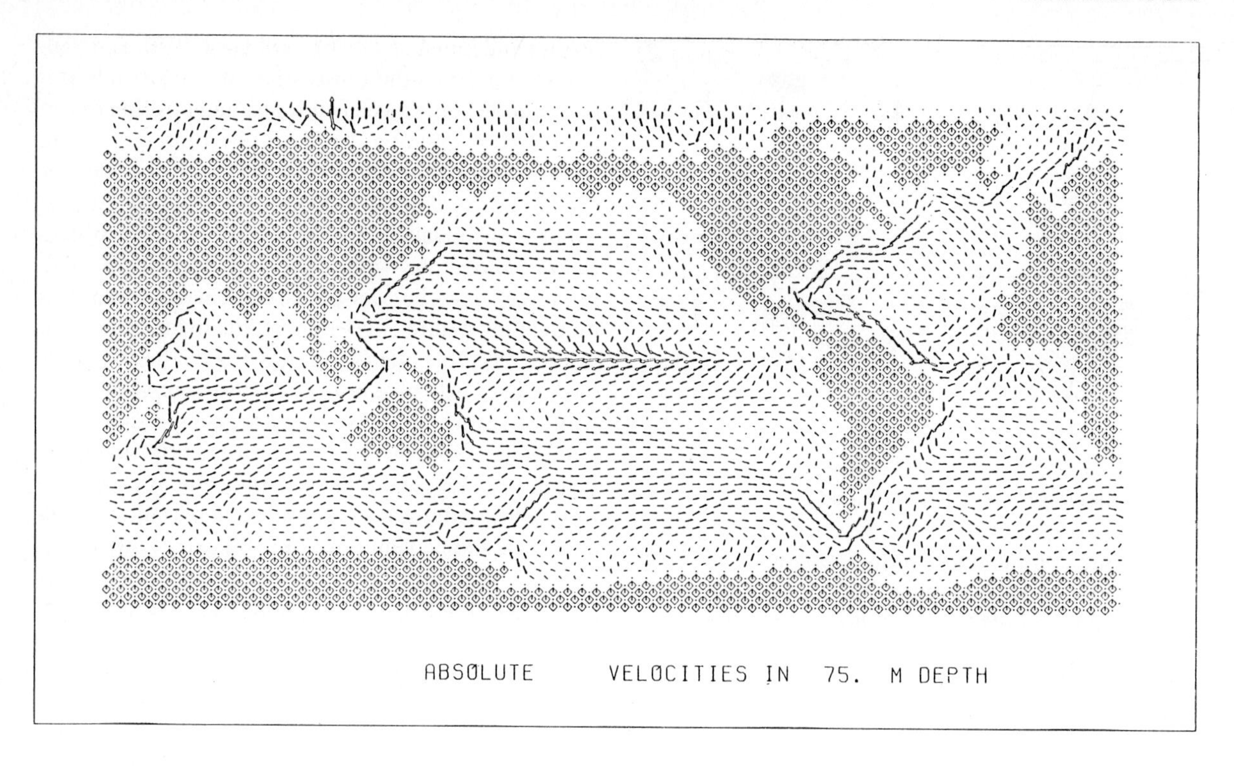

Fig. 8.30. Surface ocean circulation as simulated by the model of Maier-Reimer and Hasselmann (1987)

Rates of sedimentation in the deep sea ordinarily range from a few millimeters to a few decimeters per 1,000 years. But there are events caused by turbidity currents or mud slides which locally can yield much higher deposition rates; or take the opposite case, the minimal accretion rate of manganese nodules which is commonly measured in mm per million years. Much of this information stems from research conducted by the deep-sea drilling research vessel GLOMAR CHALLENGER and more recently by the JOIDES RESOLUTION (see Box 8.4). In essence, we have a fair knowledge on the pelagic facies of the last 180 million years, but everything before that date is sketchy to say the least. Deep-sea sediments of pre-Jurassic time are occasionally found in association with ophiolite sequences and a few often highly questionable occurrences are reported from other settings. For example, the more than 500 million year old Saint Petersburg blue clay in the vicinity of Leningrad is supposed to be of deep-sea origin.

The majority of marine sediments now exposed on land are of a shallow-water facies. They were once generated in epi-continental shelf seas on carbonate platforms or in geosynclines (see Box 8.5). Their deep-sea counterparts became mostly subducted and thus vanished from the sedimentary rock scene. This has serious consequences, because we have – in a way – only marginal information about the geologic past. It is almost like analyzing the ring in a bath tub and trying to reconstruct from that deposit some features of the water, the body, or a person's bathing habits. Take the fossil record. Sedimentation gaps in ordinary sediments and the scarcity of pre-Jurassic deep-sea sediments jointly comprise an equivalent of 95 percent of all potentially available information. From the few percent actually left as sediment, paleontologists delineate the path of evolution, geochemists outline the chemical history of air, sea and earth, and geologists reconstruct the paleogeography through time. I do not want to question the validity of the results, but only to remind everybody of how small our data bank actually is (Box 8.6).

BOX 8.4

Initial Reports DSDP, 1–96, 1969–86 (American Geological Institute, 1975, 1976)

Research on pelagic sediments has for a long time been hampered by the limitation on lengths of sediment cores obtainable. The practical length has increased from less than one meter in 1935 to about 10 meters in 1960; exceptionally, cores 20 meters long have been recovered.

In spring 1961, marine sedimentologists took advantage of a geophysical pilot project aimed at drilling through the oceanic crust down to the Mohorovičić Discontinuity. A tentative site was chosen off Puerto Rico, where a shallow depth for the Moho was indicated. As a kind of feasibility study an "Experimental Mohole" was drilled near Guadalupe Island, Mexico, at 28°59'N; 170°30'W; water depth: 3,566 m (Rittenberg et al., 1963). The core penetrated about 2.5 m of red clay followed by 172.5 m of hemipelagic ooze, both calcareous and siliceous representing about 15 million years of ocean history. Below the ooze, the drill encountered basalt.

The Mohole itself never materialized because during the lengthy discussion the costs rose sharply year by year with no foreseeable end in sight. All the same, Experimental Mohole showed the feasibility of drilling at water depths of several kilometers. In a way, the Deep-Sea Drilling Project (DSDP) established by the U.S. National Science Foundation in 1964 under the heading JOIDES (Joint Oceanographic Institutions for Deep Earth Sampling) may be viewed as an outgrowth of this experimental venture.

Following a successful drilling program in the early summer of 1965 in the Blake Plateau region (off Jacksonville, Fla.) by the motor vessel CALDRILL I, a new drilling vessel capable of drilling in water depths up to 6,000 m and with a penetration of 1 km into the seafloor was designed. The keel for the GLOMAR CHALLENGER was laid on October 18, 1967; she was launched on March 23, 1968 (400 feet long; 65 feet in beam; displacement: 10,500 tons when loaded). A new era of marine geology began (American Geological Institute, 1975, 1976).

Leg 1 of the GLOMAR CHALLENGER's first cruise commenced August 11, 1968. During its 15 years of operation the vessel completed 96 cruise legs, each lasting about 60 days, and drilled 624 holes in various parts of the world ocean. Lengths of cores retrieved ranged from a few centimeters to more than 1,000 meters. The Initial DSDP Reports alone fill roughly 10 m of bookshelf and the scientific output across all fields of the earth sciences comprises thousands of articles. It is beyond the scope of this section to paint a picture of the achievements of DSDP. In a summary of the Conference on Scientific Ocean Drilling (JOIDES, 1982) it says, "The continuous and detailed record of microfossils preserved in ocean sediments may give the best data for describing evolutionary changes and for understanding their causes. Sediments bear the imprint of ocean temperatures and currents, information critical to the reconstruction of oceanic circulation of the past and hence to the reconstruction of ancient climates. Drilling provides access to the rocks of the oceanic crust and thus helps to unravel their structures and motions, information required to understand the phenomena of seafloor spreading and continental drift, and, more broadly, the structure of the earth as a planet. Deep-sea sediments record the contributions of the rivers and winds of the past and thus the history of the continents, records otherwise lost by erosion of the land. In addition to greatly increasing our knowledge of earth history in general, the scientific information gained by drilling is basic to the search for mineral and petroleum resources both on land and beneath the seas. As the ocean is the last frontier for these resources, the importance of a thorough understanding of its geologic history and framework cannot be overstated."

With the new and versatile flagship JOIDES RESOLUTION, the Ocean Drilling Program (ODP) continues. At the moment of writing, I received a copy of a telex from the co-chief scientists Bill Bryan and Thierry Juteau aboard JOIDES RESOLUTION: "Site 669 is located on the gently sloping summit of a mountain forming the western rift valley wall of the Mid-Atlantic Ridge, immediately to the south of the Kane Fracture Zone..... fine-grained aphyric basalt rubble was recovered, with distinct alteration halos and orange clays on the outer surfaces. The basalt derives from the higher slopes of the mountain; the nature of the underlying basement remains unknown." JOIDES RESOLUTION departed Site 669 at 0200 hours June 2, 1986, to the next drilling site, to begin logging operations there.

BOX 8.5

Geosynclines (Dott and Shaver, 1974; Hallam, 1981; Mitchell and Reading, 1978)

Elongated sediment troughs bordering continents in a zone 100 to 200 km wide are generally considered the final resting site for the bulk of the detritus eroded on land. These troughs had been named geosynclines by Dana (1866), but it was actually James Hall (1859) who first discovered that rates of basinal subsidence and sedimentation are controlled by the same tectonic events that shape the nearby mountain belts.

For more than 100 years this concept remained a hallmark in geology. Flysch and molasse are familiar names even to a first year freshman in geology. Flysch, a colloquial lithological term from Simmental/ Switzerland introduced by Studer in 1827, is a geosynclinal facies composed of alternating sequences of poorly sorted sandstones, mudstones and clays. They were generated during a synorogenic phase by means of turbidity currents or debris flow sliding down the continental slope. Frequently, flysch facies commences with black shales.

Molasse facies follows the main orogenic event. It is dominated by well-sorted sandstones and conglomerates that formed in shallow-water marine environments. In case sedimentation exceeds subsidence, a freshwater molasse facies develops with intercalations of chemical sediments, red beds, or coal seams.

Through the years the classical concept of a geosyncline became extended and refined. For instance Hans Stille, about 60 years ago (Stille, 1924), distinguished between geosynclines exhibiting volcanic activity, which he named eugeosynclines, and others that were free of igneous rocks, which he termed miogeosynclines (see also Aubouin, 1965; Dott, 1974; Dott and Shaver, 1974).

At present, the general concept of a geosyncline is only maintained in its rudimentary form in that sedimentation in a trough is controlled by ongoing tectonism. According to Mitchell and Reading (1978), it is more appropriate to relate sedimentary facies to individual tectonic settings within the global structural frame of plates, such as spreading-related, sub-

Fig. 8.31. Facies succession in the Rhenish Slate Mountains (right bank, Ebbe-Anticlinorium and vicinity) during the Paleozoic. Starting with carbonate- free marine black shales of Cambrian to Silurian age, a lagoonal restricted marine environment became established in the early Devonian with alternating sections of carbonate and black shale. This facies was gradually replaced by a highly productive and ventilated shallow sea as indicated by the abundance of fossiliferous shales. A shoaling led in time to the deposition of deltaic sandstones, and eventually the sea retreated. Coarse-grained freshwater sediments, that is sandstones and conglomerates, dominate the scene for the entire lower Devonian, and acidic volcanism was widespread. A marine transgression terminated this facies and first shales and subsequently massive marine limestones characterize the facies during the middle and upper Devonian. Basic volcanism and associated siliceous limestones and shales represent the main facies during the lower Carboniferous. A massive greywacke (sandstone) facies terminates this succession. The Upper Carboniferous is typified by the wide occurrence of coal seams.
This development can largely be seen as a response to the migration of the Rhenish Geosyncline from the polar region to the equator and the complex tectonic happenings in the course of the Caledonian and Hercynian orogenies

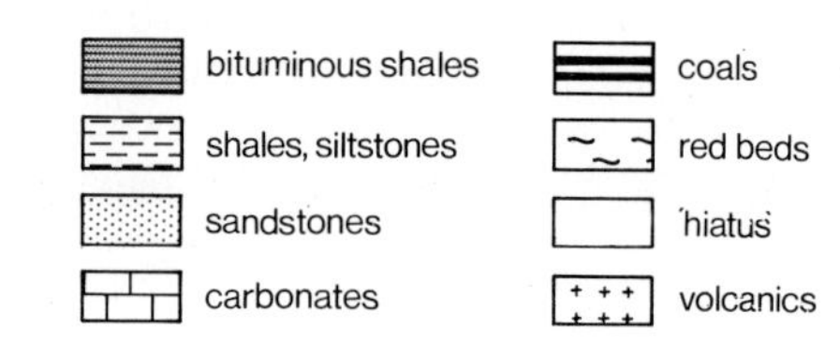

duction-related, strike-slip fault-related and continental collision-related settings (see also Hallam, 1981). However, complication arises because the lifetime of a sedimentation trough can range from several million to a few hundred million years, during which tens of kilometers of sediments are deposited. In the course of this extended period, a trough attached to a continent or a plate margin can move across the sea and tectonic, climatic, or hydrographic settings may change accordingly. Thus, interpretation of sedimentary facies requires a more mobilistic view.

In the following, the global Paleozoic wanderings of a small region, the Sauerland, about 100 km northeast of Bonn (FRG) are briefly sketched quasi as an argument against fixistic models.

The Sauerland region is part of a sedimentation trough that started to form in the late Precambrian to early Paleozoic time and which was attached to Baltica, one of the continents of that period (see Fig. 7.18). At the end of the Cambrian, carbonate-free deep-water black shales were laid down and this facies was maintained for about 50 million years. Following a hiatus of about 40 million years, shallow water calcareous black shales took over, eventually grading into a hypersaline facies dominated by alternating sequences of carbonates and black shales. Another hiatus commenced, and after a few million years a highly ventilated and fertile shallow sea developed, resulting in the deposition of fossiliferous shales and mudstones. The facies gradually changed to sandstones and conglomerates, first of a deltaic and subsequently of a freshwater facies. Volcanic ash layers, ignimbrites and keratophyrs indicate a change from a miogeosyncline to a eugeosyncline in the sense of Stille (1924). Volcanism eventually ceased, brackish-marine conditions became established and in Middle Devonian time a fully marine facies developed with coral reefs, beach sands and a massive limestone deposit. Subsequent stages are again characterized by volcanic events as one of the expressions of ongoing Hercynian mountain building.

Facies successions and some relevant details are shown in Fig. 8.31.

It is concluded that in relating sedimentary facies to global plate tectonics, the classical concept of geosynclines took on a new dimension, namely that of lateral changes of a sedimentation trough as a whole. Feedbacks on climatic change, hydrography, biology and others are associated with such movements, and were hitherto not considered or else explained in a fixistic way. In the light of this development one should re-examine sedimentary facies of the Precambrian, where tectonic regimes and settings different from those of the Phanerozoic existed.

BOX 8.6

Sediment Fluxes to the North Atlantic (Van Andel, 1983; Ehrmann and Thiede, 1985, 1986; Summerhayes and Shackleton, 1986)

Sediments recovered during the Deep Sea Drilling Project give a vivid account on the dynamics of the various stages in the evolution of ocean basins in the course of the past 180 million years. Paleooceanography, as this new branch of marine geosciences has been termed, envelopes physical, chemical, biological and geological phenomena, all of which actually seem to be interrelated.

The best-surveyed ocean basin of our globe is the North Atlantic Ocean, into which in over 100 deep-sea drill sites about 170 drill holes have been "punched".

Three major depositional sites can be discerned: (i) western basin, (ii) eastern basin, and (iii) Norwegian-Greenland Sea basin. The first two are separated by the Mid-Atlantic Ridge, whereas the volcanic Greenland-Scotland Ridge running along the line Scotland-Iceland-Greenland sets apart (i) and (ii) from (iii). Except for some minor anomalies, all three basins are underlain by oceanic crust upon which a sediment cover of up to more than 1 km has piled up over the past 130–150 million years. It is of note that plate tectonic evolution of the North Atlantic from start (Middle Jurassic) to now is minutely recorded in the magnetic anomaly pattern of the crust (Hanisch, 1984; Pitman and Talwani, 1972; Talwani and Eldholm, 1977; LePichon and Renard, 1982). Furthermore, the temporal development of paleogeography and paleobathymetry is known in detail (Sclater et al., 1977; Bédard, 1985; Thiede, 1979; Rona, 1980; Emery and Uchupi, 1972, 1984). And finally paleocirculation or, as it was so nicely put, "commotion in the ocean – in relation to paleotectonics" is a well-understood phenomenon (Berggren and Hollister, 1974, 1977).

In this context only the sediment fluxes to the deep North Atlantic are of concern. Southam and Hay (1977) have calculated sediment volume versus age, and Moore and Heath (1977) have expanded on factors which enhance or lower the probability of sediment accumulation. The smooth temporal sedimentation curve of Southam and Hay (1977) found no sup-

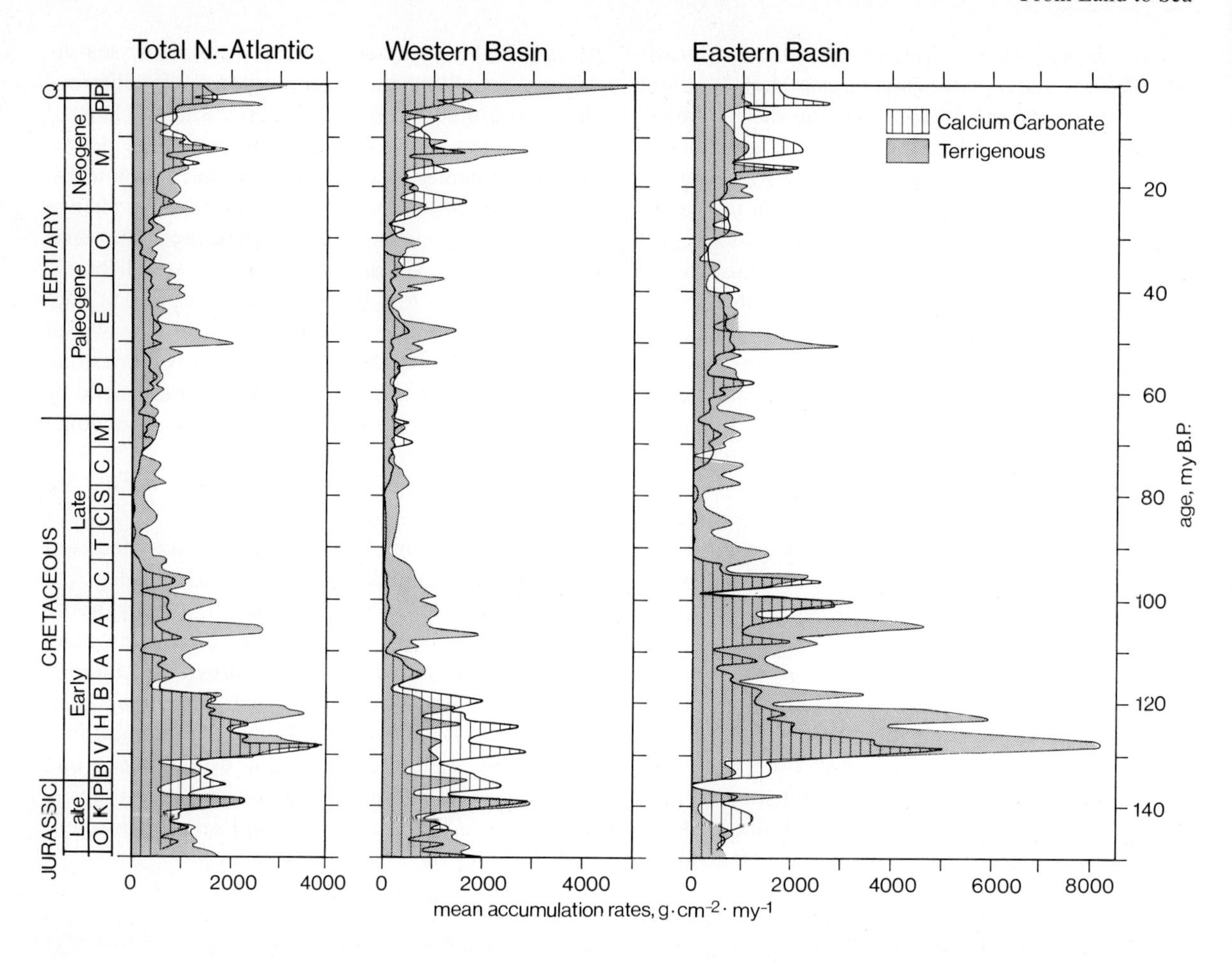

Fig. 8.32. Mean accumulation rates of terrigenous and biogenic calcareous sediment components in the three main subbasins of the North Atlantic Ocean, i.e., Norwegian-Greenland Sea, Western, and Eastern North Atlantic since Late Jurassic representing ca. 150 million years of sedimentation history. Left graph depicts the sum total of all three subbasins, whereas the data for the western and eastern subbasins are individually plotted (middle and right graph). It is of note that the amounts of terrigenous and calcareous contributions are positively correlated, emphasizing the role of plankton blooms for scavenging the fine detritus. Dissolution of deep-sea calcium carbonate in response to carbonate compensation depth (CCD; Berger, 1978) appears to be the main factor in lowering the correlation coefficient between terrigenous and biogenic calcareous sediment components (after Ehrmann and Thiede, 1986)

port in the work of Ehrmann and Thiede (1985, 1986), who could show an enormous spread in the ratio of mean accumulation rate and hiatus distribution in the North Atlantic deep-sea sediments (Fig. 8.32). Three major events could be discerned: (i) 150–110 Ma B.P. when bulk accumulation rates were high and sedimentation gaps few, (ii) 110–10 Ma B.P. when sedimentation was at a minimum and hiatus distribution high, and (iii) 10–0 Ma B.P. when sediment deposition rose by an order of magnitude, and sedimentation gaps again declined in number. All three events can be traced across all three basins of the North Atlantic and seem to exist also in other deep ocean basins (Worsley and Davies, 1979). Thus, these are events of global nature.

The younger hiatus maximum seems to be related to the acceleration of bottom-water circulation in response to climatic changes in the polar regions. A satisfactory explanation for the Mesozoic hiatus peak is harder to come by. In all probability sediment gaps are due to changes in the paleocirculation regime, perhaps related to the generation of new oceanic pathways.

The ocean floor is constantly subducted and sediments are eventually transformed to metamorphic rocks. We commonly speak of epi-, meso-, and katazonal facies to describe the pressure and temperature conditions to which a metamorphosed rock slab has been subjected. Facies indicators are specific minerals or mineral assemblages. For instance, the katazonal eclogite facies is characterized by two minerals, that is omphacite (a pyroxene) and pyrope (a garnet).

Sedimentologists are quite unhappy that their well-defined term facies was adopted by metamorphic petrologists. It was P. Eskola who, in 1920, proposed this switch of facies. I see no harm in such a transaction, because appearance, look or fabric of a rock serve here and there as diagnostic criteria of the conditions that shaped a sediment at the time of deposition or a metamorphic rock during its formative years.

Across the Sands of Time

In his book "Passage to Ararat" first published as a three-part memoir in February 1975 in The New Yorker, Michael J. Arlen describes a voyage to the region around Mount Ararat. It is a masterful story telling of a people and of a country from the moment civilized life started to appear in the valleys of Eastern Turkey about 5,000 years ago.

It was the tribes of Hurian and Hittite stock that settled here first. They mined for tin and copper and began to tend wild vines. The Kings of Nairi ruled this land. Out of Nairi came the kingdom of Urartu, and its capital stood near the shores of Lake Van.

From the remains of the tribes of Urartu came the people of Eastern Anatolia, and in Xenophon's "Anabasis" we learn much about their way of life. Actually, on his route to the Black Sea, Xenophon had visited Lake Van. The land was famous for its horses, and we hear that at the time of Xerxes as many as 25,000 horses were sent down from Eastern Anatolia to Persia every year. Maybe this was the reason that Persians, Greeks and Romans tried to control the East Anatolian Plateau in the years to come, because at that time horses and cavalry were the basis for victorious wars. The Kingdom of Armenia was founded by Artaxias I, and in 301 A.D. Christianity became the state religion. Almost 1,000 years later, however, in 1375, East Anatolia fell to the invading Mamelukes. Eventually, the Mameluke Empire was supplanted by the Ottoman Empire, which in turn vanished with the advent of Atatürk, the founder of modern Turkey.

Noah's ark, according to some recent sources that claimed to have found its remains, is supposed to have touched safe ground near the top of Mount Ararat after the 40 days flood ceased. This event – if we follow the Old Testament closely – can be placed at 4000 B.C. In summing up the biblical ages of the archfathers, the Anglican Bishop Usher arrived at a date of 4004 B.C. for the creation of the world. However, this age assignment must have already been the common belief, since the first Christians celebrated the four weeks of advent, one week each, for 1,000 years passed before Christ even was born: Meta no eite!

Back to Ararat and Noah. Close to the foot of Mount Ararat lies Lake Van, the largest soda lake on Earth, which ranks fourth in volume (607 km^3) among the closed lakes of the world; surface area is 3,574 km^2 and greatest depth is 451 m (Degens and Kurtmann, 1978). Sediment cores retrieved from the lake are as colorful as the history of mankind in this area (Fig. 8.33). An uninterrupted sequence composed of varved sediments and volcanic ash layers helped to provide a year by year account of the past 10,240 years of geological history (Kempe, 1977; Degens et al., 1984c).

With respect to the sediment layer deposited at the time Noah was supposed to have lived, there was indeed a low water level of –300 m relative to present lake level. Within a few hundred years the water surface rose by 250 m and an area of 1,500 square kilometers became flooded. Could this be the event that started the story of the biblical flood?

Varve counting as a means of absolute age dating was first proposed by de Geer in 1910. He observed in lake deposits of southern Sweden that each year, due to seasonal events, one light and one dark layer were added to the sediment column. Having varves as an absolute yardstick of time, de Geer calculated from these lacustrine deposits an age of 10,000 years as having elapsed since the retreat of the Weichselian ice shield in southern Scandinavia. This date conforms to the full length of the youngest of the geological epochs, the Holocene.

The Holocene is just a brief warm interlude within a general cold climatic setting. Such global warm interludes, or to use the proper term, interglacials, are paired with glacials, the true ice ages. One glacial-interglacial cycle has a periodicity of about 100,000 years and its duration appears to be principally governed by orbital elements of the Earth and solar feedbacks (Toon et al. 1980b; Schove, 1983). Close to 20 such cycles were measured during the Quaternary. The Plio-Pleistocene boundary or the transition from the "warm" Tertiary to the "cold" Quaternary has almost by "decree" been fixed at 2.0 Ma (Fig. 8.34).

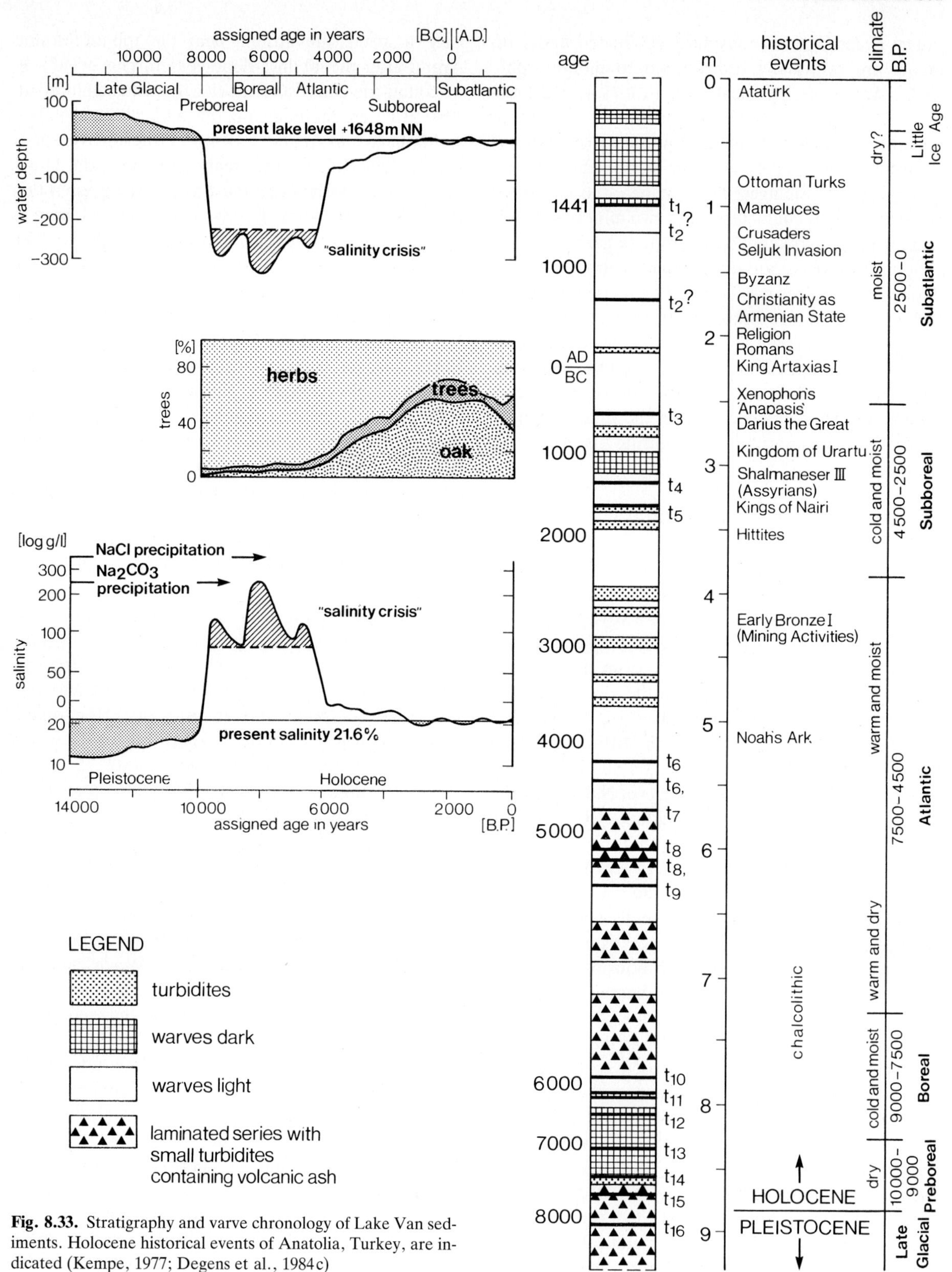

Fig. 8.33. Stratigraphy and varve chronology of Lake Van sediments. Holocene historical events of Anatolia, Turkey, are indicated (Kempe, 1977; Degens et al., 1984c)

Fig. 8.34. The youngest period in Earth history, the Quaternary, covers a time span of about 2 million years. Many attempts have been made to intercalibrate the various chronostratigraphic time scales proposed in the literature, but no common consent could be reached, except for the latest epoch, the Holocene, which envelops the last 10,000 years. The next epoch, the Pleistocene, has its base – the so-called Pliocene-Pleistocene boundary – tentatively fixed at 2 million years before present (Harland et al., 1982). However, a further subdivision of the Pleistocene encounters some problems because terrestrial and marine stratigraphic sequences can only be roughly correlated. Inasmuch as climatic change is the landmark of the Quaternary, glacial and interglacial events will serve as a means chronologically to subdivide the Quaternary. For northern and northwest Europe mean July temperatures based on pollen diagrams (Zagwijn, 1975; Hammen et al., 1971) are about 18° C during interglacials and 8° C during glacials. The terminology in usage by Quaternary land geologists from central and northern Europe as well as their proposed timing of events has been adopted. Polarity epochs are indicated; it is of note that the Réunion events are still questionable (after Lowrie and Alvarez, 1981)

Continental geologists in Europe would rather see this boundary be placed at 2.6 Ma (Zagwijn, 1975; Hammen et al., 1971) because this was about the time Greenland started to glaciate, thus introducing a bipolar ice age.

The Phanerozoic time-table as it stands today (Table 8.3) is a compromise achieved after several decades of controversial discussions in numerous international stratigraphic correlation commissions. We should bear in mind, however, that a camel, so the saying goes, is a horse defined or invented by a committee: It all started more than 200 years ago when naturalists began to collect and describe plants, animals, rocks and fossils, first at random and then more systematically. Stratigraphy from the start resembled a gigantic jig-saw puzzle to which new pieces were constantly being added. Piece by piece, be it a fossil, a stone, a structure, a relative age, an absolute age

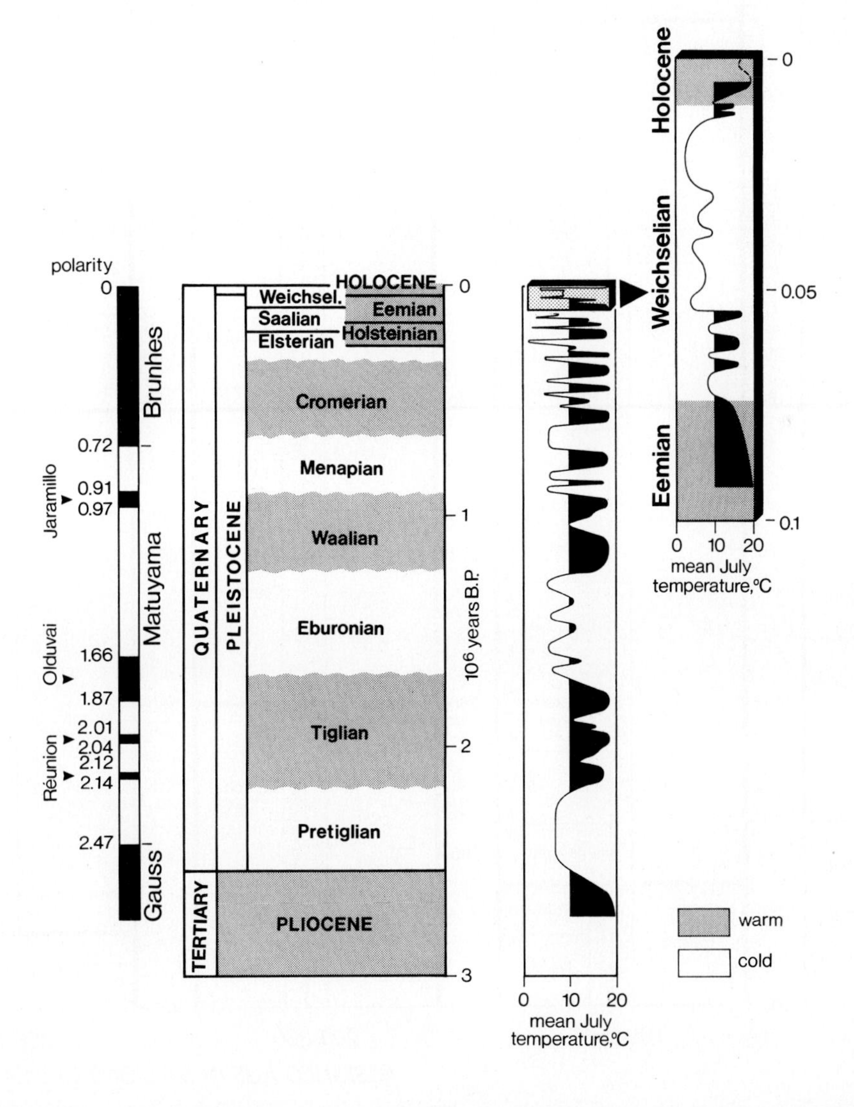

Table 8.3. Stratigraphic time table

0-100 (ASSIGNED AGE IN MILLIONS OF YEARS; scale: 0, 10, 20, 30, 40, 50, 60, 70, 80, 90, 100)

System/Series	Subdivision		Stage
Pleistocene			Glacial
			Calabrian
Pliocene			Astian
			Piacenzian
			Zanclian
Miocene	late		Tortonian-Messinian
	middle		Langhian
	early		Burdigalian
			Aquitanian
Oligocene	late		Chattian-Bormidinian
	early		Rupelian-Lattorfian
Eocene	late		Bartonian / Priabonian
	middle		Lutetian
	early		Ypresian
Paleocene			Thanetian
			Montian
			Danian
Creta-ceous	upper	Senonian	Maestrich-tian
			Campanian
			Santonian
			Coniacian
			Turonian
			Cenomanian

100-200 (scale: 100, 110, 120, 130, 140, 150, 160, 170, 180, 190, 200)

System	Series	Subdivision	Stage	Zone
Creta-ceous	lower		Albian	
			Aptian	
		Neokomian	Barrêmian	
			Hauterivian	
			Valanginian	
			Ryazanian	
Jurassic	Malm (= upper)		Purbeckian	
			Portlandian	.... ζ
			Kimmerid-gian	ε δ γ
			Oxfordian	β α
	Dogger (= middle)		Callovian	ζ
			Bathonian	ε
				δ
			Bajocian	γ β α
	Lias (= lower)		Toarcian	ζ ε
			Pliens-bachian	δ γ
			Sinemurian	β
			Hettangian	α
Triassic	upper	Keuper	Rhaetian	
			Norian	

200-300 (scale: 200, 210, 220, 230, 240, 250, 260, 270, 280, 290, 300)

System	Series	Subdivision	Stage
Triassic	upper	Keuper	Karnian
	middle	Muschelkalk	Ladinian
			Anisian
	lower	Buntsandstein (= Skyth)	Olenekian
			Induan
Permian	upper	Zechstein	Tatarian
			Kazanian
	lower	Rotliegendes	Kungurian
			Artinskian
			Sakmarian
Carbon-iferous	upper	Silesian (= Pennsylvanian)	Stepha-nian
			West-phalian

300-400 (scale: 300, 310, 320, 330, 340, 350, 360, 370, 380, 390, 400)

System	Series	Subdivision	Stage
Carbon-iferous	upper	Silesian (= Pennsylvanian)	West-phalian
			Namurian
	lower	Diantian (= Mississippian)	Viséon
			Tournai-sian
Devonian	upper		Famennian
			Frasnian
	middle		Givetian
			Eifelian
	lower		Emsian
			Siegenian
			Gedinnian
Silurian (Gotlandian)			Ludlow

0-100 | 100-200 | 200-300 | 300-400

ASSIGNED AGE IN MILLIONS OF YEARS

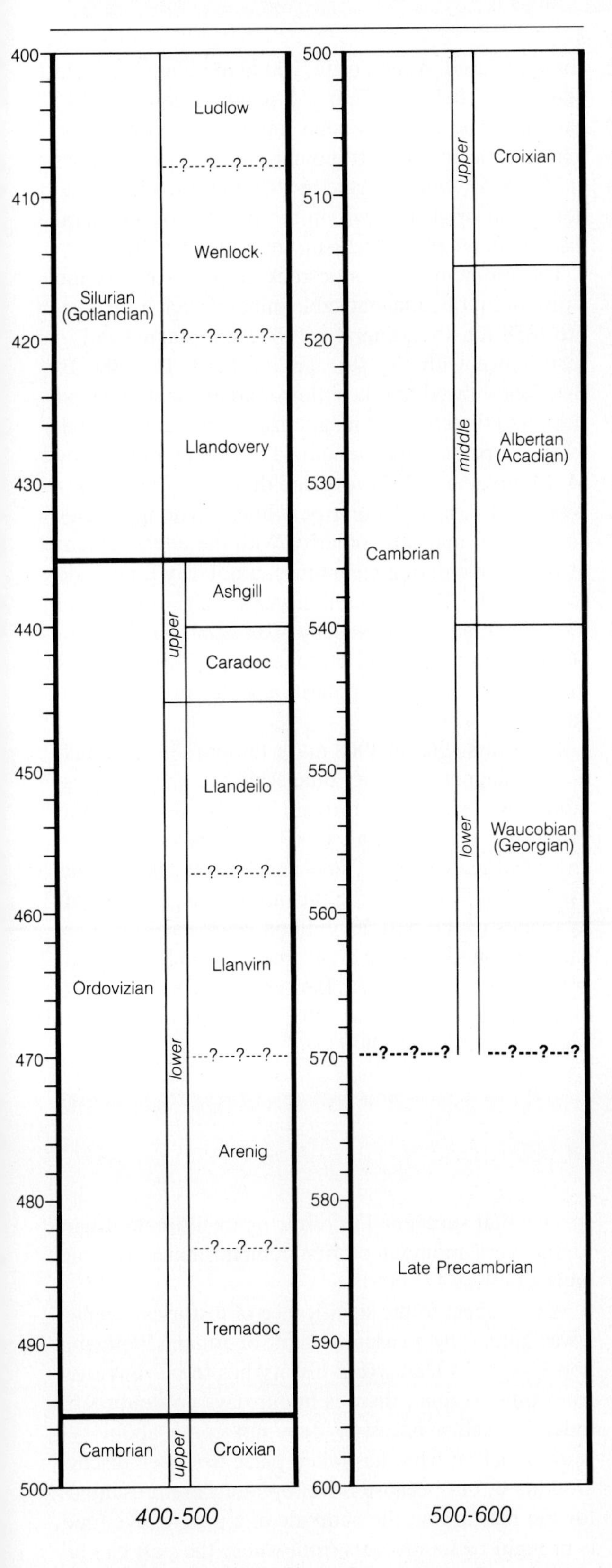

etc. became added to the system. Those pieces of information that at their time made no sense were laid aside, and quite often they became key elements years later. The naming of formations, after native tribes (e.g., Cambrian), regions (Devon), towns (e.g., Perm), or chemical features (e.g., Carboniferous) has a long-standing tradition. The 19th century European scientific community collectively came up with a biostratigraphic time-table in which national elements are deeply rooted and still form the base of a unifying concept.

Stratigraphic boundaries are quasi-congealed global events that caused extinctions or first appearance of certain species and genera or the occurrence of distinct types of sediments, such as the Cretaceous (Kreide) or the Buntsandstein, or the onset of biomineralization across all taxa close to the Precambrian/Cambrian boundary, or a reversal in the Earth's magnetic field (Box 8.7). Unfortunately, the Phanerozoic time-table gives no clues as to the precise nature of the global events responsible for many of these boundary phenomena. Quite often "deus ex machina" explanations are invoked, for instance a bolide impact, the first appearance of molecular oxygen in the air, or a sudden rise or fall of sea level. Yet it could well be that what we witness as an event is just the last drop in a global "acid-base" titration scheme, to which organisms responded by becoming extinct or by starting to evolve.

Many species of high global abundance cover a time slot of only a few hundred thousand years at the most. These are the index fossils, the true backbone of biostratigraphy. In my student days I had to memorize about 500 of them by shape, Latin name, and stratigraphic horizon. It was too bad that nobody at that time could give us a good answer as to why index fossils "suddenly" show up in the stratigraphic record in large numbers and worldwide, and vanish shortly thereafter, while coexisting species freely traverse the stratigraphic interface and prevail morphologically unchanged for hundreds of millions of years. Considering the fact that the bulk of index fossils are represented by their shells, bones or teeth, and organisms without hard parts become rarely fossilized, we – as far as representative sampling is concerned – are only looking at the tip of an iceberg.

Much attention is presently paid to interfaces where mass extinctions happened such as at the Cretaous/Tertiary boundary. The days may be not too far off, when the physical, chemical or geological factors behind the extinction or emergence of species are brought to light. Should this happen, then memorizing hundreds of index fossils by heart will make sense,

BOX 8.7

Geomagnetic Polarity Time Scale (Regan et al., 1975; McDougall, 1979)

It is common knowledge that the Earth has a magnetic field. Even though a giant bar magnet at the center of the Earth could readily explain the observed phenomenon of a dipole field at the Earth's surface, mineralogy rules out such an explanation because heat would erase permanent magnetism once temperatures exceed about 500° C (Curie point; named after Pierre Curie, 1855–1906). Instead, the Earth's magnetic field is assumed to be a result of current flow in its metallic liquid core (dynamo principle).

Strength of the Earth's magnetic field changes. For instance, the field has declined by about 6 percent since the beginning of the nineteenth century, when the first precise measurements became available. Expressed differently, should this trend continue, the Earth's magnetic field might vanish in 2,000 years. This would have considerable consequences, particularly to life, because the full force of high energy cosmic radiation, now deflected by the dipole field, would strike the Earth's surface and possibly induce mutations. Other secular variations, i.e., inclination and declination, have been reported the world over for several centuries. The explanation is offered that fluid motions in the core generate not only the main field, but account for its temporal changes originating perhaps in small eddies within the large-scale convective motion (see Press and Siever, 1986). The idea has also been advanced that earthquakes, or bolides, keep the magnetic dynamo going. Energies can be temporarily stored in the liquid core and gradually become released. A meteorite, 300 m in diameter, could generate a field of 50 to 150 gauss (1 gauss = 10^{-5} gamma). The unit has been named in honor of the mathematician and astronomer Karl Friedrich Gauss (1777–1855), who asserted in 1839 that the main magnetic field originated within the planet; for a general discussion on geomagnetism, see Lilley (1979).

The moment a volcanic rock cools off below the Curie point of a magnetizable mineral such as magnetite (858 K), the mineral will become magnetized in accordance with the surrounding field. In 1906, B. Brunhes showed that lava flows and associated baked clays had directions of magnetization about antiparallel to the present magnetic field. Subsequent work by M. Matuyama in 1929 revealed that early Quaternary lavas had reversed polarity, whereas younger lavas had normal magnetic polarity. With the advancement of radiogenic dating techniques, a polarity time-scale covering the last 3.5 million years of geologic time was eventually proposed by Cox et al. (1964a, b). Gauss, Matuyama and Brunhes were introduced as the main geomagnetic polarity epochs of the Quaternary (Fig. 8.34).

The years 1963 to 1968 mark the era during which the geomagnetic polarity time-scale became firmly established, principally through the work of marine geophysicists: Cox et al. (1963a, b, 1964a, b), Vine and Matthews (1963), Vine and Wilson (1965), and Vine (1966, 1968). In the meantime, magnetic polarity events of the past 200 million years have been accurately dated and have become useful chronostratigraphic indicators (e.g., Berggren and Van Couvering, 1974; Berggren et al., 1985; Harland et al., 1982; Lowrie and Alvarez, 1981).

because it will allow students to relate Latin names to processes formerly alive in the environment and the crust.

In "A la recherche du temps perdu", Marcel Proust traces his life back – in episodes – to the moment of first remembrance. Our Earth, an individual member among a group of possibly a billion other similar planetary bodies scattered across the universe, can only be understood in the context of its ancestral roots and the events that accumulated up to this day. That *Homo sapiens* is regarded as the last major episode in the 15 to 20 billion year history of the universe is a logical consequence of the fact that it is this species that succeeded in retracing its ultimate origin to the very moment of first remembrance, the big bang (Table 8.4).

With respect to the significance of historical events, I was guided by a casual remark of Johann Wolfgang von Goethe, "That world history has to be rewritten from time to time, there is in our days no doubt left, indeed. Such a necessity does not come about because much of what has taken place has been discovered anew, but because novel opinions are presented, for the reason that the comrade of a progressive time is brought to standpoints from where the past can be overlooked in his own way."

Table 8.4. Major events in the universe's history. Time scale is tentative because the precise age of the universe is not known. It is better to refer to such events in terms of the *red shift* which is a measure of the degree of compression of the expanding universe (Barrow and Silk, 1980)

Cosmic Time	Epoch	Red Shift	Event	Years Ago
0	Singularity	Infinite	Big bang	20×10^9
10^{-43} sec	Planck time	10^{32}	Particle creation	20×10^9
10^{-6} sec	Hadronic era	10^{13}	Annihilation of proton-antiproton pairs	20×10^9
1 sec	Leptonic era	10^{10}	Annihilation of electron-positron pairs	20×10^9
1 min	Radiation era	10^9	Nucleosynthesis of helium and deuterium	20×10^9
1 week		10^7	Radiation thermalizes prior to this epoch	20×10^9
10,000 a	Matter era	10^4	Universe becomes matter-dominated	20×10^9
300,000 a	Decoupling era	10^3	Universe becomes transparent	19.9997×10^9
1–2 $\times 10^9$ a		10–30	Galaxies begin to form	18–19 $\times 10^9$
3 $\times 10^9$ a		5	Galaxies begin to cluster	17×10^9
4 $\times 10^9$ a			Our protogalaxy collapses	16×10^9
4.1 $\times 10^9$ a			First stars form	15.9×10^9
5 $\times 10^9$ a		3	Quasars are born; population II stars form	15×10^9
10 $\times 10^9$ a		1	Population I stars form	10×10^9
15.2 $\times 10^9$ a			Our parent interstellar cloud forms	4.8×10^9
15.3 $\times 10^9$ a			Collapse of protosolar nebula	4.7×10^9
15.4 $\times 10^9$ a			Planets form; rock solidifies	4.6×10^9
15.7 $\times 10^9$ a			Intense cratering of planets	4.3×10^9
16.1 $\times 10^9$ a	Archeozoic era		Oldest terrestrial rocks form	3.9×10^9
17 $\times 10^9$ a			Microscopic life forms	3×10^9
18 $\times 10^9$ a	Proterozoic era		Oxygen-rich atmosphere develops	2×10^9
19 $\times 10^9$ a			Macroscopic life forms	1×10^9
19.4 $\times 10^9$ a	Paleozoic era		Earliest fossil record	600×10^6
19.55 $\times 10^9$ a			First fishes	450×10^6
19.6 $\times 10^9$ a			Early land plants	400×10^6
19.7 $\times 10^9$ a			Ferns, conifers	300×10^6
19.8 $\times 10^9$ a	Mesozoic era		First mammals	200×10^6
19.85 $\times 10^9$ a			First birds	150×10^6
19.94 $\times 10^9$ a	Cenozoic era		First primates	60×10^6
19.95 $\times 10^9$ a			Mammals increase	50×10^6
20 $\times 10^9$ a			*Homo sapiens*	1×10^5

References

Chapter 5-8

Addison WE (1961) Structural Principles in Inorganic Compounds. New York, John Wiley & Sons, Inc, pp 176

Allègre CJ, Shimizu N, Treuil M (1977) Comparative chemical history of the Earth, the Moon and parent body of a chondrite. Philos Trans R Soc Lond A 285:55–67

American Geological Institute (1975) Deep Sea Drilling Project. Legs 1–25. AGI Reprint Ser 1, pp 93

American Geological Institute (1976) Deep Sea Drilling Project. Legs 26–44. AGI Reprint Ser 2, pp 82

Ampferer O (1906) Über das Bewegungsbild von Faltengebirgen. Jahrb Geol R-A 56:539–622

Anders E (1977) Chemical compositions of the Moon, Earth, and eucrite parent body. Philos Trans R Soc Lond A 285:23–40

Anderson DL (1977) Composition of the mantle and core. Annu Rev Earth Planet Sci 5:179–202

Anderson DL (1981) Hotspots, basalts, and the evolution of the mantle. Science 213:82–89

Anderson OL (1986) Properties of iron at the Earth's core conditions. Geophys J R Astron Soc 84:561–579

Anderson RN, Honnorez J, Becker K et al. (1982) DSDP Hole 504B, the first reference section over 1 km through Layer 2 of the oceanic crust. Nature 300:589–594

Anhaeusser CR (1971) Cyclic volcanicity and sedimentation in the evolutionary development of Archaean greenstone belts of shield areas. Geol Soc Aust, Spec Publ 3:57–70

Anonymous (1984) Global Ocean Flux Study. Washington, DC, National Academy Press, pp 360

Arndt NT (1977) Ultrabasic magmas and high degree of melting of the mantle. Contrib Miner Petrol 64:205–221

Arndt NT, Naldrett AJ, Pyke DR (1977) Komatiitic and iron-rich tholeiitic lavas of Munro Township, northeast Ontario. J Petrol 18:319–369

Aubouin J (1965) Geosynclines. Dev Geotectonics 1, Amsterdam, London, New York, Elsevier Publ Co, pp 335

Baas Becking LGM, Kaplan IR, Moore D (1960) Limits of the natural environment in terms of pH and oxidation-reduction potentials. J Geol 68:243–284

Badgley C, Tauxe L, Bookstein F (1986) Estimating the error of age interpolation in sedimentary rocks. Nature 319:139–141

Bäcker H (1976) Fazies und chemische Zusammensetzung rezenter Ausfällungen aus Mineralquellen im Roten Meer. Geol Jahrb D17:151–172

Bäcker H (1979) Metalliferous sediments in the Red Sea, occurrence, exploration and resource assessment. In Proceedings of the International Seminar of Offshore Mineral Resources, Germinal and BRGM, 319–338, Document BRGM No 7, Orléans

Bäcker H, Lange K, Richter H (1975) Morphology of the Red Sea central graben between Subair Islands and Abul Kizaan. Geol Jahrb D13:79–123

Ballard RD (1979) Tauchfahrt ins Erzgebirge. Geo 12 (Dec 79):6–32

Ballard RD, Van Andel TN (1977) Morphology and tectonics of the inner rift valley at lat 36°50'N on the Mid-Atlantic Ridge. Geol Soc Am Bull 88:507–530

Bally AW (1983) Seismic expression of structural styles. Vols 1–3. Am Assoc Petrol Geol, Stud Geol Ser 15

Bally AW, Snelson S (1980) Realms of subsidence. Can Soc Pet Geol Mem 6:9–94

Bally AW, Bernoulli D, Davis GA, Montadert L (1981) Listric normal faults. Proc 26th Intern Geol Congr Paris, July 7–17, Oceanologica Acta No SP: Geology of Continental Margins, 87–101

Bardossy G (1981) Karst Bauxites. Bauxite Deposits on Carbonate Rocks. Amsterdam, Oxford, New York, Elsevier Publ Co, pp 441

Barrow JD, Silk J (1980) The structure of the early universe. Sci Am 242/4:98–107

Barth TFW, Newton CR (1982) Metamorphic Rocks. In "McGraw-Hill Encyclopedia of Science and Technology", 5th ed, vol 8, New York, McGraw-Hill, Inc 426–431

Basaltic Volcanism Study Project (1976–1979, 1981) Basaltic Volcanism on the Terrestrial Planets. New York, Pergamon Press, Inc, pp 1286

Baumgartner A, Reichel E (1975) The World Water Balance. München, Wien, R Oldenbourg Verlag, pp 179

Bédard JH (1985) The opening of the Atlantic, the Mesozoic New England igneous province, and mechanisms of continental breakup. Tectonophysics 113:209–232

Beloussov VV (1970) Against the hypothesis of ocean-floor spreading. Tectonophysics 9:489–511

Ben-Avraham Z, Nur A (1982) The emplacement of ophiolites by collision. J Geophys Res 87/B5:3861–3867

Ben-Avraham Z, Nur A, Jones D, Cox A (1981) Continental accretion: from oceanic plateaus to allochthonous terranes. Science 213:47–54

Berger WH (1978) Sedimentation of deep-sea carbonate: Maps and models of variations and fluctuations. J Foraminiferal Res 8:281–302

Berger WH, Winterer EL (1974) Plate stratigraphy and the fluctuating carbonate line. In "Pelagic Sediments: On Land and Under the Sea" (eds Hsü KJ, Jenkyns HC), Oxford, Blackwell, 11–98

Berggren WA, Hollister CD (1974) Paleogeography, paleobiogeography, and the history of circulation in the Atlantic Ocean. In "Studies in Paleo-Oceanography" (ed Hay WW), Soc Econ Paleontol Mineral Spec Publ 20:126–186

Berggren WA, Hollister CD (1977) Plate tectonics and paleocirculation – commotion in the ocean. Tectonophysics 38:11–48

Berggren WA, Van Couvering JA (1974) The late Neogene. Paleogeography, -climatology, -ecology 16:1–216

Berggren WA, Kent DV, Flynn JJ, Van Couvering JA (1985) Cenozoic geochronology. Bull Geol Soc Am 96:1407–1418

Blatt H, Middleton GV, Murray RC (1972) Origin of Sedimentary Rocks. Englewood Cliffs, New Jersey, Prentice-Hall Inc, pp 634

Böger H (1983a) Eine Lithostratigraphie des Unterdevons im Sauerlande und im östlichen Bergischen Lande (Rheinisches Schiefergebirge). I. Das Gebiet entlang dem Nordsaum des Siegerländer Sattels. N Jahrb Geol Paläont Abh 165:185–227

Böger H (1983b) Eine Lithostratigraphie des Unterdevons im Sauerlande und im östlichen Bergischen Lande (Rheinisches Schiefergebirge). II. Das Ebbe-Antiklinorium. N Jahrb Geol Paläont Abh 166:294–326

Böttcher AL, Burnham CW, Windom KE, Bohlen SR (1982) Liquids, glasses, and the melting of silicates to high pressures. J Geol 90:127–138

Bolin B (1980) On the exchange of carbon dioxide between the atmosphere and the sea. Tellus 12:274–281

Borradaile GJ, Bayly MB, Powell CM (eds) (1982) Atlas of Deformational and Metamorphic Rock Fabrics. Berlin Heidelberg New York, Springer-Verlag, pp 551

Borucki WJ, Pollack JB, Toon OB, Woodward HT, Wiedman DR (1980) The influence of solar UV variations on climate. In "The Ancient Sun" (eds Pepin RO, Eddy JA, Merrill RB), Geochim Cosmochim Acta, Suppl 13:513–522
Bosworth W (1985) Geometry of propagating continental rifts. Nature 316:625–627
Brewer PG, Spencer DW (1974) Distribution of some trace elements in Black Sea and their flux between dissolved and particulate phases. In "The Black Sea – Geology, Chemistry, and Biology" (eds Degens ET, Ross DA), Am Assoc Petrol Geol Mem 20:137–143
Brindley GW (1980) Crystal Structures of Clay Minerals and their X-Ray Identification. Mineral Society, London, pp 495
Brinkmann R (ed) (1972) Lehrbuch der Allgemeinen Geologie, Bd 2, Tektonik. Stuttgart, Ferdinand Enke Verlag, pp 579
Brinkmann R (1974) Geologic relations between Black Sea and Anatolia. In "The Black Sea – Geology, Chemistry, and Biology" (eds Degens ET, Ross DA), Am Assoc Petrol Geol Mem 20:63–76
Broecker HC, Petermann J, Siems W (1978) The influence of wind on CO_2-exchange in a wind-wave tunnel, including the effects of monolayers. J Mar Res 36:595–610
Broecker WS (1983) Tracers in the Sea. New York, Eldigio Press, pp 690
Brosche P, Sündermann J (eds) (1982) Tidal Friction and the Earth's Rotation. Berlin Heidelberg New York, Springer-Verlag, pp 345
Brown C, Girdler RW (1982) Structure of the Red Sea at 20°N from gravity data and its implication for continental margins. Nature 298:51–53
Browne SE, Fairhead JD, Mohamed II (1985) Gravity study of the White Nile Rift, Sudan, and its regional tectonic setting. Tectonophysics 113:123–137
Bruneau L, Jerlov NG, Koczy F (1953) Red Sea. Natl Board Fish Inst Hydr Res, Göteborg, Rep Swedish deep-sea expedition, 3, 4 Appendix, Tab 2:14–30
Brunhes B (1906) Recherches sur la direction d'aimantation des roches volcaniques. J Phys 5:705–724
Bryan WB, Moore JG (1977) Compositional variations of young basalts in the Mid-Atlantic Ridge rift valley near lat 36°49'N. Geol Soc Am Bull 88:556–570
Bullen KE (1963) An introduction to the Theory of Seismology (3rd ed). Cambridge, pp 381
Bullen KE (1975) The Earth's Density. Chapman and Hall, London, pp 420
Burke K (1977) Aulacogens and continental breakup. Annu Rev Earth Planet Sci 5:371–396
Burke KE, Dewey JF, Kidd WSF (1976) Precambrian paleomagnetic results compatible with contemporary operation of the Wilson Cycle. Tectonophysics 33:287–299
Cadée GC, Laane RWPM (1983) Behaviour of POC, DOC, and fluorescence in the freshwater tidal compartment of the River Ems. Mitt Geol-Paläont Inst Univ Hamburg 55:331–342
Cameron W, Nisbet EG, Dietrich VJ (1979) Boninites, komatiites and ophiolitic basalts. Nature 280:550–553
Canuto VM, Levine JS, Augustsson TR, Imhoff CL, Giampapa MS (1983) The young Sun and the atmosphere and photochemistry of early Earth. Nature 305:281–286
Carroll D (1959) Ion exchange in clays and other minerals. Geol Soc Am Bull 70:749–780
Chapman DS, Pollack HN (1975) Global heat flow: A new look. Earth Planet Sci Lett 28:23–32
Charnock H (1964) Anomalous bottom water in the Red Sea. Nature 203:591
Clark BC, Baird AK (1979) Is the Martian lithosphere sulfur-rich? J Geophys Res 84:8395–8403
Clark SP, Ringwood AE (1964) Density distribution and constitution of the mantle. Rev Geophys 2:35–88
Clarke FW, Washington HW (1924) The Composition of the Earth's Crust. US Geol Surv Prof Pap 127:pp 117
Cloos H (1936) Einführung in die Geologie. Berlin, Gebr Borntraeger, pp 503
Cloos H (1939) Hebung, Spaltung, Vulkanismus. Geol Rdsch 30:401–527
Cloos H (1955) Geologische Strukturkarte der Mittelgebirge. Geol Rdsch 44:480
Cochran JR (1981) The Gulf of Aden: Structure and evolution of a young ocean basin and continental margin. J Geophys Res 86 (B1): 263–287
Coleman RG (1977) Ophiolites, ancient oceanic lithosphere? In "Minerals and Rocks" (eds Wyllie PJ, Engelhardt W v, Hahn T), vol 12, Berlin Heidelberg New York, Springer, pp 229
Cox A (ed) (1973) Plate Tectonics and Geomagnetic Reversals. San Francisco, WH Freeman and Company, pp 702
Cox A, Doell RR, Dalrymple GB (1963a) Geomagnetic polarity epochs and Pleistocene geochronometry. Nature 198:1049–1051
Cox A, Doell RR, Dalrymple GB (1963b) Geomagnetic polarity epochs: Sierra Nevada II. Science 142:302–305
Cox A, Doell RR, Dalrymple GB (1964a) Geomagnetic polarity epochs. Science 143:351–352
Cox A, Doell RR, Dalrymple GB (1964b) Reversals of the earth's magnetic field. Science 144:1537–1543
Cox KG, Bell JD, Pankhurst RJ (1979) The Interpretation of Igneous Rocks. London, pp 450
Craig H (1969) Abyssal carbon and radiocarbon in the Pacific. J Geophys Res 74:5491–5506
Crisp JA (1984) Rates of magma emplacement and volcanic output. J Volcanol Geotherm Res 20:177–211
Däppen W, Gilliland RL, Christensen-Dalsgaard J (1986) Weakly interacting massive particles, solar neutrinos, and solar oscillations. Nature 321:229–231
Daly RA (1933) Igneous Rocks and the Depth of the Earth; containing some revised chapters of "Igneous Rocks and their Origin". New York London, McGraw-Hill Book Company Inc, pp 598
Dana JD (1866) Observations on the origin of some of the earth's features. Am J Sci 2/42:205–211, 252–253
Danckwerts PV (1970) Gas-liquid Reactions. McGraw Hill, pp 276
Davies TA, Gorsline DS (1976) Oceanic sediments and sedimentary processes (2nd ed). In "Chemical Oceanography" (eds Riley JP, Chester R), London New York San Francisco, Academic Press, pp 1–80
Deer WA, Howle RA, Zussman J (1972) An Introduction to the Rock-Forming Minerals. London, J Wiley & Sons Inc, pp 528
Degens ET (1965) Geochemistry of Sediments. Englewood Cliffs, New Jersey, Prentice-Hall Inc, pp 342
Degens ET, Ittekkot V (1986) Ca^{2+} stress, biological response and particle aggregation in the aquatic habitat. Neth J Sea Res 20/2–3:109–116
Degens ET, Kurtman F (eds) (1978) The Geology of Lake Van. The Miner Res Explor Inst Turkey 169, pp 158
Degens ET, Ross DA (eds) (1969) Hot Brines and Recent Heavy Metal Deposits in the Red Sea. Springer Verlag, Berlin Heidelberg New York, pp 600
Degens ET, Ross DA (1970) The Red Sea hot brines. Sci Am 222/4:32–42

Degens ET, Ross DA (eds) (1974) The Black Sea – Geology, Chemistry, and Biology. Am Assoc Petrol Geol Mem 20, pp 633
Degens ET, Ross DA (1976) Strata-bound metalliferous deposits found in or near active rifts. In "Handbook of Strata-bound and Stratiform Ore Deposits", vol 4:165–202, Tectonics and Metamorphism (ed Wolf KH), Amsterdam, Elsevier Publ Co
Degens ET, Stoffers P (1977) Phase boundaries as an instrument for metal concentration. In "Time and Strata-bound Ore Deposits" (eds Klemm DD, Schneider H-J), Berlin Heidelberg New York, Springer-Verlag, 25–45
Degens ET, von Herzen RP, Wong HK (1971) Lake Tanganyika: water chemistry, sediments, geological structure. Naturwissenschaften 58:229–241
Degens ET, Paluska A, Eriksson E (1976) Rates of soil erosion. In "Nitrogen, Phosphorus and Sulphur – Global Cycles", SCOPE Report 7 (eds Svensson BG, Soderlund R), Ecol Bull 22, Stockholm, 185–191
Degens ET, Timm J, Wong HK (eds) (1980) Rheinisches Schiefergebirge: Ebbe-Antiklinorium. Fazies, Stratigraphie, Tektonik. Mitt Geol-Paläont Inst Univ Hamburg 50, pp 284
Degens ET, Kempe S, Spitzy A (1984a) Carbon dioxide: a biogeochemical portrait. In "Handbook of Environmental Chemistry" (ed Hutzinger O) vol 1 pt C, Springer Verlag, Berlin Heidelberg New York: 127–215
Degens ET, Kempe S, Ittekkot V (1984b) Monitoring carbon in world rivers. Environment 26:29–33
Degens ET, Wong HK, Kempe S, Kurtman F (1984c) A geological study of Lake Van, Eastern Turkey. Geol Rdsch 73:701–734
Degens ET, Wong HK, Wiesner MG (1986) The Black Sea region: sedimentary facies, tectonics and oil potential. Mitt Geol-Paläont Inst Univ Hamburg 60:127–149
Degens ET, Izdar E, Honjo S (eds) (1987) Particle Flux in the Ocean. Mitt Geol-Paläont Inst Univ Hamburg 62, pp 480
Denman KL (1973) A time-dependent model of the upper ocean. J Phys Oceanogr 3:173–184
Densmore CD (1965) Missing ingredients found. Oceanus 11/3:6
DePaolo DJ (1980) Crustal growth and mantle evolution: inferences from models of element transport and Nd and Sr isotopes. Geochim Cosmochim Acta 44:1185–1196
Deuser WG, Brewer PG, Jickells TD, Commean RD (1983) Biological control of the removal of abiogenic particles from the surface ocean. Science 219:388–391
Dewey JF (1976) Ophiolite obduction. Tectonophysics 31:93–120
Dewey JF (1977) Suture zone complexities: a review. Tectonophysics 40:53–67
Dewey JF, Burke KCA (1973) Tibetan, Variscan, and Precambrian basement reactivation: products of continental collision. J Geol 81:683–692
Dietz RS (1961) Continent and ocean evolution by spreading of the sea floor. Nature 190:854–857
Dimroth E (1985) A mass balance between Archean and Phanerozoic rates of magma emplacement, crustal growth and erosion: implications for recycling of the continental crust. Chem Geol 53:17–24
Dimroth E, Rochelean M, Mueller W (1984) Paleogeography, isostasy and crustal evolution of the Archean Abitibi Belt: a comparison between the Rouyn-Noranda and Chibongamau-Chapals areas. In "Chibongamau – Stratigraphy and Mineralization" (eds Guha J, Chown EH), Can Inst Min Spec 34:73–91
Dobretsov NL (1980) Introduction in Global Petrology. Nauka, Novosibirsk, pp 200 (in Russian)
Dobretsov NL, Kepezhinscas VV (1981) Three types of ultrabasic magmas and their bearing on the problem of ophiolites. Ofioliti 6:221–236
Dott RH (1974) The geosynclinal concept. In "Modern and Ancient Geosynclinal Sedimentation", (eds Dott RH, Shaver RH) Spec Publ Soc econ Paleont Miner 19:1–13
Dott RH, Shaver RH (eds) (1974) Modern and Ancient Geosynclinal Sedimentation. Spec Publ Soc econ Paleont Miner 19, pp 380
East Pacific Rise Study Group (1981) Crustal processes of the Mid-Ocean Ridge. Science 213:31–40
Eddy JA (1977) Historical evidence for the existence of the solar cycle. In "The Solar Output and Its Variation" (ed White OR), Univ Colorado Press, Boulder, 51–71
Edmond JM, Von Damm K (1983) Hot springs on the ocean floor. Sci Am 248/4:70–85
Edmond JM, Von Damm KL, McDuff RE et al. (1982) Chemistry of hot springs on the East Pacific Rise and their effluent dispersal. Nature 297:187–191
Ehrmann WU, Thiede J (1985) History of Mesozoic and Cenozoic Sediment Fluxes to the North Atlantic Ocean. Contrib Sedimentology 15, E Schweizerbart'sche Verlagsbuchhandlung (Nägele und Obermiller) Stuttgart, pp 109
Ehrmann WU, Thiede J (1986) Correlation of terrigenous and biogenic sediment fluxes in the North Atlantic Ocean during the past 150 my. Geol Rdsch 75:43–55
Eisma D, Bernard P, Boon JJ et al. (1985) Loss of particulate organic matter in estuaries as exemplified by the Ems and Gironde estuaries. Mitt Geol-Paläont Inst Univ Hamburg 58:397–412
Emery KO, Uchupi E (1972) Western North Atlantic Ocean: Topography, rocks, structure, water, life and sediments. Am Assoc Petrol Geol Mem 17:1–532
Emery KO, Uchupi E (1984) The Geology of the Atlantic Ocean. Springer-Verlag, Berlin Heidelberg New York Tokyo, pp 1050
Emery KO, Hunt JM, Hays EE (1969) Summary of hot brines and heavy metal deposits in the Red Sea. In "Hot Brines and Recent Heavy Metal Deposits in the Red Sea" (eds Degens ET, Ross DA), Springer Verlag, Berlin Heidelberg New York, 557–571
Engel AEJ, Itson SP, Engel CG, Stickney DM, Cray Jr EJ (1974) Crustal evolution and global tectonics: a petrogenic view. Geol Soc Am Bull 85:843–858
Engelhardt W von (1977) The Origin of Sediments and Sedimentary Rocks. Stuttgart, E. Schweizerbart'sche Verlagsbuchhandlung (Nägele und Obermiller), pp 359
Fairbridge RW (1971) Upper Ordovician glaciation in Africa? Reply. Geol Soc Am Bull 82:269–274
Faulkner J, Gough DO, Vahia MN (1986) Weakly interacting massive particles and solar oscillations. Nature 321:226–229
Fisher RV, Schmincke HU (1984) Pyroclastic Rocks. Springer Verlag, Berlin Heidelberg New York, pp 472
Flothmann D, Lohse E, Münich KO (1979) Gas exchange in a circular wind/water tunnel. Naturwissenschaften 66:49–50
Folk RL (1974) Petrology of Sedimentary Rocks. Austin, Texas, Hemphill's Book Store, pp 182
Fonselius SH (1974) Phosphorus in Black Sea. In "The Black Sea – Geology, Chemistry, and Biology" (eds Degens ET, Ross DA), Am Assoc Petrol Geol Mem 20:144–150
Frakes LA (1979) Climates Throughout Geologic Times. Amsterdam Oxford New York, Elsevier Publ Co, pp 310
Frey H (1977) Origin of the Earth's ocean basins. Icarus 32:235–250

Fuchs K, Gehlen K v, Mälzer H, Murawski H, Semmel A (eds) (1983) Plateau Uplift, The Rhenish Shield – A Case History. Intern Lithosph Progr Publ no 0104, Berlin Heidelberg New York Tokyo, Springer-Verlag, pp 411
Füchtbauer H, Müller G (1977) Sedimente und Sedimentgesteine (Sediment-Petrologie, Pt 2) – Stuttgart, E Schweizerbart'sche Verlagsbuchhandlung, pp 783
Fuglister FC (1960) Atlantic Ocean Atlas of Temperature and Salinity, Profiles and Data from the International Geophysical Year of 1957–1958. WHOI, Woods Hole, Mass, pp 209
Fujii H (1975) Material and energy production from volcanoes. Bull Volcanol Soc Japan 20:197–204
Fyfe WS (1976) The evolution of the Earth's crust: modern plate tectonics to ancient hot spot tectonics? Chem Geol 23:86–114
Fyfe WS (1981) The environmental crisis: quantifying geosphere interactions. Science 213:105–110
Gaposchkin EM (1974) Earth's gravity field to the eighteenth degree and geocentric coordinates for 104 stations from satellite and terrestrial data. J Geophys Res 79:5377–15411
Garrels RM, Mackenzie FT (1971) Evolution of Sedimentary Rocks. New York, WW Norton & Co, Inc, pp 397
Gaustad JE, Vogel SN (1982) High energy solar radiation and the origin of life. Origins of Life 12/1:3–8
Gee DG, Sturt BA (eds) (1985) The Caledonian Orogen – Scandinavia and Related Areas. New York, John Wiley and Sons
Geer G de (1910) A geochronology of the last 12,000 years. Congr Geol Int XI, Stockholm 1910:241–253
Girdler RW, Styles P (1978) Seafloor spreading in the western Gulf of Aden. Nature 271:615–617
Girdler RW, Brown C, Noy DJM, Styles P (1980) A geophysical survey of the westernmost Gulf of Aden. Philos Trans R Soc Lond Ser A 298:1–43
Glikson AY (1976) Stratigraphy and evolution of primary and secondary greenstones: significance of data from shields of the southern hemisphere. In "The Early History of the Earth" (ed Windley BF), New York, 257–277
Global Ocean Flux Study (1984) Proceedings of a workshop, September 10–14, 1984. Washington, DC, National Academy Press, pp 360
Goldich SS (1938) A study in rock-weathering. J Geol 46:17–58
Goldschmidt VM (1926) Geochemische Verteilungsgesetze der Elemente. VII. Die Gesetze der Kristallochemie. Skrift. Norske Vidensk-Akad Oslo I. Math-Nat Kl 2, pp 117
Goldschmidt VM (1954) Geochemistry. Clarendon Press, Oxford, pp 730
Goodwin AM (1981) Precambrian perspectives. Science 213:55–61
Green DH (1975) Genesis of Archean peridotitic magmas and constraints on Archean geothermal gradients and tectonics. Geology 3:15–18
Green DH, Nicholls IA, Viljoen M, Viljoen R (1975) Experimental demonstration of the existence of peridotitic liquids in earliest Archean magmatism. Geology 3:11–14
Green DH, Hibberson WO, Jaques AL (1979) Petrogenesis of Mid-ocean ridge basalts. In "The Earth: Its Origin, Structure and Evolution" (ed McElhinny MW), Academic Press, London New York San Francisco, 265–299
Green JC (1981) Pre-Tertiary continental flood basalts. In "Basaltic Volcanism", Pergamon Press, New York, 30–77
Grim RE (1968) Clay Mineralogy (2nd ed). McGraw-Hill Book Company, New York Toronto London, pp 596
Gutenberg B (1941) Changes in sea level. Postglacial uplift and mobility of the earth's interior. Geol Soc Am Bull 52:721–772
Gutenberg B (1959) Physics of the Earth's Interior. Academic Press, New York, pp 240
Hall J (1859) Description and figures of the organic remains of the lower Helderberg Group and the Oriskany Sandstone. Nat Hist New York; Paleontology, Geol Surv, Albany, NY 3, pp 544
Hallam A (1981) Facies Interpretation and the Stratigraphic Record. WH Freeman and Company, Oxford and San Francisco, pp 291
Hamblin WK, Howard JD (1986) Exercises in Physical Geology (6th ed). Burgess Publ Co, Minneapolis, pp 191
Hammen TVD, Wijmstra TA, Zagwijn WH (1971) The floral record of the Late Cenozoic of Europe. In "Late Cenozoic Glacial Ice Ages" (ed Turekian KK), Yale Univ Press, New Haven, London 391–424
Hanisch J (1984) The Cretaceous opening of the Northeast Atlantic. Tectonophysics 101:1–23
Harland WB, Cox AV, Llewellyn PG, Picton CAG, Smith AG, Walters R (1982) A Geological Time Scale. Cambridge, Cambridge Univ Press, pp 128
Hartline BK (1979) In search of solar neutrinos. Science 204:42–44
Hasse L, Liss P (1980) Gas exchange across the air-sea interface. Tellus 32:470–481
Head JW, Solomon SC (1981) Tectonic evolution of the terrestrial planets. Science 213:62–76
Head JW, Wood CA, Mutch TA (1977) Geologic evolution of the terrestrial planets. Am Sci 65:21–29
Heath DF, Thekaekara MP (1977) The solar spectrum between 1,200 and 3,000 A. In "The Solar Output and Its Variation" (ed White OR), Boulder, Colorado, Colorado Assoc Univ Press, 193–212
Heezen BC, Hollister CD (1971) The Face of the Deep. New York, Oxford Univ Press, pp 659
Heezen BC, McGregor ID (1973) The evolution of the Pacific. Sci Am 229/5:102–112
Heirtzler JR, Dickson GO, Herron EM, Pitman WC, III, Le Pichon X (1968) Marine magnetic anomalies, geomagnetic field reversals, and motions of the ocean floor and continents. J Geophys Res 73:2119–2136
Henderson-Sellers A (1983) The Origin and Evolution of Planetary Atmospheres. Monogr Astron Subj 9 (Gen Ed AJ Meadows), Adam Hilger Ltd, Bristol, pp 240
Henderson-Sellers A, Meadows AJ (1977) Surface temperature of early Earth. Nature 270:589–591
Hendricks SB (1945) Base-exchange of crystalline silicates. Ind Eng Chem 37:625–630
Hermann AG, Blanchard DP, Haskin LA et al. (1976) Major, minor, and trace element compositions of peridotitic and basaltic komatiites from the Precambrian crust of South Africa. Contrib Miner Petrol 59:1–12
Hess HH (1962) History of ocean basins. In "Petrologic Studies – A Volume in Honor of AF Buddington" (eds Engel AEJ, James HL, Leonard BL), New York, Geol Soc Amer 599–620
Hickey RL, Frey FA (1982) Geochemical characteristics of boninite series volcanics: implications for their source. Geochim Cosmochim Acta 46:2099–2115
Holeman JN (1968) Sediment yield of major rivers of the world. Water Resour Res 4:737–747
Holler P (1986) Fracture activity: a possible triggering mechanism for slope instabilities in the Eastern Atlantic? Geo-Mar Lett 5:211–216
Honjo S (1976) Coccoliths: production, transportation, and sedimentation. Mar. Micropaleont 1:65–79
Honjo S, Spencer D, Farrington JW (1982) Deep advective transport of lithogenic particles in Panama Basin. Science 216:516–518

Hunt JM, Hays EE, Degens ET, Ross DA (1967) Red Sea: detailed survey of hot brine areas. Science 156:514–516

Illies JH, Fuchs K (1983) Plateau-uplift of the Rhenish Massif – introductory remarks. In "Plateau Uplift, The Rhenish Shield – A Case History" (eds Fuchs K, Gehlen K v et al.), Springer-Verlag, Berlin Heidelberg New York Tokyo, 1–8

Irving E (1977) Drift of the major continental blocks since the Devonian. Nature 270:304–309

Irving E (1979) Pole position and continental drift since the Devonian. In "The Earth. Its Origin, Structure and Evolution" (ed McElhinney MW), Academic Press, London 567–593

Jähne B, Münnich KO, Siegenthaler U (1979) Measurements of gas exchange and momentum transfer in a circular windwater tunnel. Tellus 31:321–329

Jahn BM, Gruau G, Glikson AY (1982) Komatiites of the Onverwacht Group, S. Africa: REE geochemistry, Sm/Nd age and mantle evolution. Contrib Miner Petrol 80:25–40

Jannasch HW, Mottl MJ (1985) Geomicrobiology of deep- sea hydrothermal vents. Science 229:717–725

Jenkyns HC (1978) Pelagic environments. In "Sedimentary Environments and Facies" (ed Reading HG), 314–371, Oxford, Blackwell

JOIDES (1982) Report of the Conference on Scientific Ocean Drilling, Nov 16–18, 1981. JOIDES, Washington, DC, pp 112

Jung D, Vinx R (1973) Einige Bemerkungen zur Geochemie der Magmatite des Saar-Nahe-Pfalz-Gebietes. Ann Scien Univ Besançon 3/18:197–202

Kanwisher J (1963) On the exchange of gases between the atmosphere and the sea. Deep-Sea Res 10:195–207

Keller WD (1956) Clay minerals as influenced by environments at their formation. Am Assoc Petrol Geol Bull 40:2689–2710

Keller WD (1957) Principles of Chemical Weathering. Columbia, Mo, Lucas Brothers, pp 486

Kempe S (1977) Hydrographie, Warven-Chronologie und organische Geochemie des Van Sees, Ost-Türkei. Mitt Geol Paläont Inst Univ Hamburg 47:125–228

Kempe S (1983) Impact of Asuan High Dam on water chemistry of the Nile. Mitt Geol-Paläont Inst Univ Hamburg 55:401–423

Kempe S, Degens ET (1985) An early soda ocean? Chem Geol 53:95–108

King-Hele DG, Brookes CJ, Cook GE (1980) The pear-shaped section of the Earth. Nature 286:377–378

Klein C, Hurlbut CS Jr (1985) Manual of Mineralogy (after JD Dana) (20th ed). New York, Wiley, pp 596

Kossinna SE (1921) Die Tiefen des Weltmeeres. Veröff. Inst Meereskunde, Univ Berlin, NFA, Geogr, Naturwiss 9:1–70

Kranck K (1980) Experiments on the significance of flocculation in the settling of fine-grained sediment in still water. Can J Earth Sci 17:1517–1526

Kraskovski SA (1961) On the thermal field in old shields. Izv Akad Nauk Arm SSR, Geol Geogr Nauki, Engl Transl 247–250

Kraus EB, Turner JS (1967) A one-dimensional model of the seasonal thermocline. II. The general theory and its consequences. Tellus 19:98–106

Krebs W (1976) Wiederholter Magmenaufstieg und die Entwicklung postvariszischer Strukturen in Mitteleuropa. Nova Acta Leopoldina, NF 45:23–36

Krebs W (1978) Die Kaledoniden im nördlichen Mitteleuropa. Z dt geol Ges 129:403–422

Krebs W, Wachendorf H (1974) Faltungskerne im mitteleuropäischen Grundgebirge. Abbilder eines orogenen Diapirismus. N Jb Geol Paläont Abh 147/1:30–60

Kröner A (1985) Evolution of the Archean continental crust. Ann Rev Earth Planet Sci 13:49–74

Kröner A, Hanson GN, Goodwin AM (eds) (1984) Archaean Geochemistry. Springer-Verlag, Berlin Heidelberg New York, pp 286

Krumbein WC, Sloss LL (1963) Stratigraphy and Sedimentation (2nd ed). WH Freeman and Company, San Francisco and London, pp 660

Lean J (1987) Solar ultraviolet irradiance variations: a review. J Geophys Res 92:839–868

Lee WHK, Uyeda S (1965) Review of heat flow data. In "Terrestrial Heat Flow", Geophys Monogr 8, Amer Geophys Union, 87–190

Leeder MR (1982) Sedimentology: Process and Product. London Boston Sydney, George Allen & Unwin, pp 344

LePichon X (1968) Sea-floor spreading and continental drift. J Geophys Res 73:3661–3697

LePichon X, Francheteau J (1978) A plate-tectonic analysis of the Red Sea-Gulf of Aden area. In "Structure and Tectonics of the Eastern Mediterranean" (ed Oren OH), Tectonophysics 46:369–406

LePichon X, Renard V (1982) Avalanching, a major process of erosion and transport in deep-sea canyons: evidence from submersible and multi-narrow beam surveys. In "The Ocean Floor" (eds Scrutton RA, Talwani M), Bruce Heezen commemorative volume, New York, John Wiley and Sons, 113–128

LePichon X, Sibuet J-C (1981) Passive margins: a model of formation. J Geophys Res 86, no B5:3708–3720

LePichon X, Francheteau J, Bonnin J (1973) Plate Tectonics. Amsterdam, Elsevier Publ Co, pp 300

Lerch FJ, Wagner CA, Richardson JA, Brownd JE (1974) Goddard earth models (5 and 6) NASA-GSFC Doc X-921-74–145. Greenbelt Md, Goddard Space Flight Center, pp 100

Lewis BTR (1983) The process of formation of ocean crust. Science 220:151–157

Lilley FEM (1979) Geomagnetism and the earth's core. In "The Earth: Its Origin, Structure and Evolution" (ed McElhinny MW), Academic Press, London New York San Francisco 83–111

Liss P (1973) Processes of gas exchange across an air-water interface. Deep-Sea Res 20:221–238

Liss P, Slater PG (1974) Flux of gases across the air-sea interface. Nature 247:181–184

Livingstone DA (1963) Chemical composition of rivers and lakes. US Geol Surv Prof Paper 440-G, pp 64

Lowrie W, Alvarez W (1981) One hundred million years of geomagnetic polarity history. Geology 9:392–397

Lupton JE, Weiss RF, Craig H (1977) Mantle helium in the Red Sea brines. Nature 266:244–246

MacDonald GJF (1964) Dependence of the surface heat flow on the radioactivity of the earth. J Geophys Res 69:2933–2946

McDougall I (1979) The present status of the geomagnetic polarity time scale. In "The Earth: Its Origin, Structure and Evolution" (ed McElhinny MW), 543–566, Academic Press, London New York San Francisco

McElhinny MW (ed) (1979 a) The Earth: Its Origin, Structure and Evolution. Academic Press, London New York San Francisco, pp 597

McElhinny MW (1979 b) Palaeogeomagnetism and the core-mantle interface. In "The Earth: Its Origin, Structure and Evolution" (ed McElhinny MW), 113–136, Academic Press, London New York San Francisco

McGregor VR, Mason B (1977) Petrogenesis and geochemistry of metabasaltic and metasedimentary enclaves in the Am'itsoq gneisses, West Greenland. Am Mineral 62:887–904

McKenzie D (1978) Some remarks on the development of sedimentary basins. Earth Planet Sci Lett 40:25–32

McKenzie DP (1967) Some remarks on heat-flow and gravity anomalies. J Geophys Res 72:6261–6273
McLennan SM, Taylor SR (1982) Geochemical constraints on the growth of the continental crust. J Geol 90:347–361
Maier-Reimer E, Hasselmann K (1987) Transport and storage of CO_2 in the ocean – an inorganic ocean circulation carbon cycle model. Climate Dynamics 2:63–90
Mareschal J-C (1983) Mechanisms of uplift preceding rifting. In "Processes of Continental Rifting" (eds Morgan P, Baker BD), Tectonophysics 94:51–66
Martin AK (1984) Propagating rifts: crustal extension during continental rifting. Tectonics 3/6:611–617
Martin H, Eder FW (eds) (1983) Intracontinental Fold Belts. Springer-Verlag, Berlin Heidelberg New York, pp 945
Martin JM, Burton D, Eisma D (1980) River Inputs to Ocean Systems. UNEP-UNESCO, Switzerland, pp 384
Masursky H, Boyce JM, Dial AL, Schaber GG, Strobel MB (1977) Classification and time of formation of Martian channels based on Viking data. J Geophys Res 82:4016–4038
Matheja J, Degens ET (1971) Structural Molecular Biology of Phosphates. Gustav Fischer Verlag, Stuttgart, pp 180
Matsui T, Abe Y (1986) Evolution of an impact-induced atmosphere and magma ocean on the accreting Earth. Nature 319:303–305
Matuyama M (1929) On the direction of magnetization of basalt in Japan, Tyôsen, and Manchuria. Japan Acad Proc 5:203–205
Meade RH (1982) Sources, sinks, and storage of river sediment in the Atlantic drainage of the United States. J Geol 90:235–252
Meeresforschung in Hamburg (1977) Illustrated documentation of the "Sonderforschungsbereich 94", FLEX 76. Univ Hamburg, FRG
Meissner R (1983) Evolution of plate tectonics on terrestrial planets. Ann Geophysicae 1/2:121–127
Meissner R, Bartelson H, Murawski H (1981) Thin-skinned tectonics in the northern Rhenish Massif, Germany. Nature 290:399–401
Merlivat L, Memery L (1983) Gas exchange across an air-water interface: Experimental results and modeling of bubble contribution to transfer. J Geophys Res 88:707–724
Meybeck M (1982) Carbon, nitrogen and phosphorus transport by major world rivers. Am J Sci 282:401–501
Meyer W, Stets J (1975) Das Rheinprofil zwischen Bonn und Bingen. Z dt geol Ges 126:15–29
Meyer W, Alber HJ et al. (1983) Pre-Quaternary uplift in the central part of the Rhenish Massif. In "Plateau Uplift. The Rhenish Shield – A Case History" (eds Fuchs K, Gehlen K v et al.), 39–46, Springer-Verlag, Berlin Heidelberg New York Tokyo
Middleton GV (1973) Johannes Walther's law of correlation of facies. Geol Soc Am Bull 84:979–988
Miller AR (1964) Highest salinity in the world ocean? Nature 203:590–591
Miller AR (1969) Atlantis II account. In "Hot Brines and Recent Heavy Metal Deposits in the Red Sea" (eds Degens ET, Ross DA) 15–17, Springer Verlag, Berlin Heidelberg New York
Miller AR, Densmore CD, Degens ET et al.(1966) Hot brines and recent iron deposits in deeps of the Red Sea. Geochim Cosmochim Acta 30:341–359
Milliman JD, Meade RH (1983) World-wide delivery of river sediments to the oceans. J Geol 91:1–21
Mitchell AHG, Reading HG (1978) Sedimentation and tectonics. In "Sedimentary Environments and Facies" (ed Reading HG), 439–476, Blackwell, Oxford
Miyashiro A (1978) Metamorphism and Metamorphic Belts. Boston Sydney, George Allen & Unwin, pp 492
Moore JG (1960) Curvature of normal faults in the Basin and Range Province of the western United States. US Geol Surv Prof Paper 400, B409–B411
Moore TC, Heath GR (1977) Survival of deep-sea sedimentary sections. Earth Planet Sci Lett 37:71–80
Morgan WJ (1968) Rises, trenches, great faults, and crustal blocks. J Geophys Res 73:1959–1982
Mulugeta G (1985) Dynamic models of continental rift valley systems. Tectonophysics 113:49–73
Murawski H (1964) Fazies. Geol Rdsch 54:585–595
Murawski H, Albers HJ et al. (1983) Regional tectonic setting and geological structure of the Rhenish Massif. In "Plateau Uplift. The Rhenish Shield – A Case History" (eds Fuchs K, Gehlen K v et al.) 9–38, Berlin Heidelberg New York Tokyo, Springer-Verlag
Mutter JC (1985) Seaward dipping reflectors and the continent-ocean boundary at passive continental margins. In "Geophysics of the Polar Regions" (eds Husebye ES, Johnson GL, Kristoffersen Y), Tectonophysics 114:117–131
Mysen BO (1977) The solubility of H_2O and CO_2 under predicted magma genesis conditions and some petrological and geophysical implications. Rev Geophys Space Phys 15:351–361
Nakamura K (1974) Preliminary estimate of global volcanic production rate. In "The Utilization of Volcano Energy" (ed Colp JL), Albuquerque, NM, Sandia Lab, 273–286
Naldrett AJ, Smith IEM (1981) Mafic and ultramafic volcanism during the Archean. In "Basaltic Volcanism on the Terrestrial Planets" (eds Members Basaltic Volcanism Study Project), New York Oxford Toronto Sydney Frankfurt Paris, Pergamon Press, 5–29
Natterer K (1901) Expedition SM Schiff "Pola" in das Rote Meer, Südliche Hälfte, Sept. 1897–März 1899. 15: Chemische Untersuchung von Wasser und Grundproben. Denkschr Österr Akad Wiss, Mathem- naturwiss Kl, Abt I, 69, Wien 1901:297–309
Neugebauer HJ (1983) Mechanical aspects of continental rifting. In "Processes of Continental Rifting" (eds Morgan P, Baker BH), Tectonophysics 94:91–108
Neumann AC, Densmore CD (1959) Oceanographic data from the Mediterranean Sea, Red Sea, Gulf of Aden and Indian Ocean. Woods Hole Oceanogr Inst, Ref No 60–2. (unpubl manuscript)
Newkirk Jr G (1983) Variations in solar luminosity. Annu Rev Astron Astrophys 21:429–467
Newman MJ, Rood RT (1977) Implications of solar evolution for the earth's early atmosphere. Science 198:1035–1037
Newsom HE (1980) Hydrothermal alteration of impact melt sheets with implications for Mars. Icarus, 44:207–216
Newton RC, Smith JV, Windley BF (1980) Carbonic metamorphism, granulites and crustal growth. Nature 288:45–50
Nisbet EG, Walker D (1982) Komatiites and the structure of the Archaean mantle. Earth Planet Sci Lett 60:105–113
Norton DL (1984) Theory of hydrothermal systems. Annu Rev Earth Planet Sci 12:155–177
Noy DJ (1978) A comparison of magnetic anomalies in the Red Sea and the Gulf of Aden. In "Tectonics and Geophysics of Continental Rifts" (eds Ramberg IB, Neumann E-R), 279–287, D Reidel Publ Co, Dordrecht
Nur A, Ben-Avraham Z (1982) Oceanic plateaus, the fragmentation of continents, and mountain building. J Geophys Res 87:3644–3661

Owen T, Cess RD, Ramanathan V (1979) Enhanced CO_2 greenhouse to compensate for reduced solar luminosity on early Earth. Nature 277:640–642
Parkinson JH, Morrison LV, Stephenson FR (1980) The constancy of the solar diameter over the past 250 years. Nature 288:548–551
Parsons B, McKenzie DP (1978) Mantle convection and the thermal structure of the plates. J Geophys Res 83:4485–4496
Parsons B, Sclater JG (1977) An analysis of the variation of ocean floor bathymetry and heat flow with age. J Geophys Res 82:803–827
Patchett PJ, Bylund G (1977) Age of Grenville magnetisation; Rb-Sr and paleomagnetic evidence from Swedish dolerites. Earth Planet Sci Lett 35:92–104
Pauling L (1927) The sizes of ions and the structure of ionic crystals. J Am Chem Soc 49:765–790
Pauling L (1960) The Nature of the Chemical Bond and the Structure of Molecules and Crystals (3rd ed). Cornell Univ Press, Ithaca, New York, pp 644
Pepin R, Eddy JA, Merrill RB (eds) (1980) The Ancient Sun. Geochim Cosmochim Acta, Suppl 13, pp 581
Pettijohn FJ (1975) Sedimentary Rocks (3rd ed). New York, Harper and Row, pp 628
Pettijohn FJ, Potter PE (1964) Atlas and Glossary of Primary Sedimentary Structures. Springer-Verlag, Berlin Göttingen Heidelberg New York, pp 370
Pettijohn FJ, Potter PE, Siever R (1972) Sand and Sandstone. Springer-Verlag, Berlin Heidelberg New York, pp 618
Pitman WC, Talwani M (1972) Seafloor spreading in the North Atlantic. Geol Soc Am Bull 83:619–646
Pollack HN, Chapman DS (1977) On the regional variation of heat flow, geotherms and the thickness of the lithosphere. Tectonophysics 38:279–296
Pollack JB (1979) Climatic change on the terrestrial planets. Icarus 37:479–553
Pollack JB (1984) Origin and history of the outer planets: theoretical models and observational constraints. Annu Rev Astron Astrophys 22:389–424
Pollack JB, Grossman AS, Moore R, Graboske HC Jr (1977) A calculation of Saturn's gravitational contraction history. Icarus 30:111–128
Pollack JB, Borucki WJ, Toon OB (1979) Are solar spectral variations a drive for climatic change. Nature 282:600–603
Polyak BG, Smirnov YaB (1968) Relationship between terrestrial heat flow and the tectonics of continents. Geotectonics 4:205–213 (engl translation)
Potter PE, Pettijohn FJ (1977) Paleocurrents and Basin Analysis (2nd ed). Springer Verlag, Berlin Heidelberg New York, pp 425
Prasad G (1985) Das frühtertiäre Bauxit-Ereignis. Weinheim, Geowissenschaften in unserer Zeit 3/3:81–85
Press F, Siever R (1986) Earth (4th ed). New York, WH Freeman and Company, pp 656
Ramberg H (1963) Experimental study of gravity tectonics by means of centrifuged models. Bull Geol Inst Univ Uppsala 42/1:1–97
Ramberg H (1971) Dynamic models simulating rift valleys and continental drift. Lithos 4:259–276
Ramberg H (1978) Experimental model studies of rift valley systems. In "Tectonics and Geophysics of Continental Rifts" (eds Ramberg IB, Neumann ER), 39–40, Dordrecht, D Reidel Publ Co
Ramberg IB, Neumann ER (eds) (1978) Tectonics and Geophysics of Continental Rifts. Dordrecht, D Reidel Publ Co, pp 443
Rau A, Tongiorgi M (1981) Some problems regarding the Paleozoic paleogeography in Mediterranean western Europe. J Geol 89:663–673
Reading HG (ed) (1981) Sedimentary Environments and Facies. Oxford London, Blackwell Scient Publ, pp 569
Regan RD, Cain JC, Davis WM (1975) A global magnetic anomaly map. J Geophys Res 80:794–802
Reineck H-E, Singh IB (1980) Depositional Sedimentary Environments (2nd ed). Berlin Heidelberg New York, Springer-Verlag, pp 549
Reymer A, Schubert G (1984) Phanerozoic addition rates to the continental crust and crustal growth. Tectonics 3:63–78
Richards FA (1975) The Cariaco Basin (Trench). Oceanogr Mar Biol Annu Rev 13:11–67
Richards S, Pedersen B, Silverton JV, Hoard JL (1964) Stereochemistry of ethylene-diaminetetraacetato complexes. Inorg Chem 3:27–33
Riley GA, Stommel W, Bumpus DF (1949) Quantitative ecology of the plankton of the western North Atlantic. Bull Bingham Oceanogr Coll 12, Art 3:1–169
Ringwood AE (1979) Origin of the Earth and Moon. Springer-Verlag, New York Heidelberg Berlin, pp 295
Ringwood AE (1982) Phase transformations and differentiation in subducted lithosphere: Implications for mantle dynamics, basalt petrogenesis, and crustal evolution. J Geol 90:611–643
Rittenberg SC, Emery KO, Hülsemann J et al. (1963) Biogeochemistry of sediments in Experimental Mohole. J Sed Petrol 33/1:140–172
Rogers CA (1964) Packing and Covering. Cambridge Univ Press
Rona PA (1980) The Central North Atlantic Ocean basin and continental margins: geology, geophysics, geochemistry, and resources, including the Trans-Atlantic Geotraverse (TAG). Nat Ocean Atmosph Administr, Envir Res Labs, NOAA Atlas 3
Ronov AB (1972) Evolution of rock composition and geochemical processes in the sedimentary shell of the earth. Sedimentology 19:157–172
Ronov AB, Yaroshevsky AA (1969) Chemical structure of the earth's crust. Geokhimia, 1967 (II), 1285–1309 (in Russian); also: In "Earth's Crust and Upper Mantle", Geophys Monogr 13, Amer Geophys Union, Washington DC, 37–57
Ross DA, Whitmarsh RB, Ali SA et al. (1973) Red Sea drillings. Science 179:377–380
Ross DA, Uchupi AE, Prada KE, MacIlvaine JC (1974) Bathymetry and microtopography of Black Sea. In "The Black Sea – Geology, Chemistry and Biology" (eds Degens ET, Ross DA), AAPG Memoir 20, Tulsa, Oklahoma, 1–35
Ross DA, Neprochnov YP et al. (eds) (1978) Initial Reports of the Deep Sea Drilling Project 42, Pt 2, US Government Printing Office, Washington, DC, pp 1244
Sagan C (1977) Reducing greenhouses and the temperature history of Earth and Mars. Nature 269:224–226
Sauramo M (1955) Land uplift with hinge lines in Fennoscandia. Ann Acad Sc Fennicae, Ser A, 3, 44:1–25
Scheidegger KF, Krissek LA (1982) Dispersal and deposition of eolian and fluvial sediments off Peru and northern Chile. Geol Soc Am Bull 93:150–162
Schoell M, Faber E (1978) New isotopic evidence for the origin of Red Sea brines. Nature 275:436–438
Schoenenberg R (1979) Einführung in die Geologie Europas (2nd ed). Rombach Verlag, Freiburg, pp 300
Scholle PA (1979) Constituents, Textures, Cements, and Porosities of Sandstones and Associated Rocks. Am Assoc Petrol Geol Mem 28, pp 201

Schove DJ (ed) (1983) Sunspot Cycles. Benchmark Papers in Geology 68, Hutchinson Ross Publ Co, Stroudsburg, Pennsylvania, pp 397

Sclater JG, Christie PAF (1980) Continental stretching: an explanation of the post-Mid-Cretaceous subsidence of the central North Sea basin. J Geophys Res 85 (B7):3711–3739

Sclater JG, Francheteau J (1970) The implications of terrestrial heat flow observations on current tectonic and geochemical models of the crust and upper mantle of the earth. Geophys J 20:509–542

Sclater JG, Hellinger S, Tapscott C (1977) The paleobathymetry of the Atlantic Ocean from the Jurassic to the present. J Geol 85:509–552

Sclater JG, Royden L, Horvarth F et al. (1980) Subsidence and thermal evolution of the Intra-Carpathian Basins. Earth Planet Sci Lett 51:139–162

Sclater JG, Parsons B, Jaupart C (1981) Oceans and continents: Similarities and differences in the mechanisms of heat loss. J Geophys Res 86:11,535-11,552

Scotese CR, Bambach RK, Barton C, Van der Voo, Ziegler AM (1979) Paleozoic base maps. J Geol 87:217–277

Self S, Rampino MR, Newton MS, Wolff JA (1984) Volcanological study of the great Tambora eruption of 1815. Geology 12/11:659–663

Selley RC (1978) Ancient Sedimentary Environments (2nd ed). Chapman and Hall, London, pp 287

Sharp RP, Malin MC (1975) Channels on Mars. Geol Soc Am Bull 86:593–609

Sherman GD (1952) The genesis and morphology of the alumina-rich laterite clays. Amer Inst Min Met Eng, Problems of Clay and Laterite Genesis, 154–161

Shimkus KM, Trimonis ES (1974) Modern sedimentation in Black Sea. In "The Black Sea – Geology, Chemistry, and Biology" (eds Degens ET, Ross DA), Am Assoc Petrol Geol Mem 20, 249–278

Sloane NJA (1984) The packing of spheres. Sci Am 250/1:92–101

Smith JV (1974a) Feldspar Minerals. I. Crystal Structure and Physical Properties. Springer-Verlag, Berlin Heidelberg New York, pp 627

Smith JV (1974b) Feldspar Minerals. II. Chemical and Textural Properties. Springer-Verlag, Berlin Heidelberg New York, pp 690

Smith JV (1979) Mineralogy of the planets: a voyage in space and time. Miner Mag 43/325:1–89

Smith JV (1981) Progressive differentiation of a growing moon. Proc Lunar Planet Sci 12B:979–990

Smith JV (1982) Heterogeneous growth of meteorites and planets, especially the earth and moon. J Geol 90:1–48

Smoluchowski R (1983) The Solar System. Sci Am Library, New York, 1–174

Southam JR, Hay WW (1977) Time scales and dynamic models of deep-sea sedimentation. J Geophys Res 82/27:3825–3842

Spencer DW, Brewer PG (1971) Vertical advection diffusion and redox potentials as controls on the distribution of manganese and other trace metals dissolved in waters of the Black Sea. J Geophys Res 76:5877–5892

Sprague D, Pollack HN (1980) Heat flow in the Mesozoic and Cenozoic. Nature 285:393–395

Steckler MS (1985) Uplift and extension at the Gulf of Suez: Indications of induced mantle convection. Nature 317:135–139

Steckler MS, Watts AB (1980) The Gulf of Lion: subsidence of a young continental margin. Nature 87:425–429

Steele JH (1956) Plant production on the Fladen Ground. J Mar Biol Assoc UK 35:1–33

Steele JH (1957) Notes on oxygen sampling on the Fladen Ground. J Mar Biol Assoc UK 36/2:227–241

Steele JH (1974) The Structure of Marine Ecosystems. Harvard Univ Press, Cambridge, Mass, pp 128

Stevenson DJ (1980) Saturn's luminosity and magnetism. Science 208:746–748

Stille H (1924) Grundfragen der vergleichenden Tektonik. Berlin, Borntraeger, pp 443

Stolper E, Walker D, Hager BH et al. (1981) Melt segregation from partially molten source regions: the importance of melt density and source region size. J Geophys Res 86:6261–6271

Stosch H-G, Lugmair GW (1984) Evolution of the lower continental crust: granulite facies xenoliths from the Eifel, West Germany. Nature 311:368–370

Strunz H (1966) Mineralogische Tabellen (4th ed). Akad Verlagsges, Geest & Portig K-G, Leipzig, pp 560

Suess E (1885–1909) Das Antlitz der Erde (3 vols). Prag and Leipzig, pp 2778

Suk M (1983) Petrology of Metamorphic Rocks. Dev Petr 9, Elsevier, Amsterdam, pp 322

Summerhayes CP, Shackleton NJ (1986) North Atlantic Palaeoceanography. Geol Soc Spec Publ No 21, Oxford London Edinburgh Boston Palo Alto Melbourne, Blackwell Sci Publ, pp 473

Swallow JC, Crease J (1965) Hot salty water at the bottom of the Red Sea. Nature 205:165–166

Sykes LR (1967) Mechanism of earthquakes and nature of faulting on the mid-oceanic ridges. J Geophys Res 72:2131–2152

Talwani M, Eldholm O (1977) Evolution of the Norwegian-Greenland Sea. Geol Soc Am Bull 88:969–999

Taylor SR, McLennan SM (1985) The Continental Crust. Blackwell Scientific, Palo Alto, Calif, pp 312

Teichert C (1958) Concepts of facies. Am Assoc Petrol Geol Bull 42:2718–2744

Thiede J (1979) History of the North Atlantic Ocean: evolution of an asymmetric zonal paleoenvironment in a latitudinal basin. In "Deep Drilling Results in the Atlantic Ocean: Continental Margins and Paleoenvironment" (eds Talwani M, Hay W, Ryan WBF), Maurice Ewing Ser (AGU) 3:275–296

Toon OB, Pollack JB, Ward W, Burns JA, Bilski K (1980a) The astronomical theory of climatic change on Mars. Icarus 44:552–607

Toon OB, Pollack JB, Rages K (1980b) A brief review of the evidence for solar variability on the planets. In "The Ancient Sun" (eds Pepin RP, Eddy JA, Merrill RB), Geochim Cosmochim Acta, Suppl 13:523–531

Tozer DC (1985) Heat transfer and planetary evolution. Geophys Surv 7:213–246

Trefil JS (1978) Missing particles cast doubt on our solar theories. Smithsonian 8/12:74–83

Turekian KK, Clark SP Jr (1969) Inhomogeneous accretion of the Earth from the primitive solar nebula. Earth Planet Sci Lett 6:346–348

Turner FJ (1968) Metamorphic Petrology. McGraw-Hill, New York, pp 403

Turner FJ, Verhoogen J (1960) Igneous and Metamorphic Petrology (2nd ed). New York Toronto London, McGraw-Hill Book Company, pp 694

Turner JS (1969) A physical interpretation of the observations of hot brine layers in the Red Sea. In "Hot Brines and Recent Heavy Metal Deposits in the Red Sea" (eds Degens ET, Ross DA), Springer-Verlag, Berlin Heidelberg New York, 164–173

Turner JS, Stommel H (1964) A new case of convection in the presence of combined vertical salinity and temperature gradients. Proc US Natl Acad Sci 52:49–53

Uyeda S (1981) Subduction zones and back arc basins – a review. Geol Rdsch 70:552–569
Uyeda S (1982) Subduction zones: an introduction to comparative subductology. Tectonophysics 81:133–159
Uyeda S, Kanamori H (1979) Back-arc opening and the mode of subduction. J Geophys Res 84:1049–1061
Valeton I (1983) Klimaperioden lateritischer Verwitterung und ihre Abbildung in den synchronen Sedimentationsräumen. Z dt Geol Ges 134:413–452
Van Andel TH (1975) Mesozoic/Cenozoic calcite compensation depth and the global distribution of calcareous sediments. Earth Planet Sci Lett 26:187–194
Van Andel TH (1983) Estimation of sedimentation and accumulation rates. In "Sedimentology, Physical Properties and Geochemistry in the Initial Reports of the Deep Sea Drilling Project: An Overview" (ed Heath GR), 93–101, US Dept of Commerce, Nat Ocean Atmosph Administr, Envir Data Inform Serv, Boulder
Van Andel TH, Thiede J, Sclater JG et al. (1977) Depositional history of the South Atlantic Ocean during the last 125 million years. J Geol 85:651–698
Veizer J (1983) The Archean-Proterozoic earth. In "Earth's Earliest Biosphere: Its Origin and Evolution" (ed Schopf JW), Princeton Univ Press, Princeton, NJ, 240–259
Veizer J (1985) Carbonates and ancient oceans: isotopic and chemical record on time scales of 10^7–10^9 years. In "The Carbon Cycle and Atmospheric CO_2: Natural Variations Archean to Present" (eds Sundquist ET, Broecker WS), Geophys Monogr 32, Amer Geophys Union, Washington DC, 595–601
Veizer J, Compston W (1976) $^{87}Sr/^{86}Sr$ in Precambrian carbonates as an index of crustal evolution. Geochim Cosmochim Acta 40:905–915
Veizer J, Jansen SL (1979) Basement and sedimentary recycling and continental evolution. J Geol 87:341–370
Vine FJ (1966) Spreading of the ocean floor: new evidence. Science 154:1405–1415
Vine FJ (1968) Magnetic anomalies associated with mid-ocean ridges. In "The History of the Earth's Crust" (ed Phinney RA), Princeton Univ Press, Princeton, NJ, 73–89
Vine FJ, Matthews DH (1963) Magnetic anomalies over oceanic ridges. Nature 199:947–949
Vine FJ, Wilson JT (1965) Magnetic anomalies over a young oceanic ridge off Vancouver Island. Science 150:485–489
Vink GE (1982) Continental rifting and the implications for plate tectonic reconstructions. J Geophys Res 87, B13:10,677–10,688
Vitorello I, Pollack HN (1980) On the variation of continental heat flow with age and the thermal evolution of continents. J Geophys Res 85:983–995
Voll G (1983) Crustal xenoliths and their evidence for crustal structure underneath the Eifel volcanic district. In "Plateau Uplift. The Rhenish Shield – A Case History" (eds Fuchs K, Gehlen K v, Mälzer H, Murawski H, Semmel A) 336–342, Berlin Heidelberg New York Tokyo, Springer- Verlag
Wagner CA, Lerch FJ, Brownd JE, Richardson JA (1977) Improvement in the geopotential derived from satellite and surface data (GEM 7 and 8). J Geophys Res 82:901–914
Walker JCG (1982) Climatic factors on the Archean Earth. Palaeogeogr, -climatol, -oecol 40:1–11
Walker JCG (1983) Possible limits on the composition of the Archaean ocean. Nature 302:518–520
Walther J (1894) Einleitung in die Geologie als historische Wissenschaft. 3 vols. Jena, Fischer Verlag, pp 1055
Warren PH (1985) The magma ocean concept and lunar evolution. Annu Rev Earth Planet Sci 13:201–240
Wassen JT, Adler B, Oeschger H (1967) Aluminum-26 in Pacific sediment: implications. Science 155:446–448
Wedepohl KH (1975) The contribution of chemical data to assumptions about the origin of magmas from the mantle. Fortschr Miner 52:141–172
Wedepohl KH (1981a) Der primäre Erdmantel (Mp) und die durch Krustenbildung verarmte Mantelzusammensetzung (Md). Fortschr Miner 59, Beih 1:203–205
Wedepohl KH (1981b) Tholeiitic basalts from spreading ocean ridges. The growth of the oceanic crust. Naturwissenschaften 68:110–119
Wegener A (1912) Die Entstehung der Kontinente. Geol Rdsch 5:276–292
Wells AF (1978) Structural Inorganic Chemistry (4th ed). Clarendon Press, Oxford, pp 1095
Werner CD (1981a) Outline of the evolution of the magmatism in the GDR. In "Ophiolites and Initialites of Northern Border of the Bohemian Massif", Guide Book of Excursions, PK IX, UK 2, Potsdam, Freiberg I:17–68
Werner CD (1981b) Sächsisches Granulitgebirge – Saxonian Granulite Massif. In "Ophiolites and Initialites of Northern Border of the Bohemian Massif", Guide Book of Excursions, PK IX, UK 2, Potsdam, Freiberg I:129–161
Werner CD (1984) Globale Entwicklung des basischen Magmatismus. Z geol Wiss Berlin 12:537–562
Westphal M (1977) Configuration of the geomagnetic field and reconstruction of Pangaea in the Permian period. Nature 267:136–137
Wetherill GW (1980) Formation of the terrestrial planets. Annu Rev Astron Astrophys 18:77–113
Whitmann WS (1923) Chem Metall Engng 29:146
Wickman FE (1954) The "total" amount of sediments and the composition of the "average igneous rock". Geochim Cosmochim Acta 5:97–110
Williams H, Turner FJ, Gilbert CM (1982) Petrography (2nd ed). WH Freeman and Company, New York, pp 406
Wilson JT (1965) A new class of faults and their bearing on continental drift. Nature 207:343–347
Wilson RC (1978) Accurate solar "constant" determination by cavity pyrheliometers. J Geophys Res 83:4003–4007
Windley BF (ed) (1975) The Early History of the Earth. London, Wiley, pp 619
Windley BF (1977) The Evolving Continents. London, Wiley, pp 385
Winkler HGF (1979) Petrogenesis of Metamorphic Rocks (5th ed). New York Heidelberg Berlin, Springer-Verlag, pp 348
Woerner G, Schmincke HU, Schreyer W (1982) Crustal xenoliths from the Quaternary Wehr Volcano (East Eifel). N Jb Miner Abh 144/1:29–55
Wong HK, Degens ET (1981) Geotektonische Entwicklung des variszischen Faltungsgürtels im Paläozoikum. Mitt Geol-Paläont Inst Univ Hamburg 50:17–44
Wong HK, Degens ET (1983) Effects of CO_2- H_2O and oblique collision on orogenesis – the European Hercynides as an example. Tectonophysics 95:191–200
Wong HK, Degens ET (1984) The crust beneath the Red Sea-Gulf of Aden-East African Rift System: a review. Mitt Geol-Paläont Inst Univ Hamburg 56:53–94
Worsley TR, Davies TA (1979) Sea level fluctuations and deep-sea sedimentation rates. Science 203:455–456
Worthington LV, Wright WR (1970) North Atlantic Ocean Atlas of Potential Temperature and Salinity in the Deep Water Including Temperature, Salinity and Oxygen Profiles from the ERIKA DAN Cruise of 1962. WHOI, Woods Hole, Mass, pp 6, plates 58, tables 24

Wurster P (1958) Geometrie and Geologie von Kreuzschichtungskörpern. Geol Rdsch 47:322–359
Wyllie PJ (1971) The Dynamic Earth. New York, Wiley, pp 416
Wyllie PJ (1976) The Way the Earth Works. New York London Sydney Toronto, John Wiley & Sons, Inc, pp 296
Wyllie PJ (1977) Mantle fluid compositions buffered by carbonates in peridotite CO_2-H_2O. J Geol 85:187–207
Wyllie PJ (1981) Magmas and volatile components. Am Mineral 64:469–500
Yoder HS Jr (1976) Generation of Basaltic Magma. Washington DC, National Acad Sci, pp 265
Yoder HS Jr (ed) (1979) The Evolution of the Igneous Rocks. Princeton, Princeton Univ Press, pp 588
Zagwijn WH (1975) Chronostratigrafie en biostratigrafie: Indeling van het Kwartair op grond van veranderingen in vegetatie en klimaat. In "Toelichting bij Geologische Overzihtskaarten van Nederland" (ed Zagwijn WH, Van Staalduinen CJ), 109–114
Zenkevich LA (1963) Biologiya morel SSSR: Moscow. Izd Akad Nauk SSSR, 793 pp: Engl transl (Zenkevich), 1963, Biology of the Seas of the USSR. New York, Interscience Publ, pp 955
Ziegler AM, Scotese CR, McKerrow WS, Johnson HE, Bambuch RK (1979) Paleozoic paleogeography. Annu Rev Earth Planet Sci 7:473–502
Ziegler PA (1977) Geology and hydrocarbon provinces in the North Sea. Geojournal 1:7–32
Ziegler PA (1982) Geological Atlas of Western and Central Europe. Elsevier Publ Co, Amsterdam pp 130
Ziegler PA (1984) Caledonian and Hercynian crustal consolidation of Western and Central Europe – a working hypothesis. Geol Mijnbouw 63:93–108

Part III

LIFE-SUPPORTING SYSTEM

Life is just a thin veneer around the Earth but it draws all matter from the air, the sea, and the solid earth: carbon dioxide, mineral nutrients, essential elements, and most of all, water. So in a way it is a mirror image of the interactions going on between the organic and the inorganic world. A mini-cosmos, just 50 microns across, reveals this intimate relationship.

It all started with a green planktonic diatom that lived in the sea and built a siliceous shell (outer rim) to hold the cell content. Following death, sedimentation and burial, $Fe(OH)_3$ ooze commonly found in the top-most sediment layer seeped through the pores and pillars of the siliceous membrane into the cell interior.

Meanwhile, sulfate-reducing bacteria used the protoplasm as organic substrate. This liberated hydrogen sulfide generated a reducing micro-environment, reduced three-valent iron into two-valent iron, and precipitated an iron sulfide gel from which mineral seeds grew. In time, these seeds developed to individual pyrite crystals (FeS_2) yielding aggregates better known under the term framboids, after framboise (French for "raspberry") (drawing from electron micrograph in Schallreuter, 1984).

Chapter 9

Air

General Circulation of the Atmosphere

From Weather to Climate

It is probably no overstatement to say that the topic weather has appeal to everybody. The familiar phrase "Hello! Nice weather today" or "What a miserable day" is a good indication of the impact weather has on conversation. Tomorrow's weather forecasts on TV or radio are favored programs too, even though last weekend's weather prediction may have turned out to be wrong. In any event, these are simply the snows of yesteryear.

Weather data such as air temperature or moisture content plotted over weeks, months and years are the main elements of what is commonly referred to as climate. If such a collection were extended for centuries, systematic trends in climate variability may come to light which normally are hidden under the umbrella of day-by-day weather.

For many regions of the world we have a record of climatic change for the past decades. Going further back in time, that is centuries and beyond, there are only a few stations worldwide with data set sufficiently accurately to outline general trends in climate change. Long-term climatic records revealed by stochastic means are quite useful to suggest future trends in climate patterns on regional and global scales. Of course one may wonder how the meteorological community "dares" to say something about the climate of the next few decades if it is still unable to predict with some certainty the weather's ups and downs for the coming four weeks! Well, the feedback on weather from many directions prohibits a monthly forecast, whereas longer-scale climatic variation may appear above the background noise as distinct signals.

It would be futile to continue the talk about weather-forecasting in a more detailed fashion. In this respect many excellent texts are available: e.g., Donn (1972), Barry and Chorley (1982), Wallace and Hobbs (1977), Kondratyev and Hunt (1982), and Wells (1986). And yet some critical information on the physical nature of the atmosphere is still needed to make the general discussion on the air-sea-earth-life system more transparent.

The air is held by gravity on the rotating earth, and pressure and density accordingly decrease with altitude. At a height of about 700 km above sea level one encounters – technically speaking – an almost perfect vacuum. However, concerning the description of a perfect vacuum, we already came up against some difficulties previously (see p. 12). By definition, the boundary of the atmosphere is arbitrarily set at 1,000 km. The sphere above, the exosphere, is a region from which molecules are already escaping into space and pass on their way up at a height of about 5,000 km through the magnetosphere. Although thousands of kilometers from the Earth's surface, this sphere may profoundly influence the weather.

The internal structure of the atmosphere to about 100 km above sea level is schematically depicted in Fig. 9.1. Variations in air temperature with height are a convenient means to subdivide the air column into individual spheres, layers or shells, whatever term pleases you most. The layer with clouds and precipitation or the one where the weather is brewing, is termed the troposphere. Its upper boundary, the tropopause, has an altitude of 7 to 9 km in polar regions and increases to a height of ca. 16 km towards the equator. The next layer is the stratosphere, which has its top, the stratopause, at an altitude of 50 km. It has been found useful to speak of a lower stratosphere and an upper stratosphere simply to emphasize the difference in the way temperature behaves, namely a flat profile in the lower and a steep gradient in the upper stratosphere. The ozone layer is positioned at the boundary lower/upper stratosphere. The next layer is the mesosphere, which exhibits a sharp decline in temperature from 0°C to –90°C across an air segment of 30 kilometers. The minimum is called the mesopause, from where air temperature again rises. The thermosphere, as this layer is referred to, exhibits temperature variations in excess of 300°C. This is so because that layer cannot re-emit the incoming solar short-wave radiation as long-wave radiation.

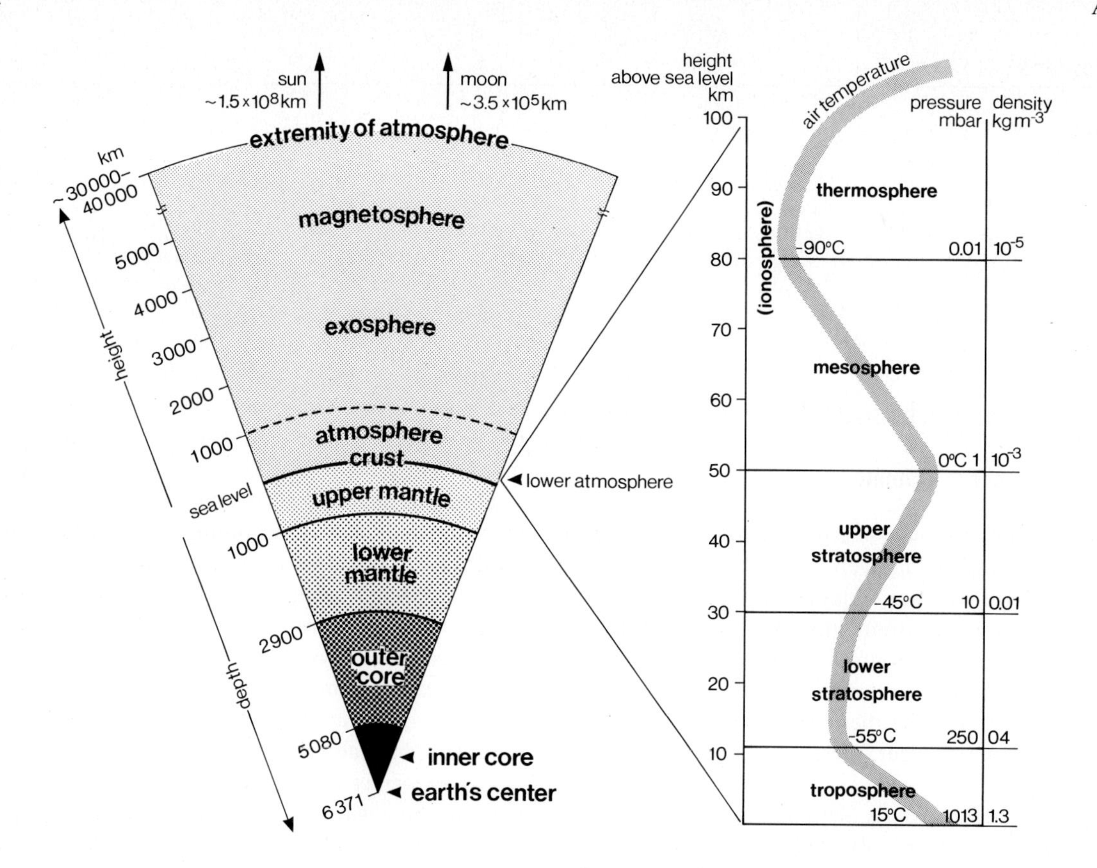

Fig. 9.1. Internal structure of the terrestrial atmosphere. Mean global changes in temperature, partial pressure, and density are depicted for the upper 100 km of the atmosphere. Thicknesses of the various "shells" of the solid Earth (Ringwood, 1979) and those of the total atmosphere are given for comparison (after Schönwiese, 1979)

At heights greater than 80 kilometers, and through the combined action of solar ultraviolet radiation and cosmic rays, electrons become separated from oxygen atoms and nitrogen molecules. This ionization is responsible for the phenomena of Aurora Borealis and Aurora Australis, the polar lights. Zones of greatest ionization in the ionosphere occur in the so-called D, E, F1 and F2 layers. Let us now turn to the motion of the air.

The atmosphere is driven by solar radiation. Polar or tropical air masses receive varying shares of heat in the course of a year. In addition, water, land, ice and plants respond differently to sunlight, leading to regional patterns characterized by a distinct atmospheric heating. In turn, a highly complex, although organized, motion is initiated that carries heat to the poles and cold to the tropics, giving rise to what is known as general circulation of the atmosphere (Fig. 9.2). As main elements we see: (i) the rain-bear-

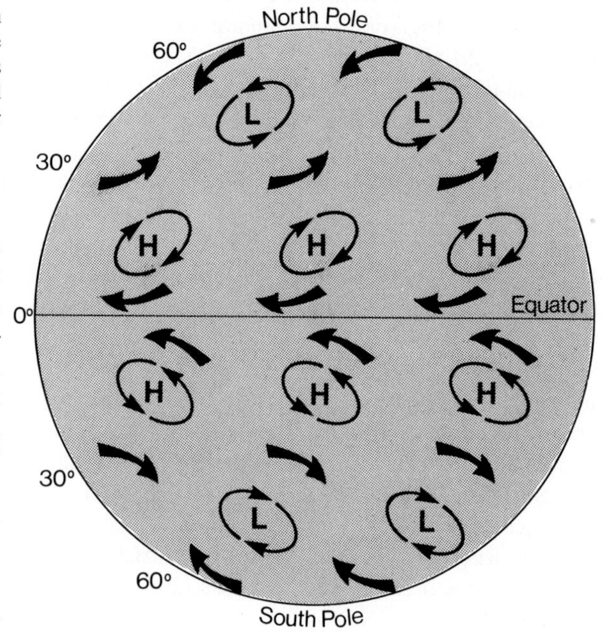

Fig. 9.2. General circulation of the atmosphere: westerlies and mid-latitude cyclones (*L*), subtropical high pressure belt with quasistationary anticyclones (*H*), trade winds and the intertropical convergence zone, which is characterized by rising motion of air masses and frequent rains (after Nicholson and Flohn, 1980)

ing cyclones of the westerlies (L), (ii) the dry subtropical high pressure belt (H), and (iii) the tropic front, with which intensive precipitation is associated. These basic features can easily be recognized on satellite imagery, taken during a northern summer (see p. 69): westerly low pressure systems are visible as cloud spirals, the subtropical high pressure belt as cloudless sky over the Sahara, and the tropic front as a zonal band of convective clouds north of the equator.

From time to time, global atmospheric circulation deviates from its "normal" behavior, producing climatic anomalies which in certain areas of the world have dramatic consequences: droughts, floods, severe winters and cold summers. Regions located along boundaries of climatic zones are especially vulnerable to a slight change in atmospheric circulation.

In the following I have chosen three case studies which well illustrate the feedback of a localized phenomenon on global weather and climate, and the coupling of air-sea-earth-life: El Niño, polar and tropic fronts, and volcanism.

El Niño

Over the past few years, and here especially during the period 1982 to 1984, abnormal climatic events were felt in various parts of the globe with considerable socio-economic impacts and losses estimated at 10 to 20 billion U.S. dollars worldwide (Climate System Monitoring, 1985). However, all of those climatic events have sprung into existence by a single factor: the El Niño – Southern Oscillation Phenomenon (ENSO).

The term "El Niño" (= Christ Child) was coined by local fishermen more than a century ago to describe the warm water that emerged around Christmas time in coastal regions off Peru, Ecuador, and Chile, roughly in the stream-line of the generally cold Humboldt Current. In some years this warming was much more intense than normally, temporarily lowering the flux of nutrient-rich upwelling waters, thus impairing phytoplankton productivity and with it anchovy fishery (see below).

The phenomenon of "Southern Oscillation" has been known to meteorologists and oceanographers for decades and is generally seen in conjunction with the pioneering work of G.T. Walker (1923, 1924) on the Indian monsoon: "When pressure is high in the Pacific Ocean, it tends to be low in the Indian Ocean from Africa to Australia; these conditions are associated with low temperatures in both these areas and rainfall varies in the opposite direction of the pressure." The switch in the sign of the pressure anomaly across the tropical ocean is commonly described as a pressure "seesaw".

Although both El Niño and Southern Oscillation were recognized quite early as major oceanographic and atmospheric events respectively, an understanding of their intimate relationship came only after the strong 1957 El Niño and following the concerted work of scientists from the atmospheric and ocean sciences during the International Geophysical Year. Bjerknes (1966) developed the concept of the Walker Circulation. In principle, convective areas of rising motion characterize the global tropical circulation, resulting in clouds and rain over Southeast Asia, the Western Pacific, the Amazon Region, and Central Africa. Warm sea surface temperatures (SST's) cause air to rise to the upper troposphere to a height of approximately 10 to 12 km, while cool SST's are commonly associated with sinking air masses. Departure of SST from its equatorial zonal mean is illustrated in Fig. 9.3 (Wyrtki, 1982). It is of note that descending branches of the overturning cells over the Atlantic and central and eastern Pacific are usually dry, and consequently those regions are often referred to as ocean deserts.

Atmospheric circulation receives much of its driving force from the large-scale tropical convection pictured here through the release of latent heat in the mid-troposphere (diabatic heat sources).

Easterly trade winds in the equatorial Pacific propel ocean circulation, leading to a sea-level lowering in the east and a rise in the west with an accompanying deepening of the thermocline. When the easterlies slacken, sea level drops and thermocline rises in the western Pacific, whereas an opposite development proceeds along South America (Fig. 9.4). This coupled atmosphere-ocean system does not stay in balance, but undergoes aperiodic oscillations of roughly 2 to 7 years as revealed by a number of observational data such as SO-index, surface wind stress, sea-level fluctuations and SST indices. It is a combination of these factors which apparently lies behind the ENSO phenomenon (see, e.g., Wyrtki, 1975, 1982; Julian and Chervin, 1978; Philander, 1983; Cane, 1983, 1986; Cane et al., 1986; Rasmusson and Wallace, 1983; Rasmusson, 1984; Salstein and Rosen, 1984; Webster, 1984; Climate System Monitoring, 1985; Ramage, 1986).

In the typical ENSO phase with a falling SO-index, pressure rises over the Indonesian Archipelago, resulting in a decline of the pressure gradient along the Equator and a relaxation in the Pacific trade winds. Eventually, the easterlies in the western equatorial Pacific may change direction, that is, blowing for sev-

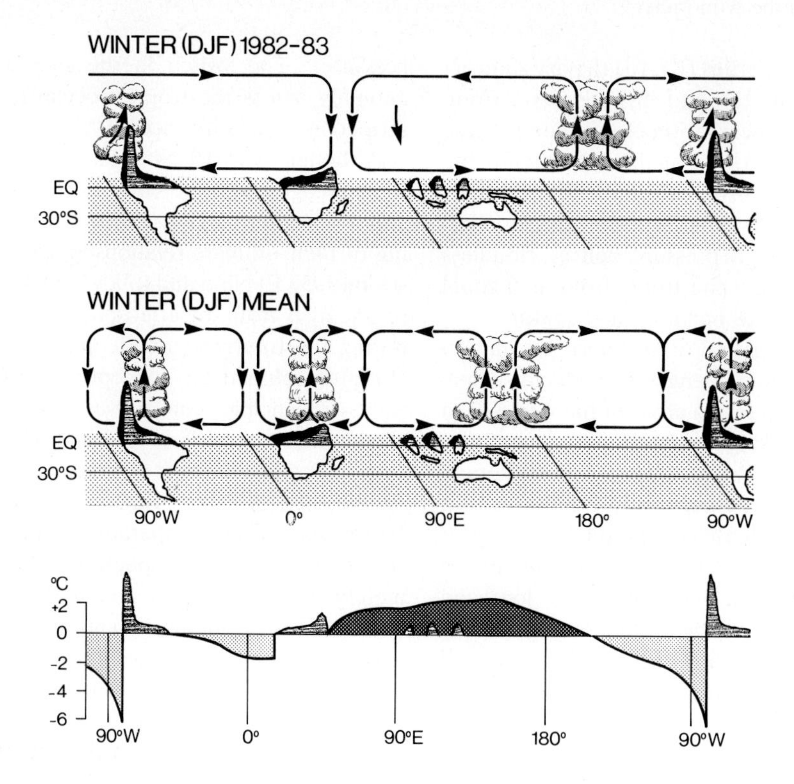

Fig. 9.3. Mean Walker circulation along the equator resulting in rising air and associated heavy rains (highly schematic). Its strongest branch over the Pacific is linked to sea surface temperature (SST). Warm temperatures induce air to rise, whereas cool temperatures lead to a sinking of air masses.
Departure of SST from its equatorial zonal mean is depicted in the lower part of the diagram. The winter (December-January-February) mean east-west overturning Walker circulation is depicted in the central part of the diagram. The top figure illustrates the ENSO Winter 1982–83 pattern (after Climate System Monitoring, 1985; Wyrtki, 1982; Ramage, 1986)

eral months from west to east. This shift in wind pattern will introduce major changes in the equatorial current system in that a rather fast subsurface wave known as the Kelvin wave spreads eastward towards the South American coastline. By definition, Kelvin waves are internal shear waves generated from relative movement between two water masses with different density. On approaching the South American continent, the Kelvin wave becomes partly deflected, while at the same time residual energy is reflected as a slow westward-moving and laterally spreading Rossby wave, which is a "planetary" wave associated with the Earth's rotation. In short, the normal wind pattern along the Equator can reverse itself completely during El Niño by a positive feedback interaction between the atmosphere and the sea surface. In the upper troposphere air flows to the west rather

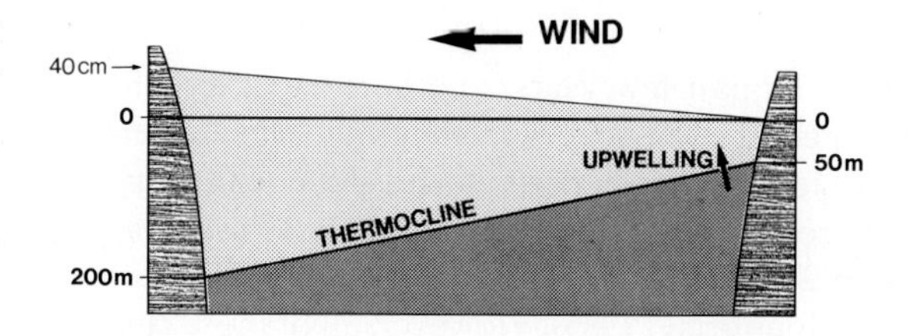

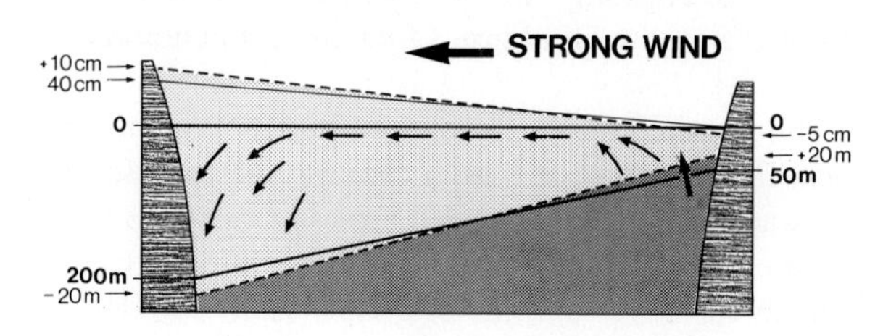

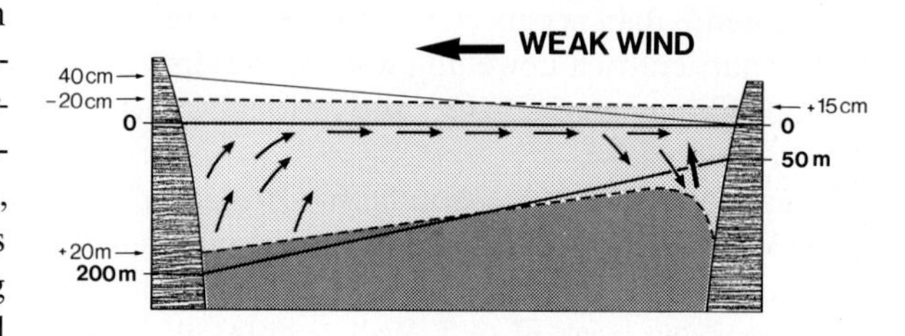

Fig. 9.4. Response of thermal structure of the equatorial Pacific to a changing wind pattern with respect to positions of sea level and thermocline, and direction of flow regime in the euphotic zone (after Climate System Monitoring, 1985; Wyrtki, 1982; Ramage, 1986)

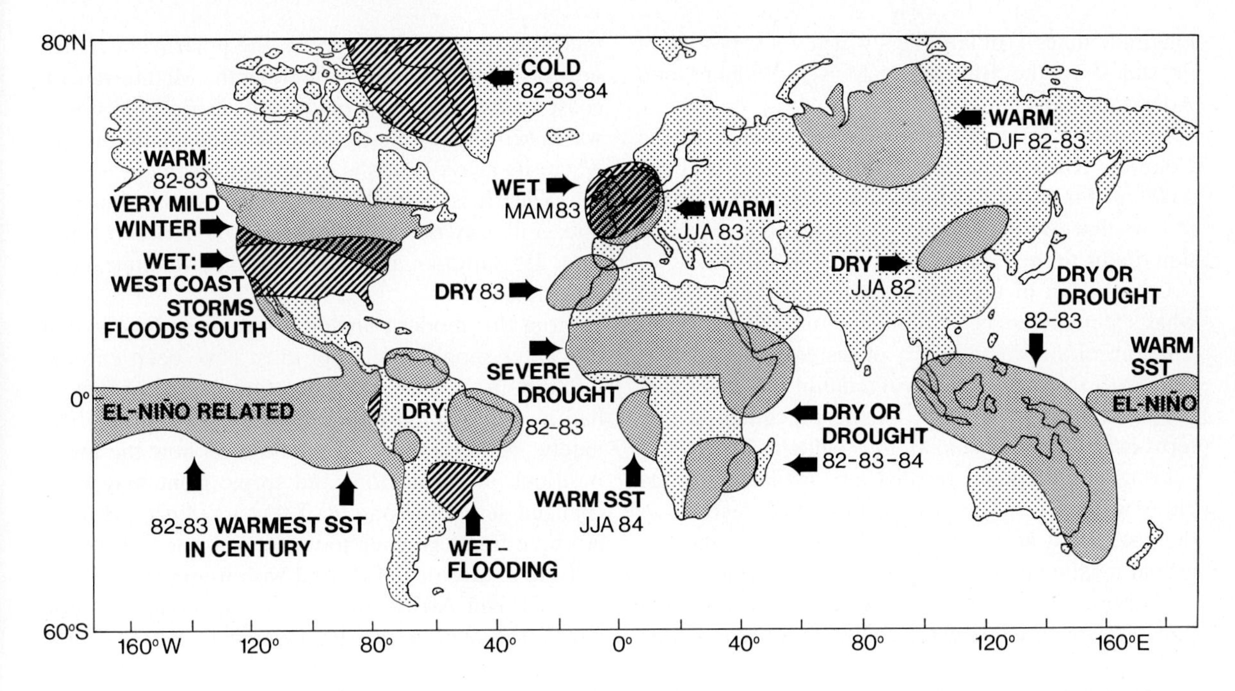

Fig. 9.5. Selected extreme climate incidents in the 1982–84 period. Dry and/or warm events are *lightly shaded;* cold or wet events are *cross-hatched* (after Global Climate System, 1985)

than to the east and sinks over Indonesia, where the weather becomes unusually dry. In contrast, torrential rain in the normally dry region of the central and eastern Pacific is a consequence of the reversal in the air motion.

The severest El Niño in a century was the one of 1982–83, with disastrous regional climatic consequences across the whole globe, some of which are depicted in Fig. 9.5. As far as general pattern and environmental impacts are concerned, the 1982–83 event seemed to resemble closely the El Niño of 1877–78 (G. Kiladis and H. Diaz: in Ramage, 1986). The 1982–83 ENSO event differed from a typical event both in timing and strength (Climate System Monitoring, 1985). By December 1982, the entire equatorial Pacific was 1 to 5°C warmer than normal and SST's reached record heights. Resulting changes in the Walker Circulation are depicted in Fig. 9.3, reflecting a pronounced shift in the convective regime towards the central and eastern Pacific.

In view of its devastating effects on the environment and society in many parts of the world, El Niño forecasts months ahead of the actual event could provide sufficient warning and lead to precautionary measures against it (Namias and Cayan, 1984; Barnett, 1984; Cane et al., 1986). Early in 1986 several groups suggested a forthcoming El Niño event, using the development of SST's in the equatorial Pacific as the prime indicator (e.g., Cane, 1986). Christmas 1986 is just two days ahead of us and El Niño is not yet in sight (Palca, 1986).

A discussion on El Niño would not be complete without a brief reference to biological response and the large reductions in plankton, fish and seabirds in the normally rich waters of the eastern equatorial Pacific (Barber and Chavez, 1983). Peruvian anchovy *(Engraulis ringens)*, once the basis of the world's largest fishery, almost vanished following recent El Niño events. The 20-fold decrease in phytoplankton biomass and productivity that took place along the coastal transect studied in 1983 (Barber and Chavez, 1983) will further decrease the growth, survival and particularly reproductive fitness of the adult anchovy. It may take a considerable amount of time before anchovy stocks are back to normal, provided, of course, that another devastating ENSO event does not again affect both reproduction and adult survival.

Polar and Tropic Fronts

Hydraulic societies (Wittfogel, 1977) that depend mainly on rivers as a water resource are in trouble if a drop in rainfall occurs in the river's catchment area. Egypt is a classical example because its water resources are principally provided by the River Nile, which is fed to almost 80% by summer rains falling in the Ethiopian mountains, a tiny catchment area – 4° of longitude times 4° of latitude – which delivers 40% of its runoff to the Blue Nile (M.A.J. Williams and Adamson, 1980).

In analogy, pratically the entire discharge of the Colorado River stems from a very limited area above 2,000 m. From the Colorado discharge in Arizona it follows that – as in Ethiopia – 40 percent of precipitation drains to the Colorado (Riehl et al., 1979).

Comparison of these two regimes makes clear to what extent relatively limited mountainous terrains can control the water supply of distant cities such as Cairo or Los Angeles. The vulnerability of water supply to small regional shifts in the precipitation patterns over such restricted areas is evident.

Long observational records are available for the Nile's water level. The gross hydrological features in the catchment area have been known since the last glacial maximum about 18,000 years ago (Flohn and Nicholson, 1980). In short, the sensitive response of a large stream to climate variability on different time scales is chronicled.

The water resources of Egypt are fed by two spatially and seasonally distinct climatic phenomena: Mediterranean winter rain at the northern margin of the Sahara, and tropical summer rain in Eastern Africa. These two events are associated with two zones of pronounced meteorological gradients. They are, to the north, the polar front, and at equatorial latitudes the intertropical convergence zone (ITCZ), often referred to as "tropic front" or "meteorological equator". The tropic front is characterized by the high-reaching convective clouds, whereas in the region of sinking air masses cloud formation is suppressed.

Spatial distribution of yearly mean precipitation across Africa is rather homogeneous zonally, but shows a strong latitudinal gradient (Fig. 9.6), both indicating the overriding control of the tropic front on rainfall in the area. In winter, the polar front moves south and brings rain to the northern margin of the Sahara. In contrast, the Nile has low water, because in its catchment area precipitation is at a minimum, due to the southward-moving tropical front. During summer, the situation is reversed: the polar front retires northwards and no rain falls along the Mediterranean coast of Africa. The tropic front also moves northwards across the equator to the Ethiopian mountains, where its heavy rainfalls cover the catchment area of the Nile. It is here that the Nile's flood is born. It makes its way north, 2,700 km through desert, crossing 35° latitude and reaching the Mediterranean 2 months later.

From this modern analog it becomes obvious that the water supply for Egypt must have been entirely different in the past, when polar front and tropic front had moved around different mean positions. Fortunately, we have some information on how the mean positions of polar front and tropic front may have changed across the past 20,000 years. Since the modern time frame is much too short to bring to light the shifting nature of ITCZ, and with it atmospheric circulation over Africa, the inferred mean atmospheric circulation over Africa is depicted in Fig. 9.7. In a simplified manner, the past 20,000 years may be characterized by three distinct patterns for the general circulation of the atmosphere with the concurrent mean latitudinal positions of the polar front and the tropic front. These broad patterns characterize: (i) the period around the glacial maximum 18,000 years ago, (ii) the Holocene warm phase between 10,000 and 4,000 B.P. (before present), and (iii) today's situation, prevailing for the past 3,000 years.

Summing up, the cold and dry period that is the time slot 20,000 to 12,000 years B.P. (Fig. 9.8a) appeared to be characterized by a southward shift of polar front as well as tropic front, and Egypt got more rain but less Nile water. With the outgoing glacial and the onset of Holocene between 10,000 and 8,000 years B.P. (Fig. 9.8b), the tropic front moved further

Fig. 9.6. Vertical meridional section through the tropical atmosphere, schematically showing the high-reaching convective clouds characteristic of the tropic front. In the region of sinking air masses, cloud formation is suppressed (after Nieuwolt, 1977)

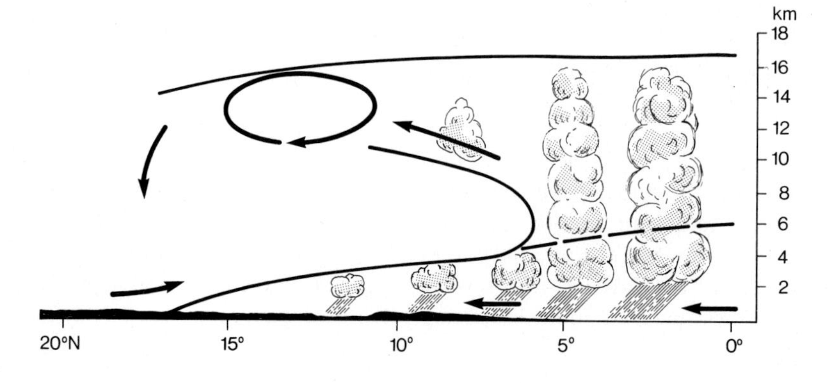

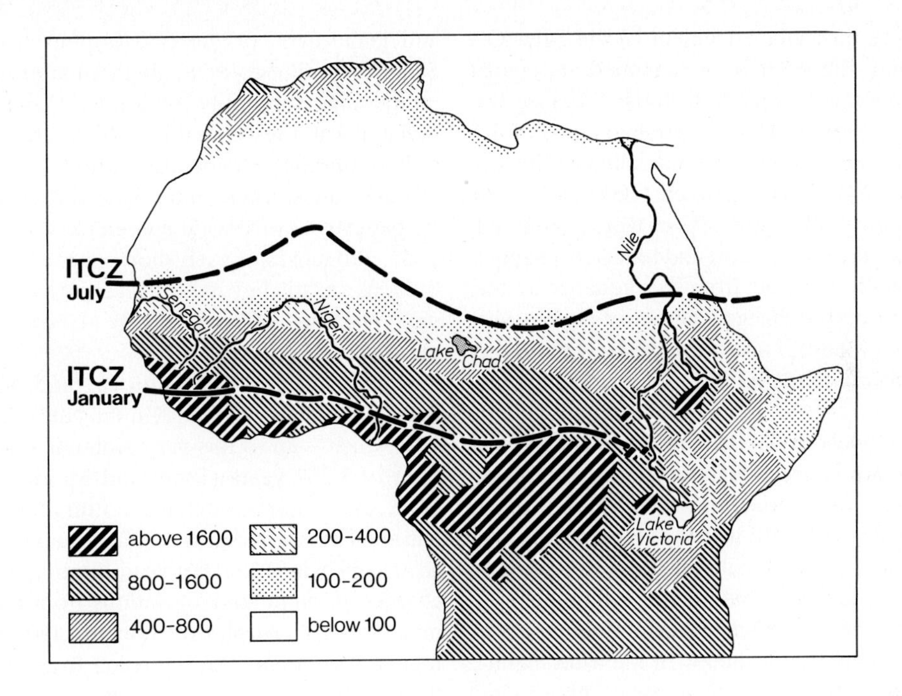

Fig. 9.7. Mean annual precipitation in Africa (from Nicholson, 1980)

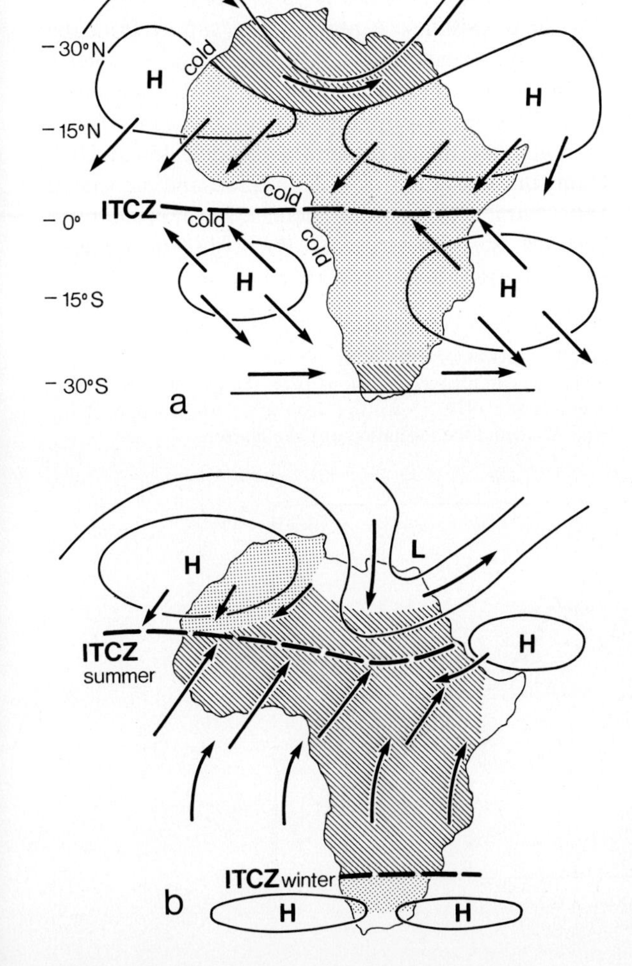

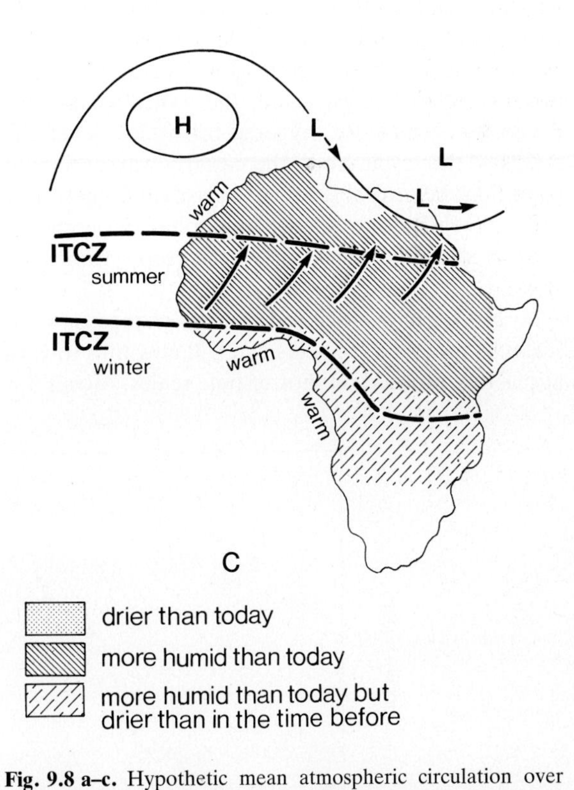

Fig. 9.8 a–c. Hypothetic mean atmospheric circulation over Africa between: **a** 20,000 and 12,000 years B.P., **b** 10,000 and 8000 years B.P., and **c** 6500 and 4500 years B.P. (after Nicholson and Flohn, 1980)

north, reaching the main catchment of the Nile. On the other hand, the polar front remained at a more southern position as compared to today, because the glaciers of the Northern Hemisphere had not yet fully retreated. The general warming that followed the last glacial reached its climax between 6,000 and 4,000 years B.P. Mean global temperatures were probably 1 to 1.5°C higher than at present, and today's arid zones were significantly reduced from the equator to the Mediterranean and a changed humidity regime was established (Fig. 9.8c). During the past few thousand years atmospheric circulation about followed the present regime.

All these models are based on fragmentary evidence and cannot be more than schematic pictures of the real events. In a series of papers (Flohn, 1980; Flohn and Nicholson, 1980) it was said that spring and autumn play an important role for the water cycle in the Sahara. One reason is that the tropical easterly jet stream – driven by the Tibetan Plateau – is only effective during summer. Associated with the wind regime in its delta is the sinking of air masses. This sinking suppresses the northward movement of rain-intensive tropical weather systems, and provides the answer to the question "Why is the Sahara so dry?" (Flohn, 1966). During spring and autumn, this mechanism is less dominant, so that tropical disturbances may easily move northward and bring rain and sandstorms as far as to the northern margin of the Sahara. They are more familiarly known under the term "Sahara Depressions", which are triggered off by low level elements of the tropic front, the easterly waves. Should conditions be favorable for an intensified triggering of Sahara Depressions, a more humid Sahara might develop without having to move the tropic front (see Box 9.1).

So far, we have discussed climatic change on time scales of thousands of years and will now turn to variations that proceed on shorter time scales. Moses' fat and lean years, or the catastrophic drought in the Sahel Zone 1968 to 1978, are familiar examples of the interannual variability of African climate and associated water resources. The Nile's catchment area as well as the Sahel Zone are located at the margin of climatic zones, where minor latitudinal shifts of climatic patterns have drastic consequences for the water cycle. Prognoses of such short- or medium-term fluctuations are still beyond the reach of modern climatology because of the highly complex feedback mechanisms.

A powerful tool to register climatic change in this part of Africa can be found in the Nile's discharge record, which – up to this very moment – covers a time span of 1,364 years (Riehl and Meitin, 1979; Riehl et al., 1979). For more than a century the Nile's runoff has been monitored at Assuan. Its mean for the years 1870 to 1976 is 91 km^3 per year. Plotting the yearly discharges as cumulative deviations from the long-term mean reveals a surplus regime until 1900 and a deficit regime since then. The transition from the surplus to the deficit regime is very abrupt, suggesting a discontinuity in climate change, which is in excellent agreement with time series of tropical rainfall around the globe (Kraus, 1955). Two lessons may be learnt from this. First, within a century, water supply from the Nile may change by 30 percent of the long-term mean. Second, the switch from fat to lean years may proceed very abruptly (Fig. 9.9). In this context, it is of note that the construction of the Assuan High Dam, the formation of Lake Nasser, and associated agricultural activities have increased evaporation and lowered the discharge of the Nile substantially (Kempe, 1983).

Fig. 9.9. Discharge rates of the Nile at Assuan since 1825, presented as cumulative deviations from the overall mean (after Riehl et al., 1979). Note the drop after construction of the High Dam and the formation of Lake Nasser

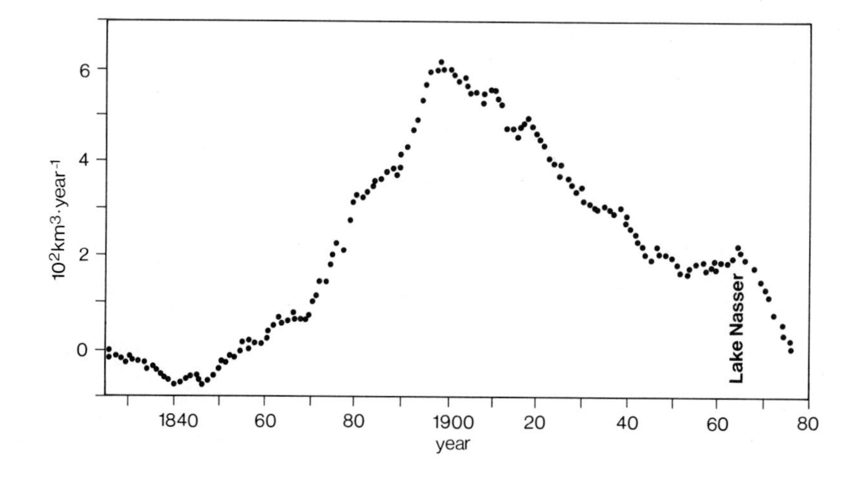

BOX 9.1

Project New Valley (Degens, 1961; Knetsch et al., 1962)

Oases are located sporadically throughout the Libyan Desert, and it is common belief that they have unlimited water resources at depth. It is further assumed that this water reservoir is continuously recharged from the south (Abyssinia-Sudan) and southwest (Equatorial Africa), where precipitation is abundant.

This belief is based largely on the fact that the oases have remained as bastions in the desert for at least the last few thousand years, and that during this time the water supply has not changed significantly.

The water is mostly artesian – brought to the surface by natural water or gas pressure. It is stored at depths from a few tens to 1,000 meters below the present surface. Geological studies reveal that there is subsurface water intercommunication between some of the oases, which could mean that water reservoirs of larger dimensions are developed at greater depth beneath the desert.

This assortment of facts and vague ideas has led to one of the most fantastic cultivation programs – Project New Valley. Basically, the project intends to irrigate land now occupied by desert by tapping the subterranean freshwater "lake" the length of 1,000-kilometer-long and up to 150-kilometer-wide meandering wadi from Wadi Halfa to Siwa (Fig. 9.10). Initially, 250,000 hectares (ha) were supposed to be cultivated. Project New Valley became a vision of a wide green belt in a glistening sand desert.

A few years following the start of the project, there was a rude awakening when it was discovered that there was no water recharge from the south and southwest, because a crystalline basement separates the Sudan from Egypt, blocking the subsurface water flow from south to north (Knetsch et al., 1962). Thus,

Fig. 9.10. Block diagram (highly schematic) showing stratigraphy and facies relationship of Egypt in a north-south profile from the three-state border: Sudan – Egypt – Lybia (Uweinat) to the Mediterranean. The uplifted Precambrian crystalline basement along the line Uweinat-Wadi Halfa is a natural barrier against the flow of water from the south (after Knetsch et al., 1962). The so-called "desert depression" from Wadi Halfa to Siwa roughly outlines the region of Project New Valley (Degens, 1961)

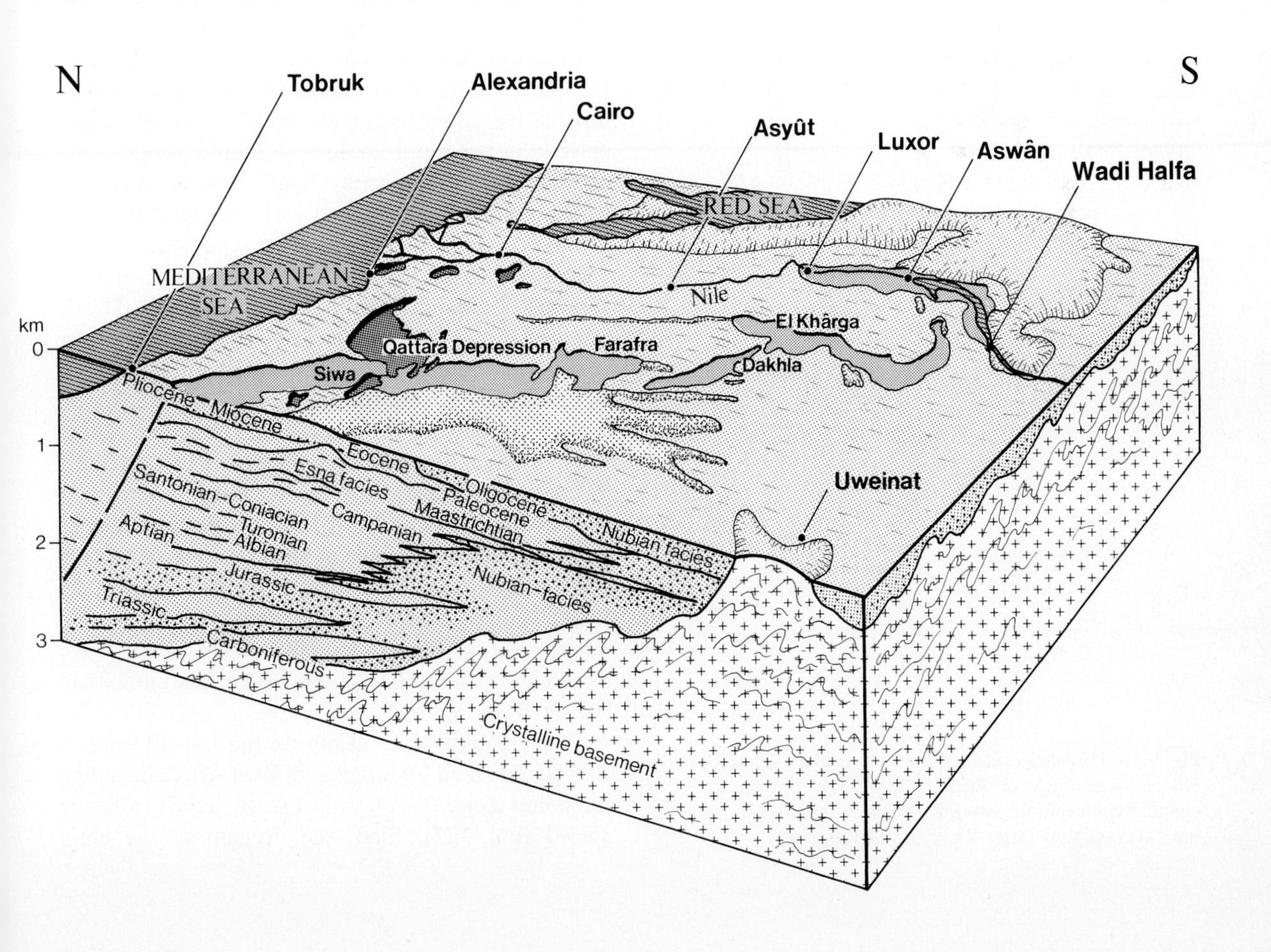

all the waters in the western Egyptian desert are fossil, that is tens of thousands of years old. In those ancient times, the region which is now the Libyan Desert having a precipitation of 1 mm per year, had clear lakes and streams and abundant rain, providing a home for a few million people. A fata morgana across the sands of time? So it seems, because just a mere residue of water is all there is left in the pore space of the Nubian sandstone.

Today less than 20,000 ha are under the plow. The water does not run so quickly as it did 20 years ago, and salt and sand are already covering the fields of the pioneers (Degens and Spitzy, 1983).

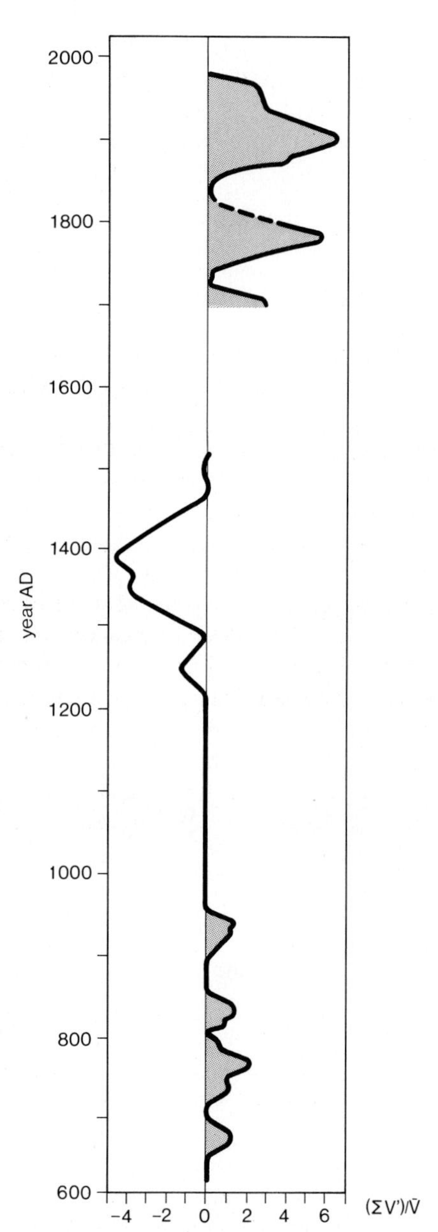

Fig. 9.11. Discharge rates of the Nile reconstructed from the Nilometer-readings at Roda, Cairo. Data are presented as cumulative deviations from the mean normalized by the long-term average flow (after Riehl and Meitin, 1979)

The records of flood from various stations along the Nile offer a unique opportunity to reconstruct the history of Nile discharge over even longer time periods. The longest record is that of "Nilometer" readings on the Island of Roda in Cairo, which starts in 622 A.D., the year in which the Mohammedan calendar begins (Riehl and Meitin, 1979). When political control passed to the Turks in 1520, about 200 years of Nile runoff remained unrecorded. Another unfortunate break of 25 years happened after the short-lived Napoleonic occupation (1800–1825).

The corresponding discharge can be computed with a transfer function that relates discharge measured at Assuan to the corresponding Nilometer reading at Cairo. The discharge history of the Nile since 622 A.D. reconstructed by this method is represented in the form of cumulative deviations from the mean in Fig. 9.11. The result is surprising. We see as characteristic features "fat" and "lean" episodes. Each episode lasts several decades. A "fat" episode is immediately followed by a "lean" episode and vice versa, indicating that there is not a long-term constant mean flow on which sporadic fluctuations are imposed. The second surprising result is the complete absence of variability from 950 to 1200 A.D. This is the time of the so-called "little climatic optimum" or "medieval warm phase" in the North Atlantic: corn was grown in Iceland, the arctic pack ice boundary retreated to the northern coast of Greenland, there were frequently warm summers and droughts throughout Europe, and the northern Sahara was more humid (Flohn, 1980). About 100 km east of Hamburg, in the province of Mecklenburg, the Cistercian Order brought Christianity to that region in the 12th Century and with it: vineyards. To produce the same quality of wine today, we would probably have to move to the Rhine Valley, which lies more than 500 km to the south.

The variability in rain during the last 40 years is also documented for a region of West Africa including the Sahel Zone. Two lessons can be learnt from the record (Fig. 9.12). First, the drought of the early

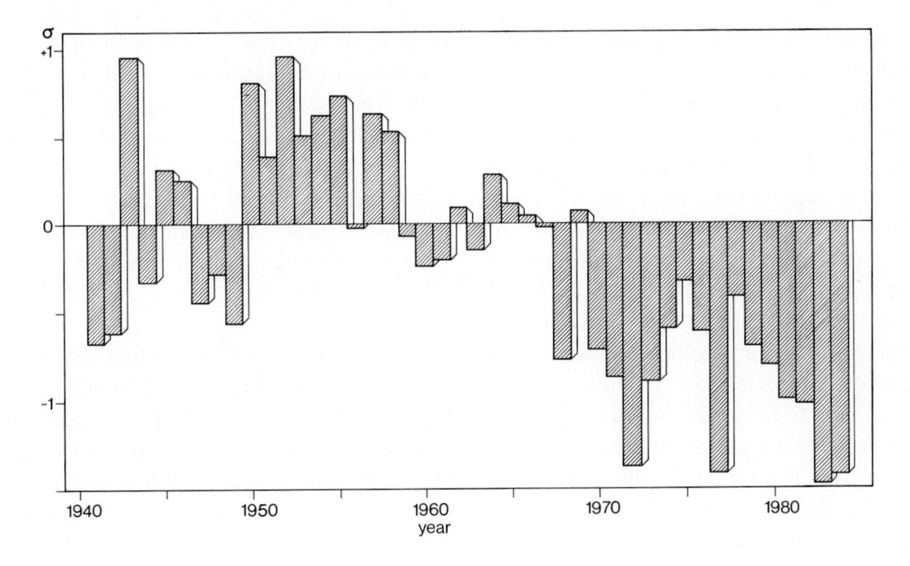

Fig. 9.12. Rainfall distribution in the Sahel Zone since 1940. Yearly average of the normalized rainfall departures (after Lamb, 1982; updated by recent measurements)

seventies is not at all unique, and second, the degree of annual variability is considerable.

In conclusion, polar and tropic fronts control the climatic happening in Africa in space and in time. However, they are critical elements of the general circulation of the atmosphere and the driving force of other important climate events; the monsoon, for instance, which dominates the weather over most of the Asian Continent, the Arabian Sea, and the extreme western position of the Northern Pacific Ocean. Except for Antarctica, monsoonal influence is felt across practically all land masses outside equatorial latitudes. In a word, the tropical belt cooks the weather for the globe, and small changes injected into that system may alter the motion of air and sea, and with it induce climatic change.

Volcano Weather

Fire, smoke, steam, torrential rain, and lightning are attributes one commonly associates with volcano weather. More hidden are emanations of climate-optimizing gases, for instance CO_2, or of climate-minimizing compounds, such as aerosols. With respect to the release of volcanic carbon dioxide, its annual discharge is far too little to change the climate in the foreseeable future. In contrast, CO_2 release to the air by weathering, respiration, deforestation or the burning of fossil fuels is in orders of higher magnitude, and consequently of major concern to the climatic development within one's own lifetime (see p. 201). Ultimately, however, CO_2 stems from the degassing of crust and mantle in the course of Earth history. It returns to the solid earth and resurfaces to the air and the sea over and over again in a never-ending cycle.

The situation is entirely different when it comes to matter and gases explosively ejected by volcanoes into the lower stratosphere. Here, at altitudes between 20 and 25 kilometers, they may substantially add to the natural aerosol layer. In the main, this layer is composed of submicroscopic droplets of diluted sulfuric acid, plus – following major volcanic eruptions – fine mineral particles. Furthermore, sulfur-containing volcanic gases can be brought up there, too, where they eventually transform into aqueous sulfuric acid and effectively enhance the aerosol layer (e.g., Labitzke, 1987, 1988). It is this topic of volcanic dust veils and climate which I will concentrate on now.

Elizabeth Stommel – she is my former piano teacher – and her husband Henry told the inspiring story of 1816: "The Year without a Summer" (H. and E. Stommel, 1979, 1983). Focal point is the eruption of Mount Tambora on the Island of Sumbawa, Indonesia, in mid-April of the year 1815, an event which preceded Napoleon's Waterloo (June 16, 1815) by about two months. As one's Waterloo stands for catastrophic defeat, the Tambora 1815 cataclysm (Box 9.2) is considered the largest explosive volcanic event of the past 10,000 years. It claimed 90,000 lives and ejected about 100 times more ash than Mount St. Helens in the State of Washington, USA, in 1980 (Lipman and Mullineaux, 1981).

The year 1816 was marked by an exceptionally cold summer especially in the New England States,

Canada, and the extreme western parts of Europe. Temperatures dropped some 1 to 2.5°C compared to the summer mean of the previous year. Snow and frost were reported from some New England areas. Harvest failures made the Germans worry about shortages of grain supply for their daily bread, whereas the French people were more concerned about the quality of their red wine because of the bad summer season. These and related economic, social and even health problems inflicted by the cold spell make fascinating reading in "Volcano Weather" by Henry and Elizabeth Stommel (1983).

Whether the Tambora cataclysm caused these events or was merely a component of them – or even a coincidence – is still a matter of controversy (Schneider, 1983). According to this source, a drop in average summer temperature by 3°C does not on its own suffice to let it snow in June and freeze in July and August 1816, as was the case in some parts of New England. However, if at the same time the jet stream dipped down unusually far to the south, the combined effects could possibly explain the few frosty days. Work by Catchpole and Faurer (1983) points to such a scenario. On the basis of data on summer sea ice severity in the Hudson Strait for the period 1751 to 1870, taken from log-books of ships sailing from London to trading posts of the Hudson Bay Company, there is clear indication of an unusual summer in 1816. That season was characterized by the frequent development of strongly meridional atmospheric circulation, allowing southward incursion of arctic air in eastern North America and in western Europe. The mean July 1816 pressure pattern over the regions adjacent to the North Atlantic Ocean, implying a blocking in the westerlies, lends further support to this supposition (Lamb and Johnson, 1966).

Since Tambora 1815, quite a number of major volcanic eruptions have occurred (see Simkin et al., 1981) that may be linked to hemispheric-scale climatic coolings by several tenths of a degree Celsius in the years – generally one or two – following a volcanic event (Landsberg and Albert, 1974; Mass and Schneider, 1977; Lamb, 1970, 1983). But it was only after modelers and theoreticians gave additional support for such a cooling trend (e.g., Pollack et al., 1976) that the climate people finally accepted the concept of volcano weather (Gilliland, 1982; Hansen et al., 1978; Robock, 1979, 1984; Oliver, 1976; Schönwiese, 1979, 1988).

First signs of volcano weather are the bright reddish colors seen in a cloudless sky at dawn and dusk. This coloration develops because aerosols disperse the blue light more effectively than the red light, a phenomenon described and physically explained by Kießling (1885) on the occasion of the catastrophic explosion of the Island of Krakatoa, Indonesia, in 1883. The Krakatoa aerosol cloud of 1883 that developed in the tropical stratosphere moved eastwards at an altitude between 20 and 25 km. Interestingly, this feature was used as an indicator that the stratosphere in tropical regions is governed by east winds. However, 80 years later it was found that the stratospheric wind regime in the tropics changes from year to year.

The drift of aerosol clouds that developed in the aftermath of the eruption of Mount Agung/Bali in March 1963 and of El Chichon/Mexico in April 1982 at an altitude of ca. 24 km has been watched closely (Matson and Robock, 1984; Rampino and Self, 1984). The temporal development of the El Chichon cloud during April 1982 is depicted in Fig. 9.13. It took the westwards-spreading cloud 3 weeks to circum-navigate the globe. During the summer it remained between 30°N and 20°S, subsequently to release substantial amounts of aerosol into the polar regions.

In principle, stratospheric aerosols lead to a warming of the stratosphere and correspondingly to a cooling of the troposphere. Using radioprobe data, Labitzke and Naujokat (1983) found a general cooling between 1962 and 1972, but a warming since then (Fig. 9.14), indicative of the volcanic emission control on stratospheric temperature trends. Since 1979 modern laser techniques rendered stratospheric aerosol measurements possible (McCormick and Trepte, 1987). It could be demonstrated that general stratospheric circulation within months following a major volcanic eruption sweeps aerosol clouds in the direction of the poles where aerosol content is rapidly building up. Since (i) residence time of aerosol above the poles is long, (ii) aerosols cause a cooling of the stratosphere during polar nights (Labitzke, 1987), and (iii) the time envelope 1955–1986 is one of highest volcanic activity, it comes as no surprise that four of the six coldest stratospheric winters recorded at the north pole over the past 31 years appear to be linked to the eruptions of Mount St. Helens and El Chichon; these are the winters of: 80/81, 82/83, 83/84, and 85/86. Temperatures at the 30-mbar level fell below –75°C (Labitzke, 1987).

Previously we have said that the ENSO 1982/83 event was unusual when compared to other El Niños. It is conceivable that volcano weather has something to do with it, inasmuch as increased aerosol levels enhance and at the same time lower the polar eddy and with it lead to a late spring warming of the stratosphere. In what way the tropospheric circulation is thereby affected is still under debate (Labitzke, 1988).

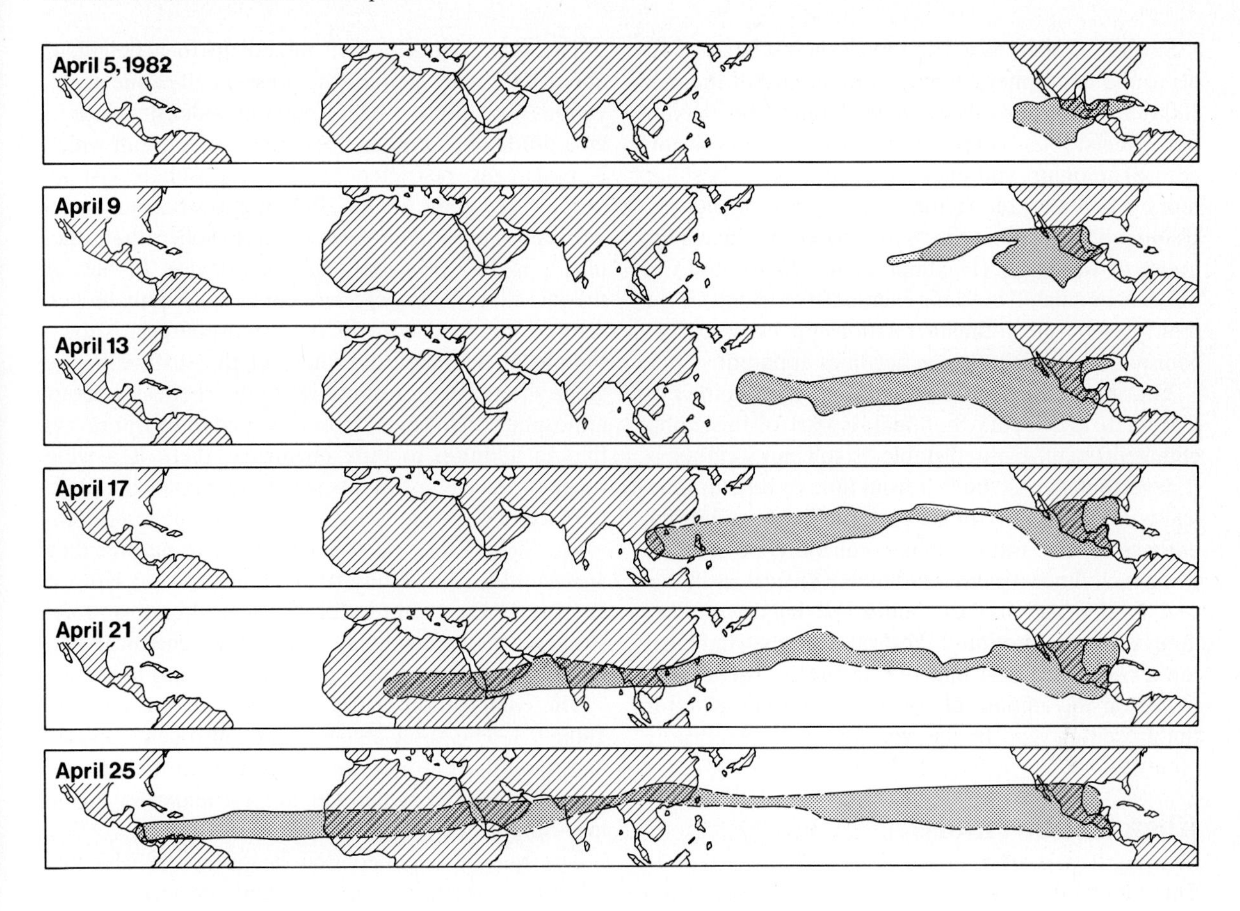

Fig. 9.13. Westward spreading of the El Chichon aerosol cloud at an altitude of ca. 24 km. The gradually enlarging cloud required 3 weeks to circum-navigate the globe (after Matson and Robock, 1984)

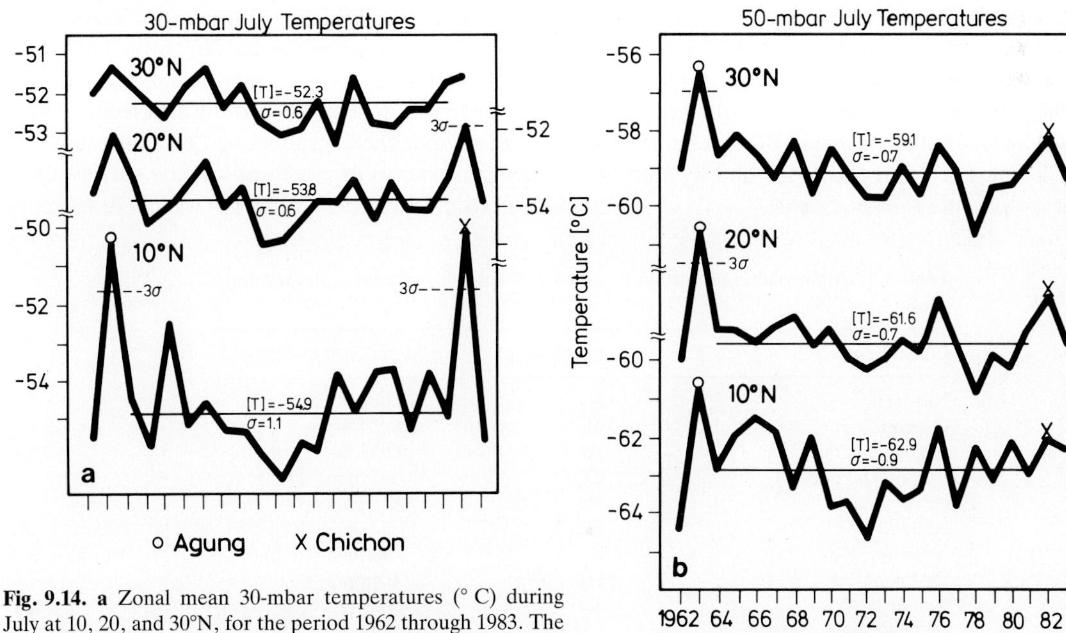

Fig. 9.14. a Zonal mean 30-mbar temperatures (° C) during July at 10, 20, and 30°N, for the period 1962 through 1983. The 18 year average (T) is for the period 1964–1981. **b** Same as **a**, but for the 50-mbar level (Labitzke and Naujokat, 1983)

Several authors have tried to fit their climate models to the hemispheric temperature record of the past 100 years. Gilliland (1982), in combining critical external effects, i.e., carbon dioxide content, volcanic aerosol amounts, and solar cycles, attains the best harmony when all three factors are properly considered. Using more extended time series based on the acidity index of ice cores (Hammer et al., 1980, 1981) or radiocarbon dated volcanic events (Bryson and Goodman, 1980), the intimate relationship between volcanism and climatic change becomes apparent.

Volcano weather, to use again the term coined by Elizabeth and Henry Stommel, is part of the global climate system. Unpredictable as volcano weather is, it nevertheless has and will from time to time dramatically influence the climate pattern. Expressed differently, past and future climate can only be assessed when including volcano weather as a critical variable.

A final question: "Do late grape harvests follow large volcanic eruptions?" Yes! say Stommel and Swallow (1983), and add that the degree of lateness depends on the amount of volcanic material ejected in the preceding year. In vino veritas!

Chemical Inventory

The atmosphere represents a mixture of gases, water droplets, ice particles and aerosols. Gases and water vapor (humidity) are the transparent constituents, whereas particles of water and ice yield fog, clouds, rain and snow. In contrast, smog and haze are principally caused by aerosols.

The total mass of the atmosphere is 5.29×10^{18} kg, which is negligible when compared to that of the solid earth (5.98×10^{24} kg) or the sea (1.35×10^{21} kg). And yet this trifling amount of air makes Earth a habitable planet by keeping it warm and moist and by storing molecular oxygen and carbon dioxide.

The atmosphere is well mixed up to a height of about 100 km, except for the presence of ozone at elevations above 20 km with a maximum density close to an altitude of 30 km, and of course water vapor which is practically restricted to the troposphere. At altitudes exceeding 100 km, lighter gases become more and more prominent because diffusion rather than mixing is their principal regulating device. In view of the fact that the troposphere contains roughly 80 percent of the mass held in the atmosphere in toto, a sample of air taken anywhere at the surface of the globe should be representative of the chemistry of our atmosphere. However, closer examination will reveal that in addition to bulk chemistry, there is a wide range of trace constituents which fluctuate from place to place in response to natural or anthropogenic emissions. Should they be of a conservative nature, they are useful for physically tracking air masses. However, should they interact or become altered and destroyed, they may help to decipher chemical and photochemical processes at work in the atmosphere.

The composition of dry air at sea level is given in Table 9.1. The three major gases: nitrogen, oxygen, argon jointly comprise almost 100 percent of the total atmosphere. The remainder are – chemically speaking – impurities.

The question immediately arises as to why just these three gases are present, and furthermore in a ratio of 78:21:1? Most straightforward is the situation concerning argon. In outer space, ^{36}Ar is the main isotope species of argon, whereas the Earth's atmosphere is dominated by ^{40}Ar (^{36}Ar: ^{40}Ar = 1:300). This is related to the former T Tauri event (see p. 42) which removed the primordial air from the proto-Earth. So all gases we presently encounter in our atmosphere were generated by the Earth itself. ^{40}Ar is the radioactive decay product of ^{40}K. Potassium contained in minerals of crust and mantle carries about 0.01 percent of the radiogenic potassium-40 which by

Table 9.1. Atmospheric constituents with small spatial and temporal variations (From Liou, 1980)

Constituent	Formula	Abundance by Volume
Nitrogen	N_2	78.084 %
Oxygen	O_2	20.984 %
Argon	Ar	0.934 %
Carbon Dioxide	CO_2	345 ppm
Neon	Ne	18.18 ppm
Helium	He	5.24 ppm
Krypton	Kr	1.14 ppm
Hydrogen	H_2	0.5 ppm
Xenon	Xe	0.089 ppm
Methane (at surface)	CH_4	1.7 ppm
Nitrous Oxide (at surface)	N_2O	0.3 ppm

beta decay yields calcium-40, and by electron capture argon-40, whereby one electron of the K-shell will become adsorbed and argon-40 is obtained by gamma-ray emission. In knowing the potassium content of our Earth and the halflife of ^{40}K (1.3×10^9 years), one can calculate how many years are needed to account for the present argon content in the atmosphere (see p. 41). Expressed differently, the atmosphere is the final sink for ^{40}Ar.

The point is now brought up: what about another noble gas helium? In the course of the radioactive decay of many heavy elements, for instance, ^{238}U, ^{235}U, and ^{232}Th, alpha particles (= helium nuclei) are emitted (see p. 24). Additional release of beta and gamma rays will yield three isotope decay series having as their respective endmembers the stable isotopes: lead-206, lead-207, and lead-208. Consequently, helium becomes abundantly generated in the lithosphere, from where it eventually will migrate into the atmosphere. Helium content in air is just 5 ppm (parts per million), because once it enters the atmosphere, its low molecular weight will promote an upward diffusion and eventually helium will escape into space, while the ten times heavier argon-40 remains Earth-bound. The reason why there is helium at all in the atmosphere is related to a steady state between emission from the lithosphere and diffusion into space.

The helium we are talking about is ^4He. There is another stable isotope of helium, namely ^3He, which is extremely rare in nature, but has been found to be a useful tracer for hydrothermal plumes emanating along active oceanic rifts. First detected on the Galapagos spreading center (Weiss et al., 1977), it has now been observed along the entire length of the East Pacific Rise and in other marine hydrothermal vent areas (e.g., Lupton, 1983). Whereas most natural gases from continental areas have a low $^3He/^4He$ ratio, those gases that appear to be linked to recent or extinct volcanism or magmatism show high $^3He/^4He$ ratios far above the atmospheric value of 1.40×10^{-6} (Sano et al., 1986). It is of great significance that the continental ^4He flux far exceeds that of the ocean so that most of the helium-4 in the atmosphere stems from the continental crust, whereas the main source of helium-3 is the upper mantle beneath the sea, from where it becomes released principally along spreading rifts. The deep mantle is also the source of ^{36}Ar, the "primordial" argon, and both ^3He and ^{36}Ar must be abundantly present at great depth.

Heavier noble gases such as neon, krypton and xenon are quite prominent in the Universe but not on Earth. They were lost together with the bulk of the hydrogen and helium at the time the solar wind swept the primitive Earth clean of many of its gases. The minute quantities found in the modern atmosphere have conservatively accumulated over aeons.

If this viewpoint of an airless proto-Earth is correct, then molecular nitrogen (N_2), too, is derived from the Earth's interior. This automatically implies that this nitrogen came to Earth in a solid form or in association with a mineral or a non-volatile compound; otherwise, nitrogen would have escaped together with hydrogen and the noble gases during the T Tauri incident.

Nitrogen can hide or be contained in: (i) metal nitrides such as siderazot, Fe_5N_2, or osbornite, TiN, (ii) silicates, e.g., the ammonium feldspar buddingtonite, $NH_4AlSi_3O_8$, and (iii) nitrogenous organic matter, for example: amino acids, amines, the bases of the purines and pyrimidines, or pyrrol complexes (Eugster, 1972). Upon thermal degradation, minerals and organics may release nitrogen in the form of molecular nitrogen or in the form of ammonia. Proponents of the ammonia concept use the atmosphere of Jupiter – where indeed ammonia and methane are major constituents – to suggest that Earth formerly had such an atmosphere too. The leading idea is the belief that this kind of atmosphere must have been reducing, a condition thought to be required for the origin of life on Earth.

At present, almost 90 percent of nitrogen contained in air, sea, crust and life resides in the atmosphere (Table 9.2). The sea carries about 1 percent of this amount and the rest is fixed in marine sediments. Less than 100 gigatons of nitrogen is all there is in the living world. And yet, it is the cell which propels the terrestrial nitrogen cycle (Box 9.2).

Organic matter will, upon decay, release its nitrogen principally in the form of ammonia. Recent sediments also carry a great portion of their nitrogen in the form of ammonia. This ammonia can be dissolved in interstitial waters or be contained in silicates either structurally or in a sorbed state. However, in the course of diagenesis, sedimentary minerals will transform ammonia in the direction of molecular nitrogen.

Table 9.2. Nitrogen content of atmosphere, hydrosphere and crust of the Earth (From Eugster, 1972)

Unit	Amount $\times 10^{20}$ g	Total Nitrogen (%)
Atmosphere	38.6	87.1
Hydrosphere	0.3	0.7
Crust, sediments	5.2	11.7
Crust, igneous rocks	0.2	0.5
Total	44.3	100.0

BOX 9.2

Nitrogen Fixation (Ludden and Burris, 1985)

A key element in cellular evolution is nitrogen. However, it is biologically useful only when combined with hydrogen, as in ammonia, or with oxygen, as in nitrite and nitrate, or in nitrogen-containing organic substrates as in proteins. When life started more than four billion years ago, a wide variety of fixed nitrogen was available in molecules such as ammonia, urea, or amino acids. As these sources became gradually depleted, organisms had to invent a machinery capable of fixing molecular nitrogen. Among extant life, only prokaryotes are capable of fixing molecular nitrogen via the nitrogenases. This enzyme system is composed of two metal-sulfur proteins with molecular weights of 200,000 and 60,000, respectively. The larger component, the iron-molybdenum protein, incorporates a cluster of unknown structure, known as the Fe-Mo cofactor, which is considered to be the site at which N_2 is reduced to ammonia at ambient temperature. The role of the second component, the iron-sulfur cluster, is to convey electrons to the iron-molybdenum protein. Simultaneous demands for protonation and electronation can be summarized:

$$N_2 + 6H^+ + 6e^- \rightarrow 2NH_3$$

and dinitrogen becomes activated by means of coordination to electron-rich metal ions. The recent finding (Robson et al., 1986) that in a variety of nitrogen-fixing bacteria molybdenum can be replaced by vanadium at the cofactor site underscores the significance of vanadium as an essential biochemical element. The close association of molybdenum and vanadium in organic-rich sediments of all ages (e.g., Volkov and Fomina, 1974) is possibly a mirror image of former nitrogenase activity.

Nitrogen fixation seems to be a difficult work assignment for an organism, because the enzyme catalyzing the transformation of N_2 to NH_3 is inactivated by molecular oxygen which in turn is generated by photosynthesis. One way out is physically to separate dinitrogen reduction and aerobic photosynthesis from each other, a strategy employed by a great number of cyanobacteria. Specifically cellular differentiation will lead to specialized cells, the heterocysts, which form at regular intervals along filaments. In heterocystous cyanobacteria the ratio of vegetative cells to heterocysts is approximately 10:1. Heterocysts have thick cell membranes which are surrounded by a mucilagineous envelope. This envelope lowers permeability of gases and functions as an oxygen barrier. In addition, respiratory enzymes will rapidly consume any free oxygen that may diffuse through the cell wall, thus keeping the internal milieu anoxic.

One remaining problem has to do with the diffusion of nitrogen into the heterocyst, because what has been said of molecular oxygen should also apply to dinitrogen gas. Heterocystous envelopes have a mean permeability coefficient of 0.3 m s^{-1} which is about 100 times slower than for vegetative cells. Mitsui et al. (1986) were able to show, however, that the rates of diffusion for the two gases are determined by their different solubilities in water. Accordingly, nitrogen influx is adequate to maintain nitrification, whereas oxygen influx is kept at a minimum. Heterocysts have been identified in rock formations as old as 2.2 billion years, which suggests that oxygen must have been around at that time (Schopf, 1978).

Non-heterocystous, that is unicellular nitrogen-fixing cyanobacteria, are known too. Their strategy is to separate nitrogen fixation and photosynthesis. Mitsui et al. (1986) report that nitrogen fixation and photosynthesis occur at different phases in the cell division cycle. Using a unicellular marine cyanobacterium, *Synechococcus* spp., they could demonstrate that it is not respiratory activity that maintains anaerobicity for nitrogen fixation as has been commonly assumed, but that nitrogenase activity appears only when the culture is in phase of net carbohydrate consumption. Endogenous carbohydrate (glycogen) appears to be the ultimate source of reductant for nitrogenase catalysis in a number of cyanobacteria, and explains the temporal separation of photosynthesis and nitrogen fixation. Of course, the easiest way out of the oxygen dilemma has been taken by all anaerobic nitrogen fixers. They simply remain in a reducing habitat.

Symbiotic nitrogen fixation between plants and bacteria is of considerable significance to agriculture. For example, species of the genus *Rhizobium* colonize the roots of leguminous plants and initiate the formation of nodules, the place where nitrogen fixation proceeds. Plant-microbe interaction involves exudation and participation of various carbohydrates and proteins stimulating, for instance, the branching of roots or enhancing nodulation by means of plant-encoded nodule enzymes. The field of molecular genetics has indeed shed new light on plant genes involved in nodulation and nitrogen fixation. Rearrangements of DNA adjacent to the nitrogen fixation (nif) genes during cellular differentiation have revealed the way the nitrogenase components are encoded (Haselkorn et al., 1983; Golden et al., 1985).

At present, there is abundant fixed nitrogen in aquatic and soil environments. Aerobic and anaerobic microorganisms are responsible for the turnover of atmospheric nitrogen into molecular forms that can be used by other organisms. As a result of this, eukaryotes lack the enzymes needed for nitrogen fixation. However, the recent claim (Yamada and Sakaguchi, 1980) that eukaryotic green algae found in hot springs can fix nitrogen shows that we still have not discovered the full extent of the presence of nitrogen-fixing organisms in the biosphere.

The biogeochemical nitrogen cycle is unusually complex, as nitrogen can occur in many valence states. The most important environmental factor regulating its cycling is the redox potential, as certain processes occur only aerobically while others only anaerobically. Furthermore, in all ecosystems, microbial processes play a paramount role in the cycling of nitrogen. Micro-organisms are the sole or major group of organisms responsible for such vital processes as nitrogen fixation, nitrification and denitrification (e.g., Rosswall, 1982).

Subsequent metamorphism or granitization will remove N_2 from the rock formation, since there is virtually no mineral of high temperature origin that will incorporate nitrogen into its lattice. So crustal nitrogen will behave almost like a noble gas and conservatively be recycled into the atmosphere. The fact that, differently from argon, nitrogen will not find its final resting place in the atmosphere, has to do with biological nitrification (Box 9.2) and subsequent transfer of organic nitrogen into the sediment column. Without organisms, all nitrogen would stay in the air. This statement is not exactly true, because lightning discharges can convert dinitrogen into various forms of fixed nitrogen. In addition, a variety of industrial processes are known which transform air nitrogen into ammonia, urea, or nitric oxide used as a base in fertilizer production. Electrosynthesis using tungsten as a catalyzing element has shown the feasibility of generating ammonia from molecular nitrogen even at room temperature and pressure (Pickett and Talarmin, 1985). This catalytic scheme helps to understand better the operation principles behind nitrogenase activity and at the same time shows new ways to synthesize N-fertilizers at low energy costs. Present fertilizer production from air nitrogen is substantial and only slightly less than the amounts fixed by bacterial processes: 2.51×10^{12} mol a^{-1} versus 3.1×10^{12} mol a^{-1} (Garrels et al., 1975).

Residence time for nitrogen in the atmosphere is estimated at close to 50 million years. This figure indicates that we should not worry that fertilizer production will exhaust the pool of nitrogen in our atmosphere.

Let us now turn to oxygen, an element said to have been discovered on August 1, 1774, by Joseph Priestley (1733–1804), an English clergyman. Actually, at least 2 years before that date, the Swedish chemist Karl Wilhelm Scheele (1742–1786) found out experimentally that air consists of two ingredients, one supporting and another inhibiting fire (see Scott, 1986). Naming the fire-loving atmospheric gas "Feuerluft", which literally means: "fire air", well describes its principal features. Priestley spoke of dephlogisticated air instead, whereas in 1778, the father of modern chemistry, the French chemist Antoine Laurent Lavoisier (1743–1794), named the gas oxygène, because of its ability to produce acids from a number of oxides.

Present oxygen content in the atmosphere is a consequence of primary productivity on land and in water. This has been known since the days of Julius Robert von Mayer (1814–1878), who concluded that sunlight provides the energy for the synthesis of plant material

$$CO_2 + H_2O \xrightarrow{\text{light}} O_2 + \text{organic matter.}$$

Photosynthesis – as this mechanism is called – is actually a more involved process than the above formula may suggest (see Box 9.3). Concerning the formation of O_2, we have to bear in mind that molecular oxygen can only accumulate in air in an amount equivalent to the sum of organic carbon stored in soils, sediments and the vegetation cover, since respiration of organic material will return CO_2 and H_2O back to the atmosphere again. A small correction factor has to be applied, because some of the oxygen is used for the oxidation of metals such as iron or manganese; rust is a typical endproduct. Furthermore, organic matter formed at some time in the past, for instance, fossil fuel, peat, forest litter or soil humus, may be combusted or start to respire, thereby lowering O_2 and raising CO_2 in the atmosphere.

The major chemical cycles of the atmosphere are governed by the plant kingdom, as this brief illustration will reveal. Let us assume for a moment that sea, air, and crust are just depots of matter to be consumed or generated. In less than 2,000 years each at-

BOX 9.3

Photosynthesis (Darnell et al., 1986; Lehninger, 1982)

Living organisms can be grouped into: autotrophs and heterotrophs. The first category of organisms can construct all matter they need from simple molecules such as water, CO_2, and dissolved minerals using solar or chemical energies. A great variety of bacteria and the green algae and higher plants belong to this group of organisms. In contrast, heterotrophs receive from other organisms all essential biochemical molecules.

The most ancient organisms were in all probability heterotrophic in nature, relying on abiotically generated organic substrates. With the gradual depletion of these food resources, life started to learn to manufacture the key biochemical molecules from scratch – so to speak – using simple inorganic molecules, e.g., H_2, H_2O, H_2S, CO_2, NH_3, PO_4, and metal ions. The first autotroph might have been an archaebacterium, using molecular hydrogen as electron donor and CO_2 as carbon source to produce organic matter and methane; accordingly, this organism is called a methanogenic bacterium.

Instead of using chemical energy directly, cyanobacteria and eukaryotic plants capture light energy and utilize chlorophyll *a* to transfer electrons to specific acceptors (Fig. 9.15). Chlorophyll *a* is the principal pigment trapping light energy and promoting photosynthesis; its peak of photon absorption lies at a wavelength of 680 nm. Chlorophyll *b*, a pigment contained in vascular plants absorbs at 650 nm. Other pigments, that is the carotenoids, absorb at still shorter wavelengths (400 to 500 nm). It is of note that the shorter the wavelength, the greater the energy yield; absorption of a mole of photons of red light (680 nm) is equivalent to about 42 kcal of energy, versus 71 kcal with blue light (400 nm). Many pigments supply energy to chlorophyll *a*, thereby extending the range of light useful for photosynthesis.

An excited chlorophyll molecule (cpl*) can transfer an electron to an acceptor and in turn becomes positively charged (cpl^+). In this capacity it is a strong oxidant and capable of removing electrons from other molecules and reconstituting itself. Four of the excited chlorophyll molecules can strip four electrons from H_2O and generate O_2:

$$2H_2O + 4cpl^+ \rightarrow 4H^+ + O_2 + 4cpl.$$

A participation of a Mn^{2+}-containing protein in the water-cleavage reaction is essential. Other two-valent ions, i.e., Co^{2+} and Ni^{2+}, have been mentioned as coordination partner. This underscores the significance of heavy metal ions in biological systems.

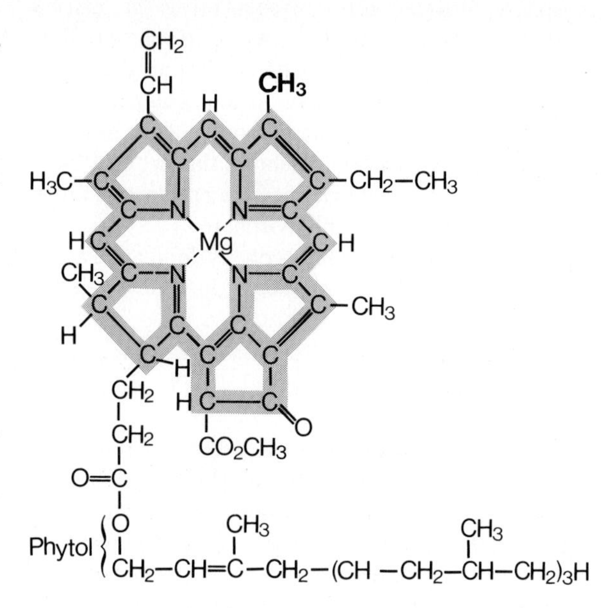

Fig. 9.15. Structure of chlorophyll *a*, the principal light-absorbing pigment. It contains four substituted pyrrole rings, one of which is reduced. The molecule has a long isoprenoid side chain and an alcohol – phytol – esterified to a carboxyl group. The four central nitrogen atoms are coordinated with magnesium. Chlorophyll *b* differs in that the $-CH_3$ group (upper right, boldface) is substituted by a –CH group

The generalized formula for oxygen-evolving photosynthesis reads:

$$12H_2O + 6CO_2 \xrightarrow{\text{light}} C_6H_{12}O_6 + 6O_2 + 6H_2O.$$

However, purple and green bacteria do not generate molecular oxygen. Instead of water, they oxidize hydrogen sulfide and produce elemental sulfur:

$$12H_2S + 6CO_2 \xrightarrow{\text{light}} C_6H_{12}O_6 + 12S + 6H_2O.$$

Molecular hydrogen from these organisms can also serve as electron donor:

$$12H_2 + 6CO_2 \xrightarrow{\text{light}} C_6H_{12}O_6 + 6H_2O.$$

The oxidized chlorophyll, cpl^+, removes electrons from H_2, thus forming protons and regenerating the original chlorophyll:

$$H_2 + cpl^+ \rightarrow 2H^+ + 2cpl.$$

We have seen that different photosystems exist among the photosynthesizing plant community. Since the H_2S based system is considered the most ancestral, it is named photosystem I. The oxygen-evolving photosynthesis is known as photosystem II. Inasmuch as respiration works the other way around, the delicate balance between oxygen generation and oxygen consumption, finding its precipitation in the molecular oxygen content in our atmosphere, comes to light.

mospheric O_2 molecule will pass through a plant cell. It takes about 3,000 years to fix one volume of air CO_2 as sedimentary organic matter, whereas it takes just 8 years to turn over the same amount via photosynthesis and respiration. The consumption of all air CO_2 by weathering requires 7,000 years. With respect to water, a full ocean passes through the biological cell in about 2 million years, whereas it takes 5 to 10 million years to cycle one volume of sea water through the crust. This comparison is simply made to show how effectively the cell recycles oxygen, carbon dioxide and water.

It is tempting to explain bulk atmospheric chemistry as a result of availability and structural disposition of a few selected metal ions in minerals and biochemical molecules. That is:

- potassium-40 contained in minerals regulates argon-40
- molybdenum, vanadium and iron in nitrogenases extract N_2
- metal ions in minerals catalyze formation of N_2 from NH_3
- magnesium in chlorophyll and two-valent manganese generate O_2
- magnesium and calcium in adenosine triphosphate (plus solar radiation of course) energize the system.

In summary, nitrogen and argon behave extremely conservatively, which explains their long residence time in the atmosphere. In contrast, molecular oxygen and CO_2 are held more or less constant at a level of about 20 percent for O_2 and a few hundred parts per million for CO_2. A higher abundance in air oxygen will trigger forest fires, and lower O_2 content. Resulting higher partial pressure in CO_2 may stimulate plant growth and induce a warming of the atmosphere. This in turn, will influence the biota . . . and so on, and so on. In short, a kind of biogeochemical "thermostat" has been operating, maintaining a steady state, since higher plants spread across the land a few hundred million years ago.

CO_2 and Other Greenhouse Gases

Atmospheric trace gases were traditionally studied individually, that is without proper consideration of their feedbacks to other atmospheric gases. Source-sink philosophy was the standard approach. Over the past two decades, however, scientists have become increasingly aware that many of the atmospheric gases exercise an influence on one another. Moreover, carbon, oxygen, nitrogen, sulfur and hydrogen are the principal elemental constituents of air, which implies that all biogeochemical cycles ought to be consulted when the global role of a particular gas is examined.

The natural biogeochemical cycling of elements has been influenced by humans ever since they took control over the management of natural resources, be it agricultural or industrial. The use or misuse of the element "fire" was the major environmental impact in the early days of human societies, when forests and brushlands were cleared by fire to make room for the plow. Principally, it is the same antique element endangering the global environment today, because of the burning of fossil fuel in addition to the continuing destruction of pristine forests by ax, saw and fire. Many of the resulting fumes and gases released from soils, auto exhaust pipes, and chimneys have altered and will alter the radiation balance of our Earth.

Solar radiation is absorbed by the atmosphere and the Earth's surface. Dickinson and Cicerone (1986) and MacCracken and Luther (1985) have graphically summarized what happens to the energy supplied by the Sun once it enters the atmosphere (Fig. 9.16). Basically, incoming solar radiation is either absorbed or reflected, whereby most of the reflection proceeds in the air on clouds or dust particles, and most of the absorption at the Earth's surface. Much of the solar energy absorbed by soils, oceans, and vegetation cover becomes quickly returned to the air in the form of sensible heat fluxes, that is heat that can be felt, for example, as dry hot desert air. The other form of heat, commonly referred to as latent heat of vaporization, is generated by plants via evapotranspiration or released in the course of evaporation and condensation of water. Eventually, all absorbed solar energy returns to space as thermal infrared radiation (Liou, 1980).

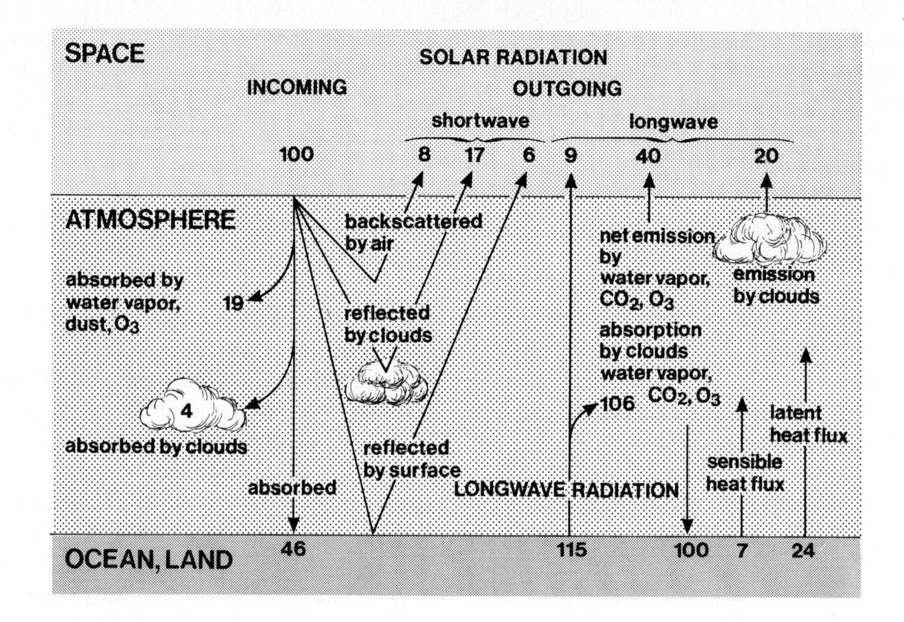

Fig. 9.16. Schematic diagram of the global and annual average fluxes of radiation within the climate system in % (Dickinson and Cicerone, 1986; MacCracken, 1985)

A great number of atmospheric trace gases, including water vapor, will lower the back flow of thermal infrared radiation. This, in essence, is the mechanism behind the warming of the global atmosphere. Gases retaining heat in the atmosphere are often called "greenhouse gases" because they operate like a glass window letting solar radiation pass through but keeping some of the reflected light in the form of heat in the lower troposphere. The possible impact of greenhouse gases on radiation balance and climate is summarized in Table 9.5.

The greenhouse effect was first described in 1863 by Tyndall, who pointed out that water vapor transmits a major fraction of the incident sunlight, but strongly absorbs thermal radiation from the Earth. The contribution of CO_2 to maintaining the Earth's heat balance was noted in 1897 by Arrhenius. He estimated that a doubling of atmospheric CO_2 would produce a global warming of about 6° C. This led Chamberlain in 1899 to propose atmospheric CO_2 as a controlling agent in the origin of ice ages. Concern about adverse climatic effects, due to fossil fuel CO_2 emissions, had been raised in 1938 by Callendar and in 1941 by Flohn.

To illustrate the combined radiative effects of water vapor and CO_2, I would like to offer the scenario of a dry and airless Earth which would behave like a black body radiator. In Fig. 9.17 the theoretical "black body" emission from the Earth's surface at 27° C is

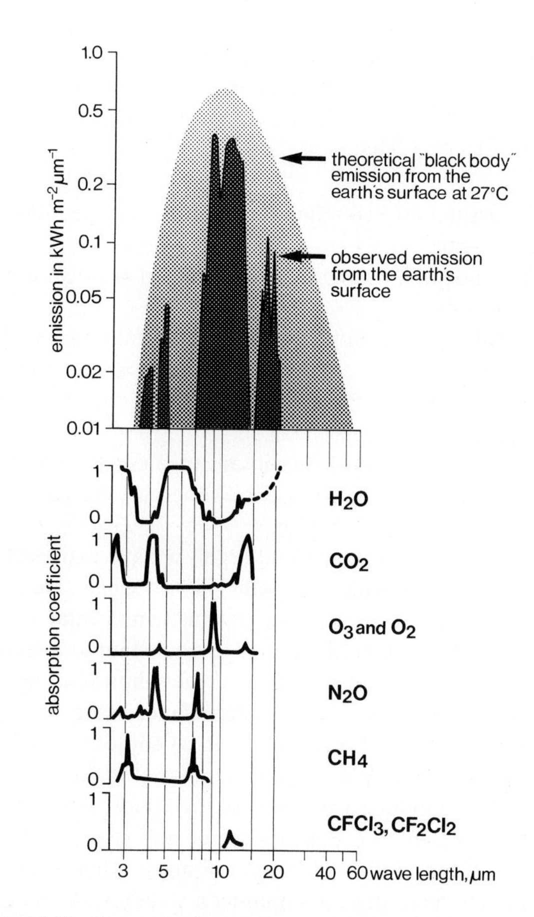

Fig. 9.17. Comparison of theoretical "black body" emission from the Earth's surface at 27° C, and the actually observed emission. Main absorber in the infrared are water vapor and carbon dioxide. The most effective absorption bands for the main CFC's is indicated (principal source: Fortak, 1971; Luther and Ellingson, 1985)

Table 9.3. The current 1985 trapping of thermal infrared radiation (ΔQ) by current tropospheric trace constituents (From Dickinson and Cicerone, 1986)

Gas	Current Concentration	$^{\Delta Q}$total (Wm^{-2})
Carbon dioxide	345 ppmv	50
Methane	1.7 ppmv	1.7
Ozone	10–100 ppbv	1.3
Nitrous oxide	304 ppbv	1.3
CFC-11	0.22 ppbv	0.06
CFC-12	0.38 ppbv	0.12

The term $^{\Delta Q}$total is the change of net radiation at the tropopause if the given constituent is removed, but the atmosphere is otherwise held fixed.

Table 9.4. Estimated pre-industrial trace gas concentrations and implied change in thermal trapping to 1985 (From Dickinson and Cicerone, 1986)

Gas	Pre-industrial concentration	$^{\Delta Q}$pre-industrial (Wm^{-2})
Carbon dioxide	275 ppmv	1.3
Methane	0.7 ppmv	0.6
Tropospheric ozone (below 9 km)	0–25 % less	0.0–0.2
Nitrous oxide	285 ppbv	0.05
CFC-11	0.00 ppbv	0.06
CFC-12	0.00 ppbv	0.12
Total		~2.2

compared to the presently observed emission from the Earth's surface. The amount of trapped infrared radiation due to tropospheric trace gases is depicted in Table 9.3. Carbon dioxide is certainly the most determining greenhouse gas. Jointly with water vapor, the two compounds make the Earth 30°C warmer. That is, without moisture and air, Planet Earth would have at its surface a mean global temperature of –15°C, as determined by its distance to the sun.

Radiative forcing of the earth-atmosphere system seems to be chiefly controlled by a few gases (Tables 9.3 and 9.4), aerosols, and surface reflectivity better known under the term: albedo. Against this background, some relevant data on the biogeochemistry of the main greenhouse gases are briefly discussed. The subject matter has been fully reviewed in: Ramanathan (1981), Ramanathan et al. (1985), R.A. Rasmussen and Khalil (1986), Callis and Natarajan (1986), Khalil and Rasmussen (1983, 1985), Crutzen and Arnold (1986), MacCracken and Luther (1985), Grassl et al. (1984), Bolin et al. (1986), Dickinson (1986), National Academy of Sciences (1984), Donelly et al. (1987).

CO_2: biogeochemical cycling: The wide attention presently paid to CO_2 has, on the one hand, to do with human activity in the area of fossil fuel combustion and deforestation of tropical forests. On the other hand, there are still many unresolved questions concerning the interplay of CO_2 within and between the four major terrestrial carbon pools air, water, earth, and life. Consequently, we are dealing with two major issues, one that is focused on anthropogenic events, and the other that pertains to the pristine global carbon cycle.

When the general CO_2 rise in the atmosphere due to human impact became substantiated by the famous Mauna Loa curve (Keeling et al., 1982; Bacastow and Keeling, 1981), tree-ring records (Jacoby, 1980; Freyer and Belacy, 1983, see Box 9.4) and related observations (Stuiver, 1978), it was felt that this trend would lead to a global warming of 2–3°C in a matter of decades. Climate watchers became alerted, research on CO_2 prospered and many models on environmental consequences were proposed (Gorshkov, 1986a, b; Gorshkov and Sherman, 1986; Wigley et al., 1980; Idso, 1982; Laurmann and Rotty, 1983).

BOX 9.4

$^{13}C/^{12}C$ Trends in Tree Rings (Freyer and Belacy, 1983)

"Es steht ein Baum im Odenwald". This is a very famous German folksong singing the praises of one tree in the Oden Forest near Heidelberg. Freyer and Belacy (1983) have taken a few trees nearby and told the epic of the stable carbon isotope ratios fixed in their tree rings. For those unfamiliar with the significance of carbon isotope ratios a little background information might be helpful.

Carbon has two isotopes, ^{12}C and ^{13}C in a ratio of 98.89 to 1.11. It has been found useful to report the carbon isotope composition of a compound as permil deviation in terms of $\delta^{13}C$ notation relative to a standard. Common reference scale is the Chicago PDB standard (Craig, 1957). A compound having a $\delta^{13}C$ of +10 would be enriched in ^{13}C, respectively depleted in ^{12}C by 10 permil relative to PDB (Peedee Formation, Cretaceous).

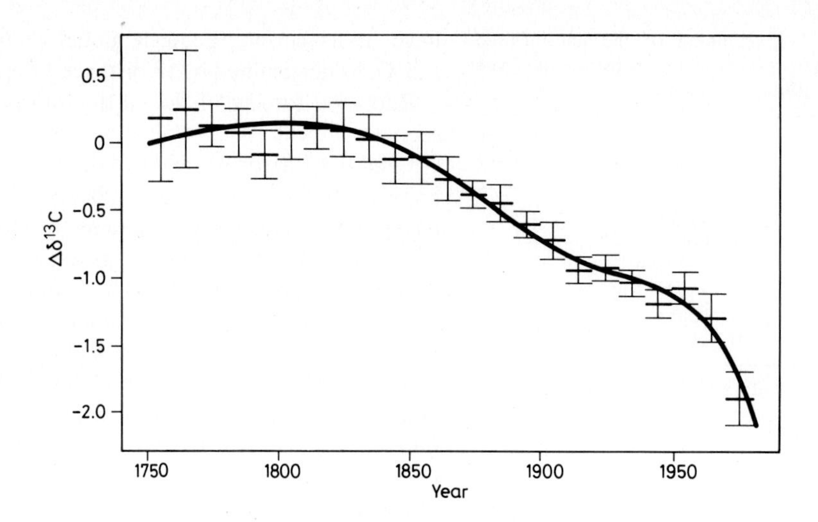

Fig. 9.18. Mean $\delta^{13}C$ variations in northern hemisphere trees (Freyer and Belacy, 1983)

It is known that for gas molecules the velocities of isotopic species are proportional to the inverse square root of their molecular weights. For carbon dioxide one can write:

$$\frac{\text{velocity}\ (^{12}C^{16}O^{16}O)}{\text{velocity}\ (^{13}C^{16}O^{16}O)} = \sqrt{\frac{45}{44}} = 1.011.$$

This implies that collisions of CO_2 of mass 44 with a photosynthesizing leaf are 1.1 percent more frequent than those of CO_2 of mass 45. In consequence, land plants will extract preferentially ^{12}C from the atmospheric CO_2 pool. Metabolic effects will further discriminate against ^{13}C, so that the final products, in the present case cellulose and lignin, are enriched in ^{12}C, respectively depleted in ^{13}C relative to atmospheric CO_2. C_3 plants (Calvin cycle) are more effectively enriching ^{12}C than C_4 plants (Hatch-Slack cycle) resulting in a mean $\delta^{13}C$ for C_3 plants of –27 permil compared to –13 permil for C_4 plants. For details on the biogeochemistry of stable carbon isotopes see Degens (1969) and Hoefs (1980).

An individual free-standing tree is expected to register in its yearly rings changes in the carbon isotopic composition of its surrounding air CO_2. Common oak and Scotch pine trees from various places in the northern hemisphere were studied for their $^{13}C/^{12}C$ record extending back as far as 500 years. Since juvenile stages of tree growth fractionate air CO_2 slightly differently, the first 50 years of record have not been considered. Furthermore, only trees which during their whole lifetime remained free-standing were included, because otherwise respiration coming from the decomposition of forest litter would have contributed isotopically light CO_2. The calculated northern hemisphere $^{13}C/^{12}C$ record derived from a total of about 50 trees reflects the recent $^{13}C/^{12}C$ change in atmospheric CO_2 observed from 1956 to 1980. This gives confidence that also the tree-ring data extending back to 1750 are a true reflection of the carbon isotope changes of air CO_2 during that time period (Fig. 9.18).

Application of a box-diffusion model (Oeschger et al., 1975) to tree-ring $\delta^{13}C$ records (Freyer and Belacy, 1983) led to the conclusion (Peng et al., 1983) that about 260 Gt C (Gt = Gigatons = 10^9 metric tons) have been liberated since the beginning of large-scale land conversion, and the present release is estimated at close to 1 Gt C a^{-1}.

Present atmospheric CO_2 in vegetated areas exhibits diurnal variations and the range in $\delta^{13}C$ is –7 to –10 permil. Accordingly, air CO_2 must have been 2 permil heavier at the beginning of the 19th century in the area where the tree grew. Fossil fuel is highly enriched in ^{12}C with a mean $\delta^{13}C$ value for coal of –25 permil, for crude oil of –26 permil, natural gas of –40 permil from Paleozoic and Mesozoic sediments and of –60 permil for methane released from Tertiary sediments (Degens et al., 1984). Thus, burning of fossil fuel and soil respiration have lowered the ^{13}C content of atmospheric CO_2.

Ecological aspects of tree-ring analysis have been summarized (Ecological Aspects, 1987).

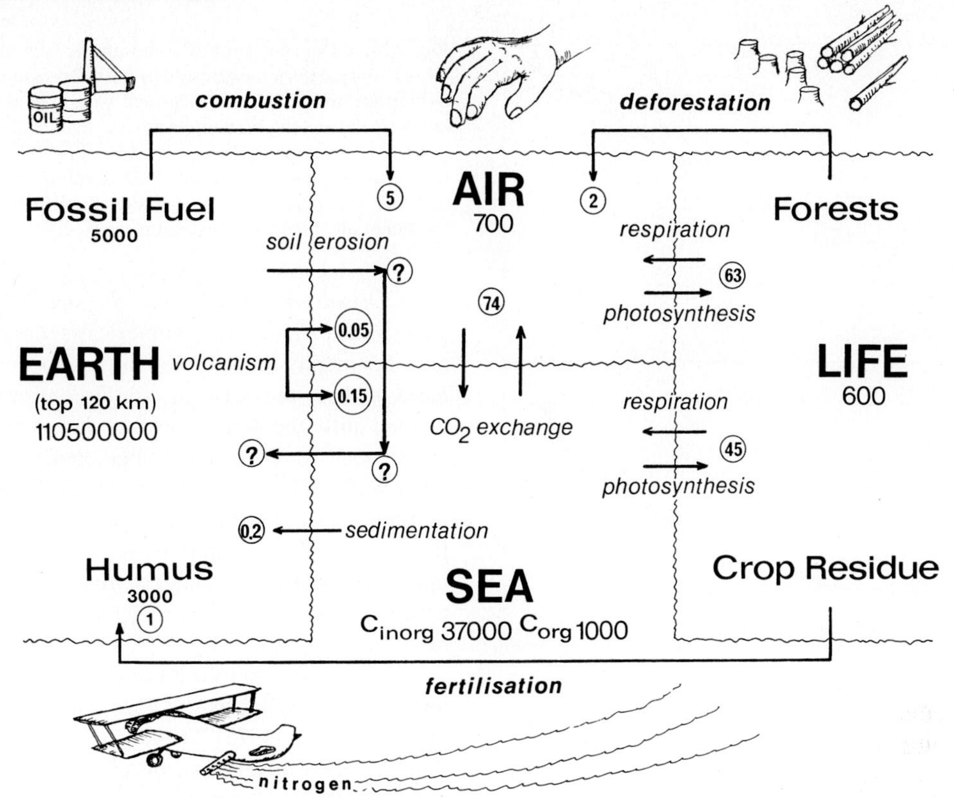

Fig. 9.19. Simplified scheme of principal reservoirs and fluxes in the natural carbon cycle. Input due to human activities is indicated as: combustion, deforestation and fertilisation. *Number in boldface* beneath AIR, LIFE, SEA, EARTH represents the size of the individual reservoir in gigatons carbon (Gt C). *Encircled numbers* are the annual flux rates in Gt C. The key role of CO_2 in maintaining the global carbon cycle becomes evident (Degens et al., 1984)

The injection of anthropogenic CO_2 into the atmosphere, even though substantial relative to volcanic CO_2 emission, i.e., 6 to 7 Gt C a^{-1} versus 0.05 Gt C a^{-1}, is minimal when compared to the amount of CO_2 annually flowing through the natural carbon cycle (Fig. 9.19).

Close examination of this simplified scheme of sinks, sources and fluxes of carbon reveals that carbon dioxide is the prime driving force of the global carbon cycle. The amount of CO_2 in the four major terrestrial reservoirs depends on the exchange kinetics through boundaries such as the air-sea interface, biological membranes, and carbonate equilibria in marine and fresh water systems. The turnover rate of the exogenic carbon cycle falls in the range of 180 Gt C a^{-1}. Where ultraphytoplankton production is considered, (e.g., Fogg, 1986), about 250 Gt C a^{-1} passes through the carbon cycle as CO_2. For a critical assessment of the published marine primary productivity figures, see Jenkins and Goldman (1985), Reid and Shulenberger (1986) and Craig and Hayward (1987). In contrast, the endogenic carbon cycle releases at the most 0.2 Gt C a^{-1} in the form of CO_2 into the atmosphere or the hydrosphere, and receives in return about the same amount of carbon as organic matter or carbonate from the terrestrial and marine environment. In short, a kind of steady state is indicated.

Superimposed on this pristine carbon cycle is the anthropogenic input which – in the case of CO_2 – manifests itself in long-term monitoring records of the types depicted in Figs. 9.20 and 9.21. Most of the CO_2 released through fossil fuel combustion into the atmosphere comes from the northern hemisphere, as can clearly be seen in Fig. 9.22. It is important to realize that less than half of the CO_2 generated by human activities remains air-borne. The rest, known as "missing carbon", must have found a temporary or permanent shelter somewhere else. Among the major sinks for that "missing carbon", the following reservoirs come to mind: (i) the dissolved carbonate pool in the sea, (ii) dissolved marine organic matter, (iii) land vegetation, (iv) soil organic matter, and (v) marine sediments. Let us find out more about the suggested possibilities.

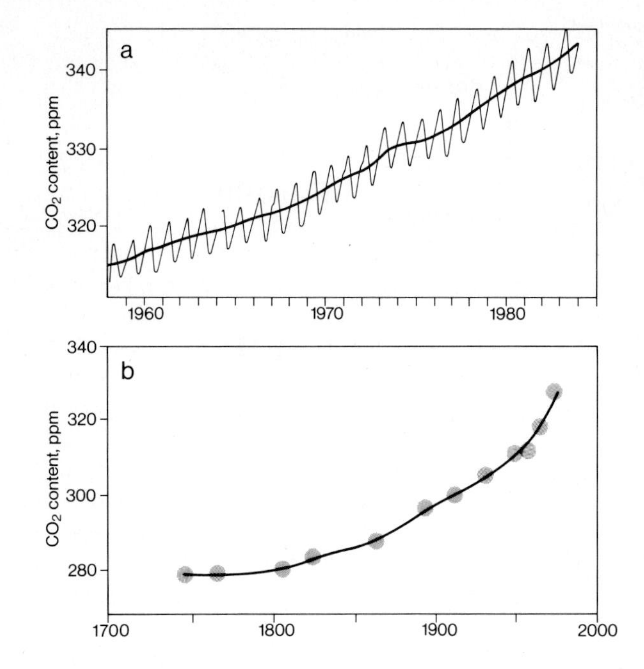

Fig. 9.20. a Concentration of atmospheric CO_2 at Mauna Loa Observatory, Hawaii, expressed as partial pressure in parts per million (ppmv) of dry air. Seasonal variations are mainly related to the extent of photosynthesis and respiration in the northern hemisphere (Bacastov and Keeling, 1981; updated by recent measurements: courtesy C.D. Keeling). **b** CO_2 content of fossil air enclosed in an ice core about 200 years of age (Neftel et al., 1985); for discussion see p. 209

(i) *Dissolved carbonates in the sea:* Many models have been proposed for numerical simulation of the ocean's uptake capacity for anthropogenic CO_2. In most of them the ocean is one-dimensional and subdivided into the boxes: surface layer, intermediate water, deep water. The exchange coefficients describing mass flux between them correspond to characteristic time scales of oceanic mixing (e.g., Bacastow and Björkström, 1981). Their basic disadvantage is that buffer factors and exchange coefficients are not known a priori, thus limiting the predictive capability of these models.

A first attempt to reproduce the continuous character of ocean mixing is the "box-diffusion-model" of Oeschger et al. (1975). They introduced vertical eddy diffusion as a mechanism of CO_2 transport into the deep sea. Further sophistications have been proposed

Fig. 9.21. a A three-dimensional perspective of the "pulse-of-the-planet": The variation of the global atmospheric CO_2 content in latitude and time. **b** Global two-dimensional map of CO_2 content for 1979/82 (Komhyr et al., 1985; Gammon et al., 1985)

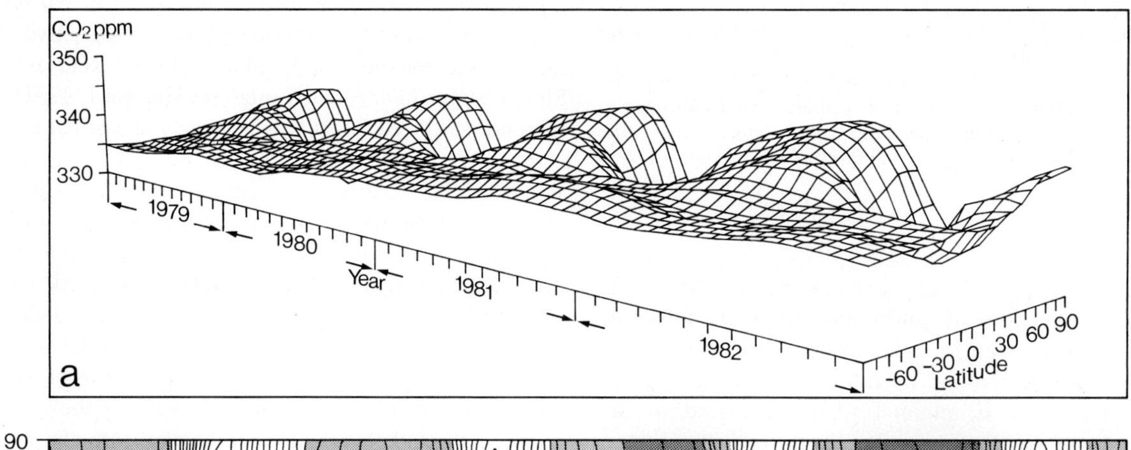

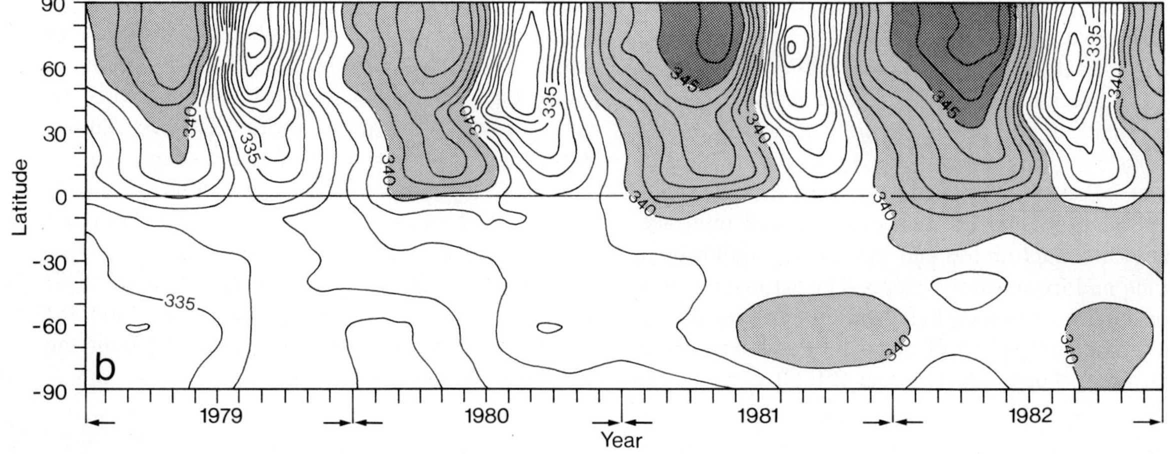

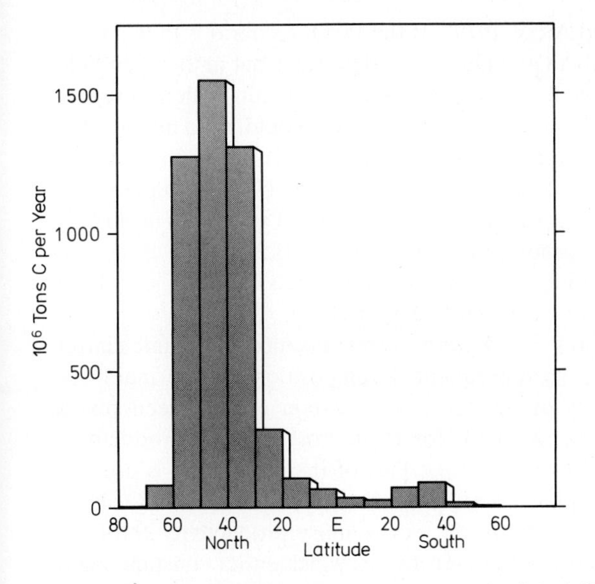

Fig. 9.22. Latitudinal distribution of annual fossil-fuel-CO_2 input to the atmosphere (Rotty, 1983, 1986)

by: Broecker and Peng (1982), Peng et al. (1983), Bolin et al. (1986) and Siegenthaler (1983, 1986).

A major step to overcome the drawbacks of oceanic box models has been taken by Maier-Reimer and Hasselmann (1987). They employed a general circulation model of the ocean, which is complex enough to reproduce the general features of oceanic circulation and simple enough to allow numerical integrations on climatic time scales. First results indicate that the simulated decay of atmospheric CO_2 following an instantaneous CO_2 input can be described as a superposition of a few exponential functions with decay times roughly 1, 10, and 500 years. They correspond to the observed characteristic time scales of mixing in the ocean: (i) equilibration with atmospheric changes occurs within a few months in the warm surface layer, (ii) within about 10 years in intermediate waters above the main thermocline, and (iii) within a few hundred years in deep waters (see also Fig. 10.28).

The decay times also depend on the magnitude of an instantaneous CO_2 input. Time constant is 160 years for a sudden doubling of CO_2, 280 years for instantly quadrupling the invasion of CO_2 into the ocean. Taking all sources of information into consideration, it appears that the ocean's dissolved carbonate pool will digest a volume corresponding to about half of the CO_2 presently released by fossil fuel burning into the air. Should the rate of fossil fuel burning either increase or decrease in the years to come, the air-borne fraction will accordingly adjust to these changes. Still left unaccounted for is the CO_2 freed by deforestation and agricultural activities in the amount of 1 to 2 Gt C a^{-1}.

(ii) *Dissolved marine organic carbon (DOC):* 1,000 Gt of organic carbon are dissolved in the sea, which is about twice the amount of all land plants taken together. Molecular size of DOC may range from less than 100 to more than 100,000 daltons (Ogura, 1974), but the bulk falls in the range of 500 to 5,000 daltons (Degens, 1970; Stuermer and Harvey, 1977, 1978). Much controversy exists as to the chemical nature and origin of the highly resistant organic matter (reviewed in Rashid, 1985). Harvey et al. (1983) suggested photolytic crosslinking of polyunsaturated lipids by hydroxide, peroxide and oxygen radicals. Accordingly, a marine source for bulk DOC in the sea was anticipated. Others (e.g., Laane and Koole, 1982) see little difference between marine and riverine DOC, which would argue for a terrestrial source. A structural model for DOC in natural environments has been proposed by Leenheer (1985).

Concerning the uptake capacity of the marine DOC pool for anthropogenic CO_2, it is important to know that DOC in the deep sea is inert to biological consumption (Jannasch et al., 1971), has an age of 3,400 years (P.M. Williams et al., 1969), and is assumed slowly to oxidize at a rate of 4×10^{-4} mg C l^{-1} a^{-1} (Craig, 1971). At this rate about 0.3 Gt of organic carbon has to be annually added to the marine DOC pool to compensate for the losses. If Harvey et al. (1983) are correct, 0.25 percent of the yearly marine primary productivity would have to enter that pool. On the other hand, if inert riverine DOC is the ultimate source for marine DOC, then today's DOC flux from land to sea amounting to 0.285 Gt C a^{-1} would be just sufficient to keep a steady state between DOC oxidation at depth and recharge from land.

During its long residence in the sea, the core of marine DOC must have been recycled several times, because the mean age of DOC is much older than the time required physically to turn over a whole ocean once (3,400 versus 500 years). In view of the small amount of land-derived DOC that is carried to the sea, and the minute quantities of marine DOC supposed to be photochemically produced, both propositions say essentially the same: the dissolved marine organic carbon pool is no major sink for anthropogenic CO_2.

(iii) *Land vegetation:* Enhanced plant growth and yield from a higher PCO_2 is widely recognized (Strain and Cure, 1985). Primarily, the rate of net photosynthesis increases, because the CO_2 gradient and hence the CO_2 flux into the leaf are enhanced. Furthermore, many important crops are C_3 species that waste some of the solar radiation they capture in respiration; and for them photorespiration is suppressed at high PCO_2. Also stomata close partially when subjected to higher PCO_2 content. In turn, a rise in external CO_2 will make more water and carbon internally available to the plant (Acock and Allen, 1985). And yet, the efficiency of CO_2 fixation in C_3 plants is in the last instance linked to an enzyme: ribulose 1,5 bisphosphate carboxylase/oxidase, abbreviated RuBisCO, probably the single most abundant protein on Earth. A PCO_2 rise will perhaps increase carboxylase specificity of RuBisCO at the expense of its oxidase specificity.

The way plants handle CO_2 is not uniform, however. Due to different metabolism between C_3 and C_4 plants, one using RuBisCO and the other phosphoenolpyruvate carboxylase in photosynthesis (Calvin cycle versus Hatch-Slack cycle), C_3 and C_4 plants show marked differences in their assimilation performance and their isotopic fractionation. C_4 plants have a much higher optimum temperature and are therefore more common in tropical areas. Many tropical grasses, corn (maize), sorghum, and sugar cane use the C_4 metabolism (for details, see Darnell et al., 1986). Nevertheless, each species responds in its own way to the CO_2 challenge. Although wide variability exists, a doubling of CO_2 concentration is likely to increase yields, on the average by one third, provided there is no shortage of water or mineral nutrient supply (Kimball, 1985).

It is a well-known fact that mineral nutrients, moisture and temperature are the determining elements which limit growth of plants to only a fraction of their photosynthetic potential (Loomis, 1976; Goudriaan and Ajtay, 1979). If the PCO_2 scenario is to work not only in greenhouses and pastures but in the open field as well, the observed increase in atmospheric CO_2 relative to pre-industrial level should have already left its imprint on land vegetation. It has been proposed that perhaps an increase in land biomass in the virgin forests of the world has occurred and still is occurring, but proof is difficult to obtain (Bolin et al., 1979). Accepting PCO_2 fertilization, plants could possibly have stored as much as 0.5 Gt C a^{-1}!

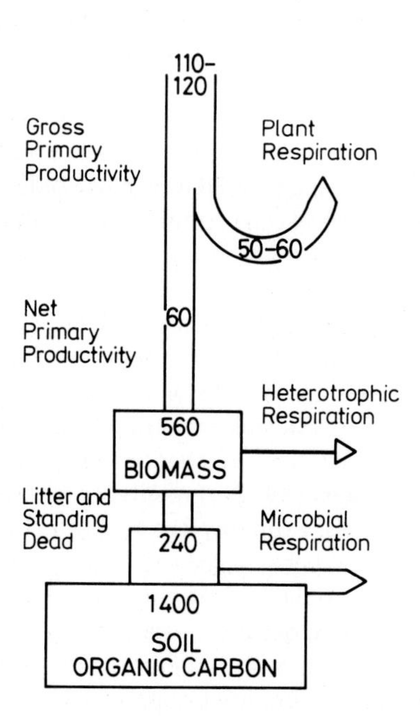

Fig. 9.23. Pool sizes and annual fluxes of terrestrial biospheric carbon cycle (Degens et al., 1984)

(iv) *Soil organic matter:* Fixation of organic matter and its subsequent decomposition are the main features of the biospheric carbon cycle. Green plants utilize air CO_2 for their gross primary productivity (GPP) (Fig. 9.23). Part of the production is used by the plant to meet metabolic demands and CO_2 is released. Only the net primary productivity (NPP) is available to heterotrophs which either consume plant matter directly or decompose plant litter gradually. Residence times of assimilated carbon range from a few hours for immediately respired carbon to several thousand years for stable humic substances accumulating in soils and peat. Only fractions of NPP persist through geologic times as lignites and organic matter in sediments.

Recent data from 2,700 soil profiles (down to 1 m depth) suggest a global reservoir of 1,400 Gt C (Post et al., 1982). Most soil carbon is contained in wetlands (14.5%), tundra (13.7%), agricultural areas (12%), wet boreal forests (9.5%), tropical woodland and savanna (9.3 %), and cool, temperate steppe (8.6%). Wet and moist tropical forests contribute only 9.9% to total soil carbon, illustrating that such soils are poor in humus. To this number are added 1,500 Gt C in the form of peat, making terrestrial humus the largest organic carbon reservoir on Earth.

Changes in land use since the beginning of large-scale land conversion have caused a loss in biospheric carbon in an amount of 260 Gt C, and present release is estimated at close to 1 Gt C a^{-1} (Fig. 9.24). Note the maximum release at the turn of the century ("pioneer effect"). The lower values for the present may be due to the fact that part of the burned vegetation is transferred to unreactive charcoal, hence representing a net sink (Seiler and Crutzen, 1980). Moreover, large parts of destroyed soils are not oxidized but rather eroded and deposited in, e.g., stream gullies, alluvial fans, behind dams, or marine sediments. Another counter-measure against a more rapid rise in atmospheric CO_2 has been invented by humans themselves. Here are the details:

World food production has doubled over the past 30 years (Brown, 1981; Barr, 1981). Total figures for grain are (expressed in million metric tons): 651 in

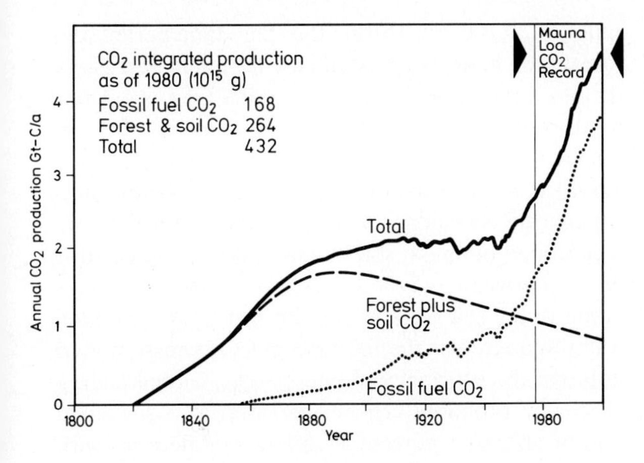

Fig. 9.24. Evolution of carbon releases by fossil fuel combustion and destruction of biosphere. Biospheric release is calculated on the basis of $\delta^{13}C$ measurements of Freyer and Belacy (1983) and by model deconvolution (Peng et al., 1983)

1950 and 1,432 in 1980. This impressive growth rate has been achieved by a combination of factors: (i) green revolution, (ii) putting about 30 percent more land under the plow, and (iii) profound changes in cultivation techniques. To keep such high yields – not even to mention further increases – will largely depend on the global condition of topsoil in terms of humus content, nutrient balance or water-retention capacity, to name but a few of the important microbiological, chemical and structural characteristics of a soil. Brown's article pictures no rosy future for the global maintenance of high-quality topsoil. Intensive cultivation has already caused massive erosion and thinning of topsoil in many parts of the world. Particularly vulnerable are the tropical countries.

Corn and grass are efficient users of solar energy. For instance in corn, about 50 percent more energy goes into roots, stems and leaves than will go into the grain directly (Loomis, 1976). Since most of the crop residue will enter the soil directly or subsequently (after being fed to animals), humus production should intensify with growing food production. This implies that some of the "missing carbon" has found temporary refuge in farmland.

The chief limiting nutrient in agriculture is nitrogen. Studies on nitrogen cycling in various ecosystems have revealed the tremendous impact this element has on the environment (Henderson and Harris, 1975; Söderlund and Svensson, 1976; Rosswall, 1979, 1980, 1981, 1982). It is thus no secret that nitrogen fertilization is the ultimate cause of the gigantic increase in world food production in recent years. Nitrate fertilization is particularly high in Europe, and ranges from 30 kg per hectare per year in Switzerland and Spain to about 240 kg per hectare per year in the Netherlands. In contrast, nitrogen fertilization in the United States is comparatively "modest", i.e., 22 kg nitrate per hectare per year.

Excessive nitrate fertilization is the main culprit behind high nitrate values in rivers and groundwaters, to the extent that in many regions of the world the suggested upper safety value of 25 mg NO_3 l^{-1} for drinking water has already been exceeded. Simply raising this level to 50 mg by decree, as done by several countries, is a solution but not necessarily the most healthy one. On the other hand, it is this nitrate that stimulates primary productivity in lakes, rivers, and oceans.

It is concluded that soil humus is a potential sink for atmospheric CO_2 in the amount of 0.5 to 1 Gt C a^{-1} largely for reasons of excess fertilization and topsoil erosion.

(v) *Marine sediments:* About 95 percent of primary productivity in the sea becomes remineralized in the euphotic zone. Only 5 percent or a total of ca. 5 Gt C a^{-1} settles to the ocean floor and feeds the benthic communities. The amount of inorganic carbon that is removed as $CaCO_3$ from the surface environment is in the order of 1.5 Gt C a^{-1}. General ocean circulation and river runoff will compensate for the losses. This is, in a nutshell, the pristine oceanic carbon cycle.

At depth, marine organic matter will be almost completely respired and the bulk of the calcareous shell material will dissolve below the carbonate compensation depth or by means of biogeochemical remineralization in the sediment. One can estimate that only 0.2 Gt C_{org} a^{-1} will eventually enter the geological column.

With regard to the "missing carbon", it is of no concern how much carbon is removed from the euphotic zone in the course of the natural global carbon cycling. A matter of interest is only how much additional carbon is extracted by the marine biota as a consequence of human activities.

Human impact on primary productivity stems largely from the pollution of estuarine, coastal and shelf environments by mineral nutrients and hazardous compounds such as heavy metals, detergents or oil. High discharge of mineral nutrients, especially from nitrate and phosphate fertilizers, fosters photosynthesis and alters aquatic ecology. Furthermore, rivers release massive amounts of clay-sized particles into the marine habitat as a result of soil erosion, due to agricultural activities. Planktonic organisms counteract this input by excreting large quantities of polysaccharides to complex interfering metal ions and to aggregate clays to make the water more transparent to light. The resulting macroflocs, rich in sugars, will settle to greater water depth where they respire. This device accelerates the removal of CO_2 from the euphotic zone, since phytoplankton, when subjected to environmental stress, may excrete more than half of its daily photosynthesized carbon into the ambience in the form of sugar polymers.

The "sugar pump" is a potential instrument to lower PCO_2, because mineral nutrients are kept in surface waters and maintain photosynthesis, whereas CO_2 fixed in sugars will be respired in the abyss.

In conclusion, a tentative assessment is made as to the whereabouts of anthropogenic CO_2 emissions into the atmosphere:

From 1860 to 1984 a total of 183 ± 15 Gt C were emitted as CO_2 by fossil fuel combustion; the present rate is 5.2 Gt C a^{-1}. During the same time period, deforestation and changing land use have released 150 ± 50 Gt C; the present rate is 1.6 ± 0.8 Gt C a^{-1}. Thus, total anthropogenic CO_2 emission is 6.8 ± 0.8 Gt C a^{-1} (Bolin, 1986). From this amount, about 2.5 Gt C a^{-1} remain airborne, 2.5 Gt C a^{-1} move into the marine carbonate pool, less than 0.5 Gt C a^{-1} increase land biomass, soil humus and riverine detritus digest between 0.5 and 1 Gt C a^{-1}, and the sugar pump in the sea accounts for the rest (Box 9.5). One way to counteract the increase in CO_2 emission is to improve the efficiency of energy use. Global energy efficiency probably can be increased at a sustained rate of at least 1 percent a^{-1} for several decades with no adverse social effects (Keepin et al., 1986).

BOX 9.5

CO_2 Scenario (Degens et al., 1984)

No mattèr what compartments we temporarily assign to account for all the "missing carbon", the final burial ground will be the sea. It is here, where the control device for PCO_2 in the air is switched on and off. The basic steps taken by the coupled air-sea-life system are:

(1) CO_2 content in air = function of: CO_2 system in ocean surface water

(2) CO_2 system in ocean surface water = function of "total CO_2" (TCO_2) plus "total alkalinity" (TA)

source		*sink*
change in TCO_2 (life-controlled), i.e.,	=	photosynthesis in the sea: deposition of C_{org}
photosynthesis on land: forest-soil humus decay		
change in TA (life-controlled), i.e.,	=	photosynthesis in the sea: deposition of C_{CaCO_3}
photosynthesis on land: forest-soil humus decay		

(3) photosynthesis in the sea = function of (apart from temperature and light flux): nutrients and essential elements

(4) nutrients and essential elements = function of: ocean circulation and continental input

(5) continental input = function of: vegetation, agricultural activities, fossil fuel burning, deforestation, pollution etc.

(6) CO_2 content in the air = function of:

(i) life processes
(ii) ocean circulation
(iii) nutrient availability
(iv) human activities

Methane: sinks and sources: The presence of methane in the atmosphere has been known for almost 50 years (Adel, 1939). That human activity was in part involved in methane release became established through the work of Seiler et al. (1978). At present we can reasonably understand sources, sinks and the tropospheric mixing ratio of CH_4.

Methane is mainly generated from organic matter by anaerobic bacteria. Water-locked soils such as rice paddy fields, swamps, lakes or marshes, and the intestines of herbivorous animals (i.e., ruminants) are the main sources. Termites have also been mentioned as a major contributor of methane. It appears, however, that this source is a comparatively minor one, and accounts for at most a few percent of the total emission. Coal mining, natural gas, and biomass burning are prominent sources too, and a summary of the CH_4 emission rates from individual ecosystems can be found in Bolle et al. (1986). Methane may also be released to the sea in the form of hydrothermal plumes (e.g., Horibe et al., 1986).

Reported current global emissions of methane differ by an order of magnitude (225 to 1210 Tg a^{-1}; 1 teragram = 10^{12}g), and a value of around 550 Tg a^{-1} is adopted which conforms with recent estimates (e.g., Seiler, 1985; Khalil and Rasmussen, 1983).

The major sink for CH_4 is oxidation by OH radicals (Box 9.6) which, via several intermediates, yields CO_2:

$$CH_4 + OH \xrightarrow[NO_2]{NO} CO_2$$

Transport into the stratosphere and biological decomposition are additional sinks. In knowing the atmospheric OH number density and assuming steady state conditions, methane turnover time is estimated at about 10 years. At present, methane content rises by ca. 18 ppbv or 1.1 percent annually.

About 200 years ago, CH_4 concentrations were less than half of present levels (Khalil and Rasmussen, 1985) and have increased rapidly since then. Going further back in time, that is 26,000 years, CH_4 level was 45 percent that of today (Craig and Chou, 1982). This may suggest that methane stayed uniform throughout the ages, but more data is needed to confirm this. The CH_4 rise since the days of the French Revolution to modern times, that is from 0.7 ppmv to 1.55 ppmv, can best be explained by contributions from anthropogenic sources; rice paddies, cattle and oil production (Khalil and Rasmussen, 1985; Crutzen et al., 1986). One cow belches a few hundred liters of methane a day. The 1.5 billion head of cattle we presently have already contribute about 25 percent to the global methane budget. These sources are likely to have increased with the population. The proposition has been made (Khalil, 1986) that 70 to 80 percent of the change of methane over the past 200 years stems from human activities and the rest from the depletion of OH radicals because of increasing levels of CO and CH_4. This relationship allows us to link growth of population directly to the increase in CH_4 emission and the decline in OH level. At the very most a doubling of atmospheric CH_4 and a 20 percent OH change per century could be expected. This projection will become invalidated if growth of population is larger than presently estimated. Furthermore, a warming of permafrost regions, the tundra, may liberate methane from clathrates, that is frozen gas hydrates.

Nitrous oxide: biogenic emissions: The significance of nitrous oxide as a greenhouse gas has become evident through the work of Weiss (1981), who measured the N_2O mixing ratios in a large number of air samples from various parts of the world. Over a period of 5 years (1976–80), the N_2O level rose from 298 to 301 ppbv and from 299 to 302 ppbv in the southern and northern hemisphere respectively. Three hundred ppbv is equivalent to 1,500 Tg N. These findings were corroborated (R.A. Rasmussen and Khalil, 1986; Khalil and Rasmussen, 1983; Bolle et al., 1986), and indicate that N_2O increases at a rate of 0.3 percent per year.

Biogenic emission is the principal source of N_2O. Its fate, however, is closely linked to nitric oxide, also liberated by biogenic processes. Once in the atmosphere, both species are key elements in the photochemistry and chemistry of troposphere and stratosphere. Chemically active NO (which is converted in a matter of seconds or minutes to NO_2) leads to the photochemical production of ozone and nitric acid (Graedel, 1985). It is also involved in the formation of the hydroxyl radical, the "detergent" of the atmosphere.

Nitrous oxide is chemically inert in the troposphere, has an atmospheric life-time of about 150 years and diffuses slowly into the stratosphere. Here, about 90 percent is photolyzed to molecular nitrogen and excited atomic oxygen $O(^1D)$, while the rest reacts with $O(^1D)$, yielding either two NO molecules or N_2 and O_2. This NO is instrumental in the destruction of ozone (Turco, 1985).

In a recent study, Anderson and Levine (1987) measured seasonal and diurnal biogenic emissions of NO and N_2O from fertilized and non-fertilized land. The rhythms of release are tuned to soil temperature, percent moisture, and exchangeable nitrate 'trite

BOX 9.6

Tropospheric OH (Hard et al., 1986; Ehhalt, 1981; Crutzen and Gidel, 1983)

The first hint as to the central role of the OH free radical in tropospheric chemistry came through the work of Weinstock (1969) and Levy (1971). Weinstock explained the enormous fluctuation in carbon monoxide as a result of the action of the OH radical. Levy related the production of excited oxygen atoms from ozone by means of ultraviolet light to the formation of OH. At present, atmospheric chemists unanimously regard the hydroxyl free radical, OH, as the sunlight's prime agent controling the trace-gas composition. At the same time it is the catalyst responsible for many of the symptoms of urban and regional air pollution (Hard et al., 1986). It is quasi the "detergent" cleaning the atmosphere from "dirt" and "dust".

The OH radical is highly labile but has no affinity to the main atmospheric gases. Thus, it exerts its full oxidative power on trace gases, most notably: methane, carbon monoxide and methyl chloride (Perner, 1986). These become converted into carbon dioxide, water and other stable compounds. For instance the gas-phase reaction:

$$OH + CH_4 \rightarrow CH_3 + H_2O$$

is the principal sink for atmospheric methane. Since methane is removed from air by this mechanism, any reduction in the level of OH radicals would seriously affect the abundance of methane. An increase by a factor of two in OH will decrease the atmospheric steady state content of CH_4 by a factor of two (Ehhalt, 1981). On the other hand, a rise in the NO_2 level leads to an increase in OH concentration.

Hydroxyl concentration in clean air varies largely in response to diurnal variation in ultraviolet light intensity. A maximum clean-air OH concentration of 6 x 10^6 cm^{-3} at summer solstice (21 June) has been reported (Hard et al., 1986), but including the measured concentrations of NO, O_3 and H_2O reduces the calculated maximum OH concentration to 2 x 10^6 cm^{-3}, which conforms with the data of Crutzen and Gidel (1983) and Arnold et al. (1986). Because of its multifarious involvements in the atmospheric chemistry of trace gases, there are indications that OH concentrations are decreasing, at least in the NO-poor regions of the lower and middle troposphere. A 20 percent or so decline of OH radicals over the past 100 years is most likely, and largely due to a rise in methane and possibly of carbon monoxide (Khalil and Rasmussen, 1985). This in turn diminishes the capability of the atmosphere to remove certain trace gases.

---

and ammonium concentration. In one case study, for instance, annual NO emission was four times higher on a fertilized relative to an unfertilized site. Of the approximately 200 kg of fertilizer added to one hectare of agricultural land, about 0.8 percent was lost as NO(N), and 1.2 percent was lost as N_2O(N). The suggestion was made that nitrification in anaerobic soils is the prime source for NO, whereas denitrification in aerobic soils yields N_2O.

In comparison to soils, the oceans emit only small amounts of N_2O. According to Seiler (in Bolle et al., 1986), average N_2O flux from the ocean into the atmosphere is about 2 Tg N per year, compared to 10 to 13 Tg N a^{-1} from all other sources. The data may also be split up differently. Natural sources, that is, oceans, lakes, rivers, and unperturbed soils, emit ca. 8 Tg N_2O(N) a^{-1}, whereas anthropogenic sources, i.e., biomass burning, N-fertilization, and acquisition of agricultural land, release between 5 and 7 Tg N_2O(N) a^{-1}.

Chlorofluorocarbons (CFC's): an avoidable hazard: CFCs are truly related to human activities and used as spray-can propellants, foams, refrigerator fluids and solvents. In the troposphere, they are practically inert, a property which made them a welcome product for industrial use in first place. The most prominent species are: $CFCl_3$(F11) and CF_2Cl_2(F12). Their atmospheric life time is about 80 and 170 years, respectively (Ko and Sze, 1982), and like N_2O or CH_4 they gradually diffuse into the stratosphere. Up to the seventies, annual increases in CFC emissions were as high as about 25 percent, but restrictions on the use of CFC's introduced by some countries led to a rapid decline (CMA, 1982). At present, 650 kt of CFC's are industrially produced per year. The mixing ratio in the lower troposphere is momentarily about 0.17 ppbv for F11 and 0.29 ppbv for F12, with slightly lower values for the southern and higher values for the northern hemisphere (± 0.01 ppbv) (Cunnold et al., 1983a, b). Rates of annual increases are about 6 percent. The

Montreal Convention, during which 43 nations agreed in 1987 to "freeze" in 1990 production of CFCs at that year's level and to further cut output in half by the end of this century, is a move in the right direction. In the light of the longevity of CFC's in our atmosphere much stronger measures were really asked for.

The stratosphere is a major sink for the industrial chlorine compounds. They become photolyzed and through the formation of chlorine atoms they profoundly influence stratospheric chemistry, especially the ozone cycle (e.g., Crutzen and Arnold, 1986).

Ozone (O_3): a gas of all traits: Near the ground in urban areas, ozone is a pollutant. You can certainly smell it, and this explains why C.F. Schönbein in 1840 chose the word "ozone" (from the Greek ozein, to smell) for this gas with a pronounced odor. In the troposphere, ozone photolysis triggers free-radical chemistry. And in the stratosphere, its main domain, it filters out high energy UV-radiation harmful to life. This will lead to a warming of the ozone belt, which in turn provides the energy to drive the stratospheric wind regime (Dickinson and Cicerone, 1986).

Ozone shares with the hydroxyl radical the property that both are not directly emitted from soils, organisms or the oceans, but that they are generated through the action of light rays from a wide array of trace compounds present in the atmosphere. As a consequence, ozone concentration is a function of natural and anthropogenic emission rates of exactly those trace compounds (for details see: WMO, 1981, 1982; Bruckmann, 1982; NASA, 1984; McPeters et al., 1984; Zerefos and Ghazi, 1985; Callis and Natarajan, 1986; Crutzen and Arnold, 1986; Bolle et al., 1986; Heath, 1986). To cut a long story short, the primary process in the troposphere leading to ozone is photolysis of NO_2 and subsequent reaction with molecular oxygen. The liberated by-product, NO, becomes the main sink for ozone in highly polluted atmospheres. However, in clean or unpolluted air, ozone concentration is proportional to the ratio of NO_2 to NO. Oxidation of organic compounds such as methane will shift the reaction in the direction of NO_2, thus favoring ozone. These degradation reactions are principally initiated by the hydroxyl free radical, and suggest that the degree of air pollution determines tropospheric ozone concentration. At low NO values (≤ 10 pptv), the hydroxyl radical and ozone will oxidize the ubiquitous methane, whereas at NO contents exceeding 10 pptv, OH and ozone should increase. This fits the global pattern. Namely, the heavily polluted northern hemisphere exhibits an ozone increase, particularly during the summer season. Since ozone's lifetime is measured in just weeks, considerable spatial variations exist. Changes by 20 to 50 percent have been measured locally, and a general rise by 1 to 2 percent seems to be the case (Bolle et al., 1986). The southern hemisphere witnesses instead a stagnation or lowering of ozone and the hydroxyl radical, except in areas of extensive biomass burning.

Stratospheric ozone follows quite a different route. We should recall that in the stratosphere ozone is spread across a vertical layer of about 30 km (see Fig. 9.1), but all O_3 molecules taken together would only produce a layer of a few mm around the surface of the globe at ambient temperature and pressure. It is hard to imagine that such a lucid veil could be so significant for protecting life on Earth. The answer can be found in Table 9.3. First, ozone absorbs in a spectral window where radiation hostile to organisms impacts the atmosphere. Second, one ozone molecule is at least 100 times more efficient than a CO_2 molecule in retaining solar energy. As a consequence, a partial destruction of the ozone layer would certainly give you a better suntan but at the same time will make you more prone to skin cancer, increase incidence of DNA damage or change your immune system profoundly. A multitude of other effects deleterious to life would result if Earth were deprived of its protective "O_3 sunglasses".

The history of ozone research and environmental implications of ozone depletion were recently reviewed (Brasseur, 1987). Following the classical work of French atmospheric scientists (e.g., Chappuis, 1880; Levy et al., 1905; Fabry and Buisson, 1913, 1921), a British group headed by G.M.B. Dobson discovered the relation between weather patterns and ozone column abundance, in that a maximum was clearly visible in low-pressure cells and a minimum linked to high-pressure events. In addition, seasonal fluctuations, that is, spring high, fall low, were noted. In recognition of this work, the atmospheric scientific community introduced the Dobson unit (1 Dobson unit = 10^{-3} atm cm), which is a measure of ozone column abundance (number of ozone molecules per unit area above the Earth's surface; Dobson, 1929). As an illustration, global ozone distribution throughout the year is depicted in Fig. 9.25. For the most recent trends in total ozone (1979 through 1986), see Bowman (1988).

Regional and global distribution of ozone is not governed by physical factors alone; actually, these are only secondary features. The crucial question is: what causes ozone to form in the first place and where does ozone eventually end? Since the work of Chapman (1930) it has been commonly assumed that ozone was generated via dissociation of molecular oxygen by means of solar UV radiation (<242.4 nm). The result-

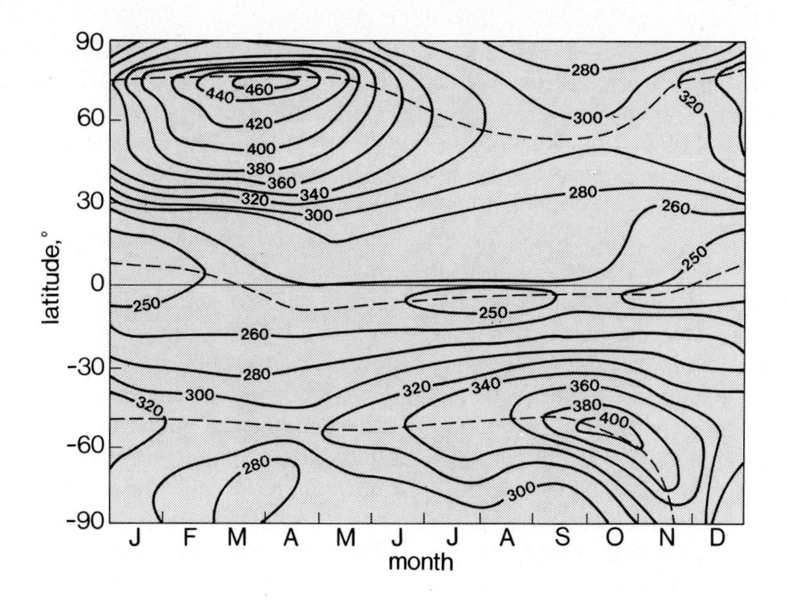

Fig. 9.25. Ozone column abundance (in Dobson units) as a function of latitude and month (London, 1980)

ing oxygen atoms recombine with O_2 by a so-called three-body reaction. The third body (M) could be either N_2 or O_2:

$$O + O_2 + M \rightarrow O_3 + M$$

ozone rapidly dissociates

$$O_3 \rightarrow O + O_2$$

and atomic oxygen will revert to ozone again. In fact, the net destruction of odd oxygen ($O + O_3$) can be described

$$O + O_3 \rightarrow 2O_2.$$

However, this recombination process, as proposed by Chapman (1930), is far too slow for the observed ozone density. A partial answer was provided by Bates and Nicolet (1950) and Hampson (1964), who attributed ozone lowerings to the activity of various hydrogen radicals formed by oxidation or photodestruction of water vapor and methane. Even by invoking such mechanisms, observed ozone densities were still too low.

Starting with the work of Crutzen (1970), Stolarski and Cicerone (1974), and Molina and Rowland (1974), we are presently witnessing a new kind of atmospheric chemistry, where most of the changes in ozone density are explained in terms of anthropogenic inputs (Crutzen and Arnold, 1986; Labitzke et al., 1986). Nitrogen oxide and stratospheric chlorine derived from CFC's are thought to be the catalysts for the depletion of the ozone layer.

At the moment the major headline is the Antarctic Ozone Hole, which is seen to be the result of a complex interplay among a variety of environmental parameters. According to Crutzen and Arnold (1986), chlorofluorocarbons diffusing into the stratosphere should, in fact, only be moderately destroyed by UV radiation. The relatively small amounts of chlorine and chlorine monoxide liberated in this fashion will react with methane and become transformed to HCl which has no affinity to ozone. However, the "methane brake" will fail once temperatures fall below –70°C. In such a regime nitrogen oxides increasingly generate nitric acid, which jointly with water vapor condenses to an aerosol of frozen particles. Removal of these gaseous N-species by freeze drying will exponentially accelerate OH content at the end of the polar night once the sun starts to rise across the antarctic horizon. Methane serves hereby as an accelerator device.

Principally, the OH radical degrades the "inactive" chlorine pool ($ClONO_2$ and HCl) and generates "active" chlorine compounds which consume ozone:

$$\begin{aligned} Cl + O_3 &\rightarrow ClO + O_2 \\ ClO + HO_2 &\rightarrow HOCl + O_2 \\ HOCl + h\nu &\rightarrow OH + Cl \\ OH + O_3 &\rightarrow HO_2 + O_2 \\ \hline \text{net effect: } 2O_3 &\rightarrow 3O_2 \end{aligned}$$

Analogously bromine, stemming from brominated hydrocarbons used in foams of fire extinguishers, could rapidly degrade ozone because "active" bromine compounds are – molecule by molecule – even more efficient than "active" chlorine molecules in destroying O_3. In essence, it is the freeze drying of

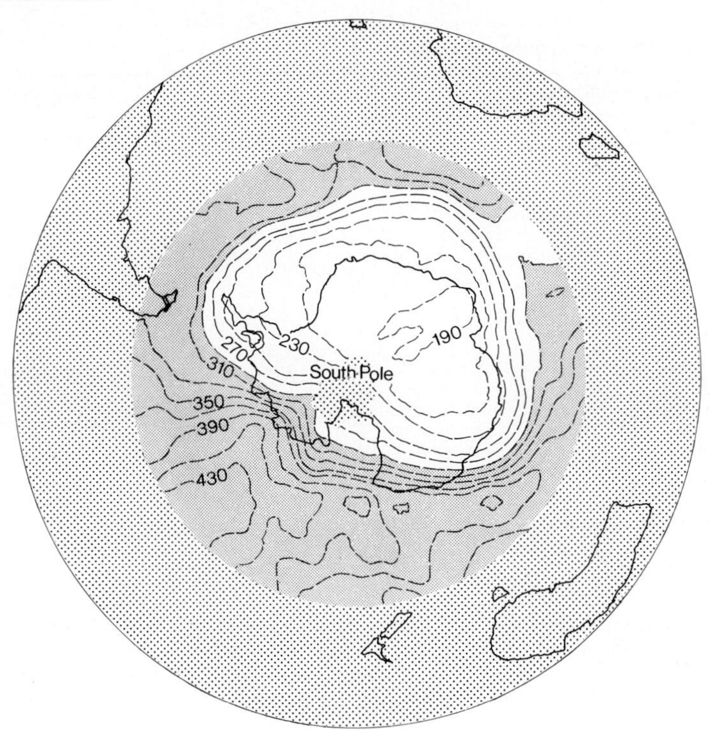

Fig. 9.26. Total column ozone abundance (in Dobson units) from the South Pole measured on October 12, 1985 by the Spin-Scan Ozone Imager aboard the Dynamics Explorer I satellite (G.M. Keating in: Brasseur, 1987)

nitric acid which promotes the spontaneous change in stratospheric chemistry so detrimental to ozone density. The widening of the springtime Antarctic Ozone Hole in recent years, now having a size comparable to the area of the United States, is a frightening sight (Fig. 9.26). Research is presently being pursued to establish whether a springtime "mini" ozone hole has already developed at the North Pole.

Longevity of CFC's in the troposphere and their slow diffusion into the stratosphere will maintain ozone destruction far into the next century, even if halocarbon emission were stopped immediately.

An entirely different outlook concerning the future development of ozone density over the South Pole has been given by Labitzke (1987, 1988). Based on the work of Stolarski et al. (1986) and Chubachi (1986), who tentatively assigned the seasonal variation of ozone to substantial reduction in wintertime planetary-scale wave activity, she was able to show that until 1981 most of the observed changes in total ozone over Antarctica are related to the delay of the Final Warmings over Antarctica which, according to her analysis, are mainly a consequence of increased aerosol content in the aftermath of recent volcanic activity (Box 9.7).

Guy Brasseur (1987) has carefully weighed the pros and cons of the chemical and dynamical theories and concludes that "because of the lack of observational as well as laboratory data, it remains difficult at this juncture to fully adopt or reject any of the theories that have recently been proposed." It is not unlikely that both schools of thought are in principle correct. That is to say, Karin Labitzke's scenario of cooling the polar regions by increased stratospheric aerosols of volcanic provenance may set the stage for the chemical mechanisms proposed by Paul Crutzen and his coworkers for the ozone thinning due to pollutant chlorofluorocarbons. As a geological observer, I can only cross my finger and hope that volcanic eruptions slow down to an all-time low. But to be on the safe side: stop CFC's, because they are essential neither for food, nor for energy, nor for civilization at all!

Smog and Haze

The number of variable constituents in air is legion, but most prominent among them is water vapor. In the troposphere – the weather sphere – concentrations may range from practically zero to a few percent, and in the stratosphere H_2O measures just a few ppmv (Table 9.5). Atmospheric condensation is achieved at very small supersaturation levels by means of condensation nuclei, tiny particles, about 0.01 to 100 micron in diameter. Highest particle den-

Table 9.5. Greenhouse gases in the troposphere (as of 1983)

Trace Constituent	Source (anthropogenic)	Sinks	Suggested Clean Air Concentration
carbon dioxide CO_2	fossil fuel destruction of vegetation and soil (humus oxidation)	ocean, biosphere	345 ppmv
water H_2O	insignificant; in stratosphere high-flying airplanes	precipitation	troposphere: 0–3 % stratosphere: 3–10 ppmv
ozone O_3	only indirect: photolysis of anthropogenic NO_2	catalytic reaction with different trace compounds such as Cl_x, NO_x, HO_x	troposphere: 25–70 ppbv (between surface + 9 km) stratosphere: up to 15 ppmv (at 25 km)
aerosols	industry; desertification, grass, bush and forest fires	sedimentation, precipitation	10 to 50 g m^{-3}
halocarbons CFC's	gas in spray cans, defrosting agents in refrigerators and insolators	no tropospheric sink and only minor one in stratosphere due to photolysis	F_{11}: 0.15 ppbv F_{12}: 0.25 ppbv ——— 0.40 ppbv Total
N_2O and NO_x	N_2O: fertilizers, NO_x: fossil fuel combustion and high-flying planes	N_2O: photolysis NO_x: precipitation dry deposition	N_2O: 0.3 ppmv NO_x: 0.5–20 ppbv
sulfur dioxide SO_2	fossil fuel, ore-processing, plants	dry and wet deposition and oxidation in atmosphere	0.2 ppbv (regionally different)
methane CH_4	rice paddy fields, fossil fuel, animals (cows etc.)	oxidation by OH radicals	1.7 ppmv
carbon monoxide CO	incomplete combustion, photochemically from hydrocarbons	oxidation by OH radicals, soil organisms	0.1 ppmv
higher C-numbered hydrocarbons	fossil fuel	photochemical destruction	ppb range
ammonia NH_3	fossil fuel, animals	oxidation to NO_x dry deposition	0.2 ppbv
krypton 85 Kr	nuclear industry	radioactive decay	1.14 ppmv

sity is observed in air masses above industrial or maritime regions, where up to several billion nuclei per cubic meter of air may be present. Such particles (excluding water and ice) are commonly referred to as aerosols. They are effective scatterers and absorbers of light rays and in this capacity can modulate climate.

Aerosols can be released from land: e.g., industry, vegetation, forest fires, desertifications, or from the ocean: e.g., sea spray, surface films, primary productivity, oil platforms. Thus, a wide variety of natural and human-produced compounds exists which may cause smog, summer haze and – last but not least – alter the radiation balance of the Earth, induce precipitation, or "bronze" trees and vegetable crops.

Reading Edgar R. Stephens' (1987) "Smog Studies of the 1950s" brought back memories of yellow photo-

Table 9.5 (continued)

Possible Influence on Radiation Balance and Climate	Comments
an increase by a factor of two (doubling) leads to an increase in global temperature by 2–4 K; in contrast, cooling of stratosphere by 10 K (these are just model calculations of equilibrium response; insufficient representation of cloud and ocean response)	preindustrial level (before 1850) 270 ppmv (tree rings), 260–280 (ice cores), since 1957 increase by 30 ppmv
direct absorption and major climate factor indirect influence via O_3 concentration	many kinds of interactions with respect to cloud formation and minor compounds in atmosphere
O_3 decomposition will result in cooling of the stratosphere and increase of UV radiation at the surface of the Earth; many types of interactions with respect to meteorological processes	tropospheric preindustrial level –12.5% of today's. Residence time 1–3 months; global effects, but regionally different. Almost all trace constituents will participate at photochemical steady state of O_3
cloud formation and precipitation processes will be regulated; absorption, reflection and scattering of solar and terrestrial radiation plus own emission in the visible range	net impact on albedo and temperature distribution cannot be evaluated at present. Recent models suggest cooling effect
increase by factor of 20 leads to an increase in global temperature of 1°C (greenhouse model calculation); increase of destruction of ozone	O_3 destruction in 100 years about 15% at present emission rate; in stratosphere present increase about 0.02 ppbv per annum, residence time F_{11}: 65 a, F_{12}: 110 a
glasshouse effect through increase of N_2O leads to reduction in ozone content; the hereby formed NO_x determines the ozone steady state in the stratosphere	residence time of N_2O a few years; residence time for NO_x a few days; contribution of fertilizers on atmospheric N_2O content unknown
through oxidation, formation of sulfate-aerosols which in turn influence radiation balance and formation of clouds (condensation centers); acid rain formation	regional effects largely through acid rain; residence time a few days
via change of tropospheric ozone concentration impact on radiation balance (indirectly) and greenhouse effect (directly)	residence time in troposphere about 5–10 a
via change of tropospheric ozone concentration impact on radiation balance (indirectly)	residence time in atmosphere about 1 month; interference with tropospheric trace compounds is most likely
via change of ozone concentration impact on radiation balance	residence time in atmosphere up to 2 a; cycles still unknown
only small influence on radiation balance	residence time a few days; cycles poorly known
change in electric field and thus impact on cloud formation	potentially important climate factor because of impact on cloud formation

chemical smog enveloping Los Angeles and Pasadena on many summer and fall afternoons. That smog not only made your eyes water, cracked rubber tires and damaged plants, but to my dismay it also oxidized spots on my paper chromatograms, whence organic compounds extracted from meteorites had migrated. It was, as we know now, ozone and PAN's chemistry (peroxyacyl nitrates) that did the job (e.g., Haagen-Smit et al., 1952a, b; Stephens et al., 1961; Leighton, 1961; Heicklen, 1976).

At about the same time, Went (1960) proposed that, due to their volatile character, isoprenoids may accumulate in the atmosphere. He calculated that about 2 x 10^8 tons of volatile plant products, predominantly isoprenoids, are annually released from living or dead organic materials to the air, where they particularize.

BOX 9.7

Polar Final Warmings (Labitzke, 1987, 1988; Muraoka et al., 1987)

Stratospheric circulation over the north and south poles differs significantly. Over the arctic, mid-winter warmings are common, whereas little interannual variability exists during the antarctic winter. This will lead to the breakdown of the arctic polar vortex in mid-winter and cause ozone-rich air to move from middle to north polar latitudes, often producing an ozone maximum already in January. In contrast, stratospheric circulation is much more symmetric over the antarctic, and the large-scale planetary waves are much less pronounced during the southern winters. In turn, air movement from middle to polar latitudes is sluggish and thus little ozone-rich air is injected into the antarctic polar vortex during mid-winter – effectively until September.

During the October-November transition time, interannual variability over the antarctic is largest. Final Warmings initiate circulation changes from the cold polar vortex to the warm polar anticyclone. These Final Warmings are linked to an intensification of the planetary-scale wave one (Fig. 9.27).

For the years 1961 to 1986 the monthly mean 50-mbar height at the South Pole shows indeed an intensification of the polar vortex in October. Furthermore, after 1979 – except for the year 1982 – the center of the polar vortex was significantly deeper, causing delayed Final Warmings. These consecutive years of late vortex breakdown (Bojkov, 1986) are seen by Labitzke (1987, 1988) as the ultimate cause of the recently observed Antarctic Ozone Hole. Chemical processes of the type outlined by Crutzen and Arnold (1986) have essentially more time to destroy the ozone. In the final analysis, increased stratospheric aerosols are responsible for a changed polar vortex and a "deep freeze" stratospheric chemistry subjected to delayed Final Warmings.

This all may suggest that once aerosol content has decreased to normal levels, early Final Warmings will recur and bring to a close the Antarctic Ozone Hole.

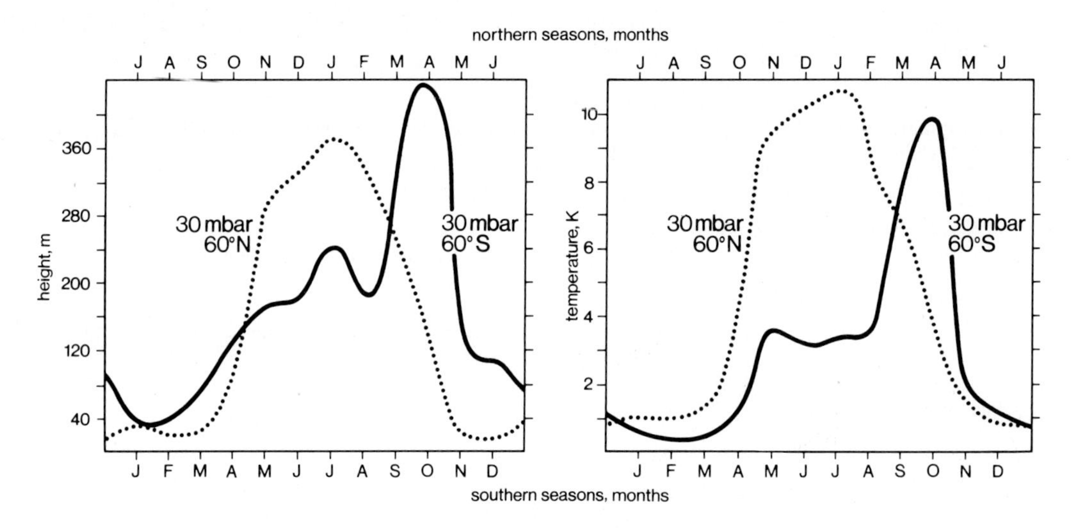

Fig. 9.27. Annual march of amplitudes of planetary-scale height or temperature wave one, for 60°N and 60°S (Knittel, 1976; Labitzke, 1987)

This condensation process was explained to be dependent upon sunlight and, partly, upon the presence of nitrogen oxides; it is the cause of "summer haze" in cultivated areas. Went considered this material, bituminous or asphaltic in nature, to be a likely source of hydrocarbons in soils and sediments. Graedel (1979), Guderian and Rabe (1981), and Kohlmaier et al. (1983) have summarized progress in this field. It is now well established that species-specific hydrocarbon patterns are emitted from forest and grassland ecosystems.

At present there is wide concern, especially in Europe and North America, about the deleterious effect of anthropogenic pollutants such as SO_2/H_2SO_4,

BOX 9.8

Sulfur Emission to the Atmosphere (Andreae, 1986; Ryaboshapko, 1983)

The atmosphere, being a highly mobile system, rapidly carries localized emissions of reduced or oxidized sulfur compounds across a wide territory. In a matter of days or weeks, sulfur dioxide released, for instance, from thermal power plants in Central Europe may be swept to Scandinavia, where it rains out. Even though the sulfur dioxide content near the emission source can reach several mg S m^{-3} air and becomes diluted on its way north by several orders of magnitude, the constant flow of sulfur dioxide to highly oligotrophic subarctic lakes, or the exposure of northern trees to acid rain will eventually take its toll. Estimates for global anthropogenic sulfur emissions are about 104 x 10^{12} g S a^{-1} (Cullis and Hirschler, 1980). The principal source is the release of sulfur dioxide from fossil fuel burning.

It came as a great surprise when it was discovered that volatile biogenic sulfur species contribute more or less an equal share of sulfur to the atmosphere as humans do via fossil fuel combustion. Most outstanding in this respect is the emission of dimethylsulfide (DMS) both on land (Adams et al., 1981) and in the sea (Ferek et al., 1986; Andreae and Raemdonck, 1983; Toon et al., 1987). Of the total biogenic sulfur flux estimated at 103 x 10^{12} g S a^{-1}, the DMS flux alone is half of it (52 x 10^{12} g S a^{-1}). The ocean is in this connection the most substantial DMS source (40 x 10^{12} g a^{-1}; Andreae and Barnard, 1984). Once in the atmosphere, DMS will be rapidly oxidized to a variety of compounds such as sulfur dioxide, sulfate, methanesulfonic acid, and dimethyl sulfoxide, all of which are part of aerosols.

A major source of cloud-condensation nuclei over the oceans is DMS released by planktonic algae, which is rapidly oxidized to sulfate aerosols. Charlson et al. (1987) have convincingly demonstrated how this mechanism has feedbacks on cloud formation and Earth radiation balance. Since clouds will reflect more light (= heat) to space than the sea does, an increase of more cloud-condensation nuclei, let us say by a factor of 2, could possibly counteract climatic warming due to doubling of air CO_2 content.

NO_x/HNO_3 or heavy metals. They are unanimously regarded as the primary stress factors on forest ecosystems. Secondary factors are insects, fungi, bacteria, and viruses, or unusual weather and climate patterns restricting, for instance, the availability of water by lowering the groundwater table. In addition, many of the pollutants lead to photochemical reactions in urban areas or on their way to forest regions. Here, ozone and other photo-oxidants produced in the atmosphere seem to bronze and kill the trees. This, however, is not the full story. It has been discovered that many reactive hydrocarbons emitted from trees such as ethane, isoprene, C_{10}-terpenes interact with O_3, $NO/NO_2/HNO_3$ or SO_2/H_2SO_4, whereby pine needles on the sunny side or the margin of a forest start to bronze first. These natural hydrocarbons can couple with the photochemical NO_x/O_3 reaction cycle, forming peroxyacyl nitrates (PAN) whose toxic effects are widely known (e.g., Grimsrud et al., 1975; Zimmermann et al., 1978; Holdren et al., 1979; Kokouchi et al., 1983; Kohlmaier et al., 1983; Graedel, 1985). Thus in a way, smog, summer haze and acid rain join in a vicious circle (Box 9.8). It has even been suggested (Crutzen, 1987) that acid rain has much to do with the extinction phenomena observed at the Cretaceous/Tertiary boundary.

Climatic Change

From Cold to Warm Epochs and Back

At present we live in a moderately warm interglacial stadial of an ice age which has little chance of persisting in time. It appears that the main factors influencing the climate within a time frame of 10,000- to 100,000 years are the orbital periodicities of the Earth (see, e.g., CLIMAP Project Members, 1976, 1984; Gates, 1976; Climate System Monitoring, 1985; Nature, 1978; Mason, 1977). Milutin Milankovitch, a Yugoslav geophysicist, was the first to suggest that astronomical cycles might account for a change from a glacial to an interglacial epoch and vice versa. From the periodicities in the tilt angle of the rotation axis (41,000 years), the precession of the Earth's axis (21,000 years), and the excentricity of its orbit (93,000 years), he plotted a curve for the radiation values averaged over the months March to September, span-

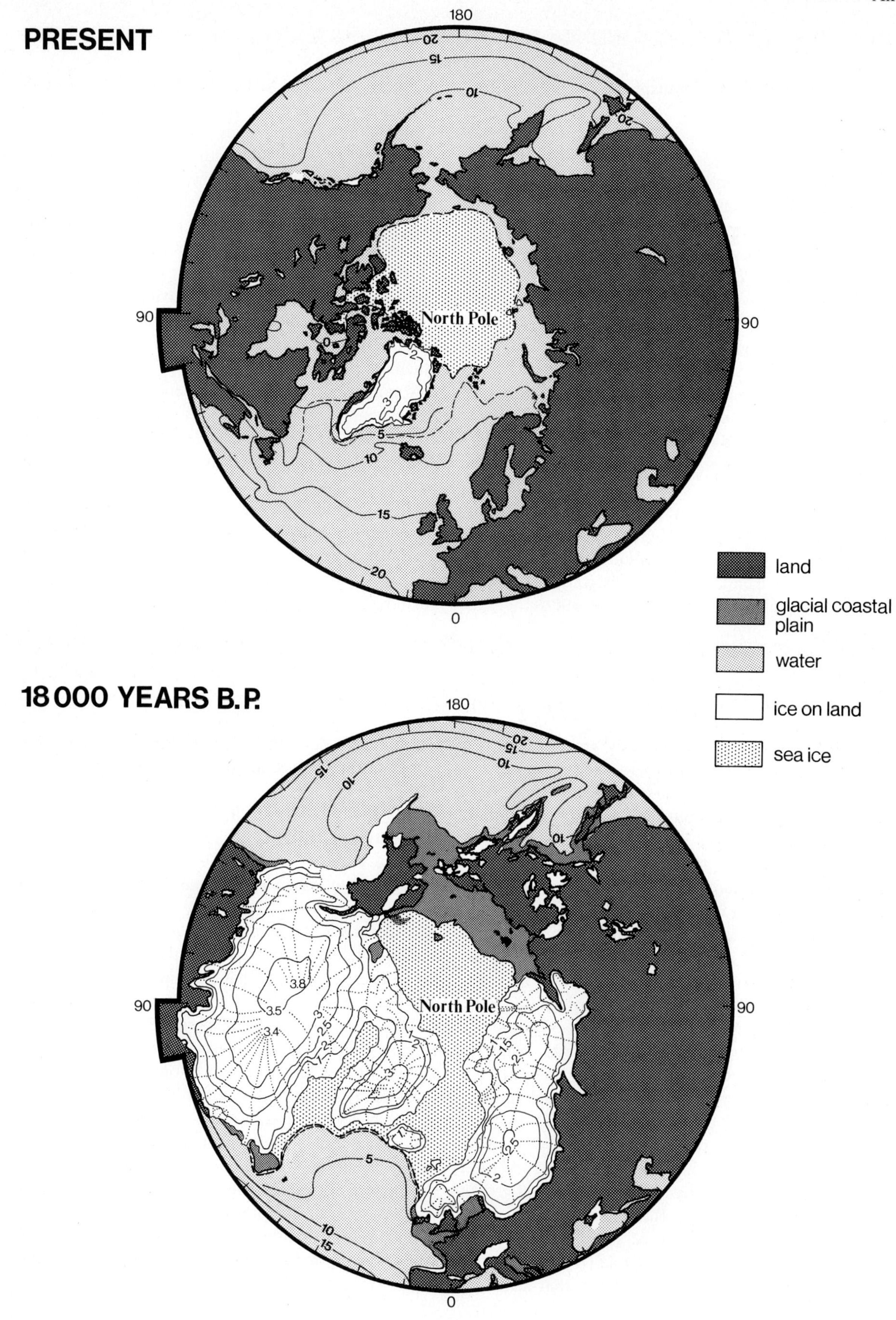
PRESENT
180
90
90
0
North Pole
18000 YEARS B.P.
180
90
90
0
North Pole
land
glacial coastal plain
water
ice on land
sea ice

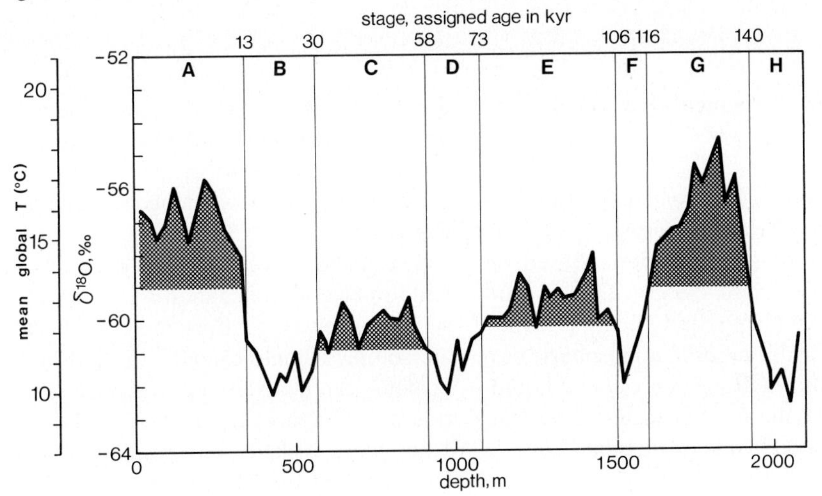

Fig. 9.29. Oxygen isotope distribution in terms of $\delta^{18}O$ in a 2.1-km-long ice core from the Antarctic (Lorius et al., 1985). This core has at its base an age of about 150,000 years and brings us back to the Saalian Glacial. Estimated differences in mean global air surface temperatures between ice age minimum and warm age maximum are around 10° C, and differences in sea level are ca. 100 m

ning the last 300,000 years. Today the solar radiation at 50°N is 847 langley day^{-1} (or 35447 kJ m^{-2} day^{-1}). The Milankovitch curve shows fluctuations between –35 and 50 ly day^{-1} with respect to the present-day value. In view of the fact that Earth has recently finished a phase for which the solar radiation incident on its surface was very high and modern incident radiation is 30 ly day^{-1} less, a cooling is expected. Weertman (1976) predicts the beginning of a new glacial epoch within the next hundred or several thousand years. It should last approximately 60,000 years, that is, its duration should match that of the Weichselian (= Wisconsinan) Ice Age. The Weertman model, however, is beset by some uncertainties, because it requires much higher precipitation rates than those prevalent in the continental polar areas today. Should Earth again revert to an ice age, the northern hemisphere could eventually look roughly the same as 18,000 years ago when Earth went through a climatic minimum (Fig. 9.28).

Ice core data seem to substantiate the validity of orbital periodicities as a controlling device for climatic change of the type recorded by the glacial-interglacial pattern of the Quaternary. Using fluctuation in oxygen-18 content with time, which is principally a measure of the temperature at the time of precipitation

Fig. 9.28. The ice situation of the northern hemisphere today and during the peak of the Weichselian Glacial, 18,000 years B.P. (GSA, 1981)

formation and the history of air masses, changes in global temperatures and ice volume changes in the course of up to about 150,000 years can be inferred (e.g., Dansgaard et al., 1971; Lorius et al., 1985; Fig. 9.29; Box 9.9). The range in mean temperature between glacial and interglacial epochs is up to about 10°C. It is relevant that from time to time glacials have warm interludes and interglacials cold spells. For instance, the Eemian Interglacial (Stage G; Fig. 9.29) and the Weichselian Glacial (Stags B, C, D, E, F; Fig. 9.29) are marked by such climatic deviations, and the observed cyclicity is striking. It matches the summer insolation curve rather closely (Lorius et al., 1985).

Milankovitch cycles have now been found in sediments of different ages and are revealed by distinct banding, colors, and the thickness of laminae and layers. Strong 100,000- and 40,000-year cycles are recorded, and observational data provide support that Milankovitch cycles are not restricted to the past one million years, but are laid down, for instance, in lake sediments 200 million years of age (Olsen, 1986; see also Arthur and Garrison, 1986).

Critics of the Milankovitch radiation curve point out that changes in orbital parameters could at the most account for a 3 to 4°C change between climatic optimum and minimum. This is insufficient to trigger an ice age, and a kind of amplification system has thus to be invoked. According to Kuhle (1986), glaciation of the Tibetan Plateau is a potential primer of a global ice age. This idea is appealing for a number of reasons. First, surface albedo of low-latitude ice sheets is higher than of polar ice caps. Second, the Tibetan Plateau is one of the critical areas regulating global atmospheric circulation. A chain reaction could conceivably come into existence involving the following steps:

BOX 9.9

Farmers' Almanac in Permanent Ice (Dansgaard et al., 1971)

Stable isotopes of hydrogen and oxygen are useful indicators of source, geochemical processes, and the meteorological history of natural waters. Based on the early work of Dansgaard (1953), Epstein and Mayeda (1953), Epstein (1959), and Craig (1961) we know that D/H and $^{18}O/^{16}O$ ratios in atmospheric precipitation normally follow a Raleigh process at liquid-vapor equilibrium. δD and $\delta^{18}O$ in meteoric surface waters are linearly related by the expression

$$\delta D = 8\,\delta^{18}O + 5,$$

which explains why researchers commonly look only for oxygen-18 in natural waters and skip deuterium analysis.

The meteorological cycle is much like a multiple-plate still with reflux. Upon evaporation, the sea preferentially loses $H_2{}^{16}O$ because of its higher vapor pressure relative to $H_2{}^{18}O$. The isotopically light vapor rises, cools, and precipitates, this time returning $H_2{}^{18}O$ rich rain back to the ocean. As long as air masses hold water vapor and permit outraining of clouds, isotopic fractionation will continue and vapor becomes lighter and lighter. Air masses, eventually reaching polar regions, have lost roughly 99 percent of the moisture they initially acquired in equatorial regions, the principal global source of water vapor. As a general rule, the colder the atmospheric temperatures and the more extended the ice sheets, the isotopically lighter the rain, snow, or ice will be. This is roughly the logic behind snow and ice stratigraphy and paleoclimatology.

Seasonal $\delta^{18}O$ variation has been observed in snow and firn layers: winter snow is isotopically lighter than summer snow (e.g., Epstein et al., 1965, 1970). Such an isotopic climate calender, a kind of "Farmers' Almanac", can be extended to the ice sheets of Antarctica and Greenland, where ice cores of more than 1 km were drilled. For results on Antarctica see Fig. 9.29 and text (Lorius et al., 1985; Jouzel et al., 1987). The Greenland data of Dansgaard et al. (1971) are briefly commented upon.

The isotope curve of the past 800 years is chronologically well integrated, in part by counting annual layers and in part by radio-carbon dating of air CO_2 enclosed in the ice. The climatic optimum of the late thirties and early forties, as well as the warm period during the middle of the 18th century can

Fig. 9.30. a Climate calendar in the permanent ice sheet of Greenland revealed by $\delta^{18}O$ variations (Dansgaard et al., 1971). Shown are the mean $\delta^{18}O$ values of 10 years. Sunspot activities are assumed to be a main factor of short-term climatic change during the Holocene; volcanic activities amplify the climate signal. During the late Pleistocene, climatic change is induced by other factors (see text). **b** Highly schematic trend of climatic change since Weichselian minimum

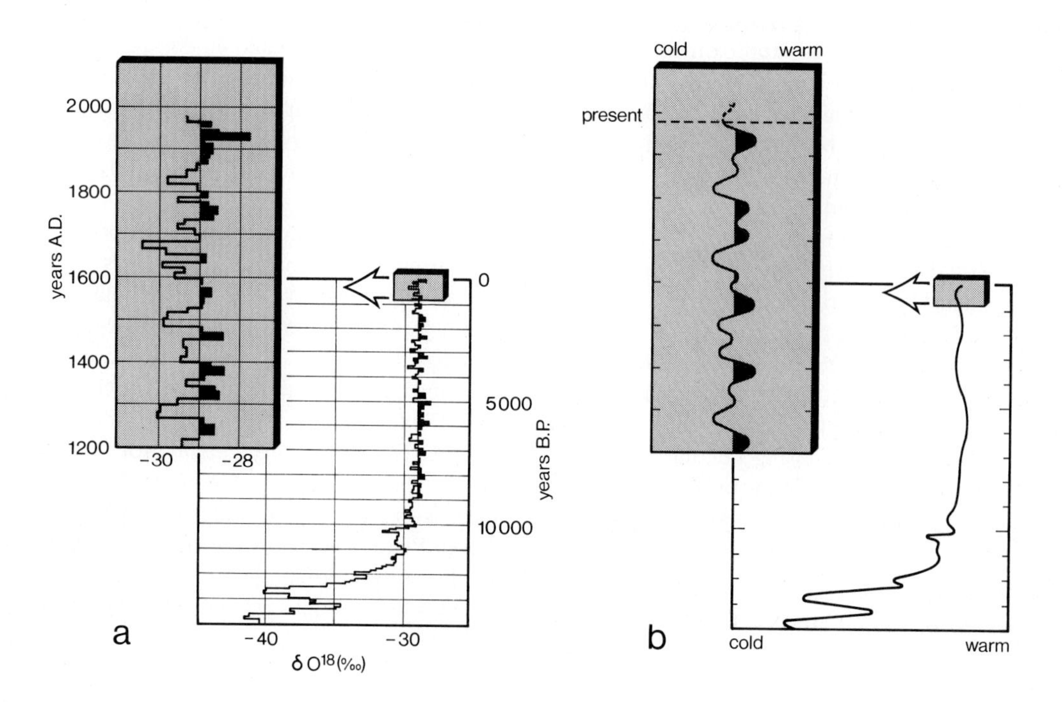

clearly be recognized. The cold spell at the beginning of the 19th century, the little ice age (Maunder Minimum), and the long enduring cold period during the 15th century come to light. This was about the time the Vikings left Greenland. Winter scenes in pictures of the old Dutch masters make you sense some of the crisp cold spread over Europe at that time. For reconstruction of the northern hemisphere temperature from the 16th to the 19th century see Grovemann and Landsberg (1979) and from 1851 to 1984 Jones (1985).

The climate calendar extends almost back to the climatic minimum of the last glacial. The general increase in temperature towards the Holocene/Pleistocene boundary is manifested in a sharp $\delta^{18}O$ rise of more than 20 permil (Fig. 9.30).

Air temperature and humidity characterize a climate. An excellent yardstick for paleoclimate is the presence or absence of specific trees and grasses as indicated by their spores and pollen found in sediments. The mean July temperature of NW-Europe deduced from pollen spectra of Holocene and Pleistocene deposits is depicted in Fig. 8.33 (Zagwijn, 1975; Hammen et al., 1971). Inferred air temperatures vary by at most 15°C between the peaks of glacial and interglacial epochs. It is worth noting that global temperature fluctuations during the Holocene only amounted to 2–3°C.

(i) orbital parameters move Earth into a threshold position for initiation of an ice age; net temperature effect: 3 to 4°C;

(ii) lowering of snow line in the Tibetan Plateau by 500 m and corresponding surface albedo change lead to a general cooling trend;

(iii) further lowering of the Tibetan snow line will generate an ice sheet about 2 to 2.4 million square kilometers in area; this ice cover causes a loss in the global heat budget roughly equivalent to a polar ice cap ca. 10 million square kilometers in size; net temperature effect: 1 to 2°C;

(iv) onset of glaciation of mountainous regions in higher latitudes; destruction of vegetation cover; substantial lowering of sea level by more than 100 m; expansion of inland ice to an areal size of 36 million square kilometers, resulting in a heat loss 3.5 times that of the Tibetan Plateau: net temperature effect: 3 to 5°C;

(v) orbital parameters move Earth into a threshold position for initiation of warm epoch; retrograde development: deglaciation, rise in sea level, return of forests at the expense of grasslands, steppes, and deserts; net temperature effect: 7 to 11°C; total time required from climatic minimum to climatic optimum: about 10,000 years. That the switch from an outgoing ice age to the beginning of a temperate climate is much more sudden (1,000 years) has been recently established (Broecker, 1987a, b).

This chain reaction scenario at the beginning and the end of an ice age is illustrated in Fig. 9.31. The critical role exercised by the subtropical Tibetan Plateau becomes evident (Kuhle, 1986). The

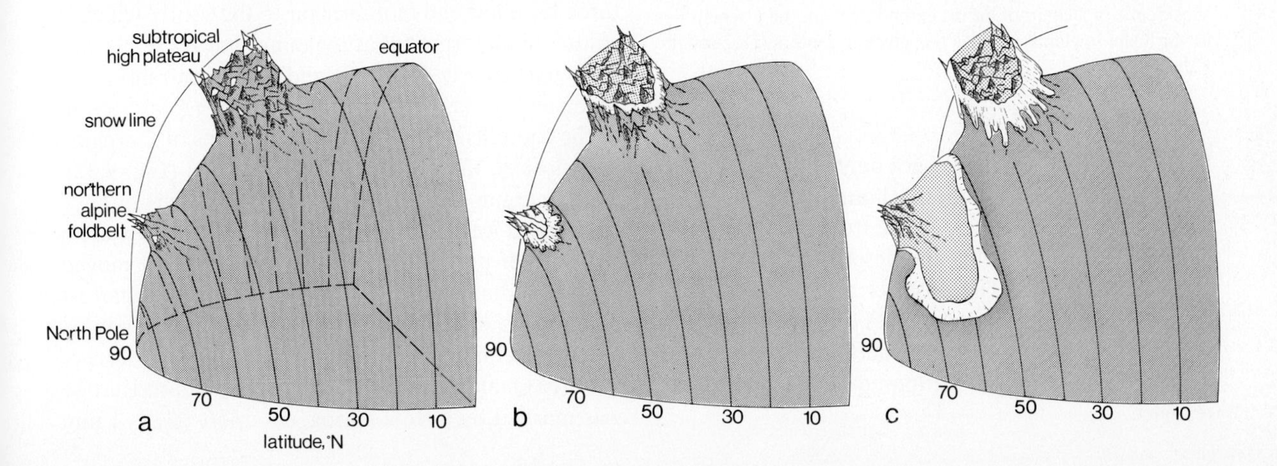

Fig. 9.31 a–c. Highly schematic representation of events in the subtropical high plateaus assumed to trigger an ice age of global dimension (Kuhle, 1986). Today's situation is depicted in (**a**); only the peaks of the mountains are ice-covered. Change in orbital parameters lead to a cooling trend and subtropical high plateaus start to glaciate (**b**). This causes a global cooling, and ice sheets on alpine foldbelts in high latitudes become extended far into the lowlands (**c**). The simultaneous advance of sea- and inland ice is not shown for graphical reasons (Kuhle, 1986)

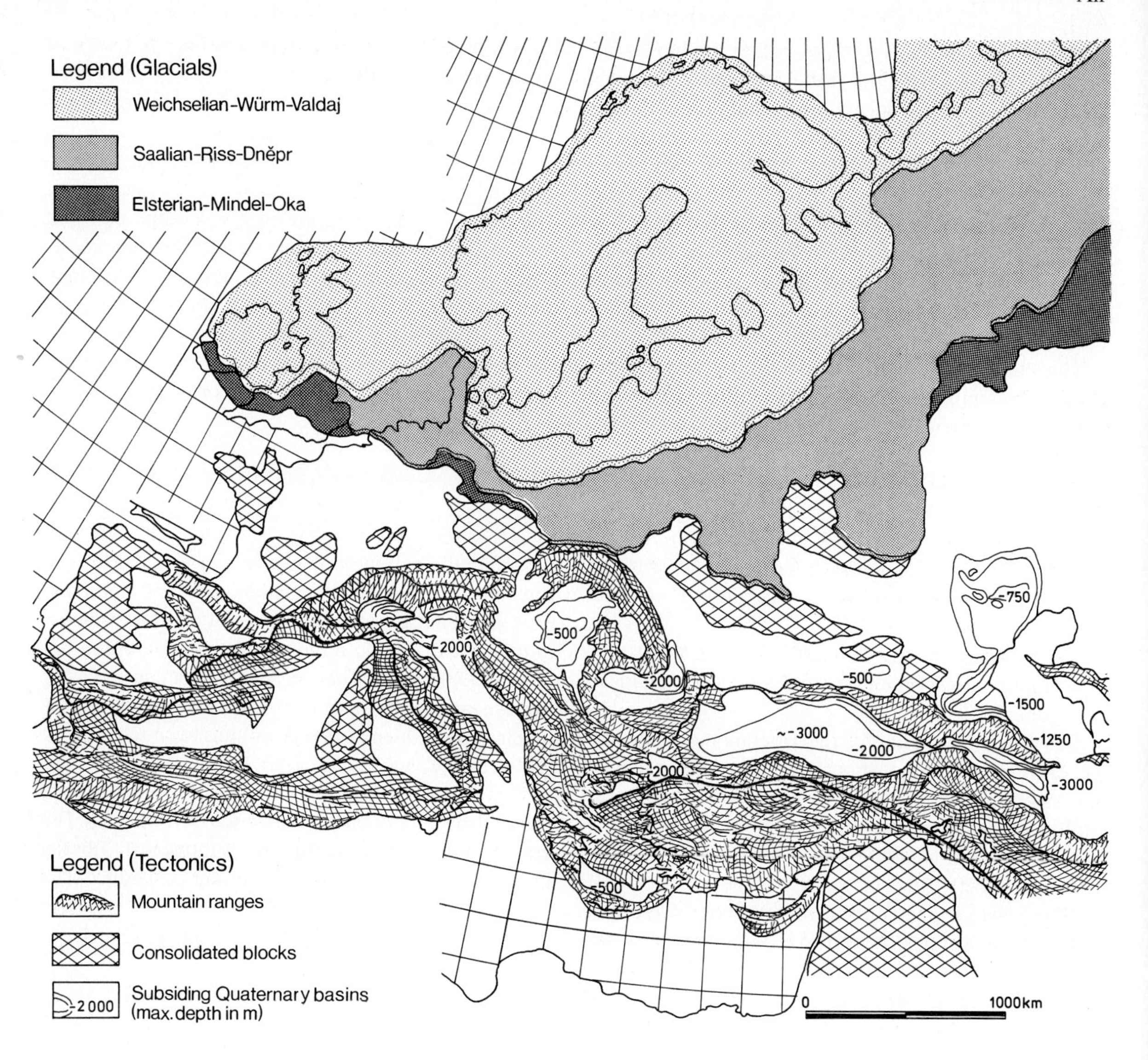

Fig. 9.32. Areal extension of the last three European glaciations: Elsterian-Mindel-Oka, Saalian-Riss-Dnepr, and Weichselian-Würm-Valdaj. Alpidic mountain ranges and associated subsiding basins are indicated. The extent of late Pleistocene subsidence in a chain of basins extending from the Caspian Sea towards the lowlands of the Po is given in meters (Degens and Paluska, 1979)

Himalayan Region triggers not only an ice age but influences the global weather profoundly. For instance, a change in cloud pattern – aside from its albedo feedback – leads to droughts and desertification in many parts of the world. The onset of an ice age may be compared to a thermostat that is lowered in discrete steps. The return to warmer conditions is simply a reversal of the thermostat settings, except at a much faster pace.

There are differences in the morphological expression of ice ages. Among the 20 or so recorded in the northern hemisphere during the Quaternary, the last three have left end moraines far to the south, which is commonly interpreted as meaning that those glacials were particularly severe. Scandinavian end moraines of Elsterian, Saalian, and Weichselian age are found at the foothills of the Central Mountains of Germany and close to the shores of the Black Sea (Fig. 9.32). Their margins are not necessarily the borders of huge land-based ice sheets. It is more likely that they are a left-over of prograding massive icebergs that moved from Scandinavia southwards across wide permafrost plains in flow after flow after flow. It is tempting to relate this ice movement to tectonic events:

Early Quaternary in Central and Northern Europe was marked by a slow epirogenic uplift (0.1 – 1 mm

a^{-1}). During glacials the flat relief kept areas of permanent ice restricted to Scandinavia proper and end moraines could not develop. About 400,000 years ago, and continuing towards the present, the regime changed and pulses of tectonic activity caused rapid subsidence of a basin chain extending from the Caspian Sea to the lowlands of the Po valley; rates of subsidence were as high as 1 to 2 cm a^{-1} with a maximum about 200,000 years ago (Degens and Paluska, 1979). Synchronous with the subsidence in the south was a general uplift in parts of Russia, Central Europe, and Scandinavia. For example, within 100,000 years Scandinavia became uplifted by 1 kilometer and more, while at the same time the Caspian and Black Sea basin floors sank by the same amount. This vertical motion provided the kinetic energy for massive erosion and formation of moraines in the aftermath of glacial events. The fact that the past three major ice ages coincide with a major tectonic pulse in Europe suggests that rapid tectonism leading to regional changes in land/water ratio, surface albedo, topography, orography and bathymetry may contribute to global climatic change due to the "weather-strategic" position of this area.

CO_2's Past

Ancient *Homo sapiens* is not quite 200,000 years old and the common ancestor of all modern humans is just 100,000 to 140,000 years of age (see p. 384). Thus, the physical and spiritual development of the human race, the migration of people across land and sea, and the foundation of nations all took shape between two glacials and two interglacials. By the way, *Homo habilis* and *Homo erectus* – our forefathers – came and vanished during the Pleistocene. Climate, certainly, has put its distinct stamp on humans.

Today, we witness the opposite development. A gigantic experiment has begun where humans, within geologic zero-time, may profoundly alter the natural environment by turning up the climatic thermostat. CO_2 and other greenhouse gases are definitively on the rise. Will they force – in a matter of 100 years or less – the climate into a situation which Earth experienced during the Eemian Interglacial or, worse even, the Cretaceous? Alternately, is Earth so resilient that it readily takes care of human foolishness without sweat? An adjustment to new conditions is only effective if societies have a certain room to move about. Unfortunately, climatic changes predicted for the next few decades may proceed at a rate 1,000 times faster than during previous millennia. Since we have no time to spare, let us immediately concentrate on CO_2.

To anticipate future developments of climate under conditions of anthropogenic perturbation, it is essential:

- to understand the degree of variability in the carbon cycle and the climate at different time scales,
- to elucidate the mechanisms bringing about this variability, and
- to integrate the knowledge from past and present observations into a quantitative understanding, that is a model which is then the tool for predictions of the future.

Air CO_2 content has been monitored since 1958 (Keeling et al., 1982) and has risen from 315 ppmv then to 350 ppmv now. Tree rings (Freyer and Belacy, 1983) and ice core data (Neftel et al., 1985; Friedli et al., 1986) have extended the record back into the middle of the 18th century. The basic message is that pre-industrial atmospheric CO_2 concentration is in the order of about 280 ppmv (Fig. 9.19). Raynaud and Barnola (1985) have also extracted air-CO_2 entrapped in Antarctic ice. The roughly 200-m-long core brought the CO_2 curve down to the beginning of the 14th century. An interesting feature is the close correspondence between CO_2 content and climate variability; the lowest values were found during the inferred global little ice age around the end of the 17th century (Maunder Minimum). Raynaud and Barnola report a mean CO_2 value of 260 ppmv for the time bracket 1650 to 1850 and of 270 ppmv for the preceding three centuries. Going further back in time (Lorius et al., 1985; Jouzel et al., 1987), ca. 150,000 years, that is the time envelope from the end of the Saalian glacial to the present, there is no noticeable lag between PCO_2 and climate change on time scales <100 years (Fig. 9.33). This close CO_2-climate relationship is of key importance to assess the CO_2 and climate problem. Is CO_2 forcing major climatic changes? Is climatic change forcing atmospheric CO_2 changes? Do we have a strongly interactive system with CO_2 providing a positive feedback to climatic change?

By taking the difference in CO_2 content between the peaks of climatic optimum and climatic minimum amounting to about 100 ppmv, and translating this into global air temperature, a 1°C change is the result, which is about 10 percent of the calculated temperature differential between glacial and interglacial epochs. This should answer the first question.

The second inquiry is of a more complex nature, but can still be reasonably resolved. A drop or rise in global air temperature will alter the flux of CO_2 between ocean and atmosphere. At present, low latitude oceans are a source, high latitude oceans are a sink for CO_2. According to Peng et al. (1979), at southern

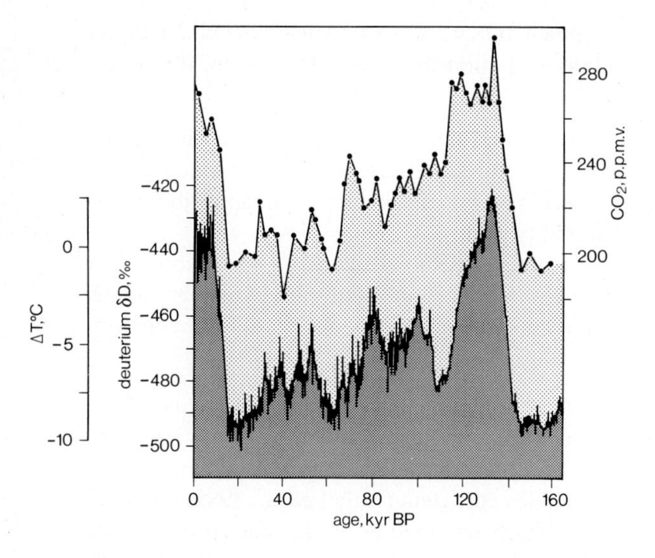

Fig. 9.33. Deuterium profile (Jouzel et al., 1987) and range in CO_2 content of air trapped in ice core (Barnola et al., 1987). Differences in PCO_2 in fossil air between glacial climatic minimum and interglacial climatic optimum are roughly 100 ppmv (180 to 280 ppmv)

latitudes higher than 40 degrees the mean exchange rate is about 50 percent above global average. CO_2 partial pressure differences between atmosphere and ocean are shown in Fig. 9.34 (Takahashi et al., 1981, 1983). Translated into a global figure it means that the natural exchange of CO_2 across the sea surface is in balance with about 390 Gt of atmospheric CO_2 entering the ocean and an equal amount of CO_2 returning to the atmosphere. For comparison, the net flux of anthropogenic CO_2 into the ocean is only of the order of 2 percent of the natural exchange, which also explains some of the difficulties associated with the CO_2 problem. A tentative answer to our original inquiry is: a cold ocean is in general a more potential sink for CO_2 than a warm ocean. However, wind pattern and deep water circulation can significantly modify the physical oceanic regime and with it introduce ocean chemistry changes (Broecker, 1982, 1987a, b; Broecker and Peng, 1982; Duplessy and Shackleton, 1985; Duplessy, 1986; Duplessy et al., 1984).

The third question is the most difficult one to address. As I have stressed throughout the preceding discussion, water and carbon dioxide are the chief molecules driving the global biogeochemical system. In my opininion there is little hope left ever to be able to understand the multifarious feedbacks of CO_2 on climate in a quantitative manner. All that is possible is an outline of general trends. These trends, however, permit a refutation of many of the existing models and limit their number to a few that can be further tested. By starting from the present situation, a rise in air temperature by 10°C will double the respiration rate. Exposing peat deposits and permafrost soils to a higher temperature regime, CO_2 and CH_4 release will accelerate. It has been estimated (Woodwell, 1984, 1986; Woodwell et al., 1987) that a temperature-enforced respiration of soil humic matter could match the CO_2 input in volume from fossil fuel burning within the next two or three decades. This may in part be compensated for by PCO_2-enhanced primary productivity. However, any rapid destabilization of climate will create biotic impoverishments, and how this might precipitate on the future CO_2 development is so far not known.

A cooling trend should work the opposite way around. Wide territories become placed under an ice cover or transform into permafrost regions. Erosion is

Fig. 9.34. Global distribution of the difference between the partial pressures of CO_2 in the atmosphere and surface ocean water. Values given in μatm. *H/L* indicate regions where the ocean is source/sink for atmospheric CO_2 (Takahashi et al., 1983)

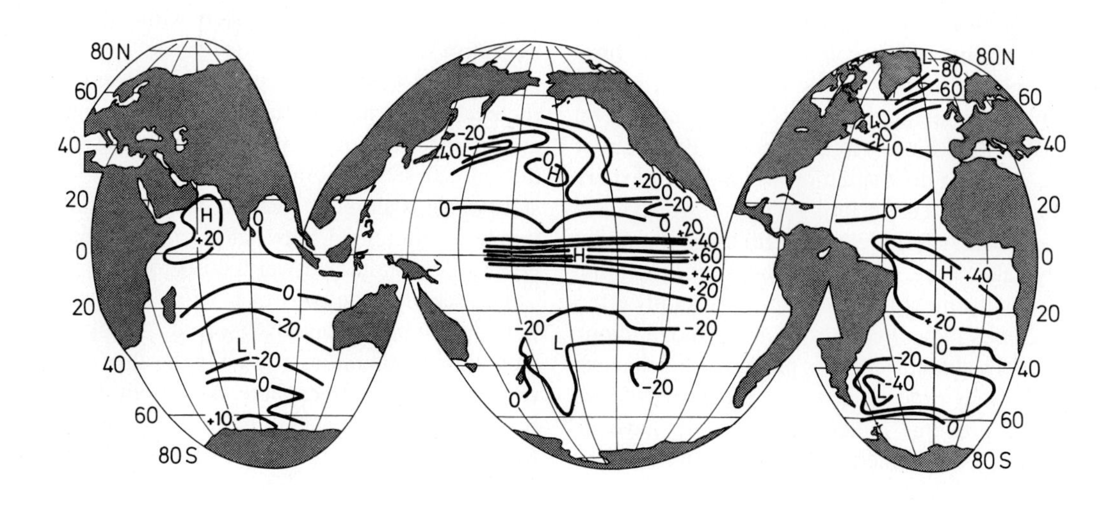

slowed down; so is respiration. Stronger upwelling along the continental margin enforces marine primary productivity. Due to the lowering of sea level, much of the organic matter produced in the euphotic zone sinks to greater water depth, where it respires in part. The deep-water circulation sets the stage for oxygen minimum zones, and organic-rich deposits (sapropels, black shales) come into existence. All listed feedbacks taken together bear responsibility for the close match between CO_2 content in air and climatic change.

On Modeling Future Climate

The farther we go back in time, the poorer becomes the resolution of factors causing climatic change, and the error bar on natural variability understandably widens (Fig. 9.35). In forecasting future climatic change, human activity has to be included as an additional variable because:

- energy use may change,
- population growth will continue or stabilize,
- worldwide strategies may be adopted to save the environment (e.g., laws, common sense) or to ruin it (e.g., event of nuclear war, human carelessness). Furthermore:
- global environmental response to human-induced climate forcing is known in direction but less in absolute extent,
- regional, local or seasonal climatic changes are difficult to assess and information is limited as to where the climatic pendulum may eventually swing, and
- the relationship between human impact and the time the environment needs to respond with a distinct climate signal are still major uncertainty factors.

This all suggests that the data bank of climate past and present offers only some help to define a realistic climate model for the decades ahead of us. Most significant is the close relationship between the amount of CO_2 in fossil air and global air temperature at that time. In essence: the sum total of climatic change generated by the air-sea-earth-life system is mirrored in atmospheric PCO_2. In addition, today's global carbon cycle is reasonably understood in terms of sinks, sources, and fluxes of CO_2 (Bolin et al., 1979; Degens et al., 1984; Sundquist and Broecker, 1985; Trabalka, 1985; Trabalka and Reichle, 1986). This in turn, gives us some idea as to the carrying capacity of the ocean and the biota for excess CO_2, and the inertia of the environment to human-induced change.

Fig. 9.36 is a scheme of how to model climate. The first requirement is a clear definition of the problem. The resulting concept is calibrated against available observations. Where validation is good, a forecast can be made, and if not, the concept has to be revised. Understanding the operation principles of today's carbon cycle, and knowing the CO_2/air temperature relationships of past millennia, will lead to a model on climatic change for the Quaternary at a resolution of a hundred to a few hundred years. Modern climate contains essentially the same elements, plus human forcings.

Fossil fuel combustion, forest clearing, and changing land use account for the bulk of anthropogenic

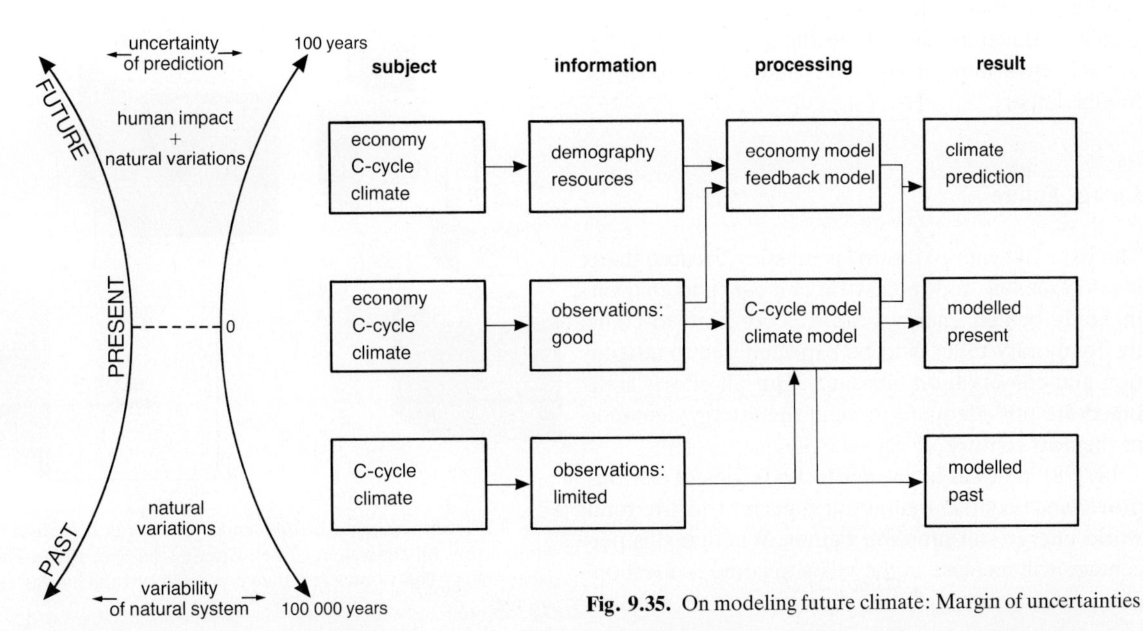

Fig. 9.35. On modeling future climate: Margin of uncertainties

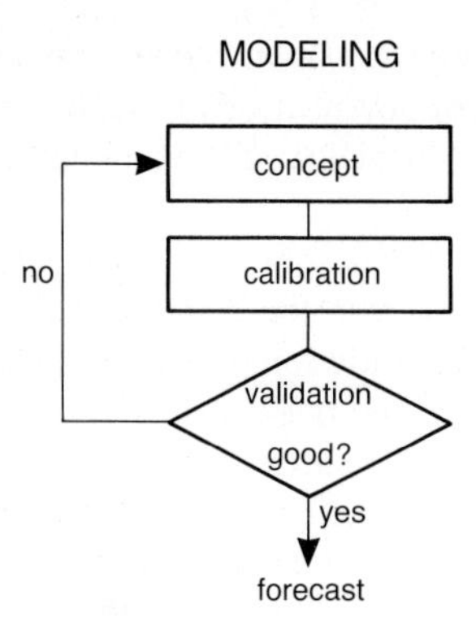

Fig. 9.36. How to model climate

CO_2 emitted into the air. Only the history of CO_2 release by the burning of coal, oil and gas is accurately known. How much the other two sources contribute may be established through modeling. That is, models are tuned to reproduce the Mauna Loa record, and the post-1850 A.D. tree-ring and ice-core data by independent variation of: (i) deforestation and agricultural activities, and (ii) the ocean's uptake capacity for human-made CO_2. The result is a tentative answer as to how much anthropogenic CO_2 is emitted, what are its sources, and what are its terminal sinks.

To perpetuate the CO_2 model far into the next century is not advisable for three reasons: (i) future energy use may differ, (ii) growth rate of world population may change, and (iii) climate forcings on ocean, soil, precipitation, or biota may vary. Yet it is not only CO_2 but additional heat that has to be sequestered by the environment. Thus, climate prediction requires an economy model, and a revised C-cycle and climate model which incorporates transient response functions describing the adjustment of the overall environment to the changing pattern of human impact.

Energy Future

The issue of "energy future" is pressing because there is a foreseeable end to easy oil and gas, and environmentally benign energy resources are hard to come by. Temporary relief is to be expected due to adaptation and conservation measures. But all efforts combined are not adequate to meet the energy demands of the next century.

In 1981 renewable energy technologies of various provenance contributed about 8 percent to the total world energy consumption figures. Whether this percentage will increase in the years to come is questionable, because even "safe" energy such as wind or water may be environmentally hazardous when used on a much wider scale than at present. In order to illustrate from what sources we presently obtain our energy, the consumption figures for the year 1981, expressed in millions of metric tons oil equivalent, are shown in Fig. 9.37. Added to the picture is the CO_2 emission rate for that year. The eminent role of oil, coal and gas in providing global energy comes to light. In contrast, nuclear energy supplies just a few percent of the total, but we have to bear in mind that up to this very moment 26 countries receive much of their electricity from nuclear power plants. In 1986, close to 15 percent of global electric energy was of nuclear origin, which is still less than 5 percent of the 1986 world energy consumption figures. Indeed, a switch from fossil fuel to nuclear energy would represent a formidable task. It is not so much the shortage of

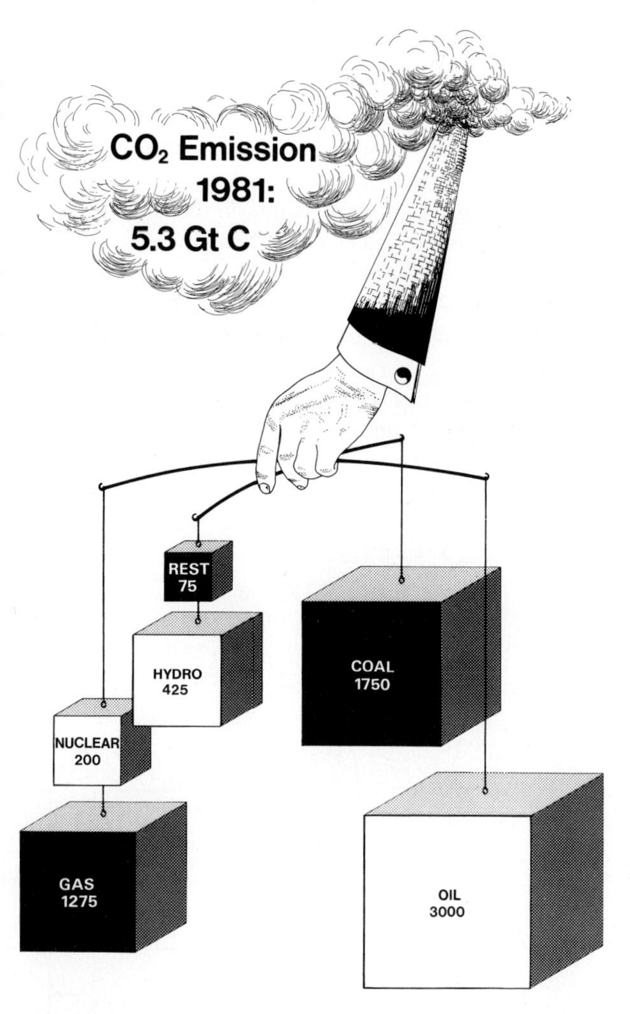

Fig. 9.37. CO_2 emission in 1981 and used energy resources in million tons oil equivalents (Shell Briefing Service, 1983). Nuclear energy contributes less than 5 percent of total energy resources

uranium, but the ten- to twenty-fold increase in nuclear reactor capacity that generates a problem.

Known reserves for various types of fossil fuels in billions of metric tons oil equivalent (Shell, Briefing Service, 1983) are depicted in Fig. 9.38. The simple message we can derive is that at present consumption figures, oil will last for 30, natural gas for 50, and coal for 250 years. Oil sands and shale oil are just a drop in the ocean. To speculate that exploration activities might lead to the discovery of new oil and gas fields, or of coal deposits that could supplement the losses, is ill-advised. New oil extraction technologies, or the exploitation of smaller fields, or going deeper into the Earth may slightly move the deadline by a few decades into the future. More is currently not in sight. By the middle of the next century, no oil or gas will be left if combustion continues at present rates, provided our assumptions are correct. Shifting to coal by that time to fill the energy gap would exhaust global reserves rather quickly.

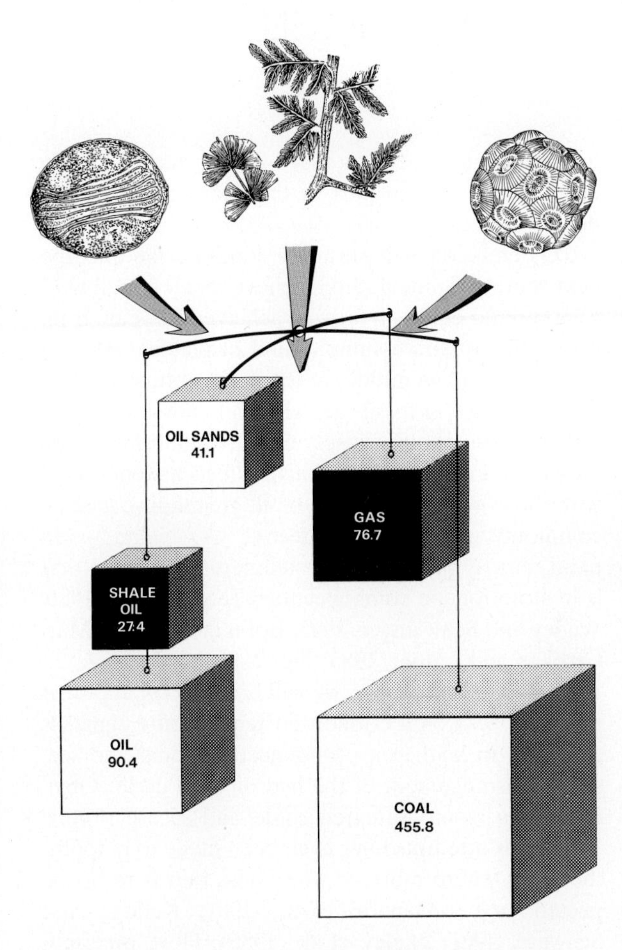

Fig. 9.38. Energy resources of fossil fuel in billion tons oil equivalents (Shell Briefing Service, 1983)

Most coal basins are located in the industrialized world and in mainland China. It is hard to imagine from where the rest of the people would get energy once the conventional resources phase out or decline and the world population continues to grow. To tap the sun could be the only long-term solution.

Feasibility studies on solar energy look promising, but many decades will pass before such a source can penetrate the market. To continue meanwhile with the present mode of energy generation and adjust to climatic change is certainly not the best of propositions. A "resilient" energy system has been advocated (Laurmann, 1986) to deal with an uncertain future. A resilient system has the advantage that it can be changed rapidly in response to new findings on energy-induced environmental damage in general and climatic change in particular. Recommended energy policies may, in principle, involve: (i) reduction in emission, (ii) CO_2 scrubbing using improved technologies (cost-reduction), and (iii) replacement of fossil fuels for renewable resources; but sufficient time should be allowed to go through this process.

Back to the Greenhouse Gases

Most models predict that a doubling of CO_2 would lead to a global warming of 1.5 to 3.5°C, with almost negligible warming in the tropics and a 7–8°C warming in polar regions. In the stratosphere the reverse is expected: slight cooling in the polar and more pronounced cooling in the tropical stratosphere.

Two positive feedbacks contribute to these results. First, the ice-albedo feedback enhances the warming at high latitudes (Budyko, 1969; Sellers, 1969). Second, the water-vapor feedback enhances warming, because CO_2-warming increases atmospheric water vapor, which then itself exerts an additional greenhouse effect (Ramanathan, 1981). Most models (e.g., Manabe and Wetherald, 1980; Washington and Meehl, 1983) compute the fast-reacting atmosphere's equilibrium response to a fixed CO_2 increase, treating the ocean merely as an evaporating swamp. However, CO_2 is increasing continuously. In addition, the transient and highly non-linear interactions between atmosphere and the more slowly varying parts of the climate system, that is the ocean and the cryosphere, are the essential characteristics of the climate dynamics (Grassl et al., 1984). Because of thermal inertia (Hasselmann, 1982) it may take the ocean several decades to reach equilibrium, so that at any given time the global-mean atmospheric temperature lags behind the corresponding equilibrium temperature (Wigley, 1985; Wigley and Schlesinger, 1985; Wigley et al., 1980; Gornitz et al., 1982).

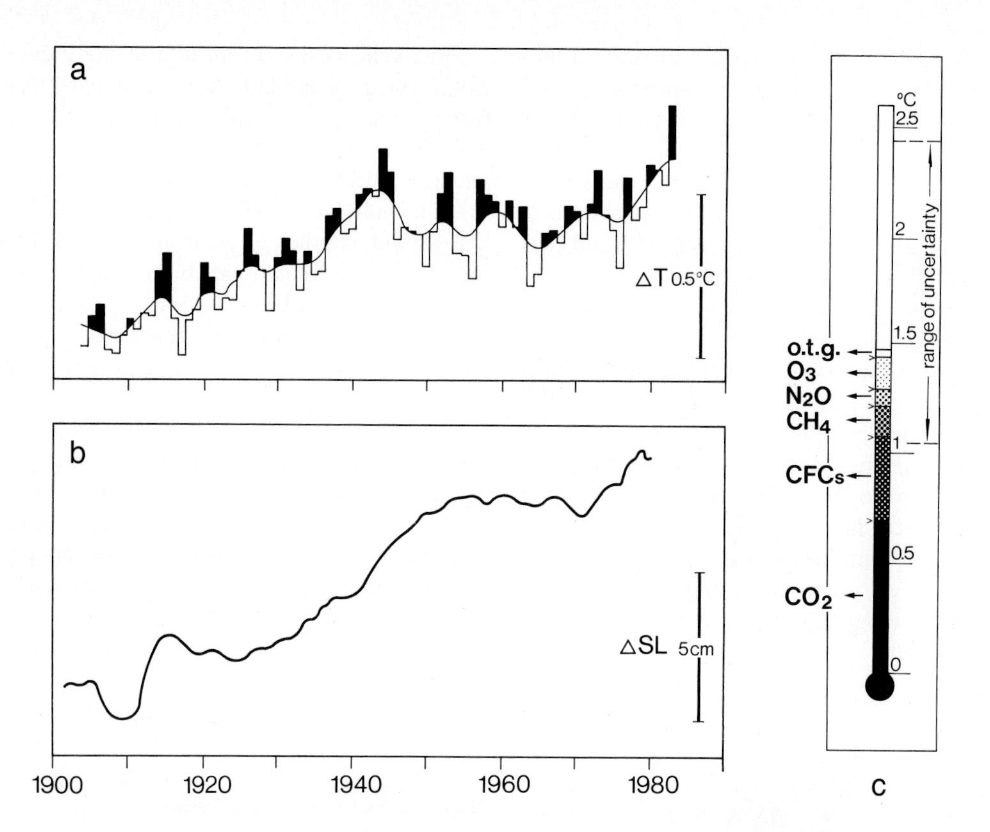

Fig. 9.39. **a** Global mean annual surface temperature changes (Wigley et al., 1986). Smooth curve shows 10-year Gaussian filtered values. **b** Global mean sea level changes since 1900 A.D. (Gornitz et al., 1982). **c** Cumulative equilibrium surface temperature warming due to all greenhouse gases for the time period 1980 to 2030 A.D. (Ramanathan et al., 1985)

Surface temperature and sea level have increased since the beginning of this century. Compared to 100 years ago, global mean air temperature has risen by about half a degree centigrade and sea level by 10 to 15 centimeters (Fig. 9.39). The rapid global warming in the years 1920 to 1940 and the cooling trend between 1940 and the mid-1960s could easily be part of natural climatic change. To ascribe these changes to human produced CO_2 is far-fetched, the more so since we know that not only CO_2 but a series of other radiatively active trace gases contribute to the greenhouse scenario, and their combined effect in the years 1960 to 1985 is about equal to that of CO_2. Ramanathan et al. (1985, 1987) have computed the cumulative equilibrium surface temperature warming due to all greenhouse gases by a one-dimensional model for the time period 1980 to 2030. A thermometer-plot (Fig. 9.39) depicts the suggested temperature contributions from the various gases. As far as rise in sea level is concerned most estimates for the next hundred years range between 0.3 and 1.6 m. Higher figures (up to 6 m) are based on the most unlikely event that a certain part of the Antarctic ice shield breaks off.

CO_2 emission will certainly continue far into the next century. Some authors believe that a doubling (2 x 280 ppmv) could already be achieved 50 years from now, while others assume a kind of plateau value of 450 ppmv for the middle of the next century. By that time, shortage in fossil fuel will limit emissions.

Other greenhouse gases will also continue to increase, and when their radiative effects are combined with those of CO_2, the sum of all greenhouse gases – commonly expressed as effective CO_2 value – will paint no rosy picture for the temperature regime that is in store for the coming century (e.g., Wigley, 1985; Wigley and Schlesinger, 1985; Bolin et al., 1986; MacCracken and Luther, 1985; Fig. 9.40).

Against this scenario, we will briefly look at potential feedbacks. A CO_2-induced temperature signal is expected to lead to a pronounced regional and seasonal rearrangement of the hydrological cycle. Obviously this would threaten some and please others. Numerous attempts have even been made to pinpoint the areas where more or where less rain is to be expected (e.g., Manabe et al., 1981; Kellogg and Schware, 1981; Wigley et al., 1986). How, precisely, such a regional pattern might evolve, cannot be credibly predicted by the models available at present.

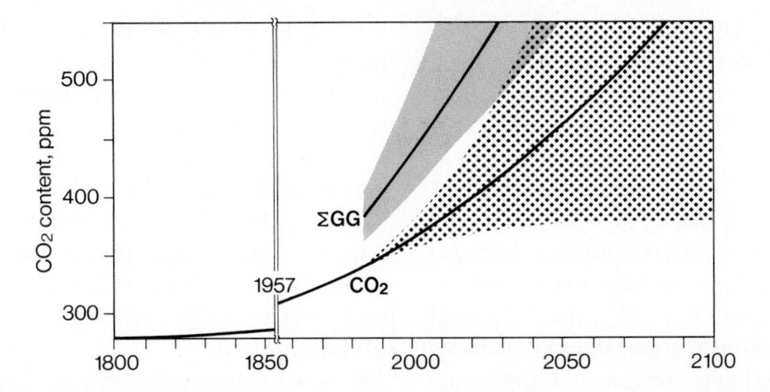

Fig. 9.40. Future development of (i) CO_2 content in air, and (ii) the sum of all greenhouse gases (in effective CO_2 values). Range of uncertainty is indicated (various sources)

Let us now turn to the global aspects and look at the clouds and the ocean. Work by Schneider (1972, 1983) has shown that a decrease in global cloud cover by 2 percent lowers global mean temperature by 2°C. Yet, cloud feedback is a key unknown in climate modeling, because the type of clouds may change, becoming wetter or drier, higher or lower, etc.

Initial appraisal of cloud feedback on climatic change has been made by Grassl (1981). Inspired by a theoretical approach proposed by Paldridge (1975), he designed a thermodynamic model in which entropy is maximized by meridional heat flow in atmosphere and ocean. Variable cloudiness is implicitly accounted for. The extremum condition by which entropy is maximized is supported by observation (Mobbs, 1982). A CO_2-doubling experiment revealed no particular sensitivity of polar temperatures but slight warmings in the tropics which, interestingly, have been noted in recent years. These surprising results appear as the combined effect of less cloudiness and weakened poleward transport of heat. On the other hand, cloud amount records from Europe and the United States indicate an increase during all seasons (Henderson-Sellers, 1986). What counts, in the final analysis, would be a prediction of cloud type in addition to total cloudiness, because the radiative properties of a cloud cover depend on cloud height, aerosol content, size of droplets etc. This has recently been done (Roeckner et al., 1987). Accordingly, cloud-albedo feedback should lead to a cooling trend that may buffer the warming trend expected from the increase in greenhouse gases.

In this context, a study by Choudhury and Kukla (1979) is of interest. They proposed that the weak absorption of CO_2 between 1 and 3 μm wavelength may lead to a weakening of the radiative energy necessary for the onset of snow melt. In high polar latitudes this may prolong the period of high snow reflectance, equivalent to a cooling, especially in early summer. The cooling effect of cloud- and ice-albedo feedbacks above the poles can be considered as one of the more potential countermeasures of nature against the warming effect of the greenhouse gases.

What about the ocean? Planktonic and benthic foraminifera that lived in the open sea during the late Pleistocene contain in their calcium carbonate tests isotope signals that are helpful to reconstruct direction and speed of deep water circulation at that time (Duplessy et al., 1984; Duplessy, 1986; Shackleton et al., 1983; Boyle and Keigwin, 1985; Siegenthaler and Wenk, 1984). Variations of the temperatures of the deep water in the North Atlantic are a proxy indicator of the deep water circulation changes. Interestingly, they correlate closely with the CO_2 record over the last 150,000 years, suggesting that the global ocean circulation has been responsible for natural variations of the CO_2 cycle. Two outstanding features emerge: (i) a much steeper slope (from Equator to North Pole, for example) for the advective transport, and (ii) indication of a much reduced speed in turnover compared to the present (about half?).

New insight into deep water circulation in the sea under conditions of doubling, quadrupling etc. of atmospheric CO_2 content has been gained through the work of Bryan and Spelman (1985) and Manabe and Bryan (1985). Without mentioning here the complex details, the crux of the matter is the following: oceans can retard surface air temperature increase considerably, and – that is new – it buffers by transferring any increment of heat to greater depth where heat dissipates without ever returning to the surface. For instance, under the conditions of a four times higher air CO_2 concentration, the deep water pole-equator convection cell moves to greater water depth, develops a flat circulation component for the advective transport, and changes direction of flow regime from south-north – that is the present situation – to north-south. This is in part due to a reduction of downwelling in the polar region. Owing to the influence of sea ice upon thermohaline circulation, the poleward heat transfer by ocean currents in a glacial ocean is very small, making the polar regions even cooler than they already are.

The most important message we can deduce from these simulation studies is that an increase of CO_2 up to eight-fold will only moderately influence the zonal mean air temperature relative to the present, whereas in a situation of half the modern CO_2 value (= glacial time), a dramatic change in heat transfer in the sea appears to be the case. Actually, the glacial scenario supports the results from the above-mentioned isotope studies of Pleistocene foraminifera. One can also infer that present climatic zones will not be shifted polewards following a substantial rise in CO_2 levels. It may help to understand the thermohaline circulation flip-flop in the ocean by realizing that the total heat contained in our air just about equals the global heat content of the upper 3 meters of sea water.

Other important negative feedbacks may have been omitted, but to be on the safe side, the worst case scenarios have to be taken seriously, until revision renders them obsolete. Our group has previously reviewed the issue of the biogeochemical cycling of carbon dioxide in the past, present and future and its impact on climatic change (Degens et al., 1984). It may be a fitting moment to close this chapter by quoting the last sentence from this paper: "We like to emphasize what the modeling community is well aware of: a model is a model is a model! Consequently it is up for revision at any time, because ingrained in a good model is – following Karl Popper – its falsification."

On Planetary Atmospheres

Mass, chemical composition, vertical structure, and general global circulation characterize the atmosphere of planets and a few moons in our solar system. Heat to carry on motion is principally provided by the sun, but internal heat sources can supplement radiant energy. For instance, Jupiter and Saturn emit far more energy than they get from the sun. Gravitational contraction accounts for the excess energy in the case of Jupiter, while for Saturn it appears to be related to precipitation of helium at the top of the metallic hydrogen zone and subsequent gravitational settling (Stevenson and Salpeter, 1977). Helium depletion in Saturn's atmosphere supports this inference. Or take the spectacular volcanic plumes of Io, a rocky moon close to Jupiter, which extend to an altitude of 300 km. Tidal forcing is the most plausible internal heat source behind Io's volcanism and tectonism.

The heat budget of an atmosphere depends primarily on: (i) distance of planetary body from the Sun, (ii) surface albedo, and (iii) internal heat flow. Over a lifetime of 5 billion years, the sun has changed in luminosity by roughly 30 percent and is emitting today less UV radiation than during initial phases. In consequence, weather, climate, and chemistry have been subjected to an evolving sun, and been tuned accordingly.

Mass, structure and chemistry of an atmosphere determine how much of the heat is retained by a parent body. Greenhouse gases and aerosols modulate weather and climate, and produce diurnal or seasonal effects. Many atmospheric gases and particles can be generated or destroyed in response to radiation. Consequently, the vertical structure may change and clouds of various compositions may originate. Take the atmospheric profiles of Jupiter and Saturn (Fig. 9.41). They exhibit tropospheric temperature structures close to the adiabatic profile. Since both planets have atmospheres of near-solar composition, temperature variations – when related to the saturation vapor pressure – suggest the formation of clouds of ammonia (NH_3), ammonium hydrosulfide (NH_4SH), and ice (H_2O) at descending levels in the troposphere (Weidenschilling and Lewis, 1973). Variations in cloud structures will locally change the temperature regime.

Mass and chemical composition of modern planetary atmospheres, including those of Moon, Titan, and Io, are known with a reasonable accuracy (Table 9.6; Pollack and Young, 1980). Models on atmospheric circulation (e.g., G. P. Williams, 1979; Ingersoll and Pollard, 1982) have helped to understand the dynamics of zonal or vertical air motions, particularly in those instances where time-series photographs from space missions could be made available. For example, rapid changes in the position of bright coloration bands surrounding the Great Red Spot on Jupiter, or interactions among clouds and zonal regimes testify to the dramatic Iovian weather pattern and conform with existing models.

Outer Planets

Atmospheres on the Giant Planets: Jupiter, Saturn, Uranus, Neptune look at first approximation chemically similar. Most prominent are hydrogen and helium which, except for helium-depleted Saturn, are consistent with the cosmic abundance. Water should follow next as the most prominent molecule of the

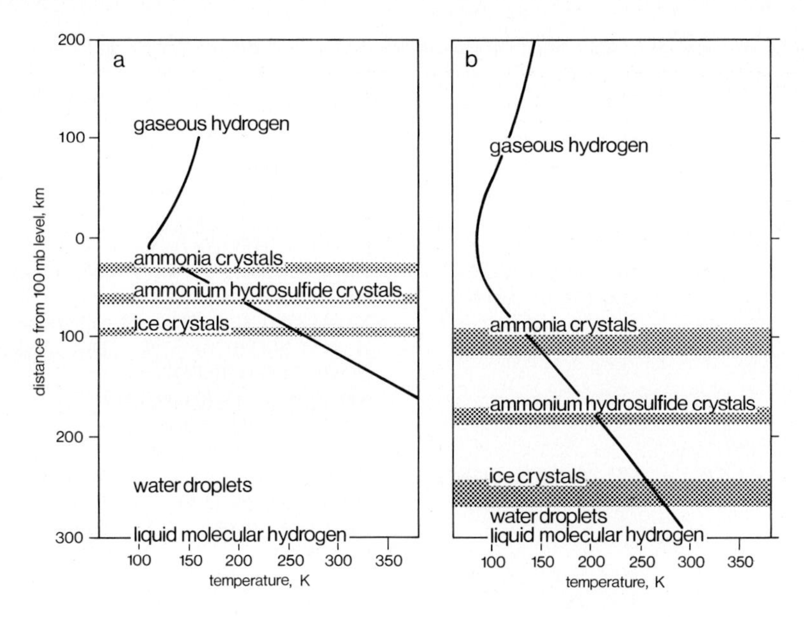

Fig. 9.41 a, b. Shown are atmospheric temperature profiles for Jupiter (**a**) and Saturn (**b**) which are consistent with the available observations. The level of the cloud layers expected to form in these atmospheres is indicated (Hunt, 1983)

solar system. However, water content in the Jupiter atmosphere is small because of cloud formation, and for the other three planets undetectable due to the very low vapor pressures in the prevailing temperature regime. Ammonia abundances on Jupiter and Saturn are modified too because of cloud formation and photochemical reactions. It should be pointed out that no ammonia has been detected on Uranus which could be related to a 100-fold ammonia depletion relative to cosmic abundance or its incorporation as NH_4SH clouds (Weidenschilling and Lewis, 1973). Methane is present in cosmic abundance on Jupiter and Saturn and appears to be somewhat enriched on Uranus and Neptune relative to the cosmic level. A methane atmosphere has been reported from Pluto (Fink et al., 1980).

One moon each of Saturn and Jupiter – Titan and Io, respectively – possesses an atmosphere which has been generated in situ by the satellites, rather than being a leftover of the primordial gas. Titan is the largest of the Saturnian moons with a substantial atmosphere which is denser than any terrestrial atmosphere except on Venus (Hunt, 1983). Its air is principally composed of nitrogen (Broadfoot et al., 1981) which makes Titan and Earth unique in this respect. However, the vertical temperature profile across a pressure range from 1 to 1,600 mbar is about 100 degrees (70 to 170 K; Maguire et al., 1981), and thus far too low to generate life as we know it. Even though life is an unlikely "commodity" on Titan, the presence of methane, its derivatives, and a wide range of other C-H-O-N gases which jointly comprise about 1 percent of the total atmosphere (e.g., Trafton, 1981; Kunde et al., 1981; Hunt, 1983), is by itself a treasure of unbelievable significance. It helps us to define a realistic primordial abiotic system from which key biochemical molecules and life on earth might have arisen. Organic compounds originate by reactions of methane and nitrogen radicals in a predominantly nitrogen atmosphere. Lack of a magnetic field and the occasional residence of Titan outside the Saturnian magnetosphere leave Titan unprotected against radiation, and cause energetic electrons and protons to ionize and dissociate certain atmospheric gases. Galactic cosmic radiation, and charged particles from Saturn's magnetosphere, could also be a contributing factor in changing atmospheric chemistry. Three-body reactions may lead to complex neutral hydrocarbons.

The discovery of HCN and CO_2 is of considerable interest, but for different reasons. This is the first time HCN has been found in a planetary atmosphere, and HCN is considered to be a convenient precursor molecule for the synthesis of amino acids and the bases of the purines and pyrimidines present in proteins and the nucleic acids respectively. Work by Khare et al. (1986) on the synthesis of organic heteropolymers by dc discharge under conditions supposed to prevail at the cloudtop atmosphere of Titan resulted in an organic residue which contained "non-biological" amino acids. This is the first hint that organic synthesis in a Titan atmospheric environment is feasible. Since HCN is quite reactive, its trace amounts (2×10^{-7} mole fraction) could signal that it is

Table 9.6. Properties of the atmospheres of solar system objects (For source references see Pollack and Young, 1980)

Object	P_s[a]	T_s[a]	Major gases[a]	Minor gases[a]	Aerosols[a]
Mercury	$\sim 2 \times 10^{-15}$	440	He (~0.98), H (~0.02)[b]		
Venus	90	730 (~230)	CO_2 (0.96), N_2 (~0.035)	H_2O (20–5000), SO_2 (~150), Ar (20–200), Ne (4–20), CO (50)[c], HCl (0.4)[c], HF (0.01)[c]	Sulfuric acid (~35)
Earth	1	288 (~255)	N_2 (0.77), O_2 (0.21), H_2O (~0.01), Ar (0.0093)	CO_2 (345), Ne (18), He (5.2), Kr (1.1), Xe (0.087), CH_4 (1.5), H_2 (0.5), N_2O (0.3), CO (0.12), NH_3 (0.01), NO_2 (0.001), SO_2 (0.0002), H_2S (0.0002), O_3 (~0.4)	Water (~5) Sulfuric acid (~0.01–0.1)[d] Sulfate, sea salt Dust, organic (~0.1)[d]
Mars	0.007	218 (~212)	CO_2 (0.95), N_2 (0.027), Ar (0.016)	O_2 (1300), CO (700), H_2O (~300), Ne (2.5), Kr (0.3), Xe (0.08), O_3 (~0.1)	Water ice (~1)[e] Dust (~0.1–10)[e] CO_2 ice (?)[e]
Moon	$\sim 2 \times 10^{-14}$	274	Ne (~0.4), Ar (~0.4), He (~0.2)	—	—
Jupiter	>>100	(129)	H_2 (~0.89), He (~0.11)	HD (20), CH_4 (~2000), NH_3 (~200), H_2O (1?), C_2H_6 (~5)[f], CO (0.002), GeH_4 (0.0007), HCN (0.1), C_2H_2 (~0.02)[f], PH_3 (0.4)	Stratospheric "smog" (~0.1) Ammonia ice (~1) Ammonium hydrosulfide (~1) Water (~10)
Saturn	>>100	(97)	H_2 (~0.89), He (~0.11)	CH_4 (~3000), NH_3 (~200), C_2H_6 (~2)[f]	Same aerosol layers as for Jupiter
Uranus	>>100	(58)	H_2 (~0.89), He (~0.11)	CH_4	Same aerosol layers as for Jupiter, but thinner smog layer plus possibly methane ice
Neptune	>>100	(56)	H_2 (~0.89), He (~0.11)	CH_4	Same aerosol layers as for Jupiter, plus possibly methane ice
Titan	$2 \times 10^{-2} \rightarrow \sim 1$	~85	CH_4 (0.1–1)	C_2H_6 (~2)	Stratospheric "smog" (~10)
Io	$\sim 1 \times 10^{-10}$	~110	SO_2 (~1)	—	—

[a] Reading from left to right, these variables are the following: surface pressure, in bars; surface temperature, in K – the numbers in parentheses are values of effective temperature; major gas species – the numbers in parentheses are volume mixing ratios; minor gas species – the numbers in parentheses are fractional abundance by number in units of ppmv; aerosol species – the numbers in parentheses are typical values of the aerosols' optical depth in the visible;
[b] these mixing ratios refer to typical values at the surface;
[c] these mixing ratios pertain to the region above the cloudtops;
[d] the sulfuric acid aerosols reside in the lower stratosphere, while the sulfate, etc. aerosols are found in the troposphere, especially in the bottom boundary layer;
[e] the ice clouds are found preferentially above the winter polar regions. Dust particles are present over the entire globe;
[f] these mixing ratios pertain to the stratosphere.

rapidly extracted from the atmosphere in the form of biochemical molecules, which – because of the cold – could not evolve further into cellular systems. The presence of carbon dioxide requires a source for oxygen, which could be provided from impacting meteorites, water photolysis, or the degassing of chlathrates (for details see Hunt, 1983). In this context mention should be made of the gas phase synthesis of organophosphorus compounds in atmospheres of Giant Planets (Bossard et al., 1986). This opens up new vistas concerning the genesis of phosphorus-containing biomolecules such as the phospholipids, adenosine triphosphate, or the nucleic acids.

As to the origin of Titan's atmosphere, it could have been captured in part or totally from the proto-Saturnian nebula. More likely, however, it was generated from crust and mantle by episodic melting events. At the surface of Titan temperatures are slightly higher (94 K) than the effective temperature (86 K) which argues for the presence of a small greenhouse effect.

Io, a close neighbor of Jupiter, has a tenuous atmosphere of sulfur dioxide, which during daytime has a pressure about a tenth of a microbar, which is equal to a ten-millionth of atmospheric pressure at sea level on the Earth (Soderblom, 1980). Volcanic emissions are the source.

Mercury and Moon have an even more tenuous atmosphere, which stems from solar wind. Helium, and to a minor extent hydrogen, is present in the atmosphere of Mercury, whereas our Moon has a few additional noble gases (Table 9.6).

Giant Planets have quasi retained most of their atmospheric gases in cosmic abundance and modified in accordance with pressure and temperature conditions prevailing at the respective parent body. Consequently, mass, chemistry and dynamics of their present atmospheres are quasi related to the planet's size, its position relative to the Sun, and the available internal heat resources. In contrast, the Inner Planets lost their primordial gases during the T Tauri event, and built up a new atmosphere ever since that incident.

Inner Planets

The air on Venus, Earth, and Mars is of complex origin and not simply a degassing product of crust and mantle. Billions of years ago all three planets must have had water galore, but only Earth has preserved it in an ocean. On Venus water has practically gone, and on Mars it is kept in a "deep freeze". Moreover, the presence of life is in some mysterious way linked to the origin and evolution of air and water here on Earth, while in turn life has profoundly altered the chemistry of these two spheres according to its needs. However, the trace quantities of noble gases – for Earth they are about 1 percent of the total atmosphere – are in a true sense "inert", whatever happens to the cycling of elements in aquatic or living system. Their presence in the atmospheres of Venus, Earth, and Mars can be used in the same fashion as fossils in a sediment which tell us something about evolution, ecosystems, or environment of deposition. Since the term chemical or biochemical fossils is already reserved for organic molecules that have preserved their molecular structure through the ages, one may call noble gases *gas fossils* to emphasize their significance in tracing the history of atmospheres and lithospheres.

Noble gases have distinct identities. They can be composed of different isotopic species of radiogenic or non-radiogenic provenance, and may even form bonds with minerals, as appears to be the case with xenon. They can be light, helium-3, or heavy, xenon-136. They can be "frozen" in minerals or clathrates, or be "free" in air. All these properties, combined with their inertness, makes them ideal gas fossils for events or processes that took shape: (i) in the gaseous solar nebula, (ii) during accretion of the proto-planet, (iii) shortly after the T Tauri incident, (iv) upon impact of comets and meteorites, or (v) during outgassing of crust and mantle in the course of a planet's lifetime.

From the wide range of possible case studies showing the usefulness of gas fossils, I will briefly give two examples just to highlight the topic. I enjoyed reading the books and reviews by Walker (1977, 1982), Pollack and Black (1982), Cameron (1983), Henderson-Sellers (1983), Holland (1984), and Budyko et al. (1987) and you may follow suit, time permitting of course.

Earth accreted in the veil of a gaseous nebula. At the beginning of proto-Planet formation, the nebula must have had a bulk chemical composition close to cosmic abundance, but with the release of volatiles from accreting small solid bodies the chemistry of the primary atmosphere changed accordingly. Even at that time, air was held by gravity on the rotating proto-Earth and an archaic global atmospheric circulation must have led to physical stratification and chemical differentiation of the atmosphere. How much and what kind of chemicals were released from the accreting Earth is difficult to judge. It is also a matter of speculation what happened to these compounds once in the atmosphere. It all depends what kind of temperature regime one envisions as prevailing at the surface of the accreting body and how hot really was the atmosphere and the crust beneath it?

I make the heuristic assumption that the T Tauri event blew away that composite of primordial gases plus the volatiles that were discharged up to this very moment from the proto-Earth. However, it is quite likely that some of the heavy gases "survived" this incident and became a kind of dowry to the newly developing secondary atmosphere which evolved in the course of more than 4 billion years to become the air we presently breathe.

The volatile inventory during accretion and afterwards is listed in Table 9.7 (Turekian and Clark, 1975; Ringwood, 1979; Walker, 1982; Fegley et al., 1986). According to their interpretations, Earth lost most of its light elements – hydrogen, carbon, nitrogen – but retained much of its noble gases contained in the primitive solar nebula – neon, argon-36, krypton,

Table 9.7. Volatile inventories on the Earth (g atom) (After Turekian and Clark, 1975, and Ringwood, 1979)

	Accreted (a)	Preserved (b)	Preservation factor (b/a)
H	2×10^{25}	3.6×10^{23}	0.018
C	2.5×10^{24}	2.3×10^{22}	0.009
N	2×10^{23}	1.0×10^{21}	0.005
Ne	9×10^{15}	3.3×10^{15}	0.4
^{36}Ar	3×10^{16}	5.7×10^{15}	0.2
Kr	4×10^{14}	2.1×10^{14}	0.5
Xe	3×10^{14}	1.7×10^{14}	0.6

xenon. Since the other terrestrial planets were subjected to "the same procedure", a comparison of noble gas content between Venus, Mars, and Earth can be quite revealing.

Because of their inertness, noble gases, with the possible exception of xenon, have no affinity to crustal rocks; actually, xenon may form weak bonds with some minerals. The lightest of them, helium, can readily dissipate to outer space, whilst the others – except for their partial loss during the T Tauri event – have no chance to escape. Thus, modern helium content in the air of terrestrial planets is a measure of helium's residence time. All the others have conservatively accumulated in the atmosphere over the entire history of the planets (Pollack and Black, 1982; Pollack, 1984; Cameron, 1983).

Heavy noble gases are partly a leftover of the primary atmosphere, and partly a degassing product of the parent body (Schubert and Covey, 1981; Anders and Owen, 1977). Cameron (1983) pointed out that collisions of planetesimals could result in atmospheric erosion which simply means a loss of atmosphere. It is, of course, difficult to quantify such impacts on atmospheric chemistry, except that the results for Mars, Venus, and Earth should be different. Alternately, the several orders of magnitude decrease in the atmospheric abundance of neon-26 and argon-36 per gram of planet from Venus to Earth to Mars, is commonly interpreted by the grain-accretion hypothesis (Pollack and Black, 1979), which essentially implies that planetary atmospheres steadily gain components. It is noteworthy that the ^{20}Ne/^{36}Ar ratio in all of them is nearly the same and comparable to that found in the more primitive carbonaceous chondrites (Pollack and Black, 1982). Whatever viewpoint is correct, non-radiogenic noble gases in the atmospheres of terrestrial planets are a potential source for unraveling many of the mysteries still attached to the early history of Venus, Mars and Earth (e.g., Lewis and Prinn, 1984).

Intercomparison of radiogenic noble gases present in the air of Venus, Earth, and Mars provide valuable information on the timing of outgassing and, by inference, thermal and tectonic histories. I will only briefly mention the case of argon-40.

As can be inferred from Table 9.8, both Venus and Mars are highly depleted in argon-40 relative to Earth. According to Hart et al. (1979) and Stacey (1980), between 25 and 75 percent of the ^{40}Ar generated from ^{40}K over the Earth's lifetime is now present in the air. As a first approximation one can say that crust and upper mantle have been flushed free of almost all of their former ^{40}Ar, while the unaccounted rest – about 50 percent or so – is still contained in the deeper mantle. At present, Earth is supposedly releasing only 7 percent of its time-averaged value; for details see Pollack and Black (1982).

Venus and Earth, being comparable in size, are a priori considered to have started outgassing at similar rates and involving about the same gaseous species. The factor of 4 deficiency in the absolute abundance of ^{40}Ar in Venus' atmosphere has the most likely explanation that Venus stopped releasing ^{40}Ar at some time in the past due to the cessation of global tectonics. It has been tentatively suggested that Earth and Venus had comparable outgassing histories for at most their first billion years, after which outgassing for Venus slowed down and eventually ceased, whereas Earth continued to be active even though at a much reduced rate relative to its formative years. For Mars, part of the factor of 16 deficiency in the ^{40}Ar abundance is accounted for by a smaller bulk endowment of potassium for Mars. A lesser fractional volume of Mars that has been outgassed and a shorter period of global degassing may explain the remaining part of the depletion. It has been proposed that the duration of global outgassing of Mars was restricted to at most its first 1.5 billion years, which does not say anything against small-scale and occasional volcanism.

Average atmospheric pressure at the surface of Mars, Earth, and Venus are: 7, 1,000, and 90,000 mbar, which represents about a 100-fold increase by going from Mars to Earth and from Earth to Venus. Mars and Venus have CO_2 and N_2 as major gases in a ratio of close to 95 to 3. Nitrogen is also the predominant gas in the terrestrial atmosphere representing 78 percent of the total. Carbon and nitrogen are highly correlated in some carbonaceous chondrites, for example, the Allende CC3V meteorite (Ott et al., 1981). Thus, N_2 and CO_2 could have its primary source in certain carbonaceous chondrites, but they could also stem from impacting comets which are

Table 9.8. Ratios of absolute abundances (g/g) between planets (After Pollack and Black, 1982)

Gas species	Venus/Earth	Earth/Mars
^{20}Ne	21(±5)	200(80–500)
^{36}Ar	72(±10)	165(±45)
^{84}Kr	2.9–80	115(40–300)
^{132}Xe	<32	35(10–100)
^{40}Ar	0.25(±0.04)	16(±3)
N_2	1.02–1.7	6.7–133
CO_2	0.6–1.8	

known to be rich in water and other volatile light elements. The remarkable correspondence in the quantity of nitrogen residing in the atmospheres of Earth and Venus strongly suggests that here and there nitrogen is an outgassing product of crust and mantle. The only difference is that part of the terrestrial nitrogen becomes constantly recycled by nitrifying bacteria, whereas on Venus all its nitrogen has found permanent refuge in the Veneran air. On Earth, CO_2 resides principally in the form of carbonates and organic matter in the lithosphere, and sea and air share only a small amount of the total. However, by taking all terrestrial carbon sources together one surprisingly finds that this sum-total equals the amount of CO_2 in the Venus atmosphere. We conclude, only by virtue of having an ocean, marine life, and ongoing global tectonics that Planet Earth was able to promote the cycling of biogeochemical elements, notably carbon and nitrogen. In contrast, Venus could only maintain a cycling of such elements as long as liquid water and active tectonism could be maintained. With the disappearance of liquid and crustal water, tectonism slowed down and eventually vanished. This led to a build-up of carbon dioxide and water vapor in the atmosphere and the establishment of a "runaway greenhouse". The bulk of the water, which must have been comparable to the amount of water we have on Earth, was lost principally through photochemical reactions leading to hydrogen and oxygen. The first molecular species vanished into space, whereas the second reacted with crustal rocks. Such a scenario, however, implies that the water must have disappeared quite early, that is at a time when tectonism was still going on, otherwise the crust could not have served as a sink for molecular oxygen. A mere trifle of water vapor is all that is left of the once huge water reservoir that Venus possessed.

For Mars, an entirely different development took place. It chiefly has something to do with its small size relative to Earth and Venus (gravitational effect; Jeans escape), and its lack of a magnetic field permitting solar wind to penetrate deeply. A clue is provided by nitrogen. Viking measurements showed that present Martian atmosphere is enriched in ^{15}N relative to nitrogen on Venus and Earth, whereas oxygen and carbon isotope ratios on Mars appear to be similar to those found on Earth (Nier et al., 1976a, b). Initially, the air on Mars should have had a nitrogen content of at least several millibars and possibly as high as 30 mbar. According to McElroy et al. (1976), the ^{15}N anomaly is attributed to selective escape of ^{14}N from Mars' upper atmosphere. Diffusive separation above the turbopause will act to enrich the upper atmosphere in ^{14}N. Escape proceeds via production of fast nitrogen atoms by electron impact dissociation of N_2 and by dissociative recombination of N_2^+. Escape flux is sensitive to CO_2 partial pressure, since N_2^+ ions are removed mainly by charge transfer to CO_2.

At present, much of the CO_2 and water on Mars is frozen near the surface or bound in some other form in the crust. In all probability, Mars has experienced a much denser atmosphere and plenty of liquid water was around. The Red Planet, as Mars is often called, contains molecular oxygen in its air, but even more oxygen is bound in its reddish soils, sediments and rocks, which may imply that in the distant past more O_2 was contained in the atmosphere, or that water reacted with "hot" metal-rich rocks which kept the oxygen and released the hydrogen.

Mars and Earth are also known as Water Planets. Earth is further characterized by its abundance of molecular oxygen in air which is tuned to the rate of photosynthesis and weathering. Photosynthesis and weathering are related to CO_2 content in air, water and soil. CO_2 content in various environmental compartments is tuned to global tectonics, hydrothermal emanations, climate, primary productivity, water chemistry and circulation, biomineralization, in one word: feedbacks. Because water and life are the principal determinants regulating the flow of CO_2, H_2O, and O_2, the origin and history of these chemical species will be discussed later.

Chapter 10

Water

"Water is the best of all things." This saying of Pindar (518–446 B.C.) has not lost its meaning over the ages. A grain seed, properly stored for thousands of years in a pottery jar, has a good chance to germinate once exposed to the humid soil. A bacterium, enclosed in a crystal of salt, is said to have come to life again after a few hundred million years of suspended animation. Some suggest, however, that the "pickled" bacterium is simply a modern contamination, but I see no reason why it could not have survived for all those years.

Water exists almost everywhere in our planetary system (Cole, 1986). On Earth it occurs not only on the surface in the form of water vapor in our atmosphere, or in lakes, rivers, and oceans, but is found in virtually all rocks down to the Mohorovičić Discontinuity; it is bound to various minerals or is present as interstitial solution. Living organisms are mainly water.

We know how water behaves in the temperature range 0 to 100°C and 44 bar, and up to 175°C at 34 kbar. It has been suggested that water at higher pressures and temperatures does not exist as H_2O but as hydronium hydroxide ($H_3O \cdot OH$). Comets, meteorites, some moons of the outer planets and even Uranus, Neptune, and Pluto are believed to carry substantial amounts of frozen water. All appears to go back to Thales of Milet (ca. 640–546 B.C.), who declared that this Earth, its air, its sea, its mountains, its life, including its gods, are just different forms of water. We shall follow this proposition and meditate on water for a moment.

Structure

Background

Water is one of the most unusual substances in the universe. Commonly, if a compound can occur in the liquid and in the solid state, the liquid state is less dense than the solid one. In turn, the solid phase would settle down and the liquid would rise. Not so for water. Ice is lighter than liquid water and floats. Without this property, no life would be established on Earth, because if water behaved "normally", the ice would sink to the bottom of the ocean and eventually all oceans would be solid. Fortunately, water is not "normal".

This property must be related to the structure of water, but so far no comprehensive molecular theory for water exists. According to Stillinger (1980), some of the more notable physical properties displayed by water are the following: (i) negative volume of melting; (ii) density maximum in the normal liquid range (at 4°C); (iii) numerous crystalline polymorphs (by now at least ten, including those that formed at elevated pressure); (iv) high dielectric constant; (v) anomalously high melting, boiling, and critical temperatures for a low molecular weight substance that is neither ionic nor metallic; (vi) increasing liquid fluidity with increasing pressure; and (vii) high mobility transport for H^+ and OH^- ions. It is striking that so many eccentricities occur together in one substance

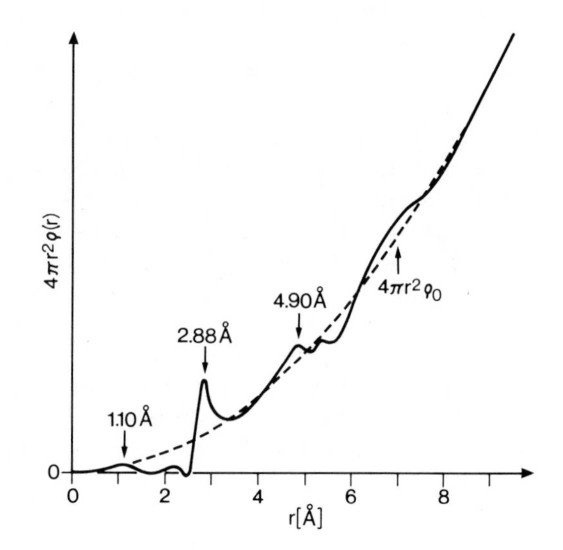

Fig. 10.1. Radial distribution function of H_2O molecules in water obtained by X-ray diffraction analysis at 25° C (after Danford and Levy, 1962). Shown is the frequency curve of distances between a randomly taken central H_2O molecule and its neighboring molecules

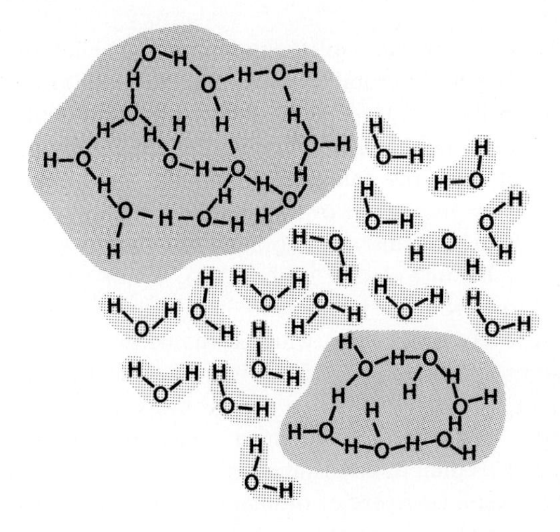

Fig. 10.2. The Frank-Wen "flickering cluster" model of liquid water. Schematic representation of hydrogen-bonded clusters and unbonded molecules (Nemethy and Scheraga, 1962)

and any one model of water's structure to be taken seriously has to explain all those features. Tables of major physical and chemical properties of water have been compiled by Kohn (1965).

What makes water so unique? An honest answer would be: we do not know yet, because water is structurally one of the least understood compounds. No equation can thermodynamically describe the reaction and properties of pure water and aqueous systems at a molecular level. The classic thermodynamic approach which is valid for ideal gases is unfortunately not applicable to water. Thermodynamic formalism operates with abstract concepts which are difficult to assess when molecular kinetic or molecular-structural interpretations are needed. The resolution of problems in the field of solution chemistry by thermodynamic means awaits future research.

Nevertheless, even if we assume for a moment that a satisfactorily thermodynamic theory for ordinary aqueous solutions exists, there would still be a problem with respect to biochemical solutions, for instance a protein molecule dissolved in water. Proteins are polyfunctional chain molecules and insight into the structural chemistry of such a solution can only be obtained by structural means. The same holds true for water; a structural treatment is essential.

Existing models on the structure of water fall into three categories: (i) crystalline, (ii) cluster, and (iii) cage:

(i) According to the classical theory of Bernal and Fowler (1933), water is tetrahedrally coordinated and forms a rigid network resembling, for example, quartz-like structures. This theory has been modified by Lennard-Jones and Pople (1951), who suggested the possibility for a "bending" of hydrogen bonds. They also pointed out that vacancies in the voluminous tetrahedral network could at increasing temperatures become filled up with water molecules. Their concept agrees with the radial distribution function of H_2O molecules in water (Fig. 10.1).

(ii) The "flickering" cluster model advanced by Frank and Wen (1957) and expanded by Nemethy and Scheraga (1962) assumes icelike microcrystalline areas termed clusters, which are separated by one or two layers composed of monomeric H_2O molecules (Fig. 10.2). This model or variants of it are presently most widely accepted.

(iii) The third category encompasses the various clathrate or "cage" models. Clathrate compounds are

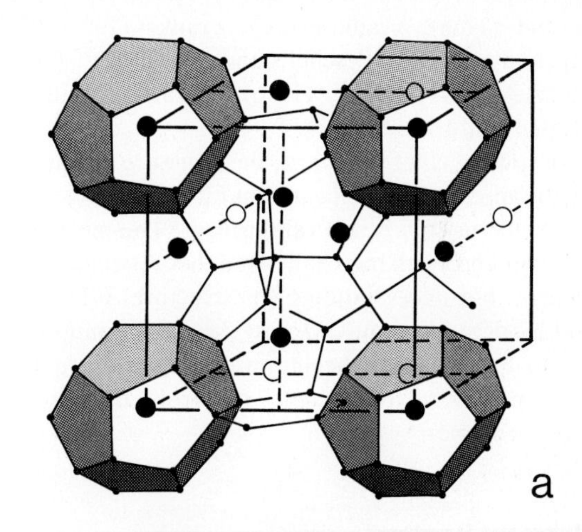

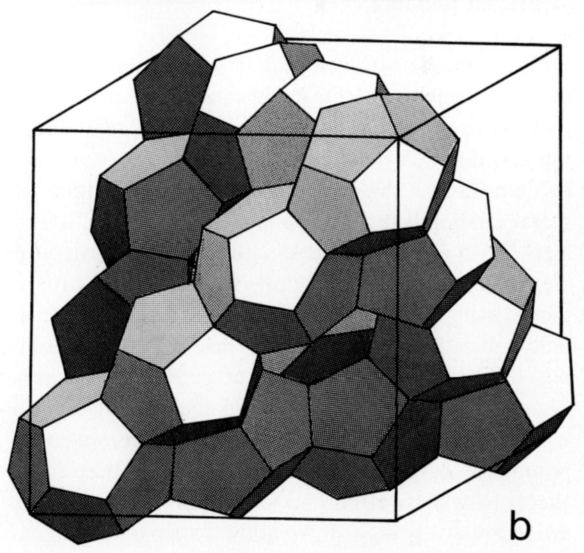

Fig. 10.3 a, b. Two kinds of cage structure found in water-clathrates (after von Stackelberg and Müller, 1954). **a** Structure composed of 12- and 14-faced polyhedra; **b** structure composed of 12- and 16-faced polyhedra. Only the dodecahedra are shown above

those in which molecules are encaged in the lattice of other molecules. Hydrates of non-polar gases such as certain noble gases or methane have the ability to form clathrates (Bhatnagar, 1962). The pentagonal dodecahedral hydrate proposed by Pauling (1961) is depicted in Fig. 10.3. His model also forms the theory of anesthesia, where water in the presence of, for example, chloroform, forms such hydrates in the cerebrospinal fluid.

Rarely has any "simple" chemical substance received so much attention as water. A small selection of review papers is listed below: Chadwell (1927), Bernal and Fowler (1933), Dorsey (1950), Pople (1951), Kavanau (1964), Danford and Levy (1962), Samoilov (1965), Drost-Hansen (1965a, b, 1966, 1967, 1972), Horne (1968a, b, 1969, 1972a, b), Kamb (1968), von Hippel (1968), Frank (1970), Franks (1972–1979), Forslind (1971), Peschel and Adlfinger (1971), Fletcher (1971), Luck, (1976, 1980a, b), Stillinger (1980).

In spite of all of this work, the structure of water remains an enigma. The comment of Walter Drost-Hansen made in 1966 is still valid today, "The most atypical liquid on earth has challenged theoreticians for decades... but understanding (its structure; ETD) may still lie decades away." Against this background I will try to extract from the literature some information that may help to elucidate the role of water in biogeochemical systems.

Hydrogen Bonding

We have previously seen that the admission of metal ions to a solution of EDTA will cause a structural rearrangement of the EDTA in giving rise to a rigid metal ion coordination polyhedron. This example was used to demonstrate the structural ordering principles of metal ion bonds at a precision of better than 1/100 of an Å. With respect to the second important structural element, the hydrogen bond, I propose an experiment, using a newspaper. The paper consists principally of cellulose, which is a macromolecule composed of roughly 2,000 units of glucose monomers joined together by glycosidic linkages. A calculation will show that per cm^2 of newspaper surface, we could potentially develop 10^{15} hydrogen bonds between two sheets of water-wetted pages. The forces of bondage are impressively high. If we allow a strip of hydrogen-bonded newspaper sheet to hang freely in the air in a kind of open noose, it would take about 50 grams of weight placed into that noose to break all hydrogen bonds of a one cm^2 structured double layer of cellulose. What actually happens is that independently of the originally random orientation of the cellulose fibers, a new structural entity is generated composed of *one* molecular layer of water and two of glucose polymers held together by hydrogen bonds. The weight does the following: it breaks the hydrogen bonds by gliding the overlapping paper sheets just 4/100 of an Å against one another. The structure will collapse. By introducing water again, a new structure will form which is identical to the previous one.

The H_2O molecule, which contains two hydrogen atoms and one oxygen atom in a non-linear arrangement, is ideally suited to engage in hydrogen bonding. It can act as a donor and as an acceptor of hydrogen. This dual ability is illustrated in vivid fashion by the crystal structure of ordinary hexagonal ice. Each H_2O molecule in the crystal has four nearest neighbors to which it is hydrogen bonded. It acts as hydrogen donor to two of the four and accepts hydrogens from the remaining two. These four hydrogen bonds are spatially arranged with local tetrahedral symmetry; that is the oxygen atoms of the neighbors occupy the vertices of a *regular tetrahedron* surrounding the oxygen atom of the central molecule. The bond angle of the free water molecule (104.5°) is only slightly less than the ideal tetrahedral angle (109.5°) that would be required by a strictly hydrogen bonding in the ice crystal.

In the vapor phase *dimers* $(H_2O)_2$ exist with a structure that displays a linear hydrogen bond (Dyke et al., 1977). The lowest energy arrangement has a plane of symmetry containing the hydrogen donor molecule to the left and the symmetry axis of the molecule to the right (Fig. 10.4). The bond length (2.98 Å) is significantly longer than the observed distance in ice, 2.74 Å. Quantum mechanical calculations reveal that starting with the most stable dimer configuration, moderate rotations of the H_2O molecules about their oxygen atoms cost little energy, provided the donated hydrogen remains essentially on axis. Thus, the quantum mechanical studies fully support hydrogen bond linearity. They emphasize op-

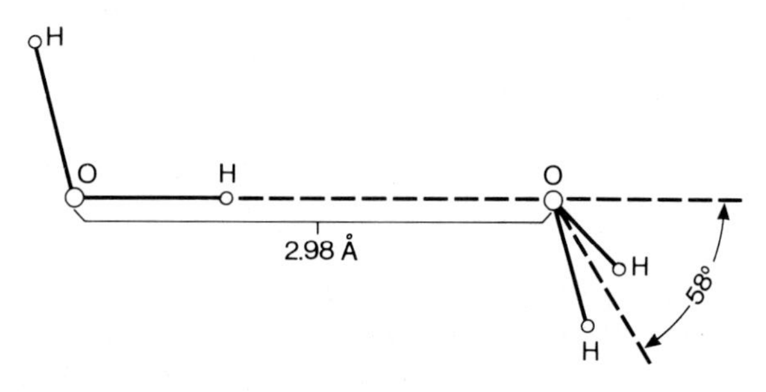

Fig. 10.4. Optimal structure of the dimer of H_2O according to molecular beam resonance studies (Dyke et al., 1977)

portunities for topological diversity in extended networks that stem from ease of rotation about linear hydrogen bonds.

After the first hydrogen bond is generated, the charge distribution within the participating monomers changes in such a way that the hydrogen acceptor molecule becomes potentially an even better hydrogen donor than before. It is capable of yielding a stronger second bond because of the existence of the first bond. Similarly, the proton donor has an enhanced ability to accept a proton on account of the bond that it has already formed. This mutual reinforcement encourages molecules to form chains of hydrogen bonds with the mean bond energy larger and the mean bond length shorter than for the simple dimer. To permit fully developed hydrogen bond order around a central water molecule, a *tetrahedral coordination* is postulated.

The melting of ice to produce liquid water obviously entails basic structural changes. *Rigidity* is replaced by *fluidity*. Molecules are much freer to change their orientation, in spite of the fact that melting has caused the density to increase by about 9 percent. Any acceptable molecular theory for liquid water must account for the topology and geometry of the hydrogen bond network after it has been altered by the melting process. The heat of melting is only 13 percent of the sublimation energy of the solid, which may be interpreted to mean that a comparable percentage of the hydrogen bonds rupture upon melting. Alternately, a structurally different alignment of non-nearest neighbors may occur which would imply that hydrogen bonds survive the "trauma" of melting (Stillinger, 1980).

Crystalline Packing Order

A clue to the structure of water is provided by its specific density changes as a function of temperature. We accept that the water molecule has basically a tetrahedral coordination; the two free pairs of electrons tend to bind hydrogen atoms, and a three-dimensional network is generated. If, on the other hand, we propose a structural order for water aggregates which conforms with the structural order of ice, we are in conflict with specific density measurements. To explain the observed densities of water, one has to accept ordered structures which are different from those established for the various types of ice structures. At the same time one has to reject the concept of a compressed van der Waals gas.

An alternative solution would be to assume that liquid water is amorphous in nature and essentially a mixture of H_2O aggregates and holes. Randomly oriented H_2O clusters composed of at most a few hundred molecules of water will disperse, aggregate, and reorientate in response to temperature-induced vibrations, and thus account for the observed changes in volume as a function of temperature.

Instead water could possibly be composed of a more regular "crystal structure"; the quotation marks are used as a reminder of its difference from solid structures. Water obeys the packing rules of crystalline materials without being an ice or a compressed gas structure. The results of neutron or X-ray diffraction studies which yield only diffuse reflexes cannot be interpreted to mean that water has a structure of a compressed gas. The superposition of scattered waves, which produce a diffraction spot, is related to constant phase differences between waves scattered and thus to a very precise repetition of water molecules across long space distances. Diffraction analysis requires a precise periodic space topology which cannot be maintained in water.

Let us assume that we can succeed in making an X-ray diffraction analysis of water or an aqueous solution at an exposure time of 10^{-12} seconds; probably we would recognize a crystalline pattern. This case is somewhat analogous to a study of the crystal structure of a metal, such as sodium, close to its melting point, except that the exposure time for the metal must be of the order of several years to make the experiments comparable to those with water. For the metal a perfect crystal structure emerges using X-ray exposure times of a few hours, which implies that each sodium atom is kept at its assigned lattice position. If, on the other hand, the observation time is extended for days or months, the atoms jump from one lattice position to another – they diffuse within the crystal structure. This conclusion is analogous to liquids, because the film which is constantly exposed to diffracted X-rays simply records all reflections, including the diffuse ones, which were generated by the disorder of the crystalline lattice. The difference between a solid and a liquid is thus determined by the average lifetime of an atom on a fixed lattice position. The difference in the time constants is as follows:

$$\frac{10^3 - 10^4 \text{ sec}}{10^{-11}\text{–}10^{-12}\text{sec}} = \frac{\text{solid}}{\text{liquid}} = 10^{15}\text{sec.}$$

It is concluded that in a crystal, an atom remains at its assigned lattice position about 10^{15} times longer than does an atom in a liquid. It is principally this difference which sets a liquid structurally apart from a solid. In Fig. 10.5, the movement of atoms or

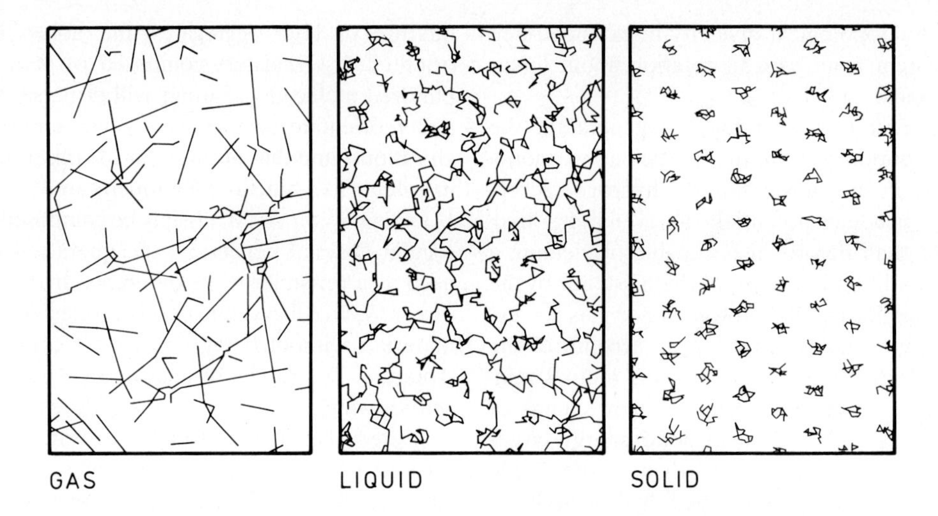

Fig. 10.5. Paths of molecular movements in gaseous, liquid, and solid phases (highly schematic; after Barker and Henderson, 1981)

molecules in gases, liquids and solids is schematically presented.

Ice

In ice structures or in hydrous minerals, water molecules are usually four-fold coordinated. The bond orbitals in oxygen tend to occupy a tetrahedral coordination (Fig. 10.6), even though the angles between the p-orbitals which bind the hydrogens are in theory not 109° but 90°, and all p_x, p_y, and p_z orbitals should be arranged perpendicular to one another. A further complication is the dipole moment of water (μ = 1.84 Debye) which allows dipole-dipole interactions. Therefore, only one hydrogen bond develops in the direction of the p-orbital when a p_x–p_y bonding state is observed. The water molecule, however, comprises:

$$sp_3 \rightarrow p_x - p_y, p_z$$

and in ice structures (Kamb, 1968; Mishima and Endo, 1980), water molecules have a tetrahedral coordination (Fig. 10.7). The water molecules can accept two hydrogen bonds and form two hydrogen bridges. This results in the formation of geometric networks that closely resemble silicate structures in which the SiO_4 unit is tetrahedrally coordinated to its neighboring SiO_4 units.

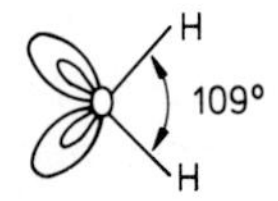

Fig. 10.6. Bond orbitals in oxygen occupy tetrahedral coordination

Water has at least ten solid phases which were given Roman numerals, i.e., ice I to ice X. Common ice, the hexagonal phase I_h, is not stable at pressures above 200 MPa. High-pressure phases were first recognized by Tammann (1900) and the latest addition is ice X, which has been synthesized at temperatures

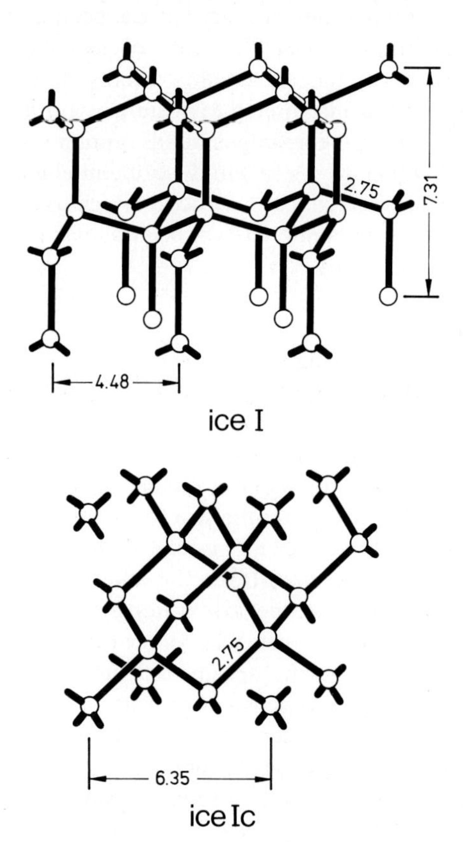

Fig. 10.7. Ice structures, ice I and ice Ic (Kamb, 1968)

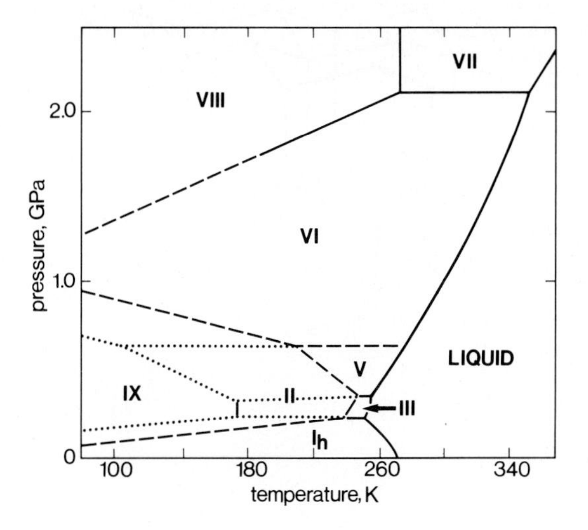

Fig. 10.8. Phase diagram of water (after Hobbs, 1974)

close to absolute zero and pressures of maximal 24 x 10^3 atmospheres. The phase diagram of water based on the work of Hobbs (1974) is shown in Fig. 10.8.

Many of the ices are expected to occur on Neptune, Uranus, Pluto, within the planetary rings, on various moons of the outer planets, and in cometary bodies. Even Mars, Earth and Venus should have carried different ices during the early stages of accretion.

Impacts into any ice surface could yield significant amounts of high-pressure forms of water ice. This situation is comparable to the production of high-pressure polymorphs of SiO_2 such as coesite and stishovite, in the case of a bolide hitting a silicate surface. Due to relatively low ambient surface temperatures on satellites in the outer solar system, the modest temperature rises accompanying the impact pressure for ice metamorphism, the high frequency of infall and the lack of tectonic activity in the majority of the moons, high-pressure ices should be ubiquitous. Fly-by missions may detect and map their distribution in the outer solar system by reflection spectroscopy at vacuum ultraviolet and mid-infrared wavelengths. This would certainly give a better understanding of histories and properties of bodies in the outer solar system (Gaffney and Matson, 1980).

Liquid State

The two principal structural elements of water are: (i) the capacity of hydrogen atoms to form tetrahedral as well as orthogonal bonding angles that allow for the construction of a great variety of possible networks, held together by hydrogen bonds, and (ii) dipole-dipole interaction, which in numerous ways can distort and even disrupt the networks generated by hydrogen bridges.

This should enable us to address the question, "What is a liquid state?" Unfortunately, as stated by W.A.P. Luck (1976), "Most university teaching courses neglect the structure of liquids, although the greater part of their students of chemistry, biology, medicine, or polymer science will later on work on questions of the liquid or amorphous state... we should know the details of the partition function of molecular distances in a liquid and of the complicated mechanism of intermolecular forces in many body systems."

BOX 10.1

Ice-Nine Excerpts from: Cat's Cradle (Kurt Vonnegut, Jr., Dell Publ. Co., Inc., New York, 1963, pp. 191)

"There are several ways", Dr. Breed said to me, "in which certain liquids can crystallize – can freeze – several ways in which their atoms can stack and lock in an orderly, rigid way (p. 38). Now suppose that there were many possible ways in which water could crystallize, could freeze. Suppose that the sort of ice we skate upon and put into highballs – what we might call *ice-one* – is only one of several types of ice. Suppose water always froze as *ice-one* on Earth because it never had a seed to teach it how to form *ice-two, ice-three, ice-four...*? And suppose that there were one form, which we call *ice-nine* – a crystal as hard as this desk – with a melting point of, let us say, one-hundred degrees Fahrenheit, or, better still, a melting point of one-hundred-and-thirty degrees" (p. 39).

"Danger! *Ice-nine!* Keep away from moisture!" (p.165).

I opened my eyes – and all the sea was *ice-nine*. The moist green earth was a blue-white pearl (p. 174).

In that bowl were thousands upon thousands of dead. On the lips of each decedent was the blue-white frost of *ice-nine* (p. 181).

At present, two principal viewpoints exist on the nature of the liquid state: (i) the gas model, and (ii) the moving crystal lattice model. According to the first model, a liquid is considered a "sticky" gas, and solutions are treated thermodynamically as are gases. Equations are introduced which incorporate interaction parameters, such as the derivation of the van der Waals equation. The proponents of the alternate model consider water a moving crystal lattice, in which the particles are driven by Brownian forces, and jump from one lattice position to another. Compared to a solid crystal, such a structure must be considered a disordered one.

Only choice or taste will determine which of the two models one should accept or reject. The selection is made the moment one decides which property of water to consider. If variability in form is chosen as the prime phenomenon, the gas model has the greater appeal. If, on the other hand, forces that hold water together are of interest, the crystalline model will be favored. For a number of reasons enumerated below, the structural model is in fact a good springboard for looking at water.

Strong support for the crystallographic model comes from studies on liquid crystals, such as soap bubbles, long-chain hetero-organic molecules or phospholipids. They all possess the featureless morphology typical of liquids, and at the same time the optical anisotropy and phase transitions exhibited in crystals. Frenkel (1957) attempts to explain this incongruity by considering the local order of molecules in liquids from a thermodynamic point of view. A crystal-chemical theory on liquids, however, does not exist.

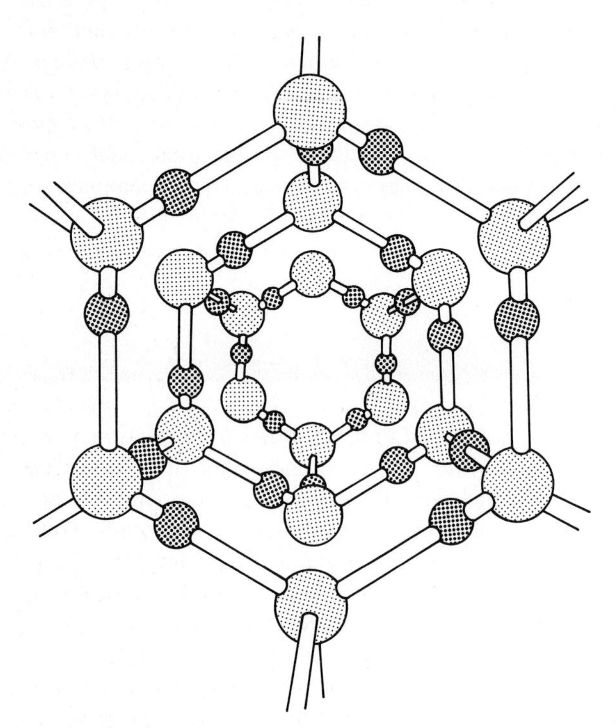

Fig. 10.9. Structural arrangement of water molecules via hydrogen bonds (Drost-Hansen, 1966). In ice models, water molecules liberated when ice melts are tucked away in voids along *c*-axis of hexagonal ice crystals. This explains how water can be more dense than ice. Model requires the existence of ice nuclei.

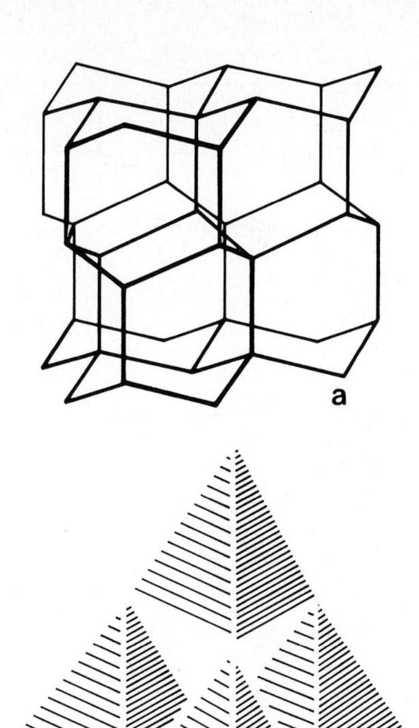

Fig. 10.10 a, b. Molecular association of H_2O molecules (liquid phase). Oxygens form OH_4 tetrahedra, which through corner-linkages condense to 6-ring units and construct a three-dimensional network similar to that of the quartz lattice. The spatial cross-linkages of the 6-ring units are shown in **a**. H-bond energy in H_2O is a polarization energy. Thus, $O_4H_9^+$ units (**b**) will form the minimum structural building block for proton trapping. The proton is located in the center of the tetrahedra complex, which is linked across corners; it is coordinated by four oxygens located in the center of the innermost tetrahedra.

The charges are the following:

4 O atoms (located in center of tetrahedra)	= (−) 8
6 H atoms (edge position)	= (+) 6
4 H atoms (located at the outer corners)	= (+) 4/2
1 H atom (center position)	= (+) 1
	(+) 1

It has been said that the presence of hydrogen bonds between two oxygens as a structural element of water has been established beyond doubt. Many layered silicates such as mica or clays which are rigid

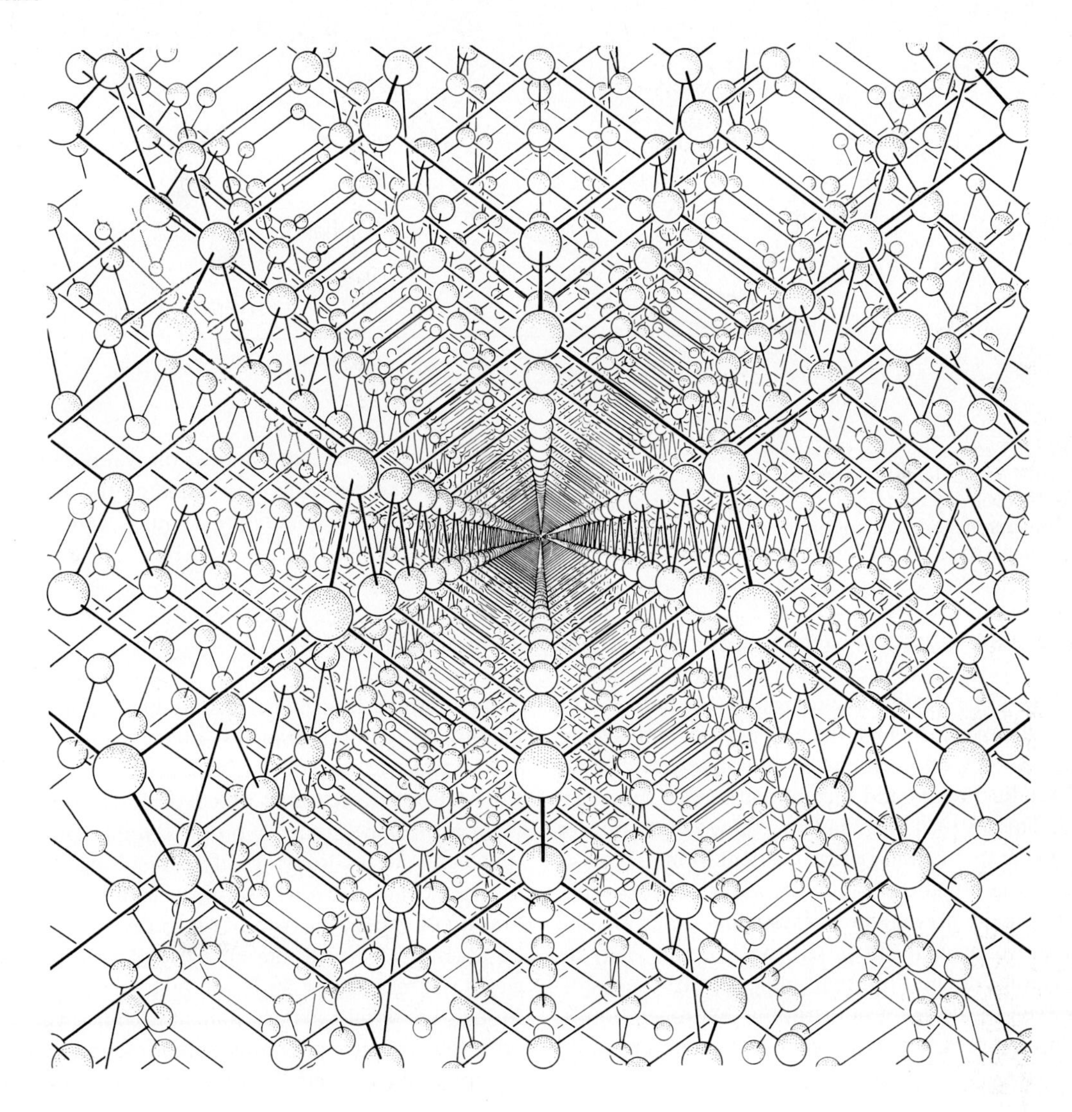

Fig. 10.11. Pointblank view down the axis of a channel in a diamond-like crystal lattice (after Brandt, 1968). Liquid water is assumed to be built according to the same structural principles. *Open circles* represent the oxygen atoms. Hydrogen atoms are located along the connecting lines. Hydrogens and the hydrogen bonds interconnecting the oxygen atoms have been omitted for graphical reasons

and stable are also held together by hydrogen bonds. The minerals and synthetic compounds whose lattices are stabilized by such bonds are legion.

The structure of water is schematically drawn in Fig. 10.9. Each water molecule has four close neighbors which are linked to one another by hydrogen bonds. The smallest structural unit of water is thus a complex composed of five water molecules (Fig. 10.10). The arrangement of these structural networks can vary. Structures in which tetrahedra are arranged in the form of six-membered rings are the most probable (Fig. 10.11). The six-membered boat-and-chair-forms developed by organic molecules are the most likely ring forms also to be generated by water. In general, a combination of the two types of ring forms characterizes the structure of water, but apparently the boat type (crystobalite structure) is most common.

At this point it should be stressed that ideal liquids and solids hardly exist in nature. There are holes or lattice defects all along which leave quasi "open ends" here or there. These, however, appear not to be randomly orientated in the case of liquid water, but are concentrated by fissure plains of defects. The percent of non-H-bonded OH groups in liquid water obtained by the overtime IR methods give an approximation of the disorder that exists (Luck, 1976). In the temperature interval 0 to 100°C, the number of non-

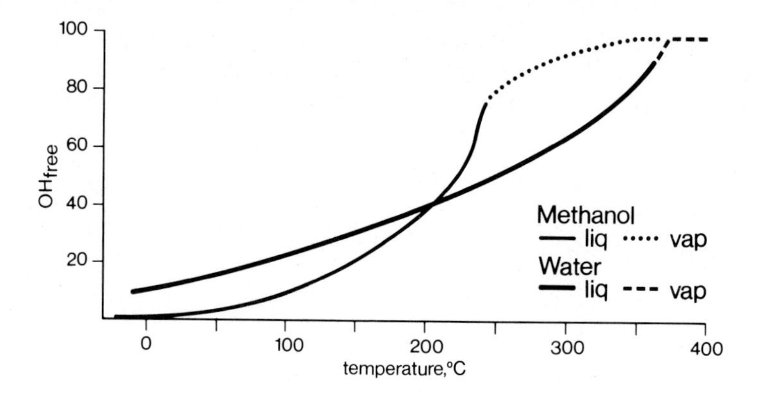

Fig. 10.12. Percent of non H-bonded OH groups in liquid water and methanol under saturation conditions as determined by the IR overtone method (Luck, 1974). Approximation of water structure requires extension of experimental work up to critical temperature

bonded OH groups is about 10 to 20 percent (Fig. 10.12). Since the relaxation time in water is 10^{-12} sec, this will be the lifetime of orientation defects too. They are moving through the water lattice similarly to a zone-melting and refining process and cause structural readjustments and recombinations as they pass along. This in turn will lead to a new generation of structured, that is non-random, orientation defects.

Accepting: (i) the existence of water networks that are principally arranged in a crystalline order, (ii) the cooperative mechanism of hydrogen bonds, (iii) dipole-dipole interaction that imposes lattice distortions even to the point of disrupting hydrogen bonds, and (iv) the presence of lattice defects in non-random distribution, a structural model on water can be constructed that conforms and explains the anomalous properties of this unique liquid called: water (Fig. 10.13).

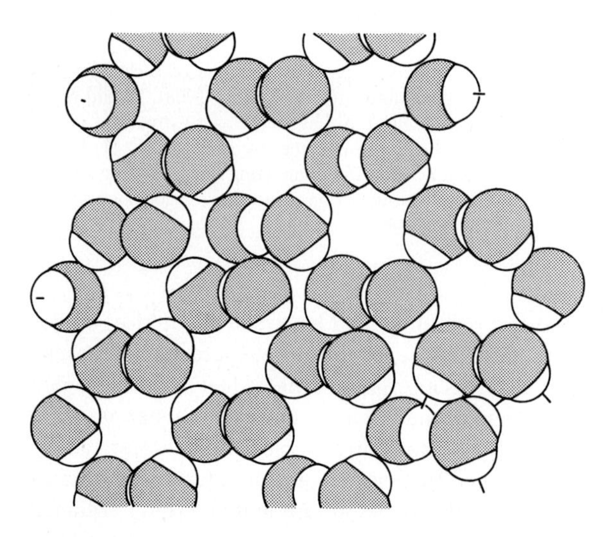

Fig. 10.13. Idealized two-dimensional model for water. In addition to H-bonds typical for ice, H-bond defects with a denser packing order are assumed (Luck, 1980 a, b)

Proton Transfer

Why is water fluid? The answer can be inferred by examining proton transfer mechanisms in liquid water. Electrical conductivity of water is based on proton transfer, which may proceed along two avenues (Gurney, 1966), that is protons jump from water molecules to hydroxyl groups, resulting in proton transfer chains; this will show up as an apparent migration of OH^-. This type of proton movement has to be assumed, because the mobility of an OH^- ion exceeds the mobility of an F^- ion by a factor of four, although we would expect similar mobility values for both ions because of comparable ionic radii and masses for hydroxyl and fluorine ions. For comparison, the equivalent conductance values in an aqueous solution at 18°C in [Ohm^{-1} cm^2] are $H^+ = 316.55$; $OH^- = 176.6$; $F^- = 46.65$; $Li^+ = 33.28$.

A second mode of proton migration involves proton jumps via water molecules. The mobility of this proton transfer is about twice that of the apparent OH^- mobility. Potential energy profiles of molecular level and distance, as are commonly used in solid state physics, illustrate that more energy has to be extracted from the thermal surroundings in case of proton movement along the OH^- level when compared to the migration along the H_2O bond. Such a proton migration is reflected by the equivalent conductance, and we describe it by saying that the binding of H^+ to OH^- is stronger than that of H^+ to H_2O. These proton migrations as described are proceeding in the form of charge transfer. Empirical data suggest that the determining step in proton migration in water is the reorientation of the water molecule. The vital step is when a water molecule receives a proton from an adjacent hydronium. The most significant aspect of proton transfer thus depends on: (i) the location – geometry – of protonic change, H^+, within a $H_9O_4^+$ complex, and (ii) the structural diffusion of such complexes by formation and breaking of hydrogen bonds with the surrounding water structure (Franck et al., 1965).

The structural model for a proton transfer complex is represented by four water tetrahedra joining in such a manner that a tetrahedral "superstructure" is built (Matheja and Degens, 1971a; see Fig. 10.10). This $O_4H_9^+$ supertetrahedron is formed easily by switching the O...H–O bond directions within the trigonal water lattice. This switching is always present in the water structure and accounts for its fluidity.

The liquid at large is communally connected. These network pathways, incidentally, provide natural routes for rapid transport of OH^- and H^+ (a proton "hole") by a direct sequence of exchange hops. It is a moving lattice and corresponds to the nature of mineral lattices, except that the structural positions occupied by hydrogen and oxygen are switched at a speed about 15 orders faster than is the case for solids.

On Physical-Chemical Properties

Water is distinguished in many ways from other liquids. The most obvious is its density anomaly with a density maximum at 3.98°C, which implies that the density of its solid form is less than that of its melt. Other properties of water are listed at random. Water has a high specific heat capacity. It can act as a strong base or acid simultaneously or consecutively. Water forms and accepts hydrogen bonds in a way unmatched by any other liquid. It is this capacity which enables water to generate hydrogen bridges, which in turn are responsible for the specific three-dimensional network of water molecules with the high dielectric constant of $\varepsilon = 80$ even though the H_2O molecules exhibit only a dipole moment of 1.84. Because the density of water decreases as the temperature rises from 4 to 100°C, the number of water molecules surrounding any other water molecule increases in the same direction. This molecular crowding is the opposite to the behavior of most other liquids when subjected to increasingly higher temperatures.

In brief, water is a medium of many unusual properties. Its excellent solvent characteristics are related to some of these qualities, though a number of these are incompatible, because they are of opposing natures. Water is a good solvent for solid compounds such as salts, but it cannot dissolve alkali metals, soluble in liquid ammonia.

When we take advantage of these properties of water which make it an outstanding solvent medium, we are usually interested only in some specific ones. By dissolving materials such as metal salts, alcohols, acids, or urea in water, we create a solvent system of well-defined characteristics and purposely suppress all other properties indigenous to water. Plasma proteins can be dissolved in water; the addition of ammonium sulfate will cause their deposition. The classification of plasma proteins is based on the way they respond towards different concentrations of ammonium sulfate. Such a classification is arbitrary, but it serves a pragmatic purpose.

A solvent medium is thus something particular and a class in itself, even though it is traditional to treat an ammonium sulfate solution as a two-phase system. The ammonium and sulfate ions are simply considered as charged particles thermally driven by Brownian movement in a zig-zag fashion through the water. This viewpoint does not properly describe the quality of a solution system according to experience. A salt solution of a specific concentration must be considered to be another species of water, and the specific two- or more phase system should be completely ignored.

When a compound is dissolved in water or another solvent medium, the resulting solution is a new product; it is no longer water, and there is no longer a dissolved compound. Granted, some resemblance to the original materials still exists. However, these molecules as we know them, such as water and salt, have lost their structural and chemical identity.

Can we predict the physical and chemical properties of water? The answer is yes, as far as the physical parameters are concerned. The predictions become more tentative and speculative when our question is aimed at the chemical properties. Most solvent media have been properly described from a physical-chemical point of view, but this description does not take care of many properties which are essential to biochemical research.

A common solvent medium is ordinarily described by the following terms: (i) chemical formula, (ii) molecular weight, (iii) dissolution properties, (iv) electrical resistance, (v) density, (vi) pH, (vii) Eh (= oxidation-reduction potential), (viii) ionic dissociation, (ix) electrical conductivity, etc. These are mostly dimensions describing the physical characteristics of the solvent medium. In addition, we must always keep in mind that a solution has well-defined chemical properties. The physical properties are predictable to a certain degree; for instance, knowledge of the equilibrium constant of the oxygen dissociation makes it possible to calculate the pH of a solution.

However, the chemical properties of a given solution can be predicted only in a qualitative fashion. The imagination of the researcher determines the selection of a solvent system for the analytical separation of, for example, an enzyme fraction in its original structural configuration.

Aqueous Environment

All natural waters contain salts in an amount as low as a few milligram per liter in some freshly fallen rain to more than 300 grams per liter in a great number of oil-field brines and evaporation pans. A river usually carries a few hundred milligrams per liter, whereas the sea has a constant value of 35 g per liter, that is a salinity of 35 permil. Thus, the amount of salt dissolved in natural waters may vary by a factor of about 100,000.

The kind of salts found in fresh waters is largely a function of rock type, weathering conditions, vegetation and human activities, and it is probably no exaggeration then to say that each creek, small river or stream has its unique "salty" personality. In contrast, bulk sea water is globally uniform as far as amount and nature of salt is concerned.

Adding salt to water will cause a structural change in the ambience near the ion. The coulombic hydration envelope appears to be of a dual nature. Next to the ion the coulombic field will crowd the polar water molecules in closely about the ion. This is the "primary" hydration zone, which itself is emplaced in the "total hydration sphere" (Fig. 10.14). The short-range and longer-range ordering effects may explain the discrepancies in hydration numbers reported in the literature for individual ions: Li^+ 3–150; Na^+ 2–70; K^+ 2–29; Cl^- 1–20 (see table in Rutgers and Hendrikx, 1962). The dependence of the viscosity of water on type and amount of ions is in support of the long-range water structure-altering properties of ions (Horne, 1972a, b). Some electrolytes, such as lithium or sodium chlorides, enhance water viscosity, presumably by increasing the amount of structure. In contrast, potassium and caesium chlorides make water less viscous and must be regarded as structure-breakers (Gurney, 1966; Horne, 1968a, b).

Natural aqueous solutions carry in general numerous ionic species, and in turn an impressive number of pair correlation functions is needed to describe their structural design (see, e.g., Barnes et al., 1979; Narten et al., 1982). In the case of silicates, where ions are present in solid solution, we have seen that the large-sized O^{2-} provides the structural frame principally in the form of tetrahedra or octahedra, into which a small-sized cation such as Si^{4+}, Al^{3+}, Ca^{2+}, K^+ etc. is placed. Within a certain ionic size range the central cation can be substituted without changing the oxygen coordination in the respective polyhedron.

According to recent findings (Cummings et al., 1980), water molecules around chlorine ions are rigidly coordinated and Cl^- hydration appears to be

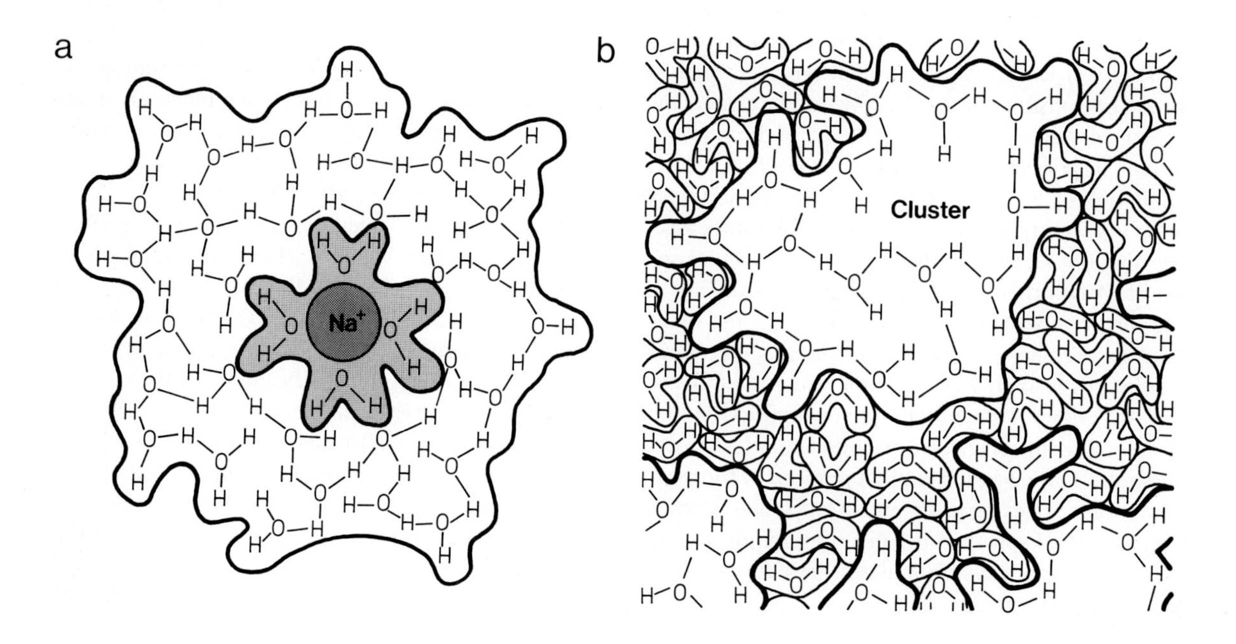

Fig. 10.14. a Primary hydration sphere in the vicinity of a cation. Clustering will structure water around hydration sphere. For comparison, the Frank-Wen flickering cluster model of liquid water is shown **b** (Frank and Wen, 1957)

unaffected by size and charge of coexisting cations and ionic strength as well. This is the more remarkable in that water molecules must be entering and leaving the hydration sphere around Cl^- on a time scale of 10^{-12} sec, whereas cation-water residence times are more in the neighbourhood of 10^{-6} sec. Caution is therefore needed when trying to equate hydration number with coordination number.

From all this it appears that size and charge of a cation is the most determining factor restructuring a water molecule, whereas Cl^-, the most ubiquitous anion in the hydrosphere, exhibits a stable hydration geometry in spite of molecular perturbations in its aqueous ambience. As pointed out by Cummings et al. (1980), these two features combined should lead to a major simplification in the theory of solutions.

The issue again becomes complicated by the fact that many transition metals form complexes in aqueous solutions. Research in the field of hydrothermal ore deposits has shown, for instance, that the majority of heavy metal ions are transported as chloro- or fluoro-complexes having stabilities determined by temperature and type of ambience. In addition to these, dissolved carbonate ions or certain organic molecules can act as ligands for a great variety of ions present in lakes, rivers or the ocean. Carbonate equilibria in the sea appear to be controlled by the existence of dissolved complexes (Pytkowicz, 1983).

To emphasize the significance the addition of a second solvent may have on the structure of water and the chemical properties of the newly generated aqueous system, the behavior of (i) a hydrocarbon chain and of (ii) a protein when added to water are briefly outlined:

(i) *Hydrophobic effect:* The adjustment of water towards non-polar solutes or non-polar side groups attached to biopolymers has long been recognized as unusual. The popular phrases "hydrophobic effect" and "hydrophobic bond" were coined (Ben-Naim, 1965; Franks, 1972–1979; Clifford, 1965; Stillinger, 1980). Typical non-polar solutes are the noble gases, hydrocarbons and sulfur hexafluoride. None can readily hydrogen-bond to water, and all are sparingly soluble. Nevertheless, non-polar solutes occupy space. If a non-polar solute molecule is to be placed in liquid water, the hydrogen bond network must reorganize around it in such a way that sufficient room is available to accommodate that molecule while at the same time not causing too much damage to the hydrogen bond order. The crystal structure of the clathrate hydrates for many of these non-polar substances shows that such rearrangements are possible in principle. The cage is usually far from perfect and has to be

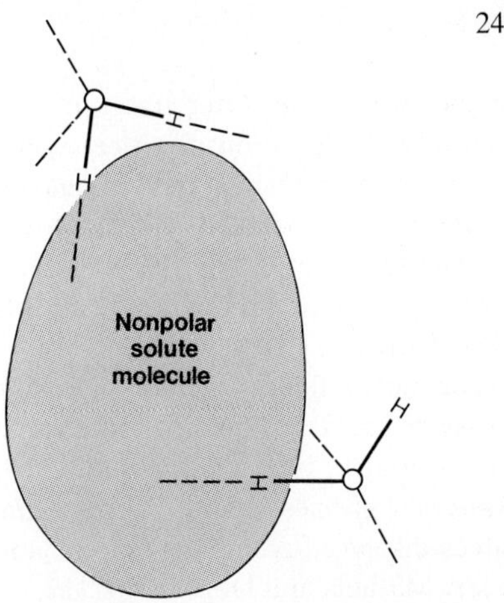

Fig. 10.15. Water molecules next to a non-polar solute. To preserve maximum number of hydrogen bonds, water molecules tend to straddle the inert solute, pointing two or three tetrahedral directions tangential to the surface of the occupied space (Stillinger, 1980)

shaped according to the structural outlay of, for instance, a long-chain hydrocarbon. Strengthening of hydrogen bonds in the cage relative to the case in pure water may partially compensate for structural irregularities. Each solvation sheath water molecule strongly prefers to place its tetrahedral bonding directions in a straddling mode (= pointing two or three tetrahedral directions tangential to the surface of the occupied space, Fig. 10.15). This arrangement obviously permits bonding to other solvation layer water molecules and always avoids pointing one of the four tetrahedral directions inward towards the region occupied by the inert solute, which would "waste" a possible hydrogen bond.

Pairs of non-polar solutes in water experience an entropy-driven net attraction for one another. This is the essence of the so-called "hydrophobic bond". These hydrophobic pair interactions are generally thought to be an important determinant for the native conformation of biopolymers containing both hydrophobic and hydrophilic side groups. Membrane structures are a good example of such interplay. The accepted qualitative explanation for the attraction is that, when two non-polar molecules are close together, their joint solvation cage entails less overall order and hence less entropy reduction than when those solutes are far apart. There is, in other words, a thermodynamically driven tendency towards the sharing of orientationally inhibited solvation sheath molecules.

In contrast, ionic solutes, because of their strong electric field, are known to twist water molecules out of orientation that can bond easily to the surrounding network. This competition of non-polar and ionic species within the water structure will lead to expulsion and precipitation if critical concentration levels are reached. The whole field of emulsions, foams and soaps, and the so-called critical micelle concentration is in effect governed by the adjustment of the structure of water to various kinds of solutes (Stillinger, 1980; Yariv and Cross, 1979).

The most refined structure on Earth – the living cell – may be viewed as the final outcome of a few million years enduring interplay among the various polar and non-polar species in the ambience and the ordering forces displayed along air-, water-, and mineral interfaces (Matheja and Degens, 1971 a).

(ii) *Aggregation and dissolution of a protein:* Myosin is a protein which can be isolated from muscle fibers. In organisms, myosin occurs in an aggregated form (polymer) and builds up a myosin filament. The aggregation and dissolution of muscle proteins is highly dependent on the kind and concentration of metal ions in the system. For myosin, the dependence on metal cations and H^+ in the actual aggregation process leading to the myosin filament has been shown by Kaminer and Bell (1966). The size of the growing myosin crystalloids is controlled by the H^+ concentration as well as by the number of potassium ions present. Many properties of myosin are known, yet the ionic properties of the solvent medium which determines the size of the myosin crystallites cannot be calculated. One can only hope that progress in this field of solution chemistry will eventually come to the point, where the chemical characteristics of solvent media can be reduced to simple physical principles and dimensions. The present trend to employ physical terms in the description of a solvent medium without considering its chemical properties is unrealistic. The different solutions used in the myosin experiments should be considered as separate, i.e., discrete systems, and not as a continuum. The behavior of myosin in a solution is affected by more than the metal ion and proton content. To define different solutions in terms of molar concentrations, i.e., 0.01, 0.1, 1 molar, and to pretend that this represents a continuum is absolutely false; each solution has its individual characteristics and cannot necessarily be looked upon as a member of an electrolyte solution series. This is analogous to the number system, which is also thought of as being a continuum; but each number retains individual properties which are uncommon to the other numbers.

Studies of biochemical systems are generally performed in aqueous solutions. Certain solvent media are applied and the results are, in sensu stricto, applicable only for the particular solvent system used. The mass of information gathered in the field of analytical biochemistry is quite impressive; unfortunately, the treatment of basic problems occupies only a small part of the actual research output recorded in the literature, and the majority of papers deal with the characterization of properties of biochemical molecules in solvent media. Several thousand articles have been published on the cleavage effect of cholinesterases on acetyl-choline. Most of these papers are concerned with the characterization of properties in the presence or absence of a great variety of salts, differences in molarities and H^+ concentration, in the presence or absence of hydrogen bond breakers with or without serotonine, adrenaline, denaturated or not denaturated and so forth.

Here and there insight into basic problems can be achieved. Polymer physical chemistry is just starting to formulate laws, and the thermodynamics of irreversible processes represents a remarkable breakthrough in this respect (e.g., Prigogine, 1979).

Interfacial Water

Air-Sea Interface

The common expression "To pour oil on troubled water" refers to the calming effect of oil upon a rough sea. Actually, oil cannot dampen larger waves already generated, but it will soften associated wind chop riding on the backs of those waves. This phenomenon had already been described by Aristotle (389–322 B.C.) and Plinius Secundus (23/24–79 A.D.). Yet the molecular aspects of organic surface films such as oil slicks in the ocean have only in recent years received wide attention with the growing awareness of marine pollution (e.g., Brockmann et al., 1976; Hühnerfuss, 1983, 1986; Hühnerfuss et al. 1982, 1983, 1984).

For more than 50 years it has been known that the structure of water near phase boundaries is different from that of bulk water (Hennikyar, 1949). Special attention should be given to the remarkable study of Deryagin (1933), who succeeded in demonstrating that a water layer about 1,500 Å thick placed between two glass surfaces develops a disjoining pressure in the order of 10^{-5} dyn cm^{-2}, and thus behaves like a solid. Peschel and Adlfinger (1971) extended Deryagin's work by showing that the disjoining pressure is a function of temperature (Fig. 10.16). The existence of four maxima in the temperature range 0 to 75°C can be reconciled only by assuming structural transitions, that is distinct crystalline phase changes associated with a regrouping of water molecules. Furthermore, interfacial water structure must extend for hundreds of Ångströms.

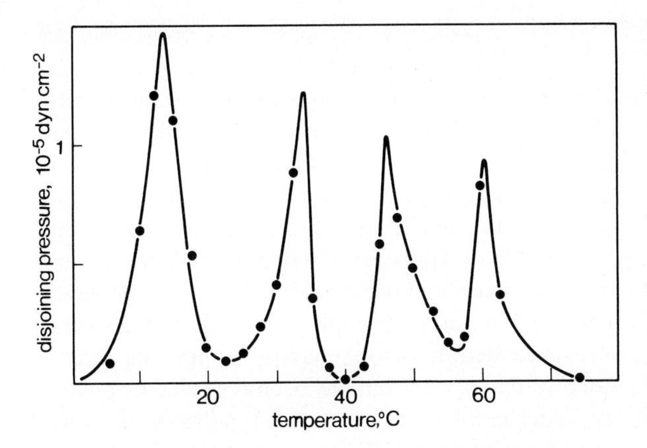

Fig. 10.16. Disjoining pressure as a function of temperature within a water layer 500 Å thick, placed between two quartz-glass surfaces (Peschel and Adlfinger, 1971)

Experimental data on organic films in sea water provide further evidence of long-range ordering effects within the interfacial zone (summarized in Garrett and Zisman, 1967; Hühnerfuss, 1983, 1986). For a boundary layer between water and Na_2SO_4/l-butanol the penetration depth amounted to 300 μm (Rosano, 1967). A schematic presentation of the interaction between a monomolecular surface film and the adjacent water layer is depicted in Fig. 10.17. In the presence of an oleyl alcohol surface film, increased hydrogen bonding and formation of ice-like clathrate structures were observed within a penetration depth of $d \leq 190$ μm (Alpers et al., 1982).

The long-range ordering effect is related to the hydrophobic part of the system, and for distances $d \leq 0.05$ μm the well-known London equation describes the nature of these "hydrophobic" interactions. However, for penetration depth beyond that and up to a few hundred μm, the chemical approach of London dispersion forces is hardly a solution. It has been suggested by Alpers et al. (1982) that, in addition to the powerful forces operating at shorter ranges, polarizing effects of this oriented layer upon the water molecules immediately adjacent to it have to be taken into consideration. The hydrophobic influence sphere becomes quasi extended and transmitted by a successive polarization of neighboring molecules to an impressive depth via long wavelength oscillations. The phenomenon as such has already been noticed by Schufle et al. (1976) and explained as a cooperative long-range structuring effect, due to "low-energy interactions" between the water molecules and the surface.

The picture which emerges of water near a water-air, or a water-hydrocarbon interface is thus one of considerable structure (Drost-Hansen, 1965a, b). Actually (as a sideline) this may have been the starting point for the proposition of a true polymer of liquid water (Box 10.2). Back to reality. Anomalies in the temperature dependence of many properties of water, for instance, surface tension, solubility, viscosity, compressibility, specific heat or magnetic susceptibility, owe their existence to the structural transition of the nature of higher-order phase transitions. The temperatures around which the anomalies – often referred to as "kinks" (Symons et al., 1967) – appear to be centered are roughly 15, 30, 45 and 60°C, and extending over ± 1 to 2°C on either side of the temperatures listed. According to Drost-Hansen (1965a, b), the large entropy of surface formation near 30 to 31°C is undoubtedly due to the "melting" of structural clusters of "cages". The minimum in the entropy at a slightly higher temperature corresponds in turn to a state of greater ordering such as found in a closer packing of spherical molecules.

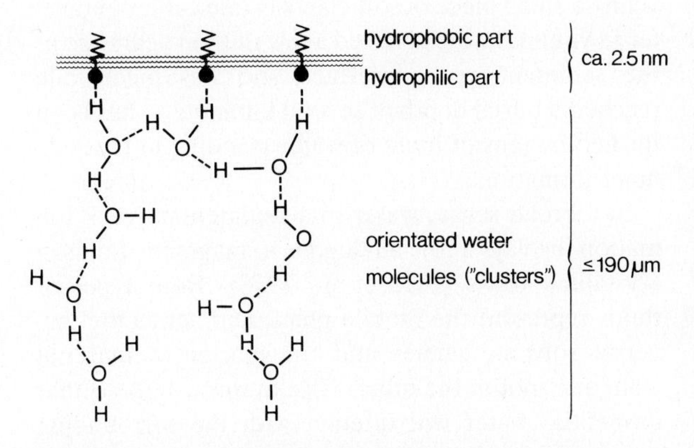

Fig. 10.17. Interaction between a monomolecular surface film, composed of a hydrophobic and a hydrophilic part, and their adjacent water molecules (Alpers et al., 1982)

Water-Mineral Interface

A great number of minerals need water to balance their structure. For instance, the presence of two molecules of water in $CaSO_4$ makes the difference between gypsum ($CaSO_4$ x 2 H_2O) and anhydrite ($CaSO_4$). Some of the water associated with minerals can be released upon heating the mineral to 105°C, whereas the rest is structurally more tightly bound and will come off only by raising temperatures to several hundred degrees centigrade. The way, in which a mineral dehydrates may thereby serve as a structural fingerprint.

In a sediment, water is contained in the pore spaces and along intergranular films. Pore diameter may range from less than a micron to a few millimeters, de-

BOX 10.2

Polywater (Franks, 1981)

"That would make a beautiful Nature paper", said my microbiology friend, when he found out that the Red Sea hot brines were sterile (see p. 137). As a matter of fact, such waters are hard to come by in nature. "Make sure you are right". And I reminded him of what happened to another famous bacteriologist from California, who in the early thirties discovered a bacterium in a meteorite, a medium supposed to be sterile. The biological establishment reacted to the finding of a "cosmic bug" in a rather funny way. They suddenly questioned the validity of his earlier and outstanding work on bacteria in marine sediments, as well as the benchmark paper on micro-organisms in ancient rocks published in Science (Lipman, 1928).

It is true that researchers love to have their papers accepted in Nature or Science. But this must have been different for B.V. Deryagin and V.V. Churaev when they placed a note in Nature on August 17, 1973 in which they openly withdrew from the concept of a new structural form of water B.V. Deryagin had jointly proposed with N.N. Fedyakin in 1962. This apparently stable form of liquid water, termed variously "anomalous water", "modified water", "orthowater", "water II", or "polywater", had indeed made a big splash in science (Franks, 1981). It could be condensated within fine silica capillaries, 1–30 μm in diameter, suspended in a close chamber containing unsaturated water vapor. What has been regarded by many a distinguished scientist as a true polymer of water turned out to be an artifact. And yet this intermezzo on polywater has possibly advanced our knowledge more than many of the papers heralded 20 years ago and forgotten in the meantime. Through the work of B.V. Deryagin across a lifetime of research, including his model on "anomalous water", we became increasingly aware of the intricate nature of water structures near gaseous, liquid, or solid interfaces. As a final note: the Red Sea hot brines continue to be sterile (Watson and Waterbury, 1969)!

pending on shape and size of mineral grains and the extent of matrix mineralization (von Engelhardt, 1960). Groundwater is stored in the more porous sediments such as sands and silts.

Large quantities of waters are continuously extracted from the hydrosphere during the deposition of sedimentary materials. Recent muds may incorporate up to 95 percent water by volume. The moment new clay-sized particles arrive at the sediment-water interface, a kind of suspension is generated which – because of its gelatinous appearance – is often termed the "fluffy" layer. It serves as a substrate for benthic life and bacteria. Much of the water becomes associated with biomolecules and mineral aggregates; in short, it recrystallizes from a liquid to an interfacial state.

The speed at which the water is released from the original bed rock in the course of diagenesis is not only a function of overburden pressure; it also depends on parameters such as mineral composition, texture, and structure of the sediment. The hydraulic gradient controls the direction of flow.

In this context, it is important to know that compaction from a loose sand to a sandstone, or from a soft clay to a shale is usually accompanied by a substantial loss in water. And yet, a deeply buried shale still contains water to about one fourth or one fifth the amount it had at the time of deposition. However, it is not the original water but some older fluid the shale happened to sink into during its downward voyage. Thus, compaction of a marine clay will not cause a recycling of interstitial solutions back to the sea. Instead, pressure will push the clay deeper and deeper into a quasi non-compactible "rigid" water column full of mineral grains and organic matter. Compaction – to speak with K.O. Emery (1960) – is nothing but a slowing-down of sedimentation. In the course of this long-enduring process, lasting perhaps for millions of years, a small piece of soft clay, say one cubic centimeter in volume, which started a few million years ago at the sediment-ocean interface and has meanwhile reached a burial depth of several kilometers, has been flushed by tens of liters of water standing in the sediment formation.

In a broad sense, water in a sedimentary rock formation displays a role analogous to magmatic fluids in crystalline rocks (Fyfe et al., 1978). Both types of fluids represent the mobile phase and act as the universal solvent, carrier and transporter of material from one spot to the other. This, in turn, suggests that interstitial water will interact with the surrounding mineral and organic matter and will not preserve indefinitely its chemical composition as fixed at the time of deposition.

The old saying by Plinius (23/24–79 A.D.) that "Tales sunt aquae, qualis terra per quam fluunt" (waters take their nature from the strata through which they flow) carries indeed a profound meaning.

On the continents, in areas where precipitation is abundant, the turnover of migrating meteoric water in subterranean water basins or groundwater tables is relatively rapid. This implies that the time of reaction with the sediment or soil is kept at a minimum. In contrast, in desert regions, where recharge and subsurface migration of water is comparatively slow and over long distances, water will stay longer in contact with the surrounding rock strata. Consequently, the chemical alteration is more severe. Very often, if time allows and if the rock formation is favorable, a chemical stratification pattern can develop in the sense that the more saline water will occupy the deepest, and the less saline water the highest level of a subterranean water reservoir (Chebotarev, 1959).

Early work on interstitial waters in pelagic sediments by Bruevich (1957) in the Okhotsk Sea, Bruevich and Zaytseva (1960) in the NW Pacific and Rittenberg et al. (1963) off Guadelupe Island, Mexico, indicated no significant differences from standard sea water either in salinity or chemical composition back to the Middle Miocene. With the advent of global ocean drilling during the seventies and eighties, a growing body of information has become available on interstitial solutions of marine sediments including those deposited under restricted conditions such as the Black Sea, Red Sea, or the Mediterranean (e.g., Manheim, 1976; Manheim and Schug, 1978; Gieskes, 1975; Michaelis et al., 1982). Accordingly, the former concept implying no temporal changes has to be modified.

Ion-Filtration by Charged-Net Clay Membranes

Experiments by von Engelhardt (1967) and Rieke et al. (1964) have shown that marine clays upon compaction (0–200 psi) release water somewhat enriched in electrolytes relative to the original interstitial water, whereas a gradual increase in pressure from 200 psi up to 200,000 psi yields water that exponentially decreases in electrolytes. Considering that the waters expelled at the initial stages of diagenesis are only slightly enriched in salinity (10–20 percent) relative to mean ocean water, this mechanism cannot account for the high salt level up to values in the order of 30 percent as found in petroleum brines and deep formation waters above crystalline basements. It has to be emphasized that all of these waters are derived from highly permeable rocks such as sandstones and certain limestones. In contrast, shales intercalated with these brine-producing sandstones are supposed to be low in electrolytes in the light of the high-pressure work on marine clays by Rieke et al. (1964).

This contrasting salinity pattern between high- and low-porosity rocks is brought about by the ability of shale beds in situ to act as ideal membrane electrodes. The most suggestive argument for this assumption is the observed constancy of a "shale baseline" on "spontaneous potential" logs found in drill holes in every part of the world (Wyllie, 1955). A quantitative theoretical treatment of the electrochemical properties of clays is given by the theory of membrane behavior of Meyer-Sievers-Teorell (Davis, 1955), which, according to the calculations based on the spo-curve (spontaneous potential) of e-logs, approximates the behavior of shales in situ in the Earth. A comprehensive mathematical treatment of this phenomenon has been presented elsewhere (Wyllie, 1955, 1958; Bredehoeft et al. 1963; Degens and Chilingar, 1967).

The filtration of salt solutions through charged-net clay membranes has been suggested as a mechanism for producing fresh water from the ocean (Ellis, 1954). The general consensus is that during compaction of clay-containing sediments salts are filtered and accumulate in the formation water retained in the strata. Thc process of salt removal or concentration depends on the large excess charge permanently attached to the clay membrane which prevents the passage of like-charged ions. In other words, the separation is effected because of the electrical properties rather than the size of the electrolyte. No such restriction is placed on the water molecules. They move and will pass the electrolytes. This process, therefore, should yield a lower salt level in the filtrate as compared to the original solution. Thus, the salt is filtered by virtue of its electrolytic dissociation and the electric properties of the clay-membranes (Samoilov, 1965).

Compacted clays and shales commonly have pore spaces which measure less than 1,000 Å in diameter. Interstitial water should thus behave as if placed between smooth mineral platelets a few 100 Å thick. Now to say that the structure of this water is of an interfacial nature and consequently of high crystalline order would be too simple a statement. We have to realize that the system is far more complex: water is salty, and polar and non-polar organic matter coexists in the mineral aggregate. Furthermore, part of the organic matter is water-soluble, another part is only sparingly soluble, and a third part clings tightly to the mineral surface. We now make the heuristic assumption that (i) the bulk of the organic matter has become amalgamated with the clay surfaces, (ii) the interstitial aqueous solution is low in salt and polar organic molecules as a consequence of ion-filtration in the course of compaction, and (iii) non-polar solvents, for example hydrocarbons, are present only in quantities soluble in the aqueous ambience. This scenario roughly describes the situation petroleum geochemists believe to exist in a petroleum source bed, prior to the migration of hydrocarbons (e.g., J.M. Hunt, 1979).

Starting from this model, interfacial water in a shale is supposed to be structured by the short-range "high-energy effects" and the long-range "low-energy effects". The first category of forces involves hydrogen bonding, and ion-dipole or dipole-dipole interactions. They are also frequently referred to as "structural" or "hydration" forces, and arise from the energy needed to dehydrate interacting surfaces which contain ionic or polar species. The second category which concerns the hydrophobic species, is governed by London dispersion forces and interactions of long wavelength oscillations involving successive polarization of neighboring molecules. To what extent diffusional interactions proceed between the various interfaces is not clear yet. However, in view of the temperature dependence of interfacial water structure and as shown by Israelachvili and Pashley (1983), the oscillatory and repulsive character of hydration force between molecularly smooth clay surfaces, temporal changes of the micro-environment across "flickering" boundaries ought to be expected.

Ancient fine-grained sediments of all environments yield considerable quantities of hydrocarbons. Ancient sediments of non-petroleum-producing regions have an average hydrocarbon content of ca. 300 ppm, which is about five to ten times higher than that observed in recent sediments of the same lithology. Taken together, this amount in hydrocarbons exceeds the world's crude oil reserves by more than 100 times (see J.M. Hunt, 1979). It is only necessary to extract small quantities of sediment hydrocarbons (though over a large area) to form petroleum pools. One may then assume, following Baker's theory (1960), that during compaction and dehydration, hydrocarbons are extracted from a sediment only in those quantities which are soluble in the interstitial aqueous solution.

Solution experiments with alkanes in water (reviewed by Wicke, 1966) revealed that the interaction of these exclusively hydrophobic compounds with water induced major structural adjustments around an alkane solute, as shown by a negative entropy term and an exothermic reaction. Moreover, the degree of hydrocarbon solubility in water is different for various hydrocarbon species. As a general rule, increase in molecular weight will decrease the solubility of hydrocarbons in the aqueous medium; and increase in salt concentration up to a level of a few percent will increase their solubility. Inasmuch as only a tiny fraction of the potentially available hydrocarbons present in a petroleum source bed is associated with interfacial water, a natural selection of molecular species will proceed in the course of time. Only those species that best fit the structural frame provided by interfacial water will survive when it comes to the extraction of interfacial fluids and the generation of crude oil.

On thermodynamic grounds, caged-in non-polar solutes tend to move closer together and share their solvation sheath molecules. Since shale beds move through a sediment column during compaction, there will be a constant flushing of new water from below and in consequence a release of earlier acquired water on top of the shale. Should the overlying stratum be a sandstone or porous limestone, a transition from subcapillary to capillary to supercapillary pores will lead to a restructuring of the interfacial water in the direction of liquid water. This feature, and the often highly salty ambience in the porous rocks, will cause a desalting of the "interfacial" emulsion and the generation of oil droplets. Because of their buoyancy, oil drops will rise, join others and – tectonic settings permitting – accumulate in the form of oil pools. The situation is just the opposite to that described earlier, when iron droplets formed in the Earth mantle and sank downwards to form an iron pool in what we refer to as the iron core.

The generation of petroleum and natural gas deposits will only proceed during late stages of diagenesis and catagenesis (Box 10.3). Elevated temperatures not only promote the cracking of organic matter in the presence of catalyzing clays, but they may induce a structural rearrangement of the interfacial water which is known to exhibit a sort of "kinkiness" at certain temperatures. This, in turn, should effect the structural order of the non-polar guest molecules in clathrates and enhance their mobility.

In conclusion, the system clay-electrolyte-water-organic matter as it represents itself in shale beds in situ may be viewed as an aggregate of layered crystals having a basal spacing less than 1,000 Å. As a function of temperature, water content and chemical speciation, individual layer units may expand or contract even to a point of changing the crystalline order irreversibly. Interfacial water is the key element of structural metamorphosis. These dynamic properties are different from those established in common minerals such as quartz, feldspar or mica, but are reminiscent of structures that prevail in biological membranes, where they function as dynamic molecular sieves (Matheja and Degens, 1971a). Here as well as there

BOX 10.3

Diagenesis of Organic Matter

The biological cell contains about 80 percent water by weight which is principally in the bound rather than in the free state. Upon burial and diagenesis, cellular water will be lost to the sediment while in turn organic matter undergoes structural and chemical changes at rates largely being controlled by microbial activities, temperature, mineral composition, and rock fabric.

Organic materials embedded in sediments can be grouped into compounds that are survivors of diagenesis and products of diagenesis. Diagenesis can be defined as the all-embracing term to describe post-depositional transformation processes, both physical and chemical, that alter sedimentary matter under low-temperature and low-pressure conditions. The first group includes all those organic molecules which are chemically similar or identical to living matter. In contrast, compounds that arose during diagenesis from the organic debris but are generally not part of plants and animals belong to the second group.

Survivors of diagenesis: The organic pigments from plants and animals show a surprising stability in a wide range of diagenetic habitats. There is ample evidence for the long-term preservation and antiquity of fossil porphyrins as old as the Precambrian. Similarly, intact peptides have been isolated from molluscan shells of the late Paleozoic. Amino acids and other biologically interesting monomers, such as sugars or the bases of the purines and pyrimidines, are common biochemical fossils in ancient sediments. This attests to the perfect preservation of many an organic compound throughout Earth history.

Products of diagenesis: Although most carbon in the Earth's crust has cycled through organisms, the bulk of the organic matter has lost its biological identity in the course of diagenesis. Polymers gradually break up into their monomeric building blocks, and these in turn can be modified severely by elimination of functional groups: for example, $-COOH$, $-OCH_3$, $-OH$, $-C=O$, $-NH_2$, hydrogenations and isomerizations, cleavage reactions, and, in general, processes which destroy the ordered building pattern of biochemical compounds.

These breakdown products can reorganize and polymerize into geochemically stable configurations; typical examples are the humic acids, coals, and kerogens. Humic acids are hereby defined as the materials that can be extracted from a sediment with 0.3 N NaOH and that subsequently can be precipitated upon acidification. Kerogen, on the other hand, is the so-called insoluble organic matter that is left in a sediment after organic solvent extraction and acid and base hydrolysis. Thus, both terms refer to a heterogeneous organic complex rather than to a well-defined organic constituent. Other constituents are stable on their own; hydrocarbons belong to this group of compounds.

In the early states of diagenesis, the bulk of the biochemical macromolecules – for instance, proteins, polysaccharides, fats, or the nucleic acids – will be rapidly eliminated. This is largely a result of microbial activity. Breakdown products, such as amino acids, sugars, bases of the purines and pyrimidines, fatty acids, and phenols, can interact and give rise to constituents commonly termed humic acids, fulvic acids, humins, and ulmins. Inasmuch as most of these complex humic materials are no longer of nutritional value, they may accumulate as organic residues even in the zone of microbial activity. Interactions with the surrounding mineral matter may stabilize the organic complex or may catalyze certain reactions.

Although plants represent the main precursor for the organic matter in sediments, microorganisms have extensively modified the organic debris in the early stages of diagenesis. This is reflected in the high abundance of sterols and hopanes, many of which are biomarkers of bacterial cell membranes. By the way, hopanes contained in kerogens are the most prolific single class of organic compounds present on Earth and they may well comprise half of all terrestrial organic matter (Ourisson et al., 1979). It is thus a matter of opinion whether plants or microbial metabolic and decay products should be considered the chief source of the organic matter in sediments. After termination of bacterial action, largely as a result of the depletion in available food materials, the diagenetic history of the organic matter is that of slow inorganic maturation and redistribution (migration of hydrocarbons). Loss in functional groups may result in the formation of hydrocarbons, phenols, amines and so on, and may cause a reduction in aliphatic side chains in former humic acids. The resulting coal or kerogen-type material will become more and more aromatic in nature. The number-one factor causing the diagenetic alteration is thermal degradation.

It is inferred that practically all former biochemical matter, independent of chemical nature, origin, or environment of deposition, is diagenetically reduced to either coal, kerogen, or the various crude oil components. If given sufficient time, the organic residues will gradually acquire the structural characteristics of graphite, whereas the petroleum will become more

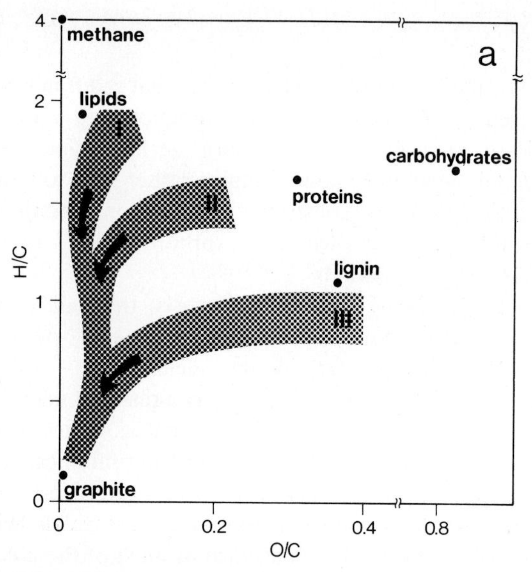

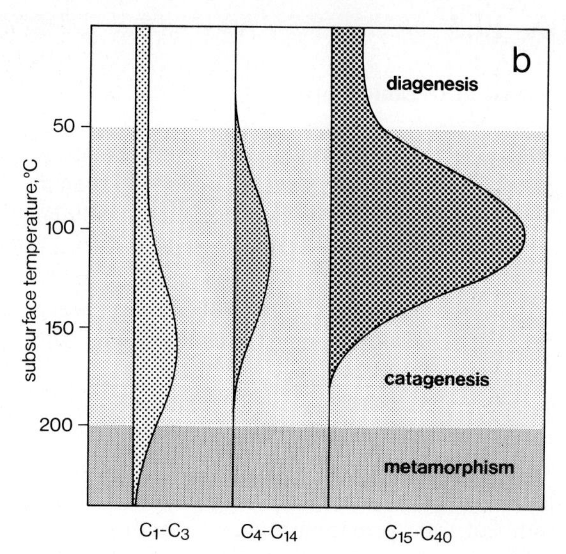

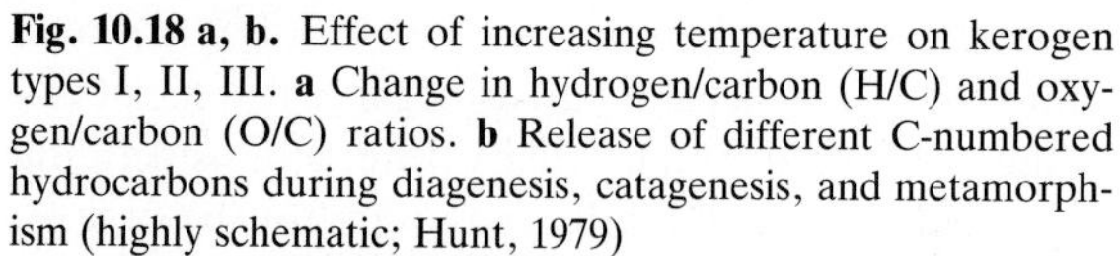
Fig. 10.18 a, b. Effect of increasing temperature on kerogen types I, II, III. **a** Change in hydrogen/carbon (H/C) and oxygen/carbon (O/C) ratios. **b** Release of different C-numbered hydrocarbons during diagenesis, catagenesis, and metamorphism (highly schematic; Hunt, 1979)

and more paraffinic. In its final stage, all organic matter will largely end up as graphite, methane, carbon dioxide, ammonia, and water. These are essentially the same products from which the organic precursor materials for life were synthesized on the primitive Earth under favorable environmental circumstances.

An illustration of the main path of diagenesis of the principal organic constituents is presented in Fig. 10.18 using O/C (oxygen-carbon) and H/C (hydrogen-carbon) ratios as a convenient plot scheme. One commonly distinguishes between three kerogen types. According to J.M. Hunt (1979), Type I is mostly normal and branched paraffins, with some naphthenes and aromatics. Type II is predominantly naphthenes and aromatics, whereas Type III contains a high percentage of polycyclic aromatic hydrocarbons and oxygenated functional groups plus some paraffin waxes.

Oil and gas obtain their hydrogen from kerogen and H/C decreases to a minimum of about 0.45 corresponding to nine condensed aromatic rings per plane (J.M. Hunt, 1979). Beyond this, no oil and only small amounts of gas are generated as H/C values drop to about 0.3. It is evident that the hydrogen-rich algal Type I kerogen is more oil-prone than the hydrogen-poor woody Type III kerogen which prefers to evolve towards coal and anthracite. In Fig. 10.18, temperature windows for oil and gas generation in sediments are schematically drawn showing the response of organic matter to different temperature regimes. Diagenesis covers the temperature range up to 50°C, whilst catagenesis envelops the range 50 to 200°C, from where on we speak of metamorphism. The subject matter has been fully covered by Tissot and Welte (1984) and J.M. Hunt (1979).

the so-called hydrophobic interactions play a fundamental role in structurally shaping interfacial water. For a more detailed account on the structural and functional role of water in biological systems see: Pullman et al. (1985), Luck (1976), Narasinga-Rao (1967) and Matheja and Degens (1971a).

Modern Ocean

Water Distribution – Chemistry – Currents

The ocean has a volume of 1.37 x 10^9 km^3, which represents about 70 % of all waters contained in crust and hydrosphere. A compilation of the size of near-surface water reservoirs is given in Table 10.1. It has been found useful to express individual water reservoirs in terms of sphere depth, which is the depth a given volume of water would yield if it were evenly

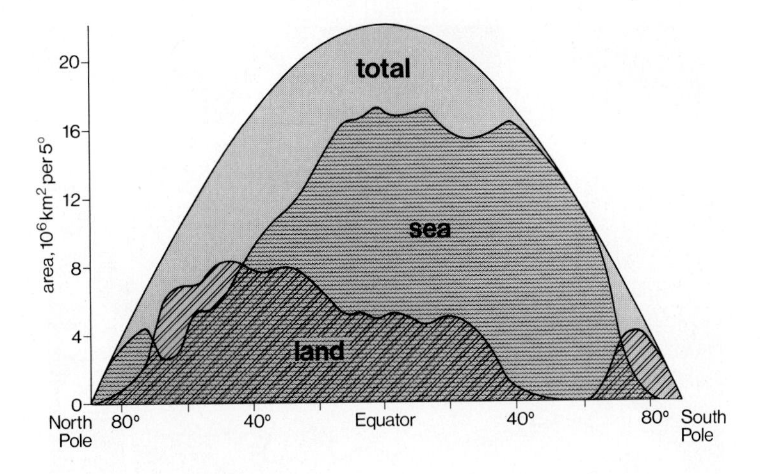

Fig. 10.19. Distribution of water and land over the earth (Baumgartner and Reichel, 1975). Note the skewed distribution between northern and southern hemisphere

Table 10.1. Distribution of the Earth's water by volume and sphere depth

Type of reservoir	Volume (km^3)	Sphere depth (m)
Water vapor and condensate in the air	$\leq 2 \times 10^4$	ca. 0.03
Organisms	1.6×10^5	0.3
Rivers and lakes	5.1×10^5	1
Groundwater	5.1×10^6	10
Glacial and other land ice	2.3×10^7	45
Crust (free + silicate-bound)	6×10^8	1,180
Sea	1.37×10^9	2,685

spread as a layer around a smooth sphere having a surface area the size of our globe, i.e., 5.1×10^8 km^2. For comparison, the sphere depth of solid land currently above sea level would amount to 244 m (Duxbury, 1977).

Mean ocean depth is about 3,800 m, and for percent ocean area between various depth zones see Fig. 8.15 (after Kossinna, 1921; and Sverdrup et al. 1970). The distribution of water and land over the Earth is uneven, as a glance at Fig. 10.19 will reveal. Obviously, the northern hemisphere is more continental than the southern hemisphere. All in all, land covers 29 percent and the sea 71 percent of the globe.

It appears that only six major elements occur in marine salt: sodium, magnesium, calcium, potassium, chlorine, sulfur and carbon. Accordingly, the bulk of the oceanic salts are Na-Mg-Ca-K chlorides, sulfates and carbonates (Table 10.2). They comprise totally 35 g of anhydrous salt per one liter of sea water or close to 5×10^{19} kg of salt for the entire ocean.

Independently of whether we take a sample of sea water from the central part of the Pacific, the Indian Ocean, or the Atlantic, the salinity is always close to 35 per mil and the chemical composition is constant. In other words, World Ocean is chemically uniform. A more refined analysis, however, will reveal that this statement is not true. Today there is an arsenal of other sensitive analytical instruments and techniques which make it possible to measure minute changes in water chemistry. This, in turn, helps to trace water masses from one location to another and to decipher the intricate interactions within the biogeochemical system (e.g., Bolin et al., 1983).

Chemical input into World Ocean can come from all directions. In the immediate vicinity of river outlets, salinity is lowered; and yet it is surprising to see that this influence is felt only in a comparatively small area. Wind and waves plus tidal movements allow for a rapid dissipation of the fresh water additions. Or let us take the example of a major inland sea, the Mediterranean, where, due to high evaporation rates combined with restriction of water exchange with the

Table 10.2. Major constituents of sea water (Standard Chlorinity = 19‰) (After Sverdrup et al., 1970, and Braitsch, 1962)

Ion		‰	Calculated molecular composition (Weight percent)	
Sodium	Na^+	10.56	78.03	NaCl
Magnesium	Mg^{2+}	1.27	0.01	NaF
Calcium	Ca^{2+}	0.40	2.11	KCl
Potassium	K^+	0.38	9.21	$MgCl_2$
Strontium	Sr^{2+}	0.013	0.25	$MgBr_2$
Chloride	Cl^-	18.98	6.53	$MgSO_4$
Sulfate	SO_4^{2-}	2.65	3.48	$CaSO_4$
Bicarbonate	HCO_3^-	0.14	0.05	$SrSO_4$
Bromide	Br^-	0.065	0.33	$CaCO_3$
Fluoride	F^-	0.001		
Boric acid	H_3BO_3	0.026	100.00	

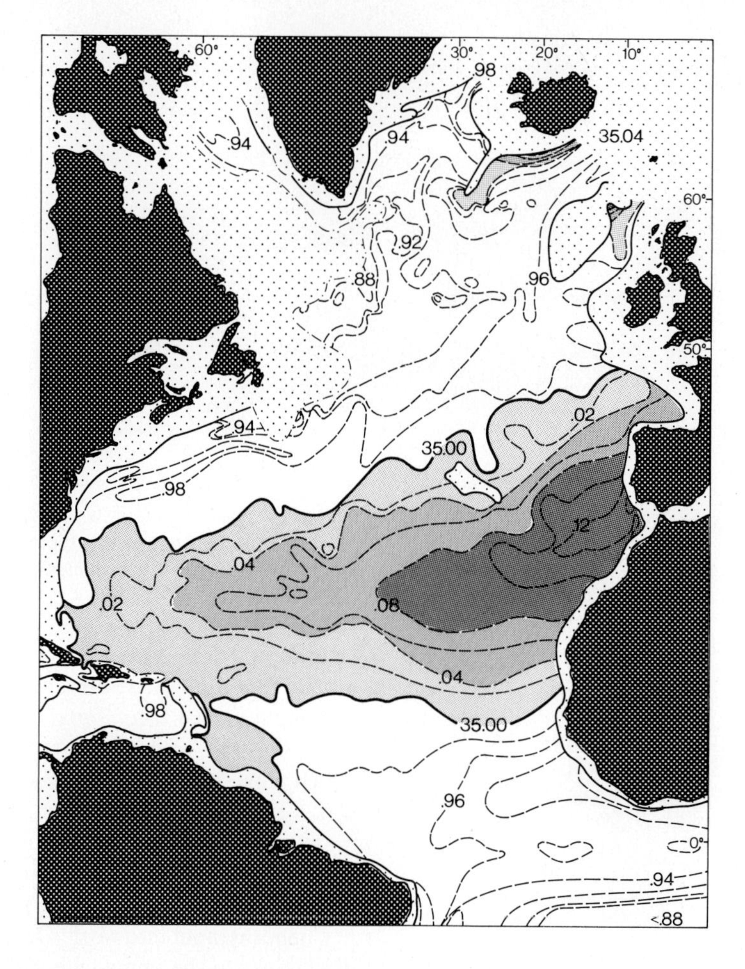

Δ
Fig. 10.20. Saline spill from the Mediterranean through the Straits of Gibraltar. Salinity pattern in permil at the 4° C potential temperature surface (Worthington and Wright, 1970)

Fig. 10.21. Schematics of water circulation at surface and depths in the Atlantic (Defant, 1961; Duxbury, 1977)
∇

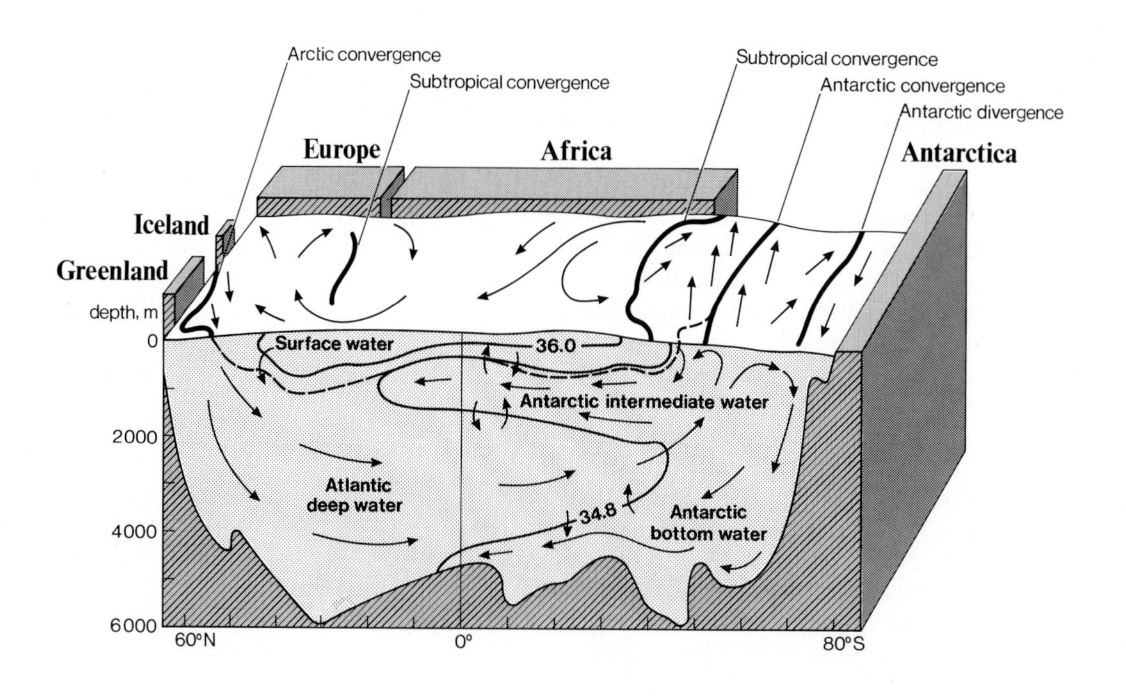

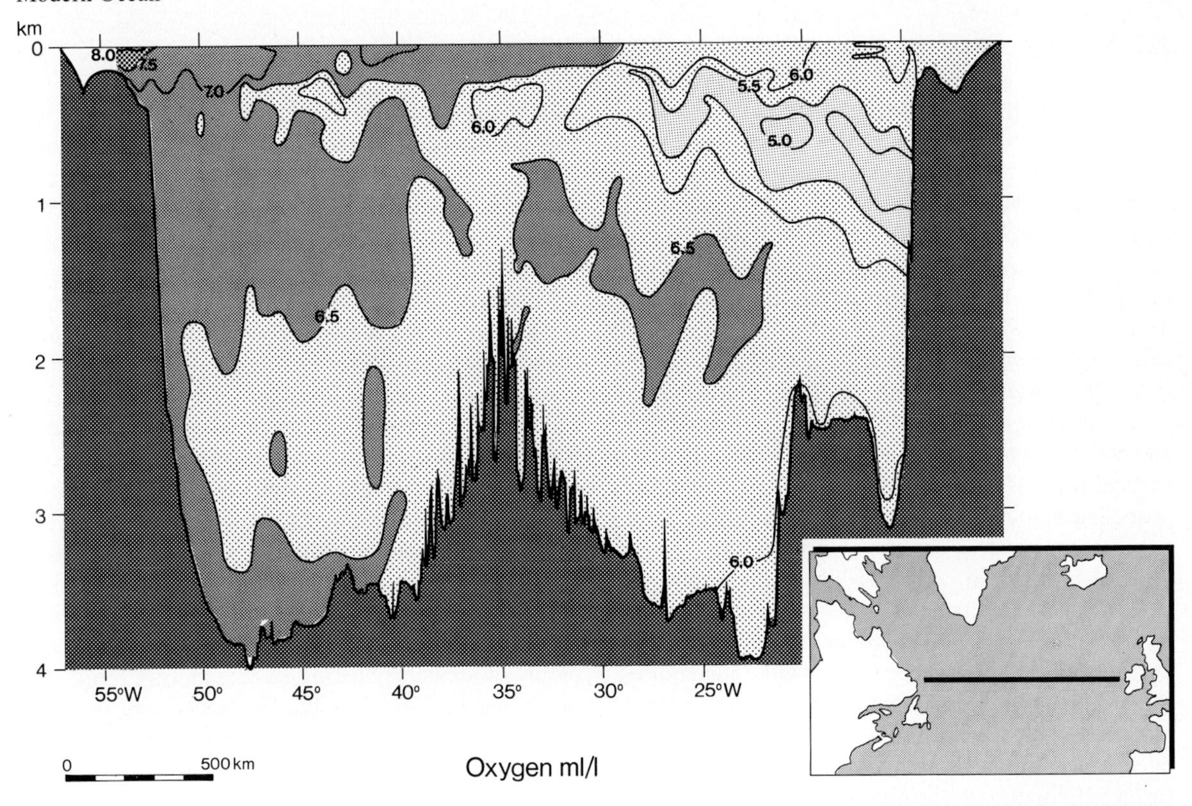

Fig. 10.22. Oxygen profile across North Atlantic, in mg O_2 per liter (Worthington and Wright, 1970)

open ocean, the salinity can be raised to values of up to 38 permil. This dense Mediterranean water will flow through the Straits of Gibraltar into the Atlantic. Because of its higher density, relative to mean ocean water, it will sink to a density compensation depth and, once there, it will rapidly spread across the whole Atlantic (Fig. 10.20).

Actually, density-driven circulation, due to thermal and salinity changes are the principal means of mixing water masses in the sea (Box 10.4). Take the Atlantic Ocean: Surface currents bring water together at the major convergences to mix and sink to the appropriate density level (Fig. 10.21). By this mechanism large volumes of water are slowly carried to a greater ocean depth and layered structures emerge. Eventually the water upwells slowly over the entire ocean surface as it becomes displaced by new and colder water masses. It is this downwelling of polar waters which supplies dissolved molecular oxygen to the deep sea (Fig. 10.22) and that is responsible for ventilation and mixing. A slowing-down of this oxygen pump, for instance, as a result of climatic change, may severely impair benthic life in deep water.

The ocean is also fed from below. Hydrothermal waters charged with gases, metal ions and organics emanate into the sea along rift zones, while at the same time water is drawn elsewhere into the oceanic crust, where it will circulate and interact with the basaltic rock for up to a few million years to surface finally back to the sea again. In a way, water present in marine sediments and the basaltic crust is simply a vertical appendix of world ocean that happens to be filled with sediment and rock. It has to be remembered, however, that most of that water is bound in hydrous silicates. This situation is comparable to the living cell, where the bulk of water is in the hydrated form.

Finally, air-sea interactions will cause exchange of gases, liquids and particles which can lead to transfer or fractionation of elements (Buat-Ménard, 1986).

Summing up, physical forces are the principal means to distribute changes that take place along the air-sea, land-ocean, cell-water, or sediment-water interfaces across the ocean at large. Two case studies may further illustrate the biogeochemical relationships that arise within or on the periphery of the sea.

Upwelling – Mineral Nutrients – Primary Productivity

The inverse relationship between dissolved CO_2 and molecular oxygen at certain depths in the ocean will find its precipitation in the distribution of all those ele-

BOX 10.4

Circulation in the Sea (Duxbury, 1977)

It is often said that ocean currents, with the exclusion of tidal movements, are essentially driven, either directly or indirectly, by the force of wind. Wind, on the other hand, is generated by solar radiation and resulting variations in heating over the Earth's surface. This, in turn, will exert a drag on the sea surface and cause surface water to flow from one climatic regime to another. Along that path, cooling or heating and evaporation or precipitation may change temperature and salinity of the water and initiate a density-driven vertical flow. Accordingly, thermohaline circulation is propelled by both internal and external forces. Internal forces arise from the mass field (density distribution) within the sea, and act whenever constant density surfaces are inclined to level surfaces. Water movement along the inclined density surfaces is gravity controlled. On the other hand, wind, the major external force on thermohaline circulation, becomes modified once the circulation has started along the line: winds promote currents, currents redistribute surface thermal patterns, thermal patterns alter wind fields, and wind fields alter currents.

The moment currents are set in motion, forces such as the Coriolis effect modify them by deflecting the moving water and sending it off in a slightly new direction. Gaspard Gustave de Coriolis (1792–1843) was the first to observe that an object bearing across a turning surface veers to the right or left, depending on the direction of rotation (McDonald, 1952). With respect to the moving surface water of the central oceans, they are deflected to the right in the northern hemisphere, and to the left in the southern hemisphere yielding respectively clockwise or counterclockwise gyres (Fig. 10.23). Continents, as well as a rugged sea floor, will deflect currents into curved paths. Inasmuch as the mean circulation of the oceans is a function of the internal mass force field and the external wind force field, both of which operate on different time scales, surface currents will generally spread more rapidly across the seas when compared to the slowly moving circulation flow of water masses in the deep sea. However, current speeds of 1 knot or more do occur in the deep sea, but they are commonly of oscillatory nature only. As a very rough estimate one can say that the euphotic zone becomes fully mixed in about 50 years, whereas the deep waters are turned over in a matter of 500 years.

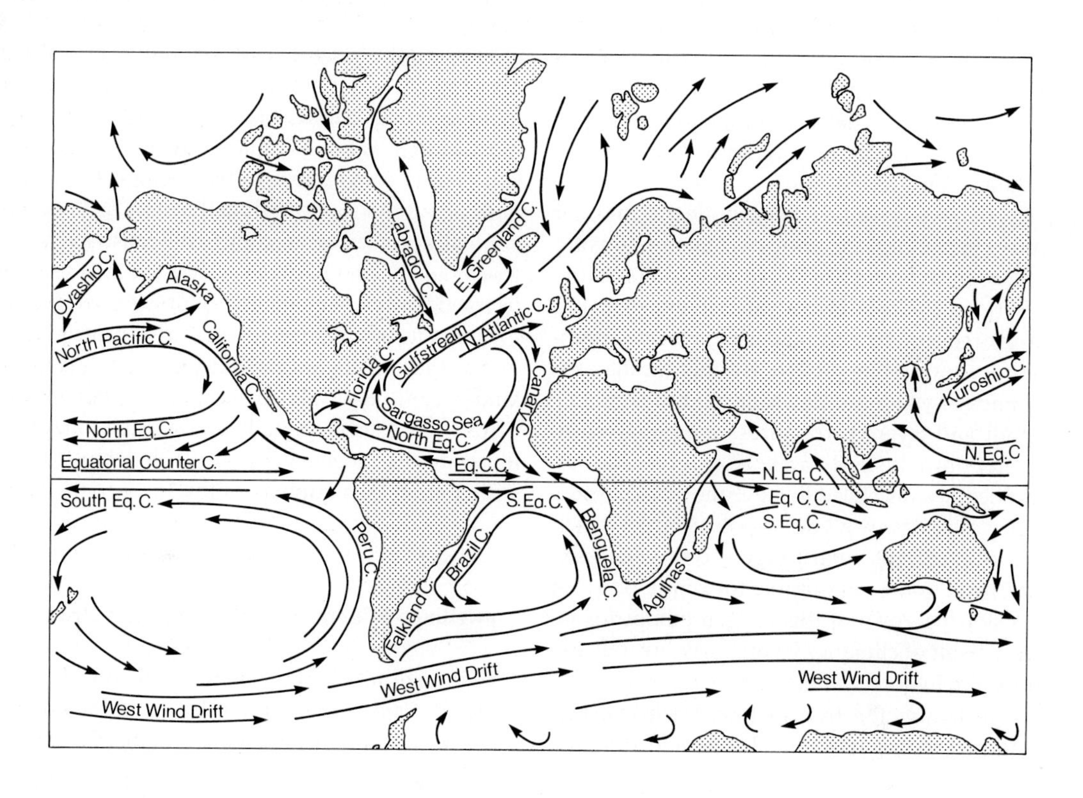

Fig. 10.23. The major surface currents of the ocean (Duxbury, 1977). *Arrows* indicate direction of principal flow but not the speed

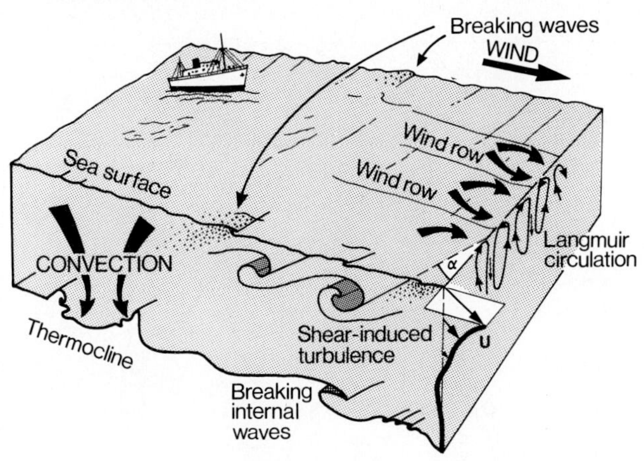

Fig. 10.24. Mechanisms affecting turbulence in the upper ocean boundary layer (after Thorpe, 1985). The spiral-like structure of the mean current (U) is due to the Earth's rotation. The sign of direction in the northern and southern hemispheres is reversed. Breaking surface waves and Langmuir circulation are processes that redistribute floating particles or biota. Breaking internal waves may affect the flux of mineral nutrients. The wakes of ships contribute little to the net fluxes, except for low frequency sound. Note the bubbles which are important in the transfer of gases, surface films, and the formation of aerosols via sea spray

Small-scale processes in the upper ocean boundary layer have a significant impact on the rate at which the water is turned over. Moreover they determine the transfer of floating particles and mineral nutrients in the sea, as well as regulate air-sea exchange of gases, organic molecules and salts. Bubbles are a convenient vehicle to accelerate the interactions (Thorpe, 1985; Fig. 10.24).

ments needed for the maintenance of aquatic life and what we commonly refer to as the nutrients or dissolved nutrient minerals of the sea. Quantitatively most important among them are soluble phosphates and nitrates, but obviously lack of any other essential trace compound will inhibit cellular growth.

Within the euphotic zone, mineral nutrients (MN) propel photosynthesis and molecular oxygen is generated at the expense of CO_2 (schematic):

$$MN + CO_2 + H_2O \xrightarrow{h\nu} (CH_2O(MN)) + O_2,$$

whereas respiration will work the other way around. Organic matter of marine plants has an O:C:N:P ratio of 212:106:16:1 by atoms or 109:41:7.2:1 by weight. Thus, one atom of phosphorus derived from the uptake of dissolved phosphate is needed to produce 106 atoms of organic carbon (Redfield ratio, named after the eminent oceanographer Alfred Redfield). Inasmuch as primary productivity removes mineral nutrients from surface waters in accordance with the Redfield ratio, any one of the essential elements may become a limiting factor in cell growth once too much tripton (organic detritus) leaves the euphotic zone. In the shallow sea, mixing of water masses and recharge from rivers or ocean currents may compensate for the losses. However, in a number of coastal environments and the equatorial currents, a plankton bloom can be triggered by upwelling of water masses from below the euphotic zone, that is from a few hundred meters depth at the most.

Present space craft technology makes it possible to measure a wide number of ocean variables critical to climate, e.g., primary productivity, ocean color, sea surface topography, ocean wave height, or sea surface winds. Numerous projects, for instance, Global Habitability or the International Geosphere-Biosphere Programme (IGBP), aimed at better understanding human impact upon the environment in a holistic fashion, take advantage of remote sensing techniques of the type presented in: e.g., B.S. Williams and Carter (1976), Short et al. (1976), Mather et al. (1979), Elachi (1980), Dixon and Parke (1983), JOI (1984), Szekielda (1986, 1987), Botkin (1986).

A good example of the usefulness of oceanography from space are satellite scanning pictures of the Somali current off East Africa, which allow the mapping of changes in surface water temperature across part of the Arabian Sea and reveal zones of upwelling (Fig. 10.25). Inasmuch as shipboard measurements were made at the same time, the remote sensing data could be calibrated and the study of the upwelling phenomenon be extended to greater water depth (Fig. 10.26). It is noteworthy that the temperature gradients during the early formation stage are directly proportional to wind speed. The block diagram (Fig. 10.27) delineates the flow paths that occur in the process of upwelling. Since intermediate waters are rich in mineral nutrients, upwelling is a means to mineralize the euphotic zone. Surface currents will subsequently carry the dissolved mineral species away from their point of emergence.

In conclusion, physical forces involved in the generation of a pycnocline or in upwelling are principal fac-

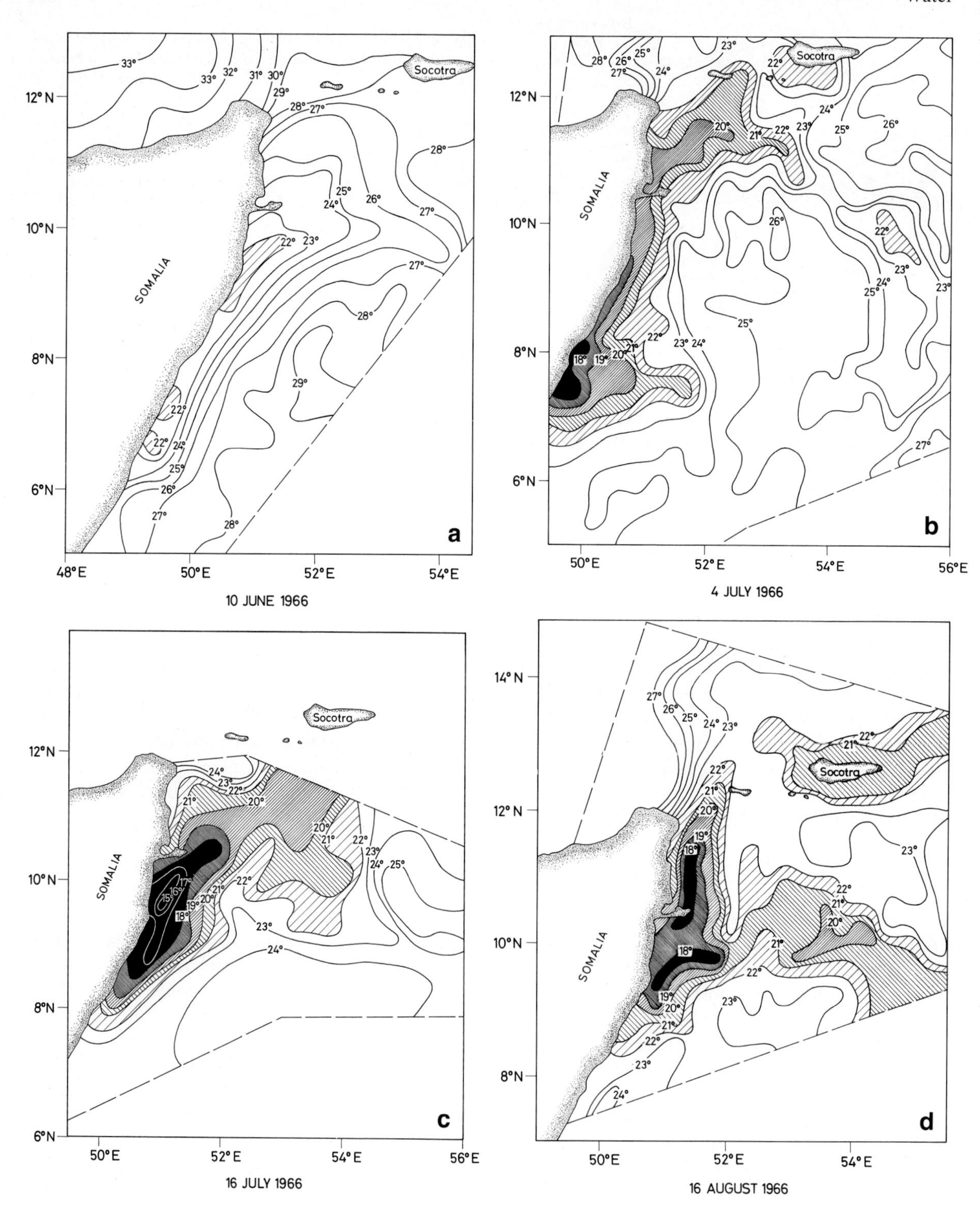

Fig. 10.25 a–d. Changes in surface water temperatures in NW Indian Ocean due to upwelling in the time period 10 June to 16 August 1966 (Szekielda, 1987). The various shapes of windows shown are due to different cloud covers on the days of satellite scanning

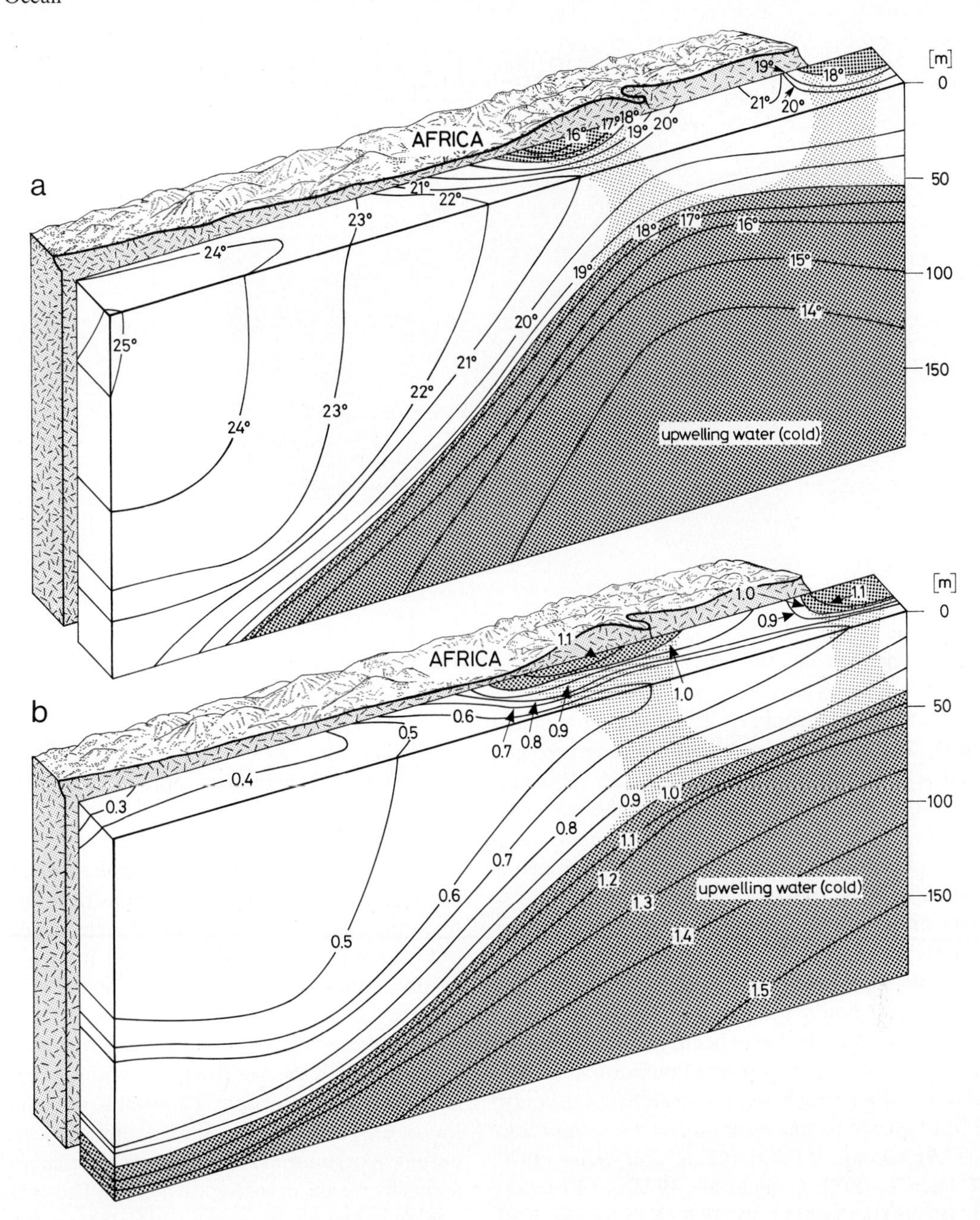

tors for maintaining optimal mineral nutrient conditions in the euphotic zone. Surface currents will counteract by dissipating the dissolved mineral matter.

Fig. 10.26 a,b. Block diagrams, depicting flow paths of water masses in the NW Indian Ocean at the time of upwelling (based on shipboard measurements and satellite data). **a** shows temperature relationships in °C, and **b** phosphate content in μg at l^{-1} (Szekielda, 1987). Note the relatively shallow water depth of the cold, nutrient-rich water mass

Carbon in the Sea

The ocean is the principal carbon reservoir on the surface of the Earth, but in comparison to the abundance of carbon in various crustal compartments, carbon in the sea represents just a trifling amount. Marine carbon is partitioned into an oxidized and a reduced pool. Bicarbonate (HCO_3^-) is the principal oxidized species and dissolved organic carbon (DOC) the major reduced form. The sizes of the two reservoirs are 39,000 and 1,000 Gt C, respectively. These figures are substantial considering that the estimated total

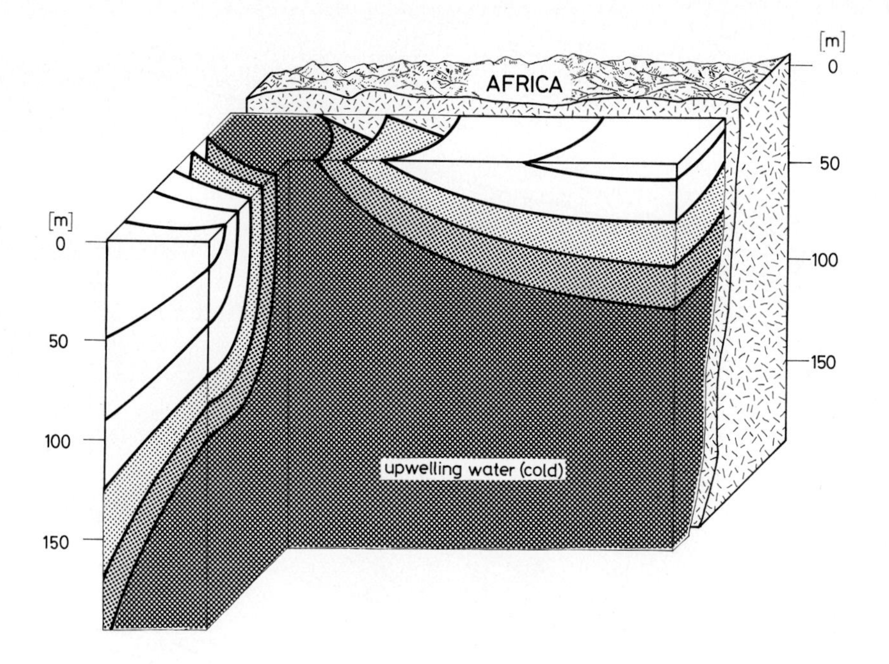

Fig. 10.27. Block diagram illustrating the flow paths that may occur in the course of upwelling

biomass on earth is only 560 Gt C (Fig. 9.23). Recent estimates for the dissolved organic carbon pool go as high as 3,000 Gt C (Suzuki et al., 1988; Sugimura and Yoshimi, 1988) and intercalibrations presently performed seem to corroborate this number. This also concerns the dissolved nitrogen pool in the sea which is substantially higher than formerly thought. High polymer organic matter ($4 \times 10^3 - 2 \times 10^4$ daltons) is the principal DOC species.

The concentrations of inorganic carbon compounds CO_{2aq}, HCO_3^-, and CO_3^{2-} in the ocean arise from thermodynamic and kinetic balances. To avoid the cumbersome calculation of ion pair influence in sea water, equilibrium constants have been measured in sea water separately. The largest set of measurements was obtained in the "Geochemical Ocean Section Study" (GEOSECS) and its successor project "Transient Tracers in the Ocean" (TTO) (Craig and Weiss, 1970; D.W. Spencer, 1972; Craig et al., 1972a,b; Broecker et al., 1979; Takahashi et al., 1981; Kroopnick, 1980; Chen-Tung et al., 1986).

The subject matter at large has been reviewed in several recommendable treatises. First, I would like to pay tribute to Lars Gunnar Sillén, who more than a quarter of a century ago, on the occasion of the International Oceanographic Congress in the home of the United Nations, gave a presentation on the physical chemistry of sea water. Actually, it was this memorable talk that triggered modern ocean chemistry. I have thoroughly read Sillén's article of 1961 many times and still find it an inspiring source (Box 10.5). Stumm and Morgan (1981), Broecker and Peng (1982) and Pytkowicz (1983) have highlighted aquatic chemistry and carbonate equilibria in the hydrosphere.

The ocean is virtually saturated with respect to $CaCO_3$ and cannot enlarge its reservoir size significantly. River input and contributions from subaquatic hydrothermal activities are therefore balanced by output from the sea. River input results from weathering of limestones, during which one mole of one biospheric CO_2 combines with one mole CO_2 from the rock. Release of CO_2 along active rifts or submarine volcanoes feed the sea from below. There are two net outputs from the ocean: a constant (very small) net loss of CO_2 to air to close the weathering cycle, and an output of organic and inorganic carbon to marine sediments by means of biological activity. The average carbonate ion would have to wait 165,000 years before it came to rest in the sediment. During that time interval dissolved carbonates play an active role in the biogeochemistry of the ocean and will travel around the globe, and from the warm surface layer to the deep sea many times over and over again.

During the GEOSECS program, samples from 122 stations in all oceans were analyzed for ^{14}C content. Radiocarbon has a half life of 5,740 years, and is thus a suitable tracer to measure time scales up to a few thousand years. Furthermore, it is naturally generated only in the atmosphere by cosmic rays and enters the ocean through its surface, which is in isotopic equilibrium with air. Stuiver et al. (1983) compiled

BOX 10.5

Buffer Factor

Dissociation of sea water, ion composition, and amount of dissolved CO_2 gas are dependent on the total amount of inorganic carbon in the sea. Quite early in the study of carbonate equilibria, it was observed that the CO_2 partial pressure increases much more rapidly than the total amount of the dissolved inorganic carbon. The resulting ratio has been referred to as the Revelle factor, evasion factor, or simply buffer factor. However, for reasons of clarity, I would like to point out that these three terms describe the same phenomenon, but are different in their precise definition, and minor correction factors have to be applied to obtain the buffer factor in sensu stricto. The primary buffer factor is realized through the process:

$$CO_{2(aq)} + H_2O + CO_3^{2-} \rightarrow 2\ HCO_3^-$$

plus a correction factor due to the participation of the borate system:

$$H_3BO_3 \rightarrow H^+ + H_2BO_3^-.$$

The numerical value for the CO_2 buffer factor varies widely among different authors. Range of values for normal surface sea water goes from 8 (Wagener, 1979) to 16 (Kanwisher, 1960), but the majority opinion is 10. The amount of atmospheric CO_2 that can exchange with the dissolved oceanic carbonates per unit time may thus differ by a factor of two, depending upon which of the buffer factors one considers to be the correct one. Furthermore, there is considerable controversy as to what extent the buffer factor changes upon increase in CO_2 content. It is important to realize that higher buffer factors imply lower buffering capacity. If CO_2 partial pressure really causes such a rise in the numerical value of the buffer factor, the predicted increases in the CO_2 content in the atmosphere would then make the ocean progressively a less potential sink for anthropogenic CO_2. Others suggest that this will not happen because any change in the carbonate/bicarbonate ratio will be registered as a change in pH too (Whitfield, 1974), and the calculated pH changes for 1958 and 2010 are:

at 313 ppm CO_2 = 8.24; at 453.5 ppm CO_2 = 8.16.

We may ask the question why sea water behaves in such a strange way, since normal fresh water exhibits none of these buffering problems. It must have something to do with the salinity of sea water and the interference of, for instance, metal-carbonate pairing of organic and inorganic metal ion complexes. The limitation of the buffering capacity is given by the restricted supply of CO_3^{2-}, which under normal conditions amounts to about 13 percent of the inorganic carbon present. Due to the extraction of CO_2 by primary producers, surface water becomes low in bicarbonate but high in carbonate ions. As a consequence, that water is supersaturated with respect to $CaCO_3$, and supersaturation may reach values as high as 300 percent.

The discussion would not be complete if the controversy concerning the pH of sea water were not at least briefly mentioned. For a long time there was no question that the pH of the ocean was not principally determined by the combined action of carbonate and borate equilibria. However, work by Lars Gunnar Sillén has cast doubt on the validity of this concept. Instead, he proposed that the pH of sea water is primarily controlled by silicate rather than carbonate equilibria. In essence, acid volatiles released from the upper crust and mantle become neutralized by bases produced during weathering of igneous rocks. Indeed, the buffer capacity of aluminum silicates is quite substantial. This can readily be seen by placing a drop of pH 7 water on the surface of an aluminum silicate. For the common aluminum silicates, the water will rapidly adjust to a pH of 8. Thus, Sillén (1961, 1967) concludes, perturbation in the environment leading to a change in pH such as a sudden escape of CO_2, are compensated for by silicate equilibria.

---

measurements for deep ocean water (below 1,500 m) as shown in Fig. 10.28. The pattern of ocean circulation becomes apparent from this graph: recent water sinks at the northern end of the Atlantic and travels south. The uniform ^{14}C-age of Antarctic circumpolar waters indicates rapid mixing of the Atlantic input with the cool, south polar surface water. These sources then feed the abyssal Indian and Pacific oceans which contain the "oldest" waters. Eventually, the water upwells slowly over the entire ocean surface as it becomes displaced by new and colder water masses.

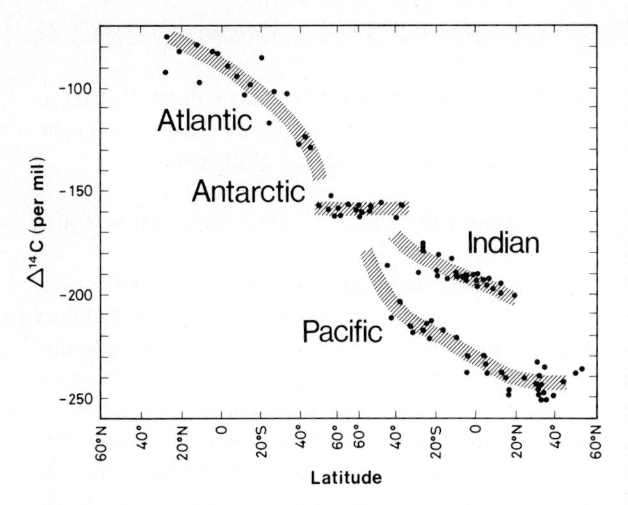

Fig. 10.28. ^{14}C (permil) variations in water samples below a depth of 1,500 m from various regions of world ocean (Stuiver et al., 1983)

Applying a box model to the ^{14}C-data, the average displacement times of ocean waters read: Atlantic, Indic and Pacific waters change every 275, 210, and 510 years respectively (Stuiver et al., 1983). The rate of upwelling inferred from these data is 4, 10, and 5 m a^{-1} for Atlantic, Indic, and Pacific respectively. Should our carbonate ion choose to move downward with the North Atlantic polar water, and continue its journey through the circumpolar Southern Ocean into the Pacific, it would stay below surface between 500 and 1,000 years. The more rapid surface currents would eventually return it to the North Atlantic. Other isotope tracers, for instance tritium or radium-226, have improved temporal and spatial resolution of residence time, mixing and circulation of water masses. Noteworthy is the detection of bomb-produced tritium (half-life for the decay of ^{3}H is 12.26 years) at a water depth of a few kilometers. For instance, the deep convective mixing in the Greenland Sea is reflected in substantial bomb tracer penetration to all depths. Residence times of 20 to 30 years for the deep cold water masses could be discerned. Roether and Münnich (1972, 1974) and Roether et al. (1974) show tritium-water depth profiles west of the Mid-Atlantic Ridge (Fig. 10.29) which reveal that deep bottom water layers close to 5 km have a turnover time of less than 8 years, whereas the intermediate water mass is substantially older. Radium-226 measurements (e.g., Chung, 1974) indicate that Ra-concentration in the deep water increases from the North Atlantic through the Antarctic to the Pacific by a factor of three. In the Pacific it increases from the southwest to the northeast, a trend which is similar to abyssal water temperature changes. This supports the inference that Ra, like heat, is continually added to deep water masses as they circulate. Isotope tracer techniques are presently widely used to follow the motion of the sea in time and in space (Bolin et al., 1983; Broecker and Peng, 1982; Broecker, 1987a, b; Box 10.6).

While residing in surface water, dissolved CO_2 can be sequestered by phytoplankton and reduced to organic matter or transformed into shell carbonates which are principally composed of aragonite or calcite. Phytoplankton is consumed by zooplankton and – after some recycling in the euphotic zone – carbon settles to the sea floor as particulate debris in the form of fecal pellets, macroflocs, or carbonate shells. At depth most of the infalling organic material is remineralized by the action of benthic organisms and bacteria. This causes $CaCO_3$ to dissolve, thereby recharging the marine inorganic carbon pool. Only a minute fraction will eventually be transferred to the geological column in a ratio of carbonate carbon to organic carbon equal to about 4 to 1 (see Table 10.3).

Throughout this book, I have given many more examples of the dynamics and biogeochemistry of the ocean, especially as they relate to the coupled air-sea-life system. Key words such as: Stratified Waters, El Niño-Southern Oscillation, Fladenground, Global Carbon Cycle, Climatic Change, or Primary Productivity may guide you further. Moreover much excellent work on the ocean as an environment is available, e.g., Ross (1970), Schlee (1973), Duxbury

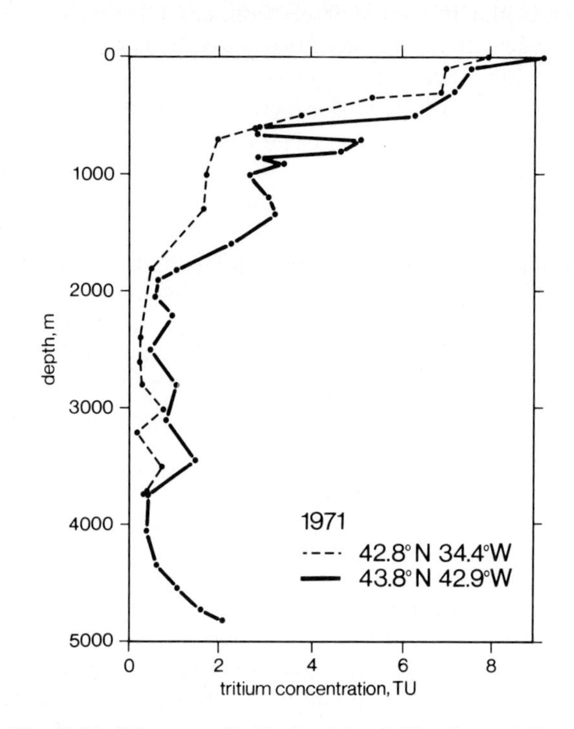

Fig. 10.29. Tritium profiles in the Atlantic (Roether and Münnich, 1972)

BOX 10.6

Conveyor-Belt-Like Ocean Current (Broecker, 1987 a, b)

An effective means to reveal climatic changes both regionally and globally is the study of temporal trends in the oxygen isotope composition of the calcitic tests of foraminifera preserved in deep sea sediments (e.g., Savin, 1977; Imbrie et al., 1984; Ruddiman, 1985; Shackleton and Opdyke, 1977; Shackleton and Pisias, 1985; Saltzman, 1987). Since foraminifera come principally in two varieties, that is as planktic or benthic species, coexisting forms living either close to the surface or on the seafloor will register in their $\delta^{18}O_{CaCO_3}$ the temperature difference in the water column at that site (Fig. 10.30 a, b). For instance, in the Subantarctic Pacific (Campbell Plateau and Macquarie Ridge), the paleoclimatic record is characterized by a general decrease in water temperature by 15°C since the Early Tertiary; noteworthy is the sharp downward slope from the Middle Miocene on which commonly is seen with the beginning of glaciation of the Antarctic continent (Fig. 10.30 b; Shackleton and Kennett, 1975; see also Vincent and Berger, 1985).

Imbrie et al. (1984) present a climatic curve for the past half million years which is used by Saltzman (1987) as basis to define a model for the late-Quaternary global variations of $\delta^{18}O$, mean ocean surface temperature fluctuations, ice-volume changes, and temporal atmospheric carbon dioxide contents. The best-fit solution for $\delta^{18}O$ with respect to the $\delta^{18}O$ record of Imbrie et al. (1984) is depicted in Fig. 10.31. A cyclicity of approximately 100,000 years between two climatic minima comes to light. Furthermore, sudden jumps over a time span of less than 1,000 years signal the beginning of a warm period. It is of note that changes in ice volume and water temperature contribute to the down-core ^{18}O signal (Mix and Pisias, 1988).

Wallace S. Broecker (1987 a, b) has come up with an interesting scenario to account for this climatic behavior by postulating a global conveyor-belt-like ocean current carrying cold and salty deep waters from the Atlantic along the Antarctic continent and South of Australia to the Pacific where it rises and returns as a warm shallow current to the northern Atlantic (Fig. 10.32). The cold deep current carries 20 times more water than all world rivers combined.

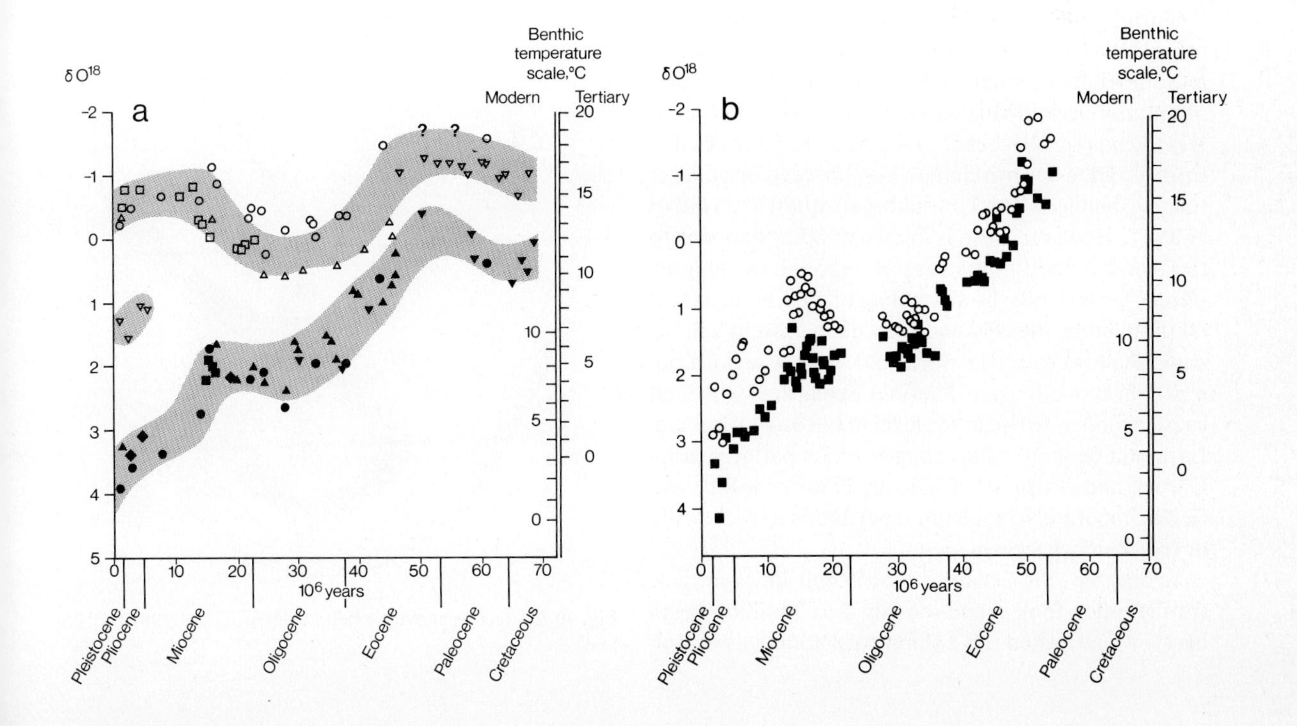

Fig. 10.30. **a** Isotopic paleotemperature data for Tertiary planktic foraminifera (*open symbols*) and benthic foraminifera (*closed symbols*) primarily from the North Pacific; for details concerning the position of stations see Savin, 1977. The "Modern" and "Tertiary" benthic temperature scales were calculated assuming water $\delta^{18}O$ values of –0.08 permil and –1.00 permil respectively. To estimate isotopic temperatures of planktic foraminifera add approximately 2.5° C to appropriate benthic temperature scale. **b** Same as **a** but from stations in the Subantarctic Pacific (Shackleton and Kennett, 1975)

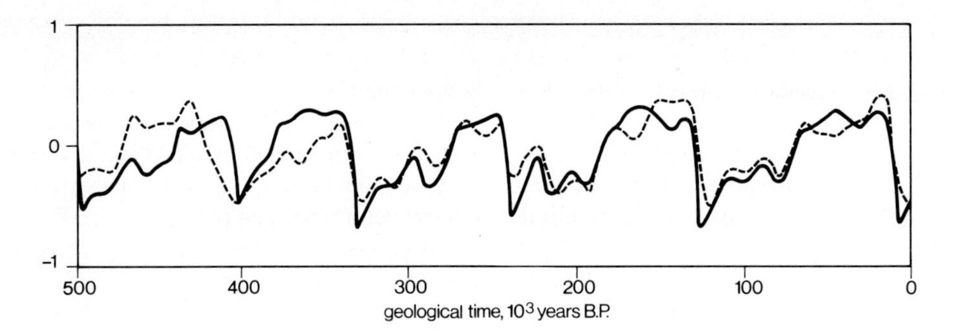

Fig. 10.31. Variations of $\delta^{18}O$ over the past 500,000 years in non-dimensional units normalized to a range of unity; *dashed curve* represents $\delta^{18}O$ record of Imbrie et al. (1984); *solid curve* is best-fit solution for $\delta^{18}O$ according to Saltzman (1987)

Circulation is principally driven by the salt of the sea. Because of today's climatic situation, the Atlantic loses more water through evaporation than it receives from rain and rivers, whilst the opposite is true for the Pacific. Trade winds carry moisture from the Atlantic to the Pacific creating a deficit in either salt (Pacific) or water (Atlantic). The amount of water leaving the Atlantic via Central America is in the order of 0.4 Sverdrup (Sv), that is about four times the water volume annually discharged by the Amazon. The extra density given to the waters of the northern Atlantic by the increment in salt gained through enforced evaporation and the removal of water by trade winds to the Pacific and rain clouds to central and eastern Europe, keeps the conveyor belt running.

During glacial time, the conveyor belt is not in operation, but starts moving again once Earth returns to a more favorable orbital configuration for the initiation of an interglacial (Milankovitch scenario; see p. 219). It is the merit of Broecker to show clearly that a switch from a cold to a warm climate may proceed in a matter of a few hundred to a thousand years quasi in a kind of "jump". However, little is known of what happens to the conveyor belt in the case of a global warming induced, for instance, by a doubling of CO_2 in the air. In all probability, the cold and salty deep current will become shallower, and the warm and shallow surface current will move deeper. Thermal expansion will then cause a moderate rise in sea level in the order of 10 cm. Little can be said on the changes in rain pattern across Europe and Africa, or of how much water will be carried by the trade winds from the Atlantic to the Pacific or of how much ice will melt.

To sum up, the ocean conveyor belt in its present configuration took shape roughly 2 to 3 million years back in time, when the Isthmus of Panama emerged,

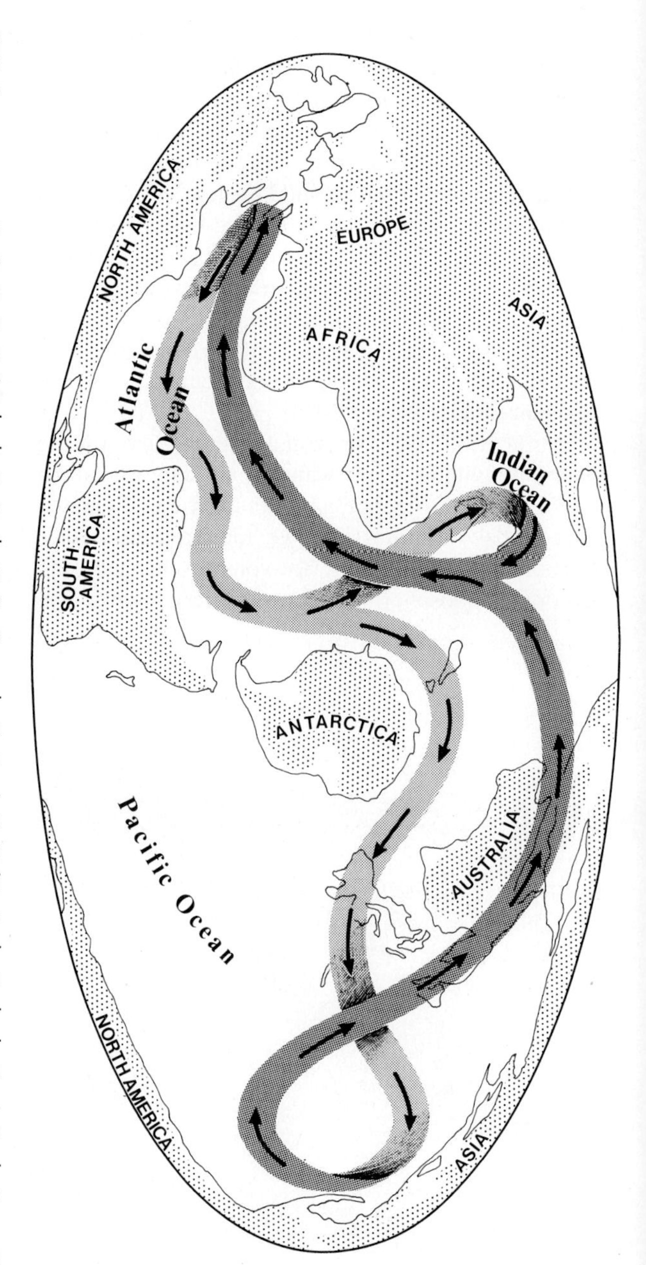

Fig. 10.32. Ocean conveyor belt (after Broecker, 1987a, b; see text)

thus closing the free exchange of Atlantic and Pacific waters. This was the starting signal for a bipolar ice age which commenced 2 million years ago at what geologists now call the Plio-Pleistocene boundary. Priming of the ocean pump enforced climatic change in the region of the emerging hominids. It is not unlikely that the stop-and-go of the Quaternary ocean conveyor belt and the associated climatic changes carried a profound stimulus to the origin and evolution of *Homo habilis* (2–1.6 Ma B.P.), *Homo erectus* (1.6–0.2 Ma B.P.), and *Homo sapiens* (0.2 Ma B.P.–? Ma A.P.).

(1977), M.G. Gross (1977), Bhatt (1978), Sears and Merriman (1980), Wells (1986). A comprehensive treatment on biogeochemical cycles in rivers, lakes, estuaries and oceans is given by Degens (1982) and Degens et al. (1983, 1985, 1988).

Origin and Evolution of Sea Water

Setting the Stage

According to Isaak Asimov, certain conditions must be met to obtain an ocean on any planet:

(i) a large volume of liquid is needed;
(ii) elements of the ocean must be plentiful;
(iii) substance must have a prominent liquid phase at STP (= standard temperature and pressure);
(iv) there should be a depression to contain it; and
(v) gravity must be sufficient to keep the ocean on the planet.

It was previously stated that the four most abundant elements in the Universe – not counting helium and neon for the reason that they are inert – are:

hydrogen – oxygen – carbon – nitrogen.

It is conceivable to create a large body of surface liquid on a planet by one or a combination of these four elements. Ordinary hydrogen, for example, melts at –259.2°C and boils at –252.8°C (1 atm) which obviously makes H_2 a poor candidate to fill an ocean. Hydrogen, however, combines readily with oxygen, nitrogen and carbon: H_2O, NH_3, CH_4. So why do we have an ocean of water, rather than one of ammonia or of methane? The answer is: our planet has an optimal climate to liquefy water, be it ice or water vapor, which is not the case for ammonia and methane (Fig. 10.33):

H_2O solid up to 0°, liquid for 100°, gas > 100°C
NH_3 solid up to –77°, liquid for 44°, gas > –33°C
CH_4 solid up to –187°, liquid for 23°, gas > –164°C.

Thus, H_2O is the logical ocean maker, because water complies with all our earthly requirements asked for in (i) to (iii). Concerning condition (iv), the present surface on Mercury or Moon bears witness to how the primordial landscape on Earth might have looked if mountain building and sediment cosmetics had not rejuvenated its surface: Mare galore! With the advent of global tectonics, either micro, intra, or macro plate, the Earth eventually generated the ocean basins we presently know of. With respect to condition (v), Earth is definitely "heavy" enough to avoid Jeans escape for water molecules.

Having set the structural frame, we can now address liquid water. First, we have to find out when the ocean started to form and how long it took to assemble all of its water. Second, we have to account for the salts in the sea. Third, we must outline its physical and chemical changes in the course of time. Questions like these are easier to raise than to answer, because modern ocean is quasi a chemical compromise

Fig. 10.33. Water, a suitable medium to form an ocean

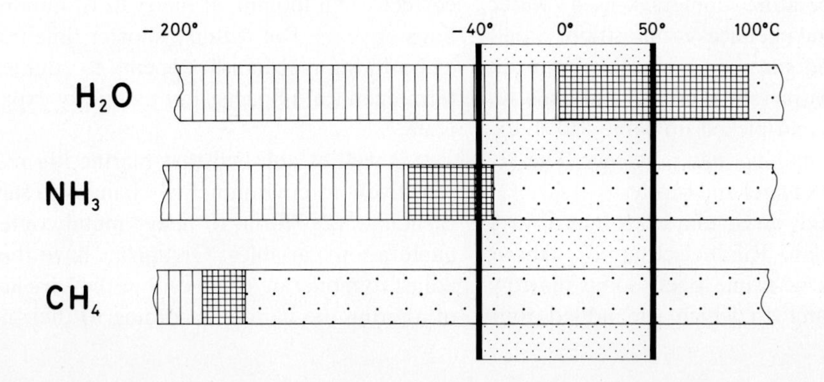

achieved after a long enduring battle between the classical elements: air-water-fire-earth, and a fifth element, the biological cell. To be frank, the sea – from start to finish – is by no means a steady state system and should therefore not be treated that way, as is, unfortunately, common practice. Like life, the ocean system is governed by irreversible thermodynamics, and chemical equilibria or steady states persist only for moments, geologically speaking.

Almost everybody agrees that the primordial sea was released from crust and mantle, but an interesting alternative has recently been offered (Box 10.7). Conflict exists when it comes to timing and rate of water discharge, as well as to the kind and proportions of co-evolving chemical species (e.g., Rubey, 1951; Wedepohl, 1963; Ronov, 1968; Fanale, 1971; Garrels and Mackenzie, 1971; Li, 1972; Degens, 1973; Holser, 1977, 1984; Maisonneuve, 1982; Walker, 1983; Towe, 1983; Veizer, 1985). Older concepts assumed that hydrothermal waters, at present discharge rates, could readily top up an ocean in a few billion years. Such ideas are no longer en vogue and rapid degassing scenarios are in. In this context I would like to mention the work of Thomsen (1980), who advocated that partitioning of ^{129}Xe (a decay product of ^{129}I having a half-life of 17 million years) between today's atmosphere and mantle-derived gases makes a substantial "juvenile" degassing after 3.3 Ga unlikely. Accepting this viewpoint, it is proposed that ongoing hydrothermal activities on land or beneath the sea are chiefly related to the recycling of ancient sea water or of meteoric water, that is water that at least once in its lifetime has gone through the meteorological cycle. Water that surfaced for the first time after being trapped in the deep mantle for ages is termed *juvenile water*. Such waters are exceedingly rare, as stable isotopes will tell.

Summing up: crust and upper mantle have released the bulk of the water contained in modern ocean during the early Archean. Ever since sea water has percolated through the basaltic crust down to the Moho and up the hydrothermal vents perhaps a few hundred times. During its few million years residence in the world of high-temperature minerals, sea water changed its isotopic and chemical composition. Small increments of ions and gases were leached from the country rock, whereas in return crustal minerals received salt and water and picked up some ^{16}O in exchange for ^{18}O, thus making water heavier. The relationship reminds me of beef and broth:

During my childhood, on Sundays, Mother always made her famous "klare Rindfleischsuppe". About three to four pounds of prime beef, some marrow bones, 3 liters of water to which she added four spoons of salt plus at least five different spices, all of which was boiled for a few hours. When we came home from church, the beef was blessed with salt and spices, the bones were partly demineralized, and the water had turned into a broth. Specks of fat were vivid proof that hydrophobic forces had done the job expected of them: separating salt from grease. The analogy to sea water and basalt cooked for a few million years at a few hundred degrees centigrade is evident (Fig. 10.34).

Fig. 10.34. Kettle-crust relationships

Once sea water has returned to the ocean again, river inputs, air-sea exchange, volcanism, and last but not least, the biota will modify its chemistry. The facies of marine sediments should reflect some of the changes, because sediments are the "metabolic" leftovers of these biogeochemical alterations. Truly, sea water has a history.

All that has been said so far could make one believe that the chemistry of sea water did in fact profoundly change in the course of geologic time. This may be correct when looking at intervals of hundreds of millions of years. But within a shorter time frame a kind of buffering mechanism appears to counteract short-term chemical impacts. Let us briefly expand on this issue.

It is well established that marine life reacts rather sensitively to environmental changes in salinity, PO_2, carbonate equilibria, or heavy metal content, just to name a few variables. Organisms have the ability to adjust to moderate changes, but they are also capable of altering the habitat according to their needs. Any

BOX 10.7

Comet-Like Objects (Frank et al., 1986a, b)

Oceans came from the sky in the form of comet-like objects. Such cosmic icicles, about 100 tons in weight and about 10 meters in diameter, arrive at the top of the atmosphere as a piston of gas with bulk speed >20 km sec^{-1} at a global rate of 20 icicles per minute. This is in essence what Frank and his coworkers infer from images (High-Altitude Satellite Dynamics Explorer 1) of the Earth's dayglow emissions at ultraviolet wavelengths. These reveal regions of transient intensity decreases to about 5 to 20 percent of surrounding values over areas estimated to be 2,000 km^2; in other words: atmospheric holes are indicated! Roughly 30,000 atmospheric holes were identified in the images taken during an observational period of ca. 2,000 hours from late 1981 through early 1985. The suggested mass flux of water is 10^{12} kg a^{-1}. For comparison, meteoric dust arrives on the Earth in an amount of 10^5 to 10^7 kg a^{-1} (Hughes, 1978), and the masses of comets are in the order of 10^{10} to 10^{16} kg a^{-1} (B. Donn and Rahe, 1982).

According to the above concept, comet-like objects have their source of origin apparently in the Oort cloud, and the frequency of impact rates on terrestrial as well as Giant Planets may be controlled by the changing position of the solar system in relation to the galactic arms. This could trigger ice ages. Moreover, the ocean as a whole could be generated this way, rather than being largely expelled from crust and mantle as the standard theory suggests. In any event, comets and planetesimals were the primary vehicle for water to come to Earth.

The idea of comet-water, associated organics and metals bound in icicles is quite intriguing and deserves further scrutiny. For the moment I would like to close the subject by quoting Louis A. Frank: "If my opinion is correct, a lot has to be changed. However, if I am wrong, I will find out soon!"

---

event which will remove the sea from its chemical ground state will switch on the biological buffering device to counteract this impact. In a sense this device is operating like a thermostat. It is conceivable that a slight departure from the ground state can be smoothed out that way. However, should changes be introduced that are too fast, or too substantial, they may trigger extinction or promote evolution of new species.

Addition of small quantities of rock salt, for example 10 mg, to one liter of distilled water will alter the system profoundly and make the water useless for analytical chemistry. Quadrupling the NaCl content in a freshwater lake from 125 to 500 mg l^{-1} could wipe out the local fish population. Adding the same amount of NaCl to sea water will not noticeably influence fauna and flora. Marine biota is simply adjusted to more salt, which over billions of years has gradually accumulated in the sea.

Present chlorinity levels, 19,000 mg l^{-1}, were most likely reached at the end of the Precambrian. Since then, ocean's chlorinity might have varied only slightly and in phase with major tectonic events or climatic changes. I speak of chlorinity only and say nothing about other chemical changes, because they could be more substantial, as we will see later.

Going further back in time, that is into the Proterozoic and the Archean, the data bank becomes progressively scanty. Unfortunately, the first 700 million years of Earth history, equivalent to the time that elapsed since the late Precambrian, have left no sedimentary record from which to deduce something about facies or paleoenvironment. Consequently, meteorites, mineral equilibria, geochemical balance calculations, or information from Moon, Venus and Mars is all there is to define a model on origin and prebiotic evolution of the sea. Nevertheless, these sources place certain boundaries within which any geologically or biochemically plausible scenario has to be discussed (Degens, 1973; Walker, 1982; Holland, 1984; Holland et al., 1986; Broecker, 1985; Budyko et al., 1987).

Soda Ocean Concept

Most authors assume that the proto-ocean resembled modern ocean in that sodium and chloride were the major ions, followed by potassium, magnesium, calcium and bicarbonate. Sulfate was absent because the sea was supposedly reducing; pH was neutral or slightly acidic. Ronov (1968) suggested a composition dominated by magnesium and calcium ions. Some postulated high atmospheric CO_2 partial pressure and a strong greenhouse effect to compensate for a faint early Sun (e.g., Hart, 1978; W.R. Kuhn and Kasting,

1983). Accordingly, the sea would be acidic. Henderson-Sellers and Cogley (1982) noted, however, that the envisaged high PCO_2's could not persist with water around, but would react with silicates and yield carbonates. The other extreme, a low PCO_2, was modeled by Henderson-Sellers and Cogley using cloud feedback for a solar flux of 0.8 its present value and today's PCO_2. Lack of land masses and ocean-albedo feedback compensated for a faint early Sun. Surface temperatures stayed above freezing, contradicting the hypothesis of a frozen Earth. Other greenhouse gases such as methane or ammonia are frequently mentioned as a temperature regulator in the early Archean. This supposition can be dismissed on various grounds. Principally, photodissociation and mineral catalysis inhibit a substantial build-up in the atmosphere.

What kind of geological evidence may conceivably support either high or low PCO_2 in the early atmosphere? We have previously seen that one volume of CO_2 presently contained in air will be consumed by weathering in about 7,000 years. For example, a granite or a basalt may chemically decompose to clay minerals and all liberated ions that cannot be accommodated in the newly formed mineral will, jointly with bicarbonate, enter the aqueous phase. Today the vegetation cover on land extracts CO_2 from the air and releases much of it into the soil, and soil-air PCO_2's of a few percent are common.

We will now assume that all of today's water amounting to 1,350 x 10^{21} g, was present from the start, and that all crustal carbon had been spontaneously released as CO_2, thus totaling 65.5 x 10^{21} g C. Weathering will immediately commence at unbelievable rates. It was calculated (Kempe and Degens, 1985), that all of this CO_2 became sequestered via surface weathering in less than 100 million years at the most. Throughout this time, the "runaway greenhouse" threshold was never reached, which contrasts with the situation on Venus, where all liquid water was lost due to the prevailing high greenhouse temperatures. Alkalies, most notably sodium and potassium, served as counter ions for the bicarbonate and chloride, whereas calcium and magnesium set free by the weathering precipitated as carbonates due to a shift in pH from acid to highly alkaline conditions. The final product is an *early soda ocean* chemically resembling the world's fourth largest closed lake: Lake Van in eastern Anatolia, and hundreds of volcanic and active rift lakes. Liter by liter, such lakes may hold dissolved carbonates in an amount exceeding sea water bicarbonate by more than a thousand times before sodium carbonates precipitate (see Fig. 8.32).

For calculation purposes, we assume that sodium is the counter ion for all of the dissolved carbonate in the early sea. Moreover, we presuppose that sodium carbonate is the principal salt species, although others may coexist, for instance, $NaHCO_3$. At the solubility maximum of 35°C, 320 g Na_2CO_3 dissolves in 1 kg solution corresponding to 470 g Na_2CO_3 per kg water or 53.3 g carbon per kg water. This implies that the primordial ocean could store any amount of sodium carbonate without ever becoming supersaturated. Soda properties, that is high pH above 9, are acquired roughly above alkalinity concentrations of 0.05 to 0.1 meq l^{-1}, which for carbon means 0.5 to 1 g per kg water. Such concentrations are quite realistic in view of the enormous pool sizes for carbon and sodium in the crustal compartment (Kempe and Degens, 1985).

I have thought a long time what kind of independent criteria could possibly be listed in support of an early soda ocean. Gas fossils – you recall the heavy noble gases in air – were helpful to unravel the transition between primary atmosphere and the air that followed. Finally it dawned on me that the temporal variation in ^{18}O content of ancient cherts and carbonates may be viewed just as *isotope fossils* of the early sea.

Cherts are microcrystalline quartz that supposedly formed in contact with sea or lake water from an amorphous silica precursor (SiO_2 x nH_2O). Their $\delta^{18}O$ is expected to reflect the environmental conditions in terms of the $^{18}O/^{16}O$ ratio of the water and its temperature. The large fractionation between quartz and water at low temperature makes cherts a potential tool to reconstruct environmental conditions of the past, particularly if analyzed jointly with co-existing minerals such as calcite or magnetite that follow different isotope fractionation curves (Clayton and Epstein, 1958). Cherts and carbonates show temporal isotopic variations and the further we go back in time the lighter the two become (e.g., Degens and Epstein, 1962; Perry, 1967; Perry and Tan, 1972; Knauth and Epstein, 1976; Kolodny and Epstein, 1976; Perry et al., 1978; Becker and Clayton, 1976). As a rough approximation one can say that the curve for cherts in the time envelope 3 Ga to the present increases in $\delta^{18}O$ by about 5 permil per 1 Ga. For the shape of the carbonate $\delta^{18}O$ curve the same suggestion was made by Weber (1965).

Students of stable isotope geochemistry are well aware that such changes can come about along various routes. For instance, using oxygen isotope ratios in cherts as old as 3 billion years, Knauth and Epstein (1976) suggest the possibility that Earth surface temperatures may have reached about 52°C at 1.3 Ga and about 70°C at 3 Ga. In contrast, Perry (1967) and

Perry and Tan (1972) interpret the $\delta^{18}O$ pattern of Precambrian cherts to be related to differences in the isotopic composition of ocean water. In the light of the data of Muehlenbachs and Clayton (1976), isotopic composition of the ocean may be held at its present value for ages (Box 10.8). This finding would on first sight be in line with the interpretation offered by Knauth and Epstein (1976). Data on coexisting phosphates and cherts may perhaps overcome this problem (Karhu and Epstein, 1986). And yet a word of caution should be voiced, because the tectonic and weathering regimes in the early Archean are difficult to evaluate with respect to their impact on the ^{18}O content in the ancient sea. In addition, metamorphism is a convenient tool to explain any observed variation, and the older a rock, the better its chance to isotopically equilibrate to a higher heat flow regime.

As is the case with ^{18}O or ^{16}O which can move in and out of minerals, water, air, and living matter in response to changes in temperature or environmental conditions, so will chemical elements do. A clue to the origin and chemical evolution of sea water can thus be provided by examining the way minerals respond when subjected to volatiles and liquids of the type released by oceanic crust, and what kind of impressions this will leave on the participating water. We have done precisely this (Degens, 1973; Kempe and Degens, 1985; Kempe et al., 1987a).

Geochemical Balance Calculations

Since the days of Viktor Moritz Goldschmidt (e.g., 1933) many attempts have been made to quantify the partitioning of chemical elements between igneous rocks, sediments, and sea water. This exercise served as a basis to calculate the total mass of sediments in the Earth that has accumulated over the past 3 to 4 billion years. Without going into detail, a wide number of authors (see Pettijohn, 1975, pp. 7–12) came to the conclusion that a sediment layer of approximately 1,000 meters around the Earth's surface has been laid down. The material is thought to be principally composed of clays and carbonates. Roughly 2 percent of the layer, namely 20 meters, is organic, the rest is inorganic in nature. About 200 meters of only mechanically disintegrated rock material, such as detrital quartz, breccias etc., have to be added.

Chemical weathering of igneous rocks and the formation of soil minerals is accompanied by massive $\delta^{18}O$ extraction from surface water. Work by J.R. Lawrence and Taylor (1971, 1972) has revealed that clay minerals and hydroxides in their oxygen and hydrogen isotope relationship about follow a line parallel to the meteoric water line (Fig. 10.35). Clays in modern marine sediments reflect in their $\delta^{18}O$ the soil environment, and diagenesis lasting for a few million years will not noticeably alter the original $\delta^{18}O$ as fixed on land (e.g., Yeh and Savin, 1977). Shales, however, will experience moderate changes in accordance with the regional geological settings. By taking together all sedimentary materials, excluding mechanical debris, a $\delta^{18}O$ of 20 permil can be assigned to the

Fig. 10.35. Meteoric water line (after Lawrence and Taylor, 1971)

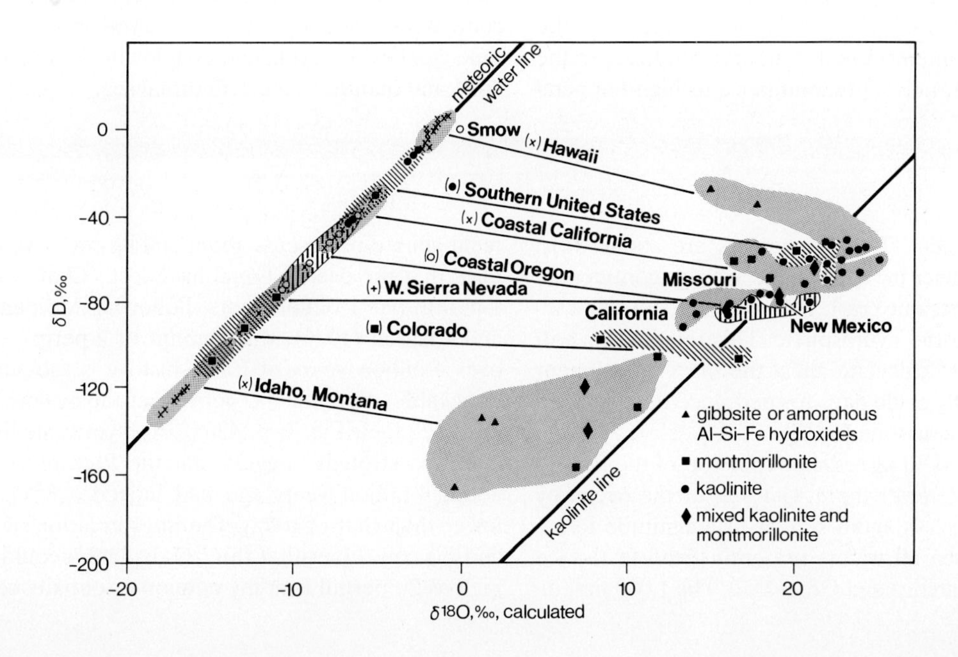

BOX 10.8

^{18}O Budget of the Sea (Mühlenbachs and Clayton, 1976)

It is common practice to report changes in the $^{18}O/^{16}O$ ratio of a mineral in terms of permil variation relative to standard mean ocean water which is arbitrarily set at: $\delta^{18}O$ equal zero. Unaltered oceanic crust would have a $\delta^{18}O$ of 5.7 permil, a "normal" granite one of 8 permil, a typical low-grade metamorphic rock one of 16 permil, a modern marine carbonate one of 28 permil, and a quartz formed at about room temperature today would have a $\delta^{18}O$ of 36 permil. As a general rule rocks get heavier as we move from basalts to granites to shales to limestones. The heaviest of all are young marine cherts that differ from oceanic basalt by about 30 permil.

The oxygen isotope record of any rock can be altered by flushing hydrothermal waters through the rock formation. It is thereby of special relevance that low temperature changes will carry a different signal than high-temperature ones. For example, sea floor spreading and subduction promote cycling of water through the oceanic crust. This will induce low-temperature weathering of basalts causing ^{18}O enrichment of the basaltic material by a few permil and a corresponding ^{18}O depletion in the participating sea water. In contrast, high-temperature alteration of the deep crust and upper mantle, for example in the vicinity of active rifts or along subduction zones, decreases the ^{18}O content of crustal rocks by a few permil and enriches sea water in ^{18}O. If low-temperature crustal weathering is the predominant factor in the diagenesis of sea water, then the ocean's ^{18}O content should decrease with time. In the opposite case, i.e., a predominance of high-temperature post-solidus exchange with sea water, the ocean would progressively gain ^{18}O.

Estimates of volumes of materials involved in: (i) formation and alteration of oceanic crust, (ii) cycling of water through the lithosphere, and (iii) continental weathering, led Mühlenbachs and Clayton (1976) to suggest that sinks and sources of ^{18}O are matched in size, thus keeping the oxygen isotope composition of sea water at about its present level through the ages. They also reported that the $\delta^{18}O$ of modern ocean is principally determined by the size of equilibrium isotope fractionation factors between water and igneous rock in the temperature range of 200 to 300°C.

Mühlenbachs' and Clayton's model is based on slow percolation rates. At present much higher convection rates are assumed, whereby one volume of sea water is supposed to flush through the crust in a matter of 5 to 10 million years, all depending of course on rate and type of global plate tectonics.

What has been said so far on the ^{18}O budget of sea water deserves a new look at the temporal ^{18}O variation in cherts. The oldest ones from Isukasia, West Greenland (3.8 Ga) have been subjected to low-grade metamorphism (Oskvarek and Perry, 1976; Perry et al., 1978), which according to the above discussion would mean a release of ^{16}O to the water in exchange for ^{18}O. On the other hand, the upward trend in ^{18}O in deep sea cherts from the late Jurassic towards the present seems to be chiefly related to global climatic change (Kolodny and Epstein, 1976). Although climatic, diagenetic, and low-temperature isotope shifts cannot be ruled out and certainly have modified the picture, the almost monotonic decrease in the ^{18}O content of ancient cherts from Jurassic to Archean time can best be explained as a dowry of ^{16}O given by crust and mantle to the primordial sea.

average sediment. Thus, sediments are about 10 to 15 permil heavier in ^{18}O than the average continental, respectively oceanic crust. This excess in ^{18}O stems ultimately from the hydrosphere. This relationship will be the basis to calculate anew the mass of sediment that potentially could have formed since the first solid crust spread across the Earth.

Formation of 50 kg sediment per cm^2 of the Earth's surface would lower the $\delta^{18}O$ value of the ocean by 1 permil, which is about similar in magnitude to adding to the sea all waters presently fixed in the ice sheets of Antarctica and Greenland. The 1,000 m sediment sphere represents about 200 kg cm^{-2}, which implies that the ocean should have lost ^{18}O at a rate of 1 permil per 1 billion years. Following Mühlenbachs and Clayton (1976), this amount of 4 permil spread over 4 billion years of Earth history is too small to noticeably affect the ^{18}O content of the oceans of the past (see their Fig. 1, p. 4368). And yet some lines of evidence strongly suggest that the Precambrian sea about 4 billion years ago had indeed a $\delta^{18}O$ much lower than that of today. The interpretation of Perry and his colleagues that this ^{18}O depletion could be as high as 20 permil is in my opinion essentially correct.

The question still remains: what caused such an exorbitant ^{18}O depletion in the Archean sea?

According to the *early soda ocean* hypothesis, large volumes of CO_2 became sequestered by crystalline surface rocks in a matter of a few million years. The sediments so derived must have been enriched in ^{18}O relative to the parent rock by about 10 to 15 permil. We have now two ways to determine the sediment mass so produced. First, we can accept a $\delta^{18}O$ value of –20 permil as derived from the chert data and assume that the ^{18}O content of early ocean was – like today – determined by water and igneous rock interactions at 200 to 300°C giving rise to "standard mean ocean water". Neglecting temperature effects, the difference of 20 permil could then signal an "explosive" weathering event during pre-Isukasia time (before 3.8 Ga). Second, magmatic water was the starting point, that is a fluid isotopically adjusted to the igneous system at magmatic temperatures. Unaltered magmatic water is supposed to have a $\delta^{18}O$ close to 6 permil (reviewed in Hoefs, 1980). Judged from the conditions assumed to have existed at the surface of the Earth prior to Isukasia time (reviewed in Kröner, 1985), the "first drop" of water released by crust and mantle had probably an ^{18}O signal of a magmatic fluid, namely ca. 6 permil. Please bear in mind that silicate melts at water pressures of a few kilobars may hold 5 percent water by weight or 30 percent by molar concentration.

To lower the ^{18}O content of the primordial sea to the assumed Isukasia level of –20 permil, a total of about 1,300 kg sediment per cm^2 Earth surface had to form. This would represent a volume of chemically weathered sediment equivalent to a layer about 6 to 7 km thick around the globe. The only mechanically disintegrated rock debris is not included in this figure. The sphere might have been stratified, having layers of calcite, dolomite, hydrous sodium silicates, bedded cherts, illite, chlorite and large amounts of interstitial solutions. In addition, massive intercalations of volcanic tuff layers are to be expected. Siliceous volcanic glass shards were abundant and must have become rapidly changed to palagonite under the extreme weathering conditions established at the time. Palagonites are known to be enriched in ^{18}O with $\delta^{18}O$ values as high as 25 permil (see Hoefs, 1980). Illuminating in this respect is the article of Eugster (1969) that expands on the mineral facies developing in some sodium carbonate lakes in East Africa associated with active rift volcanism.

Of course, such a thick section of "wet" and "muddy" sediment cannot persist in time and will constantly be worked over. Granitization is one of the principal mechanisms at depth, whereas hydrothermal activity and associated volcanism operate closer to the surface. Comets and meteorites will further complicate the matter, not only by adding extraterrestrial water, organics, minerals, and gases, but by releasing volatiles from the growing planet upon impact (Matsui and Abe, 1986). Whether an ocean existed at all, or whether mud and water were a kind of non-Newtonian fluid, resembling toothpaste or oil paint, is difficult to decide. The same is true for the atmosphere into which volcanism and bolides injected dust, gas and water. Halos and dust veils of different sorts must have formed at various altitudes and extended far into the stratosphere. Sunlight may not even have reached the surface of the Earth. The generation of different kinds of gas molecules at the time was probably more a function of mineral-organic matter-water-gas interactions than radiative forcing of atmospheric gases.

This scenario of a substantial alkali-rich sediment-water layer provides a meaningful interpretation for the origin of a continental, that is a granitic crust prior to the occurrence of the oldest terrestrial rocks from Isukasia. It is commonly assumed that following rapid cooling of the originally melted surface of the Earth,

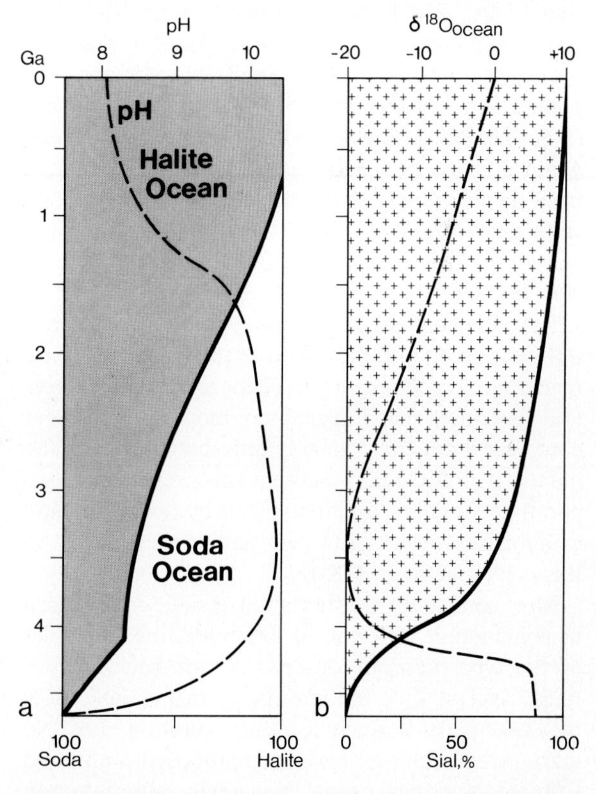

Fig. 10.36 a, b. Evolution of the soda ocean towards the halite ocean (**a**); evolution of continental crust (heavy line) and synchronous change of $\delta^{18}O$ in seawater (stippled line) (**b**); (highly schematic)

a thin crust of ultramafic rocks developed which in due course thickened, and basic volcanism must have been widespread. The early ultramafic crust was recycled back into the hot mantle. At about 4.3 Ga ago, mantle temperatures and heat flow had been lowered sufficiently to permit the establishment of a komatiite-dominated crust. One should emphasize that komatiitic rock formations rest on sial and not on sima (see p. 128). This would call for large-scale magmatic differentiation to obtain granitoid intrusives. Granitization of a thick sediment sphere would be an interesting alternative to generate sial. Accepting this idea and translating that mass of sediment into a granitic crust, a continental mass equal to about half that of the continents of today could potentially be generated from such a substrate (Fig. 10.36).

Only a few remnants exist from the continental crust older than 3.8 billion years, and examples are: the Isua sequence, West Greenland (Hamilton et al., 1983; Allaart, 1976), and the Limpopo belt, southern Africa (Barton et al., 1983). A number of geochemical anomalies in the most ancient mantle-derived rocks clearly speak for more sialic crust prior to 4 Ga ago. The other half, for which there is more direct support, came with the cratonization in the course of the Precambrian (Box 10.9).

For Venus, the situation might have been the same (Box 10.10). On Ishtar Terra there is some evidence of granites; and mountain chains do also exist. On Venus, however, tectonics stopped rather early, once the lubricant, that is water, vanished into space. Some water vapor is all there is left from the ancient Venerean sea. Water on Mars is a different story (Box 10.11).

BOX 10.9

Crustal Growth (Kröner, 1985; Reymer and Schubert, 1984)

The oldest dated rocks slightly exceed 3.8 billion years, and are thus only a little younger than the event of "terminal cataclysm" by comets and planetesimals, which in wide parts of the Earth has reset the radiogenic clock of the solid crust to zero age about 3.9 Ga ago (Wasserburg et al., 1977). Judged from the main rock types associated with the most ancient pieces of crust, craton-building processes were already well under way 3.8 billion years ago (Goodwin, 1981).

Reymer and Schubert (1984) have summarized different viewpoints published on the volume of continental crust in the Archean. There appears to be general agreement that 70 percent or more of the present continental crust had already differentiated from the mantle by 2.5 Ga ago. Subsequent growth was comparatively minimal and dominated by reworking and cannibalistic recycling (Veizer and Jansen, 1979; Allègre, 1982; Kröner, 1985).

Most commonly, the formation of sialic crust is seen in conjunction with crustal differentiation processes of the type presently observed in arc-back arc settings, except that, for reasons of higher heat flow, they take place at much faster arc accretion rates. No matter what scenarios have been proposed, water and CO_2 are the critical agents in anatexis, differentiation or cratonization. These two elements are also the main agents of weathering, and we may simply consider the alteration processes of surface or near surface rocks as an outgrowth of low temperature hydrothermal weathering. The sequence of events is as follows:

To come to a soda ocean one needs first CO_2 and H_2O to weather the top part of the crust. Unlike Mars, where the surface material was initially frozen, the situation on Earth and Venus was different in that steam or liquid water must have been abundantly present at or close to the surface. On all three planets, including Mercury and the Moon, intensive bombardment by comets and meteorites prior to 3.8 Ga produced heavily cratered areas. On Earth, a several kilometer thick megaregolith composed of terrestrial and extraterrestrial rock debris blanketed the solid crust. Low-temperature hydrothermal leaching and weathering of this breccia fostered the generation of sedimentary minerals. This was accompanied by continuing: (i) shock metamorphism of impacting bolides, (ii) rising of magma and volcanic fluids, (iii) regional metamorphism, (iv) anatexis, and (v) tectonism. At the same time liquid water and detritus accumulated and aggregated in the thousands of basins created by impacts or tectonic events. All of this became the substrate of the continental crust. In essence: Sial was born.

Further development involved the disaggregation of clay-sized minerals from associated water. Electrolytes and organic molecules are known to be rather efficient in this respect. A liquid ocean was generated also. This ocean had soda properties and must have been highly depleted in ^{18}O, due to the huge mass of sediments produced during just a few hundred million years at the most.

Assuming that the low ^{18}O content of the Archean cherts does indeed mirror the $^{18}O/^{16}O$ ratio in the early sea, the isotopic difference to magmatic water is a result of the preferential ^{18}O extraction by sedimentary minerals. Thus the degree of ^{18}O depletion in the Archean sea is a measure of the mass of sediment formed at that time. The hydrosphere, principally composed of ocean and crustal water in a ratio of about 3 to 1 (Table 10.1) is potentially capable of generating a sediment layer 8 to 10 km thick around the surface of the Earth. This represents about 70 percent of the volume of present-day continental crust, and agrees with estimates based on geochemical and petrological evidence (e.g., Veizer, 1983).

BOX 10.10

The Landscape of Venus (Pettengill et al., 1980 a, b; Colin et al., 1980; Phillips et al., 1981; Phillips and Malin, 1984; Bowin, 1983; Prinn, 1985)

The surface of Venus has been probed by radar signals from the Earth for the past 25 years outlining the main topographic features. Occasional visits by Soviet and U.S. spacecraft in more recent years have considerably improved our understanding of geological and chemical processes that shaped our sister planet in the course of Venerean history. Hallmark is the few hours' call by the two Soviet landing crafts Venera 13 and Venera 14, during which a series of photographs from the surface of Venus could be shot, and basic measurements on the composition of rocks and the atmosphere be accomplished. Listed below are a few points of information pertaining to topography, gravity and crustal evolution of Venus.

Different from Earth with its distinct bimodal distribution in the hypsometric curve of topography representing continental heights and oceanic depths, the frequency of occurrence of elevation on Venus reveals just one single peak above the median plains (Fig. 10.37). Ocean basins so familiar on Earth are lacking and the two continent-like features, that is Ishtar Terra and Aphrodite Terra, account for 5 percent of the total area, whereas on Earth close to 30 percent of the surface area is land. Maxwell on Ishtar Terra is the highest mountain of Venus and stays more than 11 km above average elevation, thus resembling Mount Everest in

Fig. 10.37. Computer-drawn block diagram of Venus' topography based on data by Pettengill et al. (1980 a, b); (courtesy of A. Paluska, Hamburg). Shaded area are regions below the median plain

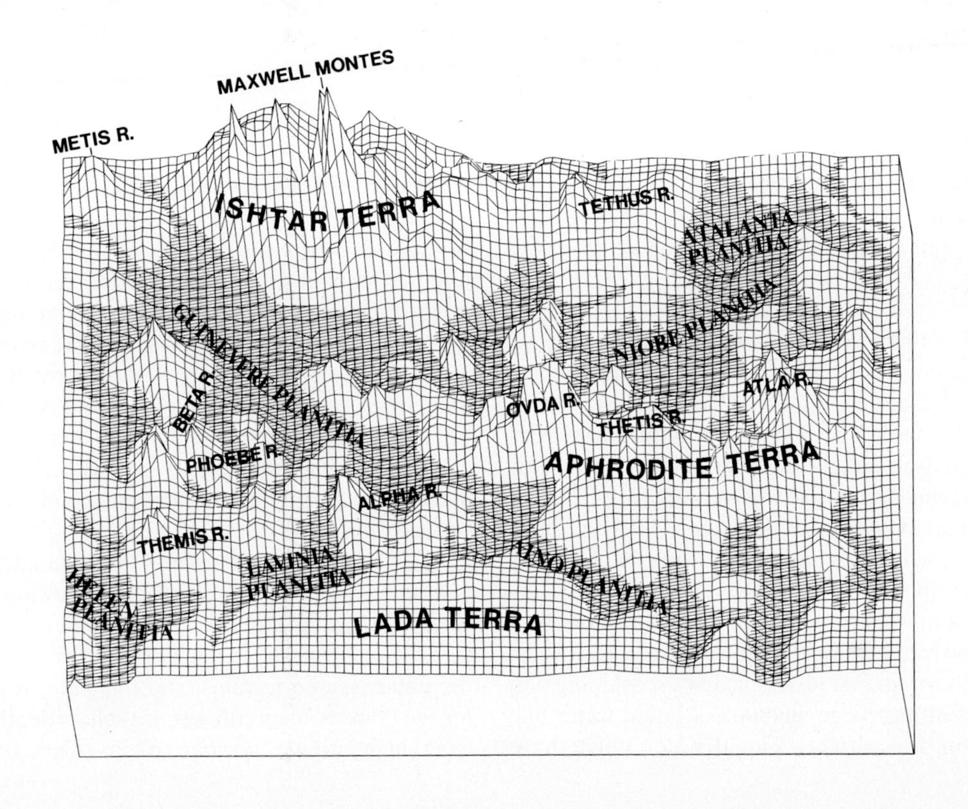

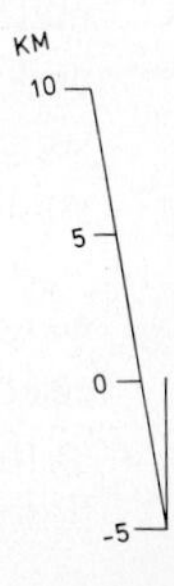

height. And yet Everest is part of a young alpine foldbelt and sialic in nature, whereas Maxwell is thought to be of volcanic origin, even though extinct. In contrast, Beta Region and a topographic high in eastern Aphrodite are believed to be still active, since these areas are associated with frequent lightning, a phenomenon also commonly observed during eruptions of terrestrial volcanoes. Broad positive gravity anomalies are closely linked with major topographic highs, which contrasts with the situation on Earth, where such anomalies show little correlation with continental masses but instead appear to be caused by deep mass anomalies. It has been suggested (Bowin, 1983) that crustal thickness, for instance in the region of Aphrodite, measures 70 to 80 km. This thickness is over twice that for continental crust on Earth and about ten times that of oceanic crust. The much thicker crust of Venus may be composed predominantly of basalt. Only a few places could conceivably contain continental (sialic) crust. Using Venus topography as indicator, the Lakshmi Plenum of the Ishtar Region is tentatively considered to be underlain by a "granitic" crust. The explanation is offered that the greater thickness of Venerean basaltic crust relative to terrestrial crust could possibly be linked to the higher surface temperatures for Venus (460° C), which would depress eclogite transition to greater depth. The scarcity of sialic crust on Venus can only be reconciled by assuming that Venus and Earth had their own distinct differentiation history.

One of the principal causes behind the different routes of Venerean and terrestrial crustal evolution seems to be related to the fate of water. Namely, loss of water from the interior of Venus in the course of the Archean may remove an asthenosphere needed to develop terrestrial-style plate tectonics (Philips et al., 1981). Please bear in mind the crucial role displayed by water when it comes to the melting of crustal rocks (depolymerization effect), and the opposite effect of CO_2 admission leading to cratonization (polymerization effect). Destabilization of water at the surface and generation of carbon dioxide would eventually bring to a halt crustal movement, thus making Venus a one-plate planet. The volcanoes on Venus are probably isolated hot spots similar to those found along the Emperor Chain with its still active member, the Hawaiian volcanoes. Actually, ongoing volcanic activity is needed to account for the currently observed cycling of elements in the Venerean atmosphere.

In conclusion, water and CO_2 are essential ingredients shaping crust and landscape of Earth and Venus but in a different fashion. On Earth, water is a kind of lubricant propelling plate movement. Furthermore, it is a critical element of weathering and life processes. On Venus, destabilization of water has stopped plate tectonics, and slowed down chemical weathering. With respect to CO_2, its high crustal abundance favors terrestrial cratonization, whereas high PCO_2 in the Venerean atmosphere initiates a "run-away greenhouse" effect. This results in surface temperatures close to 500° C and in this way depresses eclogite transition to greater depth.

BOX 10.11

Water on Mars (Carr, 1986, 1987; Fanale et al., 1986)

Published estimates on the amount of water that has been released from the Martian crust and mantle could, depending on author, yield a water layer from a few meters to more than 500 meters in thickness. Post-Viking appraisals and the discovery that some basaltic meteorites could have come from Mars (e.g., Becker and Pepin, 1984; Dreibus and Wänke, 1984), provide support for the higher figures by showing that Mars is a volatile-rich planet.

Today's atmosphere carries only minute amounts of water. However, erosional features such as deeply incised channels can best be reconciled by assuming that in the distant past large amounts of liquid water had flowed over the surface. Liquid water might have globally existed only for a brief interlude of a few hundred million to at most a billion years, and a climate different from today has to be envisaged. Impact melt sheets, which are layers of quenched molten rock and melt-rich breccias overlaying fractured basement rock, could provide steam and hydrothermal water (Newsom, 1980). Large-scale hydrothermal circulation and weathering is considered to be the mechanism for the generation of much of the Martian surface fines (soils).

At present, water is stored as groundwater and ground ice well below the surface. Following terminal cataclysm, water has been redistributed in that surface water formerly at low latitudes has migrated to the polar layered terrains, and the water responsible for the "floods" may still rest at shallow depths in low-lying, high latitude, northern plains (Carr, 1987). Ob-

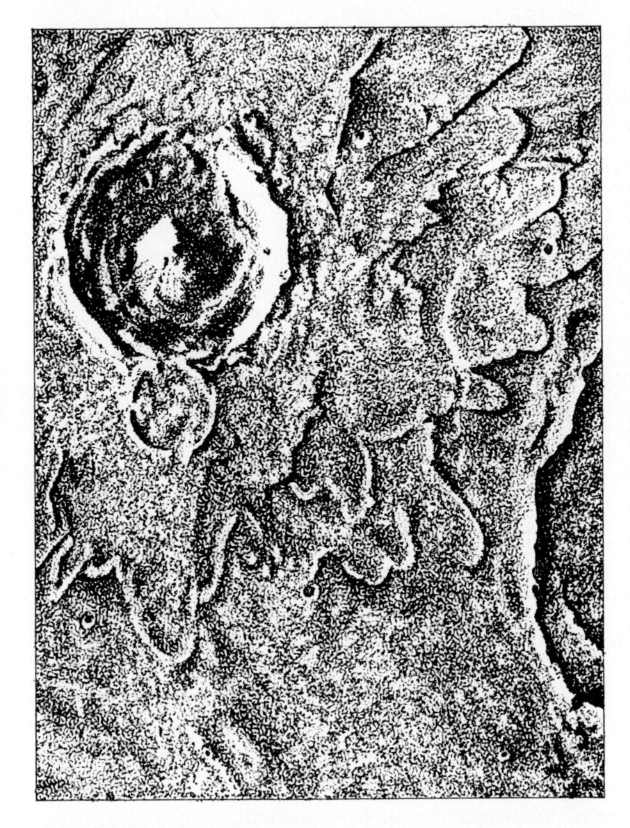

Fig. 10.38. The 18-km diameter crater Yuty (22°N, 34°W) revealing discrete lobes of ejected material that appears to have flowed outward away from the crater (Carr, 1986)

served seasonal and latitudinal distribution of water on Mars is a result of sublimation and condensation of arctic and antarctic surface ice, and the meridional transport of water vapor. The presence of a permanent H_2O ice cap in the north but not in the south is related to current climatic conditions (Davies, 1981). In addition, water can be bound in the form of clathrates or hydrous weathering products (e.g., clays, zeolites, palagonite, various hydroxides). Witnesses for the presence of subsurface water are lobate ejecta flows associated with large impact craters (Fig. 10.38). The fluidity of the debris has been ascribed to entrained water. Because of the large amount of CO_2 assumed to have outgassed early in the history of Mars, one commonly ascribes the relatively warm climate to a former greenhouse event. This could have been the case only if Mars had accreted ices, as was tentatively suggested (Berkley and Drake, 1981), otherwise CO_2 would have been rapidly consumed by the weathering of crustal rocks.

Not considering ice-albedo feedbacks, a warmer climate could have been the final outcome of a high CO_2 partial pressure and the presence of other greenhouse gases, including water vapor. Please be reminded that the Sun had already started to increase in luminosity. It was also proposed that the fixation of CO_2 as carbonates caused the atmosphere to thin and surface temperatures to fall very early in the planet's history (Carr, 1987). This scenario would be analogous to Earth in that CO_2 and water are the main agents of crustal weathering, except that Mars could not retain liquid water on its surface for long. The high sulfate to chloride ratio found in the Martian soil is of interest in the light of the soda ocean concept (see text). Terrestrial analogs of the surface rocks of Mars have recently been proposed (R.G. Burns, 1986), and data on Mars are summarized in J.A. Burns (1981).

Loss of the Soda Ocean

Crucial to the early soda ocean model is an answer as to how and when soda vanished, and why modern sea is dominated by halite. The most plausible scenario involves a gradual removal of dissolved sodium carbonates from the ocean via: (i) pore waters in sediments, (ii) migration of sea water into the crust, (iii) subduction, and (iv) biogenic reduction to organic carbon and subsequent burial (Kempe and Degens, 1985). In the course of many revolutions, the alkalies became part of the continental crust, whereas solid carbonates gave rise to carbonatites at depth, or controlled the liquidus-solidus of mineral equilibria. At the same time, chlorine, which is present in oceanic basalts in small quantities only, i.e., a few tens of ppm, was leached and became part of a rising hydrothermal salt flux to the sea. On the basis of geochemical balance calculations (Wedepohl, 1963), only 0.5 percent of the chloride content in the sea is derived from surface weathering of igneous rocks, the rest is crustal leaching. Sinks of hydrothermal chlorine include recent and ancient sediments as well as crystalline rocks derived from granitization and metamorphism of former sediments. Cl^- fixed in sediments in the form of interstitial waters, salt deposits and ordinary minerals constitute about 25 percent of the amount present in the sea. Excess of Cl^- in granites, diorites, syenites and metamorphic rocks above the amount released from the hydrothermal weathering of the oceanic crust yield another 25% (Ronov, 1968; Wedepohl, 1963). The sum total of hydrothermal Cl^- is thus about 50 percent above the present sea water chlorinity.

Injection of HCl and other acids into the surface environment steadily transformed the oceans's pH value from alkaline to acid. About 1 billion years ago the erstwhile soda ocean was thus transformed into today's salty sea. Except for small evasion in salinity due to salt deposition, or changes in the rate of hydrothermal leachings, the ocean must have stayed reasonably constant in NaCl content. On the other hand, changes in pH and possibly Eh have altered the marine environment quite substantially. This concerns in particular the carbonate system which switched from sodium- to calcium- and magnesium-dominated mineral equilibria about 1 Ga ago, as well as the sulfate system which became controlled by molecular oxygen, calcium content in the water, evaporation and, last but not least, the biota. It is here that an explanation for a distinct geochemical, sedimentological and biological facies can be found.

The occurrence of the banded iron formation (James, 1954; Gole and Klein, 1981; Simonson, 1985) in the early Proterozoic is linked to: (i) the abundance of molecular oxygen, (ii) a drop in pH below 9.3 as part of the acid-base titration scheme in the soda ocean, and (iii) microbial consumption of organic matter originally produced by photosynthesizing organisms leading to a release of CO_2 or CH_4.

The first occurrence of massive halite deposits at the end of the Proterozoic signals that prior to this time NaCl had not yet accumulated sufficiently to yield a massive salt sequence. Large halite casts from shallow water and hypersaline lake sediments from the Middle Proterozoic have been described (Neudert and Russell, 1981), but their presence indicates only that NaCl was around but not to the extent of forming salt deposits. A similar case can be made for gypsum ($CaSO_4$ x 2 H_2O; Kempe and Degens, 1985). Here it was the low Ca^{2+} content in the soda ocean that inhibited calcium sulfate deposition. Finally, the onset of biomineralization and the subsequent evolution of marine invertebrates and nanoplankton is in the final analysis determined by the evolution of carbonate equilibria in the sea. Benchmarks of geochemical events are plotted against the chemical development of the ocean in the course of 4 billion years of Earth history (see Fig. 12.26).

The rather involved interplay between air, water, crust and eventually life, combined with the unknown rates of reactions, does not warrant a thermodynamic treatment nor a mathematical formalism to decipher origin and evolution of the system at large. The only meaningful approach in my opinion is to define a model that is at least consistent with the limited observational data and that outlines and takes care of feedbacks among the individual compartments. This was the avenue I have tried to follow in the above discussion.

Other scenarios on the genesis of air and water have been proposed and the interested reader is referred to: Walker (1982), Henderson-Sellers (1983), Holland (1984), Holland et al. (1986), and Budyko et al. (1987).

Global Sea Level Changes

Before closing the subject water a few words should be addressed to causes and chronology of sea level changes. It is not so much the "biblical flood" which stands in the foreground of interest, but the fact that the sea from time to time floods continents or retreats to the outer shelf margins in what geologists refer to as transgression or regression. The phenomenon of sea level rise or fall for which the general term eustacy has been coined (Suess, 1906), has commonly been explained, for example, by suboceanic subsidence, epirogenetic uplift of large cratons, or glaciation and deglaciation.

Today, we look at the fluctuating sea level in a more refined way ever since Hays and Pitman (1973), Vail and colleagues (1977) and Pitman (1978) proposed a relationship between eustacy and stratigraphic sequences of passive margins. Basically, seismic reflection data, drilling results, and outcrop studies of sediment sections found along coastlines the world over reveal by means of stratal geometries a kind of geologic benchmark indicative of either transgressions or regressions. Terms such as onlap, downlap, hiatus, truncation, and basinward shifts of coastal onlap have become familiar expressions.

For about 10 years, the subject has been discussed in special meetings, articles, and books and continues to be debated (Sleep, 1976: Hallam, 1981, 1984; Schlanger, 1986). The timely publication by Haq et al. (1987) has brought to a climax the issue of relative sea level changes which is a combined effect of subsidence and eustacy. A condensed version of their eustatic curve in relation to magnetostratigraphy and chronostratigraphy is shown in Fig. 10.39. The long- and short-term eustatic curves are plotted. The scale (in meters) represents the best estimate of sea level

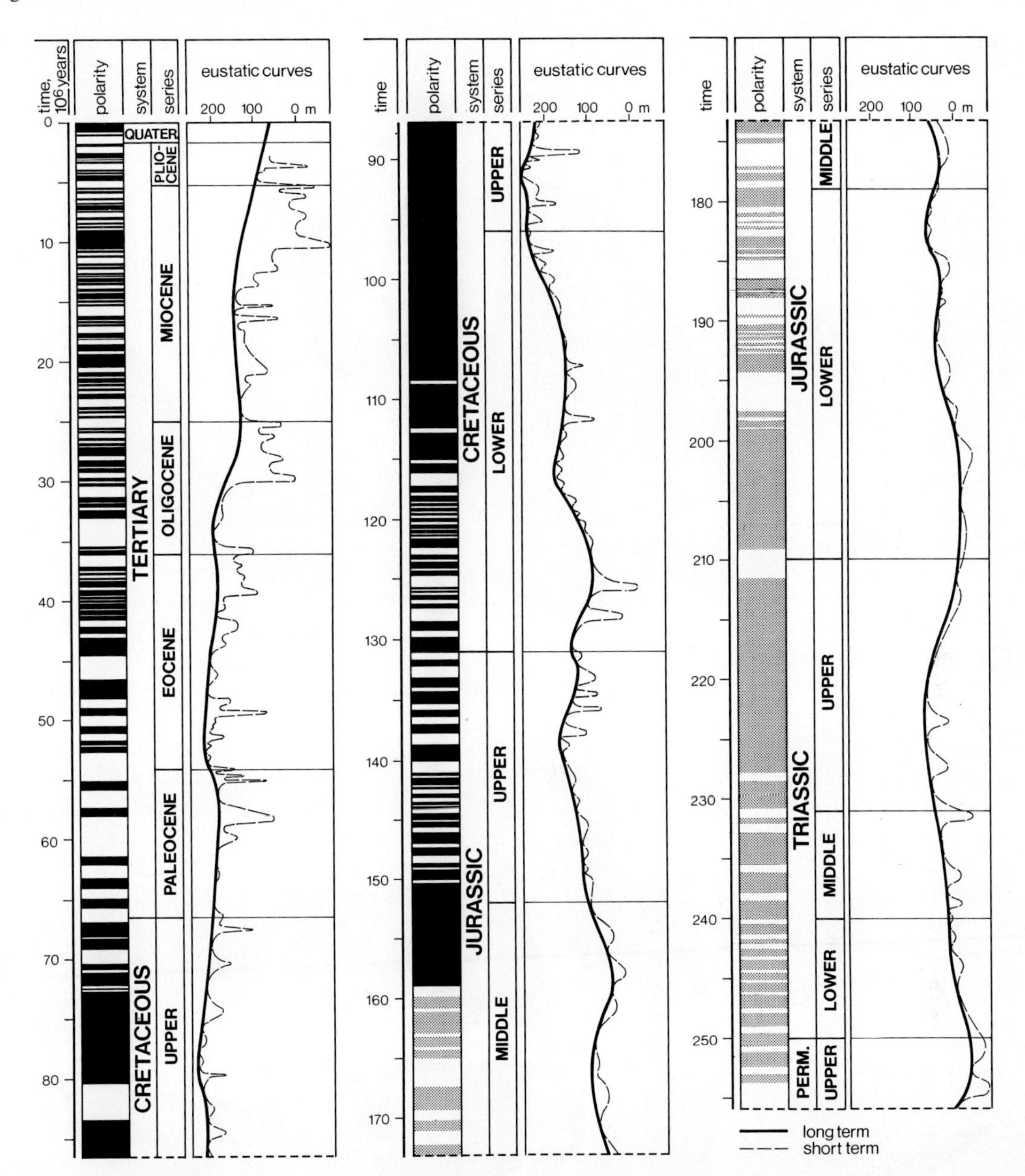

Fig. 10.39. Sea level changes in the course of the Mesozoic and Kaenozoic (after Haq et al., 1987). Magnetic polarity epochs are indicated

rises and falls compared with the present-day global mean sea level since the beginning of the Mesozoic. All in all, a total of 119 Triassic through Quaternary sea level cycles have been identified. Of these, 19 began with major sequence boundaries. A hypothetical sea level curve for most of the Paleozoic is depicted in Fig. 8.31.

To illustrate the significance an oscillating sea level has on paleogeography, a world map of the Upper Cretaceous and one of the Lower Permian are shown in Fig. 10.40. The graph reveals that by Late Cretaceous time, i.e., during a highstand, approximately 35 percent of the present land surface was covered by a shallow sea. The opposite picture, i.e., world ocean at a lowstand, is taken during the early Permian, when Pangea was still intact.

Pitman and Golovchenko (1983) have summarized general causes, magnitudes, and rates of sea level change. Aside from short-term glacial-interglacial

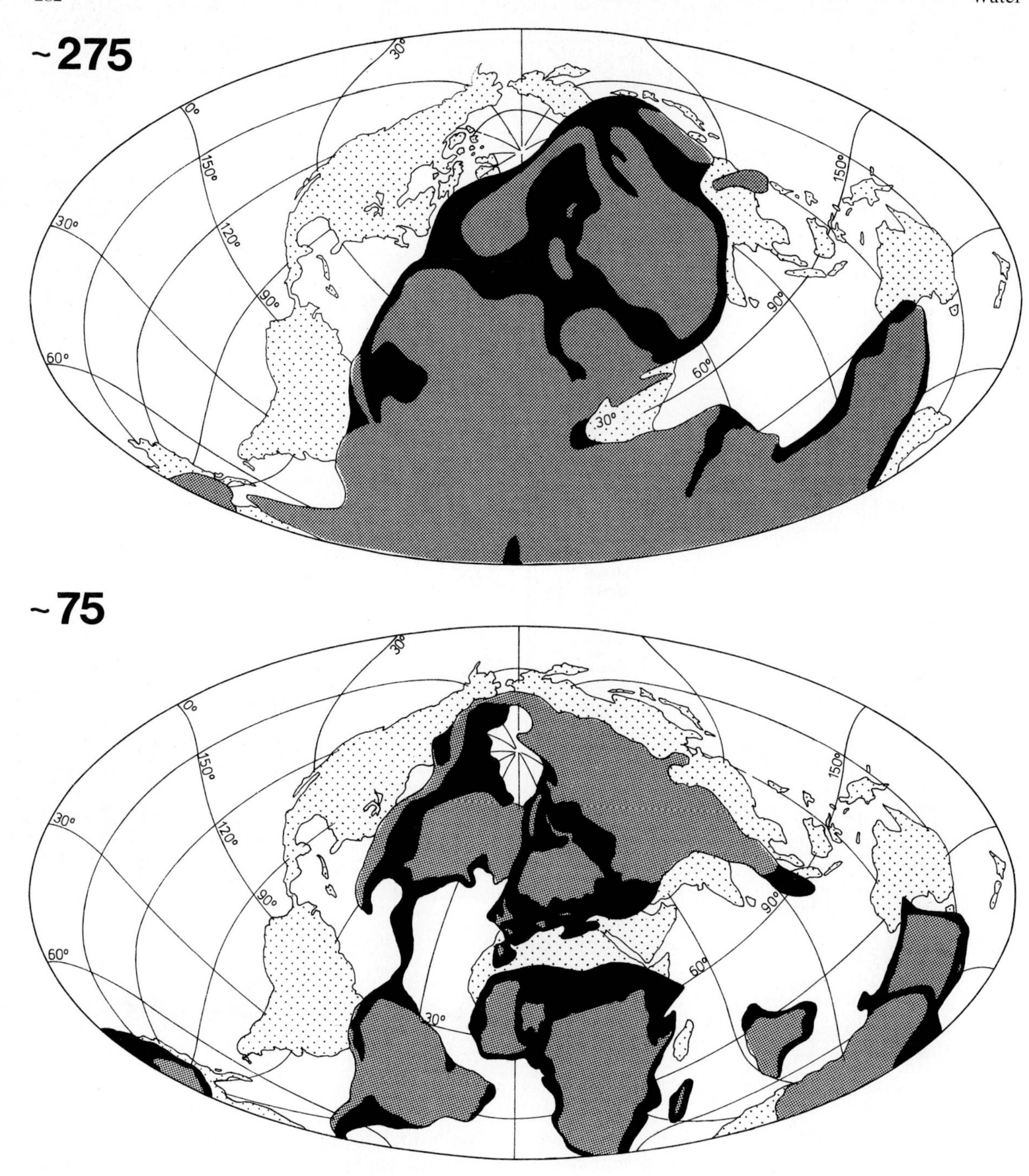

changes, which can be attributed to removal or formation of ice sheets plus expansion or contraction of water as a result of temperature effects, major sea level oscillations in the past are now ascribed to seafloor spreading in terms of ridge-crest length or ridge-crest volume. Hallam (1981, 1984) and Schlanger (1986) report major short-term eustatic changes measuring at the most a few million years during the

Fig. 10.40. Paleogeography about 275 and 75 million years ago. *Gray areas* are continents, and *black areas* epicontinental seas. Present position of continents is indicated (*stippled areas*)

Jurassic and Cretaceous and which cannot be explained by mechanisms offered so far. Perhaps rapid tectonic subsidence and uplift along active margins and associated with crustal adjustments both vertically and horizontally may in part account for this phenomenon. Alternately, the large pool of water contained in the crust which by volume represents about 40 percent of the ocean, may have been filled and drained episodically, that is in a non-steady state fashion; as a result, sea water oscillated in phase. It is also conceivable that a number of positive and negative feedbacks among a series of possible processes including sporadic additions of cometary water (see Box 10.7) could explain the observational data. However, an overall satisfying interpretation is still lacking.

Chapter 11

Life

The Essence

Definition

To quote J.D. Bernal (1959): "There is only one dominant chemical pattern of life. If more than one exists the others must be obscure, as are some of the red algae, and unnecessary to the survival of the dominant", and "We may define life, for the sake of this discussion only, as the embodiment within a certain volume of self-maintaining chemical processes". Others state (Miller and Horowitz, 1966): "An organism, to be called living, must be capable of both replication and mutation; such an organism will evolve into higher forms".

Living forms are known which through generations have remained unchanged. The attribute of mutation is not essential. Mutation is an independent event and can be regarded as a discrete thermodynamic disorder phenomenon. Moreover, mutation implies alteration but does not tell us anything regarding the direction of that change. It is better to speak of *evolution,* which is a directed and oriented development, because a highly advanced (derived) organism will never mutate back into an ancestral form.

The central problem of the origin of life deals with the minimum requirements for the formation and maintenance of a workable living system. Questions regarding the controlling elements of evolution and mutation, or the mechanisms of perception and computation (cognogeny) in the sense of Lederberg (1965), are at this point, where origin is asked for, of no relevance.

Starting from the axiom that life is derived from ordinary chemical reactions that took place on the surface of the Earth some 4 billion years ago, it is concluded that life as an autoreplicating system is an outgrowth of its environment and thus can be described and understood only in relation to that environment.

In the preceding chapters I have outlined the conditions that in my opinion led to the origin of air, water, and solid earth. From that "compost" life arose. Inasmuch as living matter appears to be a product of these environmental elements, it should reflect that heritage in structure and function. I agree with Pirie (1937) that in the present context a dispute as to at what point one should consider a form living or nonliving represents an unfruitful scientific "Scheinproblem". Whether or not an isolated chromosome or nucleus is alive, who knows? Or take a virus. A person having the "flu" would certainly call it alive, but is this truly the case? A bacteriophage – that is a bacterial virus – substitutes, following infection, for the monitoring part of the bacterial system, because no emulation between the bacterial cell and the virus takes place. The viral DNA replaces the cellular program. It is logical that a monitoring capability does not represent the essential difference between a bacterium and a virus. The basic difference is that a bacterium is capable of producing its own boundary layer, the membrane, whereas the virus cannot do this. We are thus forced to state that the real physical-chemical foundation of the cellular system is based on the genetic instruction for membrane synthesis. This implies that the codon holding the instruction for the synthesis of phospholipids is the vital biochemical part of living organisms which distinguishes them from the rest of universal matter.

All living and solid matter on Earth is governed by the same coordination principles among atoms and ions; they control the generation of identical structural entities of the type we already know of: tetrahedra, octahedra, polyhedra. Structure determines function, and function may lead to structural change, in one word: *evolution.* The movements of electrons and protons along and within polyhedra associations keep the air-sea-earth-life system going and evolving.

At this point there is no necessity to expand on molecular cell biology or the principles of biochemistry. For that several outstanding texts are available (Lehninger, 1982; Alberts et al., 1983; Darnell et al., 1986; Starr and Taggart, 1987; Streyer, 1988). Instead, the forthcoming discussion is intended solely to draw attention to the interaction between metal ions, organic molecules, and water as they relate to the synthesis of biophosphates, the true structural backbone of life. Inasmuach as their structure had apparently been shaped by mineral templates, a comparison of inorganic and cellular phosphate "crystals" may ultimately lead to an improved model on the origin of life.

The Cell

The term "cell" was invented by Robert Hooke in 1665 to identify the small units or compartments he saw in sections of plant tissues. About 200 years later, in the year 1858, Rudolf Virchow recognized that each cell originates from another cell. The invention of the electron microscope in the 1930's started a new era of investigation, and much of what we know today about the structure of cells is derived from that source. Of particular significance was the finding that a cell is subdivided into numerous compartments by various types of membranes which represent natural boundaries within and between cells. Ribosomes which are cellular units actively participating in the synthesis of proteins were recognized. The electron microscope revealed the cell in a structural and functional complexity hitherto unknown.

To us, the concerted action of the various molecular constituents of a cell appears confused; the harmony escapes us. This statement is valid in spite of the enormous body of published data in the field of biosciences. To illustrate what we know for sure, here is a check-list of the principal chemical ingredients of a cell:

(i) about 15 bases of the purines and pyrimidines which are components of the nucleic acids;
(ii) about 30 species of amino acids, the monomeric building blocks of peptides and proteins;
(iii) about 50 sugar monomers which compose the polysaccharides; their derivatives – organic acids, aldehydes ... – amount to about 70 species;
(iv) circa 20 different coenzymes which are needed for the activity of numerous enzymes (also referred to as cofactor or prosthetic group);
(v) about 50 types of lipids which contain as their basic element long-chain aliphatic hydrocarbons;
(vi) about 20 different phospholipids, the cornerstones of membranes.

To this number of about 250 molecular species should be added circa 50 phenols, which are major components of the lignins in higher plants.

At first sight, a sum total of 300 simple organic molecules seems to be manageable, chemically speaking. The first complication arises the moment monomers transform to polymers with molecular weights ranging from 10^3 to 10^9 dalton. A dalton (named after John Dalton, 1766–1844) is the mass of a single hydrogen atom. For instance, the DNA of *Escherichia coli* is about 2 x 10^9 daltons, which is equivalent to about 10^{-19} g. Even if we were able to cross the first hurdle of complexity, for instance, by unravelling the sequential order of amino acids in a protein and eventually the molecular arrangement of all essential macromolecules in a cell, we would still be faced with hurdle after hurdle after hurdle, when we are seeking the cell's logic. Please bear in mind that a cell is not just composed of carbon-containing molecules – abbreviated C_{org} – but of two other key compounds:

$$\text{water} - C_{org} - \text{mineral salt}$$

in a ratio (for the protoplast) of about 80:18:2. Water and salt conduct an electric charge. A cell, therefore, represents an electrolytic system. Moreover, all three classes of compounds rarely occur in a cell in their pure state. Instead, they are combined to a legion of structural entities. Free water is almost non-existent.

In summary, the amount and relative proportions of the various organic and inorganic molecules or ions in various cell types is known in great detail. Furthermore, the physical distribution of a number of macromolecules or cellular bodies is reasonably understood. The sequential order of monomers in close to 1.5 million peptides, proteins and nucleic acids across the plant and animal kingdom is established, but no unifying concept has yet emerged. It is the overall physiological interplay in the living cell that still remains a puzzle, even though the operation principles of many biochemical cycles, such as the urea or citric acid cycle, are well comprehended.

Unfortunately, analytical biochemical research starts frequently with the homogenization of the cellular material. Another common practice in electron microscopy is the staining of organic tissues with electron-dense salts of heavy metals to get a better contrast (Box 11.1). In both instances, the delicate fabric of a cell becomes eliminated or grossly distorted. The situation reminds me of "Humpty Dumpty" the personified egg in an old English nursery rhyme. "Humpty Dumpty sat on a wall, Humpty Dumpty had a great fall!". Well, from a scrambled egg you might reconstruct some details, but "all the King's horses and all the King's men" will never be able to put Humpty Dumpty together again and let a live chicken hatch out of it. To apprehend life in its wholeness, a close look at the most dominant structures of the genetic and metabolic apparatus – the esters and anhydrides of phosphoric acid – is helpful.

BOX 11.1

Staining Techniques

Prior to transmission electron microscopy, organic tissues are frequently stained with heavy metal solutions (e.g., salts of Pb, U, Os, W) to provide electron-dense areas. Biologists are commonly unaware of what this contrast medium might structurally do to their specimens. Staining patterns are often interpreted to reflect true images of the structural make-up of studied tissue. This is not necessarily so, as the molecular crowding of tungsten and oxygen around two phosphate tetrahedra will illustrate (Figs. 11.1 and 11.2).

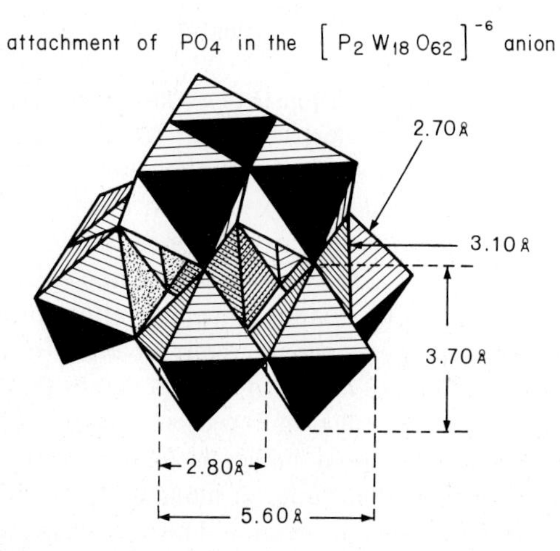

Fig. 11.1. Crowding of oxygens around tungsten and phosphorus; octahedra dimensions in Å

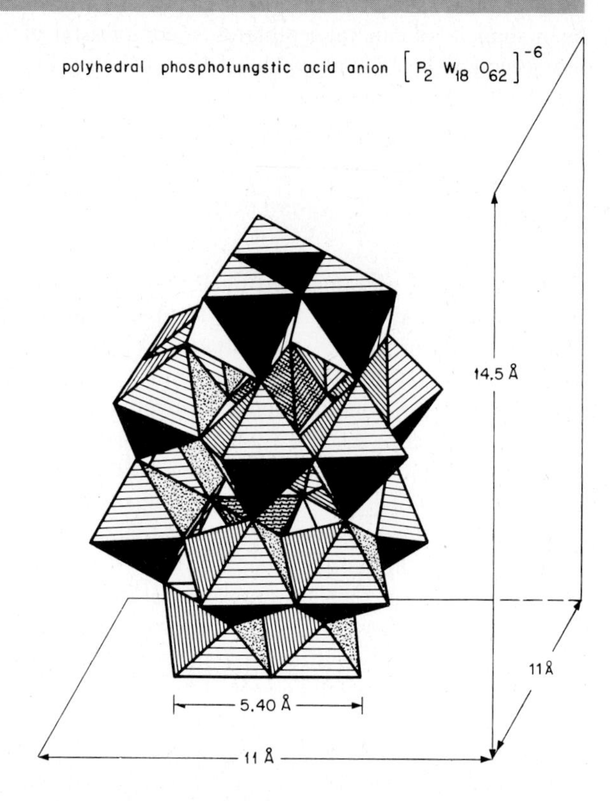

Fig. 11.2. Crowding of oxygens around tungsten and phosphorus; dimension of unit cell in Å

Assuming that a phospholipid bilayer membrane provides the oxygens for the heavy metal ions, a series of patterns may come into existence which may have little to do with the original conformation of the bilayer and their protein cover. Superimposing two linear patterns of dots and lines at different angles will give you a variety of structural designs which in the case of the phospholipid bilayer can be related to the way the heavy metal ions have reshaped the substrate (Degens et al., 1970; Fig. 11.3).

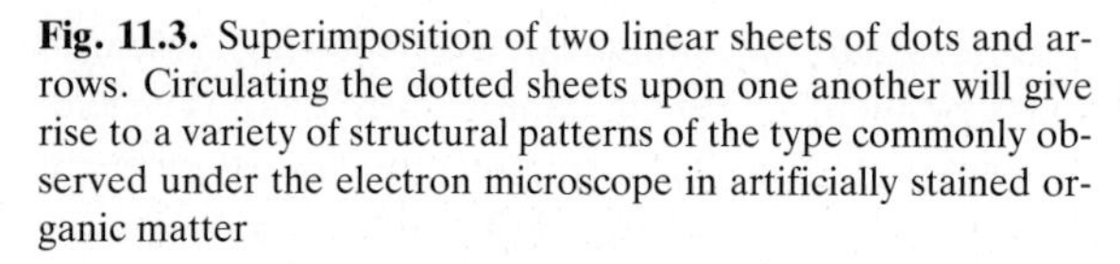

Fig. 11.3. Superimposition of two linear sheets of dots and arrows. Circulating the dotted sheets upon one another will give rise to a variety of structural patterns of the type commonly observed under the electron microscope in artificially stained organic matter

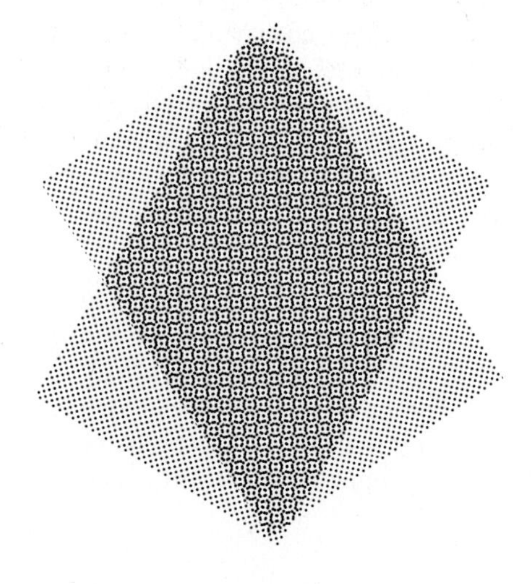

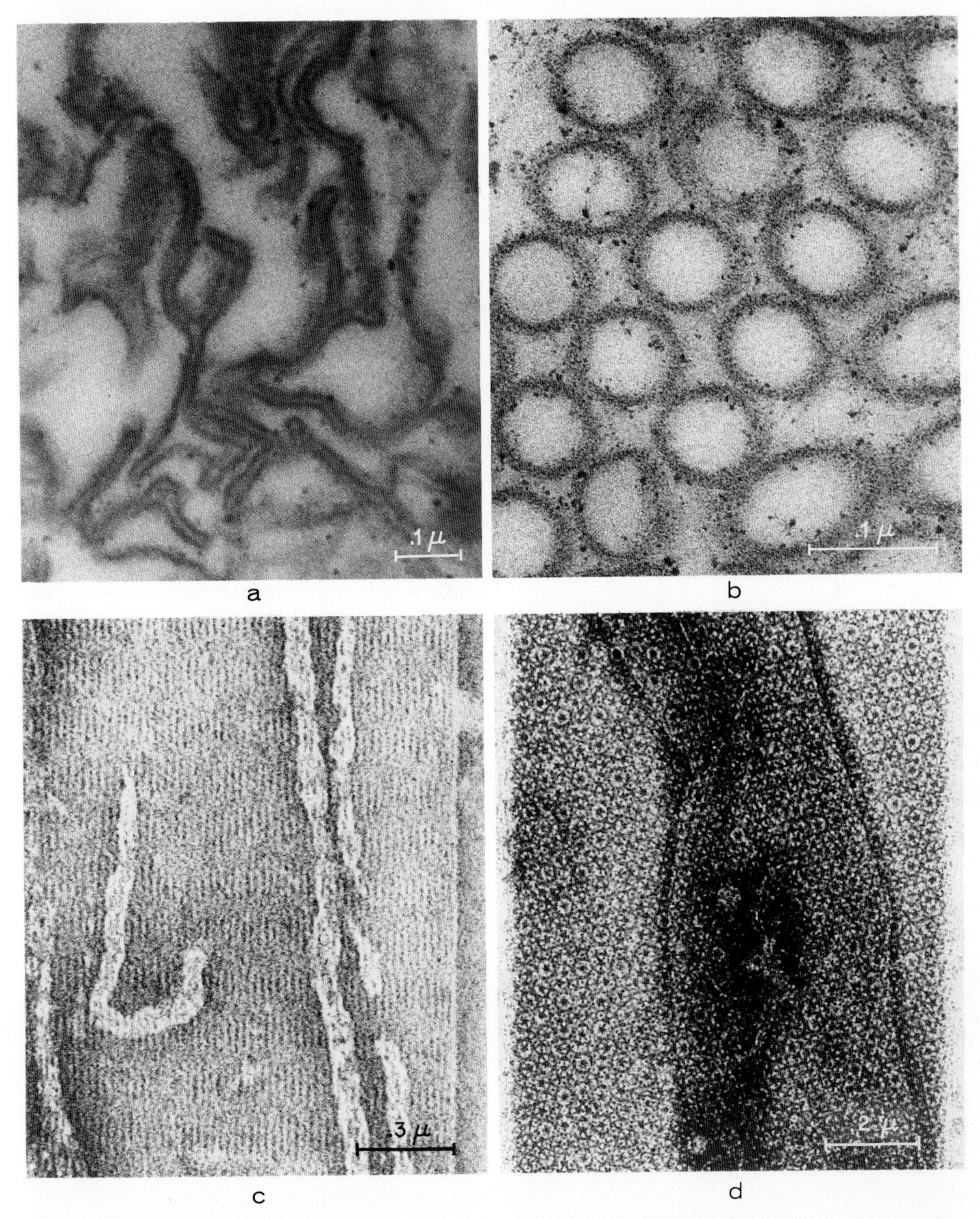

Fig. 11.4 a–d. Electron micrograph of: **a** branched tubular membranes aggregated together and bound by limiting membrane; **b** large tubular membranes having diameters of 700 to 800 Å and consisting of unit-membranes having a width of 80 Å; **c** organic crystal consisting of unique patterns of subunits resembling bacterial cell walls; and **d** organic crystal showing crystalline arrangement of subunits resembling pattern in Fig. 11.3

Biological membranes exposed for about 3,000 to 5,000 years to anoxic conditions of the Black Sea exhibit a staining pattern which is caused by the uptake of heavy metal ions from interstitial waters (Fig. 11.4). This time period was marked by a gradual transition from a fresh water to a marine environment. Resulting salinity gradients promoted natural chromatography, dissolution, redeposition and preservation of organic molecules. Noteworthy in particular is the high abundance of branched, tubular membranes which occur in the cristae of some mitochondria and a variety of bacteria but which here occur as fine sediment layers. It has been shown that configurational changes in mitochondrial membranes may occur which do not depend on electron transfer or hydrolysis of ATP. Rather, these configurational changes which can lead to tubular membranes are due to fluctuations in the ionic environment and in metal ion concentration within the cell.

Of the six biochemical endmembers in our check list, three have the PO_4 unit as an integral part of the molecule. The remaining three biochemicals require for their synthesis the active participation of at least one of the other phosphate-containing molecules. Thus, it appears that some unique properties must be ingrained in the element phosphorus.

Uniqueness of Phosphorus

In his book "Historia inventionis phosphori", G.W. Leibniz (1646–1716) stated that the element phosphorus "became known around the year 1677". Its discoverer, the physician Hennig Brand, extracted the element "from the spirit of urine". The long-time use of the alchemistic symbol for fire △, and the selection of the name:

phosphorus φωσφόρος = carrier of light

for this element had to do with the fact that phosphorus ignites when exposed to air and glows when left in darkness.

For about 100 years urine was the only source for the manufacture of phosphorus. This changed when in 1770, J.G. Gahn and C.W. Scheele found this element concentrated in the ashes of bones. Phosphorus' peculiar properties were initially thought to be a result of its biological origin and promoted its widespread use as a medicine in the early years. Since there was no obvious cure to health problems by administering phosphorus, this treatment was soon abandoned.

Surprisingly, for more than 200 years following the discovery of this element, the physiological significance of phosphorus received no scientific attention at all, except for the fact that it was a major element in bones and teeth and occurred there in the form of apatite, a calcium phosphate mineral. The situation changed in 1906, when H. Harden and W.J. Young recognized that phosphate is an essential nutrient for the fermentation of sugar in cell-free yeast juice. This paper marked the beginning of a scientific breakthrough by demonstrating that inorganic orthophosphate is a key factor in cellular dynamics.

It was soon discovered that wherever cells or organelles represent a natural thermodynamic system, phosphate compounds do actively participate in chemical reactions. The presence of phosphates in the genetic coding apparatus was considered further evidence for the wide use of phosphorus in biological systems. However, the real significance of the phosphate units in, for example, the various nucleic acids, came to light only gradually. This slow development is all the more surprising because desoxyribonucleic acid (DNA) was discovered as long ago as 1869 by F. Miescher.

In the biological sciences a quiet revolution is presently going on, namely the recognition that the architectural principles observed in biophosphates are identical to those established in the inorganic phosphates. The early adoption of the name phosphorus, the carrier of light, has not lost its true meaning over the centuries. Phosphorus, in the form of phosphate bonds, is the carrier of energy in the living system. It is the "energy currency" as George Wald (1962) so nicely put it, which becomes printed, exchanged and converted at various rates in the perpetuating cycle of life. It is phosphorus which controls the structure and shape of the cellular material and which thus selects the energy transfer sites.

The reason for the uniqueness of phosphorus is rather trivial. Phosphorus has an electron configuration where there is one electron too many for its various functional and structural work assignments. This statement can perhaps be better understood by presenting silicon as a case study:

In its pure form, the element silicon behaves as an *insulator.* When subjected to pressures exceeding about 100 kilobars it transforms into a metal and *con-*

ducts. There are indications that at higher pressures silicon atoms become stacked in the form of hexagons which are anisotropic with regard to electronic charge distribution, and *superconduct*. The use of silicon (Si^{4+}) and of structurally related germanium (Ge^{4+}) crystals as *semiconductors* in modern electronic appliances such as transistors, rests on added impurities of, e.g., arsenic (As^{5+}) at concentrations of 10^{-2} to 10^{-6}. This element has one electron that cannot become accommodated in the four-valent bond structure in the Si-crystal which hosts the arsenic. Alternatively, the addition of gallium (Ga^{3+}) will cause a depletion in one electron in the silicon or germanium lattice. A substitution of this kind will result in distinct physical properties of the silicon or germanium crystals. The use of silicon chips, for instance in computers, or of gallium arsenic crystals for TV communication systems by means of glass fiber cables, is based on the way electrons are configurated and moving in the adopted crystalline compounds.

The example of the transistor does not, of course, fully explain the functionality of phosphorus in biological systems. The contamination of crystals with "dirt", that is traces of impurities, should only serve as an analogy for the operation principles of cellular phosphate. For a given stereochemical configuration a phosphate has one electron too many. This explains why phosphates are incapable of constructing simple networks of the type observed in SiO_4 structures, in which the four electrons participating in the bonding are equally balanced and compensated for. In contrast, PO_4 structures can readily adjust to changes in the biochemical milieu simply by shifting the fifth electron around (Box 11.2). It is this flexibility in functional and structural behavior which makes phosphorus the key element in living systems.

There is currently some talk of taking advantage of this property developed in some biomolecules by fabricating what is often referred to as biochips. But solid state physics of silicon is so well understood, predictable and inexpensive, that its possible replacement by organic molecules, although challenging, is so far only a vision, nothing more. Nevertheless, it is quite remarkable how life, in its mysterious past, has come up with an electronic appliance system that outranks any human-made electronic device in both speed and efficiency.

Homo sapiens has come a long way from the stone ages, to various metal ages – please do not forget "aurea aetas..." in Ovid's Metamorphosis – and now to the nuclear age, which appears already transitory to the silicon age. Question: What kind of intelligent technology was employed by the primordial cell to generate a network of minichips to keep the system alive and evolving? Jointly with my friend Janek Matheja (1934–1974), I came up with an answer 20 years ago (Degens and Matheja, 1967, 1968): "Adoption of oxygen-coordinated metal ions associated with PO_4 groups". This statement is still valid and not refuted by new data.

BOX 11.2

Phosphorus and Its Next Neighbors

Phosphorus differs from its neighboring elements carbon, nitrogen, and oxygen, in that one of the 3 s or 3 p electrons can be transferred to the 3 d orbital (Fig. 11.5). This means that phosphorus, four-fold coordinated to oxygen (sp^3), has bonds in addition to the four bonds. The problematics of the bond order and the bond-energy relationships in the phosphate group are reviewed in Matheja and Degens (1971 a).

It is important to realize that the homologous elements N and P have complementary properties in relation to hydrogen and oxygen:

(i) affinity of group V elements to hydrogen decreases with increasing atomic number; the formation of NH_3 represents a strong exothermic reaction, and that of PH_3 a weak exothermic one (values of bond energy):

$$(94\ \text{kcal})\ N-H > P-H\ (77\ \text{kcal});$$

(ii) affinity to oxygen increases with increasing atomic number; the formation of nitrogen oxides represents an endothermic reaction, whereas that of phosphorus oxides is an exothermic one:

$$(60\ \text{kcal})\ N-O < P-O\ (100\ \text{kcal}).$$

Empirical data (Table 11.1) underline this behavior of phosphorus. For phosphorus and hydrogen, electronegativity values are identical and related to their competitive role in capturing oxygen:

$$H \rightarrow O \leftarrow P.$$

Bond energies between two atoms a and b include the electronegativities (Pauling, 1948). On the other hand, nitrogen and phosphorus resemble each other with respect to their electron affinity, which is a measure of the energy released by electron capture. Val-

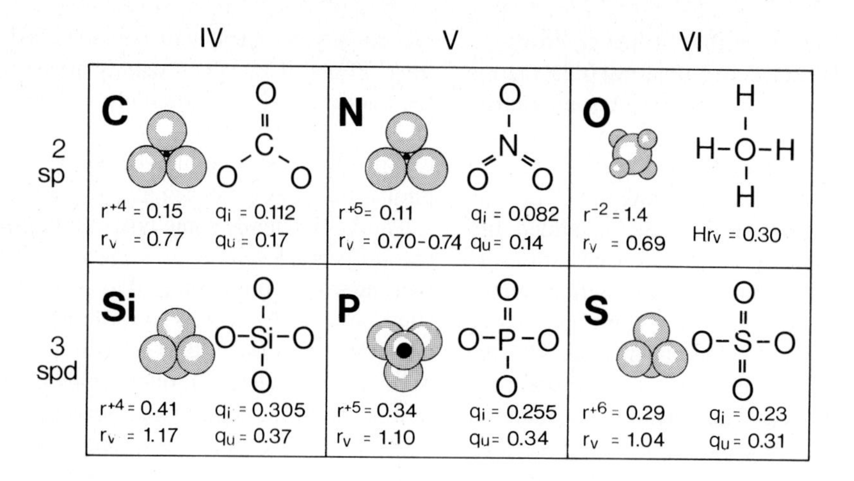

Fig. 11.5. Segment of Periodic Table of Elements. The ionic radii, r^{n+}, for the six-fold coordinated state, are given in Å units. The regular valence ratio, r_v, exists when the number of bonds are equal to their normal valence state. The ionic radii quotient for an oxygen polyhedron is q_i, and the ratio of univalent radii is q_u (= cation/oxygen). Size of the cations determines oxygen coordination. Phosphorus and silicon are too large for a triangle coordination, whilst nitrogen is too small for it

ues for N and P correspond to the electron affinity for the alkali metals; with regard to chemical reactions, low electron affinities have significant consequences.

Excitation potential and ionization energy are additional evidence for the complementary nature of phosphorus and nitrogen to the elements oxygen and hydrogen. They also indicate that in aqueous environments phosphorus occurs in the form of rather stable PO_4 tetrahedra. Because of the relationship between bond energy and length, the PO_4 group exhibits a flexible tetrahedron geometry in terms of variable bond lengths and angles. As a consequence, PO_4 polymers are able to bind metals of different valences and sizes; they simply adjust to the rigid spheres of the cations. This property of phosphorus is unmatched by any other biogenic element, and PO_4 represents an optimal coordination partner for metal cations.

Table 11.1. Electronegativity and electron affinity for major biogeochemical elements (from Preuss, 1962)

Element	Electro-negativity (X)	Electron affinity (kcal)	I. Excitation potential (eV)	I. Ionization potential (eV)
O	3.5	88	9.0	13.6
C	3.0	31	7.5	11.3
N	2.5	1	10.2	14.5
P	2.1	3	6.4	10.9
H	2.1	16.4	10.1	13.5

The importance of being ionized, a thesis developed by Davies in 1958, implies that a cell must keep its metabolites, and ionized species, different from electrically neutral ones, become restricted in moving around because of their huge size and by barriers such as cell walls which operate quasi like an electric fence. Westheimer (1987) has expanded on this issue and developed an interesting scenario as to why nature chose phosphates. It is because phosphates with multiple negative charges can react by way of the monomeric metaphosphate ion PO_3^- as an intermediate. The ionizing capacity of phosphoric acid makes it an ideal partner to stabilize the nucleic acids – which are diesters of phosphoric acid – and at the same time maintain a third ionizable group that keeps the material within the cell.

I fully agree with Davies (1958) and Westheimer (1987) that the conditio sine qua non for a living system is to be *ionic,* and that the ideal group for the perpetuation of cellular activity is the negatively charged phosphate. My viewpoint, however, differs in so far that I consider metal-ion coordination to the phosphate unit, which determines structure and regulates function, the essence of life. It has inherited this property from mineral templates, ages ago, and refined it in response to environmental changes in the course of Earth's history.

Structure and Function

Metal Ion Coordination

When it comes to metal ions in living things, one commonly associates them with molecules such as chlorophyll or hemoglobin in which magnesium, or iron respectively, are the coordinating partners. Nitrogen from the four pyrrole rings serves as ligands to the metal ions:

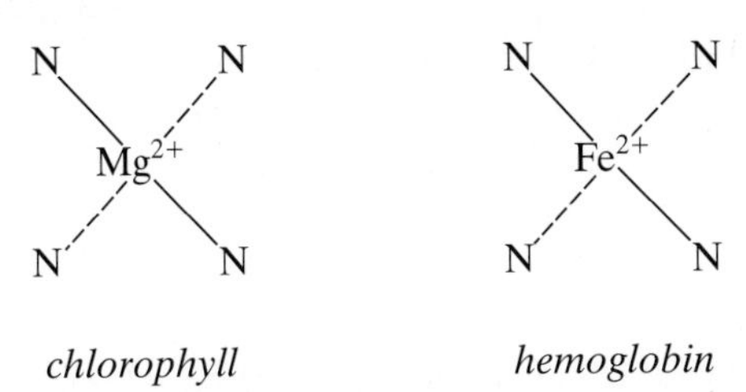

Chlorophylls are green pigments which function as receptors of light energy in photosynthesis, whereas hemoglobin is a heme protein of the red blood cell mediating the transport of molecular oxygen. It is also well known that organisms can deposit Ca^{2+} in the form of phosphates in bones and teeth or as carbonates in shell structures. To suggest that the same metal ions, plus a few others, give structural rigidity and at the same time functionality to the nucleic acids and to the phospholipids in membranes will not necessarily find unanimous approval. That these structures generated by metal ions are identical with those found in minerals and rocks is even harder to understand. A brief explanation in support of this thesis is thus needed.

Nitrogen and phosphorus are the two most prominent structural elements in biogenic compounds. Nitrogen has the ability to generate hydrogen bonds which give rise to well-defined geometric networks – just think of Linus Pauling's α-helix – and the potential to exist in the three- and four-fold coordination state:

$$\mathrm{N} \rightleftharpoons -\mathrm{N}^{+}-$$

Analogous phosphorus, in the form of the PO_4 unit, can generate metal ion bonds which determine shape and configuration of phosphate-containing bio-

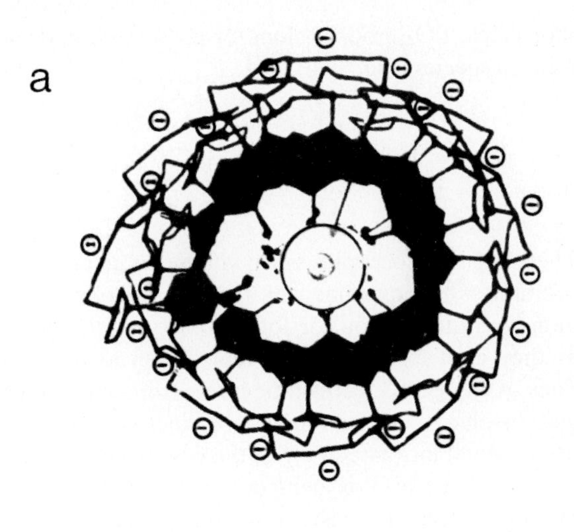

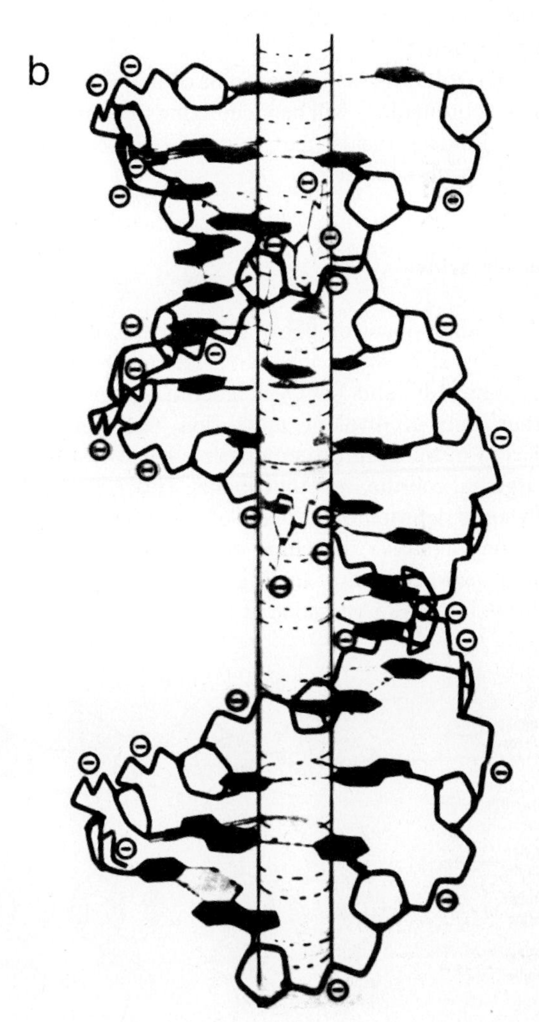

Fig. 11.6 a, b. Double helix model of DNA shown down the helical axis (**a**) and in upright position (**b**) (after Crick, 1954; Crick and Watson, 1954). The distribution of negative charges due to the presence of the phosphate groups is indicated

molecules. PO_4 groups, alone or combined, exist as ionized species (Fig. 11.6):

$$HPO_4^{2-}; \quad {}^{\ominus}O-\underset{\underset{O}{\|}}{\overset{\overset{O^{\ominus}}{|}}{P}}-O-\underset{\underset{O}{\|}}{\overset{\overset{O^{\ominus}}{|}}{P}}-O-\underset{\underset{O}{\|}}{\overset{\overset{O^{\ominus}}{|}}{P}}-O-R$$

Their large size occupies much volume in an aqueous solution, which explains why they do not obey common laws of diffusion, or follow space requirements as they exist for neutral molecules such as *n*-paraffines. As a result, electrostatic forces bring metal ions and dissolved PO_4-containing molecules near to each other. Metal ions interact with the negatively charged PO_4 units and form oxygen-coordinated polyhedra of the same type as exist in minerals. The aqueous medium will structurally adjust to these changes in coordination.

In the following a case in point of metal ion/phosphate coordinations will be made using three pillars of life: the nucleic acids, the phospholipid membranes, and the triphosphates.

Nucleic Acids

A two-dimensional model of DNA in Na^+ coordination (Fig. 11.7) shows the spatial dimension for the PO_4 tetrahedra and Na^+O_6 octahedra. A monovalent rather than a divalent metal ion was purposely selected in the graph to emphasize that it is not just charge but coordination that counts. That is to say the polyhedra determine the shape of the macromolecule and may induce a stretching of the polymer chains. These polyhedra are forcing two single-stranded DNA' s into a double helix. It is of special note that the kind of metal ions administered determines whether a single, double or even triple helix will form. The base-base hydrogen bonds cannot account for the number of stereotypes that exist. Metal ion coordination also inhibits the folding of a single stranded DNA; folding of a single stranded DNA and generation of branched structures takes place only when metal ions are removed from the system.

It is true that divalent metal ions are more effective – generally in the order of 10^4 – than monovalent metal ions in the preservation of the helical structure. In the case of the ribonucleic acid (RNA) of the tobacco mosaic virus (TMV), Mg^{2+} was found to be 2.5 x 10^4 more efficient than Na^+ in stabilizing the helical regions of RNA (Boedtker, 1960). A comparison of the valences and ionic radii of the two cations shows that the smaller sized $Mg^{2+}O_6$ octahedra has a stronger bonding strength to the nucleic acids compared to Na^+O_{6-8} polyhedra, because one third of the oxygen is furnished from at least two PO_4 groups. A Ca^{2+} solution is expected to yield a 5 x 10^3 more effective stabilization of the helical region of RNA (TMV) relative to Na^+ solutions, even though calcium and sodium have comparable ionic radii (Table 11.1).

Nucleic acids exhibit a double chain backbone:

Fig. 11.7. Model of DNA (schematic) showing metal ion coordinated phosphate groups. A sodium ion is bonded by ionic forces between two phosphate tetrahedra. Not only do metal ions determine the structural order, but they considerably reduce photo damage caused, for instance, by UV radiation. In contrast, presence of water molecules enhances photo damage. The cell takes advantage of metal ions for protection of the genetic apparatus, whereby magnesium is the most critical ion (see text). The amount of water linked to DNA is negligible and does not change during denaturation

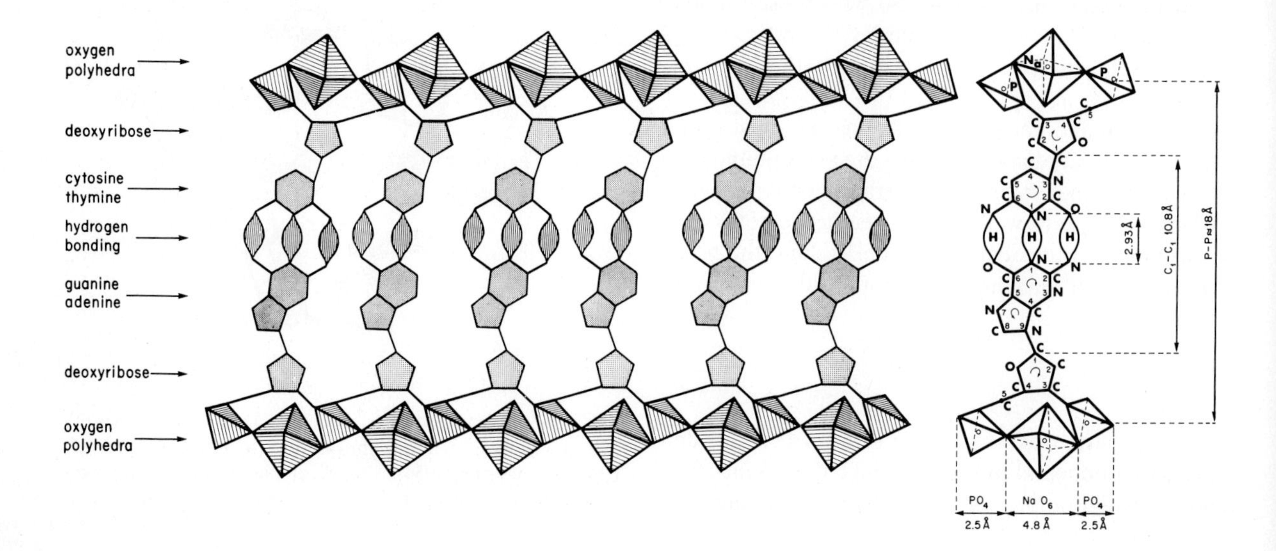

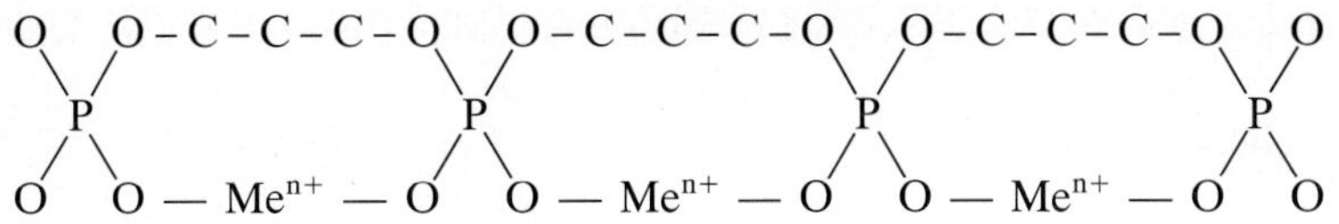

```
O   O-C-C-C-O   O-C-C-C-O   O-C-C-C-O   O
 \ /         \ /         \ /         \ /
  P           P           P           P
 / \         / \         / \         / \
O   O — Meⁿ⁺ — O   O — Meⁿ⁺ — O   O — Meⁿ⁺ — O   O
```

one that is mobile and covalent, another that is negatively charged and geometrically stable. It is suggested that in the sequence of base decoding processes the interactions of the individual enzymes with the nucleic acids proceed by exchange and elimination of the polyhedra metals. Stereospecificity is determined by these polyhedra.

Experiments on the binding of messenger RNA to ribosomal particles elucidate the action principles involved. Attachment of polynucleotides to ribosomes requires a Me^{2+} concentration exceeding the 10^{-2} molar level (e.g., Moore and Asano, 1966). It is conceivable that the low cytosolic Ca^{2+} level of $10^{-7.5}$ molar (Kretsinger, 1983) has to be maintained by the cell in part so as not to interfere with the work assignment of Mg^{2+} in the genetic apparatus.

All experimental data point in a direction indicating that: (i) a critical Mg^{2+} content of 10^{-2} M is essential, (ii) Mg^{2+} can be substituted for by either Ca^{2+} or Mn^{2+}, and (iii) maximum attachment exists when all PO_4's of the nucleic acids are associated with Mg^{2+} (e.g., Felsenfeld and Huang, 1959; Goldberg, 1966). These factors seem to imply that enzyme systems of the protein synthesis apparatus utilize the structural order provided by the metal ions associated with the PO_4 groups in the nucleic acids. The molecular "recognition" first demonstrated on ribosomes as "spatial fit" is a pattern frequently observed in enzyme systems (Box 11.3). The field of biochemical topology (Wassermann and Cozzarelli, 1986), in which physical and geometric features of DNA structure and reaction mechanisms are used to elucidate, for instance, enzymic processes, is in essence an outgrowth of such relationships.

Summing up, in cellular reactions where DNA and RNA participate, PO_4 and $Me^{n+}O_{6-8}$ polyhedra are: (i) instrumental in organizing randomly distributed compounds and forcing them into specific 3D-structures, (ii) control interaction mechanism and kinetics by introducing a specific reaction pattern, due to the stereo-order, and (iii) provide intramolecular bridges by incorporating oxygens from different molecules into the polyhedra.

Phospholipid Membranes

For many years, proteins and the nucleic acids stood in the limelight of molecular biology. The α-helix, the double helix, and the genetic code became the structural shrine of biochemistry. From this platform, researchers expanded the field in many directions, but emphasis still remained on proteins and the nucleic acids and the way they interacted. The physiological significance of, for instance, the phospholipids which are present in membrane systems, received comparatively little attention. It was commonly assumed that the sole purpose of phospholipids was to give membranes physical structure and that they acted as a valve system for the passage of matter.

In recent years, the situation has profoundly changed. We know that phospholipid membranes possess dynamic properties and participate in the metabolic activity of the whole cell. Selective ion exchange appears to be the critical element behind the dynamics displayed by membrane systems. More specifically, phospholipids are thought to serve as distinct ion exchange "resins" channelling the flow of metal ions inward, outward, and within membranes. Harris (1972) has shown that the addition of alkali salts to a mitochondrial suspension liberates H^+ ions, and fixes metal ions. It is of note that uptake of metal ions proceeds at very low concentrations and even against a concentration gradient. It was also discovered that divalent ions become more strongly adsorbed than monovalent ions; an addition of Ca^{2+} will release both H^+ and K^+.

Inorganic polyphosphates which exhibit the highest ion exchange capacity among all naturally occurring minerals (Amphlett, 1964) are an excellent analog to phospholipid ribbons. These outstanding ion exchange properties of phosphates are related to the great flexibility in the way the P–O bonds are arranged in phosphate structures. A comparison of the structural pattern of inorganic and organic polyphosphates underscores the intimate relationship between the two species (Fig. 11.10). It is concluded that coordinative bonding of metal ions with the PO_4 groups of phospholipids in cellular membranes is identical to that followed in phosphate minerals. It controls the exchange of matter not just between the cell and its ambience but between the various intracellular compartments. The decisive mechanism employed is the specific exchange of certain metal ions for others or for protons causing structural adjustment for the entire membrane units. This explains the concerted action of enzymes and transport molecules along the periphery of membranes. To visualize the change in

BOX 11.3

The Structuring of Metal Ions

Inorganic phosphate crystals form metal ion bridges:

$$\gtrless P - O - Me^{n+} - O - Me^{n+} - O - P \lessgtr$$

and typical linkage elements developed by ionic forces in phosphate structures are shown in Fig. 11.8. The polyhedra bridges may also be utilized in biochemical systems. For instance, the attachment of RNA to ribosomes, or the attachment of compounds necessary for the polyU-directed phenylalanine polymerization appears to be controlled by such linkages. Sharing of oxygens may proceed at corners, edges, or polyhedron surfaces. Oxygens may come from peptides. For example, attachment of ϕX 174 to *E. coli* C and its cell walls suggests participation of peptide oxygens in the generation of polyhedra. It was shown as an irreversible process with an encounter efficiency close to 1 in the presence of Ca^{2+} (0.1 M $CaCl_2$). No linkage takes place in the presence of Na^+ (0.1 M NaCl), nor does the addition of 0.1 M sodium chloride solution affect the bonding in the presence of Ca^{2+} (Fujimura and Kaesberg, 1962).

In Fig. 11.9, formation of a polyhedra network is schematically illustrated. Randomly coiled chains transform into an ordered polymer structure via association of PO_4 tetrahedra and metal ions. Increase in entropy due to this coordination is significant and comparable to that of solid crystalline phases.

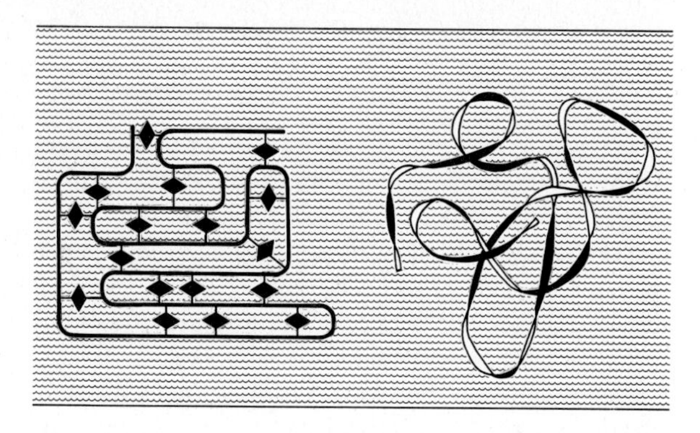

Fig. 11.9. Randomly coiled chains in a solvent medium transform into an ordered polymer structure by linking the oxygens of their PO_4 groups to metal ions

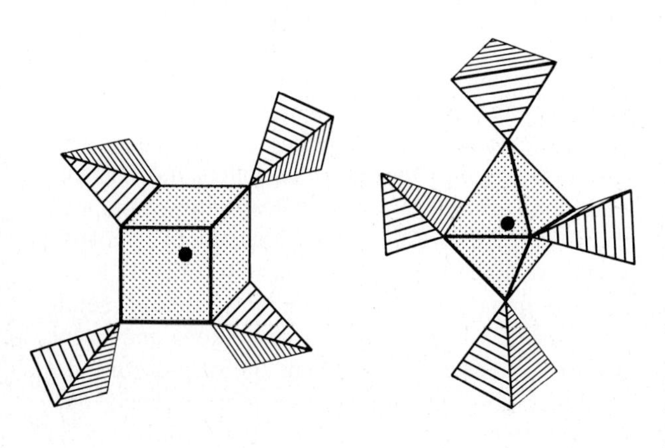

Fig. 11.8. Linkage elements (bridges) established by ionic forces in inorganic phosphate structures

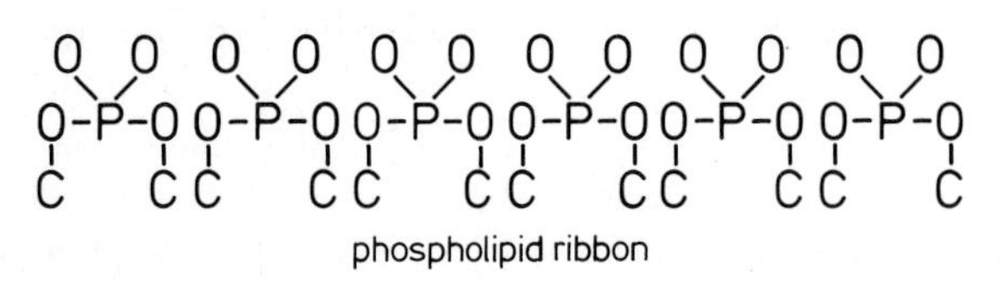

Fig. 11.10. Comparison of inorganic and organic polyphosphate structures

structural order that goes along with the removal or deposition of metal ions, we shall examine the way a solid PO_4 surface is generated.

Since the beginning of physical chemistry it has been established that metal ion-oxygen polyhedra promote the ordering process of fatty acids on a water surface in the direction of hexagonal lattices. The polyhedron formation and the organization of phospholipids differ only in the type of forces operating. Dipole-dipole interactions and hydrophobic forces are involved in the system: fatty acids – liquid water. Coordinative bonding, which determines the association of metal ions and oxygen atoms, is responsible for the formation of polyhedra networks composed of PO_4 tetrahedra linked to $Me^{n+}O_n$ polyhedra.

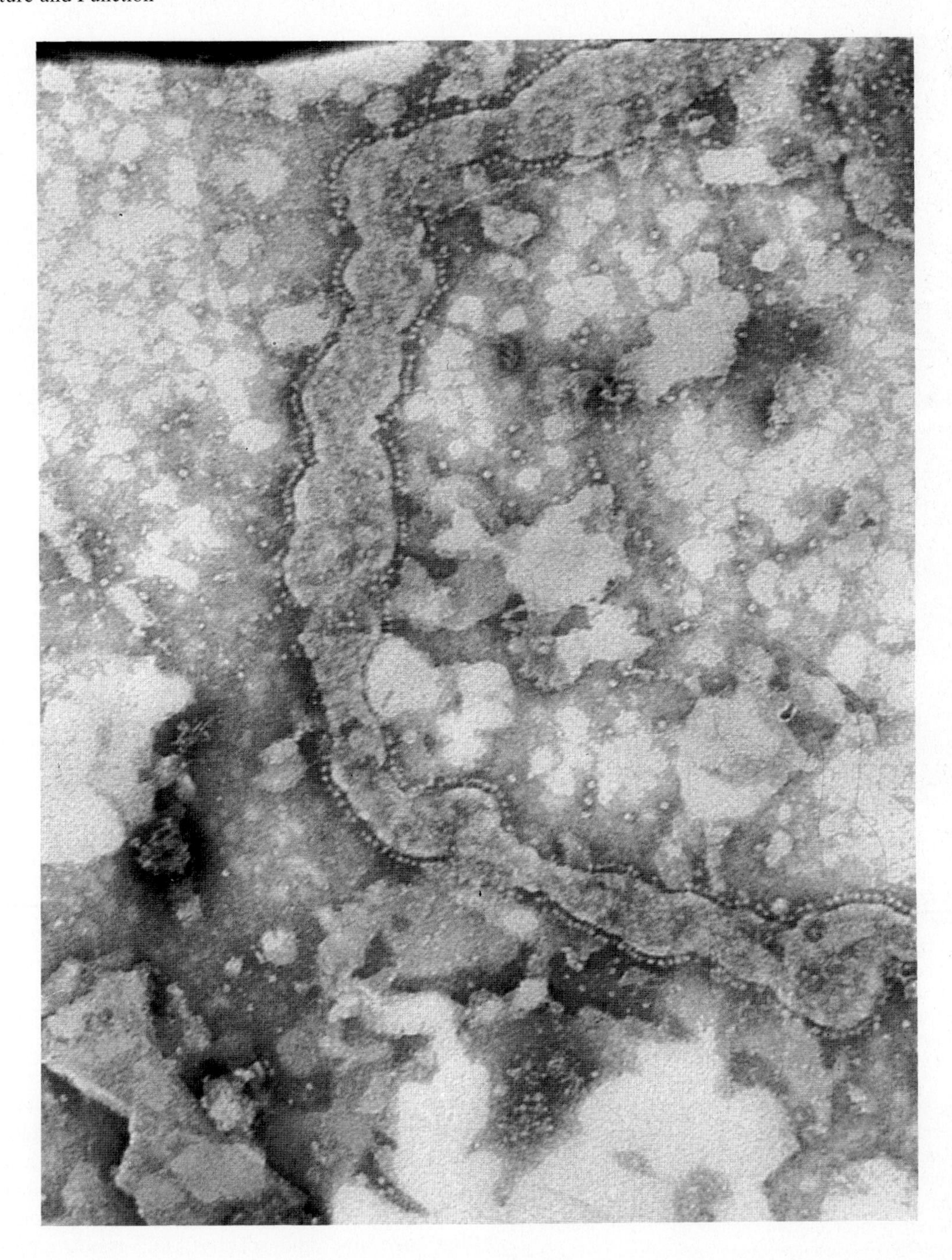

Fig. 11.11. Electron micrograph of membrane segment of beef heart mitochondria; negative staining technique (PTA) (courtesy of C.C. Remsen)

How can we possibly imagine that such networks look alike? Electron micrographs of bilayer mitochondrial membranes reveal two continuous bands composed of granular subunits 40 Å in size which are lined up like a "string of beads" (Fig. 11.11). Such a resolution, however, does not suffice to delineate the structural outlay of membranes at the molecular level. On the other hand, electron diffraction patterns of isolated membranes (Fig. 11.12) clearly show the crystalline order of biological membranes in a suitable Å range but give no clues concerning the three-dimensional arrangement of the various membrane constituents.

To overcome difficulties in explaining order phenomena in biological membranes, examples of known layered phosphate structures are presented for

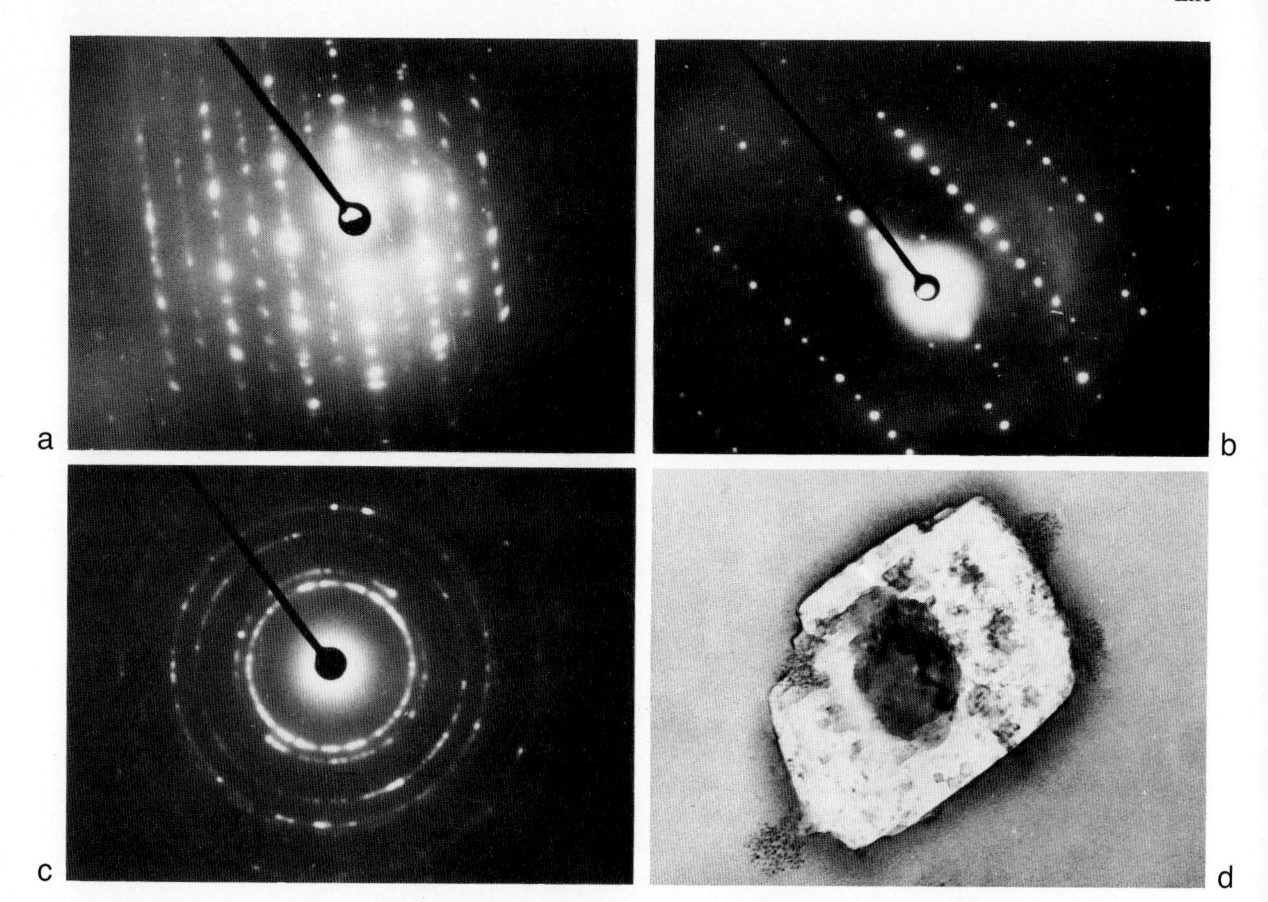

Fig. 11.12 a–d. Electron diffraction patterns of photoreceptive disc membranes from rod cells isolated from a frog's eye (**a, b, c**). The transmission picture (140,000) of the freshly prepared disc membranes is shown in (**d**). The material is untreated and for heat protection embedded in a low-vapor pressure oil immersion; during exposure to the electron beam the diffraction pattern could only be seen for a few minutes and then became gradually extinct. It is of note that the dot pattern (**a, b, c**) fluctuated in light intensity almost like a city at night seen from an airplane through haze or thin clouds. The membrane (**d**), exposed for two minutes or so, started to disassemble and some of the membrane content dispersed sideways or from the top into the oil immersion (Matheja, 1971). Rod cells are unique in that they can detect a single photon. Voltage-gated Ca^{2+} and Na^+ channels are present and the opening and closing involves two messengers: (i) a transmembrane protein (opsin) and (ii) in all probability Ca^{2+}. Light apparently releases Ca^{2+} from the disc into the cytoplasm, and the raised cytoplasmic Ca^{2+} causes the Na^+ channels to close (Alberts et al., 1983)

illustration (Fig. 11.13). Phosphate tetrahedra and metal ion oxygen polyhedra can combine to a variety of geometries including undulating surfaces, or concave/convex perforated surfaces. These loosely arranged space fabrics exhibit selective molecular sieve properties and ion exchange characteristics.

Phospholipid membranes are expected to exhibit identical properties and structures as they exist in inorganic phosphate minerals, even including holes and surface granularities. Due to changes in the intra- and extracellular milieu, one structural type can readily be transformed into another one. It is proposed that the interchangeable nature of metal ions causes membranes to act as *dynamic molecular sieves.* Their pore size and shape must be quite variable as a function of type and availability of metal ions which in the last instances are enzyme controlled. Assuming adenosine triphosphate (ATP) is capable of trapping metal ions but adenosine diphosphate (ADP) is not, a periodic pulsation of the membrane lattice is the consequence.

Membranes are natural boundaries between a cell and the ambience, among neighboring cells, as well as between different compartments of a single cell. The primary function is to regulate the traffic of nutrients, waste material, essential ions and molecules, and energy in or out of the system. A double layer of lipid molecules gives membranes a duplex nature: one end is hydrophilic and the other end hydrophobic. Phospholipids are the most prominent members of lipids in membranes.

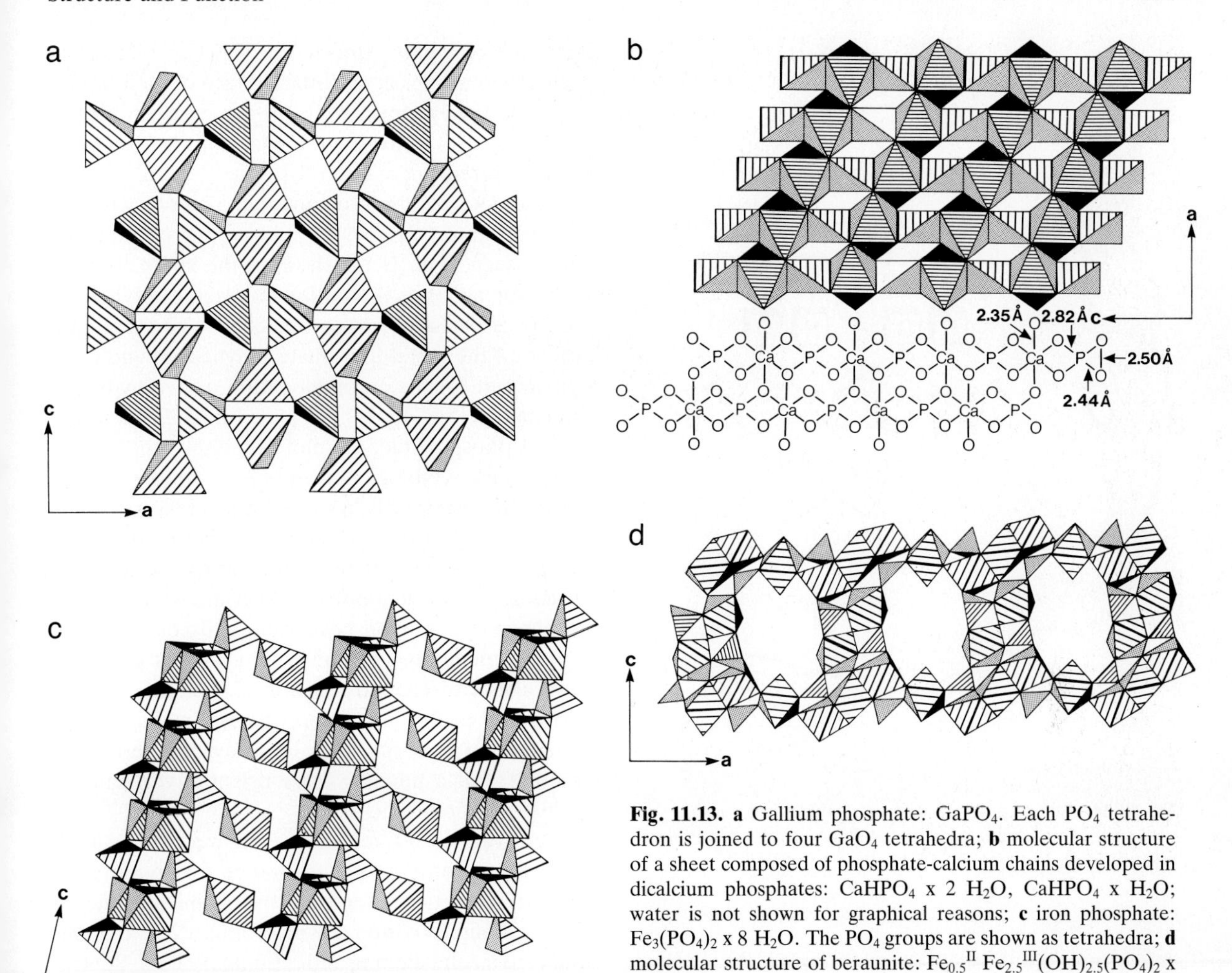

Fig. 11.13. a Gallium phosphate: $GaPO_4$. Each PO_4 tetrahedron is joined to four GaO_4 tetrahedra; **b** molecular structure of a sheet composed of phosphate-calcium chains developed in dicalcium phosphates: $CaHPO_4$ x 2 H_2O, $CaHPO_4$ x H_2O; water is not shown for graphical reasons; **c** iron phosphate: $Fe_3(PO_4)_2$ x 8 H_2O. The PO_4 groups are shown as tetrahedra; **d** molecular structure of beraunite: $Fe_{0.5}{}^{II}$ $Fe_{2.5}{}^{III}(OH)_{2.5}(PO_4)_2$ x 3 H_2O

Permeability of membranes is highly selective. Channel systems are provided by distinct proteins. The sites where the channels develop are termed gap junctions, which open and close in response to changes in the intracellular conditions. This gating effect is most critical for the maintenance of cellular function. Each junction is composed of 12 protein subunits which are rooted in the membrane (Unwin and Henderson, 1984; Unwin, 1986; Bretscher, 1985).

Opening and closing of gap junctions has been shown to depend on the intracellular level of calcium ions. Ca^{2+} appears to straighten the subunits and to constrict the channel. This structural rearrangement is considered as regulating the passage of matter between the interiors of live cells.

It was pointed out that the interchangeable nature of metal ions causes membranes to act as dynamic molecular sieves. Inasmuch as gating of voltage-sensitive ion channels is in the last instance controlled by enzymes that remove or deposit ions along the membrane surface, conformational changes and a periodic pulsation of the crystalline lattice is the consequence (Matheja and Degens, 1971a). Oscillations of ionic constituents have been observed in mitochondria (Chance and Yoshioka, 1966). Electron density distribution perpendicular through a membrane is in support of the presence of metal ions between the lipid bi-leaflets (Finean, 1966).

Intracellular level of Ca^{2+} determines the opening and closing of gap junctions (Unwin and Henderson, 1984). Ca^{2+} shapes the geometry of channels and regulates transfer of molecules and ions. At normal cytosolic Ca^{2+} levels of 10^{-7} M, channels are highly permeable to a wide range of molecular sizes with an upper size limit for peptide molecules of about 1,200 to 1,900 dalton. A rise in Ca^{2+} to 5 x 10^{-5} M in the junctional locale will drastically lower permeability (B. Rose et al., 1977). A schematic cross-section (Fig. 11.14) exhibits structural details of a bilayer membrane and the coordination of Ca^{2+}. In principle,

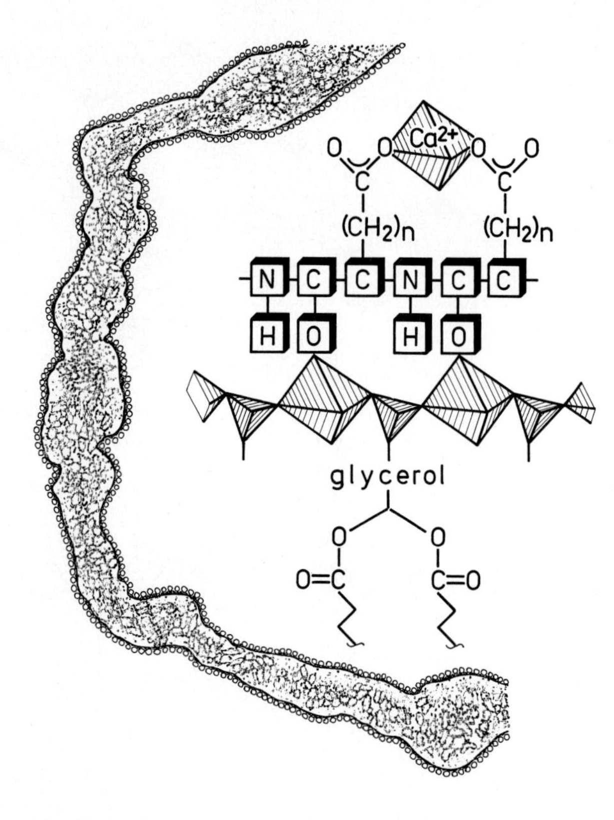

Fig. 11.14. Schematics, showing structural relationship between Ca^{2+}, membrane protein, metal ion coordinated PO_4 ribbon, glycerol, and ester bonded hydrocarbon chains

"metal-protein-phoshate gates" make the plasma membrane relatively impermeable to outside Ca^{2+}, which in the case of sea water is 10^{-2} M. The electrochemical gradient for Ca^{2+} is consequently large. Binding of Ca^{2+} to specific sites at the internal face of membranes alters conformation of membrane constituents, thus allowing transfer or passage of hydrophilic substances across the lipid barrier. High-affinity Ca^{2+} proteins and glycoproteins confer specificity for binding, removal and transportation of Ca^{2+} (Elbrink and Bihler, 1975).

The above graph is drawn after an electron micrograph of a beef-heart mitochondrial cross-section (compare Fig. 11.11). Outer grains represent structural proteins. The black lines below the protein grains reveal the position of the phosphate-metal ion oxygen polyhedra which obey a 40 Å periodicity. Glycerol and double aggregates of hydrocarbons joined to the crystalline tetrahedron-octahedron surface occupy the space in between. Structural relationships between principal molecular units of a phospholipid membrane are depicted in the center of the figure. For graphical reasons, only a cross-section through a single layer is shown and the hydrocarbon chains are only partly drawn. The way Ca^{2+} is fixed to carboxylate groups at the internal face of membranes is schematically presented. In the case of the archaebacteria, an ether bond in place of the ester bond is encountered.

These data are used to construct a cross-section of the membrane surface. A lipid-oxygen assemblage can be seen (Fig. 11.15) showing the space requirements for an oxygen-coordinated metal polyhedron and the associated hydrocarbon chains. Cross-section surface of the metal-phosphate polyhedra band gives a perfect fit for the accommodation of two hydrocarbon chains (35 to 40 Å^2) per unit. In this fashion, the logic of phospholipids or, more precisely, that of the diglycerides in biological membranes, finds its explanation. In other words, a single fatty acid is too small, and three fatty acids combined too big to fit within the space allocated for the metal/phosphate unit. An adoption of the common triglycerides would make the phosphate mineral layer rigid and compact. The resulting membrane system would lose its polyfunctional elasticity. The introduction of a metal ion about 4 Å^2 in size establishes not only the optimal areal conformation of close to 40 Å^2 in a layered phosphate structure, but a bonding force oriented vertically to the hydrocarbon chain axis.

Interaction of PO_4 tetrahedra and metal ions yields a structurally ordered surface lattice for membranes. This surface lattice determines the chemical behavior of phospholipids within the membranes in a manner similar to a zone melting and refining process. Species such as cephaline, lecithin, phosphatedylinosite, or sphingolipids will not be distributed randomly, but organized somewhat analogously to amino acids in peptides, or the bases of the purines and pyrimidines in DNA. Membranes differ, however, in that they exhibit a two-dimensional coding sequence. These views have been expressed by Lehninger (1966), who considers a two-dimensional code device on membrane surfaces.

Triphosphates (III–P)

The ability of phosphates to fix metal ions has been mentioned throughout the text. Different patterns arise depending on the type of PO_4 molecules and the kind of metal ions present. Nevertheless, there is no single law applicable to all phosphates, but a number of rules exist which apply only to orthophosphates (HPO_4^{2-}; $H_2PO_4^-$; PO_4^{3-}), or to diphosphate, or to polyphosphates with chain members of three and more. The triphosphates are thus set strictly apart from diphosphates and these from mono-phosphates

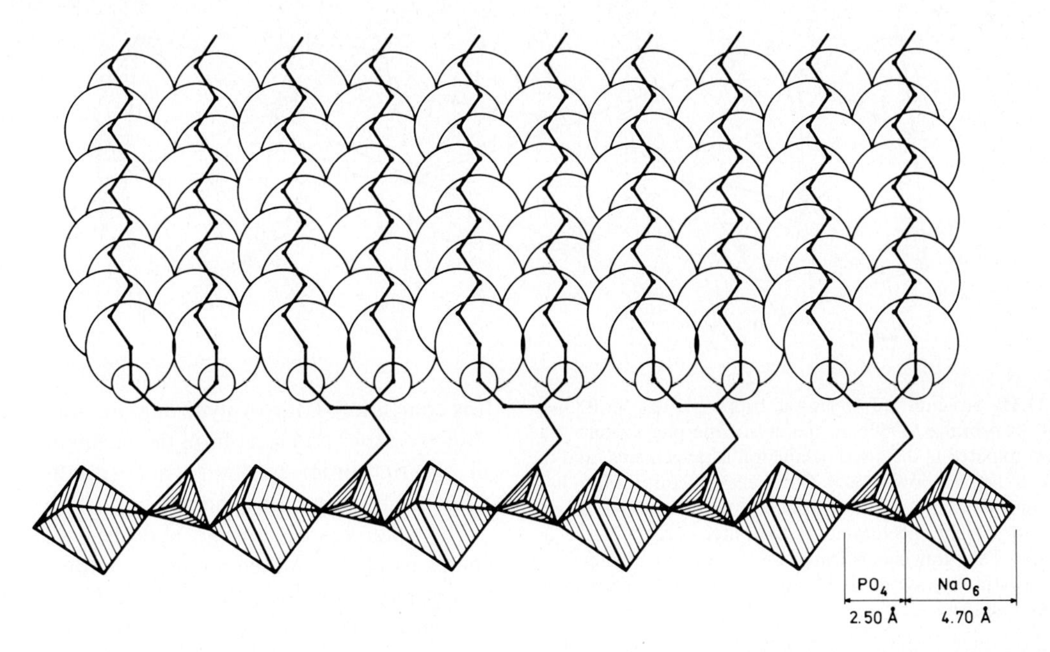

Fig. 11.15. Suggested phospholipid membrane cross-section composed of double aggregates of hydrocarbons and oxygen polyhedra. Glycerol is only schematically shown; the carbonyl groups and the PO_4 residues have been omitted

with respect to their activities in solution. Adenosinetriphosphate (ATP) is even more distinct from any of the aforementioned species, because a sugar and a heterocyclic base are part of the molecule (Fig. 11.16).

The III-P molecule can bind metal ions strongly and yield coordination complexes. In this capacity it resembles the polyphosphates. The III-P chain can yield with metal ions such monomeric units as are observed in $Mg_2P_3O_{10}^-$. In contrast, diphosphate molecules become associated in the form of dimers in the presence of polyvalent metal ions. Dimers, by the sharing of oxygens, will usually end up as insoluble salt.

Two metal ions exhibit an especially favorable bonding order in III–P: Ca^{2+} and Na^+. In their presence, $Ca_2NaP_3O_{10}$ is thermodynamically the most advantageous coordination and for that reason probably utilized in the cell for Ca^{2+} and Na^+ regulation.

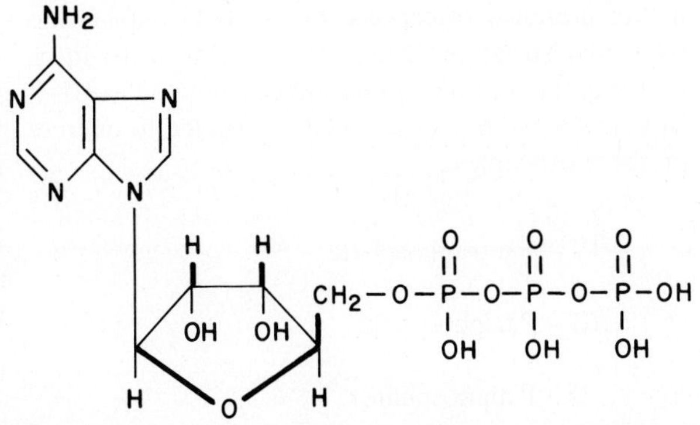

Fig. 11.16. Adenosine triphosphate (ATP)

The stability constant pK and the free energy of formation ΔF for metal ions coordinated to III–P increase for the noble-gas configurated elements in the order:

$$Cs < K < Na < Li < Ba < Sr < Ca < Mg.$$

This relationship indicates the formation of metal ion coordinated PO_4 groups because the free energy increases with growth in charge and reduction in ionic radii. Metal ions such as Fe^{3+}, Hg^{2+}, or Pb^{2+} become too strongly bonded to III–P to be of any value in cell dynamics. They simply precipitate the triphosphates.

III–P exhibits two properties: (i) release of the terminal groups and formation of the reactive P^+O_3 during hydrolysis, respectively phosphorylation, and (ii) affinity to all cations by means of coordination. These bonding forces control displacement and affinity order for metal ions. Among the phosphates, III–P is unique in that it represents a "convergence point" between (i) and (ii), both of which are a function of chain length. Order of hydrolysis is reciprocal to the strength of metal ion bonding:

$$II-P > III-P > IV-P > V-P > VI-P > VIII-P$$

order of hydrolysis

$$I-P < II-P < III-P < poly-P$$

order of metal bonding.

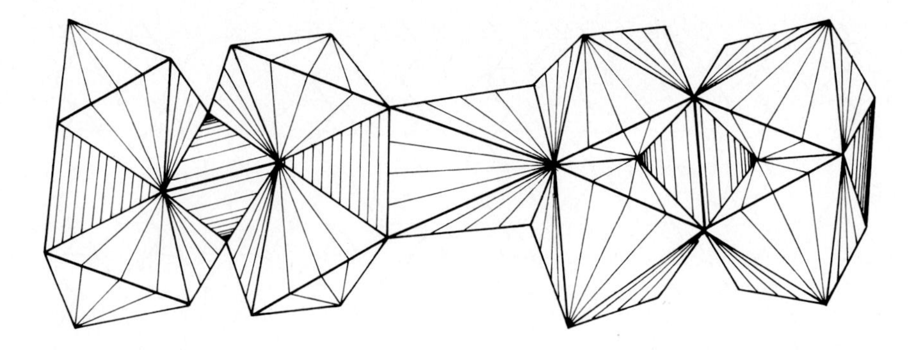

Fig. 11.17. Structure of hydrated triphosphate, $Na_5P_3O_{10}$ x 6 H_2O (Corbridge, 1960). In the crystalline phase, some Na^+ are coordinated in distorted octahedral arrangements, and the rest is four-fold coordinated. The sheet structure can be hydrated and yields hexahydrate. The oxygens of water are coordinated to sodium and located at the outer corners of the octahedra. This drawing indicates the formation of metal ion polyhedra in the case of III-P's and illustrates the interactions of metal ions and ATP

In aqueous solution the half-life of a polyphosphate chain is reduced with increasing chain length, whereas its half-life to the metal bonding increases with chain length. One might ask if III–P constitutes a compromise between maximum metal ion affinity and minimum speed of hydrolysis. In cellular systems, triphosphate molecules are apparently associated with metal ions, which gives III–P a duplex nature: a protection against hydrolysis, and a tool for its own cleavage.

Adenosine triphosphate, the most prominent energizer molecule in living systems, is an instructive example of the biochemical utilization of the III–P structure (Fig. 11.18; Box 11.4). It is to be remembered that all chain polymers of PO_4 are unstable in aqueous solution, hydrolysis starts at the terminal group, and for III–P it is a first-order reaction. Mono- and diphosphates in a molar ratio of 1:1 are the hydrolysis products. Diphosphate becomes hydrolyzed to monophosphates; again a first order reaction. For tetraphosphates, IV–P, hydrolysis yields monophosphates which are released one after the other, starting from the terminal chain. All reactions are first-order ones and proceed independently:

It is concluded: (i) hydrolytic degradations occur one after the other and start from the terminal group, (ii) all are first-order reactions, and (iii) hydrolysis is catalyzed by metal ions and pH.

The chief reason for the hydrolysis of polyphosphates is related to the change in coordination of the PO_4 group. Namely, in the II–P and III–P molecules, the phosphorus-oxygen distance within the P–O–P bridge is greater than the distance to the terminal oxygens. This implies that the π electrons at the terminal oxygens make these oxygens negative, and at the same time the phosphorus atom more positive. A configuration results which is schematically drawn:

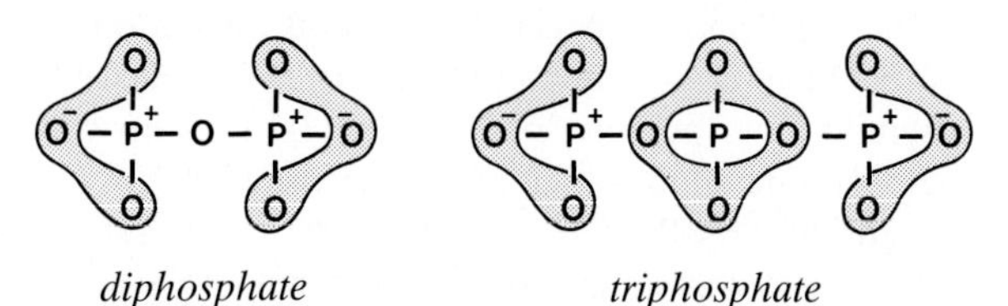

diphosphate *triphosphate*

This type of electron configuration represents a high energy bond indeed. Hydrolysis will result in a better orbital configuration, and the resulting single tetrahedra become more effectively coordinated to the solvent medium. This is accompanied by the release of a small amount of energy: 3 kcal/mol. Potential energy will be larger if a rigid triphosphate fabric deforms the network of the hydrogen bonded water structure. The rate of hydrolysis of III–P will increase in the presence of cations. Figure 11.17 depicts the molecular structure of hydrated III–P to assist in visualizing the way metal ions are coordinated to III–P. In biochemical processes, ATP is assumed to operate on these principles.

IV – P → III – P → II – P → I – P
↓ I – P ↓ I – P ↓ I – P

IV – P tetraphosphate
III – P triphosphate
II – P diphosphate
I – P monophosphate.

BOX 11.4

Adenosine Triphosphate

Conformation of the hydrated disodium ATP ($Na_2 \cdot ATP \cdot 3H_2O$) is shown in Fig. 11.18. The phosphate chain is folded back towards the purine base and a hydrophilic and a coordinative part become visible. Noteworthy is the packing of the ATP molecules into corrugated sheets which are running parallel to the *b*–*c* planes. The plane shown in the graph is cutting these sheets perpendicularly and they are extending parallel to the *b*-axis. Organization of the ATP molecules into sheets is governed by: (i) ionic bonds of phosphate oxygens with Na^+ creating polyhedra, and (ii) the base stacking which results in alternating chains of:

... purine base –
coordination polyhedron –
purine base ...

Conformation of ATP molecules has some interesting features. The terminal PO_4 group which is bent towards the adenine rings is not hydrated, and none of the water molecules is close to the terminal PO_4 unit. The terminal group is strongly bonded to the Na^+ coordination complex, composed of five phosphate oxygens and one adenine nitrogen. The sheets of ATP molecules are separated by water molecules and held together by hydrogen bonds.

ATP's functionality in cellular systems rests on its dual nature: one part is hydrophylic and composed of ribose and an adjacent phosphate, and the other, coordinative part, is composed of the purine base and the terminal phosphate. The hydrophylic part constitutes the surface of the sheet structure, whereas the coordinative part is located within the sheets. This conformation has relevance for ATP's binding by enzymes. It is to be remembered that the hydrophilic part of DNA (see Fig. 11.4), that is ribose with the adjacent PO_4, is also located at the surface of DNA and is in contact with the water molecules.

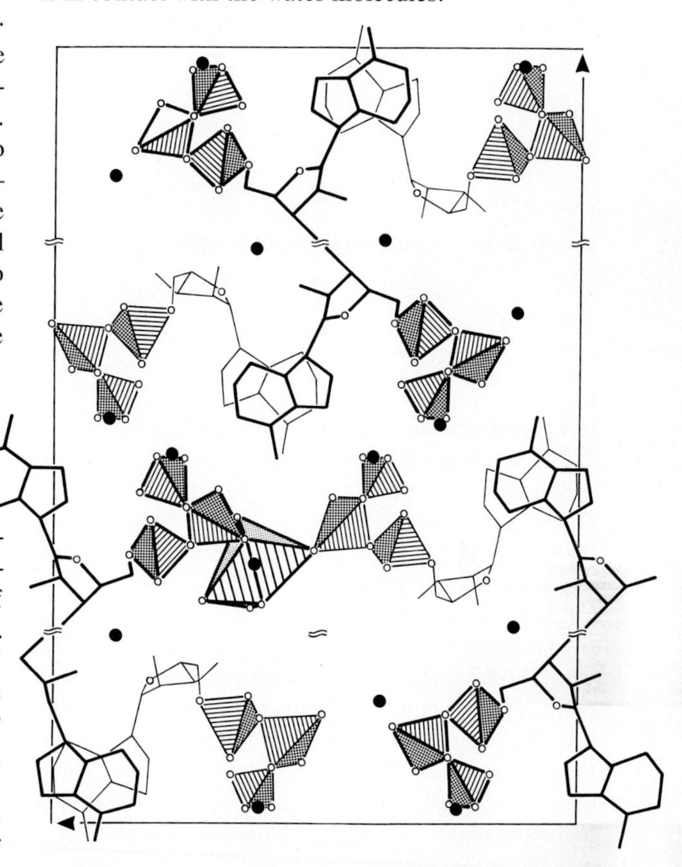

Fig. 11.18. Structure of adenosine triphosphate, Na ATP x 3 H_2O, projected down the *c*-axis (after Kennard et al., 1970). The unit cell contains 8 ATP molecules, 16 Na^+, and 24 H_2O molecules. PO_4 units are shown as tetrahedra, and the sodium ions involved in coordination polyhedra as black dots. One octahedron is depicted: $(H_2O)_4Na(O\text{-}PO_3)_2$. It is composed of two oxygens of the central phosphate groups and of four H_2O's; the remaining water molecules (not shown) are located in the blank region

Conclusion

The quiet and secret role excercised by phosphate groups in the cell will undoubtedly remain a mystery for many years to come. The problem of why pyruvic acid becomes oxidized to carbon dioxide in the citrate cycle is relatively simple: it is a combustion with a simultaneous release of water. But the involved operation principles within and between the various cellular compartments that are strictly controlled by phosphates, and which require a holistic approach for a meaningful interpretation, are only marginally known.

This statement is not intended to minimize past accomplishments in the field of molecular biology; on the contrary. To have come so far is already an achievement in view of the complexity of life. Our discussion has been concentrated on a small but essential segment of the cellular system: the phosphates. I wanted to convey the message of the uniqueness of phosphorus as a structural and functional element of life. In the element phosphorus lies the answer to the question what distinguishes life from an ordinary mineral. Life is based on PO_4 units:

```
    O
    |
O – P = O
    |
    O
```

and rock-forming minerals on SiO_4 units:

```
    O
    |
O – Si – O
    |
    O
```

It is essentially the π electron – the high energy bond – in PO_4 which maintains the animated world. Since the π bond can lie "parallel" to any of the four sigma bonds, giving rise to a variety of differently shaped tetrahedra, a dynamic and flexible tetrahedra network can be created. In contrast, the SiO_4 unit has just four sigma bonds which only permit the establishment of rigid networks.

Life's relation to the common biogenic elements was pointed out, and that PO_4 units represent forces which introduce a structural order at the molecular level in the cell. In regard to the principle of economy which applies to the biological cell and requires that only a minimum number of building blocks must be used, it is concluded that phosphates represent a meaningful solution for the life processes, which is unmatched by any other compound on Earth. In the final analysis, however, certain metal ions are the functional and structural elements of biological systems. Most outstanding in this regard are magnesium for the genetic and calcium for the metabolic apparatus. Next come sodium and potassium, for instance in the regulation of cell volume via Na^+–K^+ ATPase. The discussion on metal ion functionality in the living cell could well go on, but other subject matter is waiting.

Prolog on the Inanimate World

Elemental Part

Aside from helium, living and universal matter are principally composed of the same chemical elements:

hydrogen – oxygen – carbon – nitrogen

reading in order of abundance, as:

C – O – H – N vs H – O – C – N.

In contrast, the Earth has an entirely different bulk composition:

iron – oxygen – silicon – magnesium.

In comparing the various patterns one might suggest that living matter sprouts directly out of the ancestral universal matrix, whereas terrestrial matter must be a late derivation product of that matrix. Only oxygen ranks high in all three compartments: cosmos, life, Earth.

It is common practice to speak of carbon as *the* reference element of organic molecules and we should thus briefly examine carbon's chemistry as it relates to the abiotic synthesis of simple molecules such as carbon dioxide, methane, fatty acids, sugars, and amino acids.

We have already seen that dark interstellar clouds are excellent hunting grounds for simple carbon-containing molecules, including carbon monoxide, formaldehyde, hydrogen cyanide, methanol, ethanol etc. Observational data suggest that in the space regions accessible for studies by radio telescope all carbon is molecularly bound, that is, no atomic carbon remains. Apparently the chemistry of interstellar clouds favors the formation of carbon-containing molecules.

In transit to Earth, carbon must have been essentially present in a reduced state due to the large excess of hydrogen (Suess, 1975). Question: at what point does CO_2, the prime molecule of photosynthesis and respiration, enter the carbon scene?

Upon arrival on Earth, the reduced carbon pool became part of a differentiation process leading to core, mantle, crust, hydrosphere, and atmosphere. As a result, carbon distributed itself among the various compartments. New molecular carbon species developed in accordance with changing mineral equilibria, while other species survived the harsh treatment of the evolving Earth molecularly intact. Stable carbon isotopes are excellent tracers of the terrestrial carbon flow (Box 11.5).

Starting from a source of carbon of unspecified structure, the carbon phase diagram (Fig. 11.21) illustrates the suggested stability fields for graphite, carbynes and diamond (Whitaker, 1978). At temperatures above 3,800 K and at elevated pressures, carbon readily dissolves in melts of silicates and iron. In view of the lower melting point of iron relative to silicates, carbon can be extracted from mantle material and carried via iron droplets to the central core. The mechanism of graphite dissociation is depicted in Fig. 11.22. At high temperatures a single C–C bond can break and shift an electron into each of the adjacent double bonds, resulting in a kind of "chain reac-

BOX 11.5

Tracing the Flow of Carbon (Degens et al., 1984)

The abundance of carbon in the Earth's crust is 0.27 percent. For comparison, the mass fraction of organic carbon in carbonaceous chondrites is about 1.3 percent, and in all meteorite material collected on Earth 0.06 percent. Based on physical properties such as albedo, spectral reflectance and density, the moons of Mars, most asteroids, interplanetary dust, and many of the moons of the outer planets suggest the presence of carbonaceous chondritic material. From all we know, there is no evidence of such material on the surface of any object closer to the Sun than Mars. The high abundance of carbon-containing molecules in comets was verified during the recent encounter of Comet Halley. Even our Moon holds carbon in materials exposed at its surface. Here, sources of carbon are solar winds and impacting micrometeorites. Carbon, so it appears, is omnipresent in our solar system.

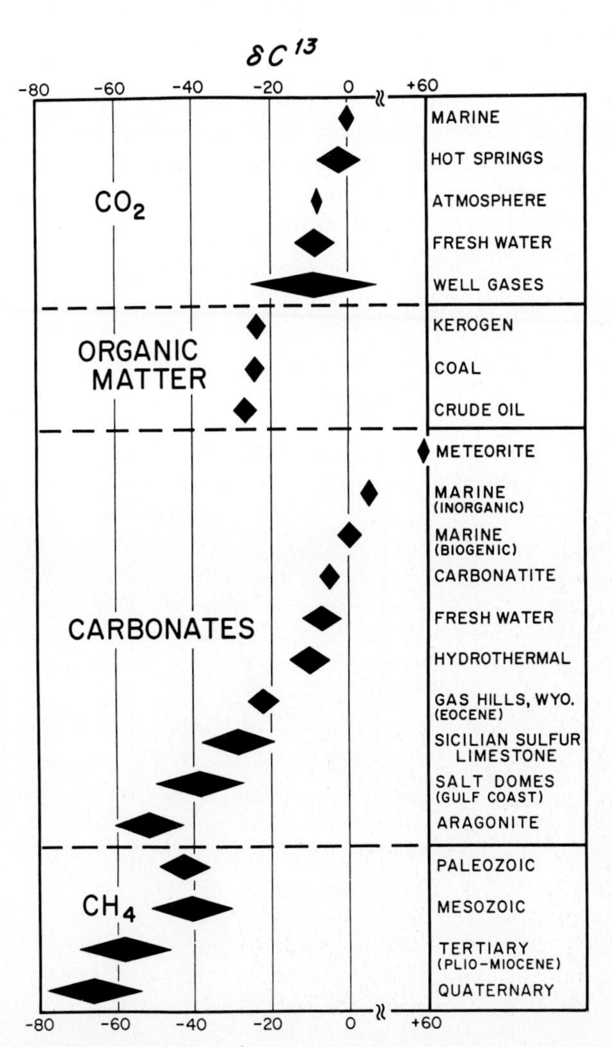

Fractionation of carbon isotopes in the course of planetary and terrestrial differentiation of matter is a valuable tool to reconstruct the flow of carbon through the various compartments of our Earth. Natural $^{13}C/^{12}C$ variation covers a range of 150 permil, heaviest being a carbonate from a meteorite ($\delta^{13}C$ + 60 permil) and lightest a bacterial methane ($\delta^{13}C$ –90 permil) (Fig. 11.19).

Meteorites most enriched in carbon have in general the highest $\delta^{13}C$ values. Carbonaceous chondrites (Type I), the assumed precursor material for much of the earthly crust, have a mean $\delta^{13}C$ that corresponds with terrestrial magmatic carbon (Fig. 11.20). The magmatic carbon pool should have a mean $\delta^{13}C$ close to –6.5 permil, the assigned value for primordial respectively juvenile carbon.

At temperatures prevailing in the upper mantle, carbon isotope fractionation between graphite, carbonate and diamond is insignificant. Even modern oceanic rift basalts more or less reflect this value. Carbon dioxide emanating from a magmatic source is about 2 permil richer in ^{13}C than the cogenetic carbonate. This is the expected fractionation between CO_2 and calcite in the temperature range 500 to 700°C. Juvenile CO_2 has an assigned $\delta^{13}C$ value of –5 permil.

The partitioning of carbon in the course of its subsequent biochemical cycling can be documented and summarized by means of $\delta^{13}C$ ranges for various carbon-bearing materials (Fig. 11.20). The two most outstanding endmembers of carbon isotopic fractionation are CO_2 and CH_4, the first one being generally depleted and the second one enriched in ^{12}C. Photosynthesis shifts $\delta^{13}C$ towards more negative values by about 15 to 25 permil relative to its CO_2 precursor. In contrast, chemical carbonate precipitation moves $\delta^{13}C$ into the more positive range by 4 to 5 permil relative to the starting bicarbonate ion. Where methane becomes oxidized to CO_2 or a contribution of biologically generated CO_2 is indicated, carbonates derived from such sources will be isotopically light. These contrasting features of isotopic fractionation between the

Fig. 11.19. $\delta^{13}C$ ranges for various carbon-bearing materials. Width of diamond-shaped symbols equals two standard deviations for the data. Individual samples have been grouped into systematic classes of compounds and are plotted in form of cumulative frequency diagrams (range: 2 sigma). The aragonite "diamond" represents the 2 sigma range of 15 Pleistocene beach rocks from the eastern United States and the light carbon stems from the oxidation of microbially generated methane (Hathaway and Degens, 1969)

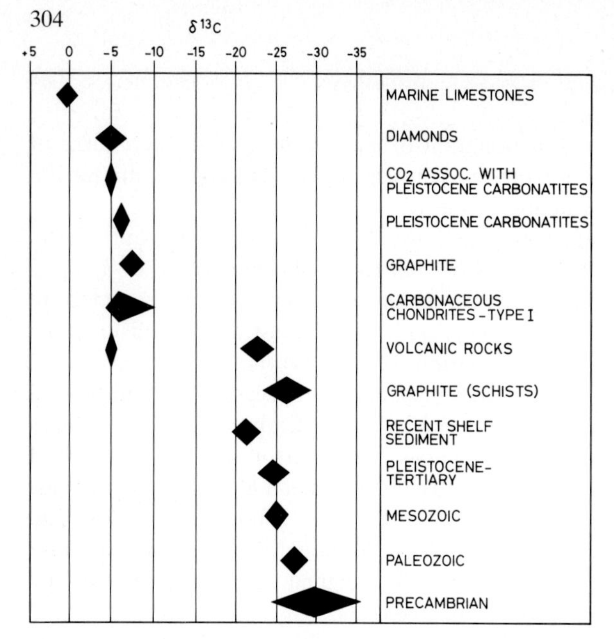

Fig. 11.20. $\delta^{13}C$ in various geological materials. The majority of volcanic rocks have a $\delta^{13}C$ in the range of common plants, whereas oceanic basalts freed of surface contaminants have a $\delta^{13}C$ in the magmatic range. Of note is the apparent age effect for organic carbon in sediments. This is principally related to differences in carbon isotope fractionation characteristics of organisms. For instance, marine chemoautotrophic bacteria produce a $\delta^{13}C$ in their organic matter which is depleted by 7 permil relative to the $\delta^{13}C$ of phytoplankton grown under the same circumstances. Part of the wide range of the Precambrian sediments is accounted for by microbially related fractionation processes; for a recent summary see Schidlowski (1987). The data bank used for the construction of this figure is based on more than 5,000 analyses taken from various literature sources

reduced and the oxidized form of carbon enable us to calculate the ratio of reduced to oxidized carbon contained in the crust. A value of 1:4 is obtained. This figure is reached by calculating the difference in ^{13}C content between the magmatic carbon pool and the two principal endmembers of the oxidized and reduced species, carbonate and kerogen, the principal C_{org} residue in sediments. It is of note that particles settling to the seafloor and collected in mid-water by means of sediment traps have instead a predominance of organic matter over calcareous debris in a ratio of 4:1. This discrepancy can best be explained by assuming a preferential loss of organic matter during the early stages of diagenesis.

At present, the bulk of carbon dioxide released from the lithosphere through either volcanism or hydrothermal vents is secondary in nature, i.e., recycled. Carbonate, sedimentary organic matter, and sea water bicarbonate are the principal progenitors of carbon dioxide emitted by volcanism and hydrothermal activities. It is this CO_2 which drives the exogenic carbon cycle. Carbon will return after being worked over by organisms and the environment to the solid Earth in different organic and inorganic forms. From this complex mixture, CO_2 will arise over and over again in a never-ending cycle.

tion" until the graphite basal plane sheet is completely transformed into $(-C \equiv C-)_n$ chains.

Within the graphite stability field, graphite may also react with hydrogen gas:

$$C + 2H_2 \rightarrow CH_4$$

or with water vapor:

$$C + H_2O \rightarrow CO + H_2.$$

The subsequent reaction of carbon monoxide with water vapor:

$$CO + H_2O \rightarrow CO_2 + H_2$$

will yield carbon dioxide.

Carbon dioxide can also be generated along a different route, from complex organic matter, which will be abbreviated to CH_2. At depth, a series of reactions may take place, for instance:

$$CH_2 \rightarrow C_{(diamond)} + H_2 \text{ or}$$
$$CH_2 + 3\,Fe_3O_4 \rightarrow CO_2 + H_2O + 9\,FeO.$$

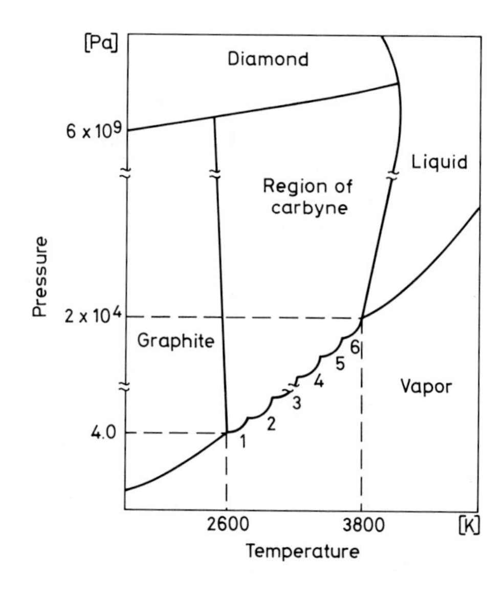

Fig. 11.21. Carbon phase diagram (Whitaker, 1978)

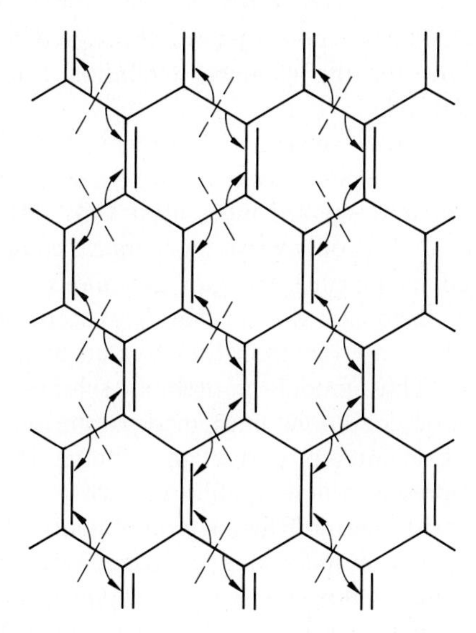

Fig. 11.22. Transformation of graphite into carbyne chains $(-C \equiv C-)_n$

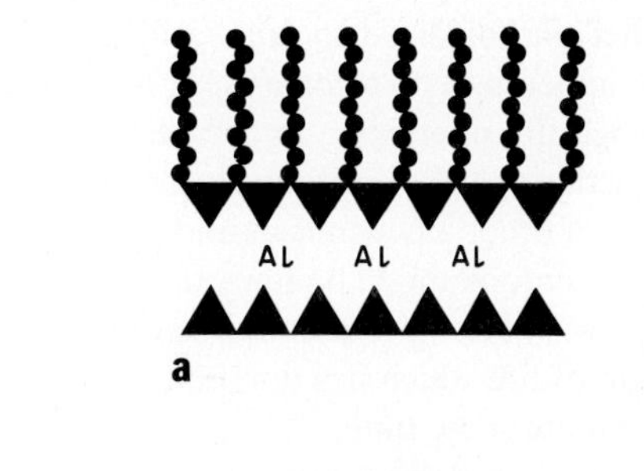

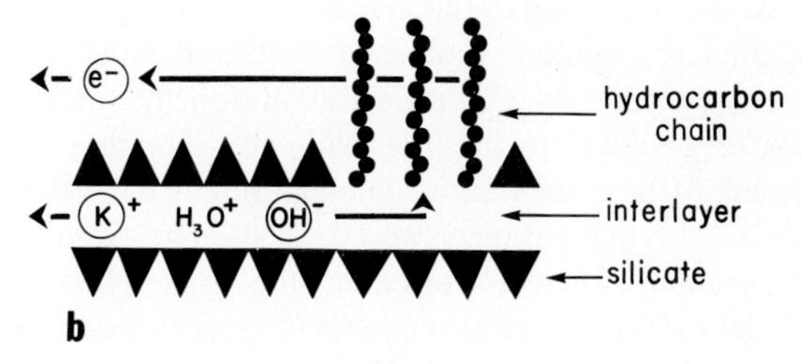

Fig. 11.23. Oriented attachment of hydrocarbon chains on layers of montmorillonite

Starting from graphite, a variety of hydrocarbons and oxy-hydroxy compounds, such as graphitic acid $C_8O_2(OH)_2$, may be generated. Reactions will commence from the outer layers and thence from atoms at the edges of the benzenoid layer, thus yielding chain and layer fragments (Dawson and Follett, 1963). That the reactions start from lattice defects accounts for the formation of higher molecular weight aromatics and chain hydrocarbons (e.g., *n*-paraffins), because lattice defects promote catalysis.

In summary, graphite is a compound which readily combines with both hydrogen and oxygen, and equilibrium should exist between these molecular bonding states:

$$[C-C] : [C-OH] : [C=O] : [CO_2] : [H_2O] : [H_2].$$

Oxidation of graphite, promoted for instance by the dissociation of water, may lead to *n*-paraffins, which by way of methyl radicals can associate with mineral surfaces. The coexistence of hydrated minerals, water, and hydrocarbons in carbonaceous chondrites can be viewed as evidence of the simultaneous hydration of silicates and the hydrogenation of carbon. Subsequent oxidation of hydrocarbons will introduce the formation of fatty acids. The presence of fatty acids in such meteorites (Nagy and Bitz, 1963) could simply be a consequence of this reaction scheme.

Based on the calculation of the density of surface films (Weiss, 1963), an oriented attachment of paraffins on layers of montmorillonite as shown in Fig. 11.23 will result. Oxidation will proceed, because the alkaline ion leaves its lattice position upon thermal activation, and is replaced by a hydronium ion, H_3O^+, at the AlO_4^- tetrahedron:

$$K^+(AlO_4)^- \rightarrow K^+ + AlO_4^-; AlO_4^- + 2H_2O \rightarrow (AlO_4)^-H_3O^+ + OH^-.$$

Hydroxyl ions can be transformed by redox reactions into free oxygen and water molecules. The monoatomic oxygen stripped of its negative charge will diffuse more easily through the silicate lattice and oxidize hydrocarbons attached to the surface of silicates. This process may result in the formation of fatty acids:

$$3\,O + H_3\text{–}C\text{–}R \rightleftharpoons R\text{–}COOH + H_2O.$$

The reaction is catalyzed by the silicates themselves due to their metal oxide nature. Oxidation of hydrocarbons does not necessarily stop at the fatty acid level; it may continue until it reaches its final end-member: CO_2. It is well to remember that in an open thermodynamic system we are always dealing with chain reactions where the entropy increase of the whole system is the determining factor. In such a system, for instance the lithosphere, the free energy of a single reaction is of *no* importance because a system is involved and not an isolated reaction in a glass tube.

Primordial Soup

The elementary presentation on carbon and CO_2 has already demonstrated the feasibility of interactions between clay minerals and certain hydrogenation products of carbon, namely paraffins, which may subsequently become oxidized to fatty acids. Since this mechanism appears to be a convenient tool to generate physiologically interesting molecules, mineral-organic interactions might conceivably be involved in the synthesis of other biochemicals and perhaps life itself, and a closer look might be rewarding. But before doing this, I wish briefly to mention another concept on the origin of life which has dominated our way of thinking for quite some time.

At the beginning of this century one assumption remained undisputed: Biological evolution must have been preceded by a chemical evolution that in due course gave rise to the first living cell. The way and means of how such an assemblage might have been achieved remained unanswered. In the late twenties, Aleksander Oparin presented the first plausible model on the chemical synthesis of critical biochemical molecules. He proposed a reducing proto-atmosphere composed of small inorganic molecules which by lightning and high energy radiation led to the synthesis of simple organic molecules such as amino acids or sugars. These products accumulated in the sea, where they interacted, formed polymers, and led to the familiar "primordial organic soup". According to Oparin (1953), life eventually arose from that substrate.

For several decades, Oparin's model remained uncontradicted, the more so because experiments simulating prebiotic atmospheric conditions and applying electric discharges, ionizing particles, or UV-radiation were able to synthesize amino acids, small peptides, sugars, heterocyclic compounds, among others (e.g., Horowitz, 1945; Urey, 1952; Miller, 1953, 1959; Calvin, 1959, 1969; Oró and Guidry, 1961; Ponnamperuma and Kirk, 1964).

Up to this very moment, the flow of articles showing the ease of prebiotic synthesis of numerous organic chemicals by means of irradiating different types of atmospheric systems has not ebbed. And yet, all these experiments, remarkable as they are, do not necessarily prove that nature has come up with this or that biochemical compound exactly along the route experimentally followed. It would be the same as proposing that primordial urea has been generated from ammonium cyanate – simply because Friedrich Wöhler synthesized urea in this fashion in the year 1828 – although such a likelihood cannot be dismissed a priori.

To narrow down the wide range of scenarios that have been suggested to account for major biochemicals or even the first living cell, three criteria should be applied to verify or refute a model. First the system must be thermodynamically feasible, second it must have a high yield in organic molecules to provide a sizeable organic pool to draw from; and third, it must allow for the chemical evolution towards a primordial cell.

Amino acids, sugars, or nitrogen heterocycles present in atmospheric systems will decompose upon irradiation into smaller molecules: CO_2, H_2O, N_2, NH_3, H_2, C. It is only when biochemical compounds generated by ionizing energies are allowed to find shelter in an aquatic system, that they escape destruction by the same radiation that created them in the first place. Thus, models on prebiotic synthesis under atmospheric conditions must incorporate the hydrosphere as an integral part of the system. This then would represent a non-equilibrium environment and conform with thermodynamic requirements. Oparin's model, therefore, meets our first condition.

At the energy flux estimated for chemical synthesis in the atmosphere/hydrosphere system (Degens, 1979), almost 15 million years are needed to match the primary productivity of a single year. To obtain the level of organic matter presently dissolved in the ocean, amounting to 1 mg organic carbon per liter, photochemically produced molecules would have to accumulate in the ocean without loss to sediments or oxidation for about 1.5 billion years, geologically an unlikely proposition. Expressed differently: there is no chance whatsoever to create a prebiotic soup on a global scale. In turn, yield of reaction in the air/water system is far too little to provide a huge quantity of organic matter along such avenues. Furthermore, polymerization of amino acids in water, a highly polar medium, cannot proceed because of dipole-dipole interaction. Quite to the contrary, any peptide or polysaccharide introduced to the sea will ultimately be hydrolyzed. With respect to our final condition, the creation of a cellular envelope composed of phospholipid membranes, there is no conceivable way to achieve this goal in either air or sea unless a mediating substrate is provided.

Prevital Monomers

Starting with the pioneering work of A. Treibs in the early thirties on porphyrins in coal and oil as old as the Cambrian (Treibs, 1936), reports on the presence of this or that organic molecule in sediments and soils grew more and more in number. By 1960, almost all principal monomeric building blocks of the living cell were found to survive diagenesis for millions of years even though only in trace quantities (first reviewed in Abelson, 1959). Today, organic geochemistry has advanced to one of the most challenging scientific fields in the earth sciences (e.g., Eglinton and Murphy,

BOX 11.6

Another Type of Suess Effect

Working with K.O. Emery and A. Prashnowsky in the late fifties on the distribution of amino acids and sugars in Santa Barbara basin sediments, off California, we could demonstrate not only that they are present in large quantities, but that during compaction and early diagenesis they must have separated in the few-meter sediment column by chromatographic means. Mentioning this to a friend and colleague at the Scripps Institute of Oceanography, Hans Suess, provoked the comment: "Egon, you are demonstrating the obvious. I would predict the presence of amino acids and sugars in recent sediments without even analyzing the samples." Then Hans opened the drawer of his desk and took out small fragments of some unpretentious looking rocks: "Well, these are samples from the worlds beyond the Earth, carbonaceous chondrites. To find amino acids or sugars in them would be something." And so we looked at the Orgueil, Murray, Bruderheim – unfamiliar names then – and soon enough discovered amino acids, sugars and quite a number of other biochemicals.

The first comments on this discovery that we were analyzing the sweat from the palms of our hands could be discarded when finding out that we were essentially dealing with racemic mixtures, where left- and right-handed optical isomers were present in equal proportions (I.R. Kaplan et al., 1963). The problems concerning the interpretation of amino acids in meteorites have been reviewed by Kvenvolden (1974). To us, most surprising was the fact that the bulk of the organic matter was present in a highly polymerized form, materials we commonly refer to as humic matter or kerogen. Since carbonaceous chondrites have a low-temperature history, judged by the high abundance of organic matter, clay minerals, and water, it dawned on us that clays were actually responsible for the observed polymerization. From that discovery the examination of clays and amino acid mixtures subjected to low-temperature reactions was imperative. We reported on these findings first at a Gordon Research Conference in the mid-sixties. I am deeply grateful to Hans Suess for sending me on this quest.

1969; Kvenvolden, 1974, 1975; Hare et al., 1980; J.M. Hunt, 1979; Duursma and Dawson, 1981; Tissot and Welte, 1984; Sundquist and Broecker, 1985; Degens et al., 1986b). It can safely be said that molecular markers of cellular provenance, so-called "biomarkers", have been discovered in sediments as old as Isukasia (3.8 Ga). The most prominent ones are the hopanes, which, "frozen" into kerogen, may constitute more than half of all fossil organic matter. Since hopanes are ultimately derived from biological membranes, their presence bears witness to the stability of the cellular envelope (Ourisson et al., 1979). All these data support the notion that organic matter in terrestrial rocks has a biogenic source, but some materials associated with crystalline rocks, including some hydrocarbon patches, might well have been formed inorganically (Robinson, 1964).

From the world beyond Earth, we may presuppose that simple organic molecules are more abundant than ordinary mineral matter. However, it is difficult to prove to what stage of complexity they have advanced: perhaps amino acids, sugars, peptides, polysaccharides ...? The only cosmic markers we are certain of are those contained in carbonaceous chondrites (Box 11.6). All lines of evidence point to the conclusion that organic matter in meteorites is of extraterrestrial origin and of abiotic nature (e.g., Lawless et al., 1972; Nagy, 1975; Field et al., 1978).

We are now faced with the problem of the point at which to draw a line between organic matter of prevital and vital origin. Probably the most elegant stand taken on this subject is the one of Cairns-Smith (1982): "It is proposed that life on earth evolved through natural selection from inorganic crystals." The idea of a wholly inorganic (non-carbon) genetic material is at the heart and at the start of his thesis. Organic molecules became introduced much later into a living entity which already existed in the form of clay organisms. Cairns-Smith's book makes fascinating reading.

My approach differs in that I consider clay minerals only as a mediating force but not already as a manifestation of archaic life. Consequently, the genesis, chemical evolution, and structural assembly of monomers, polymers, salts and water into a living cell is, in my opinion, the pivotal question.

Two molecules are of special relevance in the dark interstellar clouds: formaldehyde and hydrogen cyanide, because of their ability to yield by self-condensation or in the presence of a mineral catalyst physiologically interesting organic molecules. According to A.H. Weiss et al. (1970), formaldehyde in the presence of $Ca(OH)_2$ leads to: (i) self-condensation of formaldehyde (e.g., glycol-aldehyde), (ii) aldol condensation (e.g., pentoses and hexoses), (iii) Cannizzaro reaction (the simultaneous reduction and oxidation of two aldehyde groups in the presence of hydroxyl ions):

$$4\ CH_2O + Ca(OH)_2 \rightarrow (HCOO)_2Ca + 2\ CH_3OH$$

and (iv) isomerization of hydroxy aldehydes and ketones. The complexity of the system, particularly with regard to the competitive nature of formose and Cannizzaro reactions, makes it difficult to predict equilibrium conditions. The varieties of molecular species that may arise in the process of such reactions are legion, and the resulting carbohydrate chemistry can certainly be as complex as that of crude oil.

A similar case can be made for hydrogen cyanide with respect to the synthesis of urea, amino acids or the bases of the purines and pyrimidines (e.g., Mizutani et al., 1975). These data strongly support the inference that at the dawn of time a wide spectrum of organics must have already been present, and in unbelievable quantities (e.g., Yoshino et al., 1971).

Once on Earth, such reactions could still continue because of the omnipresence of potential mineral catalysts and the ongoing supply of fresh starting material in the form of comets, planetesimals or even de novo synthesis in Earth's primitive atmosphere (e.g., Pinto et al., 1980). Experiments reveal that in the presence of clays, CO_2 and NH_3, a synthesis of urea, amino acids and pyrimidines can be achieved (Harvey et al., 1972). Sugars arise from mixtures of clays and formaldehyde (Fig. 11.24ab), but addition of urea will shift equilibrium in the direction of Cannizzaro reactions (Degens, 1974). Adding lime to the clay-formaldehyde system will almost instantaneously yield a suite of organics, not only monomers, but complex polymers too (Box 11.7).

Lime or calcium hydroxide solutions must have regionally been quite abundant on the prebiotic Earth, even though only for brief moments. We have seen that carbonates were the main sinks for much of the CO_2 released on the early Earth. In contact with hot lava, and impacting bolides, decalcification of limestone should have produced plenty of lime. Blown into the air or released to the water, lime plus H_2O promoted $Ca(OH)_2$-catalyzed reactions, thus filling the horn of plenty from which life could emerge.

These few examples should suffice to reveal the potential of organic matter synthesis on a prebiotic earth. High yield of reaction, one of the three pre-

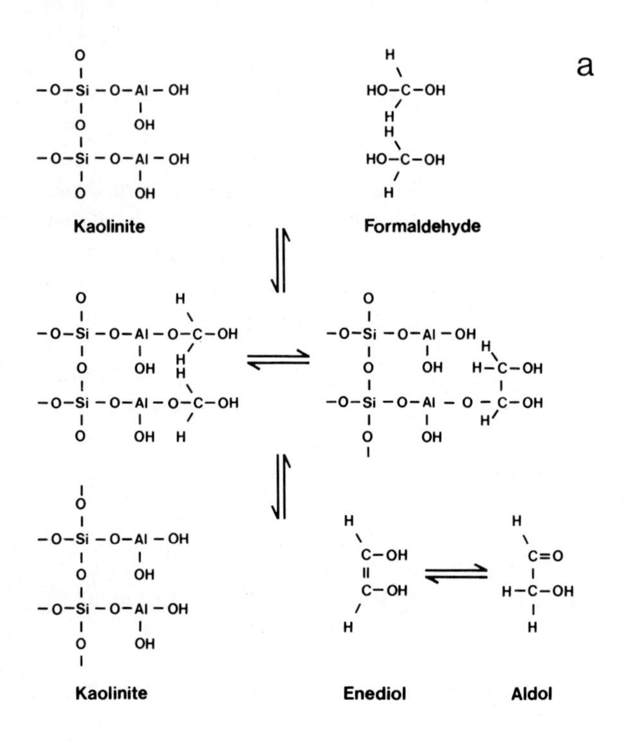

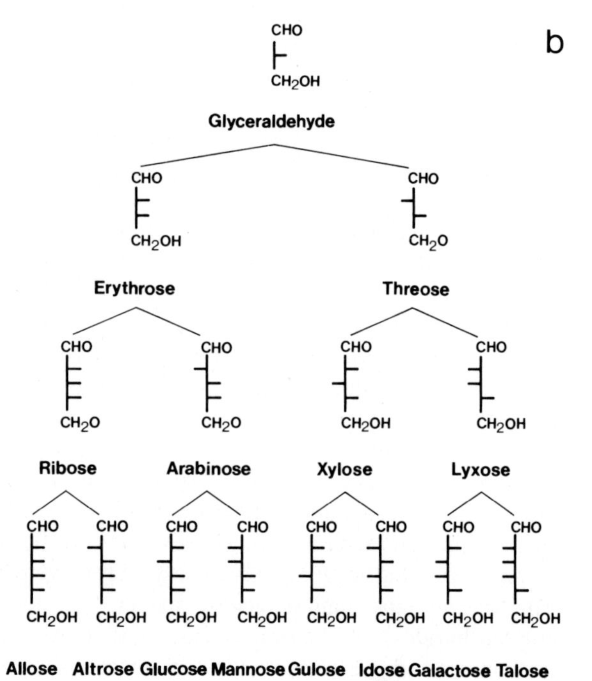

Fig. 11.24. a Possible sequences for kaolinite-catalyzed reactions; **b** addition of one unit of formaldehyde at a time to D-glyceraldehyde in the presence of kaolinite would result in the distribution of the sugars illustrated. Thermodynamic factors, such as steric repulsions of hydroxyl groups, play an important role in the distribution of sugars (after Harvey et al., 1972)

BOX 11.7

He is One of Us

In the mid-sixties, the Woods Hole Oceanographic Institution where I happened to work at that time, was visited by five sisters from the Vatican. A few Catholic families in our community were asked to have them as their guests over night.

One of the sisters stayed with us and one with our friends and neighbors, the Johnsons. All Sisters were scientists and so it was natural that they visited my laboratory, where at that time I was trying to synthesize organic molecules with the help of clays. Our nuns were startled virtually to "see" the essence of life coming out of a test tube.

That evening, the sisters were invited for supper by the Johnson's. When asked by Mrs. Judy Johnson how their day had been, one of them said, "Well, Judy, we have visited the laboratory of Professor Degens, and he showed us how easily one can get molecules of life from clays. This really shook us up in our faith." Judy Johnson moved closer to her, patted her kindly on the shoulder and remarked, "Don't worry, sister, he is one of us!"

---

requisites in a realistic model on the origin of life, is certainly met. An interesting alternative has been proposed by Hoyle and Wickramasinghe (1986) elaborating on the old idea of panspermiology according to which Earth has become inoculated from interstellar matter – a "cosmic bug" – which softly landed on Planet Earth and which continues to do so even up to this very moment.

Crystalline Blueprints

Rock-forming minerals are principally composed of oxygen ions which have as their main coordination partner silicon, aluminum, iron, calcium, magnesium, sodium and potassium ions. Crust and upper mantle may thus be viewed as an ionic oxysphere. The fact that most minerals of the solid Earth are composed of charged ions has been known since the turn of this century. Crystals may contain charge deficiencies, structural irregularities, lattice defects and, in hydrated varieties, may even develop hydrogen bonds.

In the mid-twenties, while studying the crystallography of clay minerals, Linus Pauling dealt with hydrogen bonding, a device needed to balance the clay structure (Pauling, 1930). And it was the same Linus Pauling who, 20 years later, looking at peptides and proteins presented to us the alpha helix whose critical structural element is again hydrogen bonding (Pauling et al., 1951). At the same time, George Brindley (see Brindley, 1980) opened up the field of clay-organic complexes which lead to the recognition of clays as an ideal substrate for fixing and transforming organic matter. But it was John Desmond Bernal (1954a, b, 1959) who injected into the discussion on the origin of life the idea that minerals were instrumental in the synthesis of the key organic molecules assembled in the biological cell. This proposition was especially courageous because at that time – and even today – the textbook doctrine on the origin of life always asks for a synthesis of organic molecules by high energy radiation in the atmosphere and its subsequent polymerization in a prebiotic "organic soup". The picture would not be complete without mentioning Erwin Schrödinger (1945) who put into the discussion on the origin of life the phenomenon of "aperiodic crystals" and the way of how they structure macromolecules.

To structure our discussion, I will begin with a process commonly described under the heading "epitaxis", a term derived from the Greek *tassein* meaning: to arrange or to organize.

The growth of crystalline material on other crystal surfaces is a well-studied subject in the field of crystallography. Epitaxis can also proceed on organic templates with the resultant formation of biominerals. Furthermore, organic polymers can promote the synthesis of other organic polymers, and the living cell is vivid proof of that. Finally, mineral surfaces may provide sites for activation and protection of functional groups displayed by organic molecules, and may accordingly serve as polymerization matrices. Thus, one can distinguish between four systems in which one partner represents the template and the other partner the epitaxial product:

Template	*Product*
mineral	mineral
biopolymer	mineral
biopolymer	biopolymer
mineral	biopolymer

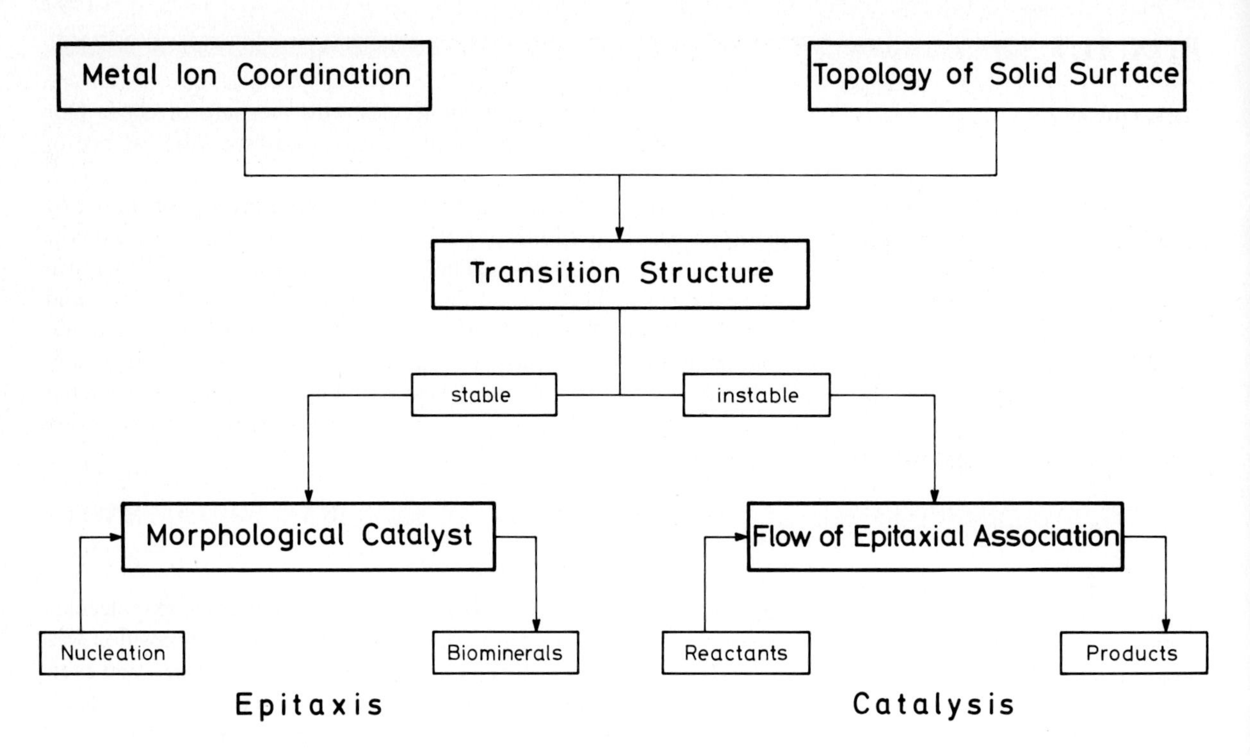

Fig. 11.25. Flow diagram of epitaxis and catalysis (Matheja and Degens, 1971a)

Epitaxis on solid-state surfaces should be viewed in relation to catalysis because both processes follow a similar reaction path. Figure 11.25 outlines the main steps. Based on the molecular configuration of catalytic processes proceeding along solid state substrates, catalysis represents a process in which a solid state surface "tries" to establish a thermodynamically favorable phase transition structure with the adsorbent. Crystal habitus is catalyzed by means of the solution partners through creation of new boundary surfaces (Neuhaus, 1952a, b). Phase transition structures emerge which, when chemically stable, lead to epitaxis or oriented intergrowth. In contrast, should transition structures introduce a chemical change of the adsorbent such as polymerization, hydrogenation, dehydrogenation etc., we are dealing with catalysis. Principles of cellular catalysis, as for instance executed by enzymes, are identical to those observed in mineral systems. A decision whether an epitaxial or catalytic association exists is thus structurally determined. Catalysis constitutes a flow of epitaxial associations, whereas epitaxis involves a "frozen-in" transition structure provided by a morphological catalyst. Such catalysts are always characterized by a distinct crystalline order, and colloidal or amorphous phases will never lead to organized growth. A few critical studies on mineral-organic interactions will follow, in order to elucidate structural principles of catalysis and epitaxis obeyed in natural environments as they relate to the origin of biomolecules.

Ultrastructural patterns that arise when amino acids or small peptides interact with mineral surfaces are depicted in Fig. 11.26. A poly-L-alanine solution evaporated at 40° C on a rhombohedral plane of D-quartz, deposits the peptide principally in β-conformation. Chain-folded helices are aligned in the form of lamellae which exhibit a sharp phase boundary at the organic-mineral contact zone. Frequently the lamellae are split along the direction of their fold axis ("zipper effect") (Fig. 11.26a). Insertion of β-pleated sheets running perpendicularly to the long axis of the lamellae act as dispersion forces and cause the formation of cross-β-structures, as indicated by IR-spectra. It is noteworthy that an increase in temperature during evaporation of the peptide solution will progressively favor β- over α-conformation.

L-alanine deposited on a rhombohedral surface of quartz exhibits two preferential directions. A rough chain-folded surface is developed which is a result of multiple nucleations (Fig. 11.26b). Below a smooth chain-folded surface (a) and a rough chain-folded surface (b) are depicted:

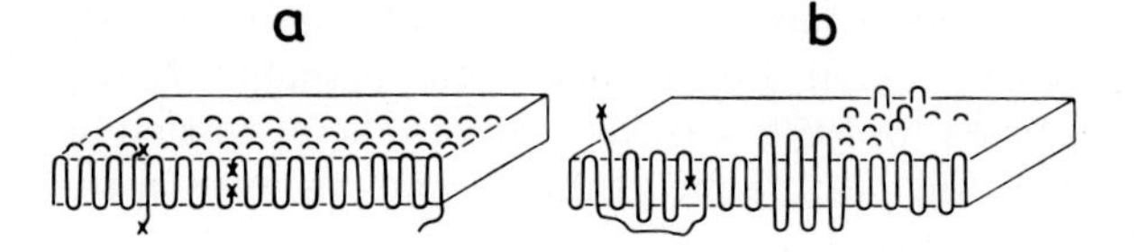

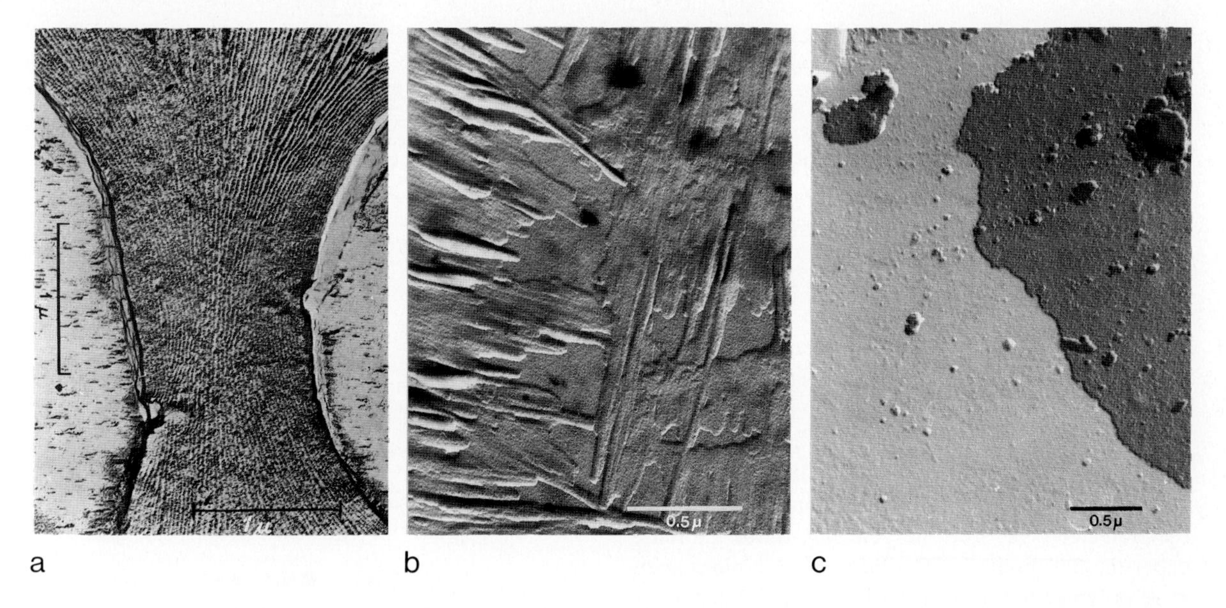

Fig. 11.26 a–c. Electron micrographs (Pt/C replica) showing epitaxial association of: **a** poly-L-alanine and quartz (Seifert, 1968); **b** L-alanine and quartz (Degens, 1976); and **c** L-aspartic acid and quartz (Degens, 1976)

Start and termination of a peptide chain is indicated by an x. In contrast to L-alanine, L-aspartic acid produces a rather smooth layer (Fig. 11.26c). The different ultrastructural patterns obtained for alanine and aspartic acid are a result of the ability of alanine to become adsorbed to the mineral substrate by means of hydrogen bonds, whereas aspartic acid is principally bound via carboxyl groups. Both alanine and aspartic acid, which were administered as monomers, require 6 N HCl hydrolysis for their recovery which implies that the mineral substrate has served as both epitaxial and catalytic agent. On the assumption that poly-aspartic acid is generated, the carbonyl groups should be aligned in a hexagonal array for steric reasons, whereby each corner of the hexagon is occupied by one oxygen. Such an arrangement represents a rather stable configuration and explains why aspartic acid was adopted, for instance, as key element in biological calcification (Degens, 1976).

With the help of clay minerals, chemical synthesis of a number of biologically interesting polymers has been successful (e.g., Solomon and Loft, 1968; Neuman et al., 1970a, b; Burton et al., 1969); this particularly concerns the formation of peptides (Degens and Matheja, 1967, 1968; Paecht-Horowitz et al., 1970; Theng and Walker, 1970). The mechanism involves carboxyl activation and the inactivation of functional groups not participating in the formation of the amide bond by so-called protective groups displayed along mineral surfaces. The relationships for a kaolinite-amino acid system are schematically illustrated in Fig. 11.27. In the presence of kaolinite, amino acids will be picked up from an aqueous solvent and brought into solid solution. Amino groups

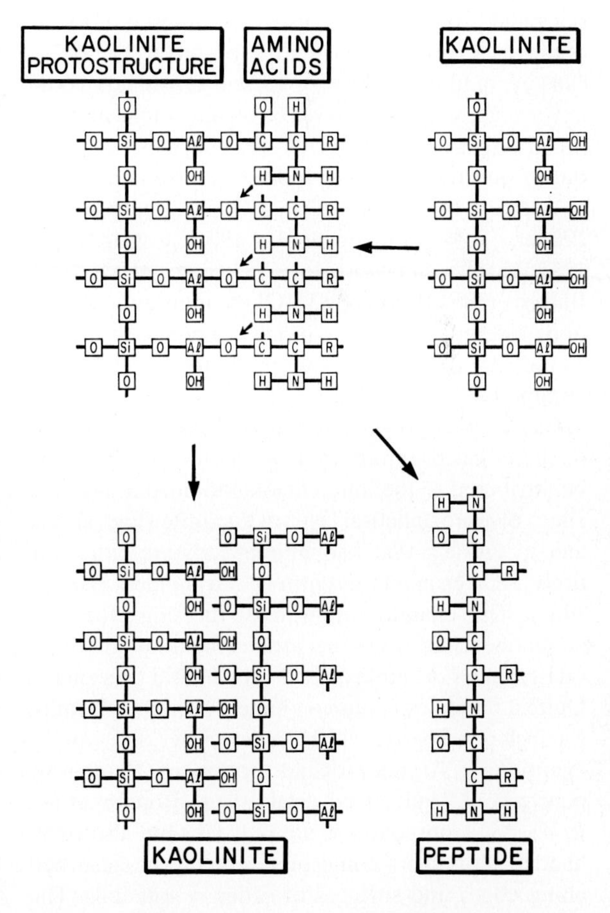

Fig. 11.27. Polymerization of amino acids along clay templates

become hydrogen bonded to the structural oxygen, or in the case of basic amino acids, occur as positive ions. They are tightly fixed to the silicate surface, and thus rendered inactive. Carboxyl groups associate with charged Al-oxy-hydroxy groups by means of ionic bridges, or become directly attached to the aluminum. The high number of oxygens coordinated to aluminum will promote either discharge of hydroxyls or uptake of protons. In contrast, silica in the clay structure is negatively charged because hydroxyls exposed at the surface will release protons; the tight bonding of silicon to oxygen prohibits OH^- removal.

In water, amino acids cannot polymerize because of dipole-dipole interactions. In solid solution, however, amino acids will polymerize, because the solvent medium does not interfere, and because this reaction step is favored energetically. In the kaolinite experiment, about 1,000 times more amino acids were polymerized to peptides than could conceivably become adsorbed to the clay surface. In consequence, a flow of freshly polymerized molecules across the catalytically effective mineral surface has to be postulated.

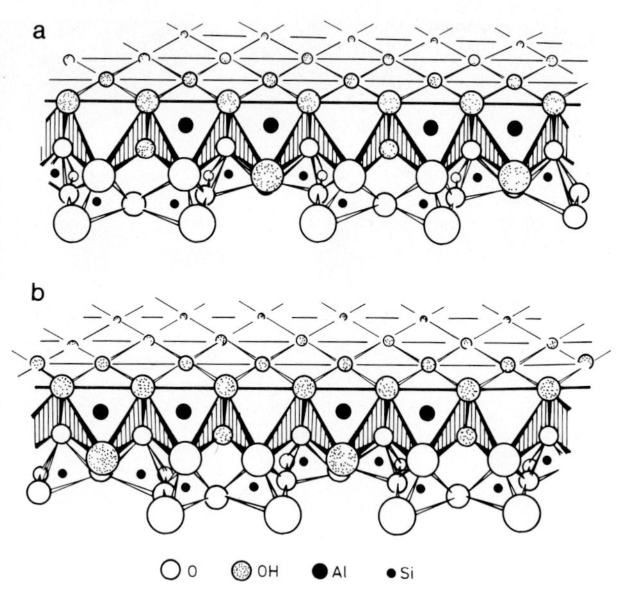

Fig. 11.28 a, b. Schematic representation of the edge of a kaolinite crystal of ideal composition (**a**), and its mirror image (**b**) viewed along the *a*-axis (after Jackson, 1971)

A remarkable characteristic of life is that all peptides and proteins are exclusively composed of the L-optical isomers of amino acids. This has to be so, because a random mixture of L- and D-isomers could never lead to a higher level of organization, and stable orderly structures, such as the enzymes, would be out of question. Many propositions have been made to account for chirality (Wald, 1957; Bouchiat and Pottier, 1984; Bonner et al., 1974; Cairns-Smith, 1982; Mosher and Morrison, 1983; Joyce et al., 1984; Barton, 1986; Mason, 1985). Of the numerous suggestions offered, minerals – in my opinion – should be considered a rather suitable promoter for chiral discrimination (Greek *chiros:* "hand").

Preferential polymerization of L-amino acids on kaolinite has been reported by Jackson (1971). It can be attributed to the inherent enantiomorphism of the edges of the octahedral layer of kaolinite (Fig. 11.28), and to the fact that kaolinite crystals are either entirely right-handed or entirely left-handed (Bailey, 1963). The enantiomorphism of the edge faces of kaolinite arises from the arrangement of O-atoms, OH-groups, Al-atoms, and octahedral vacancies. Quite a number of common minerals exhibit chirality, for instance quartz, which comes either as L- or D-enantiomer. Asymmetric adsorption of alanine by quartz could indeed be established (Bonner et al., 1974). As a consequence, not only left- but also right-handed monomers could preferentially be adsorbed along edges and surfaces of ordinary minerals. The fact that common carbohydrates such as cellulose, starch, or cane and beet sugars (sucrose) are composed of D-configurated monomers can also be related to the optical activity of the archaic mineral matrix.

The structural shape of biomolecules, which is a key element of cellular function, is asymmetric. It is conceivable that on the prebiotic Earth left- and right-handed polymers were generated by mineral printing machines having either D or L block letters. Once the first organism had chosen the L-configurated amino acid polymer, or the D-configurated sugar polymer, their mirror images had no chance to evolve further. Since free amino acids racemize at rates ranging from a few thousand (e.g., aspartic acid) to a few million years (e.g., valine), extraction of L-stereoisomers from a hypothetical amino acid pool will not noticeably enrich the D form in such a reservoir across geological time scales.

Nature is the great master of stereospecificity (Prelog, 1976). Chiral molecules are ingrained in the organic constituents of all living organisms from the most ancestral to the most highly advanced forms. A great number of minerals also exhibit chirality, but unlike organic matter they will not racemize. A D quartz will remain a D quartz for ever. Should an organic molecule, however, become stabilized by heavy metal ions, be adsorbed to minerals or enter the kerogen structure, it may keep its original chiral identity through the ages too. So in a way, universe and life are asymmetric: the *cosmos* by being matter-dominated and *life* by having selected chiral molecules, for example enzymes. Whether minerals such as chiral clays were at the base of this development is suggestive simply because of their omnipresence in the environment.

Kaolinite is also instrumental in preferentially synthesizing pentoses and hexoses from formaldehyde and transforming them into polysaccharides. Kaolinite, as we have seen, can also generate fatty acids and entertain esterification reactions leading to glycerides (Harvey et al., 1972). Furthermore, addition of calcium phosphate to an aqueous mixture of kaolinite, glycerol and palmitic acid produces compounds which appear by electron micrographs to be phospholipid monolayers which are deposited in epitaxial order with a 40 Å periodicity on the crystal surface of kaolinite.

For fatty acids there is a considerable increase in the surface adsorption relative to hydrocarbons, because hydrogen bonding will attach these molecules to the SiO_4 tetrahedra; such bonds are not developed in the case of hydrocarbons. Moreover, the melting temperature of fatty acids is about 40°C higher than that of its paraffin precursor. Let us finally turn to the phospholipids.

The PO_4 bonding on a SiO_4 surface is shown in Fig. 11.29. Two routes can be followed: (i) formation of hydrogen bridges, or (ii) coordinative sharing of oxygens by P–O–Si bonds. X-ray studies reveal that the surface energy is stored in the form of structural irregularities (D'Eustachio, 1946). Also, the isoelectric point of Al and Si oxides is strongly influenced by adsorbed impurities, which indicates the presence of structural defects. To extract energy it is favorable to cover the distorted surface with a monomolecular film as shown in Fig. 11.29.

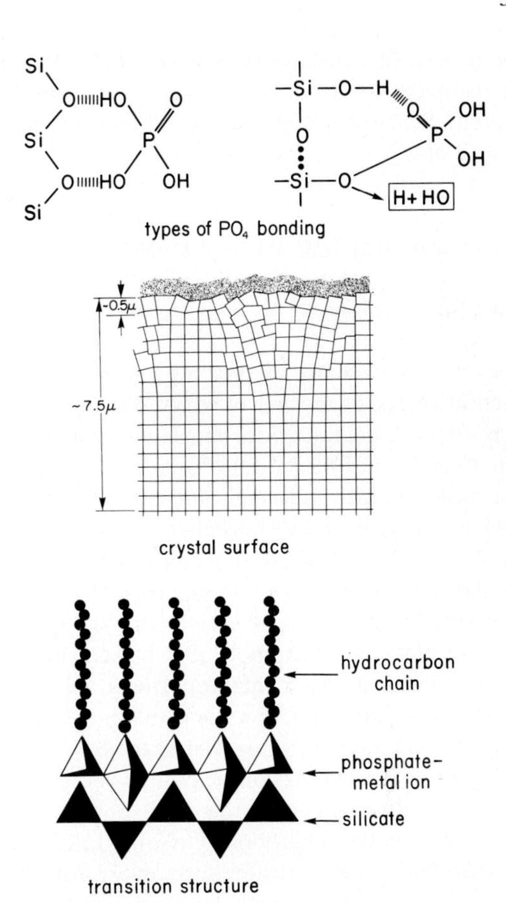

Fig. 11.29. Formation of phospholipids

Due to the structural flexibility of phosphate tetrahedra in relation to the metal ion employed, a strong bonding in the form of a smooth surface film can be created on top of a layered silicate. Indeed, experiments on polyphosphates indicate that phosphates are capable of forming transition structures on solid surfaces, thereby inducing hydrophilic properties (e.g., von Engelhardt and von Smolinski, 1957).

Phosphate minerals, for example apatite, are excellent scavengers of "almost any substance present in the solution to which they are exposed" (Burton et al., 1969; Neuman et al., 1970a, b). Phosphorylated substances such as nucleotides adsorb well, and polyphosphate compounds (pyrophosphate, ADP and ATP) readily associate with the crystalline surface. It could also be demonstrated that apatite crystals will catalyze phosphorylation at temperatures prevailing in the environment. Various biologically interesting bases react with apatite in the presence of pyrophosphate. The phosphate group of the resulting nucleotide monophosphates, as expected, arise from the adsorbed pyrophosphate (Neuman et al., 1970a, b). Since the apatite surface can change at will from an acid catalyst of hydrolysis to a dehydrating catalyst of phosphorylation as a function of pH and phosphate or pyrophosphate content, an alkaline ocean could conceivably stimulate, for instance, the phosphorylation of adenosine to AMP in the presence of a suitable mineral catalyst. In contrast, an acid ocean ($\approx$ pH 4) might yield diphosphates, but at elevated temperatures (>100°C) a stabilizing element would be Mg^{2+} (Seel et al., 1986). The occurrence of pyrophosphate in prebiotic time has been treated in detail by Miller and Parris (1964).

Summing up, crystalline blueprints are effective devices for generating leading biomolecules and for promoting chirality. Clays are outstanding in this respect, but they deliver only semi-finished products, not life itself. To accomplish this, and make a clay-based "Golem" one needs, as can be read in Gustav Meyrink (1868–1927; "The Golem", 1915), cabalistic rites. Alternately one could follow the eminent astronomer Iosif S. Shklovskii, who a few years ago commented on the universe and life, "Honestly, I wouldn't be surprised to find an old biblical god behind all of it."

Constructing the First Living Cell

Black Box Machine

Research often treats "life-making" in the form of a reaction vessel, at the top of which one injects reactants such as amino acids, expecting almost living cells to emerge at the bottom. But the plain truth is that basic research, although intensive, has not yet reached conclusive results. A belief in a direct link between the abiotic synthesis of amino acids or proteins and the generation of a cell presupposes a sudden and spontaneous generation of life. Even assuming that all molecules and macromolecules present in a cell were available in the right proportions, an infinity would still separate us from the actual molecular assemblage and the structural order of the cell.

So despite the findings of the past 50 years we find ourselves still referring to the same questions that concerned Oparin in his book "Origin of Life", "One must first of all categorically reject every attempt to renew the old arguments in favor of a sudden and spontaneous generation of life. It must be understood that no matter how minute an organism may be, or how elementary it may appear at first glance, it is nevertheless infinitely more complex than any simple solution of organic substances. It possesses a definite dynamically stable structural organization which is founded upon a harmonious combination of strictly coordinated chemical reactions. It would be senseless to expect that such an organization could originate accidentally in a more or less brief span of time from simple solutions or infusions" (first paragraph, final chapter).

To approach the problem on the origin of life, we might try to look for distinct events in a prevital environment that could lead to well-ordered sequences of chemical processes exhibiting autocatalytic properties. These should be mechanisms which provide chemical reaction chains and a catalyzing system complementary and structurally adjusted to one another. In some way this would be an extension of the concept of a "Turing Machine" (Turing, 1967; Hopcroft, 1984) and the term "black box machine" has accordingly been phrased (Matheja and Degens, 1971a).

After having all semi-finished organic molecules, essential elements and water available for constructing the first living cell, it is necessary to assemble

Fig. 11.30. Principal events for the synthesis of the first living cell

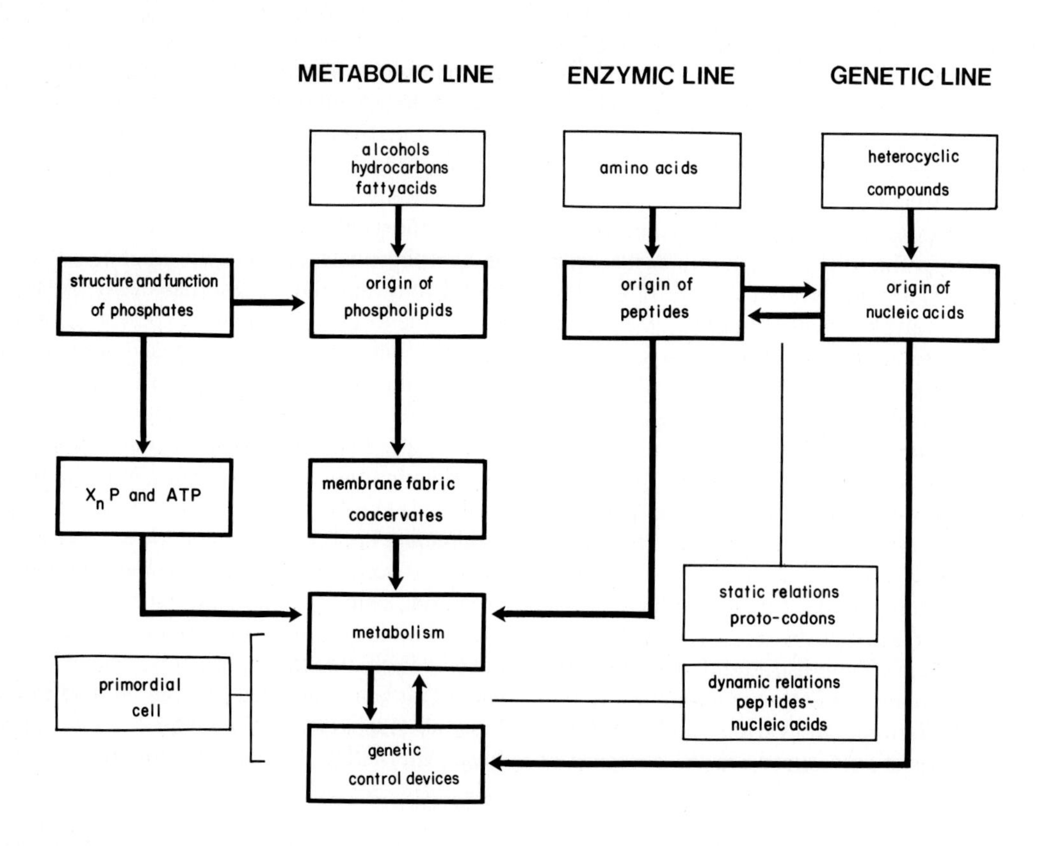

them in a certain order, one by one, almost like building a house; that is, first the basement, and finally the roof. It is only then that people can move into the individual rooms or apartments and fill the house with life.

For the creation of life we have to provide a "shell" with a series of compartments which will become *natural reaction boxes* which are frequently referred to as *black boxes*. Principal events required in structuring the primordial cell are outlined in Fig. 11.30. Three major lines can be distinguished: (i) metabolic line, (ii) enzymic line, and (iii) genetic line. Following synthesis of the monomeric building blocks, a total of 11 *black boxes* must come into operation to produce the key components of life. Arrows interconnecting individual black boxes are used to symbolize the relationships between two or more classes of compounds. To illustrate reaction mechanisms that may proceed between black boxes, we will examine one segment of the cycle more closely, that is the merger of the three main lines: Genetic, metabolic, enzymic.

Genetic Line

With respect to the origin of a self-maintaining system, the course of a familiar argument runs:

enzyme → nucleic acids → cell or nucleic acids → enzyme → cell.

The question is: did the enzymes which regulate genetic functions exist before or after the RNA or DNA event. The logic of our black box system is different in that we are searching for a chain of chemical reactions having a proper feedback acting like a single catalyst towards the environment.

It is commonly assumed that first carriers of genetic information on the prebiotic Earth were RNA molecules (e.g., Eigen et al., 1981). Information is encoded in the particular sequence of bases along the RNA strand. Participating bases are complementary in that two purine-pyrimidine pairs come into existence: adenine (A) is matched with uracil (U), and guanine (G) with cytosine (C). The successful non-enzymatic synthesis of oligonucleotides of 30–40 units in length using long polymers of the complementary nucleotide as template and Zn^{2+} as catalyzing agent has greatly advanced our understanding by showing that polymers rich in G and C offered high copying fidelity in the absence of replicases. Furthermore, such a metal-ion catalyzed template mechanism provides a clue to how polymerases would have evolved on the primitive Earth (Lohrmann et al., 1980, 1981). The first genes supposedly arose from the base sequences of a primordial RNA composed of short segments of nucleotides. Conventional routes of biochemistry were followed and nothing unusual has to be presupposed. In brief, RNA provided structure, function and information at the same time. The more elaborate and rigid DNA protein machinery must have subsequently evolved from that substrate.

Nucleic acids are linear polymers composed of four mononucleotide units. DNA is composed of the following nucleic acid bases: thymine (T), cytosine (C), adenine (A), and guanine (G), and base pairing is A = T, and G = C. In RNA, uracil (U) substitutes for thymine. The three-letter genetic code composed of 64 (4^3) code words is depicted in Fig. 11.31; each triplet is termed a codon. Bases are given as ribonucleotides, so U appears instead of T. Three codons, i.e., UAA, UAG, and UGA, are chain terminators ("stop"), and one codon, AUG (methionine), standing for "start", initiates the printing of all proteins. However, in a few instances, methionine is encoded by GUG (Darnell et al., 1986). The protein factory, the ribosomes, thus knows where to start and where to stop protein synthesis.

The "Hypercycle Model" of Eigen and Schuster (1977) and Eigen et al. (1981) rests on the RNA-instructed synthesis of proteins. Competing hypercycles must have existed in high numbers. In the end, however, rivalry among hypercycles must have led to the survival of only one of them, which in time evolved to the first living cell, the progenote. Eigen et al. (1981) state, "Presumably hypercycles still exist today as features of viral infection processes."

In this context attention is drawn to viroids, the smallest known agents of infectious disease in several plants and possibly some animals (Diener, 1981). They are nothing but short strands of RNA. The entire nucleotide sequence of the potato spindle-tuber viroid (PSTV) was worked out (H.J. Gross and Riesner, 1980), showing that the viroid is a chain of 359 nucleotides: 73 A's, 77 U's, 101 G's and 108 C's. The excess of G's and C's appears to be a characteristic of viroids in general. Perhaps an archaic element? Yet viroids are not translated into proteins; that is they do not act as messenger RNA's encoding specific proteins. Their replication relies entirely on enzyme systems of the host plant. Since primordial RNA and viroidal RNA appear to have everything in common, namely linear arrangement of the bases A, G, C, and U, predominance of G, and C, hair-pin structure and short strands, it is a puzzle why viroids are unable to encode a polypeptide acting as RNA replicase (Diener, 1981), whilst abiotic RNA is believed to do so (e.g., H. Kuhn and Waser, 1982). Back to the base oligomers!

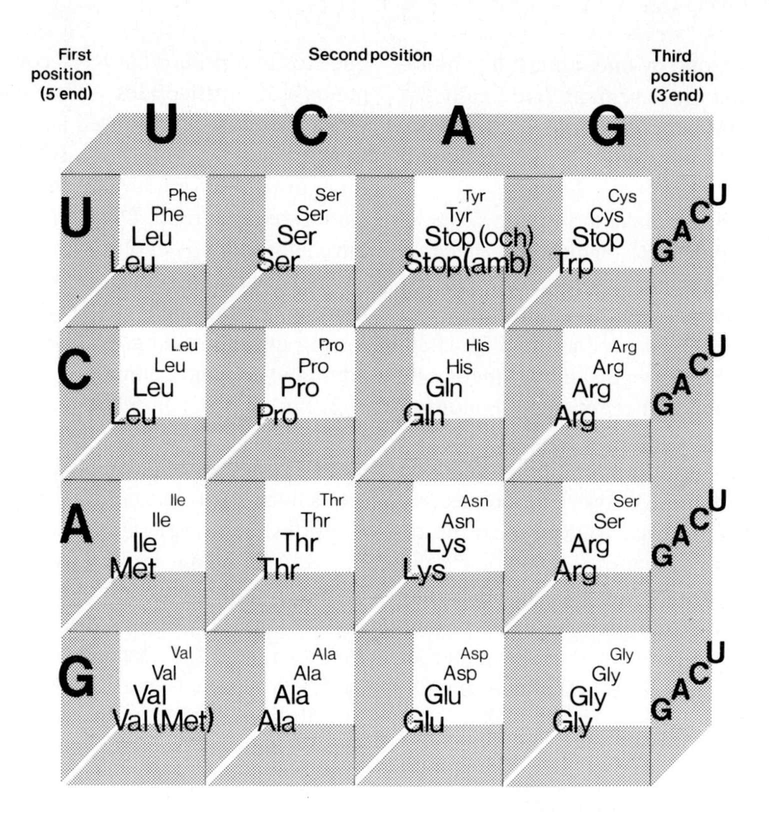

Fig. 11.31. The genetic code; bases are given as ribonucleotides

Prebiotic propagation of the nucleic acids must certainly have taken place (Eigen et al., 1981). Where, when and why is fortunately no longer everyone's guess. Outstanding in this respect were contributions of Susumu Ohno. He proposed in 1984 and 1985 that the first set of prebiotic coding sequences to be translated were repeats of base oligomers with one restriction: that the numbers of bases in such oligomeric units were not multiples of three.

Oligomeric repeats in primordial coding sequences have three virtues:

(i) if they were repeats of base nanomers, the probability of those having open reading frames of indefinite length is high;

(ii) periodical polypeptide chains encoded by them were likely to have assumed either α-helical or β-sheet secondary structures (Ycas, 1972); and

(iii) all repeats, except those where the number of bases in the primordial building block oligomer is a multiple of three, share the attribute of immortality (Box 11.8).

BOX 11.8

Immortal Coding Sequences (Ohno, 1984; Ohno and Epplen, 1983)

A few genes is all that retrovirusess possess in their genome; inherent replication error rate of reverse transcriptase is in the order of 10^{-3}/base pair/year (Gojobori and Yokoyama, 1985). For bacteria it is about 10^{-4}/base pair/year. In contrast, the number of gene loci in the mammalian genome approaches 10^{5}; inherent replication error rate amounts to about 10^{-9} base pair/year (Ohno, 1985). Thus, when it comes to DNA replication, mammalia utilize a rather refined copying and editing machinery, to keep all genes functional, whereas the genome of retroviruses can afford to be enterprising.

Non-enzymatic nucleic acid replication can be expected to have a replication error rate not less than the error rate of reverse transcriptase. This implies that an assigned polypeptide chain experiences a 100 percent amino acid sequence change every 1,000 years, a rather cumbersome situation for life to emerge.

Premature chain terminations and reading-frame shifts are the most detrimental base changes which

In the town where I was born lived a man who sailed to sea.
And he told us of his life in the land of submarines.
So we sailed up to the sun till we found the sea of green, and we lived beneath the waves in our yellow submarine.
We all live in a yellow submarine, yellow submarine, yellow submarine.
We all live in a yellow submarine, yellow submarine, yellow submarine.
And our friends are all on board, many more of them live next door, and the band begins to play.
We all live in a yellow submarine, yellow submarine, yellow submarine.
We all live in a yellow submarine, yellow submarine, yellow submarine.
As we live a life of ease, ev'ryone of us has all we need. Sky of blue and sea of green in our yellow submarine.

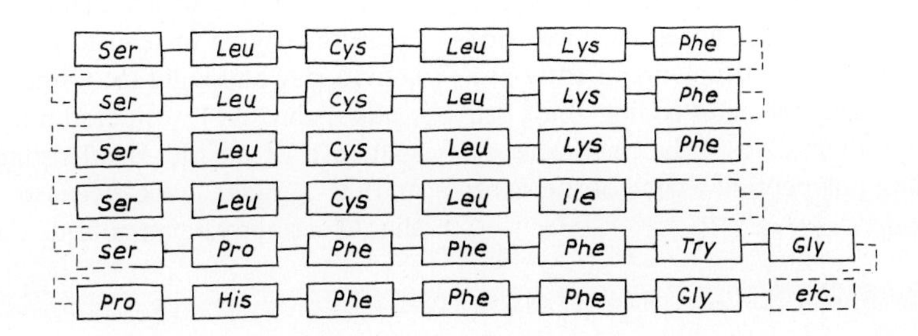

Fig. 11.32. Transcription of the Beatles' "Yellow Submarine" (John Lennon and Paul McCartney, 1966) into its base sequence. Repeats of base oligomers are evident. Striking is the abundance of thymine which occurs almost six times more frequently than adenine. It is extraordinary that no initiator nor terminator codon is employed, at least in the first critical six lines. The expression of this sequence, that is its polypeptide chain, bears interesting features. The amino acid polymer has a hydroxyl (serine) beginning and serine reappears at the start of the next four lines too. High aromaticity (phenylalanine) throughout the score is striking. Sulfur-crosslinkage via cysteine (cys) nicely connects the general theme. The rest are fillers except for a few basic amino acids which appear to give the beat. In summary, serine may stand for silk, phenylalanine for skin, and cysteine for hair, all three of which are a Beatles' trade-mark. However, the most refined phenomenon is the presence of a single tryptophane – almost at the end. This amino acid releases upon digestion indols which are known to put you nicely to sleep. This may suggest, that the air in the "Yellow Submarine" was eventually either low in oxygen, or high in carbon monoxide or both

coding sequences may sustain. They may come about by base substitution that changes an amino acid specifying codon to a chain terminator. Furthermore, deletions and insertions of bases that are not multiples of three in numbers displace the reading frame. Resulting unusual reading frames tend to be full of chain terminators and a shortening of chain lengths results, which may deprive the polypeptide of its assigned function.

However, should the number of bases in a hypothetical abiotic oligomer not be a multiple of three, coding sequences that arise by means of oligomeric repeats are quasi "immune" to the above described base change dilemma as shown for a heptameric repeat:

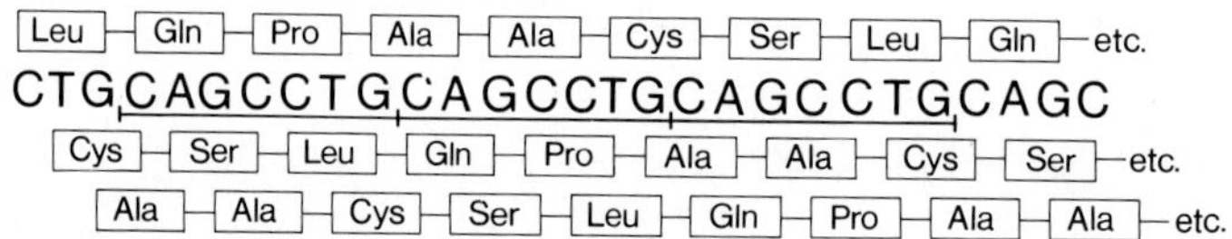

The 21-base long sequence (3 x 7) becomes the unit periodicity and encodes the heptapeptidic periodical polypeptide chain. A base change, for instance from a cystein codon TGC to a terminator TGA, can silence only one of the open reading frames, and reading-frame shifts are of no consequence because the periodicity of polypeptide chains is maintained. Hence, a good measure of immortality is ingrained in coding sequences that are repeats of $N = 3n \pm 1$ or 2 base oligomers. Such types of oligomeric repeats have the further advantage of giving longer periodicities to the polypeptide chains they encode.

In evolution, new genes come into existence from redundant copies of pre-existing genes. For instance, the adaptive immune system of vertebrates has apparently evolved by plagiarizing one ancestral gene (Mostow et al., 1984). De novo recruitments of truly new coding sequences from the non-coding base sequences have shown themselves to be near-exact repeats, thus encoding polypeptide chains of the exact periodicity.

A few months ago I had the good fortune to listen to a talk by Susumu Ohno in which he was able to demonstrate that the repetitious recurrence manifested in the coding sequences in the genome pervades many aspects of human endeavor as seen, for instance, in music and architecture. During his presentation, coding base sequences became readily transformed into musical scores through the assignment of two consecutive positions each in the octave scale to four bases in the ascending order A, G, T and C.

Chopin's Nocturne, Op. 55, No. 1, he transcribed to its base sequence. Three times the base nonamer CAACCTCCC recurred. Interestingly, the primordial building block base oligomer of the last exon of the largest subunit of mouse RNA polymerase II is the nonamer CAACCTCTC, which recurs four times. This is nothing more than a single base derived from the principal subject of Chopin's Nocturne and thus *homology* is established between the two. Transforming the first 106 codons of this "Last Exon" into a musical score for the piano, according to the modus operandi of Chopin's Nocturne, gave a fascinating concert piece. Indeed, listening to a tape of "Mouse's Last Exon" made you think of a Chopin composition even though not necessarily the Nocturne, as Susumu Ohno pointed out. "Mouses's Last Exon" has a lively dance cadence which is quite fitting for a RNA polymerase and which definitely is not a nocturnal creature, having instead to engage in transcriptional activity day and night (S. Ohno and M. Ohno, 1986).

After the talk I wondered how the transcribed base sequence of an old Beatle song, "The Yellow Submarine", might look because "the sea of green" should certainly be an antique element (Fig. 11.32).

Original periodicities have not vanished through time. Three representative segments of the human X-linked, 419 codon long coding sequence for one of the sugar metabolizing enzymes, phosphoglycerate kinase well illustrates the antiquity (Singer-Sam et al., 1983, Fig. 11.33). Such sugar metabolizing enzymes are truly ancestral in nature in the light of the fact that the amino acid sequence of this human enzyme has retained an almost 70 percent homology with the corresponding enzyme of baker's yeast, and the two are indistinguishable in their tertiary conformation. Or take cytochromes, which function as electron carriers and are the critical element in biological oxidation. Different cytochrome molecules from species covering a wide range from ancestral (baker's yeast) to highly evolved forms (man) show amino acid residues at some positions that are common to all cytochrome molecules, whereas at other positions

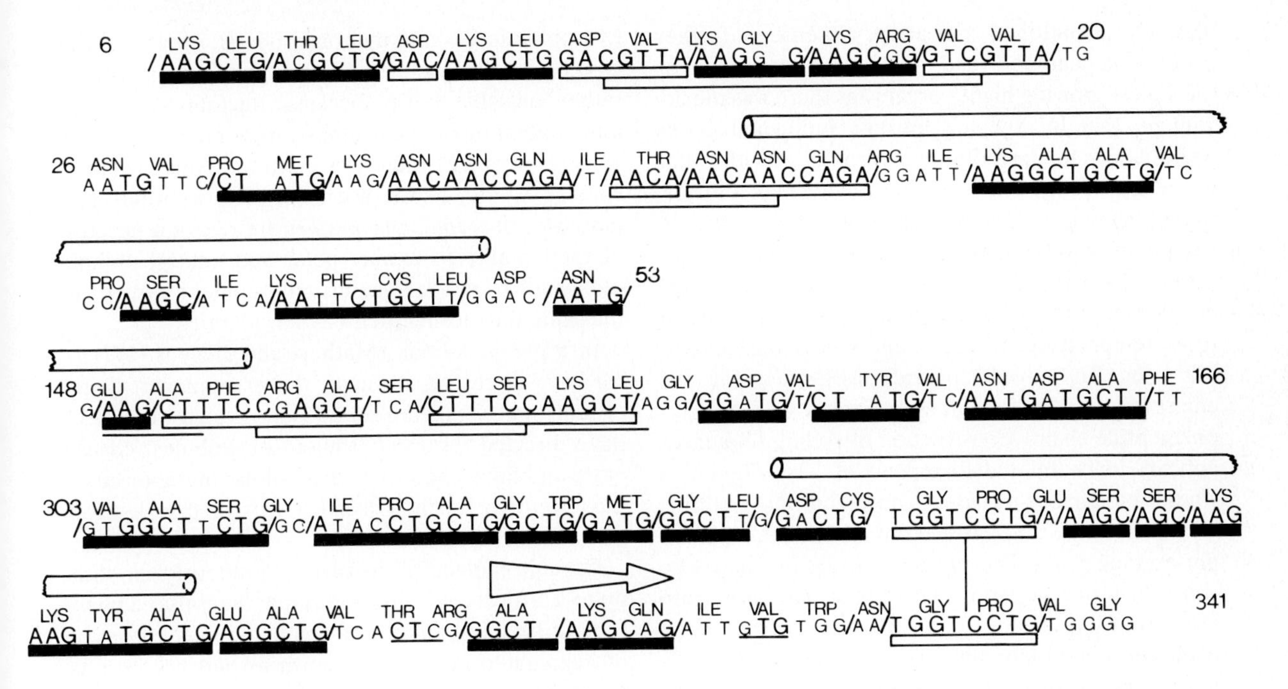

they have mutated sometimes as much as eight times (Margoliash and Fitch, 1968; Fig. 11.34).

Almost all genes of the eukaryotes are mosaic in nature, that is composed of coded segments: the *exons* and in between segments: the *introns*, which are not transcribable to polypeptide chains. According to Walter Gilbert, *introns* were originally tailored from catalytic domains of ancestral RNA' s which in the distant past allowed RNA *molecules* perhaps to operate without enzymes and permitted at the same time the assembly of larger and larger macromolecules. Eventually this exon-intron pattern became conservatively handed over to DNA. It is of considerable interest that prokaryotes and even an ancestral eukaryotic yeast possess almost no introns. A likely explanation is that for the fast-growing bacteria, synthesis of introns became an unnecessary burden and was con-

Fig. 11.33. Four widely spaced representative segments of the 419 codon long, human X-linked phosphoglycerate kinase coding sequence and their corresponding amino acids (Singer-Sam et al., 1983). *Numbers on margins* indicate the segments' position. Primordial building block hexamer AAGCTG and its parental decamer AAGGCTGCTG are depicted in large capital letters, and deviant bases are shown in small capital letters; both are underscored by solid bars. Base oligomers not directly related to the above primordial building blocks are underscored by open bars and connected to each other (Watson et al., 1982; S. and M. Ohno, 1986)

Fig. 11.34. Invariance principle observed in cytochrome *c*. Amino acids in black boxes are identical for all 25 cytochrome *c* molecules studied (Margoliash and Fitch, 1968). The number in the *white boxes* represents the number of different amino acids found at the respective amino acid residue position.

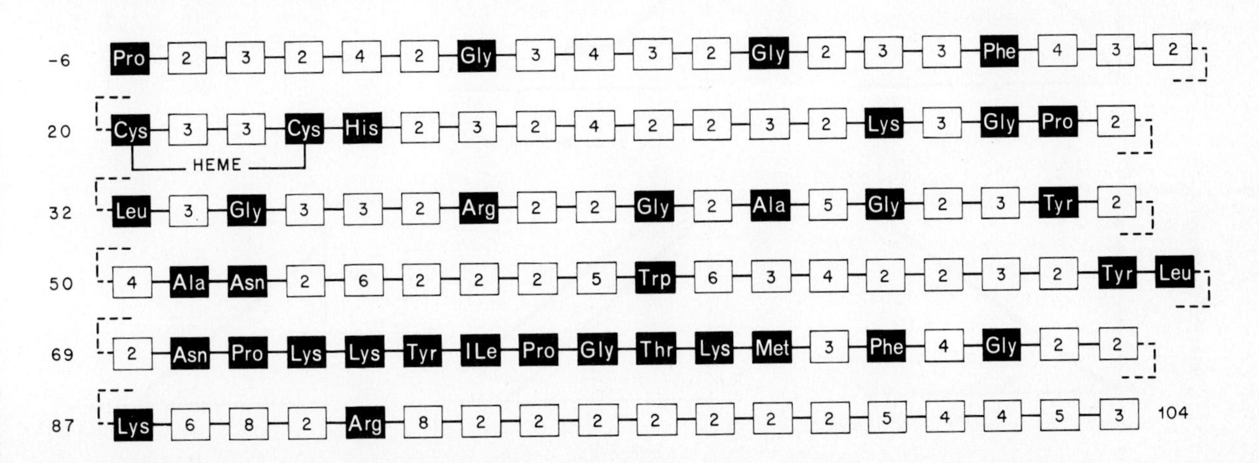

sequently abandoned, but at the expense of further progress: a dead-end road in evolutionary terms. For the slower-growing higher organisms there was the advantage that by keeping introns, individual exons could associate, reshuffle and recombine with other exons to larger genetic elements having new informational loading potential. Expressed differently, introns appear to be the key elements of evolution.

On a prebiotic Earth, the genetic line encounters no serious problems to come into existence, even at room temperature. Minerals and metal ions such as Mg^{2+} or Zn^{2+} could provide assistance for the emergence of the first RNA capable of encoding a polypeptide chain. Construction of coding sequences appears to be related to repeats of base oligomers where the number of bases is not a multiple of three. New genes, as a rule, arise from redundant copies of pre-existing genes. The mosaic structure of eukaryotic genes in terms of exons and introns may propagate exon shuffling, lead to novel proteins, and keep evolution potentially alive. Actually, my book too is nothing but a collage of work and ideas of others, introns and exons so to speak. In the course of writing they became reshuffled and restructured to provide a new genome helping to decode some of the mysteries of nature.

Metabolic Line

Several authors who contributed to the books of Oparin et al. (1959) and S.W. Fox (1965) proposed that the activity of enzymic processes in *colloidal micelles* could lead to life; such elements are held together by hydrophic forces and hydrogen bonds. However, the type of micelles sustained in this manner represents a rather unstable entity because fluctuations in the ionic milieu or of water temperature may cause such micelles to eject their cell content. Thus, when it comes to the development of a primordial metabolism, only *ionic membranes* are relevant because of their ability to extract polar and neutral organic molecules from the aqueous environment without at the same time losing their cellular identity.

In a previous book (Matheja and Degens, 1971a), we have discussed at length the principles of liquid crystals, micelles, emulsions, foams which we believe have directed the establishment of globules, coacervates and finally heterotrophic cellular metabolism on the primitive Earth. Crucial events of this development are briefly mentioned.

Organic molecules dissolve continuously in water up to a critical concentration value; thereafter a phase separation takes place, and a spontaneous formation of aggregated molecules, known as *micelles,* results.

Fig. 11.35 a–f. Dissolution properties of organics in aqueous systems: **a** micelle formation of ionic compound, sodium dodecyl sulfate, is indicated by break in electrical conductance at CMC value; **b** micelle formation of nonionic compound, decyl glucoside, is revealed by break in surface tension; **c** competitive interaction scheme between ionic and polar organics in solution concerning micelle formation; **d** interfacial tension of water-organic liquids increases linearly with the logarithm of the degree of immiscibility of water and organic liquids; **e** logarithm of the CMC (moles/liter) of potassium mono- and di-carboxylates is shown as a function of hydrocarbon chain length at 25° C; **f** effect of chain length of paraffins on CMC is depicted (for further details and references see: Matheja and Degens, 1971a)

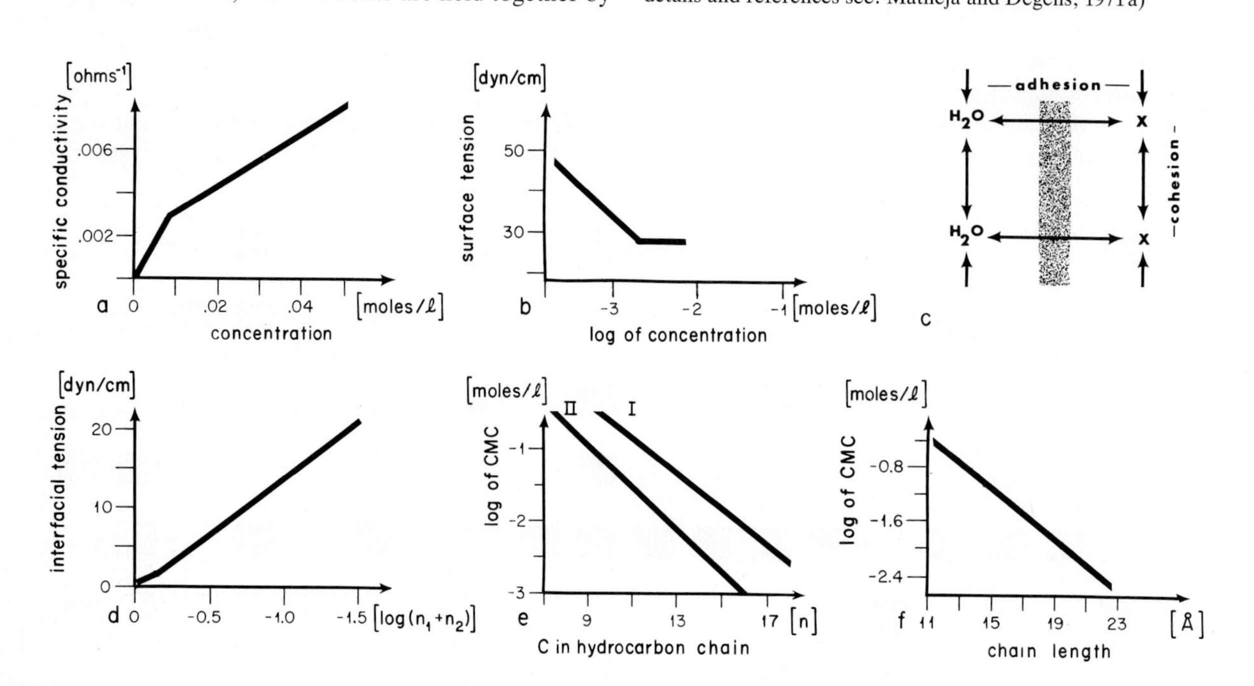

Micelle formation is reflected by a discontinuity in the physical and chemical properties of the solution (Fig. 11.35a,b). The point at which this change occurs is known as Krafft's point (Krafft and Wiglow, 1895), or the critical micelle concentration (CMC). The hexagonal water framework and its distortion by dissolved molecules shows this spontaneous phase separation to be a function of chain length and the ionic or polar properties of dissolved organic molecules. At the phase separation point the ratio for organic molecules (OM) with carbon numbers up to 20 is about:

(i) H_2O: non-ionic OM $= 10^7$–10^5:1
(ii) H_2O: ionic OM $= 10^5$–10^2:1

The number of aggregated molecules within micelles falls in the range of 10^1 to 10^5, giving micelles a radius in the range of 10 to 100 Å.

The self-propelled separation process which trips at a certain concentration level is controlled by the general chemistry of the solvent medium (Fig. 11.35e, f). Linkage forces between organic phases can be enhanced or lowered by co-solvents (Fig. 11.35c, d). Ionic and non-ionic organic molecules are clearly set apart by their behavior in aqueous solutions. *Cohesional* cross-linkage forces are predominant in polar molecules, whereas strong *adhesional* linkage forces are involved in ionic molecules. Therefore we observe a so-called clouding with increase in temperature in phases formed by polar components. The term clouding implies cross-linkage among single micelles, resulting ultimately in a complete phase separation, whereby the organic material becomes isolated from the water. In the case of high molecular compounds, clouding is frequently referred to as *coagulation.*

Clouding is caused by forces developed between individual micelles and temperature is a determining variable. Cloud temperature point is independent of concentration and pH of the solution (Fig. 11.36a, b). Radius of polar micelles increases with temperature, whereas ionic micelles do just the opposite (Fig. 11.36c). Other changes introduced by metal ions or organic additives are depicted in Fig. 11.36d–f.

Collection of matter proceeds by familiar solubilization effects where micelles act as co-solvent for water-immiscible or water-insoluble organic molecules. Metal ions lower the critical value of micelle formation, and the suppression is related to the valence of the metal (Fig. 11.37a). Critical micelle concentration (CMC) is also lowered by dissolved organic molecules (Fig. 11.37b). Thus it appears that the collection of organic molecules present in an aqueous ionic environment is founded on the nature of the ionic boundary layer of micelles which can perfectly adjust to physical or chemical perturbations in the ambience. *Question:* What compounds in an aqueous environment are best suited to yield an ionic boundary layer in the first

Fig. 11.36 a–f. Physical-chemical basis of micelle formation: **a** concentration of a solution does not affect the cloud point as shown for Triton X-100; **b** pH does not affect cloud point (*horizontal bar*), whereas increase in electrolytes will do so (*sloped curve*); **c** temperature changes control weight of micelles differently, i.e., curve I (methoxydodecaoxyethylene decyl ether), curve II (methoxypolyoxyethylene octanoate), and curve III (sodium dodecyl sulfate) which is an ionic micelle; **d** temperature dependence of micellar weight (methoxydodecaoxyethylene decyl ether) at 75°C (curve I) and 9.7°C (curve II); **e** addition of electrolytes to Triton X-100 depresses cloud point, i.e., Na_2SO_4 (lower curve), Li_2SO_4 (central curve), and $AlCl_3$ (upper curve); **f** addition of solubilizates affects cloud point, for instance, polyoxyethylene dodecyl ether (three inclined curves representing different concentrations), whereas hydrocarbons (upper curve) have very little effect (for details and references see: Matheja and Degens, 1971a)

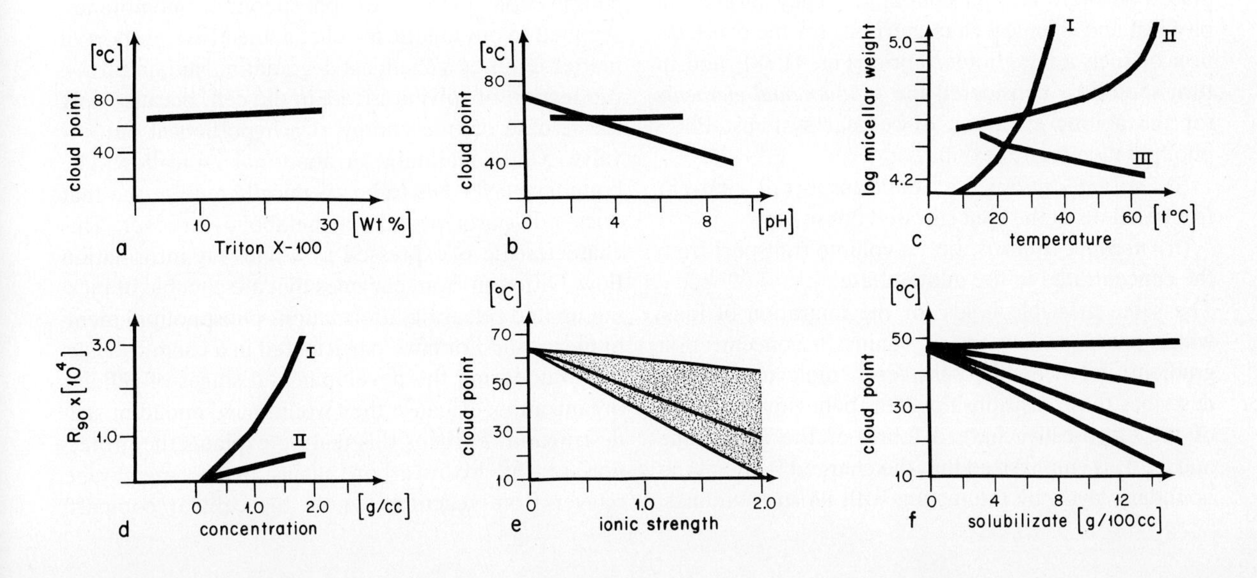

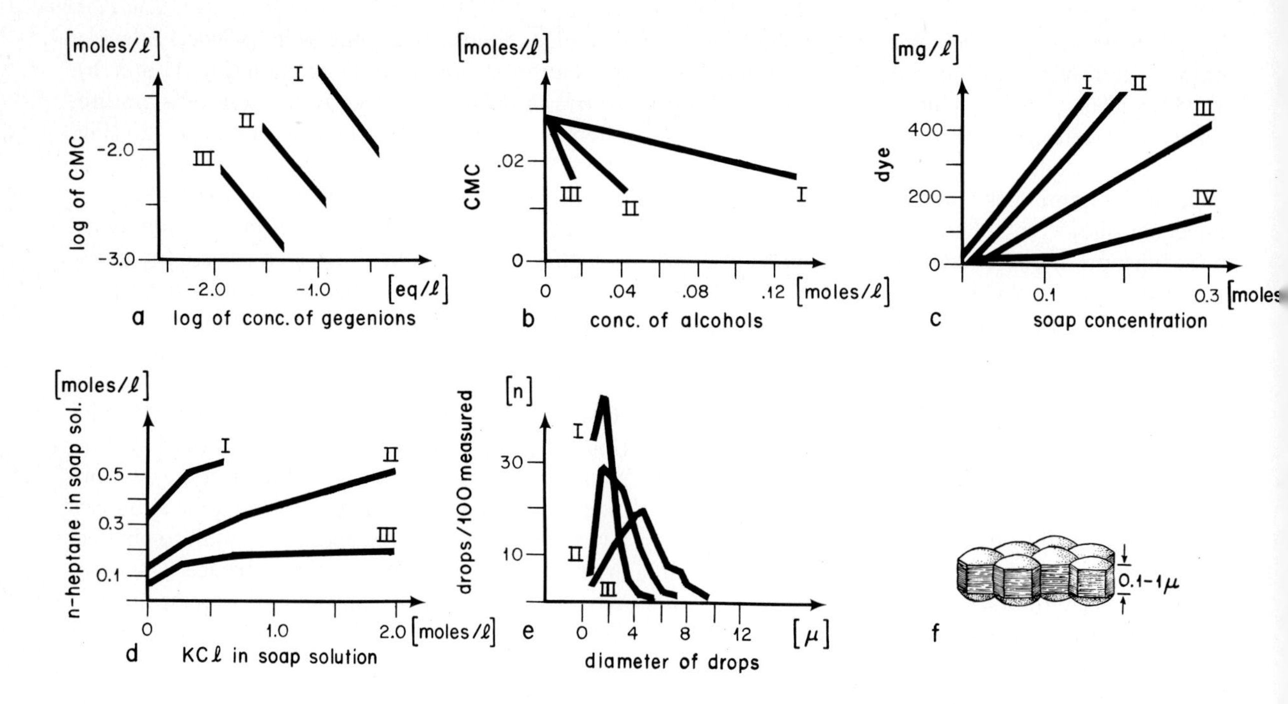

Fig. 11.37 a–f. Lowering of CMC by salt or organic molecules: **a** logarithm of CMC as a function of logarithm of gegenions, for instance R_{12}-,R_{14}-,$R_{16}CH(COOK)_2$ (curves I-III); **b** effect of isopentanol (curve I), hexanol (curve II), and heptanol (curve III) on CMC of K-dodecanoate at 10° C; **c** solubilization of dimethylaminoazobenzene in fatty acid soap solutions at 30°C: K-oleate (curve I), K-myristate (curve II), K-laurate (curve III), and K-caprate (curve IV); **d** effect of KCl on solubility of n-heptane in 0.25 M K-tetradecanoate containing 0.301,0.064, and zero mole octanol per liter (curves I–III); **e** change in number of drops as a function of aging of a 1/200 M soap solution (caesium oleate and octane): first day, third day, seventh day (curves I–III); **f** macro-cell structure of an emulsion system (for details and references see: Matheja and Degens, 1971 a)

place? *Answer:* The phospholipids. They display all physical and chemical characteristics for the construction of such a membrane fabric (Fig. 11.38), and in turn should be considered the *fundamental elements* for the abiotic evolution of cellular systems. Phospholipid membranes exhibit:

(i) *normal osmosis,* i.e., the transport of a solvent from a dilute to the concentrated phase, and

(ii) *negative osmosis,* i.e., a volume transport from the concentrated to the dilute phase.

The same principle holds for the migration of ions, which proceeds towards or against a concentration gradient; in biochemistry the term "molecular pump" describes these relationships. This behavior is a result of the electrically charged fabric of the membrane matrix. It is emphasized that the charged fabric of the boundary layer, by interacting with its surroundings, develops potential and pressure gradients which act as driving forces (e.g., Schlögl, 1956; Kobatake, 1958).

Phospholipid membranes are charged solid-state surfaces. Salts move against the concentration gradient at the expense of free energy. The coupling between the free energy of *metabolism* and ion or molecule migration across biomembranes is a routine operation in present-day life. For instance, the uptake of potassium by *E. coli* requires glucose which is burned during active transport. Data indicate that the total amount of free energy released metabolically can be fully transformed into motion against a concentration gradient. These "abnormal" operation capabilities appear to be related to the anisotropic composition pattern of phospholipid membranes. Applied to our abiotic micelle, a selective transport of matter requires a chemical degradation and, in turn, a production of solvent media in the cell, because only the release of free energy at a hypothetical outflow valve would maintain an abnormal input-flow. The boundary layer has to be chemically inert – one that does not participate in a metabolic turnover. This characteristic is expressed as a one-way information flow. Different from enzymes that are capable of picking up and releasing information, phospholipid membranes could not have participated in a chemically active way during the developmental stages of cellular organization, because they would have ended in self-destruction. Perhaps this feature explains the protection of the hydrocarbon chains in the membrane bilayer from reactions within the cellular compart-

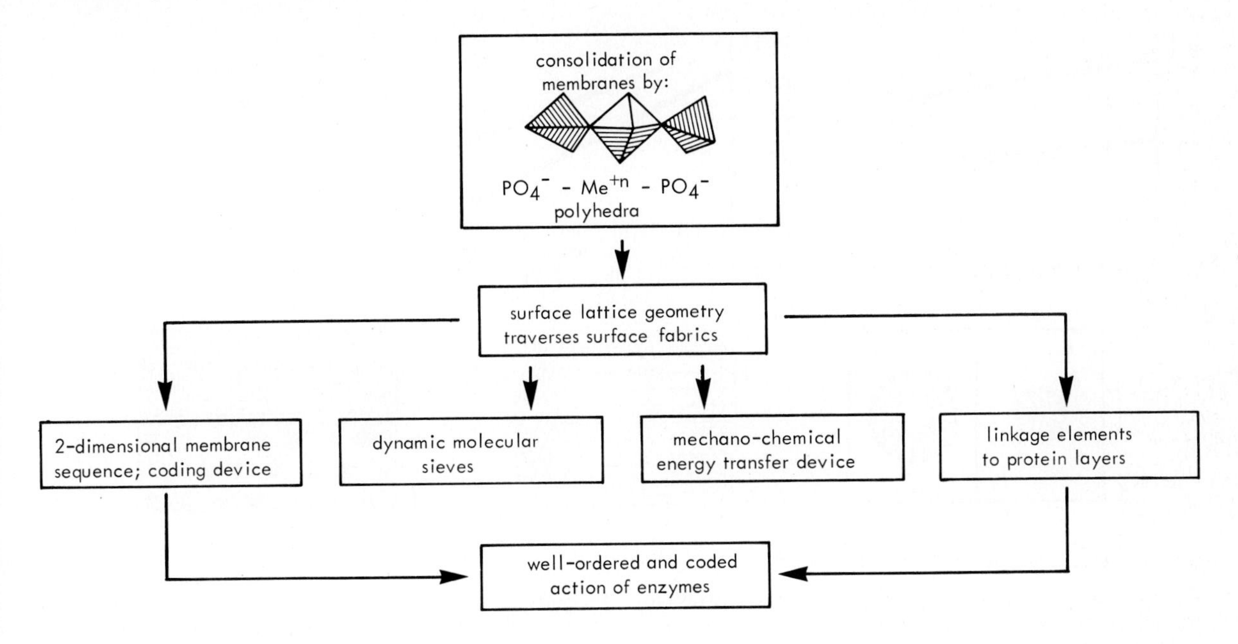

Fig. 11.38. Flow sheet depicting physical and chemical properties of phospholipid membranes and course of action

ment. Those aggregates which did not invent an effective protection for their paraffines perished by cannibalizing their hydrocarbons. Once this membrane system became established, it was not subject to further evolution, which implies that the cellular nature of organisms has been maintained ever since in its primordial habitus.

Experimental data on emulsions and foams (e.g., Berkman and Egloff, 1941) show that micelles equipped with anisotropic membranes are able to grow at the expense of other micelles by consuming them through surface attachment (= lowering in surface energy) in a process termed emulsification. These globules, soaps or emulsion particles as they are termed exhibit an *optimal critical diameter* in the order of 10^3 to 10^4 Å. A typical size distribution is shown in Fig. 11.39e and the macro-cell structure in Fig. 11.39f. Main feature of this water-organic system is an anisotropic and charged phase boundary layer. Many articles have emphasized the need for an ionic fabric for the existence of stable colloids (reviewed in Matheja and Degens, 1971a). Next on the agenda is the origin of the cell membrane proper.

The classical concept on the primordial development of a cellular membrane, also shared by Oparin (1953), involves the separation of hydrophobic chains (paraffins) from water. Boundary layers are generated, whereby the hydrophobic stacks are transformed into membranes by the uptake of polar films such as acetamide R–NH–CO–CH_3 or peptides (Alexander, 1942; Fig. 11.40). This kind of association will not lead to stable membrane systems and is therefore of no relevance for the discussion at hand. Similarly, stable soap cells formed by fatty acids and metal ions represent rather unstable fabrics which easily invert and in doing so eject their cell content (Fig. 11.41). A contraction of ionic boundary layers (hydration sphere) will proceed as a function of ionic radius and valence, so that polyvalent metal ions form water-in-oil emulsions. However, should phospholipids constitute the boundary material this reaction is unlikely, because *PO_4 tetrahedra and metal ions construct stable fabrics.*

The developmental stages of globules (macromicelles) separated from the aqueous medium by anisotropic phospholipid membranes are consequently permitted to evolve (Box 11.9). Metamorphosis, often referred to as coacervation, is accompanied by the expulsion of water via the joining of phospholipid micelles. The capability of inducing phase transformations represents the most remarkable aspect of coacervation. Water droplets become enclosed by membranes, and the original micelle content is exchanged, but according to laws different from those established in aqueous systems. That is, condensation of lipid membranes towards a rigid membrane is achieved by the uptake, for instance, of cholesterol or metal ions (de Bernard, 1958). The expulsion of water proceeds during the intercellular attachment by means of oxygen-coordinated metal bridges. Due to the ionic fabric closely attached to the coacervate, dissolved organic molecules such as peptides or carbohydrates are bounded and precipitated on the membrane surface.

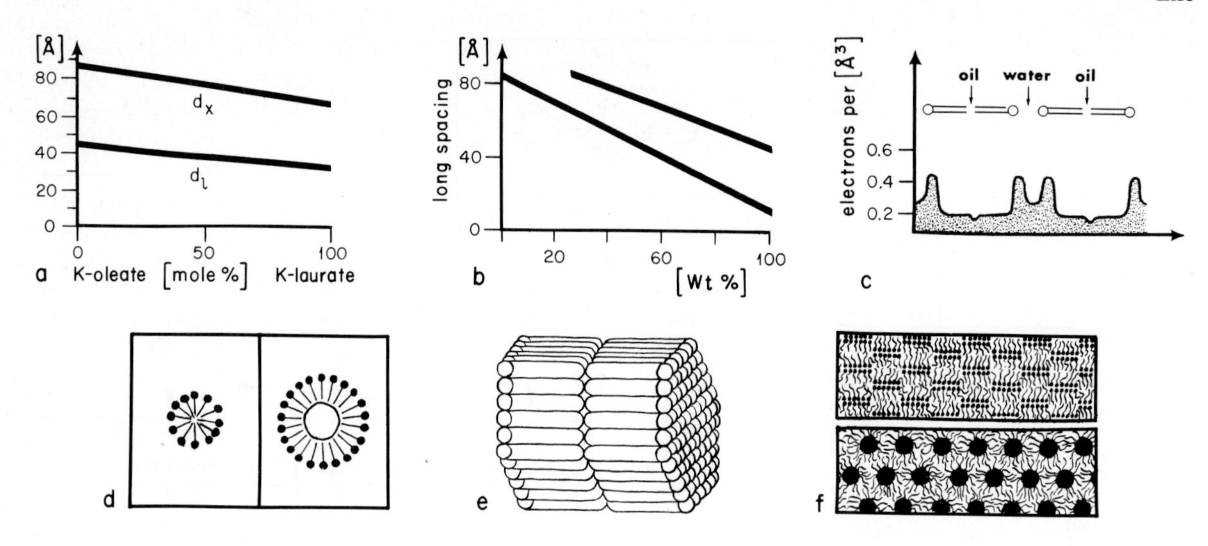

Δ

Fig. 11.39 a–f. Structural aspects of emulsification: **a** X-ray diffraction of crystals of K-oleate/K-laurate mixtures does not yield maxima established for the pure substance but intermediate maxima for the interplanar distances d_l and d_x; **b** lattice spacing as a function of concentration of Detergent X (lower curve), and Emulphor 0 (upper curve) in aqueous solution; **c** electron density distribution within a lamellar micelle of potassium laurate; highest electron density is located at the ionar group -COOK and water is positioned in central valley; **d** expansion of micelle of a 18% K-laurate solution in water before and after adding benzene; **e** structure of lamellar micelle; **f** schematics of liquid-crystalline phase of anhydrous soaps; in alkaline soaps, polar groups form ribbons (*upper graph*), and in the hexagonal phase of divalent soaps, polar groups are localized in form of cylinders which are viewed along their long axis (*lower graph*); (for details and references see: Matheja and Degens, 1971a)

Fig. 11.40 A–C. Formation of globules
∇

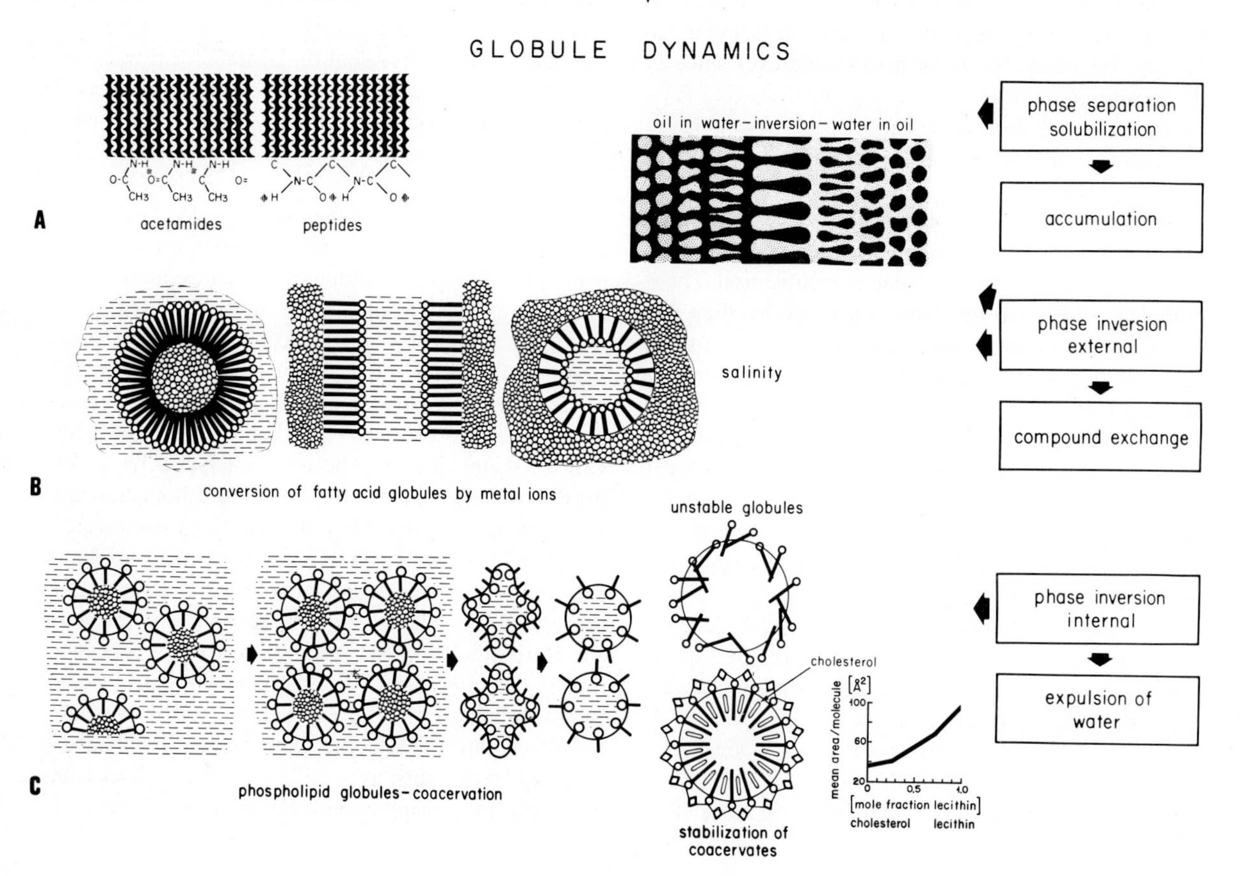

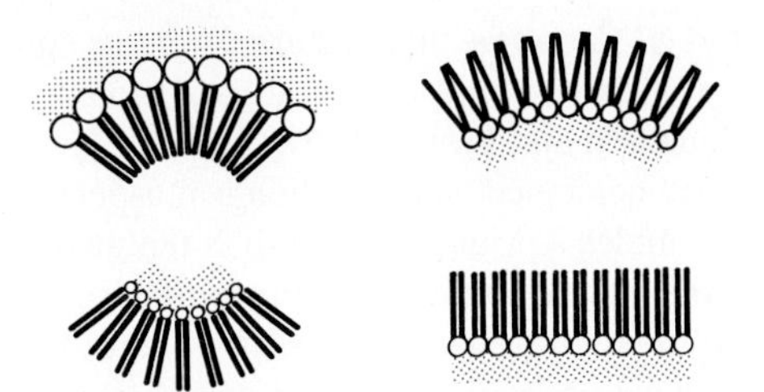

Fig. 11.41. Structure of coacervate systems based on monovalent and polyvalent metal ions

In the course of coacervate development, structures will arise which are enclosed by a double phospholipid skin (= bilayer membrane). Phosphate groups become oriented towards the aqueous phase and double layers may combine to multi-layered stacks. The ionic milieu, i.e., the effect of salt, is a regulation mechanism (Tabony et al., 1987). Membrane pouches come into existence, resulting in the formation of multi-chambered coacervates bearing striking resemblance to mitochondria. Judged by the conservative nature of mitochondria, it appears that as a system it still carries relics of its abiotic origin. The development of such a self-controlled reaction agrees with the thermodynamics of system behavior. A stable cyclic process can exist in the vicinity of a stationary state and may operate repeatedly an infinite number of times without ever passing through the stationary phase itself (Prigogine and Balescu, 1956).

BOX 11.9

The Secrets Behind Aphrodite's Bubble Bath (Degens et al., 1973)

At about the time of the Pleistocene-Holocene boundary, let us say 10,000 years ago, Aphrodite, the goddess of life, is supposed to have risen out of the foaming sea near the Isle of Cyprus. Botticelli has captured this event, and his painting: "The birth of Venus", can be admired in the Uffizi Gallery.

Wherever wind and water meet, foam is generated and up to about one million bubbles per second may come and go. In the presence of dissolved organics and phosphates, bubbles will become coated with a thin film and furthermore be stabilized in part due to their interaction with metal ions. In due course many bubbles will rupture, jetting their "cell content" a few meters high into the air, from where material is carried some distance by the wind. Other bubbles will grow via "cell lysis", thus exchanging the matter they once acquired from the sea. However, bubble formation may also proceed at depth, where hot lava, hydrothermal water, volcanic gases, or decaying organic

Fig. 11.42 a–c. Electron micrographs (carbon replicas) of globules suspended at depth in Lake Kivu waters (Degens et al., 1972); **a** largest globule measures 1.2 micron in diameter; **b** "blisters" visible on the globule surface may have been produced by the leakage of gas entrapped inside the larger spheres; subsequent solidification through resinous material fossilized this event; **c** position of sphalerite crystals (black spots up to about 50 Å) on the globule surface can be seen; notice the accretion pattern

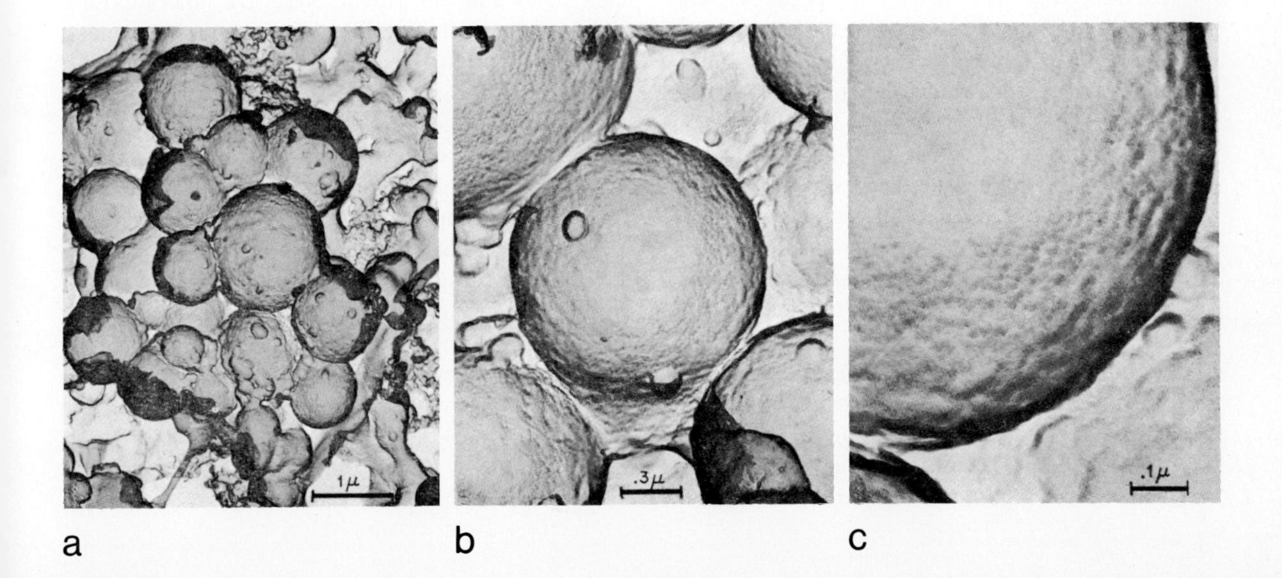

matter release bubbles into the open water column (e.g., Corliss et al., 1981). Multitudes of gas bubbles will form which upon ascent can pick up surface active compounds. A seemingly unrelated observation from the field of aquatic geochemistry may illustrate the phenomenon in question.

Lake Kivu, situated at the highest point of the East African Rift Valley, is surrounded by active volcanoes which at a depth of a few hundred meters discharge geothermal springs loaded with gases. One liter of water taken from a depth of 450 m will release at surface pressure 2 liters of a mixture of CO_2 and CH_4 gas. This implies that when deep water moves up by turbulent forces, effervescence will occur at some critical depth and tiny gas bubbles will form. Such bubbles will extract hydrophobic resinous material and generate ionic membranes that become stabilized by metal ions. It is remarkable that zinc and iron are the two heavy metals that were selected for the stabilization of the organic coat. The resulting hollow spheres are some 1 μm in size (Fig. 11.42a, b), which is about the size of a prokaryotic cell capsule. It is interesting that today deep-sea hydrothermal vents provide niches for rather unusual biological communities (e.g., Jannasch and Mottl, 1985).

In many geological and geochemical aspects, Lake Kivu resembles a primordial sea. It is tentatively concluded that the formation of a primordial cell capsid was also initiated by gas bubbles. In point of fact, gas bubbles have probably produced an infinite number of micron-sized spheres, globules, micelles, or coacervates which accumulated as thick foam at the surface of the early Precambrian ocean. Spheres have only a temporal stability. They may rupture and eject their cell content; alternately, lysis of the cell wall may take place along the line of a viral infection (Fiddes, 1977). Recombination will proceed and a transfer of metabolic, catalytic, or genetic material from one cell to the next is accomplished. This is the most critical part of chemical evolution because it relaxes evolutionary constraints required for the final assemblage of the primordial cell. Aphrodite's bubble bath in the sea off Cyprus, as drawn by Botticelli, appears to me a good scaffold for life's origin.

In conclusion, the primordial metabolism of the coacervate was in all probability maintained by means of a reversible phosphorylation cycle. In consequence, the origin of metabolism is in no way linked to the development of the genetic transcription apparatus. It must be considered an independent formation process. For this reason the abiotic origin of phosphorylation must be regarded as an equally important step towards the creation of the primordial cell. The problem, therefore, centers around the question of how to polymerize the common monophosphates into di-, tri-, or tetraphosphates, since polyphosphates are unstable in natural environments (Thilo, 1962). High-temperature synthesis can be excluded because in low-temperature aqueous environments such compounds would hydrolyze again. The only reasonable choice left is to place the polymerization effect in the coacervates. It is conceivable that phospholipid solid-state surfaces served in this capacity, because inorganic mineral surfaces also act as templates and furthermore may catalyze phosphorylation, as has been demonstrated for apatite crystals (Neuman et al., 1970a).

The establishment of an interconnected and chemical reaction pattern for the coacervate system as a whole exists when phosphorylation can be maintained. This requires a constant supply of organic molecules and metal ions that are consumed or utilized during this development. In this manner, a certain modus vivendi is established. Sources of energy were oxidizable organic compounds in the ambience. Molecules such as amino acids and sugars must have been present in heroic quantities in the environment in view of their mineral-fabricated origin.

So far, however, no vital power was involved. It is postulated that a primitive heterotrophic metabolism was established in abiotic coacervates, and that this metabolism improved progressively. Its development took shape independently of the evolution of the nucleic acids and the genetic code. By superimposing the two separately developed entities, (i) the genetic apparatus, and (ii) the heterotrophic metabolism, the primordial cell came into existence. The primitive metabolism of coacervates was kept "alive" via phosphorylation processes and became explored and embodied by the self-reproducing cycles. It is conceivable that the nucleic acids were able to encode polypeptides utilizing certain metal ions. Alternately, nucleic acids succeeded in adopting the available metalloproteins in their environs – among which must have been enzymes in the billions – for their own reproduction. In any event, the link between the two independently developed events, (i) the primitive metabolism, and (ii) the genetic reproduction apparatus, is represented by the peptides. They are the essential tool by which coacervate metabolism – for the purpose of nucleic acid replication – was utilized. The structure of phospholipid membranes and the genetic code are archaic elements – biochemists generally use the term "universal elements" – which remained steadfast in the course of evolution (Morowitz, 1967).

Enzymic Line

Proteins are by far more versatile structures than the nucleic acids. They represent the molecular force driving the evolution of cellular function from its first beginning in an archaic coacervate system up to modern *Homo sapiens.* What makes proteins so unique? Much of the reason rests on proteins' ability to adjust structurally to a selective group of metal ions which they are able to extract from the environment and utilize for specific work assignments in the biological cell: osmotic regulation, structural conformation, acid-base catalysis, electron donor-acceptor reactions, messenger events, proton conduction processes, biomineralization, and excretion mechanisms. By means of metal ions, proteins may counteract environmental changes and adjust the whole organism to the new conditions of life.

Metal binding depends on chelation, multidentate binding, because monodentate and even bidentate ligand complexes lack the stability to retain metal ions. A favorable structural setting for chelation can be provided by the common folding pattern of proteins. For instance, a helical conformation will bring into juxtaposition on any one of its sides the first and the fourth or fifth amino acid side chain.

Suitable donor groups for coordinating metal ions are: imidazole (histidine), carboxylate (aspartate and glutamate), thioether (methionine) and thiolate (cysteine). It is remarkable to see how frequently such chelation shows up in the proteins of modern organisms (R.J.P. Williams, 1981). Thiolate donor centers external to the cell might have been extremely important during the primordial stage of coacervation but today thiolates are not readily available outside of the cells.

Use of only two such chelating centers from a single helix does not suffice to keep a metal ion at low concentration fixed to the protein. At least three binding sites are required. They may be provided by two linked helices or through the association with other peptide strands. Binding of metal ions can be weak or strong, and much of the versatility of proteins is founded on the difference in bonding strength. At some sites metal ions may be readily exchanged, whilst on others they are tightly bound. For example, Ca^{2+} and Mg^{2+} like to associate with highly mobile "acidic" proteins where they become positioned in open regions on the surfaces in the influence sphere of charged amino acid residues. In contrast, iron, copper and zinc, located in the catalytic center of many enzymes, are often deeply buried in pockets of generally hydrophobic proteins. It almost looks as if they are hiding to avoid being kicked out by competitive metal ions. However, the truth of the matter is that in the course of evolution these metals were found to be the best structural fit for a certain function. Survival of the fittest?

Magnesium, a good chelator, was chosen as the key metal of the genetic apparatus and conformation of RNA and DNA are Mg^{2+}-dependent. Calcium's biochemical role is best described as that of "a jack of all trades" serving as the major driving force in the evolution of the eukaryotic cell. Copper and iron chelates provide electronic connections in mitochondrial and thylakoid membranes which are at the control center of light and oxidative energy capture. Activation of water in condensation and hydrolysis reactions, or DNA or RNA synthesis are but a reflection of the 150 zinc enzymes (Vallee, 1978). Disposition of element binding sites in proteins, and classification of metal ions on the basis of their mobility have been tabulated by R.J.P. Williams (1981). His work can be recommended as a survey on the logic of metal ions in biochemical systems in general and proteins in particular.

The symbiotic use of a dozen metal ions by organisms requires that the environment must supply them in sufficient quantities. At the same time they must be offered in a suitable chemical make-up, for example, redox-inert or -active, uncomplexed or in the ground or excited state. Furthermore, they should allow for cellular feedback. Feedback between functional value and uptake or discrimination would arise through competition. A functionally less efficient metal might eventually be replaced by a more advantageous one. Evolving proteins can be stimulants for such a development, while in turn the refinement in metal ion usage can promote molecular alterations beneficial to the organism. In addition, environmental changes in terms of ionic milieu, PO_2, PCO_2, pH, or temperature might force organisms to "rethink" their metal ion strategy over and over again, and look for better devices to manage a new situation. This reminds me of our civilization, which for a number of reasons switched from stone to bronze, to iron, then to uranium technologies.

The early ocean prior to 4 billion years ago was in many respects different from today's sea, perhaps not so much in volume as in chemical composition and fluid dynamics. Since weathering must initially have

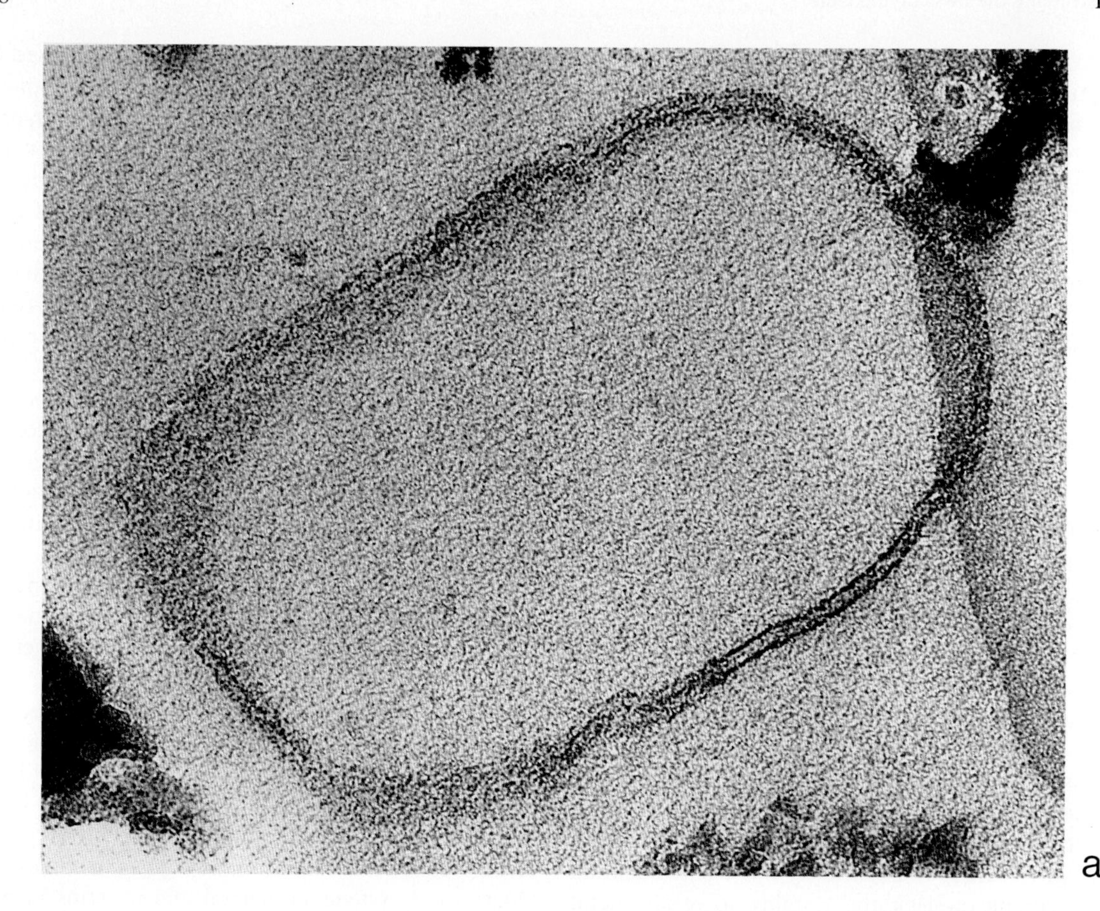

been orders of magnitude higher than at present, due to elevated PCO_2's, enforced volcanic activity and meteoritic impact, the dissolved and particulate matter in the sea must have reflected these events in kind and amount. However, in following James Hutton (1726–1797) that the "present is the key to the past", we can safely assume that the mechanism of sedimentation in world ocean remained essentially uniform through the ages. In modern ocean, planktonic organisms excrete massive amounts of low molecular weight sugars and polypeptides which stick to suspended clay-sized particles and force them to settle to the ocean's abyss. As a result, the sea is low in dissolved organics and suspended particles as well. The situation must have been the same in the prebiotic ocean, except that the low molecular weight peptides and sugars were "excreted" by minerals, causing suspended clay-sized particles to agglomerate, sink as macroflocs to the seafloor, and accumulate as organic-rich mud. Here is where the analogy ends.

Organic matter in modern marine sediments feeds benthic communities and micro-organisms and only a trifling amount of C_{org} survives oxidation and diagenesis. In contrast, the abiotic organic-rich mud, being sterile and kept under reducing condition, developed its own dynamics. Organic molecules, minerals, water and salt interacted and trillions of micelles and coacervates arose, refining and reshaping the cellular system (Box 11.10).

Fig. 11.43 a, b. Bacterial outer membranes naturally stained by heavy metals in the course of diagenesis of a few thousand year old sediment, Lake Tanganyika (water depth 1,300 m; anoxic conditions); **a** cellular outline of a bacterium; note the periodicity of the granular membrane surface and the disintegration of the membrane structure; **b** layer composed of "hash" of membrane pieces; identical patterns have been observed in other anoxic habitats, for instance the Black Sea (see Fig. 11.4)

Strongly reducing organic-rich sediments from the Black Sea and Lake Tanganyika, just about 1,000 years old, reflect some of the abiotic restructuring organic matter undergoes upon compaction and diagenesis (Fig. 11.43 a,b).

Membranes, protein crystals and other organic particulate material commonly found in anoxic sediments are usually stained by heavy metals. Under suitable conditions these metals, after becoming coordinated to functional elements in the respective organic template, can grow in epitaxial order, leading to minute crystals of sulfides, oxides and even native ele-

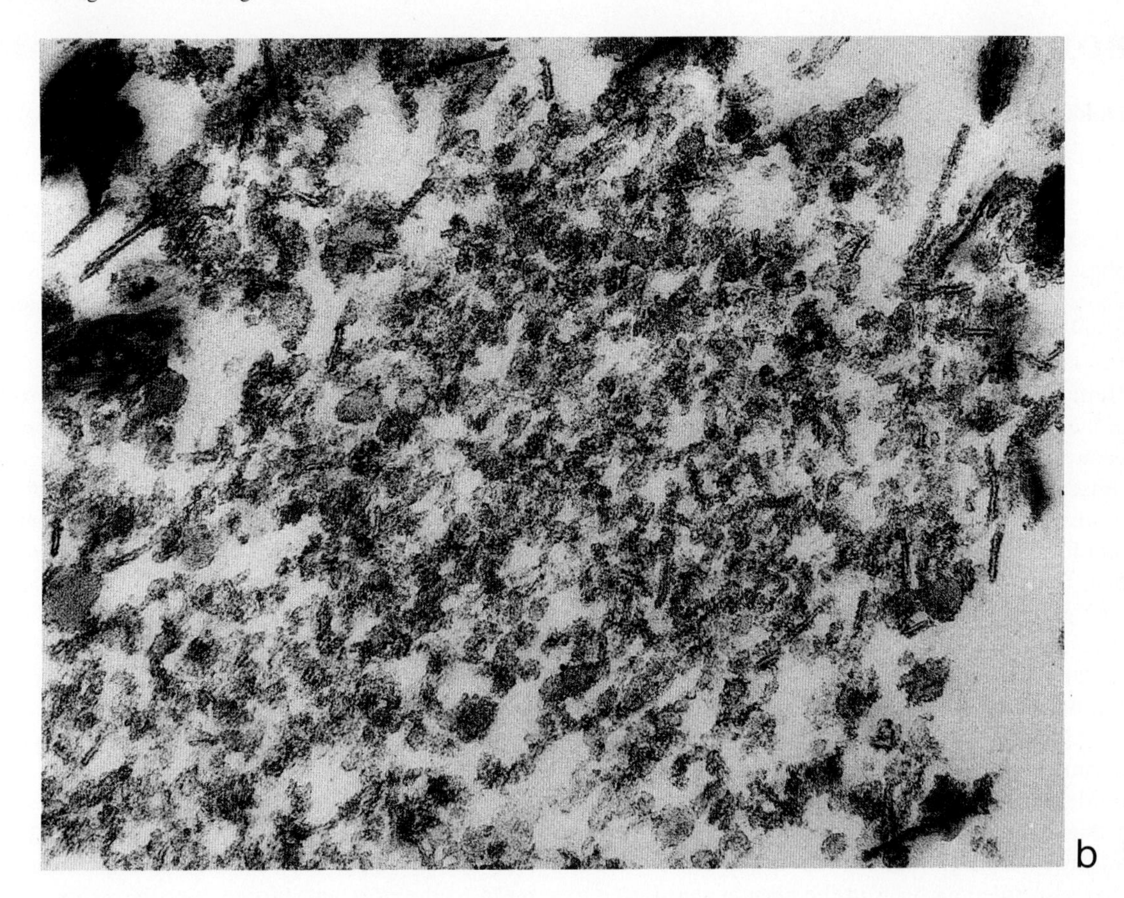

ments. However, because of its abundance, iron is the most prominent staining metal, and iron sulfides are the most common epitaxial products.

Iron is the major electron transfer agent in the biological cell. It is considered to have acted in this capacity in cellular systems from the start. A ferredoxin may have been one of the earliest proteins formed (Hall et al., 1971). This certainly argues for a reducing habitat for its point of origin. At such conditions the disulfides (FeS_2): pyrite and marcasite, the monosulfide (FeS): pyrrhotite, plus a variety of metastable iron sulfides: greigite (Fe_3S_4) and mackinavite (FeS_{1-x}), may form. Also siderite ($FeCO_3$), rhodochrosite ($MnCO_3$) and magnetite ($Fe^{2+}O{\cdot}Fe_2^{3+}O_3 = Fe_3O_4$) can come into existence under certain Eh-pH conditions (Fig. 11.44). Alabandite, MnS, requires a much lower

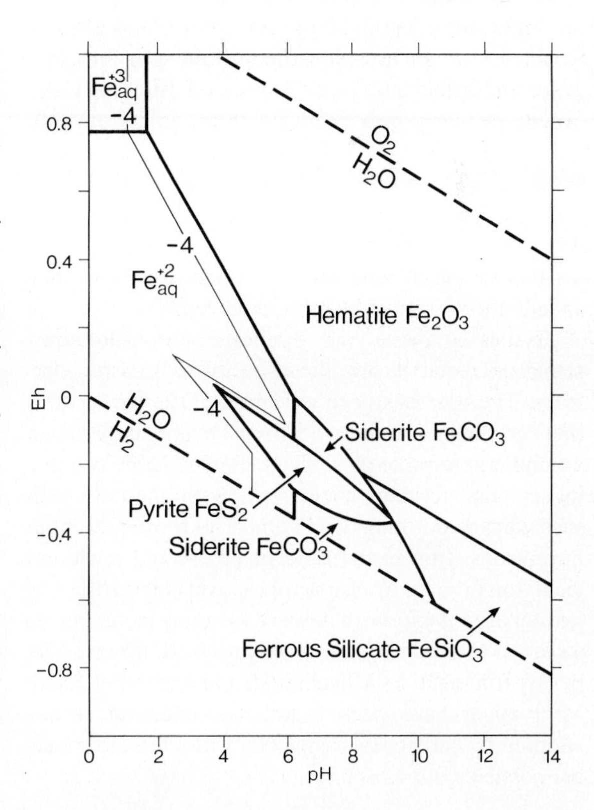

Fig. 11.44. Eh/pH stability relations among iron oxides, carbonates, sulfides and silicates at 25°C, and 1 atmosphere total pressure in the presence of water. Other conditions: Total CO_2= 10°; total sulfur = 10^{-6}; amorphous silica is present. At slight undersaturation with respect to amorphous silica, magnetite enters the ferrous silicate stability field (after Garrels, 1960)

BOX 11.10

Fluidized Beds (Yariv and Cross, 1979)

A mixture of solid particles suspended in an aqueous phase behaves in many ways like a fluid (Yariv and Cross, 1979). A recent mud having 90 percent water may be considered as in the state of suspension and has accordingly been named "fluid mud layer" or "fluffy layer". Such a system flows readily under the influence of a hydrostatic head. The overall bed flow behavior and the upward injection have considerable consequences for the transport of solids in post-depositional events. Mud volcanoes and mud diapirs that form by overburden pressure are examples of vertical transport. Mud diapirs that have ascended the stratigraphic column for several hundred meters are a common feature along the ocean margins of today.

Rates of sedimentation in modern oceanic basins can be as low as a few mm per 1,000 years (e.g., red clay facies, Central Pacific) but can reach a few decimeters per 1,000 years in productive geosynclines bordering the continents. In the prebiotic ocean sedimentation rates must have been globally at least an order of magnitude higher. Under the condition that suspended particles in the ancient sea started to aggregate about 4.2 billion years ago, in response to the build-up of dissolved organic matter and electrolytes, the former non-Newtonian fluid split into free water and a fluid mud layer. Expressed differently, the ocean, formerly in a fluidized state, in its entirety transformed into a thick section of fluffy material and salt water, i.e., the real ocean.

For the next few hundred million years, the marine fluidized beds became subjected to diagenesis, metamorphosis and eventually granitization. At the same time hydrothermal vents sprouting across the ocean's floor destabilized the fluidized beds, injected metal ion solutions, dissolved minerals, gases and heat into the system which led to the deposition of a wide spectrum of minerals, such as sulfides, oxides, carbonates, and silicates. Micelles and coacervates flourished in a warm clayey "inoculation medium" rich in essential mineral nutrients. Multitudes of bubbles that developed in this substrate could collect and refine the organic matter; they grew by coalescence with other bubbles along the same line as bubbles in liquids. The two most important features of a gas-fluidized system that distinguish it from a liquid system are the electric charge of the particles induced by mutual collisions and the large difference in the density of materials constituting the various phases (Yariv and Cross, 1979).

Density currents driven by the hydrostatic instability of a fluid mud layer and differences in temperature and salinity must have carried sediment from one basin to the next and turned over the strata. Please bear in mind that around 4 Ga the Earth surface was subjected to terminal cataclysm and associated volcanic events. By Isukasia time, the situation had normalized and life was able to take command of the environment.

Eh than found in sediments, but such conditions may arise in the vicinity of hydrothermal vents.

Crystals of pyrite and pyrrhotite are well-known semiconductors, having the capacity to transfer electrons. The idea has been advanced (Österberg, 1974) that crystalline electron carriers might have been around where proteins evolved. Polypeptides containing cysteine residues could react spontaneously with such carriers, forming bonds similar to those present in ferredoxins. Iron in the form of Fe^{2+} could react and yield iron proteins similar to rubredoxin (McCarthy and Lovenberg, 1968). With respect to manganese, its association with carboxyl functions provided, for instance, by aspartic acid, is a likelihood. On the other hand, MnS might have been a potential electron carrier adopted by early cellular systems. But alternate routes are possible (Box 11.11).

From a brief glance at the principal metal ions involved in enzymes it is interesting to see that their majority bears a striking affinity to either FeS_2 or FeS. Pyrite can readily accommodate Ni^{2+}, Co^{2+}, and Mo^{2+} in solid solution. FeS and ZnS exhibit a structural likeness to one another which is temperature-controlled. Noteworthy is the presence of sphalerite crystals – having FeS in solid solutions – in tiny organic spheres suspended in Lake Kivu. Zinc (Zn^{2+}) can be a very effective Lewis acid (electron-pair acceptor) and in $Zn^{2+}OH$ it generates a very good anionic attacking base from the major substrate of hydrolytic reactions, H_2O (Vallee, 1978; R.J.P. Williams, 1981).

The way proteins bind to nucleic acids is highly significant for understanding processes of gene expression and replication. Recently J. Miller et al. (1985) and J.M. Berg (1986) elucidated the relationship between

BOX 11.11

Manganese in the Sea

In 1891, Sir John Murray and R.F. Renard reported the discovery of manganese and iron oxide nodules on the deep ocean floor during the famous H.M.S. CHALLENGER expedition in 1873–76. Meanwhile, ferromanganese nodules are known from all major oceans (e.g., McKelvey, 1980; Halbach and Fellerer, 1980; Emerson et al., 1982; Emerson, 1984; Aplin and Cronan, 1984; Baturin, 1986; U. v. Stackelberg, 1986; U. v. Stackelberg and Marching, 1987). Growth rate proves to be roughly one millimeter per million years, which translates into an annual accretion of a few atomic layers on top of the MnO_2 deposits at a water depth of a few thousand meters.

The special attraction of such nodules lies in their content of other metals, notably nickel, copper, cobalt, molybdenum, vanadium, and zinc. The bulk of metal ions stems from infalling primary producers which decompose at depth and release their metals. Exceptions are iron and cobalt, which are chiefly of volcanic provenance; they are swept to their final burial place in the form of colloidal iron hydroxide suspensions, where they become epitaxially associated with MnO_2.

The most substantial concentration of nodules lies in the northeastern equatorial Pacific between the Clarion and Clipperton Fracture Zones. The size of this area is about 2.5 million square kilometers, and nodule concentration for half of this area averages 12 kg m^{-2}. On the assumption that 20 percent of these nodules can be harvested, about 2 billion metric tons (dry weight) of potentially recoverable nodules would contain: 25% manganese, 1.3% nickel, 1% copper, 0.22% cobalt, and 0.05% molybdenum. The global figure for all seabed nodules could be almost ten times that number. Even though expectation runs high for mining this treasure, the Law of the Sea Treaty (Ross and Knaus, 1982), and environmental hazards will probably keep these polymetallic nodules in their original place of deposition.

MnO_2 minerals are insoluble in oxic waters. Since two-valent manganese is an essential micronutrient for organisms in general and a prerequisite for the oxidation of water in photosynthesis in particular, availability of Mn^{2+} is vital. Some Mn^{2+} may be furnished to the environment and stay there temporarily due to slow oxidation kinetics. However, the abundance of reduced manganese species in marine surface waters demands a more potential source for Mn^{2+}. Sunda et al. (1983) demonstrated that dissolution of $^{54}MnO_{1.8}$ in coastal, Gulf Stream, and Sargasso sea water is strongly stimulated by light and the presence of natural humic acids. Photoactivated reduction of manganese oxides to Mn^{2+} by dissolved organic compounds has considerable impact on life by keeping this ionic species in the euphotic zone. The low Mn^{2+} content in upwelling waters may in part be responsible for the sluggish response of aquatic life to the general fertilization of surface water. Photoactivated reduction of other transition metals (e.g., Cu^{2+}, Fe^{3+}) by certain organic ligands and humic acids is documented too. Fluctuations in the availability of such metal ions in the sea may thus limit or enhance primary productivity. For a comprehensive treatment on the behavior of metal ions in aqueous solutions see Stumm and Morgan (1981).

certain metal ions, notably zinc, copper and iron, and a number of nucleic acid-binding and gene-regulating proteins. In all instances it was the coordination of the above metal ions to sulfur-rich domains in proteins, for instance Cys-X_{2-5}-Cys-X_{12}-His-X_{2-3}-His, where X may be any amino acid. It was proposed that the repeated units represent Zn^{2+}-binding domains that interact with nucleic acids via hydrophilic residues. J.M. Berg (1986) has demonstrated similar sequences in five classes of proteins involved in genetic regulations, and a possible structure for a metal-binding domain is illustrated in Fig. 11.45. It appears that a variety of nucleic acid-binding proteins contain sequences with four Cys or His residues arranged in tetrahedral coordination having a heavy metal ion in central position. They are found in regions of the proteins implied as being functionally significant.

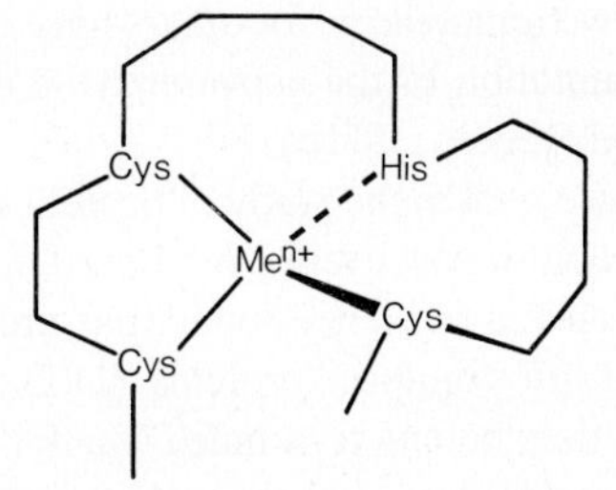

Fig. 11.45. Possible structure for metal-binding sulfur-rich domains in proteins (Berg, 1986)

Non-metals do, in fact, also participate in enzymic activity and the importance of sulfur has already been stressed. Selenium, like sulfur, can be retained by a single bond, and its adoption in biochemical systems rests principally on its close structural and functional relationship to sulfur. This by itself suggests that selenium has been with the organisms since archaic times. Since selenium's redox chemistry is largely two electron (atom) chemistry typical of non-metals, it does not have the disadvantage of redox-active metals that readily generate free radicals, for instance from dioxygen (R.J.P. Williams, 1981).

Amino acids on their own determine enzymic regulation by virtue of their side chains. Serine, for instance, is an active site in proteolytic enzymes, including esterases and in enzymes responsible for the transfer of phosphate groups. In the history of the biosciences it is rather ironical that the recognition of serine as an active part of hydrolases was a product of wartime research. During the Second World war, extremely toxic nerve gases (organophosphoesters) were developed. In England: the gas diisopropylfluorophosphate (DEP); lethal dose for humans 500 mg. In Germany: the nerve gas Tabun with a toxicity ten times that of DFP; also Sarin, which is even more deadly. Initially, organophosphoesters were used as pesticides and insecticides. The neurotoxic action has to be seen in conjunction with cholinesterase activity, where serine is instrumental in attaching the phosphate group. DFP also reacts with esterases and proteases and is widely used in biochemical research for the inhibition of enzyme activity through phosphorylation of serine residues. Serine residues appear principally in the active center of hydrolases, such as trypsin, chymotrypsin, elastase, thrombin, and subtilisin, or in phosphatases and cholinesterases. Mention should also be made of the significance of the amino acid side chains of argine and lysine in metabolic regulation mechanisms even though they are not part of the active center of enzymes. Mutations which would replace these two amino acids for others have the same effect as the mutation of the active enzyme center itself (Matheja and Degens, 1971b).

The ultimate goal in the study of protein evolution is a reconstruction of past events that have led to the vast pool of proteins in existence today that are at least in the order of 10^6 "unique" proteins (Doolittle, 1981). Fortunately, they belong to a much smaller number of classes or superfamilies – less than a thousand according to Dayhoff (1972). The problem facing reduction to protein families and superfamilies is carefully outlined in Doolittle (1981). For the time being it seems to be that the overwhelming majority of extant proteins – and certainly most enzymes – have evolved from a very small body of archetypical proteins. As is the case with genes, it is much simpler to duplicate and modify proteins genetically than to assemble them from scratch. Analogously, the metal ions involved in the enzymic machinery are ancestral elements that have only been refined in the course of time in response to the challenge of the environment.

Metal ions in biological systems are the structure-makers and -breakers, and at the same time carriers of information in the metabolic and genetic apparatus. Coordinated to PO_4 units they (i) transform phospholipid membranes into dynamic molecular sieves regulating the flow of matter into and out of the organisms (or between cellular compartments), (ii) control conformations of RNA and DNA, and (iii) are responsible for phosphate transfer. In association with proteins they maintain the daily chores of cellular activities and keep life prepared for the eventualities of environmental change. In following Lovelock (1979), Lovelock and Margulis (1973) and Schneider (1986), organisms appear to keep Earth – Gaia – a habitable planet by virtue of their cooperation and diligence (Box 11.12). Let's follow suit!

Hallelujah: Life is Born!

From Coacervate to Progenote

Coacervates, utilizing energy released by the oxidation of organic compounds during a phosphorylation cycle, were the primer of life. They "fed" mainly on sugars and their derivatives which must have been the most copious organic substrate available. After many trials and errors, the first heterotrophic biological cell, the *progenote,* capable of sustaining a primitive physiology and at the same time proficient in reproducing, emerged. Heterotrophic metabolism is emphasized by Oparin (1953), who arrived at this conclusion through a comparison of heterotrophic and autotrophic cells:

"The primitive metabolism of energy was entirely anaerobic and dependent on the interaction of organic substances with molecules of water."

Genetic considerations led Horowitz (1945) to postulate heterotrophic metabolism:

"The first living entity was a completely heterotrophic unit, reproducing itself at the expense of prefabricated organic molecules in the environment."

Glycolysis is used by Haldane (1954) as evidence that the anaerobic metabolism is the most ancestral.

The triggering event transforming a single coacervate to the progenote was the merger of the genetic line with

BOX 11.12

Weltgeist – Weltschmerz – Weltseele

Concepts and ideas like Gaia have been with us for almost a century. It was Vladimir Ivanovitsch Vernadskij (1863–1945) who first recognized life as the "Lord of Planet Earth" and founded the field of biogeochemistry. In his book "The Living Substance in the Earth's Crust and its Geochemical Role", which he started to write in 1916, he saw organisms as the main driving force in the cycling of living and mineral matter ever since life took command of crustal and environmental evolution. More than 1,000 references to works written in all European languages pay tribute to Vernadskij's linguistic skills (in striking contrast, by the way, to many texts written in English). He even envisioned the influence of human activities in restructuring Planet Earth, and by introducing the "noosphere", that is the sphere of the mind, he provided a new dimension to the environmental sciences. However, Vernadskij underestimated the acceleration and severity of human-induced changes and not everybody may share his optimism concerning the future development of our Earth, "We certainly do not witness a crisis which only horrifies weak souls, but a revolution of human thought occurring just once in a thousand years; we see the success of the natural sciences in a way never experienced before. By standing on the threshold of that revolution and looking into the opening future, we should consider ourselves quite fortunate in being invited to shape the future."

Gaia concerns Planet Earth only. Yet the entire Universe may be viewed as a biogeochemical entity in case we conform with Vernadskij that the sphere of mind – the noosphere – is simply the ultimate outgrowth of a universal biogeochemical evolution now permeating space and time. Actually, many a scholar before Vernadskij, for instance, Aristotle, Plato, Zenon, Seneca, Marc Aurel, St. Paul, Bruno, Goethe, Herder, Hegel and others, have thought along this line. What "Weltseele" to Plato, "Stoa" to Zenon, "Weltgeist" to Goethe, was God to St. Paul. In his book "Von der Weltseele" (1798), Friedrich Wilhelm Schelling (1775–1854) defined Weltseele as "nature in its entirety linked (knitted) together to an organism."

---

the two already cooperating metabolic and enzymic lines. By definition, the origin of the system of metabolism is independent of that of the genetic line, because the performance of the genetic transcription apparatus requires the assistance of a functional metabolism.

Students of biochemistry might argue against this proposition and point to the autocatalytic capacity of some RNA' s, or the propagation of viruses and viroids, all of which proceeds without their own functionally alive metabolism. However, in the past and present, such schemes had to adopt metabolically intact partners: a coacervate or a living cell.

Proteins and phospholipid membranes are the tools and driving force in metabolic processes, maintained by organic substrates available in the sediment and its energy utilization through phosphates. Considering the organization of this system, the incident of photosynthesis has only peripheral significance, although, with respect to phylogeny, the onset of photosynthetic processes is an essential turning point of evolution because of the production of the correct biochemicals and of molecular oxygen. It is of interest to note that the reductive pentose phosphate cycle is not restricted to photosynthesizing organisms. Certain anaerobic bacteria which utilize ammonia as source of energy also follow this cycle (Campbell et al., 1966).

Although no concepts have come up regarding the coupling of the genetic and metabolic lines, empirical observations leave no doubt that this integration process represents the decisive step in the formation of the progenote; a novel type of information storage device capable of information output through peptide synthesis started to act upon the metabolic system. The two properties: (i) information storage and (ii) information output controlled the metabolic reaction cycle by maintaining a precise sequence of physiological events.

The Rise of O_2

Free oxygen in today's air and sea stems from oxygenic photosynthesis. It would take all green plants about 2,000 years to generate one volume of air O_2. Of course some of it is constantly lost through oxidation in soils, water, air and sediments. However, sedimentation buries C_{org} which otherwise would respire, and lower air oxygen content. This type of biogeochemical cycling of oxygen has probably been with us for at least the past 2 billion years, starting with the deposition of banded iron formation in early proterozoic time (Cloud, 1976). It is commonly assumed that the present atmospheric O_2 level was reached some time during the Paleozoic and has stayed at around 20 volume percent ever since. In contrast, early Precambrian atmosphere has long been thought to be anoxic and highly reducing in nature, but strong evidence is gathering which casts serious doubts on the anoxic model (e.g., Towe, 1970, 1978, 1981, 1983, 1985; Clemmey and Badham, 1982; Carver, 1981).

Evolution of O_2 in hydrosphere and atmosphere has implications for the origin and development of the nucleated cell, since even the few anaerobic eukaryotes that presently exist require molecular oxygen for synthesis of some essential biochemicals (Raff and Mahler, 1972). Basically, however, all eukaryotic cells are aerobic and it is most likely that they evolved from an aerobic protoeukaryote stem cell. Critical concentration of oxygen for metazoan oxidative metabolism is placed somewhere near 3 percent.

In following the anoxic model, eukaryotic organisms are latecomers (<2 Ga); whilst by accepting the oxic model, no such restrictions are placed, and the eukaryotic cell could well be more than 4 billion years of age.

Oxidized iron deposits are reported from the oldest known sediments, the 3.8 Ga-old Isukasia formation of Greenland (Moorbath et al., 1973). Cloud (1976) revised his earlier model and explained these events on the grounds that O_2-generating photosynthesis might have been present much carlier than originally thought. But this is just one possible explanation. An alternative route involves the direct photodissociation of water vapor by solar UV photons, an idea originally advanced by Berkner and Marshall (1965a, b) but later dismissed as being too inefficient. At present, stratospheric water vapor content of $\approx 1.6 \times 10^8$ H atoms cm^{-2} sec^{-1} (Carver, 1981) is less than the current hydrogen emission rate from volcanoes estimated at $\approx 2.5 \times 10^8$ H atoms cm^{-2} sec^{-1} (Kasting et al., 1979). Thus photolysis today represents a negligible source of O_2. Under conditions of much higher water vapor content in the early atmosphere, due to a more temperate environment, atmospheric oxygen might have been present in significant quantities (Hunten and Donahue, 1976). According to a scenario by Walker (1977), a wet warm prebiotic atmosphere is potentially capable of releasing one volume of present air O_2 in about 30,000 years via the familiar Berkner-Marshall pathway. Although oxidation of surface materials is expected to be at least ten times higher than current values, still a considerable fraction could conservatively build up. This might have been sufficient to produce even a biologically effective Precambrian ozone shield, should O_2 exceed 0.5 percent (e.g., Carver, 1981; Kasting and Walker, 1981; see also Canuto, 1983; Canuto et al. 1982). But there are other sources for abiotically generated O_2 which have been largely overlooked.

In the lithosphere, oxygen fugacity controls jointly with temperature, pressure and composition the petrogenesis of mantle-derived magmas as well as the geochemistry of mantle and crustal fluids and gases (reviewed in Mathez, 1984, and Schrön, 1985), e.g.:

$$2\,Fe_3O_4 \text{ (in spinel)} + 6\,FeSiO_3 \text{ (in orthopyroxene)} = 6\,Fe_2SiO_4 \text{ (in olivine)} + O_2.$$

This O_2 can also be instrumental in regulating the C–H–O–S system. Mattioli and Wood (1986) succeeded in demonstrating that the dominant gaseous species in the shallow upper mantle are CO_2 and H_2O rather than CH_4 and H_2. The specific effect of degassing of S-species on oxidation state is not easy to predict. This is because the manner in which S dissolves in melts is a function of oxygen fugacity (Mathez, 1984). Even more speculative is the situation with regard to volcanic emanations of molecular O_2.

It is generally believed that volcanic activity is no source for O_2 in the atmosphere. The presence of trace quantities of O_2 in volcanic gases (e.g., Menyailov and Nikitina, 1980) is explained as atmospheric contaminations. Förster (1981a, b), however, found in experiments and in the field that molten basalt releases gases having a few percent lava-generated O_2 at temperatures exceeding 1,200°C. This O_2 will oxidize the melt, thereby raising temperatures to 1,300°C. The "foaming" of melts is attributed to liberated O_2:

$$6\,NaFeSi_2O_6 \rightarrow 2\,Fe_3O_4 + 3\,Na_2Si_2O_5 + 6\,SiO_2 + 0.5\,O_2.$$

An alternative explanation is a restructuring of the silicate melt as proposed by Mysen et al. (1980):

$$8\,Si_2O_6^{4-} + 4\,FeO_6^{9-} \rightarrow 4\,SiO_4^{4-} + 6\,Si_2O_5^{2-} + 4\,FeO_6^{10-} + O_2.$$

The discovery of molecular O_2 in high-temperature silicate and oxide melts may explain the horizontal movement of ignimbrites ("foaming volcanic ashes") over wide plains. The famous eruption of Mt. Pelée, in Martinique (Lesser Antilles), in 1902, killing all the islanders but one, a prisoner sitting in the dungeon, was such an event.

Extraction of gases from volcanic glass (obsidian, Pélé's Hair; Pélé's Tears) had the following bulk composition: 95% H_2O, 3.5% CO_2, 1% SO_2 and traces of CO, N_2, O_2, S_2 and CH_4 (Muenow, 1973). Heide (in Förster, 1981b) by heating rhyolitic obsidian to a temperature of 1,200°C, extracted pure O_2 gas in an amount of 0.25 to 1 percent of the total rock. The much higher formation temperatures of komatiitic lava and the wide abundance of palagonite could indeed foster the release of free oxygen.

Molecular oxygen could also be derived by meteoritic shock metamorphism of crystalline modifications of TiO_2 which are quite abundant on Earth. For instance anatase, used as a white pigment, is known to turn bluish when milled too long (Weyl, 1951). Chemically this can be explained by the ability of Ti^{4+} to trap an electron which is transferred from surface O^{2-} ions to the coordinatively unsaturated cation present at the mineral surface of anatase:

$$2\,TiO_2 \xrightarrow{\text{shock}} Ti_2O_3 + 0.5\,O_2.$$

The chemistry causing Venus night air glow (G.M. Lawrence et al., 1977) might also be applicable to the early Precambrian atmosphere and result in production of O_2.

Molecular oxygen has long been thought of as being lethal to anaerobic organisms. In an edifying paper Towe (1978) has drawn together relevant data from the literature which unambiguously demonstrate that oxygen-mediating enzymes are not the exclusive domain of aerobic organisms but are even found in a number of primitive obligate anaerobes. For instance, superoxide dismutase which provides protection against superoxide free radical toxicity, or ribulose-1,5-bisphosphatase carboxylase/oxidase (RuBisCO) the cardinal enzyme of photosynthesis both react with molecular oxygen. This supports the inference that such enzyme systems are archaic in nature and not just a late invention of life. Their presence in anaerobic bacteria can be seen as protective measures against transient oxygen toxicity.

As far as the origin of life is concerned, it is of no regard whether the early atmosphere had been reducing or oxidizing, once we understand that abiotic organic molecules can be generated both ways. However, the moment we are after the origin of eukaryotes, this is no longer the case. At present, biochemical and geological arguments strongly support the view that atmospheric photodissociation rather than photosynthesis was the primary contributor of free oxygen to the earlier Precambrian environments (Towe, 1978). Volcanism and meteoritic impacts are also likely candidates for additional O_2. The question of at what time oxygenic photosynthesis supplemented the declining abiotic O_2 contribution, or the one related to cyclic or periodic pulsations of atmospheric molecular oxygen in the course of Earth history, still remains to be answered. There is a yearning for a simple solution, but the complexity of the biogeochemical system is against it. Only one fact seems to be true: O_2 levels never dropped below 3 percent PAL from the instant eukaryotes marched in.

Four billion years ago the Earth could have readily provided the environmental niches we know of today and which give shelter to aerobic and anaerobic life in hydrothermal springs, rivers, lakes and oceans from: hot to cold(–2 to 100°C), acidic to alkaline (pH $\approx$1 to $\approx$11), fresh water to salty (<100 ppm to $\approx$300‰), deep to shallow, and oxidizing to reducing. The processes to achieve such results are manifold. For example, haloclines and thermoclines were a convenient means to stratify a water column into oxic/anoxic, warm/cold or salty/freshwater layers. Acidic hydrothermal vents could create unique metal ion- and nutrient-rich habitats in a generally alkaline sea. Formation of continents led to rivers, lakes and hydrothermal springs. Impact craters generated sizeable pools on land, or created morphological deeps beneath the sea. Into those density and turbidity currents could sweep fluidized beds. Numerous physical and chemical changes in air and water developed "pressure gradients" which forced proteins to adjust, and the metabolic and genetic apparatus of an organism to become adapted or else perish.

Taxonomic Categories

Taxonomy is a "scheme for arranging together those living objects which are most alike, and for separating those which are most unlike ... and that community of descent – the one known cause of close similarity in organic beings – is the bond, which is partially revealed to us by our classifications" (Darwin, 1859). The final outcome is a hierachical system which is often shown in the form of a phylogenetic tree, where each taxon (= taxonomic group) constitutes a unified evolutionary group.

It is customary to distinguish between two types of cells, the prokaryotic and the eukaryotic ones. Prokaryotes have no cell nucleus, while the eukaryotic cell has a well-developed nucleus. Conventional schools of biology are accustomed to viewing the relatively simple unicellular Monera (which are prokaryotic) as the evolutionary stepping stone towards the more complex unicellular Protista (which are eukaryotic), from which subsequently all multicellular organisms arose.

A revolution is occurring in taxonomy since molecular sequencing techniques have advanced to a point to permit direct measurements of genealogical relationships. Eukaryotic phylogenies have been constructed on the basis of, for instance, cytochrome *c* sequences and nucleic acid sequence data (e.g., Schwartz and Dayhoff, 1978). These "evolutionary fingerprints" have worked quite well in outlining eukaryotic phylogenies, but there is doubt as to whether the techniques can be meaningfully extended into the domain of the prokaryotes.

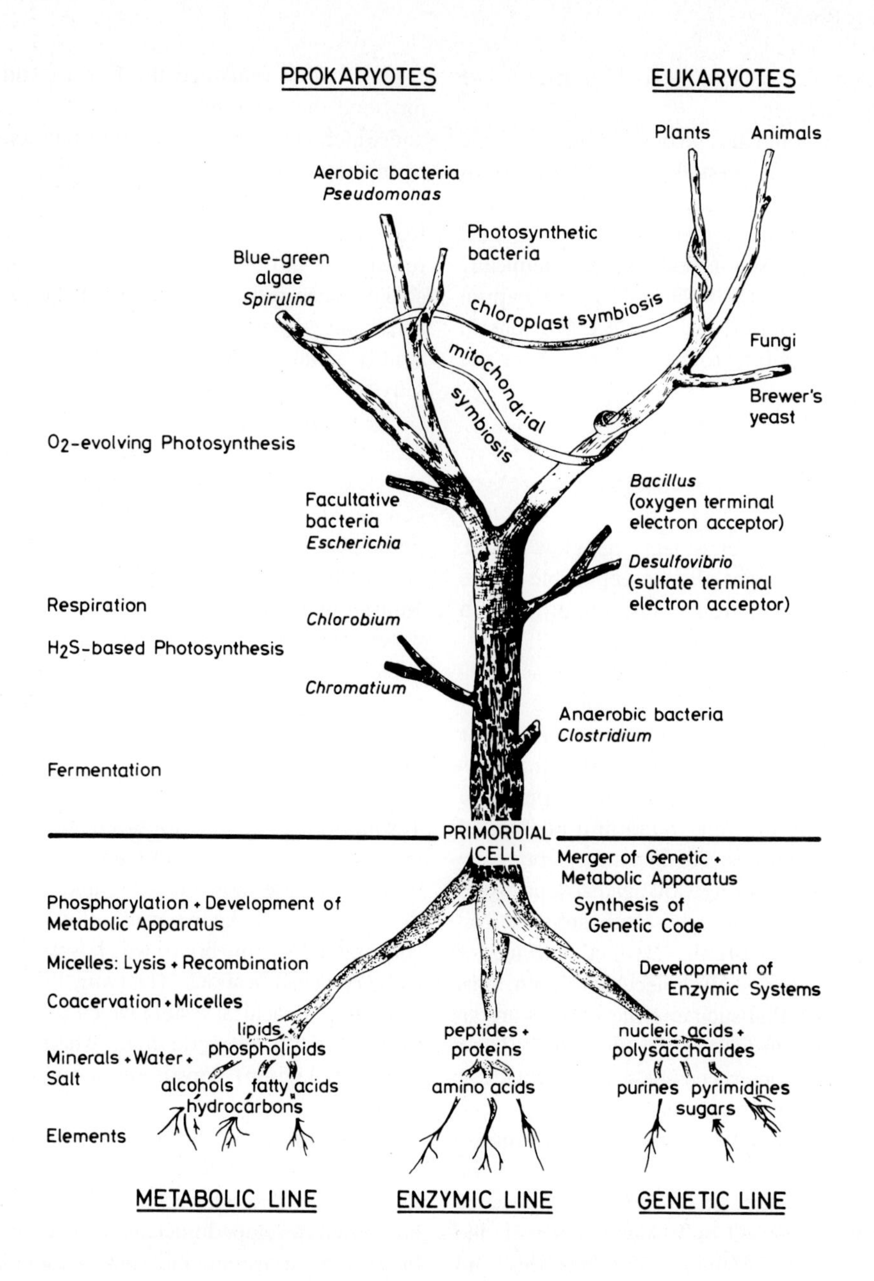

Fig. 11.46. Composite evolutionary tree (schematic) summarizing the principal steps in chemical and biological evolution. The sequence of events depicted for chemical evolution follows from the discussion in the text; the biological scheme is based on data by Schwartz and Dayhoff (1978). The upward progression from anaerobic to facultative to aerobic forms is indicated in the shading pattern. Mitochondrial and chloroplast invasions are roughly drawn between points of suggested origin and uptake, respectively

A perspective view derived from protein and nucleic acid sequence data has been constructed in the form of a "composite evolutionary tree", thus linking the eukaryotes directly to one kingdom of the prokaryotes, the true bacteria or eubacteria. The already familiar relationships between metabolic, enzymic and genetic lines leading to the merger of the genetic and metabolic apparatus are depicted as roots anchored at their base in the interstellar matter (Fig. 11.46). Although it is tempting to use such a tree in order to draw conclusions with respect to the sequence of events such as the start of: (i) photosystem I, (ii) photosystem II, (iii) respiration, (iv) sulfate reduction etc., we will soon discover that a careful analysis of the data does not warrant such inferences.

At present there is wide agreement that universally applicable sequencing techniques should involve a properly constrained molecule, which is universally distributed, exhibits constancy of function, and appears to change in sequence very slowly, that is far more slowly than most proteins (Woese and Fox, 1977; G.E. Fox

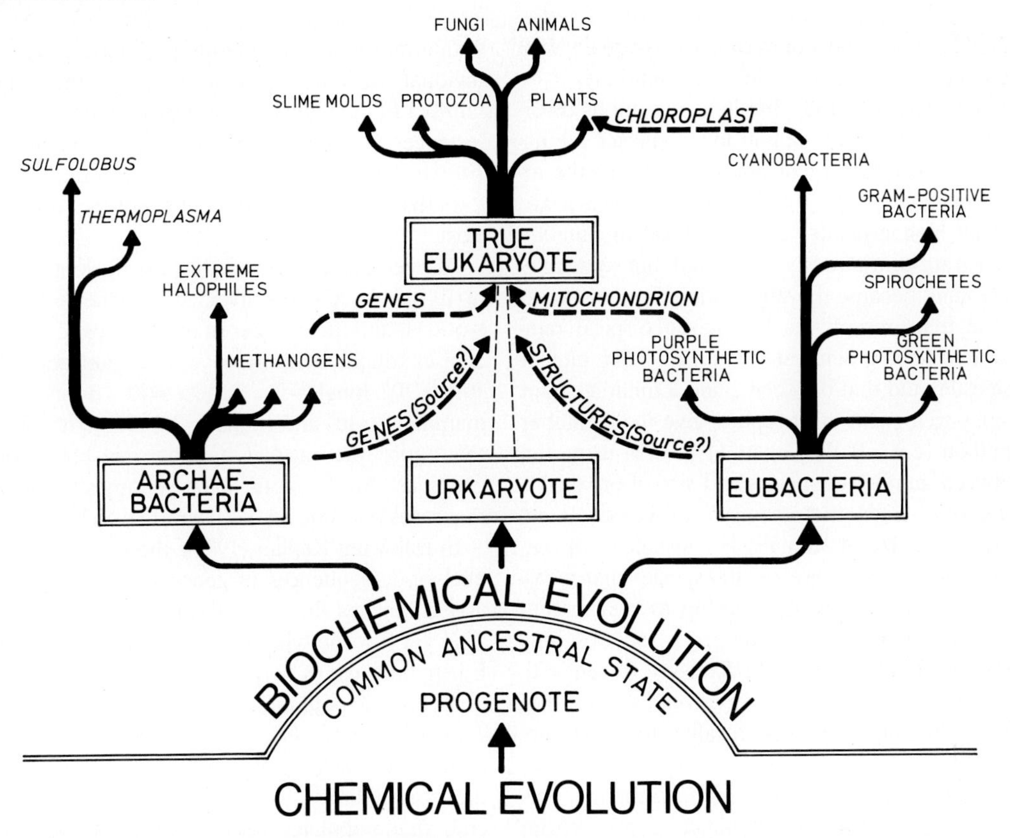

Fig. 11.47. Biochemical evolution starting from a common ancestral state: The progenote (Fox et al., 1980; highly schematic)

et al., 1980; Noller and Woese, 1981). Such molecules are the ribosomal RNA' s (rRNA' s). Their only disadvantage with respect to some proteins, for example, is the slowness of sequence change when the exact order of branching between closely related species is of interest. The 16S ribosomal RNA sequence has been found extremely helpful to establish phylogenetic relationships among the prokaryotes. Fox et al. (1980) were able to show clearly that two primary lines of descent exist as free-living forms: the true bacteria (eubacteria) and the archaebacteria, whereas a third line is represented in the eukaryotic 18S rRNA. All data indicate that the three lines of descent diverged before the level of complexity usually associated with the prokaryotic cell was reached. This level of organization has been termed the "progenote" (Fig. 11.47).

This viewpoint is not universally shared. Lake et al. (1985), using three-dimensional ribosome structure as a probe of evolutionary divergences, proposed that eubacteria and halobacteria are more closely related to each other than they are to any other known organism. They concluded that all extant photosynthetic cells are descendants from a single common ancestor that possessed a primeval photosynthetic mechanism. Cavalier-Smith (1986) rightfully straightens out some of the inconsistencies in terminology used by Woese and Fox (1977) in that one should speak of three major lineages rather than three primary kingdoms (or urkingdoms) when referring to the eubacterial, archaebacterial and eukaryotic branches of the tree of life. With respect to the work of Lake et al. (1985), Cavalier-Smith (1986) questions the validity of their arguments. For more details on the world of bacteria see Starr et al. (1981).

In summing up, organisms fall into three major lineages: archaebacteria, eubacteria, and eukaryota, all of which are independently derived from a common progenote. They share the same genetic code and operate along similar metabolic pathways. Yet genetic control mechanisms differ in numerous ways, and cell walls diverge widely in structure and composition. For instance, modern eukaryotes are set apart from bacteria by having a nuclear membrane enveloping the chromosomes, as well as about a dozen organelles such as mitochondria, chloroplasts, and the Golgi which carry out different sets of functions. Furthermore, molecular oxygen is a prerequisite of eukaryotic life, whereas numerous bacteria are known which live only in an anaerobic environment, while others are restricted to oxygenated habitats, and a third category is equally well at home under reducing and oxidizing conditions.

A characteristic feature of life is its enormous diversity. At present about 1.4 million living species are accounted for. Most prominent among them are the arthropods, mainly insects, totalling 800,000, which are followed by higher plants, about 250,000 in number. New species are continuously discovered, but some may never be found, because they are hidden, for instance in one of the treasures of this Earth, the tropical rainforest, which is being deforested at an alarming rate. It has been estimated that between 2 and 4 million extant species may well exist; most reports give their number as $\approx$3 million (e.g., R.W. Kaplan, 1985). In using the ratio between number of species and size of organisms one commonly finds that diversity progressively increases with the size of animals. For instance, for one species 1 meter in size there are 100 species that measure 10 cm. Extending this relationship to smaller and smaller-sized organisms, the number of species may reach at least 10 million (May, 1978, 1986). Recent estimates, based on current rates of discovery, suggest that as many as 40 million species are alive today (Raup, 1986).

Individuals of a species allow genetic exchange. This is the prime criterion for distinguishing one species from the next. In a number of cases, however, this test is not applicable. For instance, fossils are principally classified on the basis of morphology only, which by biological standards is a poor measure for speciation. Take the example of the famous land snail *Cerion,* believed to be divided into some 600 species. Closer examination (Gould, 1978) reveals, however, that this is nearly all invalid because any two forms interbreed when their geographical ranges overlap. This argumentation is also valid for fossil genera, families etc. (Raup, 1976). According to Flügel (1982), about 130,000 fossil species, mostly marine invertebrates, is all there are in the stratigraphic column, the history book of life. It is remarkable that paleontologists were able to come up with a phylogenetic tree, because only those species that happened to leave behind shells, bones, teeth, imprints, tracks or other forms of fossilized material became mosaic stones for the tree's construction; 130,000 fossil species – many of which may not even be true species by modern biological standards – is just a meager count of all the life forms that must have lived over the past 4 billion years. Estimates on the total number of extinct species run from 50 to 5,000 million (Hardin, 1966), 600 million (Mayr, 1975), and one to 10 billion (R.W. Kaplan, 1985). So just one in 10,000 species has "survived" as fossils in the sands of time.

The number of different genotypes exceeds by far the number of species, especially in higher organisms. Here sexual reproduction and diploidy cause nearly every individual to differ genetically from the other. The number of individuals of a species can be enormous, for instance in insects up to 10^{16}, or in the shrimp *Euphausia* about 10^{20} (Dobzhansky et al., 1977). It follows that at least 10^{22} different genotypes may presently exist.

A species is characterized by the nucleotide sequence of its genes. On the average, a gene holds 10^3 nucleotides, and the number of nucleotides of a genome increases roughly with the level of organization: bacteria 10^6 to 10^7, fungi $\approx 10^7$, insects $\approx 10^8$, fishes almost 10^9, mammals $>10^9$ and humans 2.9×10^9. It appears that the entire present-day biosphere is based on DNA genomes. Autonomous RNA genomes are found only in viruses parasitic on cells with DNA chromosomes.

In following Kaplan (1985), the number of potential nucleotide sequences of genomic size that could have arisen reaches the "unbelievable" figure of $10^{60,000,000}$. And yet the one billion or so species that ever lived on Earth made use of only an infinitesimal small fraction of that genomic potential.

Species have different life-times. Some have remained unchanged, morphologically at least, for hundreds of millions of years, like *Limulus,* the horseshoe crab that populates tidal ponds off Cape Cod, Massachusetts, and that was equally at home in some lagoons of the Paleozoic Sea. Others may come and go in a matter of thousands of years, whilst others appear suddenly to develop in a crater lake, a land-locked sea or in a lake behind a human-made dam. Charles Darwin would have been surprised, had he visited Lake Tanganyika and seen the multitude of fishes endemic to the Lake; Cichlidae alone number not less than 133 species from 38 genera, and snails show a striking resemblance to certain marine Jurassic fossils (see Degens et al., 1971).

Actually the significance of index fossils is based on a sudden and global appearance and its equally rapid extinction after a lifetime of perhaps 100,000 years. Rates of biological evolution, so it seems, can vary over a few orders of magnitude. It is even thinkable that some modern bacterial species are living fossils that kept their genome over billions of years unchanged.

A major problem facing evolution of species has to do with the second law of thermodynamics which demands that systems should decay through time, thus giving less and not more order. Although there is essentially no process within the biological cell that is not fully consistent with general theories of physics and chemistry, still the overall operation principles followed by life and the inanimate world are figuratively light years apart. Cosmic evolution is a "simple" predictable process starting with energy and matter and ending in chaos: fire or a deep freeze, as you please. Biological evolution is a complex unpredictable phenomenon which can best be defined in the context of irreversible thermodynamics as originally proposed by Prigogine (1979). Let me emphasize that such a non-equilibrium system is not unique for life and biological evolution. Air, water and earth, too, will never revert to their ancestral state but continue to evolve until the free energy runs out. Even cosmic events such as supernovae, radiation, or meteoritic impacts are an integral part of the irreversible universal scheme.

Biological evolution is just one of the multifarious outgrowths of the air-sea-earth-life system. As there are tectonic pulses, an earthquake or a volcanic eruption, there are organismic pulses which we may register as the eruption of a new species. But the analogy does not end here. In tectonics, random decay of radiogenic elements gradually accumulates energy which may become suddenly released along tectonically active zones as a volcanic explosion. In evolution, one major factor is *point mutation* which represents a single replacement of a DNA base in a gene; but the rate of change at the third position of codons is greater than the rate of change at the second position. So mutations build up at steady rates and this leads the idea of the molecular clock (Fig. 11.34) and to the discovery of a kind of genetic change known as a neutral mutation because it does not benefit or harm the organisms since protein function is maintained. However, *molecular* evolution will permit valuable insight into the branching of lineages of species (A.C. Wilson, 1985).

Regular mutation, that is any change in or nearby a gene which affects the expression of a gene and renders it active or inactive, appears to be the critical event linking the molecular and organismic evolution. Selection pressure may come from the environment but also from a species itself. According to A.C. Wilson (1985) the pressure to evolve may be generated not only by means of geological forces as there are, weathering, volcanic activity or climate, but may stem from the "brain-storming" of birds and mammals themselves (cultural drive). If a mutation occurs in the reproductive cells of organisms, it will be passed on to the next generation. Since the product of a gene is – with rare exceptions – a protein, the development of the organism depends on the particular protein specified by its genes (Box 11.13).

BOX 11.13

Old MacDonald Had a Farm ...!

Genes and chromosomes can become dissected and reconstructed (e.g., P. Berg, 1981). This is in essence the foundation on which gene technology rests. Genes determine the generation of proteins, and genetic information is ingrained in the molecular structure of DNA. The molecular design of gene structure, its organization and its function has principally been revealed in work during the 1950's and 1960's with the bacterium *Escherichia coli* and some of its host viruses (phages).

More recently, the field has expanded virulently now embracing mammalian and human cells. Initial success in the development and application of recombinant DNA techniques has fueled curiosity, and research centers have sprung up everywhere. I often wonder whether our knowledge has already advanced to the point that we really know what we are doing or even worse, what the final outcome may be. By all means: The end does not justify the means. For what can happen unexpectedly, let's have a brief look at what nature has in store for us.

Southeast Asia has long been known as a "breeding ground" of pandemic influenza A viruses (see Scholtissek and Naylor, 1988). In the "Hongkong Flu" of 1968, the hemagglutinin gene appears to have come from an influenza virus residing in water fowl. This came as a surprise, because avian influenza viruses were supposed not to be transmitted to humans. According to work of Scholtissek and co-workers (see Scholtissek and Naylor, 1988), transmission of genetic material from avian to human influenza viruses proceeds via reassortment in pigs. In essence, pigs serve as a "mixing vessel" for avian and human influenza A viruses leading to new surface antigens. The new viral strains may infect not only pigs but humans too. Since human populations have no neutralizing antibodies, a new flu may spread.

Reassortment of influenza viruses is presently seen in conjunction with age-old agricultural practices in South China. In rural areas, ducks, pigs, fish and humans live in close contact with each other. This is illustrated in a description from Ming times (1368–1644) of the brothers T'an Hsiao and T'an Chao, who farmed in T'ung-lu in Chekiang (Elvin, 1973, p. 238):

"Their ponds could be reckoned in hundreds, and in all of them they reared fish. Above the ponds they built grass huts on beams in which they reared geese and pigs. The fish fed on the manure and grew still fatter."

China has suffered in 1586–9 and 1639–44 from two of the most widespread and lethal epidemics in her recorded history, and the medical nature remains a mystery (Elvin, 1973, p. 311–312). Local gazetteers often speak of between 20 to 90 percent of the population dead in many districts. Reduction in population eased the pressure on land for a century and a half. In 1820–22 came another epidemic, the first great cholera outbreak, which spread to China by the sea-routes from Bengal. Since the symptoms of the two earlier epidemics were different from those observed during the cholera outbreak, it would not be out of line to suggest that human influenza pandemics were behind them.

The practice still continues, and duck-fish-pig cultures are widespread in Hongkong, South China, Malaysia and Nepal. By bringing together the two reservoirs of influenza viruses, a considerable human health hazard of global dimensions is created. Thus, ducks and pigs should be strictly separated from one another, so that they cannot sing in chorus like they do in the last verse of the old song: "Old MacDonald had a farm".

At present, gradualism and punctuated equilibrium are lively concerns in the discussion on the evolution of life. In plain language, everything centers around the old and new question of whether evolution of organisms proceeds in small steps almost continually or in larger steps and abruptly (e.g., Darwin, 1859; Eldredge and Gould, 1972; Gould and Eldredge, 1977; Eldredge and Cracraft, 1980; Hölder, 1983; Stanley, 1976, 1979; Gould, 1982). In this context it is remarkable that a lifetime of observation by a biologist, a paleontologist or a geologist often counts for less than a model by an overnight hero.

The fossil record is frequently used in support of long periods of stasis punctuated by bursts of rapid change ("punctuated equilibrium"). However, as pointed out by Hoffman (1982, 1985) and Schopf and Hoffman (1983), such a viewpoint is biased, because the "garden variety", larger invertebrates – the molluscs, arthropods or corals used in support of a sudden emergence of a new species – usually permits the sampling of time intervals in the order of a few 100,000 years only. One exception to this rule is the data set of Brinkmann (1929) on the famous Jurassic ammonite *Kosmoceras,* where much shorter time slots, about 10,000 years or so, could be distinguished (Box 11.14). At that resolution trends in morphology indeed showed gradualistic changes by any resasonable morphological definition of gradualism. The same can be said for many stratigraphic sections with abundant marine microfossils such as foraminifera, conodonts, and radiolaria, where gradualism appears to be a sensible interpretation of most of the data (Schopf and Hoffman, 1983).

A recent title "Neo-Darwinian Evolution Implies Punctuated Equilibria" (Newman et al., 1985) might at first sight call for a new round of discussion. The truth of the matter, however, is that by means of mathematical models it can be shown that small random variations and natural selection can lead to random genetic drift of a population between stable phenotypic states (Lande, 1985) without invoking special developmental, genetic or ecological mechanisms. On that conciliatory note I am almost ready to close this chapter, except for a final salute to Charles Darwin and Ernst Mayr:

Reading Ernst Mayr's "The Growth of Biological Thought: Diversity, Evolution, and Inheritance" (Mayr, 1983) one will come to recognize the workings and achievements of biological evolution without having to rely on mathematical formalism. Mayr implants in us a deep appreciation of the mysteries of the biological cell and gives us a feeling as reflected in the last line of Charles Darwin's "On the Origin of Species": "There is grandeur in this view of life."

BOX 11.14

***Kosmoceras:* A Documentation on Gradualism**
(Brinkmann, 1929)

No single article in the field of biostratigraphy has – in my opinion – advanced our knowledge more than one by Roland Brinkmann in 1929 on the bituminous Oxfordian clay near Peterborough, England. In this treatise, Brinkmann equated systematic alteration in shell morphology of the ammonite genus: *Kosmoceras* with geologic time. Roughly 15,000 measurements, on more than 3,000 individual specimens, composed of a total of 14 different species, spread across 13 m of sediment section, equivalent to about 1 million years of Earth history, about describe the effort and range of the study. At certain intervals abrupt changes in shell development were observed. Apparent "punctualism" was nothing but an artifact caused by many a hiatus in the sediment record. Times of sedimentation, and times of non-deposition could be readily discerned. After taking care of the gaps, lines of species evolution followed a gradual although not necessarily a linear curvature (Fig. 11.48), quote:

"The bituminous Oxfordian clay facies indicates changes from oxygenated to reducing conditions close to the sediment-water interface and vice versa, and an active and fluctuating sediment transport in a shallow basin. Sedimentation gaps, truncated horizons, and the presence of brecciated and well-sorted shell material suggest periods of strong current activities. Whereas facies changes are often abrupt and principally controlled by physical factors, the evolution of the studied ammonites progresses continuously with time, although not necessarily on a linear scale."

The question of gradualism and punctualism is indeed nothing new, but "an old hat", and was already a lively topic during the twenties (see Brinkmann, 1929, p. 226–246). For a re-evaluation of Brinkmann's data, see Raup and Crick (1981, 1982).

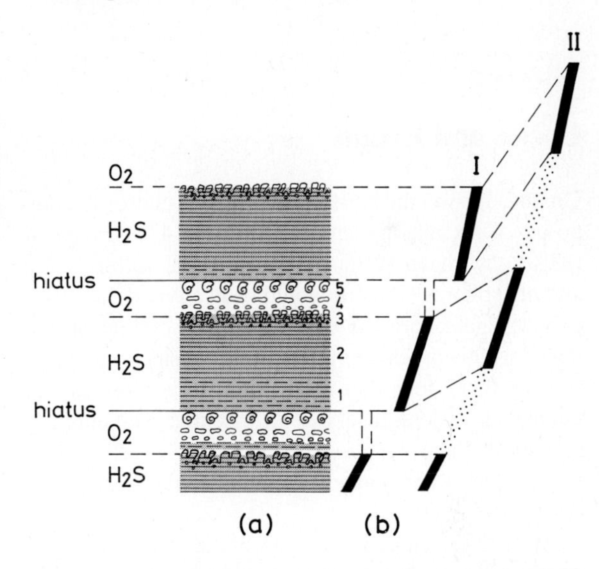

Fig. 11.48. a Small-scale cycles of Oxfordian clays: *1* greenish clay; *2* brownish bituminous clay; *3* burrows; *4* shell hash; *5* ammonite bank overgrown by oysters; **b** phylogenetic regression line of *Kosmoceras* (schematic); dashed lines in *I* mark times of erosion, active transportation, and redeposition; *stippled lines* in *II* connecting the regression lines of evolution indicate actual time elapsed between termination of anoxic conditions and start of new anoxic environment. Without properly taking care of the two sedimentation gaps – as is done in *I* – a sudden (punctuated equilibrium) speciation would be inferred. By considering both time gaps – as shown in *II* – a smooth (gradualism), although not necessarily linear evolution of the genus *Kosmoceras* can be observed (Brinkmann, 1929)

Chapter 12

Biogeochemical Evolution

Cycles and Events

On the dry weight basis, bulk organic matter is composed of six major elements: carbon (47%), oxygen (33%), hydrogen (9%), nitrogen (9%), sulfur (1%), and phosphorus (1%). It would be ill advised, however, to investigate only the biogeochemical cycling of just these six major elements and not consider the large suite of metal ions that are of vital importance in sustaining life. Close to about 25 essential elements are structurally and functionally involved in cellular activities, and a deficiency in any one of them would be deleterious to life.

All chemical elements are part of the global cycling of matter, but their majority bears no affinity to organisms, except that many of them are toxic. Accordingly, the biological cell had to develop strategies to fend them off.

In general, biogeochemical cycles describe the pathways along which organic and inorganic substances move and interact in the various compartments of our Earth. Globally combined, they can be looked upon as a complex and dynamic network of flows of matter and forces in the air-water-earth-life system. To assess their operation principle requires a constant crossing of disciplinary boundaries between physics, chemistry, and the environmental sciences. A holistic approach seems to be mandatory. This was actually a matter of course during the formative years of biogeochemistry. There is first the Russian school, and the names of V. Vernadskij, A. Fersman, M.M. Kononova, A.I. Oparin and A.B. Ronov pass revue. It was they, who more than 50 years ago recognized the significance of biogeochemical processes in the exogenic cycling of matter. In commemorating the 125th birthday of Vladimir Iwanowitsch Vernadskij (March 12, 1863 – January 5, 1945) the USSR Academy of Science and biogeochemists the world over celebrated in March 1988 this outstanding scientist and his students. Next in line is the German school and the eminent scholars V.M. Goldschmidt, R. Brinkmann and A. Treibs come to mind. Goldschmidt was the first who demonstrated that chemical elements have a logic and a likening or dislikening to other elements and feel more at home either in the dead or in the living world. Then we have the British and U.S. schools, and here I should like again to pay tribute to L. Pauling, G.W. Brindley and J.D. Bernal.

Following the "Big Bang" of organic geochemistry and biogeochemistry in the "roaring" twenties, thirties and forties, the field expanded during the second half of this century and up to this very moment at an unbelievable rate. Scholars like Philip H. Abelson, Irving A. Breger, John M. Hunt, Stanley L. Miller, Rudolph von Gaertner, Marlies Teichmüller, Alexander Lisitsin, Evgenii Romankevich, Tony Hallam, Geoffrey Eglinton, Pierre Albrecht, Bernard Tissot and Max Blumer stand for a generation of scientists who advanced the field of biogeochemistry.

Growing specialization and the explosion of scientific literature has almost brought to a halt this productive – because reflective – period of interdisciplinary synthesis. In the aftermath of scientific revolutions which arose in all areas of natural sciences from classical domains towards fields such as particle physics, cosmochemistry, molecular biology, organic geochemistry, or plate tectonics, the flow of integrated concepts across disciplinary boundaries has ebbed. Consequently, this is a moment to reflect on past achievements in biogeochemical research without becoming hopelessly outdated by the flood of new advances in these fields. And yet the wealth of already published information has reached a dimension which makes it a formidable task to a reviewer to address the biogeochemical cycling even of a single element, not to mention that of a wide range of interrelated elements. This is because physical, chemical, biological and geological processes ought to be considered that operate on very different time scales; that is from millions of years for the slow movement of the Earth's crust, weeks and days for the rapidly changing scene of the water surface of the sea, to fractions of seconds during which a high energy bond in ATP becomes generated and cleaved.

Starting with the global cycling of nitrogen, phosphorus and sulfur (Svensson and Söderlund, 1976, SCOPE Report 7), the Scientific Committee on Problems of the Environment (SCOPE) has fostered research on the global cycle of many a biogeochemical element. At the time of writing, SCOPE Report 31 has been released entitled: "Lead, Mercury, Cadmium and Arsenic in the Environment" (Hutchinson and Meema, 1987). SCOPE Report 33 entitled "Nitrogen Cycling in Coastal Marine Environments" (Blackburn and Sørensen, 1988) and SCOPE Report 36 entitled "Acidification in Tropical Countries" (Rodhe and Herrera, 1988) are two more recent products. SCOPE volumes should be consulted for an up-to-date review on the global cycling of biogeochemical elements.

Whilst global biogeochemical cycle studies help to assess modern environmental problems, it has been found more useful to speak of biogeochemical evolution when it comes to unravelling the geological past. This is because cycling of elements is tuned to events such as: mountain building, seafloor spreading, volcanism, climate, transgression, regression, onset of biomineralization, mass extinction, emergence of land plants and so on and so on... I would like to emphasize again that we are dealing with an irreversible system where not only organisms but all compartments of our Earth evolve in phase with one another. Namely, the evolution of plants and animals can only be understood in the context of the evolution of air, water and solid earth, as they in turn are affected and controlled by cellular activities.

Throughout the text, I have sketched the way this or that biogeochemical element participates in the global cycling. This time, I will select just one element, that is calcium, and follow its biogeochemical evolution as reflected in sediments and fossils. The choice of Ca^{2+} has been made, first, because it is ionic in character and thus contrasts with the behavior of covalently bonded carbon molecules. Second, Ca^{2+} is the most versatile metal ion in the metabolic and genetic apparatus. Third, it is instrumental in controlling CO_2, the basic molecule of life. And fourth, it resides in the form of calcite, aragonite and dolomite in fossils and limestones of all ages.

Ca^{2+} Cybernetics

Calcium ranks fifth among elements and third among metals in the Earth's crust. It is equally at home in magmatic, metamorphic and sedimentary rocks and accounts for 3.64 percent of the crust as a whole. It is the third most common cation in sea water, the third most prominent element of vegetation, and the most abundant metal ion in the human body (e.g., Erulkar, 1981). The high calcium content in all living cells must be seen in conjunction with its multifarious work assignments in the biological cell which requires Ca^{2+}'s easy accessibility at the right time, the right concentration, and the right place. Many articles in *Science, Nature,* biochemical journals, hand- or textbooks and a number of recent symposia volumes pay tribute to calcium as a second messenger, a structure maker or breaker, and a driving force in the evolution of the cell (see, e.g., articles and references in: Alberts et al., 1983; de Meis, 1981; Erulkar, 1981; Borle, 1981; Campbell, 1983; Franchi and Camatini, 1985; Darnell et al., 1986; Carafoli and Penniston, 1985; Gill et al., 1986; Kretsinger, 1983; Watterson and Vincenzi, 1980; Cheung, 1979; Herzberg and James, 1985; Black and Brand, 1986; Vyas et al., 1987; Drust and Creutz, 1988; Altmann, 1988).

Crustal Ca^{2+}

Enrichment of calcium relative to magnesium in average igneous rocks is related to the preferential removal of magnesium, for instance in the form of olivine, during gravitational settling of early differentiated crystallites. Furthermore, calcium-incorporating minerals such as pyroxene or amphibole follow somewhat later in Bowen's fractionation series. Especially interesting is the case of Ca-rich plagioclases (labradorite rocks), which have low specific gravity and thus may rise from heavier magmas (Goldschmidt, 1954). The closeness in ionic radii for divalent calcium and monovalent sodium is striking and accounts to a large extent for the ease of isomorphous substitution in the plagioclases.

Crustal rocks are subjected to passage of hydrothermal waters which are instrumental in leaching certain metal ions from the rock formation in exchange for elements present in solution. In due course the leached metals may surface, whereas metals picked up by the rocks will lead to new mineral assemblages at depth. It is now recognized that cold sea water can enter the oceanic crust at discrete points and circulate in convection cells down to depths of several kilometers (e.g., Edmond and Von Damm, 1983; Henley and Ellis, 1983). The idea has been advanced that the Moho may represent the level to which intracrustal water penetrates (Lewis, 1983; see Fig. 7.9). Such changes not only proceed at higher temperatures, but are a common feature of diagenesis of water and sediments. For instance, calcium chloride waters are widespread among the petroleum brines and formation waters in general. This relationship can best be linked to dolomitization processes, that is uptake of Mg^{2+} from the water phase in exchange for Ca^{2+} from solid $CaCO_3$. Magnesium may also proxy in chlorites or certain mixed-layer clay minerals for alkalies, iron and calcium. Variations established for Na/K and Ca/Na ratios between fossil and modern sea waters can also be accounted for by adsorption and exchange phenomena.

Table 12.1. Annual Ca^{2+}, Mg^{2+} and Si^{4+} input(+)/efflux(–) to/from oceans (in 10^{12}g) (From Wollast and Mackenzie, 1983)

	Ca^{2+}	Mg^{2+}	Si^{4+}
Rivers	+495	+129	+203
Hydrothermal reactions	+ 72	– 61	+ 25
Submarine weathering	+ 30	+ 27	+ 10
Pore water	+291	– 93	+185
Total flux	+888	+ 2	+423

Take, for example, recent Black Sea sediments (Shishkina, 1959). In an 8-meter sediment core, Ca^{2+} content increases three fold from top to bottom.

Wollast and MacKenzie (1983) have reviewed literature data on the biogeochemistry of submarine basalt-sea-water reactions, and calculated the elemental fluxes to the sea owing to hydrothermal reactions (Table 12.1). All data agree that percolating sea water loses Mg^{2+} to oceanic basalt, thereby yielding serpentinites and greenstones, whilst in exchange Ca^{2+} and Si^{4+} are gained as an input to world ocean. Direction and magnitude of sodium and potassium fluxes are still uncertain. Submarine weathering of basalts (halmyrolysis) are sinks or sources of cations; but with respect to Ca^{2+}, a release to the sea is indicated (Table 12.1). The same is true for the Ca^{2+} flux to the ocean across the sediment-water interface (Table 12.1). With due consideration for the uncertainties in estimating global fluxes of elements, there is still a general trend discernible in that the inputs of Ca^{2+} and Si^{4+} to the sea are substantial and in a ratio of 2:1 in favor of Ca^{2+}, whereas Mg^{2+} gains and losses come out about even.

Inasmuch as carbonate equilibria are involved in Ca^{2+} and Mg^{2+} cycling and regulation, and on the other hand PCO_2 and alkalinity are controlled by life processes, weathering and crustal leaching, the above relationships signal a constant addition of alkalinity to world ocean. This channeling of ions and gases may be viewed as a kind of umbilical cord of the living cell to the solid Earth.

It is concluded that the sea is constantly recharged from the crustal compartment with calcium ions at an annual rate of about 400 x 10^{12}g. This would be slightly less than Ca^{2+} recharge of modern ocean via rivers, which has been estimated at close to 500 x 10^{12}g per year (Wollast and MacKenzie, 1983).

Ca^{2+} in the Sea

The ocean is virtually saturated with respect to $CaCO_3$ and cannot enlarge its size reservoir significantly. At first sight river input appears to be balanced by output from the sea, but as we have learned above, there is an additional source of Ca^{2+} derived from the crust beneath the sea by hydrothermal leaching, basaltic weathering and pore water expulsion. Crustal processes will add calcium ions and carbonates to the ocean at rates specified in Table 12.1.

Ca^{2+} becomes chiefly extracted from sea water by the biological cell, largely in the form of aragonite and calcite. Even the generation of inorganic deposits such as oolites, which are small grains up to a few mm in size found in some lagoons or tropical waters, is mediated by calcifying organic templates secreted into the ambience by coralline algae. In brief, all Ca^{2+} introduced to the sea has to be sequestered by marine life. It is obvious that those organisms which live close to the source of Ca^{2+}, namely estuaries, marginal seas or hydrothermal vents, will be subjected to higher doses of Ca^{2+}, than nanoplankton in the pelagic realm.

The Ca^{2+} content in modern sea measures 10^{-2} molar. It is obvious that any outside change in the supply of Ca^{2+} will exert its influence upon marine life. Such changes may come about by: (i) climate forcings in terms of weathering, erosion, vegetation cover, sea level fluctuations; (ii) kind of rocks exposed on land, for instance ratio of limestones to shales to crystalline rocks; (iii) rates of seafloor spreading leading to transgression, regressions, bathymetric adjustments, hydrothermal vent activities; and (iv) tectonism. Since life is the final control device for Ca^{2+} in the sea, it has only two ways to cope with the situation, either to let the Ca^{2+} level increase, or remove Ca^{2+} via calcification mechanisms.

Going back in time, sea water chemistry must have undergone gradual and episodic changes in response to the dynamics of our Earth. This viewpoint is not shared by everybody. Instead, the unchanging mineral spectrum of marine evaporites (Box 12.1) is listed in support of a constant salt pattern in world ocean at least throughout the Phanerozoic. And as far as the Precambrian sea goes, a wide range of opinions has been expressed concerning its chemistry (see p. 271). Most established is the idea of a neutral to slightly alkaline ocean with a chemistry dominated by sodium chloride and calcium-magnesium carbonate equilibria.

An entirely different scenario has recently been proposed (see p. 271) in which Ca^{2+} content has supposedly been raised from 10^{-7} molar at about 4 billion years ago to 10^{-2} molar close to the Precambrian/Cambrian boundary, from where on it fluctuated in response to the evolution of the biogeochemical system. In contrast, the chlorinity curve of the sea must have stayed rather smooth in the aftermath of the cataclysmic events during the first billion years of

BOX 12.1

Evaporites

The first salt to separate upon evaporation of sea water is calcium carbonate. As evaporation proceeds to about one third its original volume, calcium sulfate appears. In the majority of cases it will be gypsum ($CaSO_4$ x 2 H_2O), but under certain conditions it will be anhydrite ($CaSO_4$). However, activating constituents in a sea water brine will bring about almost immediate conversion of anhydrite crystallites to gypsum. Thus, anhydrite found in the stratigraphic column seems to be chiefly a product of diagenesis of gypsum, which is stable only up to about 800 m depth of burial; below that, anhydrite will form.

When ocean water is reduced to one tenth of its original amount, the mineral halite (NaCl) will precipitate. Reaction products obtained under static isothermic evaporation of sea water at 25°C, assuming stable equilibrium conditions, are illustrated in Fig. 12.1. The chemical composition (in mol/x mol H_2O), and the weight of the residual solution in grams at the start of the calcite, gypsum, halite etc. precipitation are indicated. It is the nature of the sea water system, however, that frequently causes the deposition of metastable mineral phases, because stable mineral phases are not precipitated within their stability field for kinetic reasons. A good example is quartz, which should form once dissolved silica exceeds about 4 to 5 ppm, but cannot do so even though such levels are normally observed in pore waters of recent sediments. Little is known concerning the time required for the conversion of the metastable into the stable conformation. The only fact remains that concentration of participating electrolytes, chemical nature of metastable products, temperature, and overburden pressure are critical factors in controlling the inversion time. The phenomenon of metastable-stable equilibria in the sea water system has been fully discussed by Braitsch (1962) and Borchert and Muir (1964).

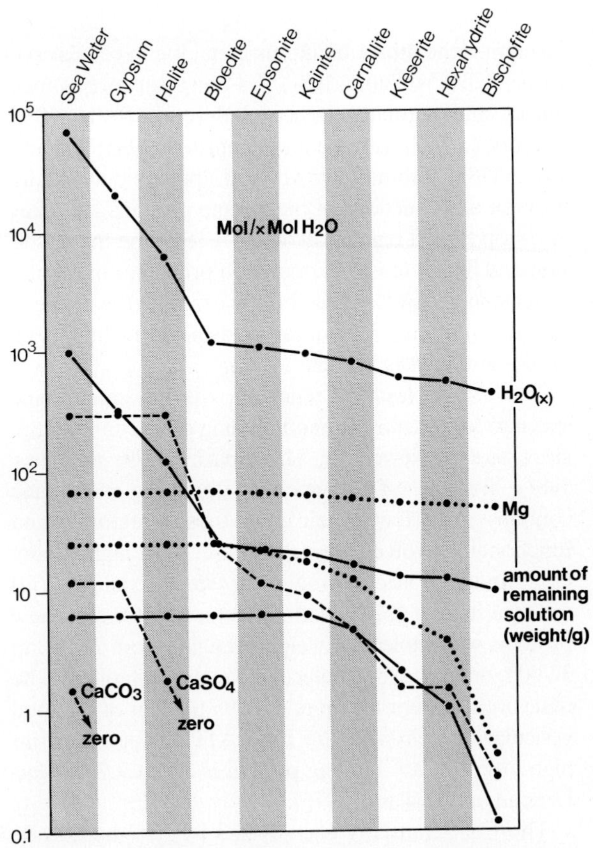

Fig. 12.1. Chemical composition of ocean water under isothermic evaporation at 25°C, and at the start of precipitation of the individual mineral species (after Braitsch, 1962)

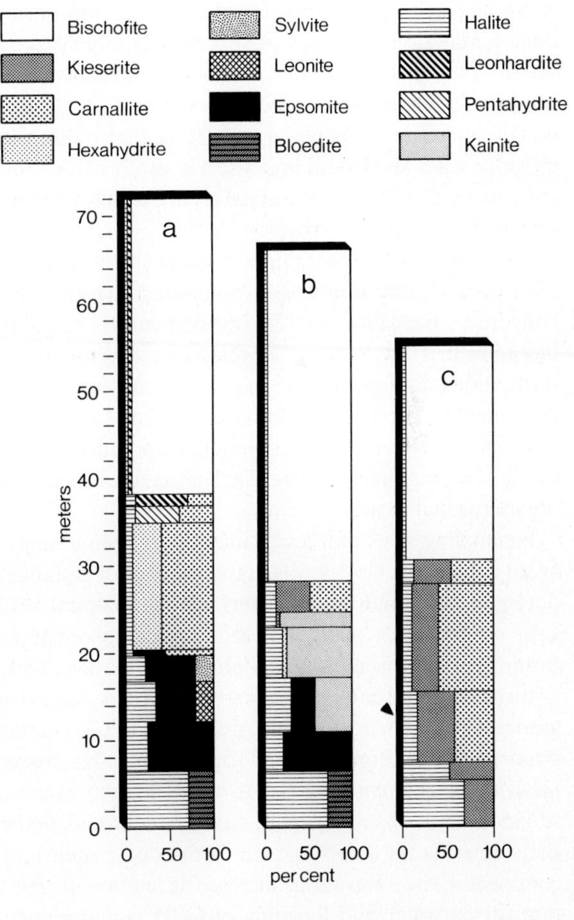

Fig. 12.2 a–b. Static evaporation of ocean water at 25°C; **a** metastable equilibrium; **b** stable equilibrium, without reaction at the phase boundaries; **c** stable equilibrium, with complete reaction at the phase boundaries (after Braitsch, 1962)

The model depicted in Fig. 12.1 may be expanded by comparing mineral sequences obtained under: (i) metastable equilibrium conditions, (ii) stable equilibrium conditions without reaction at the phase boundaries, and (iii) stable equilibrium conditions with complete reaction at the phase boundaries (Fig. 12.2). This picture illustrates that even at constant water temperature the type of salt minerals and the total amount of salts obtained during complete evaporation of sea water will differ. It is significant that a deposit of 0.4 meter limestone, 4.8 meter gypsum, and 100 meter halite will form before the salts seen in Fig. 12.2 will precipitate upon isothermic evaporation at 25°C. But to produce such a layer of salt, a water column of 8,500 meters of mean ocean water composition has to evaporate. The message we can draw from this relationship is that calcium minerals can be extracted from the sea in restricted environments without noticeably changing the chlorinity pattern of world ocean.

Earth history. The element chlorine was subsequently added to the sea only in small increments, that is in trace quantities constantly leached from crustal rocks. Even massive extraction of rock salt, for instance during the Permian, could only change ocean salinity by about 3 to 4 permil (Stevens, 1977).

The answer to the approximately 100,000-fold increase in Ca^{2+} is a consequence of a gradual shift in carbonate mineral equilibria from an early highly alkaline ocean dominated by sodium carbonates, to a late Precambrian halite ocean controlled by Ca^{2+} and Mg^{2+} carbonate equilibria. It is perhaps no coincidence that the cytosolic Ca^{2+} level in all organisms is about 10^{-7}, or the anticipated Ca^{2+} value of a soda ocean, which strongly argues for its archaic derivation.

Calcium minerals precipitate at early stages of evaporation and many limestones, dolomites, gypsum and anhydrite deposits are a consequence of this mechanism. This, in turn, suggests that episodic Ca^{2+} fluctuations in lagoons or shallow inland seas were prominent in the past. Addition of riverine calcium bicarbonate coupled with sea water evaporation could locally and regionally create Ca^{2+} anomalies to which organisms had to adjust.

Fluctuating sea water levels must have been a major agent controlling Ca^{2+} levels in the ocean. For instance, during a highstand not only terrigenous material was kept on the inner shelf, but gross extraction of lime from the shallow sea became a prominent feature. This, in turn, lowered the dissolved carbonate pool, caused a gradual rise of calcite compensation depth (CCD) in the open marine environment, and triggered massive dissolution of pelagic carbonates. At lowstand, the reverse situation took shape, namely, former highstand inner shelf deposits became removed to the outer shelf and continental rise. Resulting increase in carbonate content of sea water and lowering of CCD would reduce carbonate dissolution at depth. Climatic feedbacks associated with changing sea levels will severely impact general ocean circulation, thereby further affecting the biogeochemistry of Ca^{2+}.

These few examples on Ca^{2+} regulation in the sea may suffice to demonstrate the wide range of factors that operate on calcium distribution in a salty ambience which by itself remains comparatively constant in the course of geologic time.

Cellular Ca^{2+}

Calcium is the most ubiquitous metal ion in cellular systems (R.J.P. Williams, 1981). Its universal role as messenger and regulator is accepted (e.g., Borle, 1981; Campbell, 1983; Carafoli and Penniston, 1985; Williams et al., 1985). From all we know, it appears that the disposition and functions of membrane proteins are a key in transport and regulation of Ca^{2+}. To prime the discussion and illustrate Ca^{2+}'s operation principles in cellular systems in general, we will briefly examine the cooperative action of Ca^{2+} ions in sarcoplasmic reticulum membranes and muscle proteins.

The sarcoplasmic reticulum is a membrane specialized for transport and binding of calcium in muscle tissue (reviewed in: Hasselbach, 1979; de Meis, 1981). Because of this specialization, the membrane contains only a few proteins, and those structurally and functionally involved are: (i) ATPase, (ii) a high affinity Ca^{2+} binding acidic protein of 55,000 dalton, (iii) calsequestrin of 55,000 dalton, (iv) a set of three acidic proteins with molecular weights ranging from 20,000 to 32,000, and (v) a low molecular weight proteolipid. The first three account for more than 95 percent of the total vesicular protein and the ratio ATPase:calsequestrin: high affinity Ca^{2+} binding protein is about 7:2:1 (MacLennan and Holland, 1975).

The Ca^{2+} transport is energized by ATP hydrolysis, acetylphosphate, or carbamylphosphate through a transport ATPase which is tightly bound to microsomal

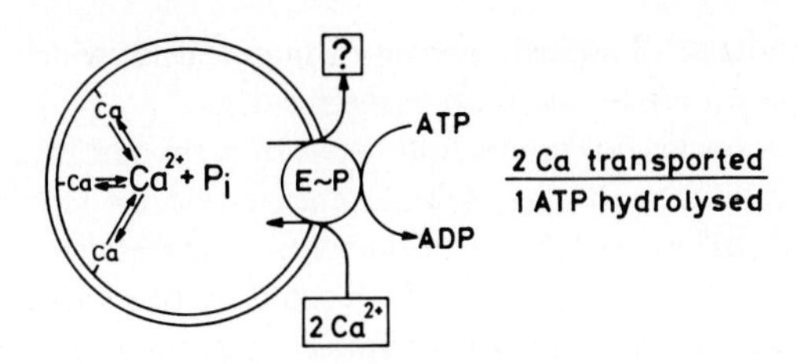

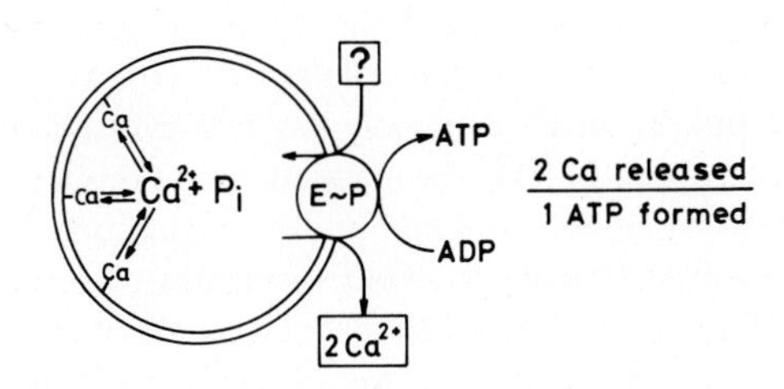

Fig. 12.3. Relationship between ATP, Ca^{2+} translocation, and phosphoprotein in sarcoplasmic reticulum (after Martonosi et al., 1974)

membrane; the relationships are depicted in Fig. 12.3; for one mole of ATP two calcium atoms are transferred across the membrane, where they are bound to membrane-linked cation-binding sites. Reversal of this process leads to hydrolysis of phosphoprotein, liberation of orthophosphate (P_i), and release of two calciums; the carrier will return to the cytoplasmic surface. In summary, the ATPase protein represents a complete transport system containing both the site of ATP hydrolysis and the Ca^{2+} carrier within a single protein.

During muscle relaxation and contraction, calcium ions are constantly on the move from one site to another. A major site of Ca^{2+} sequestration in the interior of the sarcoplasmic reticulum is calsequestrin. A minority viewpoint is that calsequestrin is located at the exterior face of the membrane and that the role of calcium binding could possibly be exercised by the high affinity Ca^{2+} binding acidic proteins. However, the weight of evidence suggests that calsequestrin is not located at the exterior, and that the functions of the two major Ca^{2+} binding proteins inside the vesicle are intertwined.

Carrier-mediated active Ca^{2+} transport is responsible for the relaxation of muscle. However, the rate of efflux from sarcoplasmic reticulum membranes during reversal of the transport process is 10^2 to 10^4 orders too low to account for massive calcium release from sarcoplasmic reticulum in stimulated muscle. Instead, passive diffusion of calcium across the sarcoplasmic reticulum membrane will proceed during excitation of muscle. The rate of calcium release observed during excitation is 1,000 to 3,000 μmol mg^{-1} protein min^{-1}, which is an increase of about 10^4 to 10^5 over the resting state.

Phospholipids, which constitute close to 40 percent of the mass of the sarcoplasmic reticulum membranes, dominate the permeability characteristics through formation of liposomes which are capable of scavenging calcium ions and carry them across the membrane (Martonosi et al., 1974). The permeability properties of liposomes depend on their phospholipid content. Rate of calcium release markedly increases with rising temperatures and in the presence of local anesthetics and microsomal proteins. When contraction is initiated by a nerve impulse, the impulse travels through transverse tubules, the T-system. Depolarization may cause a potential change across the sarcoplasmic membrane which in turn induces the release of calcium.

Contraction-relaxation processes in muscle proceed by a sliding filament mechanism, whereby actin and myosin filament move relative to one another. The reaction is energized by ATP and regulated by the level of calcium ions. In vertebrate skeletal muscle, contraction is controlled by the interaction of calcium ions with a specific protein, i.e., troponin, which is attached to the actin filaments, whereas among many invertebrates troponin is lacking and myosin and not actin is controlled. However, in a few invertebrates both types of filaments are involved in regulation (reviewed in Rüegg, 1986).

The message of calcium is conveyed from the troponin to which it binds, via tropomyosin to the actin filament. As soon as the nerve impulse ceases, the Ca^{2+} becomes quickly removed and returned to the storage sites situated in the membranes of the sarcoplasmic reticulum. When Ca^{2+} concentration in the sarcoplasm reaches 1 x 10^{-7} molar, the fiber is relaxed.

Protein sequence and structure of troponin (TN-C) and carp muscle calcium binding parvalbumin (MCBP) are known (e.g., Collins, 1974). MCBP has two calcium binding regions, each consisting of an α-helix, a loop about Ca^{2+} and another α-helix, the so-called EF hand (Fig. 12.4). The calcium binding component of troponin from rabbit muscle contains four homologous EF hands. Four Ca^{2+} are bound to TN-C. The amino acid sequence in the four calcium binding regions of TN-C and the two regions of MCBP are depicted in Fig. 12.5. Calcium ions are present in octahedral coordination – Ca^{2+} O_6 – and different oxygen ligands are incorporated in the Ca^{2+} coordination complex. In the case of TN-C, acidic amino acids provide 17 of the 24 oxygens needed for the four calcium octahedra that can become part of one TN-C molecule. Judged by the invariability of amino acids at the X (Asp) and -Z(Glu) vertices of the Ca^{2+} coordination polyhedra in all six calcium binding

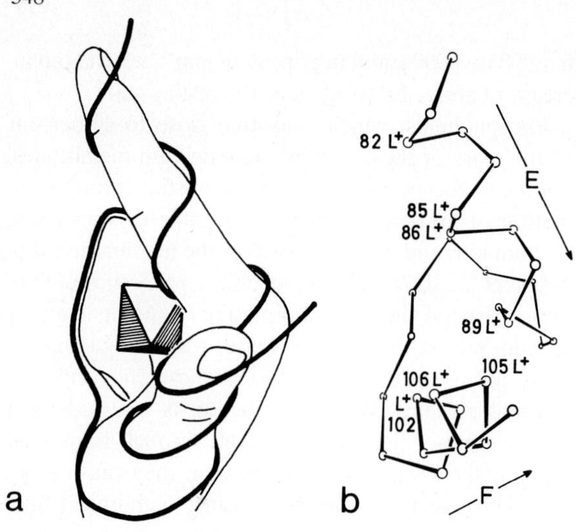

Fig. 12.4 a, b. The symbolic EF hand in **a** represents helix E (forefinger), the calcium binding loop (middle finger enclosing a $Ca^{2+}O_6$ octahedron), and helix F (thumb). The alpha-carbon skeletal models of the carp MCBP EF hand (**b**); (after Tufty and Kretsinger, 1975)

regions shown in Fig. 12.5, it appears that the presence of these two amino acids at these sites is crucial for the structural incorporation of calcium ions and may also be essential for the 3-D alignment of the two helices. Presence of EF hands appears to be a thermodynamically favorable conformation. It is the presence of carboxyl groups at specific sites which is the most probable control mechanism for Ca^{2+} fixation.

In molluscan muscles and some primitive invertebrates, contraction is mediated by the myosin and not by the thin filaments (Szent-Györgyi et al., 1973). The calcium dependence of actin-induced ATPase activity of molluscan myosin requires the presence of a specific light chain (EDTA-light chain). Removal of EDTA-light chains from myosin preparations will lower the amount of calcium by about 40 percent. The amino acid composition of the light chain is characterized by a high abundance of aspartic acid and glutamic acid, which account for about one third of the residues.

In conclusion, interaction between actin and myosin in vertebrate and invertebrate muscle systems requires a critical level of Ca^{2+}. Data indicate that interaction between myosin and actin is prevented by blocking sites on actin in the case of vertebrates, whereas in the case of molluscan muscles it is the sites on myosin which are blocked in the absence of Ca^{2+}. The acidic amino acids are of utmost importance in the regulation process.

The main element of the cellular Ca^{2+} control system is the upkeep of an extremely low free cytosolic Ca^{2+} content, about 10^{-7} M. The maintenance of this level appears to be an ancestral and universal condition of life that evolved initially to avoid formation of insoluble $Ca_2(PO_4)_3$ in the cytosol, so that cells could utilize inorganic phosphate as a source of energy (Kretsinger, 1977a, b, 1983). Ca^{2+} maintenance is a formidable task, bearing in mind that there is a multitude of intracellular and extracellular Ca^{2+} pools which exceed the free calcium content in cytosol by orders of magnitude.

Intracellular level of Ca^{2+} determines the opening and closing of gap junctions in membranes (Unwin and Henderson, 1984; Erulkar, 1981). Ca^{2+} shapes the geometry of channels and regulates transfer of molecules. At normal cytosolic Ca^{2+} levels, channels are highly permeable to a wide range of molecular sizes with an upper size limit for peptide molecules of about 1,200 to 1,900 dalton; a rise in Ca^{2+} to 5 x 10^{-5} M in the junctional locale will drastically lower permeability (B. Rose et al., 1977). Basically, "metal ion-phosphate gates" make the plasma membrane relatively impermeable to extracellular Ca^{2+}, which in sea water is 10^{-2} M, that is 100,000 times the cytosolic level. Binding of Ca^{2+}

Fig. 12.5. Comparison of EF hand amino acid sequences in carp muscle parvalbumin (MCBP) and calcium binding component of troponin (TN-C); (after Collins, 1974; Tufty and Kretsinger, 1975)

	Start	X		Y		Z		-Y		-X			-Z	End
MCBP - CD	51	ASP	GLU	ASP	LYS	SER	GLY	PHE	ISO	GLU	GLU	ASP	GLU	62
MCBP - EF	90	ASP	SER	ASP	GLY	ASP	GLY	LYS	ISO	GLY	VAL	ASP	GLU	101
TN - C1	27	ASP	ALA	ASP	GLY	GLY	GLY	ASP	ISO	SER	VAL	LYS	GLU	38
TN - C2	63	ASP	GLU	ASP	GLY	SER	GLY	THR	ISO	ASP	PHE	GLU	GLU	74
TN - C3	102	ASP	ARG	ASN	ALA	ASP	GLY	TYR	ISO	ASP	ALA	GLU	GLU	113
TN - C4	138	ASP	LYS	ASP	ASN	ASP	GLY	ARG	ISO	ASP	PHE	ASP	GLU	149

to specific sites at the internal face of membranes alters conformation of the membrane constituents, thus allowing transfer or passage of hydrophilic substances across the lipid barrier. High affinity Ca^{2+} proteins and glycoproteins confer specificity for binding, removal and transportation of Ca^{2+} (Elbrink and Bihler, 1975).

To prevent intracellular precipitation of insoluble calcium phosphates, most cells are equipped with ionic pumps invented for the speedy removal of deleterious calcium as well as other toxic metal ions. For instance, cytoplasmic Ca^{2+} level is affected by Ca^{2+}–Na^{+} exchange and by excitation-concentration coupling in the sarcoplasmic reticulum system. The calcium-sodium pump is of particular biogeochemical interest, because we have previously seen that because of nearly identical ionic radii, sodium and calcium can readily proxy for one another in plagioclase minerals. However, should the cooling of a magma be rapid, zoned crystals will form with a Ca^{2+}-rich inner core and a Na^{+}-rich outer rim; metamorphosis on the other hand may transport Na^{+} and Ca^{2+} in or out of the plagioclase crystal in response to a changing temperature regime. Cellular systems may have adopted sodium and calcium for the same structural reason as minerals did by making use of them in certain isomorphous substitution series. In particular, high affinity calcium-binding proteins serve as carriers for shipping a Ca^{2+} surplus to the external environment. The shipping system is located within the endoplasmic reticulum and the Golgi apparatus (McFadden et al., 1986). Calmodulin, an intracellular receptor protein, serves as a major regulator of cellular calcium concentrations by activating an assortment of metal ion pumps. Calmodulin has been identified in almost all eukaryotic cells (Watterson and Vincenzi, 1980; Cheung, 1979; Sasaki and Hidaka, 1982). Prokaryotic cells do not contain calmodulin but some, e.g., *Escherichia coli,* utilize a precursor protein resembling that of the EF hand structure typical for calmodulin. For recent advancements of cellular Ca^{2+} regulation see Irvine and Moor (1986) and Gill et al. (1986). The functional role of ions in the gating of calcium channels has been discussed by Saimi and Kung (1982).

Eukaryogenesis – Cell Aggregation – Multicellularity

Intracellular Ca^{2+} regulation has not only structural and functional importance in maintaining metabolic activity in eukaryotes but appears to have a profound impact concerning histogenesis and embryogenesis. It is this aspect which may help to define not only developmental stages of cells in terms of their ontogeny and phylogeny, but instruct us on the behavior of organisms in response to biogeochemical stimuli in the environment.

Jointly with Josef Kazmierczak, Venugopalan Ittekkot, and Stephan Kempe, I have expanded on the behavior aspect; for details and pertinent references see: Degens (1976), Degens et al. (1984, 1986a), Kazmierczak et al. (1985), Kempe and Degens (1985), Kempe et al. (1987a), Kazmierczak and Degens (1986). Extensive reviews on the homeostasis of intra- and extracellular Ca^{2+} levels have been prepared by: Kretsinger (1977a, b, 1983) and Lowenstam and Margulis (1980).

Of the two modern theories of eukaryotic cell origin, that is the autogenous (Taylor, 1976) and the serial endosymbiotic theory (Margulis, 1981), the latter is gaining more acceptance (Knoll, 1983). Accordingly, chloroplast and mitochondrion symbionts were acquired by means of endocytosis (Box 12.2), a process which is calcium-dependent with a maximum uptake at 10^{-4} M external Ca^{2+} level (e.g., Prusch and Minck, 1985).

BOX 12.2

Membrane Fusion Processes

Small polar molecules can be carried via transport proteins in a kind of piggy-back ride across cell membranes. However, for macromolecules and particles, cells utilize membrane-bounded intracellular vesicles into which they package matter to be transferred. The vesicles fuse with the plasma membrane in a process termed exocytosis, to open subsequently and eject the enclosed molecules to the extracellular medium.

Macromolecules and particles can enter a cell by first becoming glued to the outside of a membrane, then embraced and swallowed by a small increment of the plasma membrane, to be finally released as intracellular vesicle in a process termed endocytosis. Should the enclosed matter be fluid, one speaks of pinocytosis ("cell drinking"); where we are dealing with large particles such as bacteria or cell debris, the term phagocytosis ("cell eating") is used (Alberts et al., 1983). Ionic events are essential in triggering phagocytosis (Young, 1985). It is of note that because of the bilayer adherence step, endocytosis and exocytosis are not simply a reverse of each other, suggesting that both are regulated separately.

Secreted or ingested matter becomes generally sequestered in vesicles rather than just being trans-

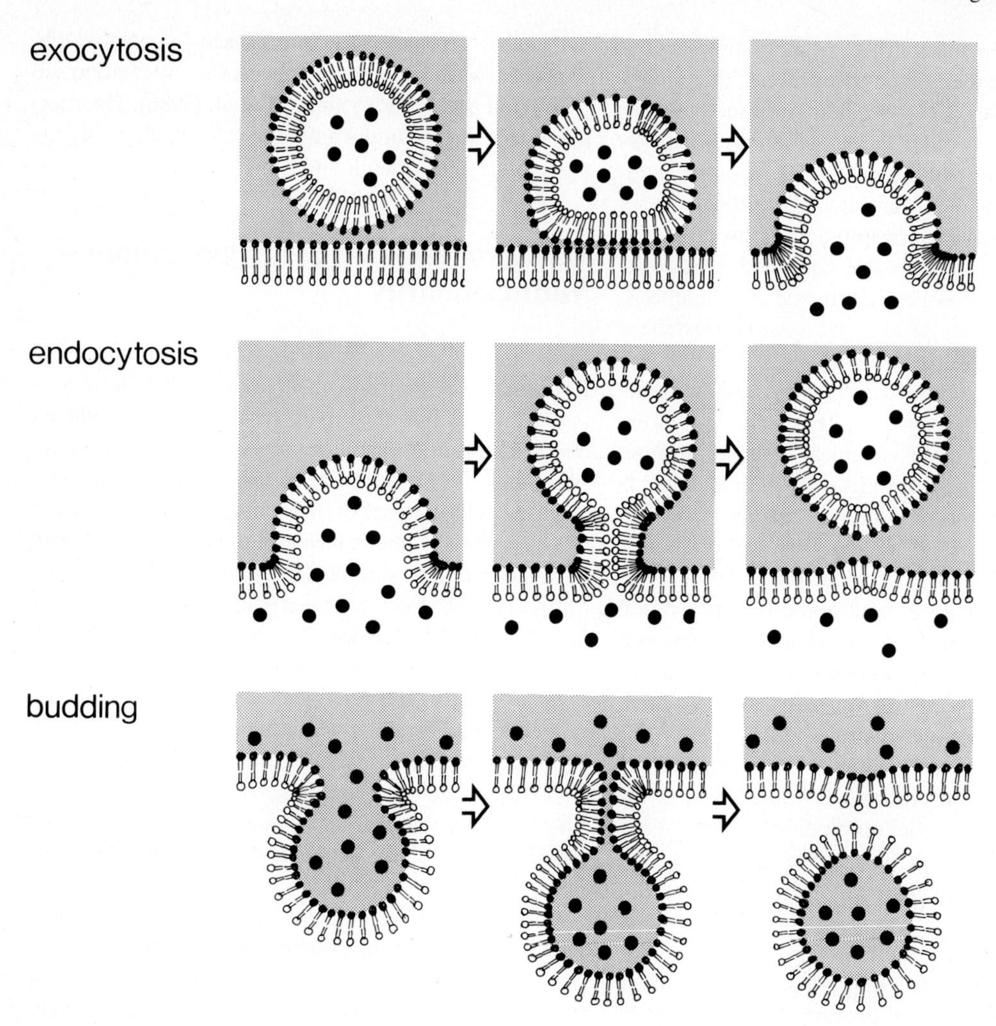

Fig. 12.6. Membrane fusion processes: exocytosis, endocytosis, and budding (after Alberts et al., 1983)

ported and conservatively handed over to macromolecules and organelles. Each vesicle fuses only with specific membrane structures and a form of topochemical coding device appears to be operative along

Fig. 12.7. Electron micrograph of the outer membranes of *Nitrosomonas sp.* a marine nitrifying bacterium; freeze etching technique. A grid pattern has been constructed to emphasize (i) the ordered arrangement of the structural proteins, (ii) the displacement of subunits along shearing zones of the two dividing cells, and (iii) the spherical nature of the bacterium. Individual subunits are spaced about 160 Å apart (Watson and Remsen, 1969)

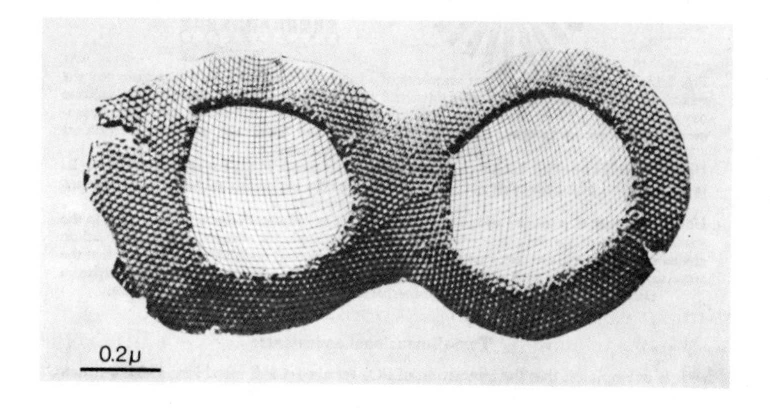

the surface of membranes, allowing an orderly transfer of matter between cell and environment.

A similar mechanism operates when it comes to the transfer of macromolecules among various compartments of the cell. Vesicles bud from the endoplasmic reticulum and Golgi apparatus (Rothmann, 1985; McFadden et al., 1986), and migrate and fuse with another membrane. An illustration on: exocytosis, endocytosis, and budding is drawn in Fig. 12.6.

Membrane fusion is a fundamental cell-membrane phenomenon, and in cell fusion and cell division bilayer adherence and bilayer joining is observed. Fig. 12.7 depicts a bacterial cell at the moment of cell division. The ordered arrangement of the structural protein becomes displaced along the shearing zone of the two dividing cells. A kind of zipper effect is indicated.

A number of nuclear activities are Ca^{2+}-mediated: DNA synthesis, regulation of cell proliferation, assembly and disassembly of microtubules in chromosome movement during mitosis, or activation of histone kinase. Chromosome breakage at times of external calcium deficiency underscores the need for Ca^{2+} to maintain chromosome configuration. High external Ca^{2+} raises free cytoplasmic calcium but not the nuclear one, indicating that nuclear Ca^{2+} is regulated independently of cytoplasmic calcium by gating mechanism in the nuclear envelope (D.A. Williams et al., 1985).

These observations may mean that the shift of the genetic material to the cell interior in proto-eukaryotes, and its membranization was an adaptive response to protect the genetic apparatus against high doses of Ca^{2+} from the ambience.

The significance of Ca^{2+} in cell aggregation, cell-to-cell adhesion, and cell fusion is fully documented in modern biochemistry texts (see: Starr and Taggart, 1987; Alberts et al., 1983; Darnell et al., 1986). For instance, a Ca^{2+} level not lower than 10^{-5} M is needed to keep slime mold cells tightly aggregated or to let yeast flocculate (Esser and Kües, 1983; A.H. Rose, 1984). Ca^{2+} appears to be the coordinating element between electronegative ligands of adjacent cells or a co-factor in activating and linking specific glycoproteins to carbohydrates.

As in algal cell aggregates, the integrity of cells in metazoan tissues too is strictly controlled by extracellular Ca^{2+} concentration. Experiments carried out on embryos and tissues of various animals demonstrate that in most of them cells fail to adhere when extracellular Ca^{2+} falls below 10^{-4} M. In the sponges *Microciona* and *Haliclona,* cells dissociate when Ca^{2+} in the medium drops below 10^{-5} M and reaggregate into a gel when Ca^{2+} is again raised above that level (Kretsinger, 1977a, b).

About 60 years ago MacCallum (1926) proposed that life must have originated in an environment with calcium concentration at the cytosolic level, grew adjusted to it and developed elaborate mechanisms to maintain it against the massive build-up of Ca^{2+} in the environment. The "soda ocean concept" (Kempe and Degens, 1985) provides a meaningful geochemical scenario which fully conforms with physiological requirements. Temporal changes in the Ca^{2+} level of the Archean and Proterozoic ocean in fact account not only for a series of cell contact phenomena but explain the evolution of Ca^{2+}-regulation and Ca^{2+}-extrusion systems in eukaryotes, resulting in the generation of a whole spectrum of calcium-binding and calcium-regulating proteins. In a way, the origin of metazoans had to await a Ca^{2+} rise in world ocean by about a thousand-fold above the cytosolic level of 10^{-7} M.

Against this background we will project the Precambrian fossil record and reassess the biogeochemical evolution of soft cellular tissue.

From Greenland to Australia

The first trace of life is found in the form of microfossils in sediments of the 3.8-billion-year-old Isukasia (Isua) Formation, Greenland (Pflug, 1978, 1982; Cloud, 1976). Since these are also the oldest rocks so far detected here on Earth, the first cell – the progenote – had to come earlier. The first metazoans are known from the 680-million-year-old Ediacara Formation, Australia (Cloud and Glaessner, 1982). Thus, evolution from bacteria to algae to metazoans took about 3 billion years.

The mode of evolution of early prokaryotes still remains an enigma (Knoll, 1983). One line of thought follows an entirely prokaryotic development for much of the Precambrian, whilst others assume quite an early origin of unicellular eukaryotes. Lack of agreement among the researchers as to the timing of eukaryogenesis has principally to do with the different types of yardsticks employed in measuring and classifying early microfossils (e.g., Kazmierczak, 1981; Awramik, 1982; Pflug, 1978, 1982). Namely, most Precambrian microfossils are characterized by spherical and filamentous shapes of very modest diagnostic value.

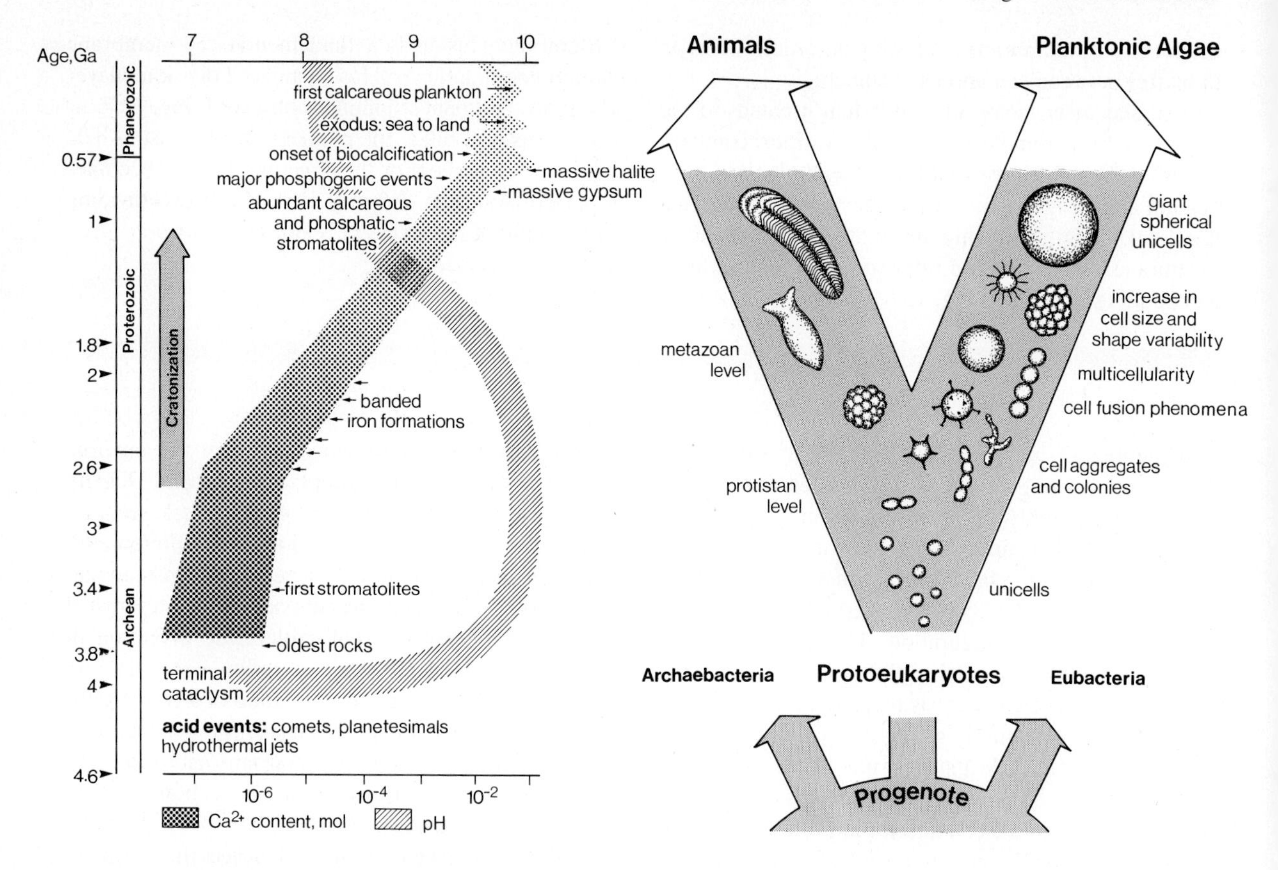

Fig. 12.8. Evolutionary lineages of animals and planktonic algae as documented by the paleontological record plotted in relation to temporal changes of Ca^{2+} concentrations and pH values in the ancient marine environment as implicated by the "soda ocean model" (Kempe and Degens, 1985). The primary splitting of the progenote cells into three superkingdoms: Archaebacteria – Urkaryotes – Eubacteria took place in an acidic environment which preceded the early soda ocean. Biogeochemical events are indicated (after Kazmierczak and Degens, 1986)

Cell size is often the only tool for separating bacteria from microbial eukaryotes. Ranges in cell size do overlap: prokaryotes – 0.5–55 μm, average 2–4 μm, largest 80 μm; eukaryotes – 1–350 μm, average 20 μm. An independent criterion has recently sprung up in favor of an early appearance of eukaryotic life, that is burrows (Kauffman and Steidtmann, 1981) and fecal pellets (Robbins et al., 1985) from rock formations at least 2 billion years old. This implies that the earliest metazoans and the first aggregates of eukaryotic algal cells are of about the same age. Thus, there appears to be no longer a restriction in placing the origin of eukaryotes close to the Archean/Proterozoic boundary, or even earlier.

Metazoan evolution is tightly linked to the presence of molecular oxygen in air and surface waters, and the biosynthesis of a structural protein, i.e., collagen, the most dominant extracellular fibrous protein in all metazoa (Towe, 1970). Collagen is the only protein having both hydroxyproline and hydroxylysine which are hydroxylation products of proline and lysine, respectively. Its precursor protein is believed to exist in chrysophytan and chlorophytan phytoflagellate algae which could possibly be the ancestral groups for the first metazoa (Aaronson, 1970; Lamport, 1977). The question comes up whether collagenous proteins were solely invented to strengthen the metazoan body, or whether they served primarily in metal ion detoxification and only later were adopted as body-building material. We advanced the idea (Kazmierczak and Degens, 1986) that Ca^{2+} build-up in the sea led to two main lineages among the protists: one characterized by cell walls rich in polysaccharides, and the other by containing collagen in addition. The first divergence generated in time the higher plants, whereas the second one evolved to the metazoans.

Synthesis of collagen may have given early metazoans a selective advantage, because hydroxy amino acids can scavenge Ca^{2+} thus counteracting Ca^{2+} stress imposed by the late Precambrian sea. The simultaneous rise in dissolved phosphates could enforce Ca^{2+} migration into cells (Cotmore et al., 1971). A highly schematic graph illustrates the biogeochemical evolution of soft tissue in the course of the Precambrian (Fig. 12.8).

As a final note, it is rather likely that Adolf Seilacher, one of the foremost paleontologists, is right in proposing that Ediacaran organisms were not the ancestors of the multicellular organisms that started the Cambrian radiation about 550 million years ago, but just a failed biological experiment (see Lewin, 1984). Ediacaran organisms may represent a short-lived fauna that relied on collagen to overcome chemical stress. Their odd "Bauplan", however, led to early extinction.

Biomineralization

Fool Canst tell how an oyster makes his shell?
Lear No.
Fool Nor I neither; but I can tell why a snail has a house.
Lear Why?
Fool Why, to put his head in.
(William Shakespeare: King Lear, Act I; Scene 5)

Organisms lay down minerals for various reasons: (i) as a deposit for physiologically essential elements, (ii) as skeletal material, (iii) to remove toxic elements, and (iv) in the course of a series of diseases such as atherosclerosis, or stone formation in the urinary tract. In short, biomineralization falls into two categories – normal and pathological. I will focus only on the beneficial aspects of biomineralization.

We know about 20 skeletal minerals in plants and animals (Lowenstam, 1973; Lowenstam and Margulis, 1980), of which four are abundant: aragonite, calcite, carbonate-apatite (dahllite, francolite) and opal. The others are either trace constituents or occur in only a few less common species. Thus, when it comes to the mineralization of biological tissues, calcium, silicon, carbonate and phosphate are the principal elements to deal with.

In 1962 it was said that "the nature of the local mechanism of calcification is one of the most important unsolved problems in biochemistry" (Urist, 1962). Much has been learned over the past 25 years on biomineralization, but a universally acceptable model is still out of reach. The reason is that researchers in the fields of biochemistry, the medical sciences, biology, chemistry, and geology define the problem in their professional way, almost untouched by progress achieved in neighboring fields. Even within a single discipline, let us say the earth sciences, the general approach is one of specialization. The viewpoint of a paleontologist on the origin of shells, bones or teeth differs greatly from that of a geochemist or that of a crystallographer.

What has been said for the earth scientists may well apply to biochemists, except that they study principally soft parts to resolve the question of calcification. In other words, the competence of the two professions stops on one or the other side of the boundary where soft and hard tissues meet. This is the terra incognita where, perhaps, the answer to biomineralization could actually be found.

Shape and color of shells have always attracted the curiosity of people, and King Lear and the Fool are no rare exceptions at all (Fig. 12.9). This, perhaps, may be the reason why the majority of articles in the field of biomineralization are descriptive in nature. Morphologies of mollusc shells are depicted in great detail, or the intricate relationship between shell organic matrix and mineral phase is examined under the electron microscope. Other papers deal with the chemical composition of the organic matrix quasi "frozen" in mineral matter, the type and shape of crystals or the trace metal content of various minerals. Only more recently has emphasis shifted in the direction of mechanisms that initiate and control mineral deposition in an organism and which relate crystal growth to distinct physiological and environmental happenings. The subject matter has been thoroughly reviewed in other places (Degens, 1976, 1979; Krampitz and Witt, 1979; Westbroek and DeJong, 1983; Degens et al., 1984; Braster, 1986).

Enzymic Activity

The two enzymes most frequently mentioned in connection with biomineralization are (i) carbonic anhydrase and (ii) alkaline phosphatase. The first one – among other physiological tasks – is regulating the carbonate system, whereas the second one is involved in Ca^{2+} transport processes.

Carbonic anhydrase is a zinc-containing enzyme with a molecular weight of approximately 30,000. It can occur in the form of multiple isoenzymes which differ in their specific activity. A schematic drawing of the molecule is shown in Fig. 12.10 (Kannan et al., 1972, 1975). Catalysis of CO_2 hydration-dehydration by carbonic anhydrase represents the starting point and two aspects have to be considered: (i) the mode of binding of the CO_2 substrate at the active site, and (ii) the physical-chemical state of ligands on the zinc ion.

A direct coordination of CO_2 to the zinc ion can be excluded because all potent inhibitors of enzymic activity become directly coordinated to the zinc atom. This implies that CO_2 binding is rather specific, perhaps involving stereochemical adjustment, such as fixation of a CO_2 molecule which is known to increase rates of hydration by several orders of magnitude.

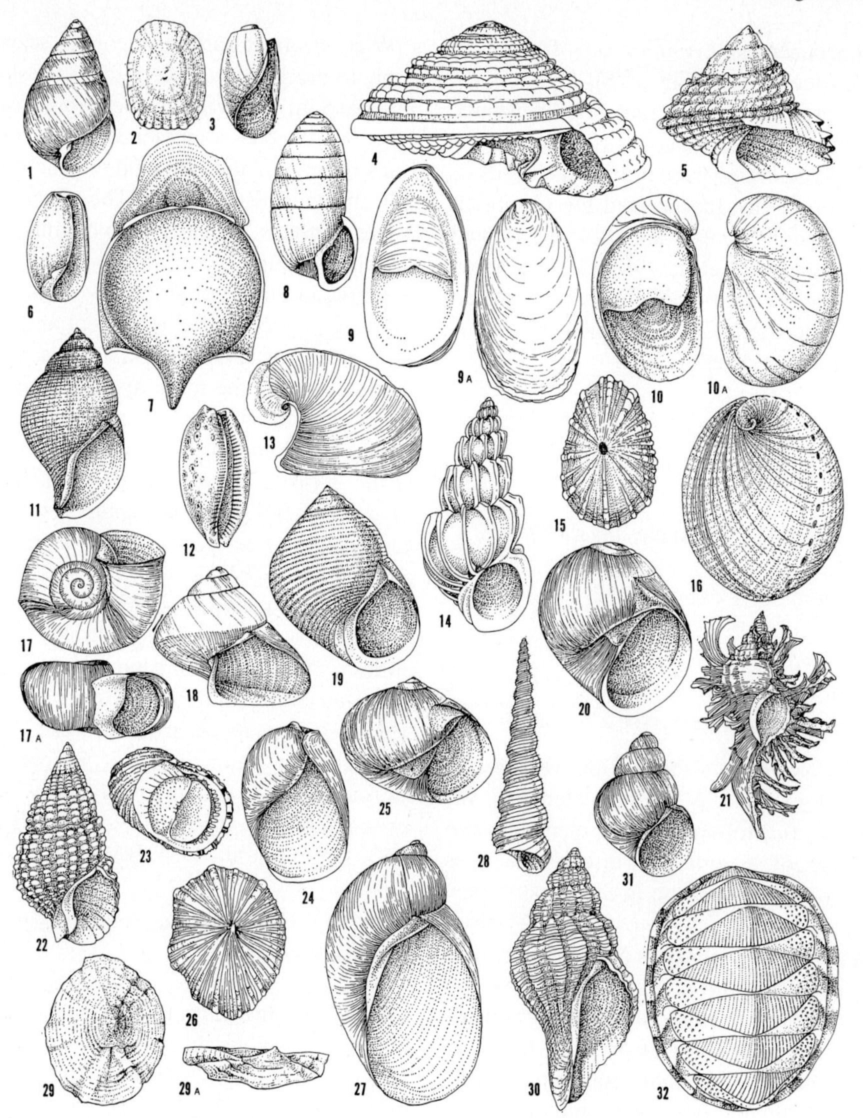

Fig. 12.9. Shells of snails: 1. *Achatinella lorata,* x 1; 2. *Acmaea pustulata,* x 2/3; 3. *Akera soluta,* x 1/3; 4. *Architectonica nobilis,* x 2; 5 *Astraea caelata,* x 2/3; 6. *Bulla striata,* x 2/3; 7. *Cavolina tridentata,* x 2; 8. *Cerion regium,* x 2/3; 9. *Crepidula fornicata,* x 1 1/3; 10. *Crepidula plana,* x 2/3; 11. *Colus trophius;* 12. *Cypraea zebra,* x 1/3; 13. *Dolabella callosa,* x 2/3; 14. *Epitonium angulatum,* x 2; 15. *Fissurella barbadensis,* x 2/3; 16. *Haliotis cracherodi,* x 1/3; 17. *Helisoma trivalvis,* x 1 1/3; 18. *Janthina janthina* x 5/6; 19. *Littorina littorea,* x 1; 20. *Lunatia triseriata,* x 2; 21. *Murex brevifrons,* x 1/3; 22. *Nassarius trivittatus,* x 2; 23. *Nerita plexa,* x 1/3; 24. *Oxynoe viridis* x 2 1/3; 25. *Polinices duplicatus,* x 1 1/3; 26. *Siphonaria alternata,* x 1 1/3; 27. *Succinea ovalis,* x 2; 28. *Turritella terebra,* x 1/3; 29. *Umbraculum indicum,* x 1/3; 30. *Urosalpinx cinerea,* x 4 2/3; 31. *Viviparus georgianus,* x 2/3; and the ancestral form (Amphineura); 32. *Chaetopleura apiculata,* x 2 (after Ghiselin et al., 1967)

There is general agreement that carbonic anhydrase activity is linked to the zinc ion and its ligands. At the active site water is bound to the zinc but the state and actual involvement of the water in the actual catalysis is a matter of controversy. According to a widely accepted opinion, the zinc-bound water is ionized and the activity is a function of pH.

This viewpoint is in conflict with proton relaxation data (Koenig and Brown, 1972), which indicate that only the high-pH form of the enzyme has a water ligand and that there is no bound water or OH^- at low pH. The idea has been advanced that only the high-pH form catalyzes both the hydration and dehydration reaction. Furthermore on kinetic grounds there is strong evidence against HCO_3^- as a substrate in the dehydration reaction. Instead it was proposed that dehydration of HCO_3^- catalyzed by carbonic anhydrase proceeds by the initial formation of a complex between the high-pH form of the enzyme and a neutral H_2CO_3 molecule which displaces the H_2O on the metal. Thus, the reaction that is reversibly catalyzed reads:

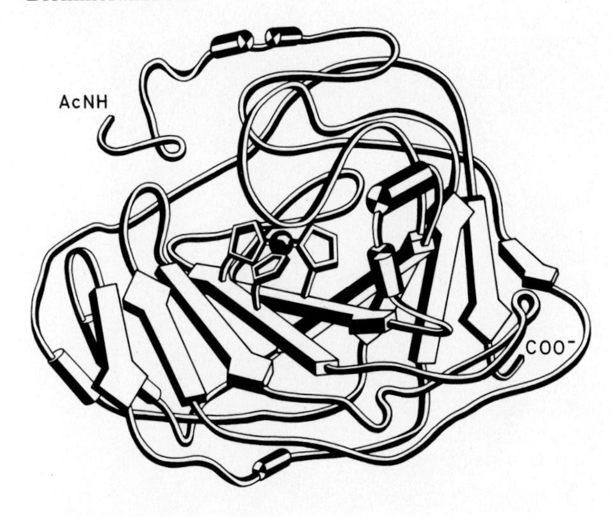

Fig. 12.10. Idealized drawing of the main chain folding of human erythrocyte carbonic anhydrase C (HCAC). The helices are presented by *cylinders* and the pleated sheet strands are drawn as *arrows* in the direction from the amino to carboxyl end. The *ball* supported on three histidyl residues represents the zinc ion (after Liljas et al., 1972; Kannan et al., 1972)

$$CO_2 + H_2O \rightarrow H_2CO_3.$$

The idea of H_2CO_3 as substrate for carbonic anhydrase is strongly supported by inhibitor binding studies. Small anion or sulfonamide inhibitors are linked to the zinc ion at high pH. The complex picks up a proton and the inhibitor is bound in a neutral form. Moreover, in the presence of inhibitors such as Diamox or sulfanilamide (Fig. 12.11), enzymic activities are strongly reduced and result in a decrease in shell thickness of eggs or malformation in calcareous tests of the kind shown in Fig. 12.12. In the presence of carbonic anhydrase inhibitors, rates of calcification are effectively lowered sometimes to the point of a standstill of mineral deposition.

These studies indicate that the actual role of carbonic anhydrase in calcification does not rest on its ability to hydrate CO_2 but to remove carbonic acid from the site of calcification:

$$Ca^{2+} + 2HCO_3^- \rightarrow Ca(HCO_3)_2 \rightarrow CaCO_3 + H_2CO_3.$$

The following conclusion can be drawn from these data:

Carbonate deposition in biological systems involves interaction of Ca^{2+} and HCO_3^- resulting in the formation of an unstable intermediary product $Ca(HCO_3)_2$. As long as calcium is not the limiting factor, the rate of formation of calcium carbonate will depend on the rate by which carbonic acid is removed from the calcification site. In the presence of carbonic anhydrase, calcification is significantly increased, due to the formation of a complex between the high-pH form and a neutral H_2CO_3 molecule, which is the substrate for carbonic anhydrase (Koenig and Brown, 1972).

Alkaline phosphatase, complementary to carbonic anhydrase in calcium homeostasis, is a dimeric zinc metalloenzyme composed of two identical subunits. The number of zinc atoms per protein molecule varies in different preparations, but only two seem to be required for catalytic activity. The molecular weight of the monomer has been reported to be 42,000, so that the natural dimer would be twice that value. Alkaline phosphatase is a phosphorylating enzyme and has 760 residues per dimer.

Action of phosphatase on glucose-1-phosphate substrate involves a breaking of the P-O bond and a release of a PO_4 group. In contrast, phosphorylase which operates on the same substrate opens the C-O bond. Apparently this affinity to specific bond sites, which in the

Fig. 12.11. Inhibition of carbonic anhydrase in a crude enzyme preparation from *Serraticardia maxima* by various compounds (after Okazaki, 1972)

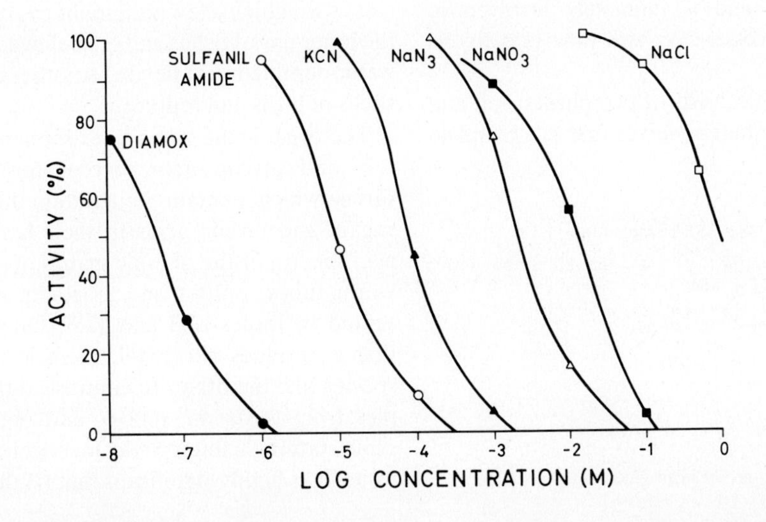

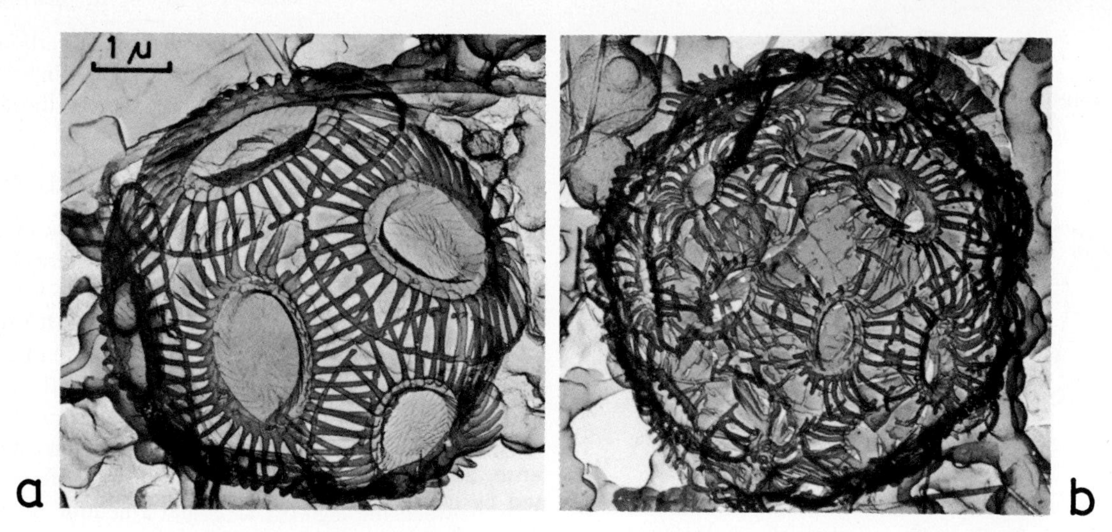

Fig. 12.12. **a** A healthy coccolithophorid alga, *Emiliania huxleyi,* and **b** a deformed coccolith from the same species (courtesy of S. Honjo). Malformations can arise through environmental stress such as abnormal temperatures, salinity or nutrient deficiency, heavy metal pollution, and chlorinated hydrocarbons

present case amounts to a difference of ca. 1.5 Å (Fig. 12.13) between the two bonds "left" and "right" from the oxygen, is caused by aspartic acid and isoleucine in the enzymes.

Alkaline phosphatase has a dual nature and exhibits phosphatase and pyrophosphatase activities. It is even quite likely that alkaline phosphatase, pyrophosphatase, and a calcium ATPase are one and the same enzyme. This enzyme has three separate functions: (i) breaking down ATP, (ii) hydrolyzing inorganic pyrophosphate and possibly other phosphate esters, and (iii) opening of C-O bonds. In this capacity it has a profound impact on transport and regulation of phosphates and calcium, and is intimately involved in biomineralization processes. We will now turn to the skeletal products.

All skeletal material consists of two phases: soft and hard tissue. Soft tissue hereby serves first as a template in the nucleation and oriented growth of biominerals, and second as a terminator of mineral deposition. The amount of organic matter in skeletal material can be as low as 0.01 percent in some mollusc shells, and as high as 20 to 30 percent in vertebrate bones or teeth; in a few isolated cases concentrations may go up to 90 percent.

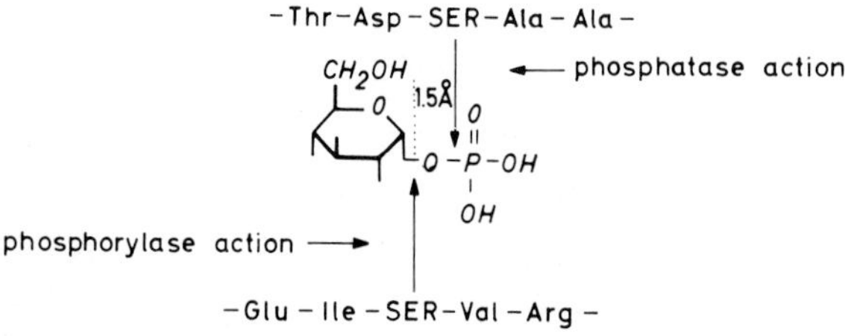

-Thr-Asp-SER-Ala-Ala-
CH_2OH ← phosphatase action
1.5Å
O-P-OH
OH
phosphorylase action →
-Glu-Ile-SER-Val-Arg-

Fig. 12.13. Release of PO_4 group from glucose-1-phosphate

Gross structure and composition of skeletal organic matrix are different in calcified (calcite, aragonite), phosphatised (carbonate-apatite), and silicified (opal) tissues which provide a logical subdivision in discussing the subject matter at large.

Calcification

Aragonite and calcite are the two calcium carbonate species most ubiquitous in biological systems. Their major domain are the lower plants and invertebrates and they rarely occur in higher organisms. One of the few exceptions are otoliths, which are small aragonitic stones weighing a few milligrams to grams deposited in the inner ear of fishes and are believed to function as a water depth control device. Another example are egg shells of birds and reptiles.

The organic matrix in biocarbonates is chiefly protein and glycoprotein, except for some ancestral forms which contain in addition high amounts of sugars and uronic acids in their hard parts. Amino acid spectra of the organic matrix in shell carbonates, worm tubes, oolites and aragonite needles are presented in Tables 12.2 and 12.3. On a relative basis, that is residues per 1,000, there is virtually no one species like the other. In contrast, different shell samples from the same species and obtained from the same natural habitat yield identical amino acid patterns. It is highly significant that (i) the mineralogy of

Table 12.2. Amino acid composition of various carbonates, in residues/1000

Amino acid	*Hydroides* (Serpulid) (Worm tubes)	*Porites* (Scleractinian coral)	*Eunicea* (Alcyonarian coral)	Oolites (Bahamas)	Aragonite Needle mud (Bahamas)
Aspartic acid	213	344	745	302	313
Threonine	97	21	13	51	50
Serine	42	29	13	46	54
Glutamic acid	120	100	37	109	121
Proline	62	35	11	38	35
Glycine	102	114	79	139	141
Alanine	90	67	60	93	79
Cystine (half)	4	2	1	2	2
Valine	72	56	14	46	43
Methionine	8	11	1	24	20
Isoleucine	41	46	5	28	24
Leucine	49	69	6	52	42
Tyrosine	18	10	1	7	7
Phenylalanine	36	33	3	38	26
Ornithine	4	11	1	2	3
Lysine	18	27	5	14	22
Histidine	3	12	1	2	2
Arginine	21	13	4	7	16
Total (μ moles/g $CaCO_3$)	8.4	2.9	17.2	3.8	21.8

Table 12.3. Amino acid composition of organic matrix in shell carbonates (geometric means in Res./1000; numbers in parentheses are 1 σ ranges)

	Amphi neura (1)	Cephalo-poda (4)	Gastro-poda (40)	Pelecypoda (51)	Brachio-poda (1)	Bryozoa (3)	Echinoidea (4)	Portunid-crabs (6)	Pisces (Otoliths)	Avian Shell (Goon)
OH-proline	–	–	.01 (0–0.1)	–	–	0.1 (0–7)	1 (0–36)		30	–
Aspartic acid	128	98 (75–128)	114 (69–185)	136 (77–240)	97	117 (100–136)	88 (78–98)	81 (73–92)	162	100
Threonine	60	34 (18–64)	46 (28–76)	28 (18–44)	39	59 (53–66)	48 (42–55)	57 (53–61)	57	66
Serine	68	95 (62–146)	85 (58–124)	66 (43–102)	70	77 (60–98)	81 (77–86)	91 (80–105)	44	101
Glutamic acid	98	65 (42–128)	87 (52–145)	50 (35–72)	77	102 (93–112)	109 (106–112)	94	170	128
Proline	93	55 (15–205)	61 (36–102)	52 (28–98)	36	50 (36–67)	78 (66–92)	103 (89–119)	50	72
Glycine	131	153 (89–262)	153 (99–238)	271 (185–398)	301	148 (127–172)	171 (121–241)	113 (101–125)	125	130
Alanine	88	126 (74–216)	89 (52–153)	66 (43–101)	76	92 (89–95)	85 (79–91)	115 (102–130)	96	86
Cystine (half)	14	7 (0–65)	9 (4–23)	14 (6–33)	8	13 (4–39)	1 (0.5–2)	3 (1–9)	17	36
Valine	45	36 (20–64)	48 (34–69)	33 (22–49)	74	48 (38–60)	38 (31–45)	65 (59–72)	70	53
Methionine	12	8 (5–13)	14 (6–31)	18 (8–39)	13	16 (14–18)	21 (18–25)	7 (3–11)	1	1
Isoleucine	30	21 (17–26)	31 (20–46)	20 (13–31)	37	31 (25–39)	26 (21–32)	28 (25–31)	30	29
Leucine	52	44 (29–67)	71 (51–100)	34 (26–44)	47	59 (46–76)	58 (47–70)	49 (44–54)	73	63
Tyrosine	31	33 (15–75)	11 (5–24)	11 (2–61)	4	23 (21–25)	20 (16–25)	25 (20–30)	1	24
Phenylalanine	37	30 (20–45)	24 (14–43)	30 (19–48)	19	28 (24–33)	25 (19–32)	30 (24–37)	12	38

Table 12.3. (continued)

	Amphi neura (1)	Cephalo-poda (4)	Gastro-poda (40)	Pelecypoda (51)	Brachio-poda (1)	Bryozoa (3)	Echinoidea (4)	Portunid-crabs (6)	Pisces (Otoliths)	Avian Shell (Goon)
OH-Lysine	–	–	.03 (0–0.2)	.03 (0–0.6)	0.1	0.4 (0–8)	0.6 (0–11)	3 (0–7)	–	–
Lysine	41	20 (5–83)	26 (17–41)	22 (13–38)	31	49 (40–61)	45 (35–57)	52 (28–97)	20	17
Histidine	11	5 (0–35)	3 (1–16)	4 (1–14)	0.3	17 (10–30)	16 (13–20)	26 (17–39)	7	19
Arginine	61	23 (10–56)	22 (9–52)	24 (6–96)	70	50 (41–60)	69 (60–80)	35 (18–68)	14	36
Protein/ hexosamines[a]	0.98	1.37	108.30	143.97	110.97	21.93	33.36	0.40	96	54

[a] From arithmetic means.

carbonates, (ii) the content in trace elements, and (iii) the stable isotope distribution are markedly affected by fluctuations in salinity, water temperature, Eh/pH conditions, and some anthropogenic factors. The same environmental parameters also seem to govern the chemistry of the shell organic matrix. This feature suggests a cause-effect relationship between the mineralogy and organic chemistry of a shell. In the final analysis, however, it is simply a reflection of the environmentally controlled dynamics of the cell.

To explain the species-specific amino acid pattern one can assume that each organism secretes one particular calcifying protein, which to a limited extent is controlled by the environment. Alternately, a series of different proteins, but each one excreted at a constant rate, can produce the observed pattern. Changes in rates of secretion, perhaps environmentally induced, could conceivably affect the ratio among the individual proteins and may account for amino acid variations.

Chemical studies confirm (Degens, 1976, 1979) that shell organic matrix is a composite of a few proteins and glycoproteins, but that the bulk organic matrix contains just three endmembers: the first being high in aspartic acid (> 50%) and glycine, the second being enriched in glycine, alanine and serine in a ratio of about 3:2:1, and the third being an EDTA-extractable glycoprotein. The first protein bears a striking resemblance to troponin, the second to silk, and the third to a Golgi-derived glycoprotein containing glucuronic acid or a similarly acting carbohydrate. Species-specific differences in the amino acid spectra of shell organic matrix depend chiefly on the mixing ratio of such proteins.

Matrix proteins are composed of two structural units: a low molecular weight member with strong affinity to Ca^{2+} ions, and a high molecular weight member with no affinity to Ca^{2+}. The first acts as organic template initiating calcification ($Ca^{2+}T$) and the second as a binding matrix (BM) for the aspartic acid-rich $Ca^{2+}T$. Attachment of $Ca^{2+}T$ to BM will activate the mineralizing substrate, and calcium carbonate crystals are laid down in epitaxial order. Crystal growth terminates upon secretion of a newly formed $Ca^{2+}T$ on the growing crystal surface, whereby a kind of organic blanket is formed. Subsequently, BM becomes attached to $Ca^{2+}T$. A sandwich is generated with the BM embedded between two sheets of the $Ca^{2+}T$ (Fig. 12.14). In this way, the organic substrate is activated and calcification may start again. Thousands of such layers can be generated in the lifetime of an organism and signal diurnal, seasonal, and

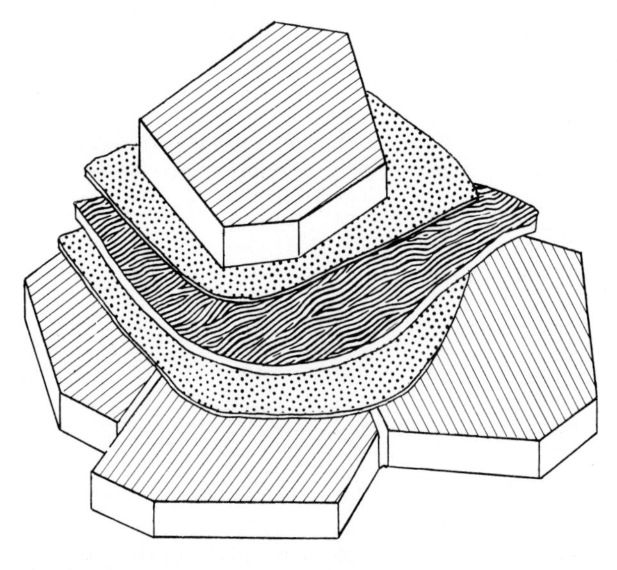

Fig. 12.14. Schematic representation of structural relationship between layers of Ca^{2+}-affine template ($Ca^{2+}T$; *dotted*), the silken binding matrix (BM; *banded*) and solid $CaCO_3$ (*block-shaped*)

other events (Box 12.3). Whereas $Ca^{2+}T$ forms a cover only a few Å thick, BM may have a thickness ranging from 30 Å to several hundreds of Å depending on species or even environmental stress. For comparison, the size of the individual $CaCO_3$ crystals is commonly of the order of a few microns.

BM's amino acid composition is identical to that described for the β-configuration of silk fibroin, where serine supplants alanine in the layer structure in concentrations up to 15 mol %. In silk, polypeptide chains are arranged in a zig-zag configuration and so-called pleated sheets are generated (Fig. 12.16). The structure is maintained by hydrogen bonding between adjacent chains. It is of note that the level of serine in shell matrix proteins is commonly around 8–10 mol percent and may reach concentrations as high as 20 mol percent. The main function of serine lies in its ability to form hydrogen bonds which increase the degree of cross-linkage among protein chains and rise their hydrophobic nature. The oxygen of the hydroxyl group can form two hydrogen bridges by supplying one H-atom and attracting another one. This process increases the dispersion bonding forces between residues by bringing these groups in closer contact. In this way the proteins become remarkably resistant to chemical influences and this may explain why organisms adopted silk-type proteins as an inert substrate in mineralization processes.

BOX 12.3

Growth Rings in Carbonate Shells

Growth patterns, particularly microstructural increments within corals, molluscan shell structures or fish otoliths have received wide attention. Marked periodicities associated with many of these structures in modern organisms and fossils as well could be attributed, for instance, to circadian rhythms (Rhoads and Pannella, 1970), physiological stress situations (Pannella, 1971), environmental events (Lutz and Rhoads, 1977) or changes in the Earth's rotation (Brosche and Sündermann, 1982). In the last instance, however, all growth rings signal times of mineral deposition followed by times of organic matrix deposition, so that interpretation of this phenomenon may assist in elucidating mechanisms of calcification and the biochemical regulation of metal ions, notably Ca^{2+}.

Corals have generally a steady growth rate throughout the year provided that environmental parameters stay constant and symbiontic algae are absent (Bak, 1973). A significant aspect of skeletal growth in corals is the existence of an internal calendar where daily, seasonal, annual, and event records are kept on file in the form of individual growth rings or series of growth bands. It is the number of bands per year in fossil corals that served in studies on the Earth-Moon system by revealing the effect of tidal friction on number or length of days per year (Brosche and Sündermann, 1982). Although in the course of the Phanerozoic the observed time-trends are not linear, paleontological data seem to suggest that about 500 million years ago we had roughly 50 more days per year than we have now.

Should corals become impacted by resuspended bottom sediment, they free themselves from the detrital load by secreting massive amounts of mucus (Dodge et al., 1974). During such events coral growth is greatly inhibited, and variability in growth decreases. Apparently the energy expenditures for removal of sediment particles and the mobilization of Ca^{2+} for muscle or energy transfer has left no excess calcium for biomineralization. Thus, during an active stage little if any calcium is exported to the cell periphery and consequently calcification is at a minimum.

With respect to mollusc shells, growth increments of 1 to 100 microns thick are common. They appear to be controlled by tidal effects since increments are in phase with a circalunadian rhythm (24 hours and 50 minutes) (Palmer, 1975). During exposure at low tide, growth stoppage occurs, giving rise to a sharp line composed of organic matrix. Furthermore, specimens from intertidal zones or a turbid subtidal mud bottom exhibit low growth rates. In contrast, calcification rates are high in specimens living in sand or in molluscs that stay elevated – half a meter or so – above the muddy bottom. Comparison of shallow and deep water forms reveals that in deep water specimens increments have a uniform thickness without sharp boundaries and no apparent seasonal or tidal indications.

A remarkable lining pattern is established in fish otoliths. Close examination reveals zones of fast and slow growth (Pannella, 1971). Each zone is characterized by alternating light and dark colored microlayers (Fig. 12.15). During fast deposition, production of organic matrix is high but calcification is even higher, whereas at times of slow deposition calcification is almost lacking, and the observed increments are due to organic matter contribution. The slow growth zone is composed of 100 to 110 bands and the fast growth zone of 230 to 260 bands, which represent winter and summer layers respectively. In addition, there are band series of weekly, fortnightly and monthly periodicities. At times of spawning, calcification is virtually nil. All this indicates that growth takes place by daily increments which are most likely related to known circadian rhythms in the vital responses of fish.

It is concluded that microgrowth patterns in calcified tissues reflect events of many sorts. And yet, the ultimate reason behind this phenomenon can be explained on physiological grounds as a result of the utilization of calcium in energy transfer and muscle activities. Namely, at times of stress these activities have higher priorities than calcification.

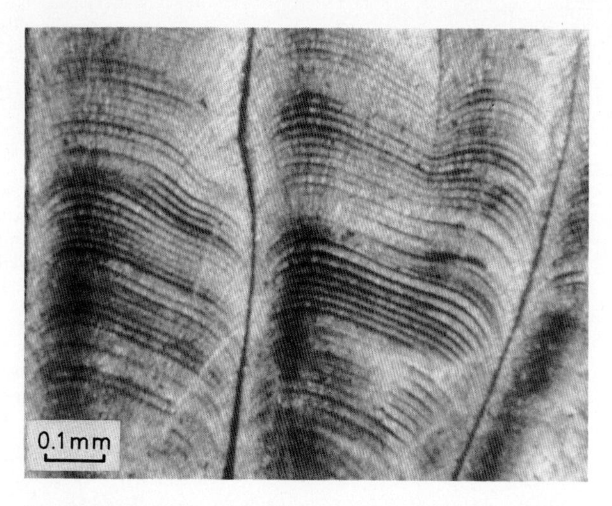

Fig. 12.15. Otolith thin-section of *Roccus lineatus* revealing banding pattern (*light:* aragonite; *black:* proteinaceous matter) (Degens et al., 1969)

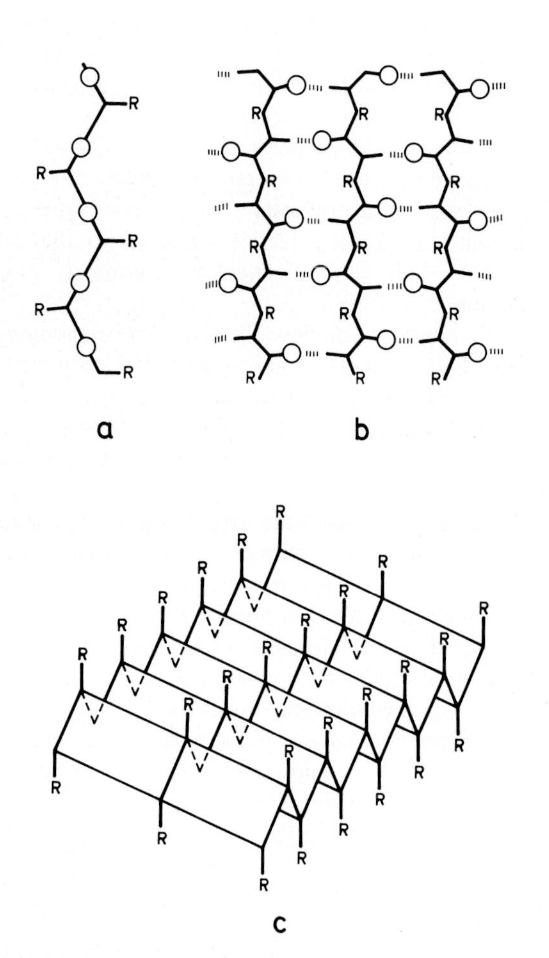

Fig. 12.16 a–c. Schematic representation of beta-pleated sheets, **a** top view, **b** edge view (hydrogen bonds are indicated), and **c** three parallel chains in beta-structure

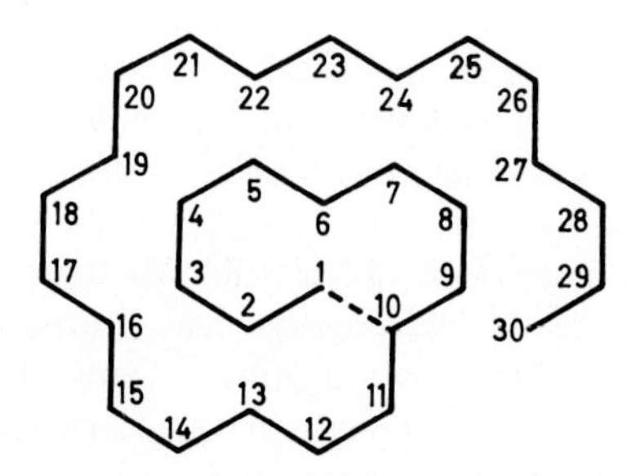

Fig. 12.17. Space-filling model of the B chain of insulin. *Numbers* indicate position of carboxyl oxygens (after Warner, 1964). The $Ca^{2+}T$ in nacreous layer of molluscs is assumed to be built according to the same principles

The $Ca^{2+}T$ has a different structural order. The most probable one involves a spiral peptide chain which gives rise to a hexagonal network. A structural analog is found in the B chain of insulin which is composed of 30 residues (Fig. 12.17). Its carbonyl oxygens are located at the corner of nine continuous regular hexagons which are arranged in a honeycomb pattern. All carbonyl oxygens are positioned in a single plane, thus producing a hydrophilic surface. The side chains are located on the opposite side and they generate a hydrophobic surface.

Attachment of $Ca^{2+}T$ to the BM may involve water which would link the two fractions via hydrogen bonds. Sterically, the water would fit perfectly and greater stabilization would be achieved. On the other hand, the high serine content could provide the necessary hydrogen bridges. In Fig. 12.18, a block model is shown in which only the organic phases appear;

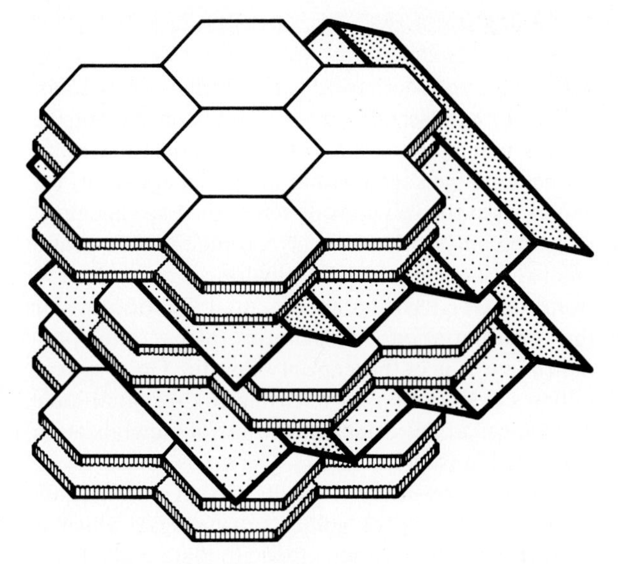

Fig. 12.18. Schematic representation of vertical cross-section of nacre showing alternating layers of $Ca^{2+}T$ (hexagons) and BM (pleated sheet conformation). The $CaCO_3$ has been omitted for graphical reasons

$CaCO_3$ has been deleted. Hexagonal plates of $Ca^{2+}T$ alternate with pleated sheets of BM. The hydrophobic sides of $Ca^{2+}T$ are facing each other and encase the mineral phase. An electron micrograph of the nacreous layer (= mother-of-pearl) of a clam further illustrates the structural relationships between organic matrix and its associated carbonate (Fig. 12.19).

There is no energy problem in $CaCO_3$ deposition, since the overall crystallization will liberate energy in the order of 2–4 kcal mol^{-1}. To allow $CaCO_3$ deposition – which must proceed rather quickly – the surface must be kept clean from interfering components. This may be done by chemical species such as sugar derivatives, and their action may be compared to that of jet gases on the crystallisation of ice at high altitudes.

The final question is of how Ca^{2+} is brought to the site of calcification. This might be mediated by glucuronic acid or a similarly acting carbohydrate. A brief operation principle is presented:

Glucuronic acid forms a soluble and inactive complex with the calcifying template. This complex is

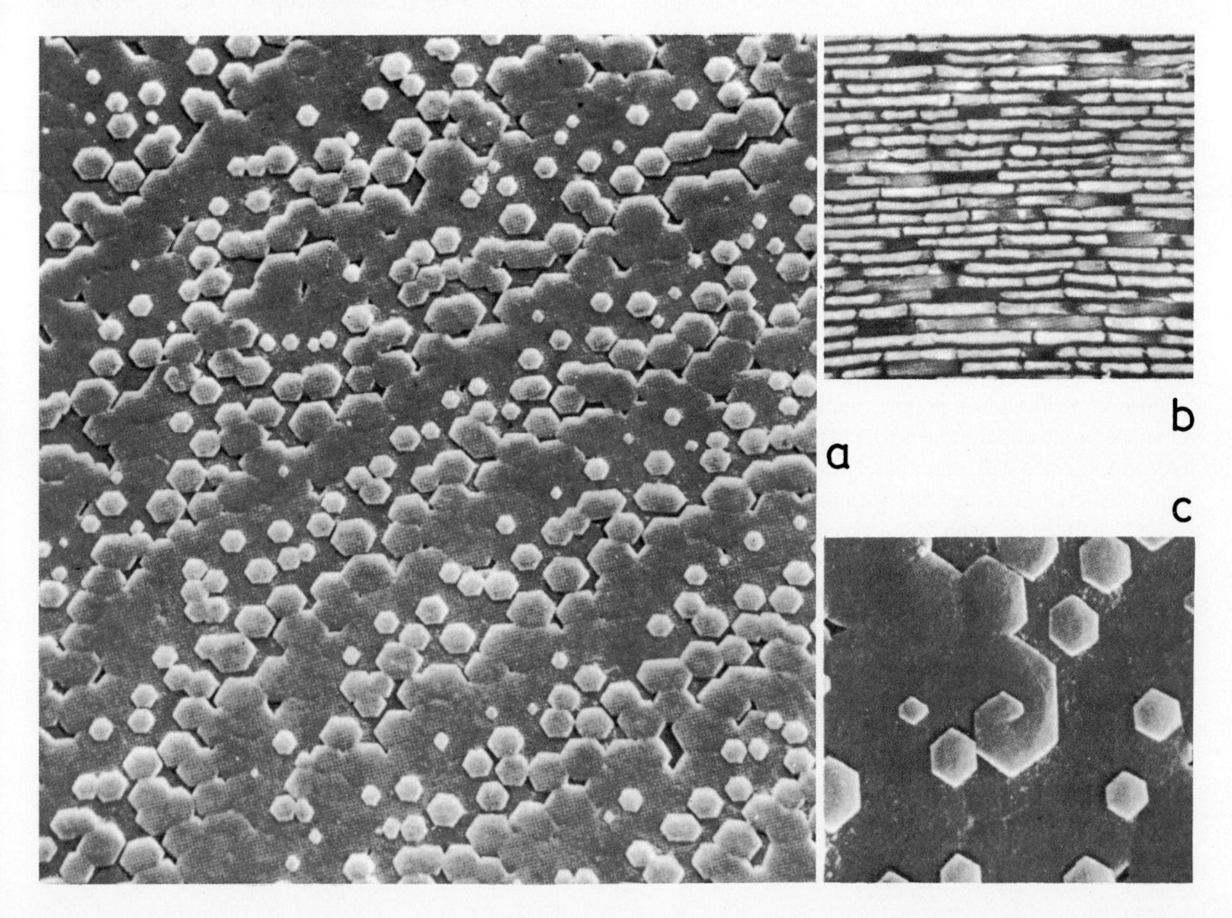

Fig. 12.19. **a** Electron micrograph of horizontal section of pelecypod nacre (mother-of-pearl) showing steps formed by overlap of mineral laminae on growth surface; **b** vertical section exhibiting brick-wall pattern; **c** terminal screw locations (after Wise and Villiers, 1971)

BOX 12.4

Golgi Complex

Eukaryotic cells contain complex membraneous organelles known variably as the Golgi complex, Golgi apparatus or Golgi body which occupy a central position in the transport system of the cell. It is this organelle which appears to be the starting point of Ca^{2+} secretion into the ambience.

The Golgi complex is in a constant state of flux (e.g., Northcote, 1971; Alberts et al., 1983; Rothman, 1985). Cisternal membranes arise from the rough endoplasmic reticulum, change to smooth endoplasmic reticulum and become the Golgi cisternae. Such cisternae then break down to vesicles which can fuse with and extend the plasmalemma.

On one surface, the Golgi complex has a fresh face and on the opposite surface a mature face. Membranes on the fresh face resemble those of the endoplasmic reticulum, and membranes on the mature or secreting face are similar to the plasmalemma. This suggests that the Golgi apparatus operates as a membrane transformer from: endoplasmic reticulum to vesicles to Golgi apparatus to vesicles to plasmalemma, a process which appears to be irreversible. Vesicles derived from the Golgi cisternae have an inner mucopolysaccharide coat which – when transferred to the plasmalemma – appears on the outside of this membrane. Many other materials are transformed and packaged within the Golgi apparatus for export from the cell interior across the plasmalemma, including polysaccharides, glycoproteins, cations and anions. It appears that carbohydrate moities derived from the Golgi system are essential "lubricants" for the transportation of material across the plasmalemma. It is in this capacity that the Golgi system initiates calcification by releasing the essential ionic and molecular ingredients to the site of mineralization (reviewed in Degens, 1976).

A particularly interesting calcification pattern is followed by a group of golden-brown algae which is characterized by a series of calcite plates called coccoliths. The individual plates are already manufactured in the Golgi complex proper and become transported and extruded via Golgi-derived vesicles to the cell surface, where they assemble in the form of interlocking structures of the kind shown for *Emiliania huxleyi* in Fig. 12.12. This group of calcareous nanoplankton known since the Jurassic is one of the most abundant $CaCO_3$ removers from tropical to polar seas ever since. For an extended review on calcification in biological systems see: Degens (1976).

---

transported from the Golgi apparatus (Box 12.4) to the nucleation site. Here, the carboxyl group might become transformed into an acetyl group, thereby rendering the molecule chemically inactive. Alternately, an enzyme – a protease – might open the polysaccharide peptide complex. Removal of glucuronic acid causes polymerization of $Ca^{2+}T$ and its subsequent attachment to BP. The freely exposed hydrophobic site will immediately function in the fixation of calcium ions and the formation of $Ca^{2+}O_6$ or $Ca^{2+}O_9$ polyhedra, which will promote the synthesis of calcite and aragonite nuclei respectively. There are indications that also amino sugars are linked to the active calcium transport. Thus, glucuronic acid together with glucosamine may not only serve as building blocks of cell walls (e.g., hyaluronic acid) but also act as a carrier in calcification.

Phosphatization

Mineralogy: The nature of mineral phases in bone, dentin, enamel and other phosphatic tissues is according to most bioscientists, hydroxyapatite, $Ca_{10}(PO_4)_6(OH)_2$, and deviation in Ca/P from common hydroxyapatite (Ca/P = 1.667) in mineralized tissues is explained by the presence of amorphous phosphate. In contrast, many crystallographers favor the idea of carbonate-apatite, i.e., dahllite, as the major crystalline phase in biophosphates.

Crystals in bone, dentin and enamel can vary greatly in size and in shape (Selvig, 1973). For example, the bulk of apatitic crystals in young bone material is small and compact, having a mean cross-sectional width of only about 90 Å; in addition, there are occasionally larger crystals (500 to 1,000 Å) which have a thin platelet-like form. The two crystal populations exhibit no continuous progression of sizes and are morphologically distinct from one another, suggesting that their formation is linked to different events.

Crystals showing a regular hexagonal outline occur in hypermineralized surface regions of cervical cementum and root dentin (Fig. 12.20). In such structures more than 100 repeat periods can be counted

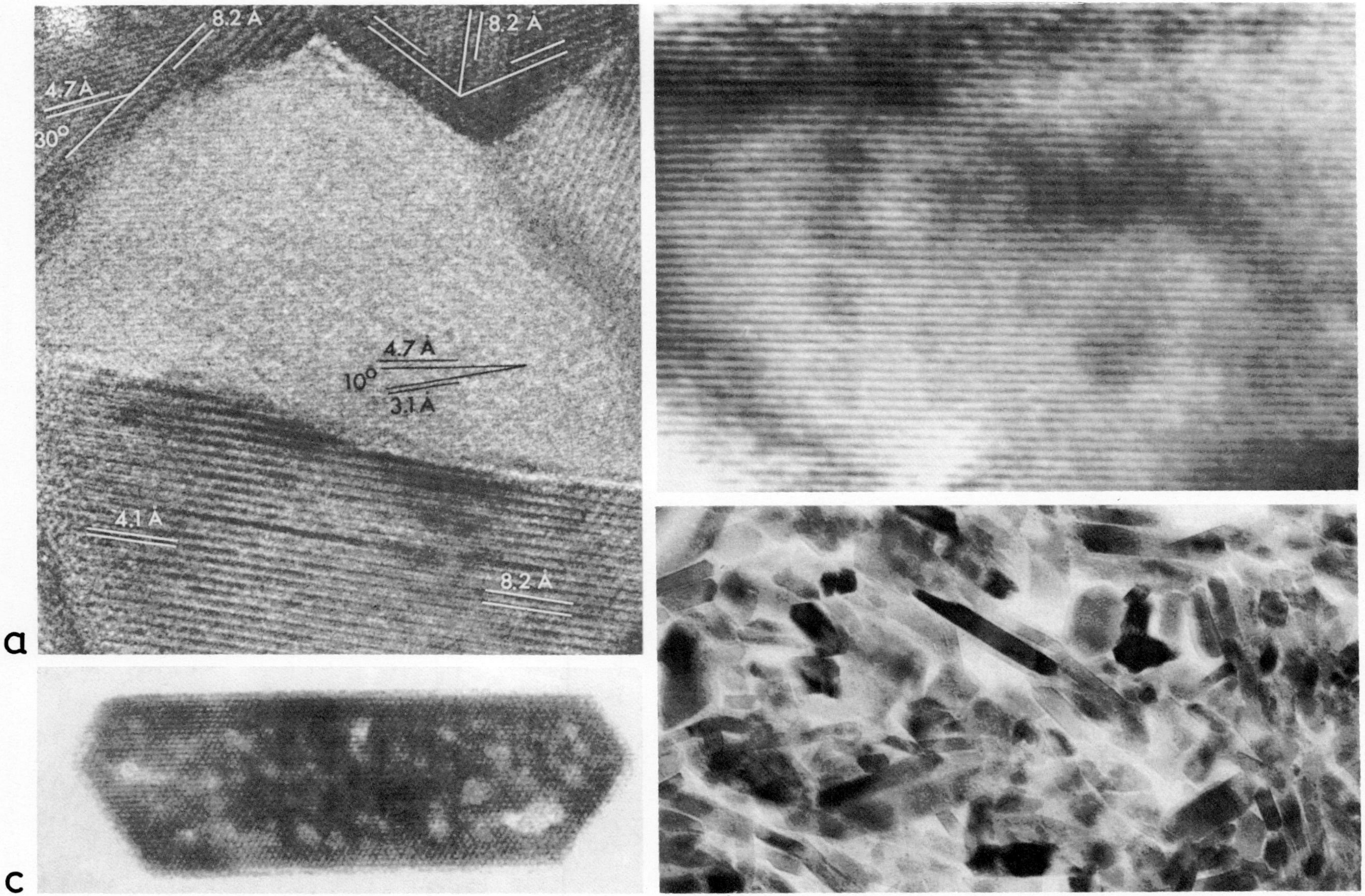

Fig. 12.20. a Electron micrograph of enamel showing lattice fringe patterns in individual apatite crystals; **b** regularity of repeating pattern; **c** apatite crystal in hypermineralized dentin, but slightly damaged by the electron beam; **d** crystals in hypermineralized dentin showing various sizes and shapes (after Selvig, 1973)

per single crystal. In normal bone and dentin, commonly less than ten and as little as four repeats are encountered. The mineral phase appears to be dahllite, and variation in Ca/P ratio is possibly related to isomorphous substitution of calcium by sodium or magnesium, and exchange of tetrahedral PO_4 groups by planar CO_3 groups.

In addition to a wide range of apatites, young bones contain calcium monohydrogen phosphate dihydrate (= brushite) which can readily hydrolyze into hydroxyapatite. Octacalcium phosphate has also been found, which is seen as a possible precursor of the small apatite crystallites in bones and teeth. Fluorine has been mentioned as an agent in the conversion to hydroxyapatite, and its therapeutical value rests on this action. In this context it is of interest that the shell of the inarticulate brachiopod genus *Lingula* is composed of francolite (carbonate fluorapatite).

The subject of amorphous phases in biophosphates is even more controversial than the nature of the crystalline phases. One source of information relegates amorphous phases to the "realm of mystics", whilst others voice strong opinions in favor of amorphous phases. Experimental evidence exists that as much as 40% of mineral matter in bone is amorphous calcium phosphate (ACP) (Eanes et al., 1973). ACP represents an agglomeration of hydrated calcium and phosphate ions with little or no three-dimensional order. Loss of excess hydration water and structural ordering may lead to epitaxial outgrowth of metastable and stable crystalline phases. This is a rapid process lasting at the most a few hours. With this in mind it is hard to understand why bones should have such a high ACP content. Thus, the concluding statement is at best tentative:

ACP-formation, most likely mediated by some organic substrate, is the initial solid phase. The process is similar to the deposition of amorphous silica ($SiO_2 \times nH_2O$) in diatoms, radiolaria, siliceous sponges and silico-flagellates (see p. 366). The two amorphous minerals differ in that metastable ACP readily transforms into a stable crystalline substance either directly or via metastable crystalline precursors such as brushite or octacalcium phosphate. Amorphous silica is metastable because the conversion into crystalline phases such as quartz, tridymite or cristobalite is kinetically slow at physiological temperatures. Crystallization proceeds generally only in the geological environment, since thousands of years are needed to alter amorphous silica into quartz. The rare presence of tiny quartz crystals up to about 100 microns in human kidney stones remains an enigma.

At the onset of biomineralization the mechanism of phosphate and silica deposition is essentially the same. Both start with a highly hydrated amorphous phase having glass-like physical-chemical properties. However, the kinetics of the two differs. ACP will rapidly alter in the direction of apatite in hours and days, whereas amorphous silica requires many years or higher temperatures to yield quartz. As a result, biophosphates may contain a variety of amorphous and crystalline phases, whereas biogenic silica is solely amorphous in nature.

Chemistry: In the region of the mineralization front, phosphorus and calcium sharply rise and the curves closely resemble first order reaction curves (Figs. 12.21 and 12.22). The difference in the behavior of phosphorus and calcium comes best to light in the

Fig. 12.21. a Calcium and phosphorus concentrations and **b** molar Ca/P during mineralization in normal rats (Wergedal and Baylink, 1974)

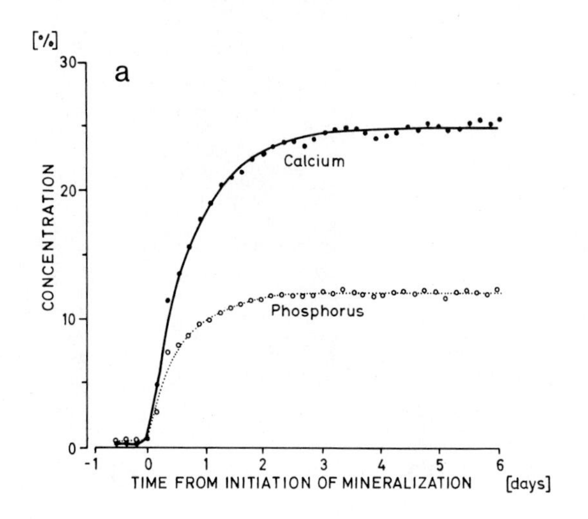

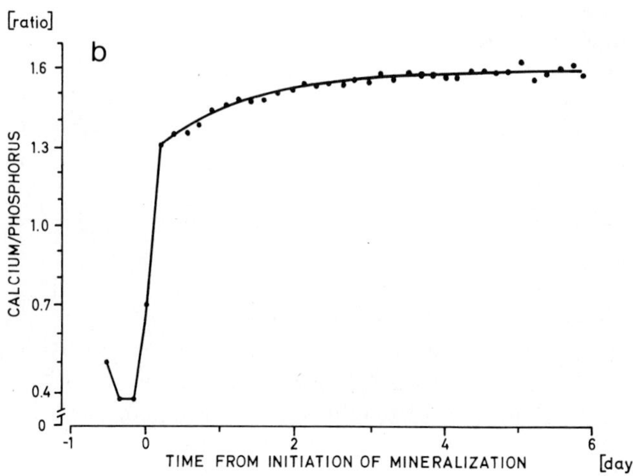

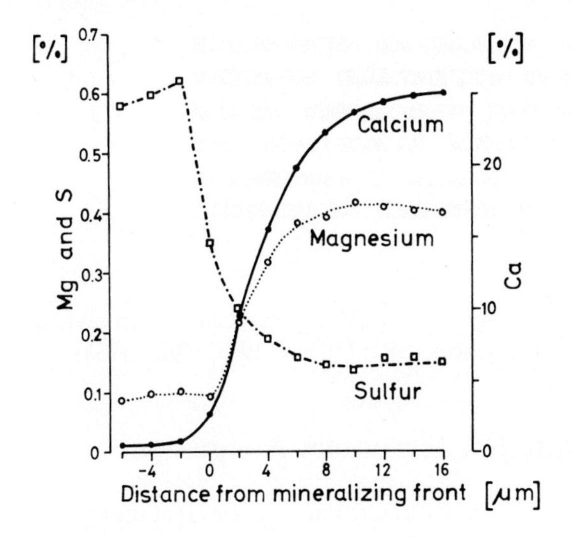

Fig. 12.22. Calcium, magnesium and sulfur concentrations during mineralization in normal rats (Wergedal and Baylink, 1974)

Ca/P curve (Fig. 12.22). On the mineralizing front and extending up to 12 percent Ca level, a value of 1.35 is recorded. From there on, values steadily increase to approach 1.60 at maximum mineralization. The shape of the magnesium curve is similar to the calcium curve, except that the enrichment factor is lower by at least an order of magnitude and the plateau level is reached sooner. In the case of sulfur, concentration drops sharply at the mineralizing front. Other elements have been suggested as being involved in phosphate deposition, most notably silicon. Coexistence of phosphates, silicates and aluminates in plaque powder of some atherosclerotic aorta tissue may support this inference. The elements Al, Si, and Ca are also found in plaques of glia cells of Alzheimer patients, old people, and humans with Down syndrom. So it appears that aging has something to do with the deposition of these three metal ions in human brain cells.

Organic matrix: In contrast to calcareous invertebrates with their species-specific shell organic matrix, apatite depositors follow a rather conservative pattern set by collagen for bone, scale, and dentin deposition, or by keratin-type protein in the case of enamel formation. The amount of organic matrix varies considerably among biophosphates; bone and dentin have values around 25 percent (dry weight), whereas mature enamel contains only about 1 percent.

Mineralization mechanisms: Ca^{2+}-accumulating vesicles, about 100 nm in diameter, have been identified in intercellular matrices of hard tissues (Anderson, 1973; Ali et al., 1970). A trilaminar membrane envelops the vesicle, which appears to be derived from plasma membranes. This relationship suggests – in a wider sense – that matrix vesicles are Golgi-elaborated structures. In a closer sense they are linked to mineralizing cells such as chondrocytes from which they became separated by budding; their de novo formation in the cell matrix can be ruled out. Isolated matrix vesicles are rich in lipids and phospholipids which show a high affinity for Ca^{2+}; at the same time, the vesicles are depositories of inorganic pyrophosphatase, ATPase, and alkaline phosphatase. Lack of acid phosphatase activity, β-glucuronidase and cathepsin D indicates that we are not dealing with lysosomal structures. The high content of phospholipids, and here of phosphatidylserine and phosphatidylinositol, cholesterol and glycolipid in almost twice the cellular amount, represents a clue to the origin of the vesicles and the mode of Ca^{2+} sequestering.

In the process of Ca^{2+} fixation to the internal face of the vesicular membrane, mucosubstances seem to act as Ca^{2+} carrier and transfer their Ca^{2+} to the membrane surface where potent acceptors are abundantly present. An especially interesting structure that may arise is one of corrugate layers of calcium monohydrogen phosphate, $CaHPO_4 \cdot 2H_2O$, which has been proposed as the initial mineral phase in biophosphates (Francis and Webb, 1971).

Following the separation of vesicles from their stem cell, the process of phosphatization continues, implying that Ca^{2+} is actively moving into the vesicle. The presence of pyrophosphatase at the vesicle membrane is needed to hydrolyze inorganic pyrophosphate (PP_i) into phosphate ($2P_i$), because PP_i is a known inhibitor of mineralization. Intravesicular phosphatization terminates with the rupture of the vesicle and the release of mineral matter and vesicle content at or close to the mineralizing front.

Subsequent to the rupture of the matrix vesicle, the phospholipid template plus mineral matter interacts with an organic substrate which is principally collagen in the case of dentin, cartilage and bone, and some keratin-type protein in the case of enamel. The ability of collagen to form biophosphates appears to be linked to the existence of "hole zones" which develop between the ends of collagen molecules (Fig. 12.23). Collagen fibrils and the rather lengthy needles of apatite crystals lie parallel to one another. It is unknown whether the collagen binds the essential elements in the form of single ions, ion clusters, amorphous phosphates, or microcrystalline material. Thus, the alignment of collagen fibrils and apatite crystals as seen in the late stages of mineralization does not necessarily imply that collagen is the nucleator of apatite. It is

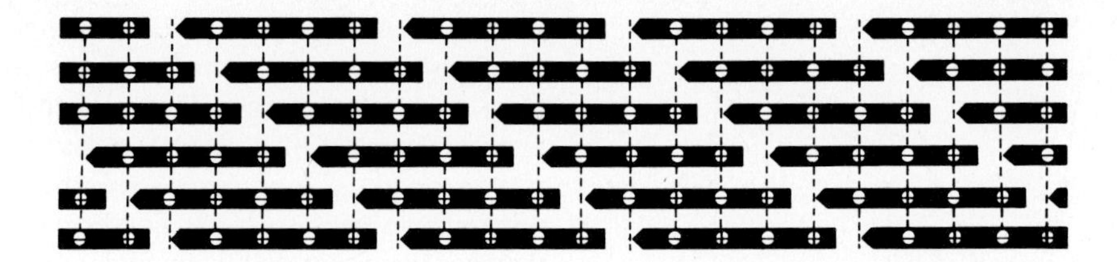

Fig. 12.23. Molecular overlap and hole zone in collagen fibrils (Höhling et al., 1971)

more probable that much of the apatite in fully mineralized tissues was formed by mineral growth upon pre-existing crystals, whereby collagen exercises structural control. In this capacity, collagen has the same physiological function in phosphate deposition as the silk-type protein in carbonate formation.

Silicification

Silicon is, next to oxygen, the most abundant element in the Earth's crust (Aston, 1983). It occurs, at least in trace quantities, in most plant and animal tissues; in a variety of organisms it is considered an essential element. Spectacular is the enrichment of silicon in mineralized tissues of unicellular organisms, e.g., diatoms and radiolaria, and a few metazoans, e.g., sponges and gastropods. Some higher plants also contain silica depots. The mineral phase is always amorphous silica in various stages of hydration: SiO_2xnH_2O.

As in the case of carbonates and phosphates, silica deposition, for instance in diatoms, can be traced back to the Golgi apparatus, where constituents of the cell wall are synthesized and ions are packed for transport across the silicalemma. Whereas glycoproteins are critical in Ca^{2+} transport, proteins or lipoproteins appear to be more suitable Si-carriers and substrates in Si fixation. Membranes of Golgi cisternae, and here in particular the surface composed of structural proteins, could provide sites for the coordination of Si^{4+} which is ultimately extracted from the endoplasmic reticulum. During transformation of vesicles into silicalemma, the sugars and uronic acids present on the inner walls of cisternae become expelled to the outside, and as a consequence protein and silica layers will turn upside down. Lipids and phospholipids will remain between the carbohydrate and protein layers and stabilize the membrane system through metal ion coordination. A hypothetical arrangement of organic layers in a diatom cell wall is depicted in Fig. 12.24 (Hecky et al., 1973). The biogeochemistry of silicon has recently been reviewed (C.P. Spencer, 1983; Calvert, 1983; Hurd, 1983).

Life is a Temporary Assignment

The origin and extinction of species is at the start and the heart of paleobiology and paleontology. The centennial of Charles Darwin's death was considered by many a scientist a welcome opportunity to elaborate on extinction in the fossil record, quite frequently in a radical way.

By definition, extinction is termination of evolutionary lineage. The matter of origin of species through time is harder to construe, whilst in living forms it can be established beyond doubt. That is, speciation hinders fertilization of members of one population by those of another, and gene flow between such populations is precluded. This has led to models which describe much of the evolving eukaryotic genome as a mirror image of molecular mechanisms of turnover processes involving repetitive DNA sequences (e.g., M.R. Rose and Doolittle, 1983). As useful as such molecular biological criteria are to distinguish one modern species from another living one, the concept obviously fails when it comes to separate clearly a given fossil as a species from the rest of the fauna or flora.

This is certainly not the place to verify or falsify paleontological nomenclature. Paleontologists have their own way of looking at speciation in terms of where to draw a line in the divergence of populations creating species, quite apart from the mechanisms that have led to the origin or extinction of the species. For the most part, paleontologists have to rely on morphological characteristics of shells, bones, teeth, spores, pollen, burrows, imprints etc., unless they are dealing with extant organisms which permit intercalibration with established biological classifications.

Calcareous, phosphatic and silicified tissues are the principal remains of life in the stratigraphic record. From a biogeochemical point of view, it is reasonable to relate observed changes in composition or structure of fossil material to biogeochemical processes controlling the cycling of calcium carbonate, calcium

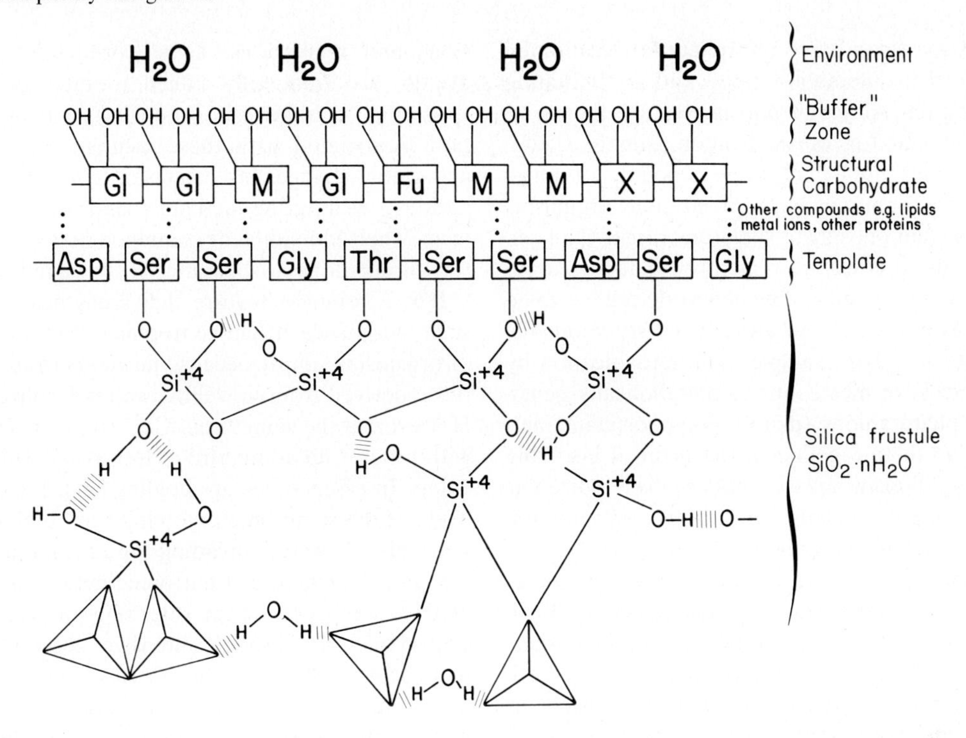

phosphate or silica in the course of time. This can be done without too much worrying about speciation.

Fig. 12.24. Hypothetical arrangement of organic layers in the diatom cell wall; glucose (*Gl*), mannose (*M*), fucose (*Fu*), xylose (*X*), serine (*Ser*), aspartic acid (*Asp*), glycine (*Gly*), threonine (*Thr*); *hatched lines* represent hydrogen bonds (after Hecky et al., 1973)

Biomineralization Through Time

In the preceding chapters I suggested that the evolution of soft tissues appears to be related to increasing Ca^{2+} content in the Precambrian sea, which promoted the development of an intricate cellular regulation system for that ion. Thus, the unfolding of eukaryotes can be viewed – in the sense of A.J. Toynbee – as a cellular response to a Ca^{2+}-mediated environmental challenge. Moreover, the establishment of hard tissues is nothing but an outgrowth of this development and might accordingly be treated in the same fashion.

Signs that organisms are involved in the extraction of lime from the ocean are found almost from the beginning of the Archean. Cyanobacterial communities laid down one microlayer of calcium carbonate after the other which eventually grew to sizeable stromatolitic structures. However, stromatolites form in tidal environments which may mean that Ca^{2+} is of fresh water origin and precipitates upon photosynthetic extraction of CO_2 according to the familiar formula:

$$Ca^{2+} + 2HCO_3^- \rightleftharpoons CaCO_3 + CO_2 + H_2O.$$

By contrast, eukaryotic calcification is almost entirely an enzymic process and unrelated to carbonate equilibria. The same holds true for phosphatization and silicification in which the respective mineral equilibria are not involved.

The first Vendian skeletal remnants of problematic derivation are principally phosphatic, whereas subsequent lower Cambrian skeletons, among which representatives of five modern phyla have been recognized, are predominantly calcareous (Lowenstam and Margulis, 1980). Should formation of biocarbonates be an adaptive response of organisms exposed to calcium stress, then the question arises, why the first wave of biomineralization led to the formation of phosphates rather than of carbonates. An answer can be found in Cook and Shergold (1984), who assign the clustering of sedimentary phosphate deposits around the Precambrian-Cambrian boundary to a global phosphogenic event. This event might have been a consequence of the ocean's overturn following a long period of anoxia during which phosphates accumulated in the deep sea. A post-glacial event in the aftermath of the Eocambrian ice age? Changes in general ocean circulation due to rise in sea level and/or land-sea configuration? Such are likely explanations.

Organisms responded to phosphate fertilization by increasing their metabolic rates and by initiating skeletonization. Although phosphates are known to be effective inhibitors of calcification (Simkiss, 1974), a simultaneous uptake of large amounts of phosphates and calcium may have forced many organisms to form calcium phosphate skeletons rather than carbonate shells. It is well to remember that regulation of Ca^{2+} and phosphate in the eukaryotic cell has to be strict, so as not to lead to unbalanced situations deleterious to life. For example, cell detoxification by binding excessive metal ions to metabolically generated phosphate anions (mainly polyphosphate) and precipitating them together in the form of insoluble granules is still maintained in modern skeletonized or non-skeletonized animals, particularly in environments enriched in heavy metals.

Once phosphate content in the sea was lowered to normal levels, priority shifted in the direction of carbonate deposition which in no time led to the famous Cambrian radiation event recorded in the stratigraphic column, and which marked the "explosion" of metazoan life. And yet a note of caution should be sounded, because all we know about the evolution of eukaryotic life at that early stage rests almost entirely on skeletal material, and the animated world composed only of soft tissue had little chance to escape anonymity.

We have studied in some detail incidents of skeletal evolution following the first wave of phosphate and carbonate biomineralization (Kazmierczak et al., 1985; Kazmierczak and Degens, 1986). They all appear to succeed geological events that triggered changes in the chemical and physical settings of world ocean and atmosphere. This in turn caused adjustments in the global cycling of biogeochemical elements. In order to illustrate how organisms might respond to a changing Ca^{2+} content in the ambience, the phenomenon of sea level rise and fall has been selected as a case study.

Density stratification in mid-water induced by salinity or temperature gradients quasi cuts the ocean into a top and a deep layer. Depending on general ocean circulation, mixing of water masses can be restricted to the point where oxygen minimum zones or even reducing conditions can temporarily be established below a pycnocline. The modern Cariaco Trench and Black Sea are excellent examples for anoxia at depth.

At times of highstand, the sea occupies much of the land, which in turn leads to a warmer climate causing mid-water stratification, yielding stagnant conditions even in shallow water habitats. In contrast, lowstand favors a mixed oxygenated ocean and a reduction in the areal size of epicontinental seas. Global transgressions and regressions, as we ordinarily call such events, are frequently linked to rates of seafloor spreading, epirogenesis or deglaciation and glaciation. Associated with these incidents are temporal and global changes in volcanism, hydrothermal emanations, weathering activities and vegetation patterns. Their individual or combined impact on marine Ca^{2+} oscillations is schematically shown in Fig. 12.25.

It is a common feature that areas near the outer shelf margin are attractive trophic niches for a variety of organisms simply because nutrients from land and those derived from upwelling waters fertilize the sea. However, at the same time, Ca^{2+} stress environments with little or no admixture of terrestrial sediment develop. In essence, we are holding back riverine Ca^{2+} while at the same time enforcing Ca^{2+} build-up from nutrient-rich waters invading the epicontinental sea. The final outcome is a eutrophic epicontinental sea, whereas the open ocean stays oligotrophic and calcium-depleted. Through intensive accumulation of biologically produced calcium carbonate such areas functioned for a long time as calcium traps removing huge quantities of Ca^{2+} from sea water. This sort of trap is often referred to as carbonate platform (J.L. Wilson, 1975).

Sea level rise might allow the migration of corrosive upwelling waters onto cratons, causing intensive dissolution of the previously deposited carbonate on the platform (Parrish, 1982). Corroded surfaces, hardgrounds, stratigraphic gaps are consequences of such processes. Spectacular examples of worldwide solution hiatuses are noticed near the boundaries Frasnian/Famennian and Cretaceous/Tertiary, which interestingly conform with times of mass extinction. Dissolution of carbonate platforms was thus a temporary significant addition to riverine and upwelling Ca^{2+} that poisoned the shelf margins.

The bulk of Phanerozoic carbonate rocks has been laid down by organisms. In the course of the Paleozoic, it was the shallow water benthic community which shared the responsibility for removing "surplus" oceanic Ca^{2+} deleterious to life in the form of $CaCO_3$. In contrast, planktonic eukaryotic algae (acritarchs) remained unmineralized but increased considerably in number and diversity near the turn of the Precambrian and Cambrian. Some of them achieved – impressive for unicellular organisms – millimetric sizes, while others developed a great variety of shapes (e.g., Vidal and Knoll, 1982; Wang, 1985). In our opinion (Kazmierczak and Degens, 1986), it represents a response of acritarch organisms to a continuous increase in Ca^{2+} concentration in their ambience, combined with synergistic action of other metal ions and high phosphate levels. Interestingly, many

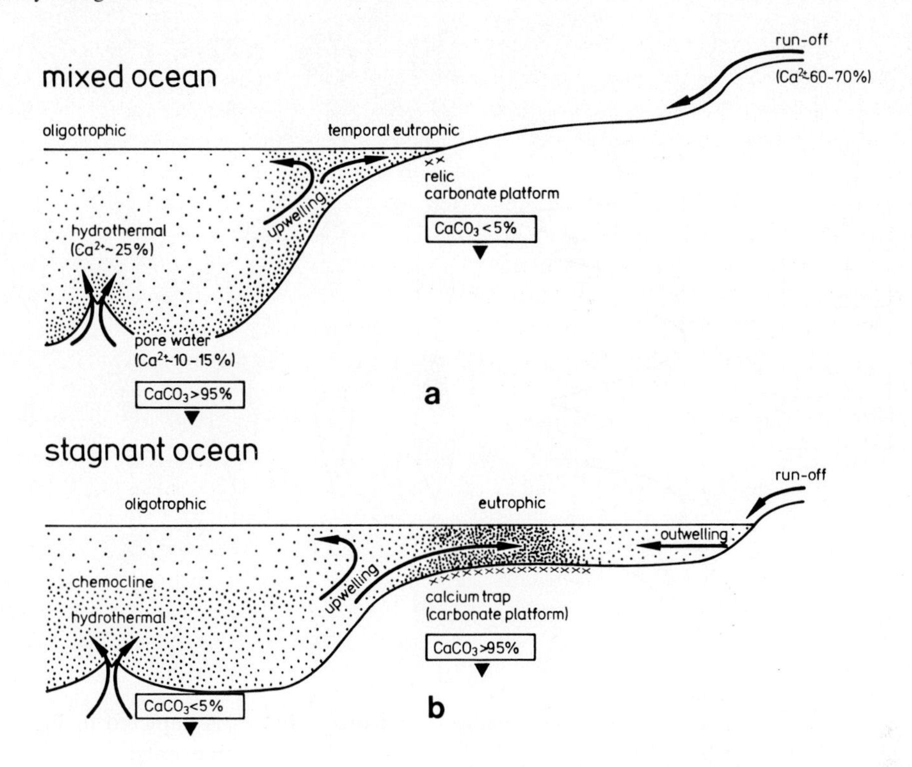

Fig. 12.25. **a** Ca^{2+} sources and sinks in a mixed (= regressive) and **b** in a stagnant (= transgressive) ocean; *dot density* indicates Ca^{2+} content (after Kazmierczak et al., 1985)

modern planktonic species respond by gigantism when exposed to excess metal ions in their environment.

With regard to Paleozoic acritarchs, they too responded to Ca^{2+} stress in a similar fashion. Specimens occurring in carbonate rocks are as a rule much bigger than similar coeval forms occurring in non-carbonate rocks. The former are often found embedded in calcareous coatings, formed post mortem by mineralization with $CaCO_3$ of thick sheets of Ca^{2+}-complexing extracellular substance released by the cell.

Until calcareous plankton appeared at the turn Triassic/Jurassic, the biotic control over oceanic calcium cycling was exerted mainly by benthic communities. The question of course comes up as to why planktonic organisms suddenly became involved in removing Ca^{2+} from the water column and still continue to do so. To some extent it is linked to the gradual decline in the size of carbonate platforms. The prime reason, however, can be seen in conjunction with tectonism. Namely, Caledonian and Variscan orogenies had recycled ancient marine carbonates so laboriously extracted by early Paleozoic life to the surface again. Land vegetation remobilized much of this material, a situation almost like that of a garbage dump returning its toxic debris into the environment.

During the remaining part of the Mesozoic much of the $CaCO_3$ extracted by the plankton came to rest in shallow water environments. The pelagic realm was subsequently occupied and today the deep sea above the carbonate compensation depth is the chief burial place of biogenic carbonate. Nature apparently has learned to dump the detrimental $CaCO_3$ in places which eventually become subducted beneath continents rather than unfold upon them and start another vicious circle. The situation is reminiscent of Archean time, during which hydrothermally released calcium became instantaneously precipitated as calcium carbonate or possibly as calcium sulfate. As a final comment, silica exhibits a similar pattern in that benthic organisms extracted the bulk of Si^{4+} from the Paleozoic ocean – except for some early radiolaria – and planktonic forms such as diatoms did not emerge before the Jurassic. Furthermore, gigantism among diatoms is a common feature in habitats loaded with dissolved silica.

In summing up, I would like to present the relative abundance of carbonate rocks since the Vendian onset of biomineralization (Fig. 12.26). Although the curve is based generally on the present-day areal and/or volumetric distribution of carbonate deposits (e.g., Garrels and Mackenzie, 1971; Ronov et al., 1980; Brouwer, 1983), it essentially reflects the extent and temporal distribution of ancient carbonate platforms

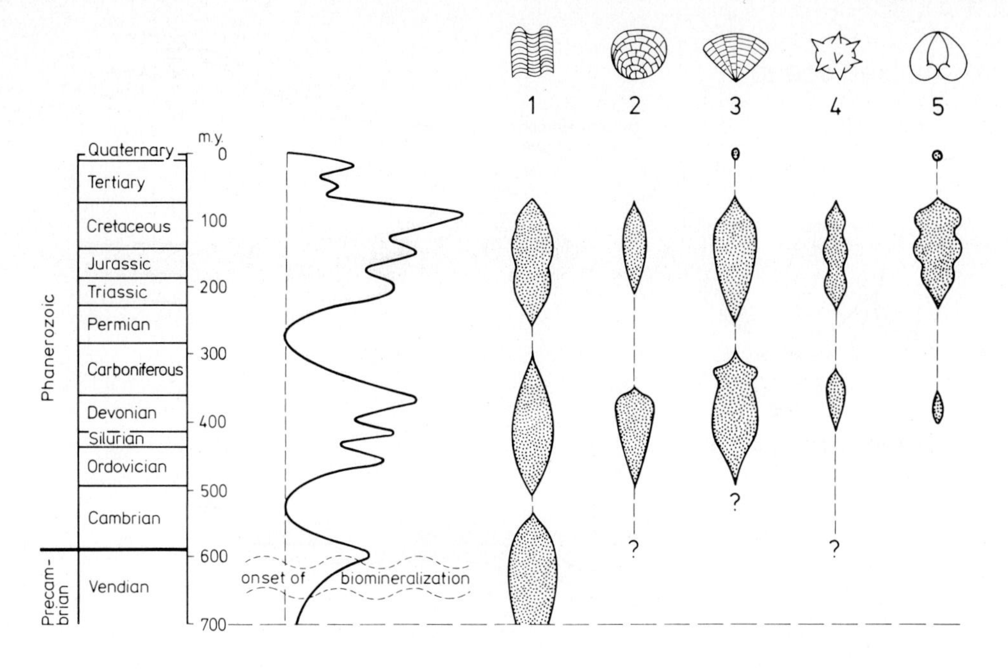

Fig. 12.26. Curve representing relative abundance of marine carbonate rocks during the Phanerozoic and occurrence of: calcareous stromatolites (1), stromatoporoid stromatolites (2), sclerosponges (3), calcispheric structures (4), thick-shelled molluscs (5); (various sources; see Kazmierczak et al., 1985)

as compared with their present-day extent. In view of the calcium trapping efficiency of the outer shelf areas of epicontinental seas it is not surprising that the $CaCO_3$ distribution pattern follows roughly the transgression-regression curve.

Extinction in the Course of the Phanerozoic

It has been found useful to distinguish between "normal or background extinction" and "mass extinction" (Raup and Sepkoski, 1982, 1984). The rate of background extinction has decreased from about 4.6 to 2.0 families per million years since the Early Cambrian. Expressed differently, the decline in background extinction rate from the Early Cambrian to the Recent implies that roughly 710 family extinctions did not occur that would have if the Cambrian rate had been sustained (Raup and Sepkoski, 1982). On the other hand, diversity increased by 680 families, which may mean that net increase in standing diversity has more to do with decrease in extinction than with increase in origination.

Mass extinctions stand out clearly above this background level. The Phanerozoic diversity curve compiled from the familial data of marine vertebrates and invertebrates is depicted in Fig. 12.27 revealing five mass extinction events:

(i) Late Ordovician (Ashgillian; 440 Ma; –12%)
(ii) Late Devonian (Frasnian; 370 Ma; –14%)
(iii) Late Permian (Guadalupian-Dzhulfian; 245 Ma; –52%)
(iv) End Triassic (Norian; 210 Ma; –12%)
(v) End Cretaceous (Maastrichtian; 65 Ma; – 11%)

Thus by all standards, the most devastating event in the history of life was the extinction of major parts of the marine ecosystem at the termination of the Permian Period.

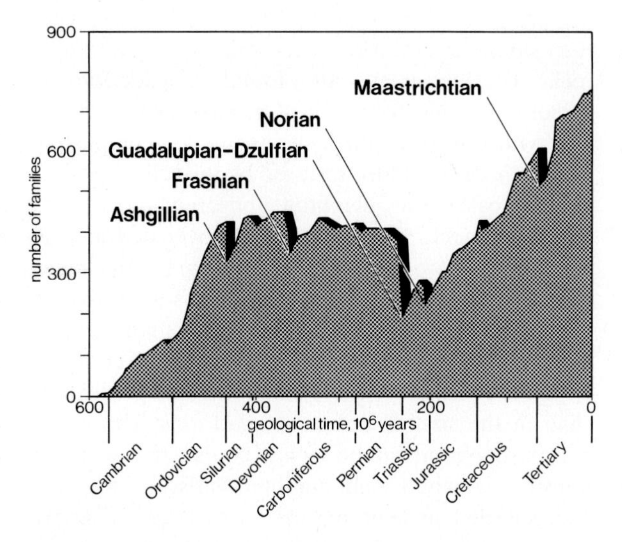

Fig. 12.27. Extinction curve of marine organisms (Raup and Sepkoski, 1982)

Explanations to account for this phenomenon are plentiful (see Stevens, 1977). A careful analysis of all the data, however, seems to support the idea advanced by Fischer (1964) that Permian mass extinction should be seen in connection with massive salt extraction from world ocean. To what extent formation of evaporites had led to a brackish sea, is worth a discussion.

Making the heuristic assumption that the salinity of the Permian ocean prior to the onset of massive evaporation had been close to modern standard mean ocean water (35 permil), rock salt (NaCl) extraction, conservatively estimated at 1.6 x 10^6 km^3, would have lowered oceanic salinity by about 10 percent and decreased salinity level to 31.5 permil (Stevens, 1977). Since this number does not include brines seeping into the sediment and salts being removed by weathering and erosion during the Triassic and possibly younger geologic periods, ocean salinity might well have been much lower at the end of the Permian.

It is the merit of Stevens (1977) to show the impact lower salinity levels might have on marine biota. In general, groups of fossils thought to be highly stenohaline marine forms, such as corals, crinoids and ammonoids, experience high extinction rates, whereas more euryhaline categories such as worms or sponges exhibit a higher survival rate (Fig. 12.28). Raup (1979) has estimated that between 88 and 96 percent of all marine species vanished at or near the Permian/Triassic boundary. This number is somewhat misleading, because the extinction event was spread over about 10 million years, that is the Guadalupian and the Dhzulfian Age, during which simultaneously new species arose. Among the foraminiferans, which include both euryhaline and stenohaline forms, all 13 families that became extinct are thought to have been stenohaline marine,

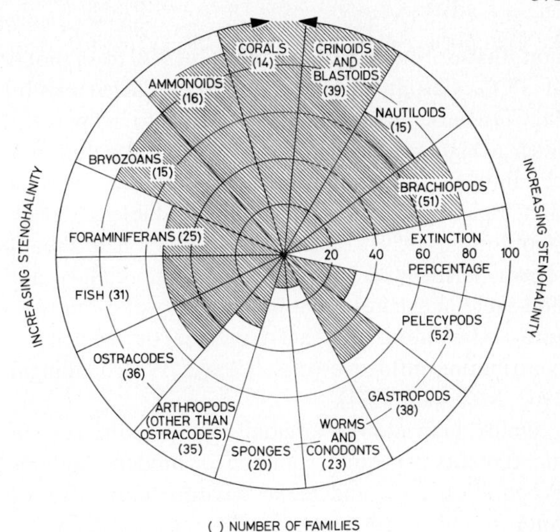

Fig. 12.28. Percentage of familial extinctions in major Permian aquatic groups (Stevens, 1977)

whereas all assumed euryhaline forms survived the trauma of salinity deprivation. Many ancestral forms, i.e., trilobites, blastoids, and certain brachiopods became extinct.

Especially revealing is the case of the corals, where the ancient forms (rugose and tabulate corals) became supplanted by the hexacorals (scleractinian corals). This change in symmetry appears to go along with a change in mineralogy, namely from the Paleozoic calcite-dominated to the Mesozoic aragonite-depositing corals (Oliver, 1980). Moreover, it has been suggested that the decline of the calcitic articulate brachiopods and the rise of the predominantly aragonitic molluscan bivalves might be linked to more favorable environmental conditions in general for aragonite depositors during the Mesozoic era (e.g., Gould and Calloway, 1980; Box 12.5).

BOX 12.5

Non-Skeletal Marine Carbonates

Carbonates are extracted from today's ocean principally by the action of organisms. Only in high saline environments or where oceans are many times supersaturated with respect to calcite will non-skeletal carbonates precipitate. Most prominent among them are calcite, low-Mg calcite (<5 mol% $MgCO_3$), high-Mg calcite (5–20 mol% $MgCO_3$), and aragonite (Burton and Walter, 1987). Protodolomite and dolomite has also been encountered, even in the deep sea, but this carbonate is considered of secondary, that is early diagenetic origin. Siderite and mangano-siderite are sporadically found in reducing habitats.

Laboratory experiments investigating growth rates of carbonate precipitates in the temperature range of 5 to 37°C and variable carbonate ion concentrations (2.5 to 15 times supersaturated with respect to calcite) showed a marked temperature dependence. Rate of precipitation of aragonite and calcite were about equal at 5°C, whereas aragonite precipitation rates rose sharply in the higher temperature regime, i.e., up to a factor of 4 relative to calcite precipitation. $MgCO_3$ content in calcite increased in the same direction, that is from less than 5 mol% at 5°C to 14 mol% at 37°C. Certain primitive calcareous algae exhibit $MgCO_3$ enrichment at a level of up to 30 percent. Such a high magnesium content in solid solution is certainly a non-equilibrium configuration, because that high-Mg calcite would only be stable at 800°C (Goldsmith, 1959). Earlier investigators had already shown that $MgCO_3$ content in calcareous algae and non-skeletal calcites are a function of water temperature and salinity, and that phylogenetic level has a contributing influence (e.g., Chave, 1954; Chilingar, 1962; Milliman, 1974).

Ooids, grapestones, aragonite needles and carbonate cements are quite common in modern lagoons, evaporation pans and reefal environments, and are often found in coexistence with one another (e.g., Lowenstam and Epstein, 1957; Gonzales and Lohmann, 1985). It is generally assumed that carbonate equilibria are the determining mechanism behind the precipitation of carbonates. However, closer examination of a variety of non-skeletal marine carbonates indicates the presence of well-defined calcifying matrices of the type recognized in calcareous algae, corals and other reef-building organisms. As an illustration, I present the amino acid spectrum of Bahama oolites (Fig. 12.29) which bears a striking relationship to the amino acid pattern of coralline algae, suggesting a direct involvement of algal-derived organic templates in the nucleation of aragonite (rather than calcite), and its epitaxial growth.

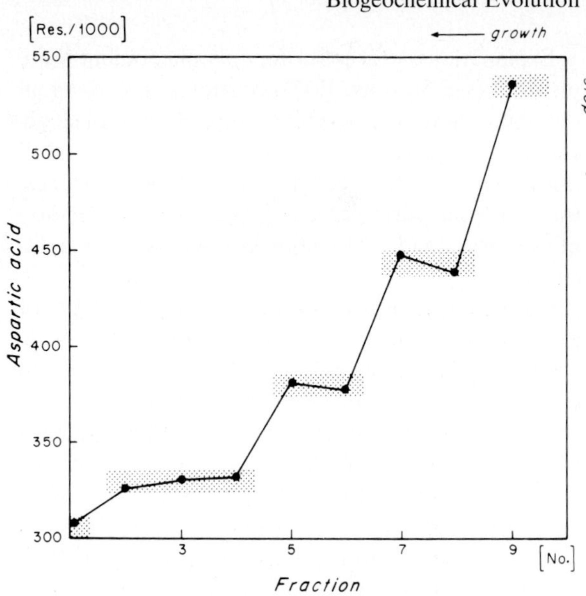

Fig. 12.29. Changes in aspartic acid concentration in consecutive HCl etchings of a 0.5 mm oolite fraction. Sample No. 1 represents the outer portion of the oolites and sample No. 9 the inner core. Steps indicate that oolites grow intermittently (Degens, 1976)

During the Precambrian and prior to the onset of Vendian biocalcification, inorganic precipitation was the principal means of Ca^{2+} and Mg^{2+} carbonate removal from the sea. After the biomineralization event, organisms controlled more and more carbonate equilibria in the open ocean, thereby limiting non-skeletal carbonate deposition to restricted marine environments.

Numerous workers have documented temporal shifts in the mineralogy of non-skeletal carbonates during the Phanerozoic, and have attributed them chiefly to atmospheric PCO_2 changes. Some authors consider low sea level stands, water temperature, Mg/Ca ratio of sea water additional factors (reviewed in Burton and Walter, 1987). In the light of the findings of Burton and Walter, temperature and degree of carbonate mineral saturation (CO_3^{2-} ion concentration) of sea water seem to be the most likely master variables in the formation of non-skeletal carbonates. Furthermore, participation of organic templates in the mineralization event ought to be considered (Degens, 1976).

An oscillating trend in Phanerozoic non-skeletal carbonate mineralogy has been reported by Sandberg (1983). Late Proterozoic non-skeletal carbonate deposition was dominated by aragonite, which became substituted by calcite at the cross-over to the Cambrian; calcite prevailed from here on until the Late Carboniferous, at which time aragonite replaced calcite and continued to do so until the end of the Triassic. Again, calcite moved in as chief non-skeletal carbonate for the entire Jurassic and Cretaceous. Close to the K/T boundary, aragonite came back and is presently the most abundant non-skeletal carbonate mineral in ooids and cements.

Much insight on carbonate mineralogy and mode of formation can be gained from lacustrine environments, and outstanding in this regard is the contribution of Müller et al. (1972). In 25 lakes with differing hydrochemistry and salinity, primary non-skeletal carbonate precipitation is chiefly dependent on the Mg/Ca ratio of the lake water. By rule of thumb one can say that low-Mg calcite precipitates at $Mg/Ca < 2$, high-Mg calcite in the Mg/Ca range from 2–12, and aragonite at $Mg/Ca > 12$. At very high Mg/Ca ratios hydrous Mg carbonates can be expected. In those instances, where lake waters fluctuated in Mg/Ca ratio, thereby crossing the various mineralogical thresholds, all three primary carbonates may coexist. The presence of dolomite signals early diagenetic changes, and only those lakes having high-Mg carbonate muds can yield proto-dolomite or dolomite. Aragonite inhibits dolomite formation.

All this suggests that in the marine realm oscillations in Mg/Ca ratio could also be a decisive element determining non-skeletal carbonate mineralogy, and might to some extent prove responsible for the observed temporal variations. The abundance of dolomite is in all probability linked to the availability of high-Mg calcites, both biogenic or abiotic, in the stratigraphic column.

Factors that control the formation of calcite and aragonite in sea water are mainly discussed in terms of $CaCO_3$ precipitation kinetics as determined principally by changes in temperature, PCO_2, salinity, Mg/Ca ratio, and sulfate or phosphate contents (reviewed in: Sandberg, 1983; Railsback and Anderson, 1987; Burton and Walter, 1987). It is well to remember, however, that organisms – as part of their metabolic activity – secrete calcifying organic templates that inoculate either calcite (CaO_6) or aragonite (CaO_9) crystal seeds which then grow in epitaxial order (see p. 362). Nobody will deny that oscillations in the physical chemistry of the environment may precipitate physiological changes leading to the generation of aragonite- or calcite-prone organic templates. It is to be emphasized that these changes are enzymically controlled and in no way linked to the kinetics of carbonate equilibria in the sea. Instead, they must be seen in conjunction with biochemical adjustments of organisms to a changing habitat.

Prior to the extraction of halite, a substantial quantity of calcium carbonate and sulfate minerals will – upon evaporation – be removed from sea water. It is difficult to assess their quantity, because many generations of calcite, aragonite, dolomite, gypsum and anhydrite layers may form without ever exceeding the solubility product for halite deposition. Borchert and Muir (1964) have estimated that for every 1 km^3 of halite an amount of 0.39 km^3 of $CaSO_4$ becomes extracted from the sea. In using the figure of Stevens (1977) for the mass of Permian halite deposits, a total of 6.2×10^5 km^3 of $CaSO_4$ is derived. This is equivalent to 1.9×10^{21} g of Ca^{2+} or about twice the present mass of marine Ca^{2+} (Railsback and Anderson, 1987). To this number, the amount of calcium bound to inorganically precipitated limestones and dolostones has to be added.

At present, 0.9×10^{15} g Ca^{2+} are annually released to world ocean (Wollast and Mackenzie, 1983). Assuming the same rate for the Permian, more than 2 million years of weathering and hydrothermal leaching are needed to match the evaporative loss in Ca^{2+}. Since regular skeletal carbonate extraction continued simultaneously, a Ca^{2+} depletion in sea water is quite plausible, the more so since sulfur isotopes signal elevated sulfate contents in Late Permian to Middle Triassic oceans (François and Gérard, 1986; Box 12.6).

BOX 12.6

Marine Sulfate and Its $\delta^{34}S$ (Claypool et al., 1980)

Residence time of sulfate in sea water is about 2×10^7 years, which is much shorter than that of chloride. Because of rapid mixing of world ocean in a matter of 1,000 years, dissolved sulfate in the sea carries a uniform $^{34}S/^{32}S$ signal commonly expressed as $\delta^{34}S = +20.0$ permil relative to a meteorite standard (Sasaki, 1970). Sulfates precipitated from modern sea water about reflect the sulfur isotopic composition of dissolved sulfate. This then can be used as a baseline to interpret age curves for $\delta^{34}S$ in marine sulfates throughout geologic time.

Sulfate evaporites, principally gypsum and anhydrite, exhibit temporal changes in $\delta^{34}S$ from +10 to +30 permil and the corresponding age curve is plotted in Fig. 12.30. Observed variations must be chiefly seen and interpreted in connection with fluxes from a reduced (i.e., pyrite) and an oxidized (i.e., gypsum-anhydrite) sulfur pool. Light $\delta^{34}S$ values imply (i) a significant contribution to the sea from weathering solutions or hydrothermal vents, or (ii) a maximum in the sulfate/sulfide burial ratio, or perhaps (iii) a combination of the above (Railsback and Anderson, 1987). Since evaporation involves almost no isotope fractionation between dissolved sulfate and sulfate minerals, heavy $\delta^{34}S$ values can only come about by the removal of isotopically light sulfur, that is in the form of sedimentary sulfides.

Many attempts have been made to link sulfur isotope age curves into one global mass balance equation (e.g., Garrels et al., 1975; Garrels and Lerman, 1981, 1984; Berner et al., 1983; Lasaga et al., 1985). The most recent attempt in this direction is the one of

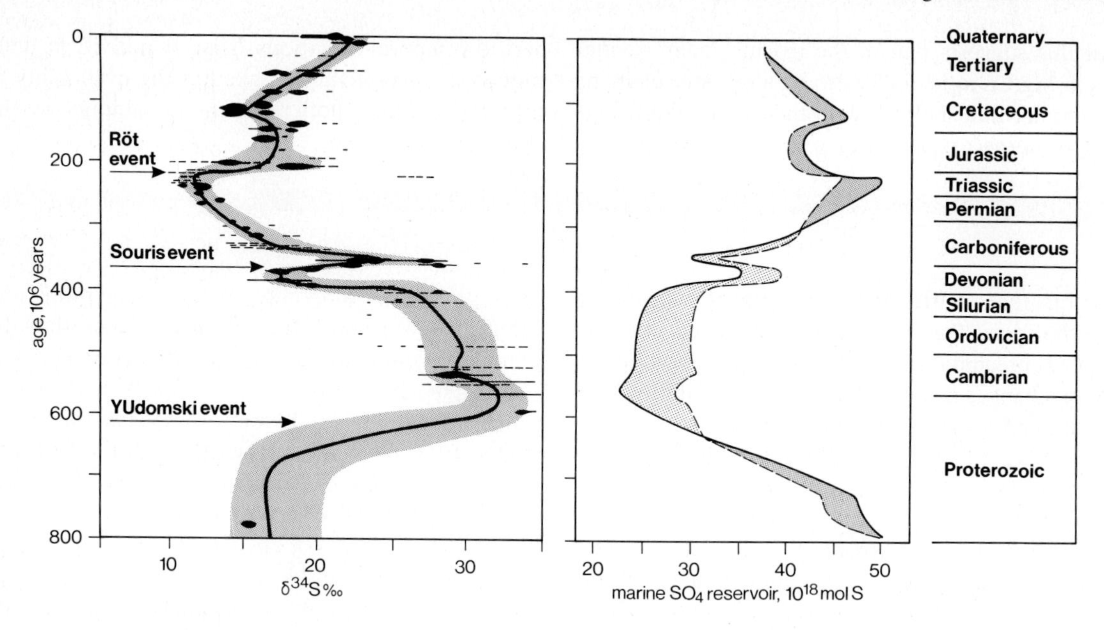

Fig. 12.30. Age curves for δ^{34}S and sulfate (Claypool et al., 1980; Holser, 1977; François and Gérard, 1986)

François and Gérard (1986) which gives a coherent picture of sea water sulfate reservoir changes with time (Fig. 12.30). Assuming relatively constant ocean volumes, SO_4^{2-} content in Late Permian to Middle Triassic oceans were 25 to 100 percent higher than during much of the Paleozoic (reviewed in Railsback and Anderson, 1987). On the other hand, sharp rises in δ^{34}S recorded in sulfate evaporites (cf. Yudomski, Souris, Röt events; Fig. 12.30) are possibly related to massive sulfide extractions in a formerly stratified sea (Holser, 1977, 1984).

Fig. 12.31. δ^{34}S distribution in finger and toe nails of selected participants of SCOPE/UNEP International Sulfur Workshop, Pushchino, USSR, 1983

δ^{34}S(CDT) of your nail sample. In commemoration of 1983
SCOPE/UNEP International Sulphur Workshop, Pushchino, Moscow Region.

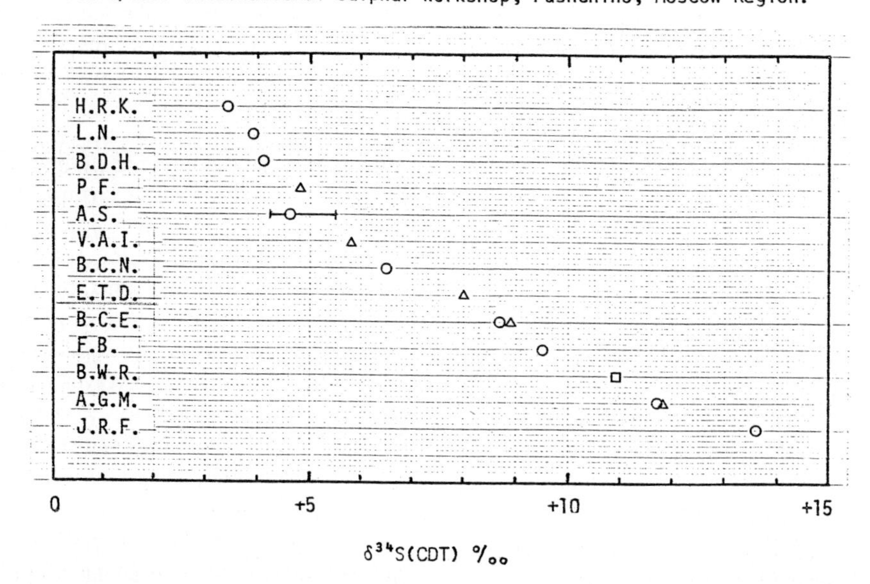

H.R.K.(KROUSE); L.N.(NEWMAN); B.D.H.(HOLT); P.F.(FRITZ); A.S.(SASAKI); V.A.I.(ITTEKKOT); B.C.N.(NGUYEN); E.T.D.(DEGENS); B.C.E.(EGBOKA); F.B. (BUZEK); B.W.R.(ROBINSON); A.G.M.(MENON); J.R.F.(FRENEY). o(Finger nails); △(Toe nails); ◻(Finger and toe nails, composite); ⊢—⊣(Variation range over 14 months). Analyst: Akira Sasaki.

Inasmuch as sulfur cycling can be linked to cycles of several other elements, especially carbon an' oxygen, and iron and calcium, understanding of the interaction of biogeochemical cycles should be a prime research objective in the field of the environmental and the earth sciences, because no one element operates in isolation (Bolin and Cook, 1983).

The global biogeochemical sulfur cycle has for many years been the focus of interest (Ivanov and Freney, 1983) and the stable isotopes of sulfur have received wide attention (SCOPE/UNEP workshop, Pushchino, USSR, 1983). During that meeting, Akira Sasaki from Japan asked participants to contribute finger and toe nails and twelve participants kindly offered nail clippings. The results are summarized in Fig. 12.31, showing that the most continental people from the middle of Canada were isotopically the lightest, whilst the more oceanic ones from Australia happened to be the heaviest. The person from Japan reflected most likely some volcanic sulfur contribution, and the man from Europe was the mean guy. Apparently, the type of food, the kind of soil, and geography regulate sulfur isotope composition in humans.

Having a skeleton as a detoxification device was by no means beneficial, if not fatal, at times when due to rapid lowering of Ca^{2+} content in the ambience biocalcification was hindered. There are many examples in the fossil record where sudden diversity declines of shelled marine biota might reasonably be linked to an ambience low in calcium (Kazmierczak et al., 1985).

It is tentatively concluded that the Permian sea was low in Cl^- and Ca^{2+}, but high in SO_4^{2-}. Loss in chlorinity might have in part been compensated by a gain in sulfate. Little can be said on the behavior of dissolved carbonates, except that calcite became replaced by aragonite as the major skeletal component. In view of the limited potential for osmoregulation of the majority of the marine invertebrates and furthermore their strong dependence on Ca^{2+}, the observed and anticipated multitudinous biogeochemical changes in sea water chemistry induced by evaporation appear to be profound enough reasonably to explain mass extinction close to the Permian-Triassic boundary. On the other hand, glaciation continued into Late Permian time, and climatic fluctuations could well have been a contributing factor in a number of extinctions. The two earlier mass extinctions, namely those of Late Ordovician and Late Devonian, share, according to Stanley (1987), the cooling trend observed in the Permian, which leads him to favor climatic deterioration as the essential controlling agent of the three Paleozoic mass extinctions. Climate certainly can be a promoter or inhibitor of life, I doubt, however, whether any one mass extinction event of global dimensions has been caused just by climatic change, or a single factor, or the same set of circumstances.

On land, life apparently survived the crisis well. The Upper Carboniferous plants composed of spore-bearing and seed-bearing varieties were both thriving well in the warm and moist environment. Climatic change during the Early Permian leading to ice ages and arid climates gave the more resistant seed plants a tremendous advantage. Gymnosperms became the predominant land plants already during the Zechstein and continued to flourish until they were gradually replaced by the angiosperms – the flowering plants – in the course of the Cretaceous. For sentimental reasons I have drawn in Fig. 9.38 a leaf of *Ginkgo biloba* – also known as maidenhair or apricot tree – because it is the only surviving member of the Ginkgoales, one of the three orders of the gymnosperms (Pinophyta). The male Ginkgo tree grows much more slowly than its female counterpart, but both appear to be quite resistant to air pollution in city streets. Survival of the fittest?

The history of non-marine vertebrates, that is amphibians, reptiles, birds and mammals, is marked by distinct events involving ecological replacements, mass extinctions and adaptive radiations (e.g., Russell, 1979; Benton, 1985). The quality of the fossil record for these organisms is variable because certain kinds of environments are poorly preserved. Furthermore, some geological periods have been studied more intensively than others, so that a kind of bias is ingrained in the data. In due consideration of these shortcomings, Benton (1985) has plotted the standing diversity with time for families of terrestrial tetrapods revealing six apparent mass extinctions (Fig. 12.32). The resolution is to the level of the stratigraphic stage (duration 2–19 Ma; mean duration 6 Ma). A total of 730 families are distinguished, of which 469 are extinct.

Three assemblages of families succeeded each other through geologic time: (i) labyrinthodont amphibians, anapsids, mammal-like reptiles, (ii) early diapsids, dinosaurs, pterosaurs, and (iii) the "modern" groups: frogs, salamanders, lizards, snakes, turtles, crocodiles, birds, mammals. Family diversity

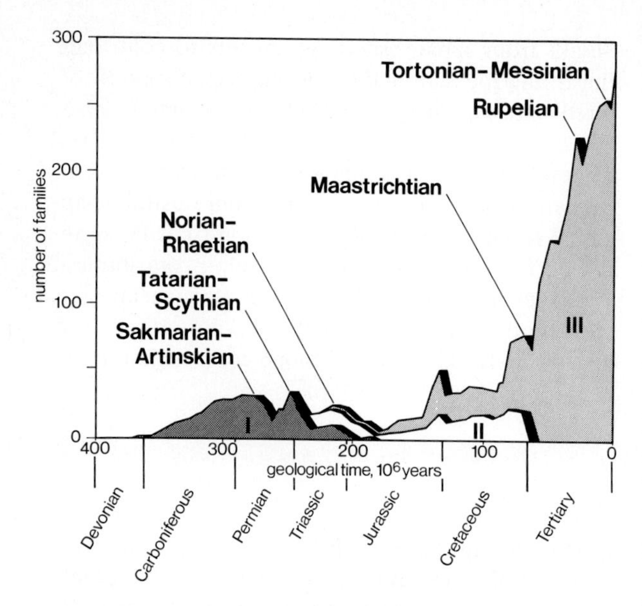

Fig. 12.32. Standing diversity with time for families of terrestrial tetrapods. Three assemblages of families were grouped together: I. labyrinthodont amphibians, anapsids, mammal-like reptiles; II. early diapsids, dinosaurs, pterosaurs; III. "modern" groups, i.e., frogs, salamanders, lizards, snakes, turtles, crocodiles, birds, mammals (Benton, 1985)

rose with time, especially from the Cretaceous on. Benton (1985) concludes that the extinction events, including the famous terminal Cretaceous extinction, resulted from a slightly elevated extinction rate combined with a depressed origination rate, and that apparent mass extinctions are statistically not set apart from background extinctions. And finally, the tetrapod record – contrary to the inferences drawn from the record of marine animals – offers little evidence of improved survival changes of organisms as time progresses. We will now turn to one of the most controversial issues in the field of earth sciences: mass extinction at the Cretaceous/Tertiary boundary and its causes.

In 1980, Walter Alvarez and his co-workers explained the iridium anomaly located in a few centimetersthick clay layer at the Cretaceous/Tertiary boundary as a result of the impact of an asteroid about 10 km in diameter. They regarded this as the triggering event of the Late Cretaceous mass extinction. Inasmuch as all dinosaurs were apparent victims of that catastrophe, it came as no surprise that the idea immediately developed popular appeal and still continues to do so. Beyond that, an avalanche of articles and books authored by many a respectable scientist fanned curiosity in a manner that often made it difficult to decide what are facts and what is fiction. Inasmuch as in science one should always be guided by a Chinese proverb, "to search for the truth in the facts", I will draw attention to a few reasonably consistent observations without becoming hopelessly entangled in a still unresolved scientific dispute (for more details see, e.g., Alvarez et al., 1980, 1984; Hsü, 1980, 1981; Hsü et al., 1982; Keith, 1982; Raup and Sepkoski, 1982, 1984; Officer and Drake, 1983, 1985; Campsie et al., 1984; Dingus, 1984; Hallam, 1984, 1987; Hoffman, 1985; Raup, 1986; Raup and Jablonski, 1986; Jablonski, 1986; Stanley, 1987; Officer et al., 1987; McKinney, 1987).

Sloan et al. (1986) present data from Hell Creek Formation (Montana, Alberta, Wyoming) that reveal not only a gradual dinosaur extinction commencing 7 million years prior to the Cretaceous/Tertiary (K/T) boundary, but indicate that of the 30 dinosaur genera living in the Hell Creek area 8 million years before K/T, the number reduced to 12 just before the K/T boundary event, whereas between 7 and 11 genera crossed the K/T boundary into the Tertiary. Argast et al. (1987) voice caution because a reworking of dinosaur teeth cannot be completely ruled out. However, reports that dinosaurs survived well into the Early Paleocene have also come in from many other regions of the world which support the Hell Creek Formation inferences (reviewed in Sloan et al., 1986). Of special interest is the observation that the final 300,000 years of the Cretaceous witnessed a sharp decline of taxa of dinosaurs and a simultaneous rise in Paleocene-like mammal taxa.

In the marine realm, dramatic changes in extinctions and speciations well before the advent of the iridium anomaly at the K/T boundary are familiar phenomena. Take, for instance, the case of the ammonites in the course of the Cretaceous, which has been so nicely presented by Erle G. Kauffman (1976; Fig. 12.33). Times of origin and extinction of species are not necessarily synchronous. Extinction episodes are invariably associated with regression of epicontinental seas (seven cases) or peak transgression (one case). In contrast, rapid evolutionary radiations are equally associated with: (i) early transgression, reflecting normal radiation into new ecological niches of the advancing sea, (ii) peak transgression to early regression leading to stress on lineage evolution due to overall environmental changes, and (iii) late regression, reflecting evolutionary response to high stress environments at the shelf edge. Regressions and transgressions are directly keyed to rifting episodes which, in turn, provide fine tuning to major evolutionary events (e.g., Kauffman, 1976; Hallam, 1984). All ammonites and many other forms of marine life suddenly vanished from the global scene at the end of the Maastrichtian, at which time sea level stood low and

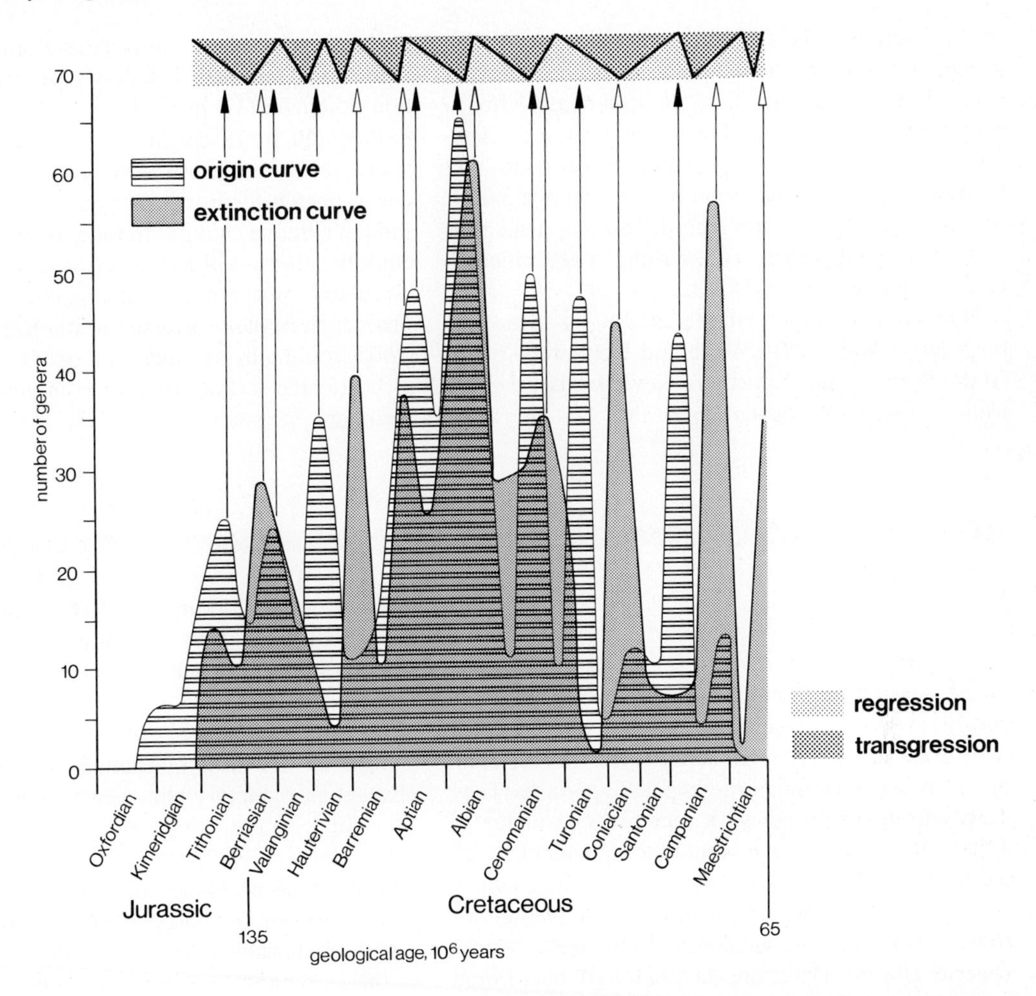

Fig. 12.33. Origin and extinction curves of ammonite genera which occur in the Cretaceous (after Kauffman, 1976)

world climate had cooled off considerably (Alvarez et al., 1984; Stanley, 1987). Back to the iridium anomaly.

Prior to 1980, when Alvarez and co-workers reported their iridium find in thin layers of K/T boundary clays, paleontologists had no clear notion as to the actual reasons behind mass extinction, except that large-scale environmental changes of complex nature were involved. Extraterrestrial causes were sporadically mentioned too but were generally dismissed by the community at large as "deus ex machina" solutions. For example, Digby J. McLaren suggested in his 1970 benchmark paper on the Frasnian-Famennian mass extinction a large body oceanic impact as a triggering mechanism. However, it took 10 years before that idea gained respectability. The present situation is almost the reverse, in that conventional views implying that mass extinctions are, for instance, just intensifications of background rates or related to earthbound mechanisms are frequently dismissed as too simplistic unless they become tied to the scenario of an impacting comet or asteroid.

The radical view runs as follows: (i) mass extinctions are clearly set apart from background extinctions, (ii) mass extinctions are catastrophic, (iii) mass extinctions are periodic and come at intervals of 26 million years, and (iv) mass extinctions are the result of extraterrestrial factors (e.g., Raup, 1986). The first two factors principally agree with viewpoints expressed by Officer et al. (1987) and Hallam (1987), but they see instead cataclysmic volcanic events at the end of the Cretaceous as promoters of large-scale environmental changes that in toto have caused mass extinction. By the way, the iridium anomaly is not in conflict with this idea, because intensive global volcanism can carry the iridium signature too, as demonstrated by Officer et al. (1987). The principal ideas are well summarized in their abstract "...transition is marked by a period of intensive volcanism. The volatile emissions from this volcanism would lead to acid rain, reduction in the alkalinity and pH of the surface ocean, global atmospheric temperature changes, and ozone layer depletion. These environmental effects, coupled with those related to the major sea level regression of the Late Cretaceous, provide the framework for an explanation of the selective nature of the observed extinction record." So it appears that "volcanoes fight comets for the right to exterminate" (The Economist, 1987) (Box 12.7).

Plants testify to the extreme conditions at the K/T boundary (Wolfe, 1987; Wolfe and Upchurch, 1986). In the Raton Basin of northern New Mexico, the K/T iridium event is registered in the 1-3 cm thick boundary clay. Late Cretaceous flora, governed by seed-bearing plants, suddenly gave way to a vegetation dominated by ferns. Eventually, after a short interlude (10 to 15 cm above boundary clay) higher plants recovered. Ferns are known to be good invaders of barren territories left after volcanic events or sudden climatic change. In time, however, a new vegetation pattern will be established best suited for a given soil, morphology and climate. Associated with this switch in plant diversity at the K/T boundary is a global cooling by as much as 4 or 5°C which appears to be related to cooling effects of aerosols. Volcano Weather?

BOX 12.7

Iridium

In the search for a common cause, the widespread occurrence of Ir anomalies at the K/T boundary in marine as well as terrestrial deposits led Alvarez et al. (1980, 1984) to the meteorite impact hypothesis. They derive from their findings a meteorite diameter of 10 km. Its kinetic energy would be equivalent to 10^8 megatons TNT, which is 10^7 times the mechanical energy of the Mt. St. Helens eruption of 1980 or 10^6 times that of the 1883 Krakatoa (Toon, 1984). Atmospheric effects of the impact-generated dust cloud were simulated by Toon et al. (1982) and Pollack et al. (1983). The results compare well with scenarios of a nuclear winter: light levels too low for photosynthesis during several months, continental surface temperatures dropping below freezing for approximately twice as long as sub-photosynthetic light levels persist. Given such meteorite impact and climate response scenarios, the idea that the extinctions were the ecosystem response to these catastrophic happenings looks plausible. Nonetheless, the question remains unanswered of why some species became extinct while others survived. The proposal by Toon et al. (1982) that small animals on land survived because they were more abundant, could more easily find food in the dark and could burrow under ground to be protected from freeze fails to explain why mouse-sized marsupials became almost as extinct as dinosaurs did (Van Valen, 1984). Independent support for the impact theory was attributed to mineralogical evidence (Bohor et al., 1984): quartz grains from an iridium-rich K/T boundary claystone show features of shock metamorphism indicating high velocity impact between a large extraterrestrial body and the Earth.

An alternative scenario involves enhanced volcanism as a causal mechanism leading to sudden changes in planetary climate and subsequent extinctions (e.g., Kent, 1981; Hinz, 1981; Officer et al., 1987; Hallam, 1987). In fact, the extensive Deccan volcanism occurred 65 million years ago; McLean (1982) suggests an explosive volcanic eruption of deep mantle material – perhaps from the initial extrusion of the Deccan traps. This idea has been expanded by Rampino (1987).

The age range of Deccan flood basalts, that span a section about 2 km thick, covers a time envelope from 66.6 to 68.5 Ma (Duncan and Pyle, 1988). This implies that roughly 1 km^3 tholeiitic flood basalt was generated per year, or more likely a few hundred to a few thousand km^3 in the course of a single event. Other well-dated flood basalt provinces, for instance the Columbia River Basalts (16 ± 1 Ma), were also erupted over a short period of time. Enhanced Ir in dust from Kilauea eruptions (Zoller et al., 1983) indicates the possibility that volcanism might as well be the source of the Ir anomaly at the K/T boundary. Also enriched in the dust are arsenic, selenium, antimony and mercury and their surprising concentration at the K/T boundary has been a difficulty for the impact hypothesis (Van Valen, 1984). Moreover, the features on quartz grains, quoted above as mineralogical evidence for an impact, have very recently been found in volcanic material (Smith, 1984).

The Late Cretaceous was a time of pronounced marine regression which in the last instance is related to seafloor spreading and plate tectonics. This implies a gradually changing climate. There is also evidence of active circumpacific volcanism (Officer and Drake, 1983). A scenario unifying impact and volcanism would imply that a state of very gradual change was suddenly struck by the bolide impact, which induced volcanic activity capable of adding to the immediate climatic consequences of the impact, the more persistent climatic and geochemical effects of volcanism on a longer time scale. About 10^{-2} of the impact energy would be transferred to seismic waves to be compared to the 100–500 megatons mechanical energy of major earthquakes (Clube and Napier, 1982). It is known from lunar research that mare volcanism was triggered by meteorite impacts during the "Terminal Lunar Cataclysm" (Wasserburg et al., 1977). According to Courtillot et al. (1986), the great Deccan flood basalts in India formed within a period of at most 3 million years at or close to the K/T boundary. Was this the place where the supposed 10 km in diameter sized bolide hit, generating a transient crater 100 to 200 km wide, excavating 20 to 40 km of crustal or even upper mantle material? This and related observations led Rampino (1987) to suggest that extinctions at the K/T boundary might be related both to impact and volcanic phenomena.

In summary: there is principal agreement that at the K/T boundary a series of different physical, chemical and biological events combined to produce irreversible ecosystem changes in a rather short time. In contrast, a wide range of speculations exists as to the nature of the mechanism triggering this chain of events. The following three scenarios seem to be consistent with the observations and only choice and taste determine which of these one personally accepts:

(i) Meteoritic impact, affecting life through atmospheric, shallow oceanic and Earth surface processes;
(ii) volcanic incidents, affecting life through mantle, crust, deep and shallow ocean and atmospheric processes; and
(iii) meteoritic impact initiating volcanism, involving all effects listed in (i) and (ii).

As in the past, when catastrophic natural causes could almost instantaneously and irreversibly upset the dynamic equilibrium of air-sea-earth-life interactions and wipe out large segments of the ecosystem, humans too would follow suit, should a nuclear impact and its aftermath become reality.

Age Curves of ^{13}C in Marine Carbonates

Variations in the stable-isotope composition of the dissolved inorganic carbon in the oceans have been studied in great detail. Following the reconnaissance studies of the early fifties and sixties (e.g., Craig, 1954; Sackett and Moore, 1966), Deuser and Hunt (1969) gave the first systematic account on $\delta^{13}C$ variations as a function of water depth at selected stations in the North and South Atlantic. As part of the Geochemical Ocean Sections Program (GEOSECS), a global coverage has now been achieved (e.g., Kroopnick, 1980) showing rather uniform $\delta^{13}C$ distribution for deep and intermediate waters in the range of about +0.5 to +1 permil relative to the PDB standard (Craig, 1957). Waters of the euphotic zone are slightly heavier due to the preferential extraction of ^{12}C by primary producers. Atmospheric CO_2 over the oceans typically has $\delta^{13}C$ values near –7 permil (Keeling, 1961) which reflects equilibrium with oceanic bicarbonate. Experiments indicate that in bicarbonate $\delta^{13}C$ is heavier by 9 permil relative to its associated CO_2 (e.g., Thode et al., 1965; Mook et al., 1974). It is that system from which skeletal and non-skeletal carbonates obtain their ^{13}C signal.

Non-skeletal carbonates from highly saline environments have $\delta^{13}C$ values around +4 to +5 permil (Lowenstam and Epstein, 1957; Degens and Epstein, 1964; Deuser and Degens, 1969; Gonzales and Lohmann, 1985). A small fractionation factor between aragonite and calcite (Rubinson and Clayton, 1969) explains the observed relationships, in that aragonite is slightly heavier where both mineral forms coexist.

$CaCO_3$ precipitation is an equilibrium process with isotopic fractionation in an environment which slowly reaches saturation through evaporation. If we assume a $\delta^{13}C$ of 0 permil for the bicarbonate in sea water, the value for the solid carbonate will come out to be +4.5 permil and that of the CO_2 –4.5 permil, because they form in a 1:1 molar ratio. While the carbonate will remain fixed, the CO_2 will quickly reequilibrate with the dissolved carbon pool and, in the case of continued evaporation and precipitation in a restricted environment, will lead to a gradual shift of the $^{13}C/^{12}C$ ratio in that reservoir (Deuser and Degens, 1969).

Some ancestral organisms, like certain calcareous algae, which precipitate carbonate in the process of photosynthesis, exhibit the expected equilibrium fractionation of +4 to +5 permil relative to dissolved marine bicarbonate. However, the majority of skeletal organisms, such as molluscs and foraminifera, lay down $CaCO_3$ with the help of enzymes and organic templates which implies that $^{13}C/^{12}C$ ratio of the

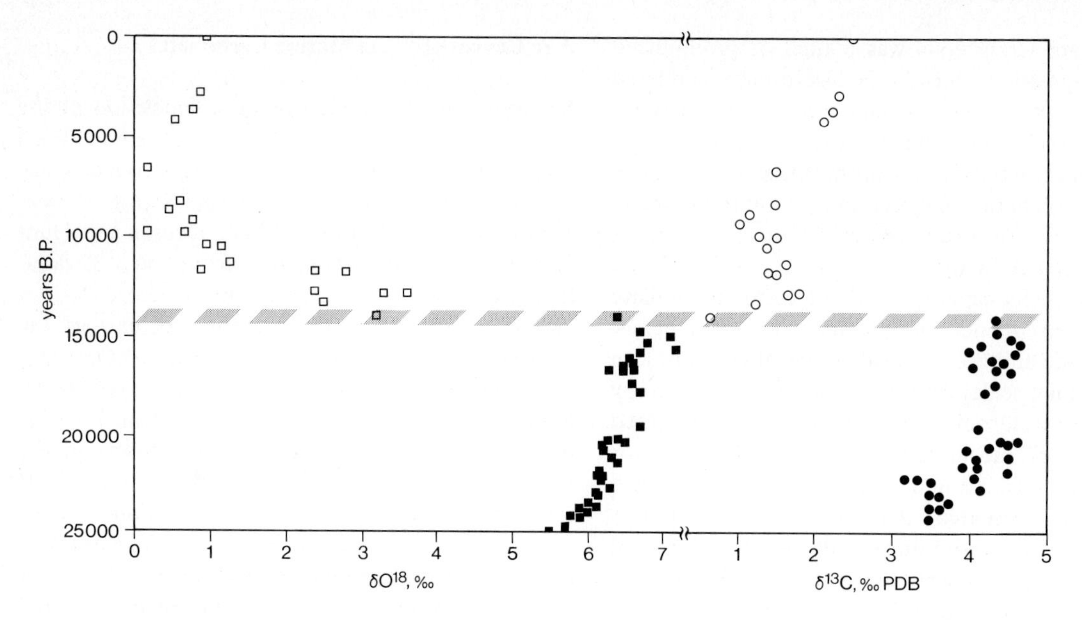

Fig. 12.34. $\delta^{13}C$ and $\delta^{18}O$ distribution in pteropod shells of core 118K, Red Sea (Deuser and Degens, 1969). *Open dots* and *squares* unaltered shells; *filled dots* and *squares* aragonite incrusted shells

dissolved carbonate is conservatively handed over to the shell carbonate. Expressed differently: (i) coexisting skeletal and non-skeletal carbonates should differ by about 4 to 5 permil in $\delta^{13}C$, and (ii) the majority of shell-forming organisms reflect in their $\delta^{13}C_{CaCO_3}$ the $\delta^{13}C$ of the dissolved marine carbonate pool. The fact that coexisting skeletal and non-skeletal marine carbonates formed in the same environment may exhibit a $\delta^{13}C$ difference by as much as five permil ought to be expected.

A particularly striking example is the isotope record of a 25,000 year sediment section from the Red Sea (Fig. 12.34). At about 14,000 years before present (B.P.), the $\delta^{13}C$ values of pteropods exhibit an abrupt change of close to 4 permil within just a few centimeters, which timewise corresponds to a few hundred years at the most. The associated $\delta^{18}O$ values undergo a net change of about 5.5 permil but with a transition zone of 30 cm, reflecting a sedimentation interval of ca. 1,000 years (13,000 to 14,000 years B.P.). Thin-layered pteropod shells prior to 14,000 B.P. are characterized by massive aragonite filling and overgrowth, which must be seen in connection with a highly saline evaporative environment which yielded non-skeletal aragonite in a $\delta^{13}C$ range corresponding to isotopic equilibrium.

The interval is also marked by the sudden appearance of foraminifera, which speaks for a rapid normalization of the Red Sea habitat at the brink of the outgoing Weichselian ice age and the beginning Holocene. This is a worldwide phenomenon (Broecker, 1987; see p. 267), and related to swift ice volume changes and the warming of the sea in much less than 1,000 years.

The fact that non-skeletal and skeletal carbonates consistently differ in $\delta^{13}C$ requires a reassessment of published $\delta^{13}C$ age curves of marine carbonates. The best way of opening the discussion is to begin with juvenile CO_2 and follow its isotopic imprint on carbonates of all ages.

Juvenile CO_2 has a $\delta^{13}C$ of –5 permil and the primordial sea should reflect this value in its dissolved carbonates. Accepting a soda ocean (see p. 271) and its enormous capacity for dissolved carbonates, calcite or aragonite could only precipitate in the vicinity of hydrothermal vents, active volcanoes or in the tidal sea, but not in the open ocean. Precambrian carbonates – largely stromatolites – have on the average a $\delta^{13}C$ of +0.39 permil (Archean), +0.43 permil (Early Proterozoic), +1.76 permil (Middle Proterozoic) and +1.50 permil (Late Proterozoic) (Veizer, 1985; Schidlowski, 1987). The uniformity in $\delta^{13}C$ across 3 Ga suggests either a large carbonate reservoir in the ocean or an effective recharge of the dissolved carbonate pool by juvenile CO_2. The $\delta^{13}C$ values in Archean and Early Proterozoic carbonates are the expected ones for non-skeletal carbonates, especially for aragonite or high-Mg calcite. The moderate increase by about 1 permil during Proterozoic time may be related to the preferential extraction of light carbon by organisms and the gradual lowering of dissolved bicarbonate.

Enormous fluctuations to the positive and negative side have recently been demonstrated in dolomites and limestones in Late Vendian time (Magaritz et al., 1986; Aharon et al., 1987; Schidlowski, 1986, 1987). Consistent in all reports, however, is the pronounced lightening close to the Precambrian/Cambrian boundary, which can best be linked to the onset of enzymic biocalcification that progressively replaced non-skeletal carbonate precipitation in the course of the Phanerozoic. Since calcareous fossils roughly reflect the $\delta^{13}C$ of the former marine bicarbonate pool, their $\delta^{13}C$ in turn provides an age curve for dissolved carbonates in the sea. Two major trends come to light: (i) a gradual rise in $\delta^{13}C$ from the Ordovician to the Permian of close to 3 permil, and (ii) a sudden drop in $\delta^{13}C$ of 2.5 permil at the K/T boundary.

Extended anoxic marine events and the emergence of land plants culminating in the formation of black shales and coal measures can readily account for the observed Paleozoic trend. For the Mesozoic and Caenozoic carbonate record, $\delta^{13}C$ oscillations are of a more complex derivation because of the rapidly changing geological and environmental settings on the continents and in the ocean. Concerning a rise in $\delta^{13}C$, the best interpretation is still the C_{org} burial mechanism. Yet, isotopic lightening can be produced by one or a combination of the following processes ($\delta^{13}C$ values of the source materials are given in brackets):

(i) oxidation of common terrestrial organic matter (–25);
(ii) discharge of fresh water bicarbonate originating from the dissolution of carbonate rocks by soil CO_2 (–10);
(iii) precipitation of non-skeletal marine carbonates (+5);
(iv) submarine emanation of hydrothermal or volcanic CO_2 (–5); and
(v) oxidation of dispersed magmatic carbon in pillow basalts (–5).

In view of the many diverging factors that influence the partitioning of carbon isotopes in the global carbon cycle, the rather modest spread in $\delta^{13}C$ values for the marine carbonate record of the past 3.5 Ga comes unexpectedly (Schidlowski, 1986, 1987), the more so because the addition of small amounts of "alien" carbon can substantially alter $\delta^{13}C/^{12}C$ ratio of the bicarbonate in the sea (Degens et al., 1986a). For instance, to lower $\delta^{13}C$ from 0 to –1 permil in a hypothetical marine bicarbonate pool holding 40,000 Gt C, one can envisage the following scenarios:

(i) 1,600 Gt C of organic carbon has to become oxidized, which is about three times the standing stock of present land vegetation or about half the carbon in all modern soils and plants; or
(ii) 4,000 Gt C of fresh water bicarbonate has to be added to the sea, representing the outflow of riverine carbon in 10,000 years; or
(iii) 8,000 Gt C of non-skeletal carbonate ooze must precipitate, which would require roughly 10,000 years on condition that all limestones and dolomites formed at a given time are of non-enzymic derivation; or
(iv) 8,000 Gt C of endogenic CO_2 has to be injected into world ocean which, at present release rates, would require 40,000 years.

For illustration, I would like to present the $\delta^{13}C$ age curve for the past 70 million years obtained from a deep sea core; each data point reflects the sum total of all carbonates in a sediment sample (Fig. 12.35; Shackleton and Hall, 1984). The location almost precludes contributions from non-skeletal carbonates or from turbidites. The observed shifts could be due to changes in burial and/or oxidation rates of organic matter. However, by focusing on the $\delta^{13}C$ of carbonates across the Cretaceous/Tertiary boundary (Fig. 12.36; Shackleton and Hall, 1984), other mechanisms seem equally plausible, as model calculations imply (Spitzy and Degens, 1985). Assuming that the observed isotope shift at the K/T boundary by about –2.5 permil is entirely due to enforced emanations of juvenile CO_2 or the oxidation of magmatic carbon, 20,000 Gt C are required to reproduce the shift by –2.5 permil in an ocean reservoir of 40,000 Gt C. Assuming further that this lightening took place over a time envelope of 100,000 years – a not unrealistic postulate – then a doubling of volcanic CO_2 emanations over present rates could achieve this result. It is to be emphasized that other source-sink combinations are conceivable, involving, for instance, lowering of primary productivity, oxidation of land plants, or destruction of soil organic matter (Spitzy and Degens, 1985).

In conclusion, the ocean's carbon isotope composition responds very sensitively to even minor changes in the flux rates of different carbon species entering or leaving the ocean. On the other hand, the carbon cycle has been severely impacted through geologic history: periods of massive volcanism, extinctions, biomineralization events, anoxia in extended ocean basins, ice ages and so on and so on... As a result one should expect a concurrently pronounced variability in the carbon isotope record. This, however, is not the

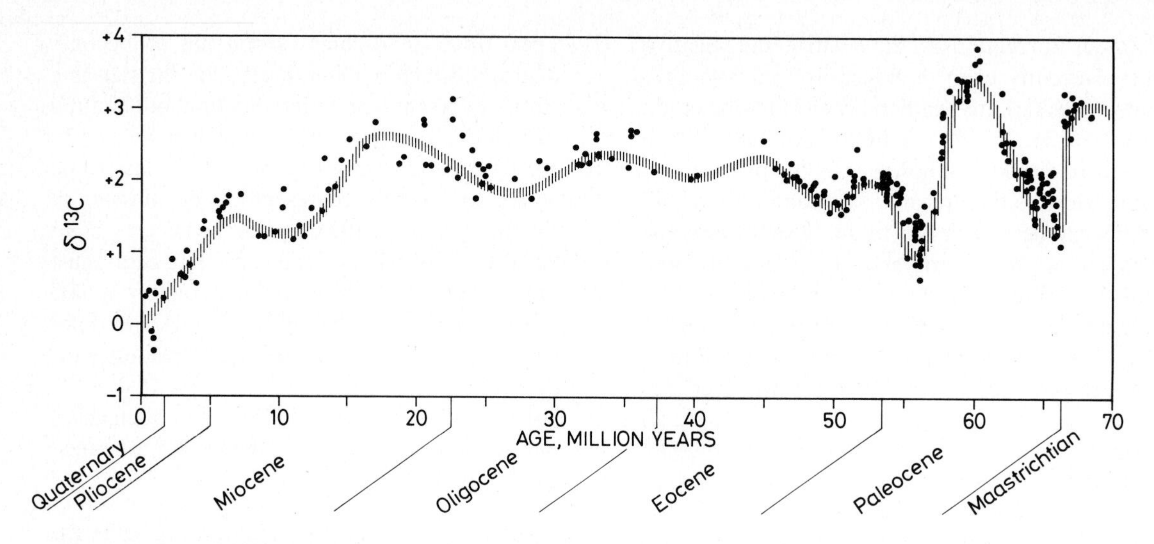

Fig. 12.35. Carbon isotopic composition of bulk sediment (carbonate fraction) from DSDP Site 525, with data from Site 528 between 25 and 45 million years (Walvis Ridge, Cape Basin); (after Shackleton and Hall, 1984)

case. Instead, $\delta^{13}C$ in marine carbonates varies within narrow bounds. Throughout the Precambrian this can partially be explained by the dissolved carbon content of the ocean, which was orders of magnitude higher than today. Since then, that is from the time for which we assume the ocean's carbon content not to be significantly different from today, counterbalancing mechanisms must have been operating as part of the ocean's biogeochemical response to changes in carbon cycle flux rates. Since organisms are actively involved in the production as well as oxidation of organic matter, and furthermore are the chief agents in the formation and destruction of limestones, they appear to be the critical factors keeping the isotope signal within narrow bounds during the Phanerozoic. Volcanism is only of secondary importance, because free CO_2 becomes readily sequestered through weathering, carbonate dissolution in the sea, enforced biomineralization, or enhanced primary productivity. This would also rule out much higher or lower PCO_2's in the air with a range of 400 ± 200 ppm for most of the recorded history.

Alas: *Homo sapiens!*

About 20 billion years ago the decision was made to create a universe. Fifteen billion years following this event our Sun, the planets, the moons, the comets etcetera arose from the collapse of a dark molecular cloud that stood near the galactic spot where our solar system now resides. A few hundred million years later Planet Earth gave birth to life. From that moment on it still took another 4 billion years of biogeochemical

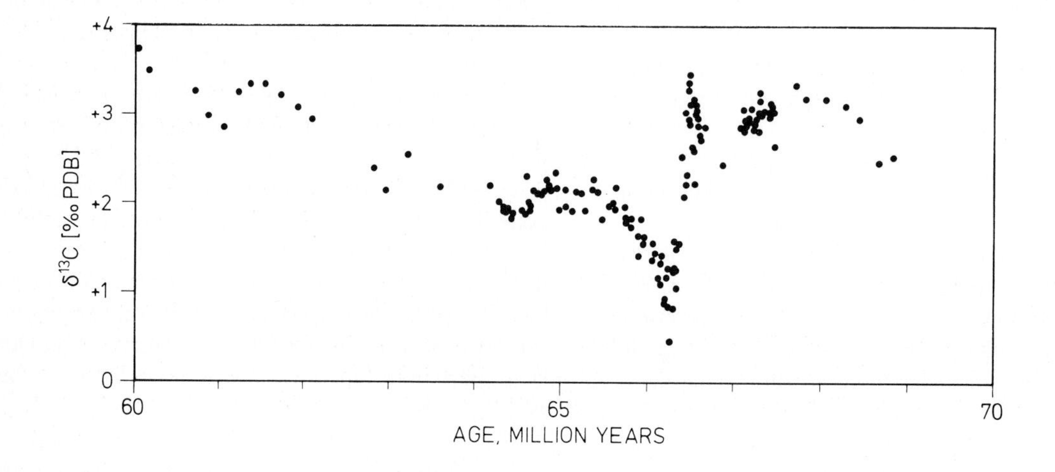

Fig. 12.36. $\delta^{13}C$ of carbonates across the Cretaceous/Tertiary boundary as registered in sediments from DSDP Site 527 (Shackleton and Hall, 1984)

evolution to shape the common mother of us all: The "African Eve" or, alternatively, the "African Lilith" who – according to the Talmud – was the first wife of Adam!

Two hundred thousand years of human history have gone by. *Homo sapiens* has meanwhile reached a level of intellectual competence which allows him or her to wonder in a scientifically meaningful way about the origin and evolution of the universe and life. Does this pondering and wondering contradict religious documents and the beliefs of people? By no means! The Bible says that God created "Heaven and Earth" in less than a week. This statement should not be taken literally. Some billions of years on a human hour-glass could well be just one day in God's life.

This chapter gives a brief account of what we presently know about the history of our immediate ancestors prior to the Holocene. We are familiar with traditional or classical history from our school days: Assyrians, Persians, Greeks, Romans, Arabs and so forth. They will not be considered, nor will human trials and errors in creating societies and nations, or cultural affairs.

Our report recommences again when humans became conscious of the harm they are doing to the global habitat. For me, environmental awareness began when reading in 1951 Rachel L. Carson's book "The Sea around Us" and a few years later "The Silent Spring". I joined the arena of research where emphasis is placed on the return to clean water and air, and food and energy for everbody.

Throughout the text, I have drawn attention to various aspects of human interference with the pristine state of the environment where people are aggressors, victims, or both. Comparatively "innocent" sins at the beginning advanced in the end to become global impacts. Take for example the release of carbon dioxide from the burning of fossil fuel; it led to interactive biogeochemical processes and phenomena endangering world climate. In closing this book I will report – sine ira et studio – on three case studies: oil pollution, radioactive fallout, and nuclear war. They may stand for countless insults *Homo sapiens* is afflicting upon Nature and the generations to come.

Roots

The claim has recently been made (Cann et al., 1987) that the first human being was an African woman who lived about 200,000 years ago. This supposition is based on high-resolution mapping of human mitochondrial DNA (mtDNA) sequences of 147 individuals in five populations. "Mitochondrial Eve" or "African Eve" became headlines but at the same time aroused many objections which particularly concerned the place and time of origin of *Homo sapiens sapiens*.

mtDNA accumulates mutations much faster than does nuclear DNA and has the further advantage that it is inherited only through the maternal line. On the other hand, mtDNA is only a co-traveller in the genetic mechanisms leading to a new species which implies that "Mitochondrial Eve" must be older than the actual *Homo sapiens sapiens*. Perhaps an archaic *Homo sapiens?* Others question the divergence rate estimated by Cann et al. (1987) at 2–4 percent per nucleotide site per million years, and assume a rate of 0.71 percent instead (Saitou and Omoto, 1987). Accordingly, "Mitochondrial Eve" could be twice as old.

At present, paleoanthropologists and biochemists are still debating this issue. Carefully weighing the available data set from fossil evidence and molecular mapping, modern humans appear to have diverged from archaic humans about 100,000 to 140,000 years ago. All present-day humans have their root in Africa and speedily populated the Old World about 50,000 years back in time, thereby gradually replacing "archaic" *Homo sapiens* (including the Neanderthals). There is some indication that in the east genetic interchange took place, leading to the continuation of a distinctive regional line. By 35,000 B.P. the pure-bred Neanderthals vanished from the global scene, and from that moment on, *Homo sapiens sapiens* was the sole "ruler" of this world.

Local races developed at about the time of massive migration 50,000 years ago. From their former homeland (Java?), the Australian aborigines crossed on natural rafts the about 100-km-wide Sunda-Sahul barrier 40,000 years B.P., bringing with them a canine companion, the Dingo (Elkin, 1974). First, they occupied only the humid-subtropical belt spreading around the margin of their continent; they moved into the central part of today's arid zone at about 22,000 years B.P. (Smith, 1987). It is not clear yet whether this migration was not promoted by pluvial episodes during a glacial high stage, and that they later adjusted to desertification. Their profound artistic skills in shaping tools and making paintings, combined with their uninterrupted historical achievements and age-old wisdom and vision, makes the Aboriginal society the oldest surviving human culture on this Earth (Elkin, 1974; Supp, 1985).

The time of human settlement of the Americas is still uncertain. Some place this event close to 32,000 years B.P., whilst others date the most ancient Indian settlement at 12,000 years B.P. Distinctive stone weapons called Clovis "fluted" points were found at

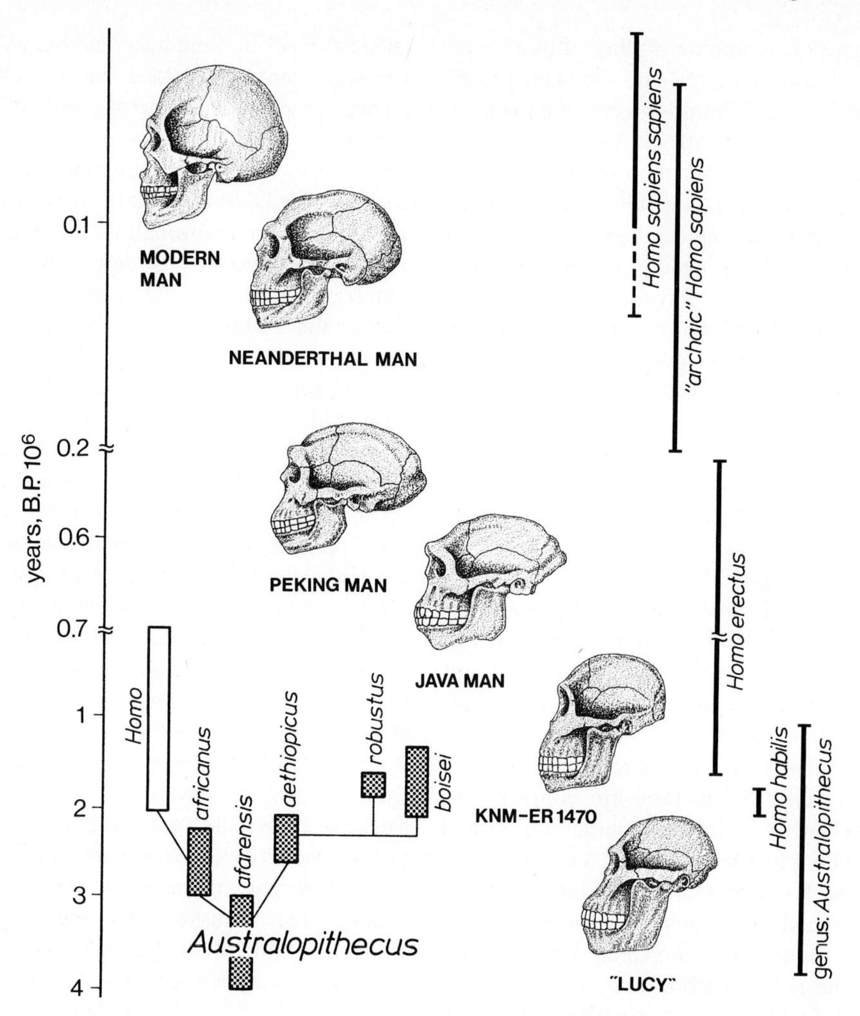

Fig. 12.37. Alternative phylogenies of the major hominid taxa (Cronin et al., 1981)

sites from Alaska to Mexico (Diamond, 1987). Clovis sites cluster in the Western United States between 11,500 and 11,000 B.P., and in the Eastern United States at 11,000 B.P.

The ancestral roots of *Homo sapiens* rest in *Homo erectus* which first appeared in Africa, where it probably originated 1.6 million years ago. From here *H. erectus* migrated to southeastern and eastern Asia, and Europe about 1 million years ago. Soleilhac, Massif Central, France (800,000 years B.P.), Java man (700,000 years B.P.), Peking man (at least 600,000 years B.P.), and Heidelberg man (500,000 years B.P.) are members of this "clan". They not only developed complex stone tools, but must – at the time of Java man – have borne Prometheus, who supposedly robbed the element fire from Zeus, brought it to Earth, and gave it to man for use. *Homo erectus* vanished about 250,000 years ago.

Olduvai Gorge in Tanzania has brought to light a variety of hominid fossils including *Homo habilis,* the immediate predecessor of *Homo erectus* (Johanson et al., 1987), having its first appearance 2 million years ago in eastern and possibly southern Africa. It was a short-lived species, considering that it disappeared in Africa 300,000 years after its first emergence.

Still older roots, going back to the earliest human-like ancestors, that is the genus *Australopithecus,* penetrate down to an age of 3.75 million years. Perhaps the gracile skeleton of "Lucy" found in 1974 and belonging to *A. afarensis,* may in some mysterious way be viewed as the mother of us all (Johanson and Edey, 1981). Phylogenetic relationships of the major hominid taxa are depicted in Fig. 12.37.

People were always on the move and the compulsion must be seen in connection with climatic change, availabilty of food resources, and perhaps simple human fears. A deep and dark tropical rainforest is not an inviting place to conquer unless you use it as a shelter to hide from other creatures. In contrast, the open savannah has many obvious advantages: game, water, good climate, and a wide sky. The coastline could have served as a highway to foreign land and the scarcity of human traces during much of pre-historic times may have something to do with the rise and fall of sea level. Eustacy restricted or allowed migration to another island or continent, while at the same time erasing the near-shore human record by either burying or eroding former settlements. For reviews on divergence and migration of hominids and hominoids as well, see: Washburn (1978), Walker and Leakey (1978), Trinkaus and Howells (1979), Cronin et al. (1981), Pilbeam (1984), Tobias (1985), Wood et al. (1986), Walker et al. (1986), Ciochon and Fleagle (1986), Delson (1985, 1987), Johanson et al. (1987).

Humans and Their Environment

Many an ancient place has fallen into oblivion, except for those centers where something historical has happened or is supposed to have happened. It could be a battle – just think of Homer's Troy or Alexander's Issos – or a natural event such as the eruption of Mt. Vesuvius in 79 A.D., which buried Pompeji and the writer Plinius Secundus, or simply the birthplace of a "Mensch": Bethlehem!

More recent natural catastrophes have remained almost unnoticed by the global community. Who would recall that in 1815 Mt. Tambora volcano on Sumbawa, Indonesia erupted, burying 90,000 people, or in 1976 the Tangshang earthquake near the city of Tianjin, China, where the death toll rose to 300,000 and possibly more. Today other names have sprung up and they are synonymous with disasters inflicted by humans upon people and the environment and this during times of peace: Bikini, Windscale, Three-Mile-Island, Chernobyl, or Bhopal, Seveso, AMOCO CADIZ, Sandoz, or Ozone Hole, Acid Rain, Smog, Itai-Itai (= outsch-outsch)... the list has no end. In most instances the names of people or institutions behind these insults to nature are almost forgotten, and there is a good chance that even the meaning of the words from Bikini to Itai-Itai will share the same fate. Renaming Windscale into Sellafield appears, at first sight, an excellent means to cover up the nuclear incident in 1957 – but in vain. Thirty years later, the full story and the not very flattering role of former politicians in playing down the event was released.

Some environmental impacts can be felt and seen, e.g., an oil spill in the ocean, or a colored chemical tide flowing down the river wiping out on its way to the sea: clams, fishes, birds. Others are more hidden: radioactive fallout or metal ion pollution. A third category still resides in the mind of some people: nuclear war. These three environmental battlefields will be briefly discussed below and may serve as a reminder of what can go wrong if all of us do not become guardians of the environment and the decision-makers.

Oil Spills

Submarine seeps are often mentioned as a major source of open ocean oil pollution, and their contribution to fouling the high sea and beaches is often considered equal to that of humans. Thus, the popular argument goes, since marine life has survived these natural seepages, it will also survive oil pollution by human activities. An ardent advocate against this reasoning was Max Blumer, who in 1972, in a one-page note in Science, clearly spelled out that we are the chief culprits polluting the sea with oil at an estimated rate of 5 million tons per year. The sheer existence of huge oil reservoirs beneath the shallow sea proves that nature had sealed them, otherwise they would have vanished long ago. At the published figures for submarine seeps, only 20,000 years would be needed to exhaust all oil deposits off continents (Blumer, 1972). Should a leak develop, it commonly heals by transforming liquid oil into solid asphalt – a kind of blood clotting so to speak.

In 1975 the National Academy of Sciences published a report on petroleum in the marine environment. Since then, an impressive amount of research has been done where even long-lasting effects on ecosystems could be evaluated. Teal and Howarth (1984) have updated the earlier report.

Some generalized findings, succeeding an oil spill, are as follows: (i) the bulk of the oil becomes speedily transferred to the bottom sediment via planktonic organisms (e.g., fecal pellets) and physical forces (e.g., wave action, aggregation), (ii) oil in anoxic sediments has a long residence time, but even in oxic habitats oil may persist for years, (iii) oil contaminates planktonic organisms, especially the zooplankton, and is deleterious to benthic invertebrates (e.g., clams, worms), (iv) oil decreases abundance and diversity of benthic communities, (v) fish are also contaminated but to a lesser degree, (vi) birds may fall "prey" to oil slicks and die miserably, and (vii) humans may not only step into "tar balls" on beaches but also be affected in numerous ways by production, shipping, and spilling of oil at sea.

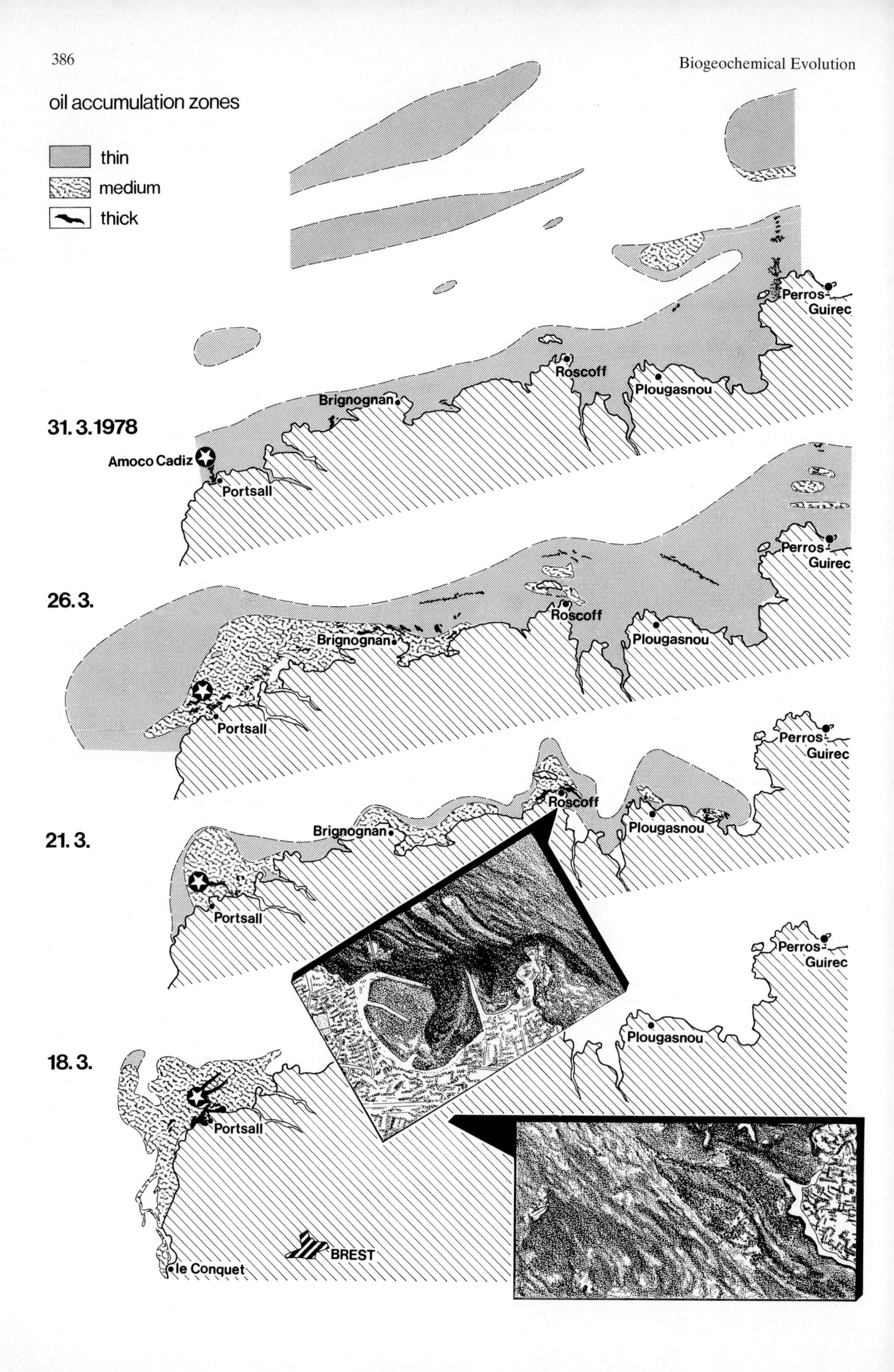
oil accumulation zones
thin
medium
thick
31. 3. 1978
Amoco Cadiz
Portsall
Brignognan
Roscoff
Plougasnou
Perros-Guirec
26. 3.
21. 3.
18. 3.
le Conquet
BREST

Table 12.4. Major offshore oil spills (from Teal and Howarth, 1984)

Event	Date	Tons of oil spilled	% Cleaned up
FLORIDA	09/16/69	630	< 1 %
ARROW	02/04/70	15,000	< 1 %
ARGO MERCHANT	12/15/76	28,000	0 %
PLATFORM BRAVO	04/22/77	20,000	< 1 %
TSESIS	10/26/77	1,100	~ 65 %
AMOCO CADIZ	03/16/78	250,000	< 8 %
ICTOC 1	06/03/79	457,000–1,400,000	< 10 %

In Table 12.4 seven particularly well-known oil spills are shown, stemming from the grounding of one barge and four tankers, the blow-out of a North Sea oil platform in the Ekofisk field, as well as that of an exploratory well 80 km from the shore in Bahia de Campeche, Mexico. Except for the tanker TSESIS, mopping-up operations was highly inefficient, considering that only a few percent of spilled oil were recovered (Box 12.8).

The best way to counteract an oil spill is rapid removal of an oil carpet by mechanical recovery. Use of sinking agents or detergents causes the toxic and undegraded oil to spread in the ocean; the biological damage is then greater than if the spill had been left untreated (Blumer et al., 1971). Bacteria normally present in the sea will slowly degrade spilled oil, sunlight will evaporate or alter the hydrocarbons, and wind and waves will do their share physically to eliminate the last visible evidence of an oil slick. This is often taken as a sign that, since no objectionable residue remains at the sea surface, nature possesses self-cleaning capacities. In reality, the problem is simply handed over to marine life, especially to burrowing organisms consuming particles that settle on the sea bed.

In the case of the diesel fuel spilled by the barge FLORIDA, sediments and bodies of surviving animals carried the hydrocarbon sign as a kind of molecular fingerprint for many months after the accident. Furthermore, bacteria for preference utilize paraffins, leaving behind the more toxic aromatic hydrocarbons, which we commonly group into the class of potent carcenogenic substances. Oil may also concentrate other fat-soluble poisons such as insecticides, thereby enforcing the downfall of the environment.

It is thus of paramount importance to reduce the inflow of oil into the marine habitat. The largest single source of oceanic oil pollution is the discharge of bunker oil from tankers and non-tanker vessels along shipping routes. This inflow will automatically come to halt once severe fines are imposed, or low-cost facilities are made available in major ports to clean up the vessel. Since all crude oils have characteristic "fingerprints", conviction of polluters should be easy.

On March 16, 1978, the tanker AMOCO CADIZ grounded off Brittany, spilling about 250,000 tons of oil (Teal and Howarth, 1984). It was the most substantial oil spill of a tanker ever. A day-to-day account of fates and effects of oiling the beaches and marine life during the first 4 weeks following this incident, has been explicitly given by Vandermeulen et al. (1978), Le Fèvre (1978), D'Ozouville et al. (1978), Bellier (1978), Berne et al. (1978), Berne and D'Ozouville (1979), and Piboubes (1979). In essence, it was a repeat of the FLORIDA spill except on a grander scale. In Fig. 12.38 the movement of the oil slick along the northern coastline of Brittany and away from it is followed over a time period of 10 days. The aerial photograph of Roscoff harbor on March 23 is – in the true sense – an oil painting of a natural catastrophe inflicted by man.

Fig. 12.38. Case study on an oil spill: AMOCO CADIZ (Berne and D'Ozouville, 1979)

Chernobyl

Nuclear energy is produced in 26 countries and close to 400 reactors are presently operating. Without nuclear energy, an additional 200 million tons of oil would have to be burned annually, to compensate for the apparent global energy deficit. The relations among the various resources, i.e., oil, gas, water, nuclear, and alternatives, are illustrated in Fig. 9.37. Although nuclear energy supplies less than 5 percent of the total global energy consumed, it contributes in some countries, for instance in France, more than 50 percent of the electricity, and worldwide almost 15 percent.

BOX 12.8

A Small Oil Spill (Blumer et al., 1971)

The 1,000-dollar advance I received from Prentice Hall for my first book "Geochemistry of Sediments" allowed me to fulfill a much longed for wish, that is to buy a small sailboat. Quite apart from fun, I hoped to get with it a better feeling for water, wind, and weather. Each year, at the beginning of May, it was anchored at West Falmouth Harbor, a small inlet in Buzzards Bay, Massachussetts (see sketch below); at the end of September, the boat was taken up again to spend the winter in my backyard.

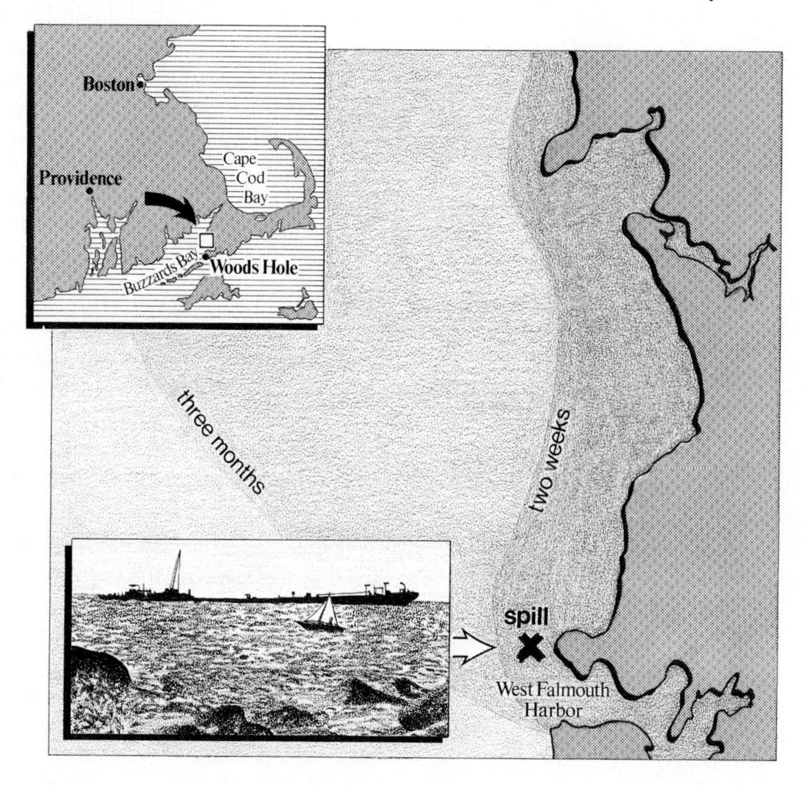

In 1969 boat retrieval commenced a month later, because on September 16, the oil barge FLORIDA came ashore releasing 630 tons of # 2 fuel oil into the coastal waters, less than a mile from my anchor place. In the 2 weeks following the incident I made many turns across the troubled waters upon which oil had been poured, and the sea played in a thousand colors.

At the same time, marine biologists from the Woods Hole Oceanographic Institution, who had worked for years in that region, embarked on an interdisciplinary study to assess the impact of a small oil spill on a formerly pristine environment (Blumer et al., 1971). The ecological effects of the spreading blanket of oil were severe. During the first few days offshore marine life in the immediate area was greatly decimated and dead animals littered the tidal flats. The oil spread northwards across the bottom of Buzzards Bay, still retaining its toxicity. Since oil destabilizes the upper sediment layer, fluid mud moved rapidly down the offshore slope contaminating benthic communities far beyond the place of the accident. Marsh sediments yielded oil even at a depth of 1 to 2 feet, ruling out shellfishing for more than 2 years.

The owner of the FLORIDA had to pay a fine of 100,000 dollars to the City of Falmouth and another 200,000 dollars to the State of Massachusetts, actually a trivial amount considering the environmental disaster and the financial loss to local fishermen. By comparison, the owner of the AMOCO CADIZ got away with a fine of about 80 million dollars, although the actual damage was estimated at 10 to 20 times that amount. The lesson we can draw from this ruling is simple. You will be fined 500 dollars when spilling a ton of oil into the sea. However, should you spill 100,000 tons of oil or more, you will get a discount of 150 dollars per ton.

Since my 1969 cruises through a sea of oil off Cape Cod, I became more alert to marine pollution.

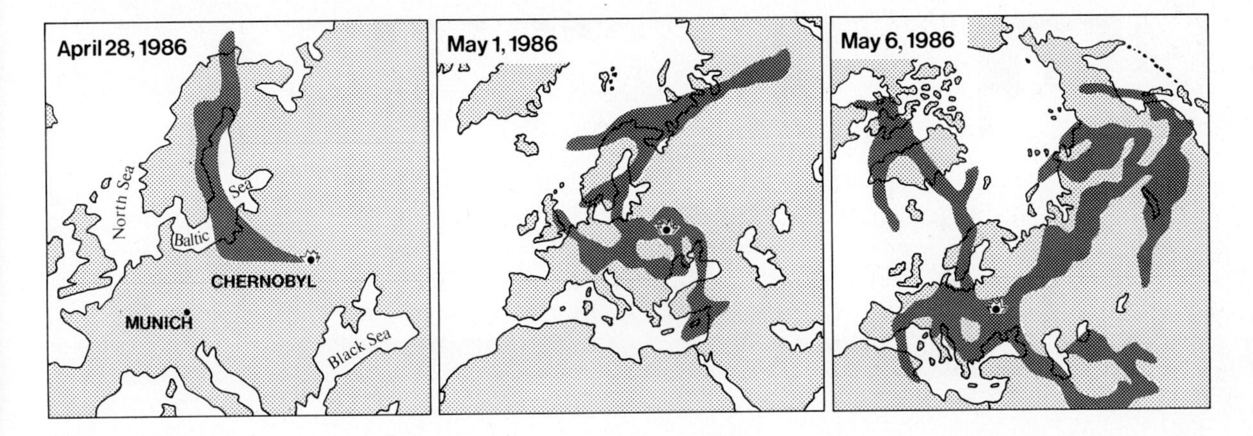

Fig. 12.39. Spread of Chernobyl cloud across Europe, April 28 to May 6, 1986

As is the case with all man-made machines, nuclear reactors, too, are programmed for some minor and major repairs due to: rust, metal fatigue, cracks in concrete foundations, or water and steam leakages. Inasmuch as these are anticipated, engineers and construction experts have built in safety devices galore. However, as regards danger number one, the human factor, there is always a good chance that something may unexpectedly go wrong. The Three-Mile-Island reactor incident is an example par excellence of that, but fortunately, it proved possible to rectify it at the last moment before it got out of control. The Chernobyl event, according to Soviet reports, also went to the account of human failure, but in this case the results were disastrous.

On April 26, 1986, at one o'clock twenty three minutes and forty four seconds, reactor block 4 exploded, sending off a fireworks display 1,000 meters high up into the dark sky. Massive quantities of radionuclides were released to the lower atmosphere, spreading a radioactive blanket first across the Baltic Sea and Scandinavia and then, in a complicated track pattern, across wide regions of Europe, Siberia, Greenland and North America (Fig. 12.39). Truly a local accident of global environmental consequences.

Irresponsible operation mistakes were at the core of the matter. Temperatures went up to 5,000°C, emitting virtually all of the noble gases, evaporating parts of the volatile core material which led to its fragmentation. Subsequently, the graphite moderator caught fire, thereby reheating the core and causing the release of a radioactive plume composed of volatiles and aerosols. The heroic efforts of firefighters brought fire and emission to a halt on May 6 at a tragic loss of 31 rescue workers. The final count gave a total emission of 1.8×10^{18} Bq (Box 12.9), corresponding to roughly 3–4 percent of the total core inventory, al-

BOX 12.9

Becquerel

Recently we have learned a new term: Becquerel. It is a measure of radioactive decay, and named after the French physicist Henri Becquerel (1852–1908), who in 1896 discovered that the element uranium radiates. Formerly, other units were used: Rem, Curie, Sievert. Actually the last two have been redefined a number of times and a great deal of confusion has resulted from this. That Rem was discarded is good, because the term is used in psychiatry as an abbreviation for "rapid eye movement".

The Becquerel (Bq) is a count of how many atomic nuclei of a radioactive substance decay per second. Its relation to the newly defined activity unit Curie (Ci) is: Becquerel times 3.7×10^{10}. Inasmuch as total radiation is not necessarily a reliable yardstick in measuring its effect upon a biological cell simply because alpha particles are about 20 times more efficacious relative to beta or gamma particles, the term Sievert (Sr) has been introduced. 1 Sievert is equal to 1 Joule (J) of energy taken up by the organism (in kg) times 20 for alpha particles and times 1 for beta or gamma particles. Actually, 1 Sievert is equal to 100 Rem. So much for that.

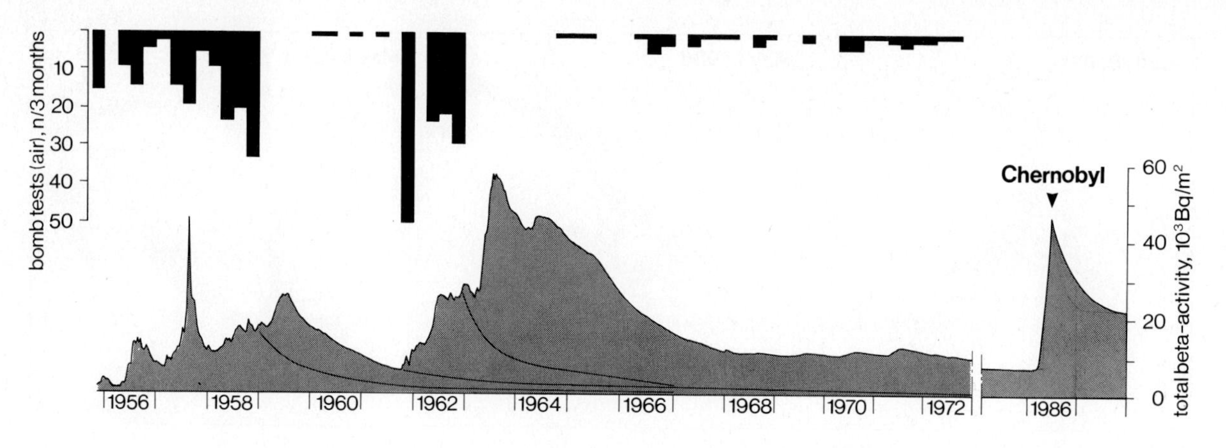

Fig. 12.40. Total beta activity (Bq m^{-2} soil) of fall-out near Munich, F.R.G. since fall 1955 (Stierstadt, 1987, and pers. comm.). The number of bomb tests at three months interval is indicated at the left margin

though the release rate for a few nuclides was much higher: I (20%), Te (15%), and Cs (10–13%) (Gesellschaft für Reaktorsicherheit, 1987).

Outside the "near-field" region with a radius of 30 km, almost all of the radiological impact is caused by the volatile radionuclides, radioiodine and radiocesium, brought in by the air (e.g., Volchok and Chieco, 1986; Goldman et al., 1987; Goldman, 1987). Of immediate concern was iodine-131 (8-day half-life), whereas worries are now chiefly concentrated on radiocesium, that is cesium-137 and cesium-134, with half-lives of 30 and 2 years respectively. How much of the actual radiation dose is absorbed into tissue, and how long it will stay there, and what its eventual effects will be on humans, and what the overall environmental risks are, is still hotly debated. A few more years will be needed to get a reasonable assessment of the latent health effects. For the time being it may thus be more meaningful to compare radioactive fallout from the Chernobyl event with that from atmospheric nuclear explosions.

Total beta-activity of fallout in the city of Munich, Germany, has been measured in a collector (cf. "Sammler") since 1955 (Stierstadt, 1987); activity is expressed in Bq per square meter (Fig. 12.40). The shape of the fallout curve is chiefly a function of number and size of nuclear bomb explosions in the air, the observance of the 1963 Treaty Banning Nuclear Weapons Tests in the Atmosphere, in Outer Space, and Under Water by the major nuclear powers, and the decay rate of the radionuclides. The fact that some nations in later years did not comply with the treaty is regrettable and accounts for the slightly elevated levels (Fig. 12.40). The Chernobyl peak is quite similar in height and decay behavior to the nuclear weapons' peak in 1963 (Stierstadt, 1987). Since ^{137}Cs is the main contributor to that peak, the collector is expected to stay at about 20,000 Bq m^{-3}; the last entry on July first, 1987 was 21,500 Bq m^{-3} (Stierstadt, pers. comm.).

Except for the immediate fatalities, there is so far no supporting evidence for health damage inflicted upon humans by the Chernobyl fallout. Years of observation will be needed to prove or disprove that this or that cancer, mental retardation, genetic effect etc. is directly caused by the radiation doses stemming from Chernobyl; but the question will be whether any future health problem can ever be linked, unequivocally, to the Chernobyl fallout. According to Goldman (1987): "On the basis of what we know now, the answer is "not likely."

With the disappearance of the radioactive cloud and after the first year of general concern, interest in Chernobyl has declined almost along the line of the short-lived radionuclides of iodine, ruthenium or cerium. By decree, new standards were set as to how many becquerel in milk, meat, vegetables or nuts were safe for human consumption, in short, everything was back to normal.

Looking at the marine environment, however, the full impact comes only gradually to light. In a study of the North Sea and Black Sea (Kempe et al., 1987b; Kempe and Nies, 1987; Buesseler et al., 1987) it could be shown that total activity was as high as 670,000 Bq per kg of suspended matter, the main food resource of marine life. This is more than a million times the radioactive load present in Black Sea and North Sea waters, or 1,000 times the upper level set by the European Community for human food (600 Bq kg^{-1}), or 100 times the becquerel counts of whey powder produced from contaminated milk and considered unsafe "even" for livestock consumption. It is ironic that 5,000 tons of radiating whey powder that came by train "Out of Rosenheim" are presently waiting to be processed as radioactive waste at Lingen, Northwest Germany, whereas at the same time, 100 km further north in the German Bight, marine life is subjected to much higher doses. O tempora! O mores!

ENUWAR

Hiroshima and Nagasaki were devastated in 1945 by comparatively small atomic bombs having an energy yield of about 15 and 20 kilotons (kt), respectively. A 1-kt explosion is equivalent in energy release to the detonation of about 1,000 tons of TNT. Physical, thermal, chemical, and biological impacts upon the air-water-earth-life system were studied in the weeks and months following these bombings; they provided a first glimpse into the after-effects of a nuclear explosion. Subsequent atmospheric nuclear tests, generally on the lower scale, yielded additional insight into the immediate environmental consequences of a nuclear war (ENUWAR).

Considering the vast arsenal of nuclear weapons stored in underground silos and bunkers east and west of the political fence, a frightening scenario took shape in the minds of some environmentally concerned scientists. Potentially, humans could bomb themselves into "the stone age", leaving the environment behind in ruins. The argument runs as follows.

At present about 24,000 strategic (long-range) and theater (middle-range) nuclear warheads with an explosive power of about 12,000 megatons are assembled worldwide. This arsenal contains in essence about 1 million "Hiroshima-size" bombs. A plausible scenario for a global nuclear war could involve in the order of 6,000 megatons divided among more than 12,000 warheads.

Starting with an article in Ambio (Crutzen and Birks, 1982), thoughtful attention has been given by numerous scientists and organisations to the environmental consequences of nuclear war. The Scientific Committee on Problems of the Environment (SCOPE), under the leadership of Sir Frederick Warner, has been spearheading ENUWAR (Warner, 1988). In a two-volume technical monograph the physical and atmospheric effects (Pittock et al., 1986) and the ecological and cultural effects (Harwell and Hutchinson, 1986) were spread out in great detail. A summary was prepared by Crutzen (1985) and Sagan (1985), and a popular account was given by Dotto (1986). According to these scenarios, our atmosphere might revert to conditions Earth had experienced at the Cretaceous/Tertiary boundary (Crutzen, 1987) where – due to bolides from above and volcanoes from below – we witnessed global fires, smoke, soot, acid rain, ozone depletion, devastation of forests, a muddy sea, freezing land, less sunlight and a change in rain pattern resulting in starvation and mass extinction at all organismic levels (Fig. 12.41). What formerly had happened to the dinosaurs might well occur to humans; indeed an apocalyptic vision which reminds me of "The Book of Revelation" supposed to have been written by the Apostle John at the end of the first century A.D.

Fig. 12.41. ENUWAR: Separation of smoke and precipitation (Malone et al., 1986)

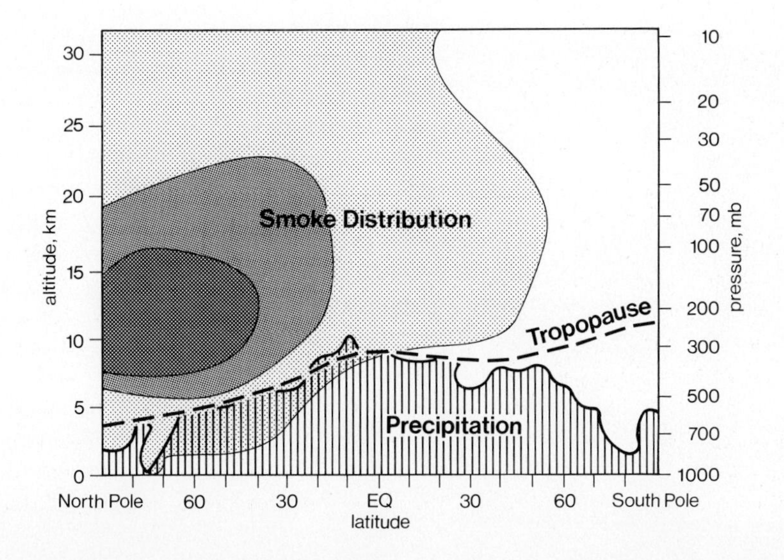

Critics of ENUWAR have voiced the opinion that anything after a nuclear exchange of that dimension is of no consequence any more – at least to human societies as we now know them – since recovery of pre-war conditions is unlikely (Box 12.10). And yet, ENUWAR has started a dialog among people from all nations and made them aware of what might be in store for us and nature in case we fail. East and west should meet to resolve some of today's involved environmental issues.

I have written this book to force myself to meditate on air, water, life and earth, and to unravel their interplay in nature. However, the wealth of natural phenomena that revealed itself to my eyes was quite intriguing but at the same time so hopelessly involved, that I gave up, many times. But, as in Hemingway's "The Old Man and the Sea", I picked up the pen, again and again, to bring to a close this biogeochemical Odyssey on air-sea-earth-life interactions.

BOX 12.10

Extinction of Big Brain (Kurt Vonnegut: Galapagos. – Grafton Books, a Division of the Collins Publishing Group, London – Glasgow – Toronto – Sydney – Auckland, 1985, pp. 269)

Kurt Vonnegut's satire describes the fate of a handful of human survivors including a few cannibal girls from the Ecuadorian rainforest, on Galapagos Archipelago in the aftermath of a crazy war game. They became the new humanity after all other forms of people had vanished from Earth:

"Meanwhile, in other parts of the world, particularly in Africa, people were dying by the millions because they were unlucky. It hadn't rained for years and years. It used to rain a lot there, but now it looked as though it might never rain again.

At least the Africans had stopped reproducing. That much was good. That was some help. That meant that there was that much more of nothing to spread around." (p. 248).

After a million years of natural selection – in the Darwinian sense – a new *Homo* species evolved, a furry creature with flippers and a small brain. All that was left of Big Brain were fossils.

References

Chapter 9–12

Aaronson S (1970) Molecular evidence for evolution in the algae: a possible affinity between plant cell walls and animal skeletons. Ann NY Acad Sci 175:531–539

Abelson PH (1959) Geochemistry of organic substances. In "Researches in Geochemistry" (ed Abelson PH), New York, John Wiley & Sons, Inc 79–103

Acock B, Allen LH Jr (1985) Crop responses to elevated carbon dioxide concentrations. In "Direct Effects of Increasing Carbon Dioxide on Vegetation " (eds Strain BR, Cure JD), Washington DC, US Dept of Energy, DOE/ER-0238:53–97

Adams DF, Farwell SO, Robinson E, Pack MR, Bamesberger WL (1981) Biogenic sulfur source strengths. Environ Sci Technol 15:1493–1498

Adel A (1939) Note on the atmospheric oxides of nitrogen. Astrophys J 90:627–628

Aharon P, Schidlowski M, Singh IB (1987) Chronostratigraphic markers in the end-Precambrian carbon isotope record on the Lesser Himalaya. Nature 327:699–702

Alberts B, Bray D, Lewis J et al. (1983) Molecular Biology of the Cell. Garland Publishing, Inc, New York London, pp 1146

Alexander AE (1942) The structure of the surfaces of solutions. Trans Faraday Soc 38:54–63

Ali SY, Sajdera SW, Anderson HC (1970) Isolation and characterization of calcifying matrix vesicles from epiphyseal cartilage. Proc Natl Acad Sci USA 67:1513–1520

Allaart JH (1976) The pre-3760 m.y. old supracrustal rocks of the Isua area, Central West Greenland, and the associated occurrence of quartz-banded ironstone. In "The Early History of the Earth" (ed Windley BF), Wiley, London New York Sydney Toronto, 177–189

Allègre CJ (1982) Chemical geodynamics. Tectonophysics 81:109–132

Alpers W, Blume H-JC, Garrett WD, Hühnerfuss H (1982) The effect of monomolecular surface films on the microwave brightness temperature of the sea surface. Int J Remote Sensing 3/4:457–474

Altmann J (1988) Ins and outs of cell signalling. Nature 331:119–120

Alvarez LW, Alvarez W, Asaro F, Michel HV (1980) Extraterrestrial cause for the Cretaceous-Tertiary extinction. Science 208:1095–1108

Alvarez W, Kauffman EG, Surlyk F, Alvarez LW, Asaro F, Michel HV (1984) Impact theory of mass extinctions and the invertebrate fossil record. Science 223:1135–1141

Amphlett CB (1964) Inorganic Ion Exchangers. Elsevier Publ Co, Amsterdam New York, pp 141

Anders E, Owen T (1977) Origin and abundances of volatiles. Science 198:453–465

Anderson HC (1973) Calcium-accumulating vesicles in the intercellular matrix of bone. In "Hard Tissue Growth, Repair and Remineralization" (eds Elliott K, Fitzsimons DW), Amsterdam London New York, Elsevier-Excerpta Medica-North-Holland, 213–245

Anderson IC, Levine JS (1987) Simultaneous field measurements of biogenic emissions of nitric oxide and nitrous oxide. J Geophys Res 92/D1:965–976

Andreae MO (1986) The ocean as a source of atmospheric sulfur compounds. In "The Role of Air-Sea Exchange in Geochemical Cycling" (ed Buat-Ménard P), Dordrecht Boston Lancaster Tokyo, D Reidel Publ Co, 331–362

Andreae MO, Barnard WR (1984) The marine chemistry of dimethylsulfide. Mar Chem 14:267–279

Andreae MO, Raemdonck H (1983) Dimethylsulfide in the surface ocean and the marine atmosphere: a global view. Science 221:744–747

Aplin AC, Cronan DS (1984) Ferromanganese oxide deposits from the Central Pacific Ocean, I Encrustations from the Line Islands Archipelago. Geochim Cosmochim Acta 48:427–436

Argast S, Farlow JO, Gabet RM, Brinkman DL (1987) Transport-induced abrasion of fossil reptilian teeth: implications for the existence of Tertiary dinosaurs in the Hell Creek Formation, Montana. Geology, 15:927–930

Arnold F, Knop G, Ziereis H (1986) Acetone measurements in the upper troposphere and lower stratosphere – implications for hydroxyl radical abundances. Nature 321:505–507

Arrhenius S (1897) On the influence of carbonic acid in the air upon the temperature of the ground. Philosoph Mag 41:237–275

Arthur MA, Garrison RE (1986) Milankovitch cycles through geologic time. Paleoceanography 1:369–586

Aston RS (ed) (1983) Silicon Geochemistry and Biogeochemistry. Academic Press Inc, London New York, pp 248

Awramik SM (1982) The Pre-Phanerozoic fossil record. In "Mineral Deposits and the Evolution of the Biosphere" (Dahlem Conference) (eds Holland HD, Schidlowski M), Springer Verlag, Berlin Heidelberg New York, 67–82

Bacastow R, Björkström A (1981) Comparison of ocean models for the carbon cycle. In "Carbon Cycle Modelling" (ed Bolin B), SCOPE Report 16, Chichester New York Brisbane Toronto, John Wiley & Sons, 29–79

Bacastow R, Keeling CD (1981) Atmospheric carbon dioxide concentration and the observed airborne fraction. In "Carbon Cycle Modelling" (ed Bolin B), SCOPE Report 16, Chichester New York Brisbane Toronto, John Wiley & Sons, 103–112

Bailey SW (1963) Polymorphism of the kaolin minerals. Am Mineralogist 48:1196–1209

Bak RPM (1973) Coral weight increment in situ. A new method to determine coral growth. Mar Biol 20:45–49

Baker EG (1960) A hypothesis concerning the accumulation of sediment hydrocarbons to form crude oils. Geochim Cosmochim Acta 19:309–317

Barber RT, Chavez FP (1983) Biological consequences of El Niño. Science 222:1203–1210

Barker JA, Henderson D (1981) The fluid phases of matter. Sci Am 245/5:94–102

Barnes P, Finney JL, Nicholas JD, Quinn JE (1979) Cooperative effects in simulated water. Nature 282:459–464

Barnett TP (1984) Prediction of the El-Niño of 1982–83. Mon Wea Rev 112:1403–1407

Barnola JM, Raynaud D, Korotkevich YS, Lorius C (1987) Vostok ice core provides 160,000-year record of atmospheric CO_2. Nature 329:408–414

Barr TN (1981) The world food situation and global grain prospects. Science 214:1087–1095

Barry RG, Chorley RJ (1982) Atmosphere, Weather and Climate (4th ed). London New York, Methuen, pp 407

Barton JK (1986) Metals and DNA: molecular left-handed complements. Science 233:727–734

Barton JM Jr, Ryan B, Fripp REP (1983) Rb-Sr and U-Th-P isotopic studies of the Sand River Gneisses, central zone, Limpopo mobile belt. In "The Limpopo Belt" (eds van Biljon WJ, Legg JH), Geol Soc S Afr Spec Publ 8:9–18

Bates DR, Nicolet M (1950) The photochemistry of water vapor. J Geophys Res 55:301

Baturin GN (1986) Geochemistry of Non-Manganese Concretions in the Sea. Publ Nauk Moscow, pp 328

Baumgartner A, Reichel E (1975) The World Water Balance. München Wien, R Oldenbourg Verlag, pp 179

Becker RH, Clayton RN (1976) Oxygen isotope study of a Precambrian banded iron-formation, Hamersley Range, Western Australia. Geochim Cosmochim Acta, 40:1153–1165

Becker RH, Pepin RO (1984) The case for a Martian origin of the shergottites: Nitrogen and noble gases in EETA 79001. Earth Planet Sci Lett 69:225–242

Bellier MP (1978) L'Expérience de l'AMOCO CADIZ, Mars-Septembre 1978. Ministère des Transports, Direction des Ports et de la Navigation Maritimes, Janvier 1979, pp 192

Ben-Naim A (1965) Thermodynamics of aqueous solutions of noble gases. I. J Phys Chem 69:3240–3245

Benton MJ (1985) Mass extinction among non-marine tetrapods. Nature 316:811–814

Berg JM (1986) Potential metal-binding domains in nucleic acid binding proteins. Science 232:485–487

Berg P (1981) Dissections and reconstructions of genes and chromosomes. Science 213:296–303

Berkley JL, Drake MJ (1981) Weathering of Mars: Antarctic analog studies. Icarus 45:231–249

Berkman S, Egloff G (1941) Emulsions and Foams. Reinhold Publishing Corporation, New York

Berkner LV, Marshall LC (1965a) History of major atmospheric components. Proc Natl Acad Sci 53:1215–1226

Berkner LV, Marshall LC (1965b) On the origin and rise of oxygen concentration in the earth's atmosphere. J Atmos Sci 22:225–261

Bernal JD (1954a) The origin of life. New Biol 16:28–40

Bernal JD (1954b) The physical basis of life. Proc Physic Soc, London B-62:597–618

Bernal JD (1959) The problem of stages in biopoesis. In "The Origin of Life on the Earth" (eds Clark F, Synge RLM), New York London Paris Los Angeles, Pergamon Press Inc, 38–53

Bernal JD, Fowler RH (1933) A theory of water and ionic solution, with particular reference to hydrogen and hydroxyl ions. J Chem Phys 1:515–548

Berne S, D'Ozouville L (1979) AMOCO CADIZ. Cartographie des Apports Polluants et des Zones Contaminées. CNEXO, Centre Océanologique de Bretagne, Brest, Mai 1979, pp 175

Berne S, Brossier R, Fontanel A et al. (1978) Télédétection des Pollutions par Hydrocarbures de l'AMOCO CADIZ. Publ du CNEXO, Série "Actes de Colloques", 6:9–26

Berner RA, Lasaga AC, Garrels RM (1983) The carbonate-silicate geochemical cycle and its effect on atmospheric carbon dioxide over the past 100 million years. Am J Sci 283:641–683

Bhatnagar VM (1962) Clathrate compounds of water. Research 15:299–304

Bhatt JJ (1978) Oceanography. New York Cincinnati Toronto London Melbourne, D van Nostrand Company, pp 322

Bjerknes J (1966) A possible response of the atmospheric Hadley circulation to equatorial anomalies of ocean temperature. Tellus 18:820–829

Black CC Jr, Brand JJ (1986) Rates of calcium in photosynthesis. Calcium and Cell Function 6:327–355

Blackburn TH, Sørensen J (eds) (1988) Nitrogen Cycling in Coastal Marine Environments. SCOPE Report 33, John Wiley & Sons, Chichester New York Brisbane Toronto Singapore, pp 451

Blumer M (1972) Submarine seeps: Are they a major source of open ocean oil pollution? Science 176:1257–1258

Blumer M, Sanders HL, Grassle JF, Hampson GR (1971) A small oil spill. Environment 13/2:1–12

Boedtker H (1960) Configurational properties of tobacco mosaic virus ribonucleic acid. J Mol Biol 2:171–188

Bohor BF, Foord EE, Modreski PJ, Triplehorn DM (1984) Mineralogic evidence for an impact event at the Cretaceous-Tertiary boundary. Science 224:867–869

Bojkov RD (1986) The 1979–1985 ozone decline in the Antarctic as reflected in ground based observations. Geophys Res Lett 13:1236–1239

Bolin B (1986) How much CO_2 will remain in the atmosphere? In "The Greenhouse Effect, Climatic Change, and Ecosystems" (eds Bolin B, Döös BR, Jäger J, Warrick RA), SCOPE Report 29, Chichester New York Brisbane Toronto Singapore, John Wiley & Sons, 93–155

Bolin B, Cook RB (eds) (1983) The Major Biogeochemical Cycles and Their Interactions. SCOPE Report 21, Chichester New York Brisbane Toronto Singapore, John Wiley & Sons, pp 532

Bolin B, Degens ET, Kempe S, Ketner P (eds) (1979) The Global Carbon Cycle. SCOPE Report 13. Chichester New York Brisbane Toronto, John Wiley & Sons, pp 491

Bolin B, Björkström A, Holmen K (1983) The simultaneous use of tracers for ocean circulation studies. Tellus 35B, 206–236

Bolin B, Döös BR, Jäger J, Warrick RA (eds) (1986) The Greenhouse Effect, Climatic Change, and Ecosystems. SCOPE Report 29. Chichester New York Brisbane Toronto Singapore, John Wiley & Sons, pp 541

Bolle H-J, Seiler W, Bolin B (1986) Other greenhouse gases and aerosols. In "The Greenhouse Effect, Climatic Change, and Ecosystems" (eds Bolin B, Döös BR, Jäger J, Warrick RA), SCOPE Report 29, Chichester New York Brisbane Toronto Singapore, John Wiley & Sons, 157–203

Bonner WA, Kavasmaneck PR, Martin FS, Flores JJ (1974) Asymmetric adsorption of alanine by quartz. Science 186:143–144

Borchert H, Muir RO (1964) Salt Deposits: The Origin, Metamorphism, and Deformation of Evaporites. London, Van Nostrand, pp 338

Borle AB (1981) Control, modulation and regulation on cell calcium. Rev Physiol Biochem Pharmacol 90:14–153

Bossard AR, Kamga R, Raulin F (1986) Gas phase synthesis of organophosphorus compounds and the atmosphere of the giant planets. Icarus 67:305–324

Botkin DB (ed) (1986) Remote Sensing of the Biosphere. Washington, DC, National Academy Press, pp 135

Bouchiat M-A, Pottier L (1984) An atomic preference between left and right. Sci Am 250/6:76–85

Bowin C (1983) Gravity, topography, and crustal evolution of Venus. Icarus 56:345–371

Bowman KP (1988) Global trends in total ozone. Science 239:48–52

Boyle EA, Keigwin LD (1985) Comparison of Atlantic and Pacific paleochemical records for the last 250,000 years: changes in deep ocean circulation and chemical inventories. Earth Planet Sci Lett 76:135–150

Braitsch O (1962) Entstehung und Stoffbestand der Salzlagerstätten. In "Mineralogie und Petrographie in Einzeldarstellungen", vol 3 (eds Engelhardt W v, Zemann J), Springer Verlag, Berlin Heidelberg New York, pp 232

Brandt W (1968) Channelling in crystals. Sci Am 218 (March 1968), 91–98

Brasseur G (1987) The endangered ozone layer. Environment 29/1:6–11

Braster M (1986) Why do lower plants and animals biomineralize? Paleobiology 12/3:241–250
Bredehoeft JD, Blyth CR, White WA, Maxey GB (1963) Possible mechanism for concentration of brines in subsurface formations. Bull Am Assoc Petrol Geol 47:257–269
Bretscher MS (1985) The molecules of the cell membrane. Sci Am 253/4:86–90
Brindley GW (ed) (1980) X-Ray Identification and Crystal Structures of Clay Minerals. Mineralogical Society, London, pp 495
Brinkman R (1929) Statistisch-biostratigraphische Untersuchungen an mitteljurassischen Ammoniten über Artbegriff und Stammesentwicklung. Berlin, Weidmannsche Buchhandlung, pp 249
Broadfoot AL, Sandel BR, Shemansky DE et al. (1981) Extreme ultraviolet observations from Voyager 1 encounter with Saturn. Science 212:206–211
Brockmann UH, Kattner G, Hentzschel G, Wandschneider K, Junge HD, Hühnerfuss H (1976) Natürliche Oberflächenfilme im Seegebiet vor Sylt. Mar Biol 36:135–146
Broecker WS (1982) Ocean chemistry during glacial time. Geochim Cosmochim Acta 46:1689–1705
Broecker WS (1985) How to Build a Habitable Planet. Eldigio Press, Palisades, New York, pp 291
Broecker WS (1987a) The biggest chill. Nat Hist 10:74–82
Broecker WS (1987b) Unpleasant surprises in the greenhouse? Nature 328:123–126
Broecker WS, Peng T-H (1982) Tracers in the Sea. Eldigio Press, Lamont-Doherty Geol Observatory, Palisades, New York
Broecker WS, Takahashi T, Simpson HJ, Peng T-H (1979) Fate of fossil fuel carbon dioxide and the global carbon budget. Science 206:409–418
Brosche P, Sündermann J (eds) (1982) Tidal Friction and the Earth's Rotation. Berlin Heidelberg New York, Springer-Verlag, pp 345
Brouwer A (1983) Global controls in Phanerozoic sedimentary record. Newsl Stratigr 12:166–174
Brown LR (1981) World population growth, soil erosion, and food security. Science 214:995–1002
Bruckmann P (1982) Cycles of organic gases in the atmosphere. In "Physico-Chemical Behaviour of Atmospheric Pollutants", Proc 2nd Europ Symp Dordrecht, D Reidel Publ Co, 336–348
Bruevich SW (1957) The salinity of the interstitial waters (sediment solutions) of Okhotsk Sea. Dokl Akad Nauk SSSR 113:387–390 (in Russian)
Bruevich SW, Zaytseva ED (1960) On the chemistry of the sediments of the northwestern part of the Pacific. Tr Inst Okeanol Akad Nauk SSSR 42:3–88 (in Russian)
Bryan K, Spelman MJ (1985) The ocean's response to a CO_2-induced warming. J Geophys Res 90/C6: 11,679–11,688
Bryson RA, Goodman BM (1980) Volcanic activity and climatic changes. Science 207:1041–1044
Buat-Ménard P (ed) (1986) The Role of Air-Sea Exchange in Geochemical Cycling. Dordrecht Boston Lancaster Tokyo, D Reidel Publ Co, pp 549
Budyko MI (1969) The effect of solar radiation variations on the climate of the earth. Tellus 21:611–619
Budyko MI, Ronov AB, Yanshin AL (1987) History of the Earth's Atmosphere. Berlin Heidelberg New York London Paris Tokyo, Springer Verlag, pp 139
Buesseler KO, Livingston HD, Honjo S et al. (1987) Chernobyl radionuclides in a Black Sea sediment trap. Nature 329:825–828
Burns JA et al. (eds) (1981) Results from the Mars data analysis program. Icarus 45:1–516
Burns RG (1986) Terrestrial analogues of the surface rocks of Mars. Nature 320:55–56
Burton EA, Walter LM (1987) Relative precipitation rates of aragonite and Mg calcite from seawater: temperature of carbonate ion control? Geology 15:111–114
Burton FG, Neuman MW, Neuman WF (1969) On the possible role of crystals in the origins of life. 1. The adsorption of nucleosides, nucleotides and pyrophosphate by apatite crystals. Curr Mod Biol 3:20
Cairns-Smith AG (1982) The Genetic Takeover and the Mineral Origins of Life. Cambridge University Press, pp 477
Callendar GS (1938) The artificial production of carbon dioxide and its influence on temperature. QJ R Meteorol Soc 64:223–240
Callis LB, Natarajan M (1986) Ozone and nitrogen dioxide changes in the stratosphere during 1979–84. Nature 323:772–777
Calvert SE (1983) Sedimentary geochemistry of silicon. In "Silicon Geochemistry and Biogeochemistry" (ed Aston SR), London, Academic Press Inc, 143–186
Calvin M (1959) Evolution of enzymes and the photosynthetic apparatus. Science 130:1170–1175
Calvin M (1969) Chemical Evolution. Oxford, Clarendon Press, pp 278
Cameron AGW (1983) Origin of the atmospheres of the terrestrial planets. Icarus 56:195–201
Campbell AK (1983) Intracellular Calcium. Its Universal Role as Regulator. Chichester, J Wiley & Sons, pp 556
Campbell AE, Hellebust JA, Watson SW (1966) Reductive pentose phosphate cycle in *Nitrosocystis oceanus*. J Bacteriol 91:1178–1185
Campsie J, Johnson GL, Jones JE, Rich JE (1984) Episodic volcanism and evolutionary crises. Eos 65/45:796–800
Cane MA (1983) Oceanographic events during El Niño. Science 222:1189–1195
Cane MA (1986) El Niño. Annu Rev Earth Planet Sci 14:43–70
Cane MA, Zebiak SE, Dolan SC (1986) Experimental forecasts of El Niño. Nature 321:827–832
Cann RL, Stoneking M, Wilson AC (1987) Mitochondrial DNA and human evolution. Nature 325:31–36
Canuto VM (1983) Oxygen and ozone in the early Earth's atmosphere. Precamb Res 20:109–120
Canuto VM, Levine JS, Augustsson TR, Imhoff CL (1982) UV radiation from the young Sun and oxygen and ozone levels in the prebiological palaeoatmosphere. Nature 296:816–820
Carafoli E, Penniston JT (1985) The calcium signal. Sci Am 253/5:50–58
Carr MH (1986) Mars: a water-rich planet? Icarus 68:187–216
Carr MH (1987) Water on Mars. Nature 326:30–35
Carver JH (1981) Prebiotic atmospheric oxygen levels. Nature 292:136–138
Catchpole AJW, Faurer M-A (1983) Summer sea ice severity in Hudson Strait, 1751–1870. Clim Change 5:115–139
Cavalier-Smith T (1986) The kingdoms of organisms. Nature 324:416–417
Chadwell HM (1927) The molecular structure of water. Chem Rev 4:375–398
Chamberlain TC (1899) An attempt to frame a working hypothesis of the cause of glacial periods on an atmospheric basis. J Geol 7:545–584
Chance B, Yoshioka T (1966) Sustained oscillations of ionic constituents of mitochondria. Archiv Biochem Biophys 117:451–464

Chapman S (1930) A theory of upper atmospheric ozone. Mem R Meteorol Soc 3:103–125
Chappuis J (1880) Sur le spectre d'absorption de l'ozone. CR Acad Sci Paris 91:985–986
Charlson RJ, Lovelock JE, Andreae MO, Warren SG (1987) Oceanic phytoplankton, atmospheric sulphur, cloud albedo and climate. Nature 326:655–661
Chave KE (1954) Aspects of the biogeochemistry of magnesium-1: Calcareous marine organisms. J Geol 62:266–283
Chebotarev II (1959) Metamorphism of natural waters in the crust of weathering, 1, 2, and 3. Geochim Cosmochim Acta 8:22–48, 137–170, and 198–212
Chen-Tung A Chen, Rodman MR, Ching-Ling Wei, Olson EJ, Feely RA, Gendron JF (1986) Carbonate Chemistry of the North Pacific Ocean. TRO34. CO_2, Washington, DC, US Dept Energy, pp 176
Cheung Wai Yiu (1979) Calmodulin plays a pivotal role in cellular regulation. Science 207:19–27
Chilingar GV (1962) Dependence on temperature of Ca/Mg ratio of skeletal structures of organisms and direct chemical precipitates out of sea water. Bull South Calif Acad Sci 61/1:45–60
Choudhury B, Kukla G (1979) Impact of CO_2 on cooling of snow and water surfaces. Nature 280:668–671
Chubachi S (1986) On the cooling of stratospheric temperature of Syowa, Antarctica. Geophys Res Lett 13:1221–1223
Chung Yu-Chia (1974) Radium-226 and Ra-Ba relationships in Antarctic and Pacific waters. Earth Planet Sci Lett 23:125–135
Ciochon RL, Fleagle JG (eds) (1986) Primate Evolution and Human Origins. Benjamin, Cummings, pp 396
Claypool GE, Holser WT, Kaplan IR, Sakai H, Zak I (1980) The age curves of sulfur and oxygen isotopes in marine sulfate and their mutual interpretation. Chem Geol 28:199–260
Clayton RN, Epstein S (1958) The relationship between O^{18}/O^{16} ratios in coexisting quartz, carbonate and iron oxides from various geological deposits. J Geol 66:352
Clemmey H, Badham N (1982) Oxygen in the Precambrian atmosphere: an evaluation of the geological evidence. Geology 10:141–146
Clifford J (1965) Properties of micellar solutions, Pt 4. Trans Faraday Soc 61:1276–1282
CLIMAP Project Members (1976) The surface of the ice-age earth. Science 191:1131–1137
CLIMAP Project Members (1984) The last interglacial ocean. Q Res 21:123–124
Climate System Monitoring (1985) The Global Climate System. World Climate Data Programme (WCDP), World Meteorological Organization, pp 52
Cloud P (1976) Beginnings of biospheric evolution and their biogeochemical consequences. Paleobiology 2:351–387
Cloud P, Glaessner MF (1982) The Ediacarian period and system: Metazoa inherit the earth. Science 217:783–792
Clube SVM, Napier WM (1982) The role of episodic bombardment in geophysics. Earth Planet. Sci Lett 57:251–262
CMA, Chemical Manufacturing Association (1982) World Production and Release of Chlorofluorocarbons 11 and 12 through 1981. Report of the Fluorocarbon Program Panel, 1982
Cole GHA (1986) Concerning the occurrence of water in the planetary bodies. Surv Geophys 8:439–457
Colin L et al. (1980) Pioneer Venus. Special Issue. J Geophys Res 85:7575–8337
Collins JH (1974) Homology of myosin light chains, troponin-C and parvalbumins deduced from comparison of their amino acid sequences. Biochem Biophys Res Commun 58:301–308
Cook P, Shergold JH (1984) Phosphorus, phosphorites and skeletal evolution at the Precambrian/Cambrian boundary. Nature 308:231–236
Corbridge DEC (1960) The crystal structure of sodium triphosphate $Na_5P_3O_{10}$, phase I. Acta Cryst 13:263–269
Corliss JB, Baross JA, Hoffman SE (1981) An hypothesis concerning the relationship between submarine hot springs and the origin of life on Earth. Oceanologica Acta 26:59–69
Cotmore JM, Nichols G, Wuthier RE (1971) Phospholipid-calcium phosphate complex. Enhanced calcium migration in the presence of phosphate. Science 172:1339–1341
Courtillot V, Besse J, Vandamme D et al. (1986) Deccan flood basalts at the Cretaceous/Tertiary boundary? Earth Planet Sci Lett 80:361–374
Craig H (1954) Carbon 13 in plants and the relationships between carbon 13 and carbon 14 variations in nature. J Geol 62:115–149
Craig H (1957) Isotopic standards for carbon and oxygen and correction factors for mass spectrometric analysis of carbon dioxide. Geochim Cosmochim Acta 12:133–149
Craig H (1961) Isotopic variations in meteoric waters. Science 133:1702–1703
Craig H (1971) The deep metabolism: oxygen consumption in abyssal ocean water. J Geophys Res 76:5078–5086
Craig H, Chou CC (1982) Methane: The record in polar ice cores. Geophys Res Lett 9:1221–1224
Craig H, Hayward T (1987) Oxygen supersaturation in the ocean: biological versus physical contributions. Science 235:199–202
Craig H, Weiss RF (1970) The GEOSECS 1969 intercalibration station: Introduction and hydrographic features, and total CO_2–O_2 relationships. J Geophys Res 75:7641–7647, together with 9 following papers, 7648–7696
Craig H et al. (1972a) The GEOSECS program: 1970–1971. Earth Planet. Sci Lett 16:47–145
Craig H, Chung Y, Fladeiro M (1972b) A benthic front in the South Pacific. Earth Planet. Sci Lett 16:50–65
Crick FHC (1954) The structure of the hereditary material. Sci Am 191 (October 1954), 54–61
Crick FHC, Watson JD (1954) The complementary structure of deoxyribonucleic acid. Proc R Soc London A 223:80–96
Cronin JE, Boaz NT, Stringer CB, Rak Y (1981) Tempo and mode in hominid evolution. Nature 292:113–122
Crutzen PJ (1970) The influence of nitrogen oxide on the atmospheric ozone content. QJR Meteorol Soc 96:320–325
Crutzen PJ (1985) The global environment after nuclear war. Environment 27/8:6–11, 34–37
Crutzen PJ (1987) Acid rain at the K/T boundary. Nature 330:108–109
Crutzen PJ, Arnold F (1986) Nitric acid cloud formation in the cold Antarctic stratosphere: a major cause for the springtime "ozone hole". Nature 324:651–655
Crutzen PJ, Birks JW (1982) The atmosphere after a nuclear war: Twilight at noon. Ambio 11:114–125
Crutzen PJ, Gidel LT (1983) A two-dimensional photochemical model of the atmopshere, 2: The tropospheric budget of the anthropogenic chlorocarbons, CO, CH_4, CH_32Cl and the effect of various NO_x sources on tropospheric ozone. J Geophys Res 88:6641–6661
Crutzen PJ, Aselmann I, Seiler W (1986) Methane production by domestic animals, wild ruminants, other herbivorous fauna, and humans. Tellus 38B:271–284
Cullis CF, Hirschler MM (1980) Atmospheric sulfur: natural and man-made sources. Atmos Environ 14:1263–1278
Cummings S, Enderby JE, Neilson GW, Newsome JR, Howe RA, Howells WS, Soper AK (1980) Chloride ions in aqueous

solutions. Nature 287:714–716

Cunnold DM, Prinn RG, Rasmussen RA, Simmonds PG, Alyea FN, Cardelino CA, Crawford AJ, Fraser PJ, Rosen RD (1983a) The atmospheric lifetime experiment 3. Lifetime methodology and application to three years of $CFCl_3$ data. J Geophys Res 8:8379–8400

Cunnold DM, Prinn RG, Rasmussen RA, Simmonds PG, Alyea FN, Cardelino CA, Crawford AJ (1983b) The atmospheric lifetime experiment 4. The results for CF_2Cl_2 based on three years of data. J Geophys Res 88:8401–8414

Danford MD, Levy HA (1962) The structure of water at room temperature. J Am Chem Soc 84:3965

Dansgaard W (1953) The abundance of O^{18} in atmospheric water and water vapor. Tellus 5:461–476

Dansgaard W, Johnson SJ, Clausen HB, Langway CC (1971) Climatic record revealed by the Camp Century ice core. In "The Late Cenozoic Glacial Ages" (ed Turekian KK), New Haven, Conn, Yale Univ Press, 37–56

Darnell J, Lodish H, Baltimore D (1986) Molecular Cell Biology. New York, Scientific American Books, pp 1187

Darwin Ch (1859) On the Origin of Species by Means of Natural Selection, or the Preservation of Favoured Races in the Struggle for Life. Reprint of the First Edition (1950) London, Watts & Co, pp XX + 426

Davies BD (1958) On the importance of being ionized. Arch Biochem Biophys 78:497–509

Davies DW (1981) The Mars water cycle. Icarus 46:398–414

Davis LE (1955) Electrochemical properties of clays. Proc First Nat Conf Clays and Clay Techn, Div Mines Calif Bull 169:47–53

Dawson IM, Follett EAC (1963) The oxidation of graphite. I. The observation of oxidation on synthetic graphite before and after irradiation. Proc R Soc Lond A 274:386–394

Dayhoff MO (1972) Atlas of Protein Sequence and Structure, vol 5. Nat Biomed Res Foundation, Washington, pp 124, and Data Section, pp 417

De Bernard L (1958) Associations moléculaires entre les lipides. II. Lécithine et cholestérol. Bull Soc Chim Biol 40:161–170

Defant A (1961) Physical Oceanography Vols 1, 2. Oxford New York, Pergamon Press, vol 1, pp 729, and vol 2, pp 598

Degens ET (1961) Project New Valley. Engineer Sci Mag, Nov 1961, Calif Inst Technol, pp 5

Degens ET (1969) Biogeochemistry of stable carbon isotopes. In "Organic Geochemistry" (eds Eglinton G, Murphy MTJ), New York Heidelberg Berlin, Springer-Verlag, 304–329

Degens ET (1970) Molecular nature of nitrogenous compounds in seawater and recent sediments. In "Organic Matter in Natural Waters" (ed Hood DW), Inst Mar Sci, Alaska, Publ No 1:77–106

Degens ET (1973) Accounting for the salts in the sea. Nature 243:504–507

Degens ET (1974) Synthesis of organic matter in the presence of silicate and lime. Chem Geol 13:1–10

Degens ET (1976) Molecular Mechanisms on Carbonate, Phosphate, and Silica Deposition in the Living Cell. Topics Current Chem 64:1–112

Degens ET (1979) Primordial synthesis of organic matter. In "The Global Carbon Cycle" (eds Bolin B, Degens ET, Kempe S, Ketner P), SCOPE Report 13, Chichester New York, John Wiley & Sons, 57–77

Degens ET (ed) (1982) Transport of Carbon and Minerals in Major World Rivers, pt 1. Mitt Geol-Paläont Inst Univ, Hamburg, SCOPE/UNEP Sonderband 52, pp 766

Degens ET, Chilingar GV (1967) Diagenesis of subsurface waters. In "Diagenesis in Sediments" (eds Larsen G, Chilingar GV), Developments in Sedimentology 8, Amsterdam London New York, Elsevier Publ Co, 477–502

Degens ET, Epstein S (1962) Relationship between O^{18}/O^{16} ratios in coexisting carbonates, cherts, and diatomites. Bull Am Assoc Petrol Geol 46:534–542

Degens ET, Epstein S (1964) Oxygen and carbon isotope ratios in coexisting calcites and dolomites from recent and ancient sediments. Geochim Cosmochim Acta 28, 23

Degens ET, Matheja J (1967) Molecular Mechanisms on Interactions between Oxygen Co-ordinated Metal Polyhedra and Biochemical Compounds. Technical Report, Woods Hole Oceanographic Institution, Ref no 67–57

Degens ET, Matheja J (1968) Origin, development, and diagnesis of biogeochemical compounds. J Brit Interplan Soc 21:52–82

Degens ET, Paluska A (1979) Tectonic and climatic pulses recorded in Quaternary sediments of the Caspian-Black Sea region. Sediment Geol 23:149–163

Degens ET, Spitzy A (1983) Paleohydrology of the Nile. In "Transport of Carbon and Minerals in Major World Rivers", Pt 2 (eds Degens ET, Kempe S, Soliman H), SCOPE/UNEP Sonderbd, Mitt Geol-Paläont Inst Univ, Hamburg 55:1–20

Degens ET, Deuser WG, Haedrich RL (1969) Molecular structure and composition of fish otoliths. Mar Biol 2:105–113

Degens ET, Watson SW, Remsen CC (1970) Fossil membranes and cell wall fragments from a 7000-year-old Black Sea sediment. Science 168:1207–1208

Degens ET, von Herzen RP, Wong HK (1971) Lake Tanganyika: water chemistry, sediments, geological structure. Naturwissenschaften 58:229–241

Degens ET, Okada H, Honjo S, Hathaway JC (1972) Microcrystalline sphalerite in resin globules suspended in Lake Kivu, East Africa. Mineral Deposita 7:1–12

Degens ET, von Herzen RP, Wong HK et al. (1973) Lake Kivu: structure, chemistry, and biology of an East African rift lake. Geol Rdsch 62:245–277

Degens ET, Kempe S, Spitzy A (1984) Carbon dioxide: a biogeochemical portrait. In "The Handbook of Environmental Chemistry", vol 1, Pt C (ed Hutzinger O), Berlin Heidelberg New York Tokyo, Springer-Verlag, 127–215

Degens ET, Kazmierczak J, Ittekkot V (1986a) Biomineralization and the carbon isotope record. Tschermaks Min Petr Mitt 35:117–126

Degens ET, Kempe S, Soliman H (eds) (1983) Transport of Carbon and Minerals in Major World Rivers, pt 2. Mitt Geol-Paläont Inst Univ, Hamburg, SCOPE/UNEP Sonderband 55, pp 535

Degens ET, Kempe S, Herrera R (eds) (1985) Transport of Carbon and Minerals in Major World Rivers, pt 3. Mitt Geol-Paläont Inst Univ, Hamburg, SCOPE/UNEP Sonderband 58, pp 645

Degens ET, Meyers PA, Brassell SC (eds) (1986b) Biogeochemistry of Black Shales. Mitt Geol-Paläont Inst Univ Hamburg, SCOPE/UNEP Sonderband 60, pp 448

Degens ET, Kempe S, Gan Wei-bin (eds) (1987) Transport of Carbon and Minerals in Major World Rivers, pt 4. Mitt Geol-Paläont Inst Univ, Hamburg, SCOPE/UNEP Sonderband 64, pp 512

Delson E (1987) Evolution and palaeobiology of robust *Australopithecus*. Nature 327:654–655

Delson E (1985) Palaeobiology and age of African *Homo erectus*. Nature 316:762–763

De Meis L (1981) The Sarcoplasmic Reticulum: Transport and Energy Transduction. In "Transport in the Life Sciences" (ed Bittar EE), vol 2, John Wiley & Sons, New York Chichester Brisbane Toronto, pp 163

Deryagin (Derjaguin) B (1933) Die Formelastizität der dünnen Wasserschichten. Z Physik 84:657–670

Deryagin BV, Churaev VV (1973) Nature of "anomalous water". Nature 244:430–431

Deuser WG, Degens ET (1969) O^{18}/O^{16} and C^{13}/C^{12} ratios of fossils from the hot-brine deep area of the central Red Sea. In "Hot Brines and Recent Heavy Metal Deposits in the Red Sea" (eds Degens ET, Ross DA), Springer-Verlag, Berlin Heidelberg New York, 336–347

Deuser WG, Hunt JM (1969) Stable isotope ratios of dissolved inorganic carbon in the Atlantic. Deep-Sea Res 16:221–225

D'Eustachio D (1946) Surface layers on quartz and topaz. Physic Rev 70:522–528

Diamond JM (1987) Who were the first Americans? Nature 329:580–581

Dickinson RE (1986) How will climate change? In "The Greenhouse Effect, Climatic Change, and Ecosystems" (eds Bolin B, Döös BR, Jäger J, Warrick RA), SCOPE Report 29, Chichester New York Brisbane Toronto Singapore, John Wiley & Sons, 207–270

Dickinson R, Cicerone RJ (1986) Future global warming from atmospheric trace gases. Nature 319:109–115

Diener TO (1981) Viroids. Sci Am 244/1:58–65

Dingus L (1984) Effects of stratigraphic completeness on interpretations of extinction rates across the Cretaceous-Tertiary boundary. Paleobiology 10:420–438

Dixon TH, Parke ME (1983) Bathymetry estimates in the southern oceans from Seasat altimetry. Nature 304:406–411

Dobson GMB (1929) Atmospheric ozone. Gerlands Beitr Geophys 24:8–15

Dobzhansky T, Ayala FJ, Stebbins GL, Valentine JW (1977) Evolution. Freeman & Co, San Francisco

Dodge RE, Aller RC, Thomson J (1974) Coral growth related to resuspension of bottom sediments. Nature 247:574–576

Donn B, Rahe J (1982) Structure and origin of cometary nuclei. In "Comets" (ed Wilkening LL), Univ of Arizona Press, Tucson, 203

Donn W (1972) The Earth our Physical Environment. New York, John Wiley & Sons Inc, pp 621

Donnelly RF, Frederick JE, Chameides WL (eds) (1987) Solar Variability and Its Stratospheric, Mesospheric, and Thermospheric Effects. J Geophys Res 92:795–914

Doolittle RF (1981) Similar amino acid sequences: chance or common ancestry? Science 214:149–159

Dorsey NE (1950) Properties of Ordinary Water Substances. Reinhold, New York

Dotto L (1986) Planet Earth in Jeopardy. Environmental Consequences of Nuclear War. SCOPE, John Wiley & Sons, Chichester New York Brisbane Toronto Singapore, pp 134

D'Ozouville L, Gundlach ER, Hayes MO (1978) Effect of coastal processes on the distribution and persistence of oil spilled by the AMOCO CADIZ – preliminary conclusions. Publ du CNEXO, Série "Actes de Colloques" 6:69–96

Dreibus G, Wänke H (1984) Accretion of the Earth and the inner planets. Proc 27th Int Geol Congr, Moscow 11:1–11

Drost-Hansen W (1965a) Aqueous interfaces. Pt 1. – Ind Eng Chem 57/3:38–44

Drost-Hansen W (1965b) Aqueous interfaces. Pt 2. Ind Eng Chem 57/4:18–37

Drost-Hansen W (1966) The puzzle of water. Int Sci Technol October 1966:86–96

Drost-Hansen W (1967) On the structure of water and some comments on water-solute interactions. Adv Chem Ser 66, pp 32

Drost-Hansen W (1972) Molecular aspects of aqueous interfacial structures. J Geophys Res 77:5132–5146

Drust DS, Creutz CE (1988) Aggregation of chromaffin granules by calpactin at micromolar levels of calcium. Nature 331:88–91

Duncan RA, Pyle DE (1988) Rapid eruption of the Deccan flood basalts at the Cretaceous/Tertiary boundary. Nature 333:841–843

Duplessy J-C (1986) CO_2 air-sea exchange during glacial times: importance of deep sea circulation changes. In "The Role of Air-Sea Exchange in Geochemical Cycling" (ed Buat-Ménard P), Dordrecht Boston Lancaster Tokyo, D Reidel Publ Co, 249–267

Duplessy J-C, Shackleton NJ (1985) Response of global deep-water circulation to Earth's climatic change 135,000–107,000 years ago. Nature 316:500–507

Duplessy JC, Shackleton NJ, Matthews RK, Prell W, Ruddiman WF, Caralp M, Hendy CH (1984) ^{13}C record of benthic foraminifera in the last interglacial ocean: implications for the carbon cycle and the global deep water circulation. Quat Res 21:225–243

Duursma EK, Dawson R (eds) (1981) Marine Organic Chemistry. Amsterdam Oxford New York, Elsevier Scientific Publ Co, pp 521

Duxbury AC (1977) The Earth and its Oceans. Reading, Mass, Addison-Wesley Publ Co, pp 381

Dyke TR, Mack KM, Muenter JS (1977) The structure of water dimer from molecular beam electric resonance spectroscopy. J Chem Phys 66:498–510

Eanes ED, Termine JD, Nylen MU (1973) An electron microscope study of the formation of amorphous calcium phosphate and its transformation to crystalline apatite. Calc Tissue Res 12:144–158

Ecological Aspects of Tree-Ring Analysis (1987) CO_2, CONF-8608144, Washington, DC, US Dept of Energy, pp 726

Edmond JM, Von Damm K (1983) Hot springs on the ocean floor. Sci Am 248/4:70–85

Eglinton G, Murphy MTJ (eds) (1969) Organic Geochemistry. New York Heidelberg Berlin, Springer-Verlag, pp 828

Ehhalt DH (1981) Chemical coupling of the nitrogen, sulphur, and carbon cycles in the atmosphere. In "Some Perspectives of the Major Biogeochemical Cycles" (ed Likens GE), SCOPE Report 17, Chichester New York Brisbane Toronto, John Wiley & Sons, 81–91

Eigen M, Schuster P (1977) The hypercycle. A principle of natural self-organization. Pt A: Emergence of the hypercycle. Naturwissenschaften 64/11:541–565

Eigen M, Gardiner W, Schuster P, Winkler-Oswatitsch R (1981) The origin of genetic information. Sci Am 244/4:78–94

Elachi C (1980) Spaceborne imaging radar: geologic and oceanographic applications. Science 209:1073–1082

Elbrink J, Bihler I (1975) Membrane transport: its relation to cellular metabolic rates. Science 188:1177–1184

Eldredge N, Cracraft J (1980) Phylogenetic Patterns and Evolutionary Process. New York, Columbia Univ Press, pp 349

Eldredge N, Gould SJ (1972) Punctuated equilibria: an alternative to phyletic gradualism. In "Models in Paleobiology" (ed Schopf TJM), San Francisco, Freeman, Cooper & Co, 82–115

Elkin AP (1974) The Australian Aborigines. Angus and Robertson Publishers, Sydney, pp 397 (first paperback edition)

Ellis CB (1954) Fresh Water from the Ocean. Ronald Press Co, New York, pp 311

Elvin M (1973) The Pattern of the Chinese Past. Stanford Univ Press, Stanford, California, pp 346

Emerson S (1984) The manganese nodule program (MANOP).

Geochim Cosmochim Acta 48:895

Emerson S, Kalhorn S, Jacobs L et al. (1982) Environmental oxidation rate of manganese(II): bacterial catalysis. Geochim Cosmochim Acta 46:1073–1079

Emery KO (1960) The Sea off Southern California: A Modern Habitat of Petroleum. Wiley, New York, pp 366

Engelhardt W v (1960) Der Porenraum der Sedimente. Miner Petrogr Einzeldarst (eds Engelhardt W v, Zemann J), Springer Verlag, Berlin Heidelberg New York, pp 207

Engelhardt W v (1967) Interstitial solutions and diagenesis in sediments. In "Diagenesis in Sediments" (eds Larsen G, Chilingar GV), Developments in Sedimentology 8, Elsevier Publ Co, Amsterdam London New York, 503–521

Engelhardt W v, Smolinski A v (1957) Die Reaktion von Polyphosphaten mit Tonmineralen. Kolloid-Z 151:47–52

Epstein S (1959) The variations of the O^{18}/O^{16} ratio in nature and some geological applications. In "Researches in Geochemistry" (ed Abelson PH), New York, John Wiley & Sons, 217–240

Epstein S, Mayeda TK (1953) Variations of the O^{18}/O^{16} ratio in natural waters. Geochim Cosmochim Acta 4:213–224

Epstein S, Sharp RP, Gow AJ (1965) Six-year record of oxygen and hydrogen isotope variations in South Pole firn. J Geophys Res 70:1809–1814

Epstein S, Sharp RP, Gow AJ (1970) Antarctic ice sheet: Stable isotope analyses of Byrd Station cores and interhemispheric climatic implications. Science 168:1570–1572

Erulkar SD (1981) The versatile role of calcium in biological systems. Interdiscipl Sci Rev 6/4:322–332

Esser K, Kües U (1983) Flocculation and its implication for biotechnology. Process Biochem 18:21–23

Eugster HP (1969) Inorganic bedded cherts from the Magadi Area, Kenya. Contrib Mineral Petrol 22:1–31

Eugster HP (1972) Ammonia, in minerals and early atmosphere. In "The Encyclopedia of Geochemistry and Environmental Sciences" (ed Fairbridge RW), New York Cincinnati Toronto London Melbourne, Van Nostrand Reinhold Company, 29–33

Fabry C, Buisson H (1913) Sur l'absorption de l'ultraviolet par l'ozone et l'extrémité du spectre solaire. CR Acad Sci Paris 156:782–785

Fabry C, Buisson H (1921) A study of the ultraviolet end of the solar spectrum. Astrophys J 54:297–322

Fanale FP (1971) A case for catastrophic early degassing of the earth. Chem Geol 8:79–105

Fanale FP, Salvail JR, Zent AP, Postawko SE (1986) Global distribution and migration of subsurface ice on Mars. Icarus 67:1–18

Fegley B Jr, Prinn RG, Hartman H, Watkins GH (1986) Chemical effects of large impacts on the Earth's primitive atmosphere. Nature 319:305–308

Felsenfeld G, Huang S (1959) The interaction of polynucleotides with cations. Biochim Biophys Acta 34:234–242

Ferek RJ, Chatfield RB, Andreae MO (1986) Vertical distribution of dimethylsulphide in the marine atmosphere. Nature 320:514–516

Fiddes JC (1977) The nucleotide sequence of a viral DNA. Sci Am 237/6:54–67

Field GB, Verschuur GL, Ponnamperuma C (1978) Cosmic Evolution: An Introduction to Astronomy. Boston Dallas Geneva Illinois Hopewell New Jersey Palo Alto London, Houghton Mifflin Company, pp 450

Finean JB (1966) The molecular organization of cell membranes. Progr Biophys Molec Biol 16:143–170

Fink U, Smith BA, Benner DC, Johnson JR, Reitsema HJ, Westphal JA (1980) Detection of a CH_4 atmosphere on Pluto. Icarus 44:62–71

Fischer AG (1964) Brackish oceans as the cause of the Permo-Triassic marine faunal crisis. In "Problems in Palaeoclimatology" (ed Nairn AEM), New York, Interscience Pubs, 566–577

Fletcher NH (1971) Structural aspects of the ice-water system. Rep Prog Phys 34:913–994

Flohn H (1941) Die Tätigkeit des Menschen als Klimafaktor. Zeitschr f Erdkunde 9:13

Flohn H (1966) Warum ist die Sahara trocken? Z Met 17:316–320

Flohn H (1980) Possible climatic consequences of a man-made global warming. Bericht RR-80-30, IIASA, Laxenburg

Flohn H, Nicholson SE (1980) Climatic fluctuations in the arid belt of the "old world" since the last glacial maximum; possible causes and future implications. In "Paleontology of Africa and the Surrounding Islands" (eds van Zinderen EM et al.), Rotterdam, Balkema, 3–21

Flügel E (1982) Palaeontologische Beiträge zur Evolution der Organismen. In "Evolution" (ed Siewing R), UTB 748, G Fischer, Stuttgart New York

Förster H (1981a) Freigesetzter Sauerstoff und seine Rolle bei der Bildung von Gesteinen und Erzen. Fortschr Miner 59/1:234–235

Förster H (1981b) Die Abgabe von Sauerstoff aus Silicatgesteinen bei hohen Temperaturen. Naturwissenschaften 68:522–523

Fogg GE (1986) Light and ultraphytoplankton. Nature 319:96

Forslind E (1971) Structure of water. Q Rev Biophys 4/4:325–363

Fortak H (1971) Meteorologie. Berlin Darmstadt Wien, Deutsche Buch-Gemeinschaft, pp 287

Fox GE, Stackebrandt E, Hespell RB et al. (1980) The phylogeny of prokaryotes. Science 209:457–463

Fox SW (ed) (1965) The Origins of Prebiological Systems and of their Molecular Matrices. Academic Press, New York and London

Franchi E, Camatini M (1985) Evidence that a Ca^{2+} chelator and a calmodulin blocker interfere with the structure of inter-Sertoli junctions. Tissue and Cell 17:13–25

Francis MD, Webb NC (1971) Hydroxyapatite formation from a hydrated calcium monohydrogen phosphate precursor. Calc Tiss Res 3:335–342

Franck EU, Hartmann D, Hensel F (1965) Proton mobility in water at high temperatures and pressures. Disc Faraday Soc 39:200–206

François LM, Gérard J-C (1986) A numerical model of the evolution of ocean sulfate and sedimentary sulfur during the last 800 million years. Geochim Cosmochim Acta 50:2289–2302

Frank HS (1970) The structure of ordinary water. Science 169:635–641

Frank HS, Wen WY (1957) Structural aspects of ion-solvent interaction in aqueous solutions: a suggested picture of water structure. Disc Faraday Soc 24:133–140

Frank LA, Sigwarth JB, Craven JD (1986a) On the influx of small comets into the earth's upper atmosphere. I. Observations. Geophys Res Lett 13/4:303–306

Frank LA, Sigwarth JB, Craven JD (1986b) On the influx of small comets into the earth's upper atmosphere. II. Interpretation. Geophys Res Lett 13/4:307–310

Franks F (ed) (1972–1979) Water – A Comprehensive Treatise, Vols 1–6. New York London, Plenum Press (for details see Vol 6, pp 455)

Franks F (1981) Polywater. Cambridge, Mass, MIT Press, pp 208

Frenkel JI (1957) Kinetische Theorie der Flüssigkeiten. Deutscher Verlag der Wissenschaft, Berlin, pp 510

Freyer HD, Belacy N (1983) $^{13}C/^{12}C$ records in northern hemispheric trees during the past 500 years – anthropogenic impact and climatic superpositions. J Geophys Res 88/C11:6844–6852
Friedli H, Lötscher H, Oeschger H, Siegenthaler U, Stauffer B (1986) Ice core record of the $^{13}C/^{12}C$ ratio of atmospheric CO_2 in the past two centuries. Nature 324:237–238
Fujimura R, Kaesberg P (1962) The adsorption of bacteriophage ϕX 174 to its host. Biophys J 2:433–449
Fyfe WS, Price NJ, Thompson AB (1978) Fluids in the Earth's Crust. Elsevier Sci Publ Co, Amsterdam Oxford New York, pp 383
Gaffney ES, Matson DL (1980) Water ice polymorphs and their significance on planetary surfaces. Icarus 44:511–519
Gammon RH, Sundquist ET, Fraser PJ (1985) History of carbon dioxide in the atmosphere. In "Atmospheric Carbon Dioxide and the Global Carbon Cycle" (ed Trabalka JE), DOE/ER-0239, Washington, DC, US Government Printing Office, 25–62
Garrels RM (1960) Mineral Equilibria at Low Temperature and Pressure. Harper and Brothers, New York, pp 254
Garrels RM, Lerman A (1981) Phanerozoic cycles of sedimentary carbon and sulfur. Natl Acad Sci Proc 78:4652–4656
Garrels RM, Lerman A (1984) Coupling of the sedimentary sulfur and carbon cycles – an improved model. Am J Sci 284:989–1007
Garrels RM, Mackenzie FT (1971) Evolution of Sedimentary Rocks. New York, WW Norton & Co Inc, pp 397
Garrels RM, Mackenzie FT, Hunt C (1975) Chemical Cycles and the Global Environment: Assessing Human Influences. Los Altos, California, William Kaufmann Inc, pp 206
Garrett WD, Zisman WA (1967) The organic chemical composition of the ocean surface. Deep-Sea Res 14:221–227
Gates WL (1976) Modeling the ice-age climate. Science 191:1138–1144
Gesellschaft für Reaktorsicherheit (1987) Neuere Erkenntnisse zum Unfall im Kernkraftwerk Tschernobyl. GRS-S-40, pp 74
Ghiselin MT, Degens ET, Spencer DW, Parker RH (1967) A phylogenetic survey of molluscan shell proteins. Bull Mus Comp Zoology, Harvard Univ 262:1–35
Gieskes JM (1975) Chemistry of interstitial waters of marine sediments. Earth Planet Sci Lett 3:433–453
Gill D, Weda T, Chueh S-H, Noel MW (1986) Ca^{2+} release from endoplasmic reticulum is mediated by a guanine nucleotide regulatory mechanism. Nature 320:461–464
Gilliland RL (1982) Solar, volcanic and CO_2 forcing of recent climatic changes. Clim Change 4:111–131
Gojobori T, Yokoyama S (1985) Rates of evolution of the retroviral oncogene of Moloney murine sarcoma virus and of its cellular homologues. Proc Natl Acad Sci USA, 4198–4201
Goldberg A (1966) Magnesium binding by *Escherichia coli* ribosomes. J Mol Biol 15:663–673
Golden JW, Robinson SJ, Haselkorn R (1985) Rearrangement of nitrogen fixation genes during heterocyst differentiation in the cyanobacterium *Anabaena*. Nature 314:419–423
Goldman M (1987) Chernobyl: a radiobiological perspective. Science 238:622–623
Goldman M et al. (1987) Health and Environmental Consequences of the Chernobyl Nuclear Power Plant Accident. Report DOE ER-0332, Dept of Energy, Washington, DC
Goldschmidt VM (1933) Grundlagen der quantitativen Geochemie. Fortschr Mineral 17:112–156
Goldschmidt VM (1954) Geochemistry (ed Muir A). Oxford, Clarendon Press, pp 730
Goldsmith JR (1959) Some aspects of the geochemistry of carbonates. In "Researches in Geochemistry" (ed Abelson PH), John Wiley & Sons Inc, New York, 336–358
Gole MJ, Klein C (1981) Banded iron-formations through much of Precambrian time. J Geol 89:169–183
Gonzalez LA, Lohmann KC (1985) Carbon and oxygen isotopic composition of Holocene reefal carbonates. Geology 13:811–814
Goodwin AM (1981) Precambrian perspectives. Science 213:55–61
Gornitz V, Lebedeff L, Hansen J (1982) Global sea level trend in the past century. Science 215:1611–1614
Gorshkov VG (1986a) Anthropogenic disturbance of the global carbon cycle. USSR Academy of Sciences, Nuclear Physics Institute, Leningrad, pp 70
Gorshkov VG (1986b) Atmospheric disturbance of the carbon cycle: impact upon the biosphere. Il Nuovo Cimento 9C/5:937–952
Gorshkov VG, Sherman SG (1986) Atmospheric CO_2 and destructivity of the land biota: seasonal variations. Il Nuovo Cimento 9C/4:902–917
Goudriaan J, Ajtay GL (1979) The possible effects of increased CO_2 on photosynthesis. In "The Global Carbon Cycle", SCOPE Report 13 (eds Bolin B, Degens ET, Kempe S, Ketner P), Chichester, John Wiley & Sons Ltd, 237–249
Gould SJ (1978) Heroes in nature. NY Rev, Sep 28:31–32
Gould SJ (1982) Darwinism and the expansion of evolutionary theory. Science 216:380–387
Gould SJ, Calloway CB (1980) Clams and brachiopods – ships that pass in the night. Paleobiology 6:383–396
Gould SJ, Eldredge N (1977) Punctuated equilibria: the tempo and mode of evolution reconsidered. Paleobiology 3:115–151
Graedel TE (1979) Terpenoids in the atmosphere. Rev Geophys Space Phys 17/5:937–947
Graedel TE (1985) The photochemistry of the troposphere. In "The Photochemistry of Atmospheres" (ed Levine JS), Orlando, Fla, Academic, 39–76
Grassl H (1981) The climate at maximum entropy production by meridional atmospheric and oceanic heat fluxes. QJR Meteorol Soc 107:153–166
Grassl H, Maier-Reimer E, Degens ET, Kempe S, Spitzy A (1984) CO_2, Kohlenstoff-Kreislauf und Klima. 1. Globale Kohlenstoffbilanz. Naturwissenschaften 71:129–136, 234–238
Grimsrud EP, Westberg HH, Rasmussen RA (1975) Atmospheric reactivity of monoterpene hydrocarbons, NO_x photooxidation and ozonolysis. Int J Chem Kinet Symp 1:183–195
Gross HJ, Riesner D (1980) Viroids: A class of subviral pathogens. Angew Chem: Int Ed Engl 19/4:231–243
Gross MG (1977) Oceanography, a View of the Earth (2nd ed). Englewood Cliffs, New Jersey, Prentice-Hall Inc, pp 497
Groveman BS, Landsberg HE (1979) Reconstruction of the northern hemisphere temperature: 1579–1880. Dept Meteor, Univ Maryland College Park, Publ No 79 181/182:1–59, 1–46
GSA (1981) Maps of Northern Hemisphere Continental Ice, Sea Ice, and Sea-Surface Temperatures in August for the Modern and Last Glacial Maximum. Geol Soc Amer, Map and Chart Ser MC-36, Map 7A & 7B
Guderian R, Rabe R (1981) Effects of photochemical oxidants on plants. Final report for the Commission of the European Communities, Directorate General for Environment, Consumer Protection and Nuclear Safety
Gurney RW (1966) Ionic Processes in Solution. Dover Publ Inc, New York
Haagen-Smit AJ, Bradley CE, Fox MM (1952a) Formation of ozone in Los Angeles smog. Proc Nat Air Pollut Symp 2nd, 54–56
Haagen-Smit AJ, Darley EF, Zaitlin M, Hull H, Noble W (1952b) Investigation on injury to plants from air pollution in the Los Angeles Basin. Plant Physiol 27:18

Halbach P, Fellerer R (1980) The metallic minerals of the Pacific seafloor. GeoJournal 4.5:407–421
Haldane JBS (1954) The origins of life. New Biology 16:12–27
Hall DO, Cammack R, Rao KK (1971) Role for ferredoxins in the origin of life and biological evolution. Nature 233:136–138
Hallam A (1981) A revised sea-level curve for the early Jurassic. J Geol Soc London 138:735–743
Hallam A (1984) Pre-Quaternary sea-level changes. Annu Rev Earth Planet Sci 12:205–243
Hallam A (1987) End-Cretaceous mass extinction event: argument for terrestrial causation. Science 238:1237–1242
Hamilton PJ, O'Nions RK, Bridgwater D, Nutman A (1983) Sm-Nd studies of Archaean metasediments and metavolcanics from West Greenland and their implications for the Earth's early history. Earth Planet Sci Lett 62:263–272
Hammen TVD, Wijmstra TA, Zagwijn WH (1971) The floral record of the Late Cenozoic of Europe. In "Late Cenozoic Glacial Ice Ages" (ed Turekian KK), New Haven London, Yale Univ Press, 391–424
Hammer CU, Clausen HB, Dansgaard W (1980) Greenland ice sheet evidence of post-glacial volcanism and its climatic impact. Nature 288:230–235
Hammer CU, Clausen HB, Dansgaard W (1981) Past volcanism and climate revealed by Greenland ice cores. J Vulcanol Geoth Res 11:3–10
Hampson J (1964) Photochemical behavior of the ozone layer. Techn Note 1627 (Can Arm Res Devel Establ Quebec)
Hansen JE, Wang WC, Lacis AA (1978) Mt Agung eruption provides test of a global climatic perturbation. Science 199:1065–1068
Haq BU, Hardenbol J, Vail PR (1987) Chronology of fluctuating sea levels since the Triassic. Science 235:1156–1167
Hard TM, Chan CY, Mehrabzadeh AA, Pan WH, O'Brian RJ (1986) Diurnal cycle of tropospheric OH. Nature 322:617–620
Hardin G (1966) Biology. Freeman & Co, San Francisco, pp 198
Hare PE, Hoering TC, King K Jr (eds) (1980) Biogeochemistry of Amino Acids. New York Chichester Brisbane Toronto, John Wiley & Sons, pp 558
Harris EJ (1972) Transport and Accumulation in Biological Systems (3rd ed). Academic Press Inc Publishers, New York, and Butterworths Scientific Publications, London, pp 454
Hart MH (1978) The evolution of the atmosphere of the earth. Icarus 33:23–39
Hart R, Dymond J, Hogan L (1979) Preferential formation of the atmosphere-sialic crust system from the upper mantle. Nature 278:156–159
Harvey GR, Mopper K, Degens ET (1972) Synthesis of carbohydrates and lipids on kaolinite. Chem Geol 9:79–87
Harvey GR, Boran DA, Chesal LA, Tokar JM (1983) The structure of marine fulvic and humic acids. Mar Chem 12:119–132
Harwell MA, Hutchinson TC (eds) (1986) Environmental Consequences of Nuclear War. Vol 2: Ecological and Agricultural Effects. SCOPE Report 28, John Wiley & Sons, Chichester New York Brisbane Toronto Singapore, pp 523
Haselkorn R, Rice D, Curtis SE, Robinson SJ (1983) Organization and transcription of genes important in *Anabaena* in heterocyst differentiation. Ann Microbiol Inst Pasteur 134B:181–193
Hasselbach W (1979) The sarcoplasmic calcium pump. A model of energy transduction in biological membranes. In "Biochemistry", Top Curr Chem 78:1–56
Hasselmann K (1982) An ocean model for climate variability studies. Progr Oceanogr 11:69–92
Hathaway JC, Degens ET (1969) Methane-derived marine carbonates of Pleistocene age. Science 165:690–692
Hays JD, Pitman WC, III (1973) Lithospheric plate motion, sea level changes and climatic and ecological consequences. Nature 246:18–22
Heath D (1986) Observations of ozone and UV solar flux variability. 26th Plen Meet COSPAR, Workshop VIII
Hecky RE, Mopper K, Degens ET (1973) The amino acid and sugar composition of diatom cell-walls. Mar Biol 19:323–331
Heicklen J (1976) Atmospheric Chemistry. Academic Press, New York, pp 406
Henderson GS, Harris WF (1975) An ecosystem approach to characterization of the nitrogen cycle in a deciduous forest watershed. In "Forest Soils and Forest Land Management" (eds Bernier B, Winget CH), Quebec, Les Presses de l'Université Laval, 179–193
Henderson-Sellers A (1983) The Origin and Evolution of Planetary Atmospheres. Bristol, Adam Hilger Ltd, pp 240
Henderson-Sellers A (1986) Increasing cloud in a warming world. Clim Change 9:267–309
Henderson-Sellers A, Cogley JG (1982) The Earth's early hydrosphere. Nature 298:832–835
Henley RW, Ellis AJ (1983) Geothermal systems ancient and modern: a geochemical review. Earth-Sci Rev 19:1–50
Hennikyar JC (1949) The depth of the surface zone of a liquid. Rev Mod Phys 21:322–341
Herzberg O, James MNG (1985) Structure of the calcium regulatory muscle protein troponin-C at 2.8 Å resolution. Nature 313:653–659
Hinz K (1981) A hypothesis on terrestrial catastrophes. Geol Jb E22:3–28
Hippel A v (1968) The dielectric relaxation spectra of water, ice and aqueous solutions and their interpretation. 5. Ice I as a proton dielectric. Laboratory for Insulation Research, Mass Inst Tech, Techn Rep 5:1–20
Hobbs PV (1974) Ice Physics. Oxford Univ Press, Clarendon London, pp 837
Hoefs J (1980) Stable Isotope Geochemistry (2nd ed). Berlin Heidelberg New York, Springer-Verlag, pp 208
Höhling HJ, Kreilos R, Neubauer C, Boyde A (1971) Electron microscopy and electron microscopical measurements of collagen mineralization in hard tissues. Ztschr Zellforsch Mikrosk Anat 122:36–52
Hölder H (ed) (1983) Evolutionsproblematik aus paläontologischer Sicht. Paläont Z 57/3–4, 175–328
Hoffman A (1982) Punctuated versus gradual mode of evolution. A reconsideration. In "Evolutionary Biology" 15 (eds Hecht MK, Wallace B, Prance GT), New York London, Plenum Press, 411–436
Hoffman A (1985) Patterns of family extinction depend on definition and geological timescale. Nature 315:659–662
Holdren MW, Westberg H, Zimmermann PR (1979) Analysis of monoterpene hydrocarbons in rural atmospheres. J Geophys Res 84/8:5083–5088
Holland HD (1984) The Chemical Evolution of the Atmosphere and Oceans. Princeton, NJ, Princeton Univ Press, pp 582
Holland HD, Lazar B, McCaffrey M (1986) Evolution of the atmosphere and oceans. Nature 320:27–33
Holser WT (1977) Catastrophic chemical events in the history of the ocean. Nature 267:403–408
Holser WT (1984) Gradual and abrupt shifts in ocean chemistry during Phanerozoic time. In "Patterns of Change in Earth Evolution" (eds Holland HD, Trendall AF), Dahlem Konferenzen 1984, Berlin Heidelberg New York Tokyo, Springer-Verlag, 123–143

Hopcroft JE (1984) Turing machines. Sci Am 250/5:70–80

Horibe Y, Kim K-R, Craig H (1986) Hydrothermal methane plumes in the Mariana back-arc spreading centre. Nature 324:131–133

Horne RA (1968a) De generatione et corruptione: the origin and fate of biomaterial in the oceans and its dependence on local water structure. Cambridge, Mass, Arthur D Little, Inc, pp 30

Horne RA (1968b) The structure of water and aqueous solutions. Surv Progr Chem 4:1–43

Horne RA (1969) Marine Chemistry. The Structure of Water and the Chemistry of the Hydrosphere. New York, Wiley-Interscience,pp 568

Horne RA (1972a) Structure of seawater and its role in chemical mass transport between the sea and the atmosphere. J Geophys Res 77:5170–5175

Horne RA (ed) (1972b) Water and Aqueous Solutions. New York, Wiley-Interscience, pp 837

Horowitz NH (1945) On the evolution of biochemical syntheses. Proc Natl Acad Sci 31:153–157

Hoyle F, Wickramasinghe NC (1986) The case for life as a cosmic phenomenon. Nature 322:509–511

Hsü KJ (1980) Terrestrial catastrophe caused by cometary impact at the end of the Cretaceous. Nature 285:201–203

Hsü KJ (1981) Origin of geochemical anomalies at Cretaceous-Tertiary boundary. Asteroid or cometary impact? Oceanologica Acta SP, 129–133

Hsü KJ, He Q, McKenzie JA et al. (1982) Mass mortality and its environmental and evolutionary consequences. Science 216:249–256

Hughes DW (1978) Meteors. In "Cosmic Dust" (ed McDonnell JAM), New York, Wiley & Sons, 123

Hühnerfuss H (1983) Molecular aspects of organic surface films on marine water and the modification of water waves. La Chimica et l'Industria 65/2:97–101

Hühnerfuss H (1986) The Molecular Structure of the System Water/Monomolecular Surface Film and Its Influence on Water Wave Damping. Habil, Hamburg University, pp 245

Hühnerfuss H, Lange P, Walter W (1982) Wave damping by monomolecular surface films and their chemical structure. Pt 1: Variation of the hydrophobic part of carboxylic acid esters. J Mar Res 40:209–225

Hühnerfuss H, Alpers W, Garrett WD, Lange PA, Stolte S (1983) Attenuation of capillary and gravity waves at sea by monomolecular organic surface films. J Geophys Res 88/C14:9809–9816

Hühnerfuss H, Lange P, Walter W (1984) Wave damping by monomolecular surface films and their chemical structure. Pt 2: Variation of the hydrophilic part of the film molecules including natural substances. J Mar Res 42:737–759

Hunt GE (1983) The atmospheres of the outer planets. Annu Rev Earth Planet Sci 11:415–459

Hunt JM (1979) Petroleum Geochemistry and Geology. San Francisco, WH Freeman and Company, pp 617

Hunten DM, Donahue TM (1976) Hydrogen loss from the terrestrial planets. Annu Rev Earth Planet Sci 4:265–292

Hurd DC (1983) Physical and chemical properties of siliceous skeletons. In "Silicon Geochemistry and Biogeochemistry" (ed Aston SR), London, Academic Press, 187–244

Hutchinson TC, Meema KM (eds) (1987) Lead, Mercury, Cadmium and Arsenic in the Environment. SCOPE Report 31, John Wiley & Sons, Chichester New York Brisbane Toronto Singapore, pp 360

Idso SB (1982) Carbon Dioxide: Friend or Foe? Tempe, Arizona, Inst Biosph Res, pp 92

Imbrie J, Shackleton NJ, Pisias NG et al. (1984) The orbital theory of Pleistocene climate: support from a revised chronology of the marine $\delta^{18}O$ record. In "Milankovitch and Climate", Pt 1 (eds Berger A, Imbrie J et al.), Dordrecht, Reidel, 269–305

Ingersoll AP, Pollard D (1982) Motion in the interiors and atmospheres of Jupiter and Saturn: scale analysis, anelastic equations, barotropic stability criterion. Icarus 52:62–80

Irvine RF, Moor RM (1986) Micro-injection of inositol 1,3,4,5-tetrakiphosphate activates sea urchin eggs by a mechanism dependent on external Ca^{2+}. Biochem J 240:917–920

Israelachvili JN, Pashley RM (1983) Molecular layering of water at surfaces and origin of repulsive hydration forces. Nature 306:249–250

Ivanov MV, Freney JR (eds) (1983) The Global Biogeochemical Sulphur Cycle. SCOPE Report 19, Chichester New York Brisbane Toronto Singapore, John Wiley & Sons, pp 470

Jablonski D (1986) Background and mass extinctions: the alternation of macroevolutionary regimes. Science 231:129–133

Jackson TA (1971) Preferential polymerization and adsorption of L-optical isomers of amino acids relative to D-optical isomers on kaolinite templates. Chem Geol 7:295–306

Jacoby G (ed) (1980) Proceedings of an International Meeting on Stable Isot. Tree-Ring Res. US Dept Energy Conf 790518, Washington, DC, pp 150

James HL (1954) Sedimentary facies of iron formation. Econ Geol 49:235–293

Jannasch HW, Mottl MJ (1985) Geomicrobiology of deep-sea hydrothermal vents. Science 229:717–725

Jannasch HW, Eimhjellen K, Wirsen CO, Farmanfarmaian A (1971) Microbial degradation of organic matter in the deep sea. Science 171:672–675

Jenkins WJ, Goldman JC (1985) Seasonal oxygen cycling and primary production in the Sargasso Sea. J Mar Res 43:465–491

Johanson DC, Edey M (1981) Lucy: the Beginnings of Humankind. Simon and Schuster, New York

Johanson DC, Masao FT, Eck GG et al. (1987) New partial skeleton of Homo habilis from Olduvai Gorge, Tanzania. Nature 327:205–209

JOI (Joint Oceanographic Institutions Incorporated) (1984) Oceanography from Space. A Research Strategy for the Decade 1985–1995. Washington DC, JOI, 1–20

Jones PD (1985) Northern hemisphere temperatures 1851–1984. Climat Monitor 14:14–21

Jouzel J, Lorius C, Petit JR et al. (1987) Vostok ice core: a continuous isotope temperature record over the last climatic cycle (160,000 years). Nature 329:403–408

Joyce GF, Visser GM, Van Boeckel CAA et al. (1984) Chiral selection in poly(C)-directed synthesis of oligo(G). Nature 310:602–604

Julian PR, Chervin RM (1978) A study of the Southern Oscillation and Walker Circulation phenomenon. Mon Wea Rev 106:1433–1451

Kamb B (1968) Ice polymorphism and the structure of water. In "Structural Chemistry and Molecular Biology" (eds Rich A, Davidson N), WH Freeman and Company, San Francisco and London, 507–542

Kaminer B, Bell AL (1966) Synthetic myosin filaments. Science 151:323–324

Kannan KK, Liljas A, Waara I et al. (1972) Crystal structure of human erythrocyte carbonic anhydrase C VI. The three-dimensional structure at high resolution in relation to other mammalian carbonic anhydrases. Cold Spring Harbor Symp. Quant Biol 36:221–231

Kannan KK, Notstrand B, Fridborg K et al. (1975) Crystal structure of human erythrocyte carbonic anhydrase B.

Three-dimensional structure at a nominal 2.2-Å resolution. Proc Natl Acad Sci USA 72:51–55

Kanwisher I (1960) PCO_2 in the sea-water and its effects on the movement of CO_2 in nature. Tellus 12:209–215

Kaplan IR, Degens ET, Reuther JH (1963) Organic compounds in stony meteorites. Geochim Cosmochim Acta 27:805–834

Kaplan RW (1985) On the numbers of extant, extinct, and possible species of organisms. Biol Zentralbl 104/6:647–653

Karhu J, Epstein S (1986) The implication of the oxygen isotope records in coexisting cherts and phosphates. Geochim Cosmochim Acta 50:1745–1756

Kasting JF, Walker JCG (1981) Limits of oxygen concentration in the prebiological atmosphere and the rate of abiotic fixation of nitrogen. J Geophys Res 86:1147–1158

Kasting JF, Liu SC, Donahue TM (1979) Oxygen levels in the pre-biological atmosphere. J Geophys Res 84:3097–3107

Kauffman EG (1976) Plate tectonics: major force in evolution. The Science Teacher 43/3, pp 6

Kauffman EG, Steidtmann JR (1981) Are these the oldest metazoan trace fossils? J Paleont 55:923–947

Kavanau JL (1964) Water and Solute-Water Interactions. Holden-Day, San Francisco, California

Kazmierczak J (1981) The biology and evolutionary significance of Devonian volvocaceans and their Precambrian relatives. Acta Paleont Polonica 26:299–337

Kazmierczak J, Degens ET (1986) Calcium and the early eukaryotes. Mitt Geol-Paläont Inst Univ Hamburg 61:1–20

Kazmierczak J, Ittekkot V, Degens ET (1985) Biocalcification through time: environmental challenge and cellular response. Paläont Z 59/1–2:15–33

Keeling CD (1961) Concentration and isotopic abundances of carbon dioxide in rural and marine air. Geochim Cosmochim Acta 24:277–298

Keeling CD, Bacastow RB, Whorf TP (1982) Measurements of the concentration of carbon dioxide at Mauna Loa Observatory, Hawaii. In "Carbon Dioxide Review" (ed Clark WC), 377–385

Keepin W, Mintzer I, Kristoferson L (1986) Emission of CO_2 into the atmosphere. In "The Greenhouse Effect, Climatic Change, and Ecosystems" (eds Bolin B, Döös BR et al.), SCOPE Report 29, John Wiley & Sons, Chichester New York Brisbane Toronto Singapore, 35–91

Keith ML (1982) Violent volcanism, stagnant oceans and some inferences regarding petroleum, strata-bound ores and mass extinctions. Geochim Cosmochim Acta 46:2621–2637

Kellogg WW, Schware R (1981) Climate Change and Society. Boulder Colorado, Westview Press, pp 178

Kempe S (1983) Impact of Aswan High Dam on water chemistry of the Nile. In "Transport of Carbon and Minerals in Major World Rivers", Pt 2 (eds Degens ET, Kempe S, Soliman H), Mitt Geol-Paläont Inst Univ Hamburg, SCOPE/UNEP Sonderbd 55:401–423

Kempe S, Degens ET (1985) An early soda ocean? Chem Geol 53:95–108

Kempe S, Nies H (1987) Chernobyl nuclide record from North Sea sediment trap. Nature 329:825–828

Kempe S, Kazmierczak J, Degens ET (1987 a) The soda ocean concept and its bearing on biotic evolution. In "Origin, Evolution, and Modern Aspects of Biomineralization in Plants and Animals" (ed Crick RE), Plenum Publ Co, New York

Kempe S, Nies H, Ittekkot V et al. (1987b) Comparison of Chernobyl nuclide deposition in the Black Sea and in the North Sea. In "Particle Flux in the Ocean" (eds Degens ET, Izdar E, Honjo S), Mitt Geol-Paläont Inst Univ Hamburg, SCOPE/UNEP Sonderband 62:165–178

Kennard O, Isaacs NW, Coppola JC et al. (1970) Three-dimensional structure of adenosine triphosphate. Nature 225:333–336

Kent DV (1981) Asteroid extinction hypothesis. Science 211:648–650

Khalil, MAK (1986) Trends of atmospheric methane: past, present and future. Abstract for "Symposium on CO_2 and Other Greenhouse Gases" of the Commission of European Communities, Brussels, Belgium, November 3–5, 1986

Khalil MAK, Rasmussen RA (1983) Sources, sinks, and seasonal cycles of atmospheric methane. J Geophys Res 88/C9:5131–5144

Khalil MAK, Rasmussen RA (1985) Causes of increasing atmospheric methane: depletion of hydroxyl radicals and the rise of emissions. Atmos Environ 19:397–408

Khare BN, Sagan C, Ogino H et al. (1986) Amino acids derived from Titan tholins. Icarus 68:176–184

Kießling J (1885) Dämmerungserscheinungen im Jahr 1883 und ihre physikalische Erklärung. Hamburg/Leipzig

Kimball BA (1985) Adaptation of vegetation and management practices to a higher carbon dioxide world. In "Direct Effects of Increasing Carbon Dioxide on Vegetation" (eds Strain BR, Cure JD), Washington DC, US Dept of Energy, DOE/ER-0238, 185–204

Knauth LP, Epstein S (1976) Hydrogen and oxygen isotope ratios in nodular and bedded cherts. Geochim Cosmochim Acta 40:1095–1108

Knetsch G, Shata A, Degens ET, Münnich KO, Vogel JC, Shazkly MM (1962) Untersuchungen an Grundwässern der Ost-Sahara. Geol Rdsch 52:587–610

Knittel J (1976) Ein Beitrag zur Klimatologie der Stratosphäre der Südhalbkugel. Meteorol Abh FU Berlin A/2:7–18

Knoll AH (1983) Biological interactions and Precambrian eukaryotes. In "Biotic Interactions in Recent and Fossil Benthic Communities" (eds Tevesz MJS, McCall PL), Plenum Publ Corp, New York, 251–283

Ko MKW, Sze ND (1982) A 2-D model calculation of atmospheric lifetimes for N_2O, CFC-11 and CFC-12. Nature 287:317–319

Kobatake Y (1958) Irreversible electrochemical processes in membranes. II. Effect of solvent flow. J Chem Phys 28:442–448

Koenig SH, Brown RD (1972) H_2CO_3 as substrate for carbonic anhydrase in the dehydration of HCO_3. Proc Natl Acad Sci USA 69:2422–2425

Kohlmaier GH, Bröhl H, Siré EO (1983) Über die mögliche lokale Wechselwirkung anthropogener Schadstoffe mit den Terpen-Emissionen von Waldökosystemen. Allg Forst- u J-Ztg 154/9–10:170–174

Kohn PG (1965) Tables of some physical and chemical properties of water. In "The State and Movement of Water in Living Organisms", Symposia of the Society of Experimental Biology 19, Cambridge, 3–32

Kokouchi Y, Okaniwa M, Ambe Y, Fuwa K (1983) Seasonal variation of monoterpenes in the atmosphere of a pine forest. Atmos Environ 17/4:743–750

Kolodny Y, Epstein S (1976) Stable isotope geochemistry of deep sea cherts. Geochim Cosmochim Acta 40:1195–1209

Komhyr WD, Gammon RH, Harris TB, Waterman LS, Conway TJ, Taylor WR, Thoning KW (1985) Global atmospheric CO_2 distribution and variations from 1968–1982 NOAA/GMCC CO_2 flask sample data. J Geophys Res 90/3:5567–5596

Kondratyev KY, Hunt GE (1982) Weather and Climate on Planets. Oxford New York Toronto Sydney Paris Frankfurt, Pergamon Press, pp 755

Kossinna E (1921) Die Tiefen des Weltmeeres. Veröff Inst Meereskunde Univ Berlin, NFA, Geogr-naturwiss 9:1–70
Krafft F, Wiglow H (1895) Über das Verhalten der fettsauren Alkalien und der Seifen in Gegenwart von Wasser. III. Die Seifen als Krystalloide. Chem Ber 28:2566–2573
Krampitz G, Witt W (1979) Biochemical aspects of biomineralization. In "Biochemistry", Top Curr Chem 78:57–144
Kraus EB (1955) Secular changes of tropical rainfall regimes. QJR Meteorol Soc 81:198–210
Kretsinger RH (1977a) Evolution of the informational role of calcium in eukaryotes. In "Calcium Binding Proteins and Calcium Function" (eds Wasserman RH, Corradino RA et al.), North Holland Publ, New York, 63–72
Kretsinger RH (1977b) Why does calcium play an informational role unique in biological systems? In "Metal-Ligand Interactions in Organic Chemistry and Biochemistry" (9th Jerusalem Symposium), Pt 2 (eds Pullman B, Goldbleum N), Lidel Publ Co, Dordrecht, 257–263
Kretsinger RH (1983) A comparison of the roles of calcium in biomineralization and in cytosolic signalling. In "Biomineralization and Biological Metal Accumulation" (eds Westbroek P, De Jong EW), D Reidel Publ Co, Dordrecht, 123–131
Kröner A (1985) Evolution of the Archean continental crust. Annu Rev Earth Planet, Sci 13:49–74
Kroopnick P (1980) The distribution of ^{13}C in the Atlantic Ocean. Earth Planet, Sci Lett 49:469–484
Kuhle M (1986) Die Vergletscherung Tibets und die Entstehung von Eiszeiten. Spektrum der Wissenschaft, Sep 1986:42–54
Kuhn H, Waser J (1982) Evolution of early mechanisms of translation of genetic information into polypeptides. Nature 298:585–586
Kuhn WR, Kasting JF (1983) Effects of increased CO_2 concentrations on surface temperature of the early Earth. Nature 301:53–55
Kukla GJ (1977) Pleistocene land-sea correlations. 1. Europe. Earth Sci Rev 13:307–374
Kunde VG, Aikin AC, Hanel RA, Jennings DE, Maguire WC, Samuelson RE (1981) C_4H_2, HC_3N and C_2N_2 in Titan's atmosphere. Nature 292:686–688
Kvenvolden KA (ed) (1974) Geochemistry and the Origin of Life. Stroudsburg PA, Dowden, Hutchinson & Ross, Inc, pp 422
Kvenvolden KA (1975) Advances in the geochemistry of amino acids. Annu Rev Earth Planet Sci 3:183–212
Laane RWPM, Koole L (1982) The relation between fluorescence and dissolved organic carbon in the Ems-Dollart estuary and the western Wadden Sea. Neth J Sea Res 15/2:217–227
Labitzke K (1987) The lower stratosphere over the polar regions in winter and spring: Relation between meteorological parameters and total ozone. Ann Geophys 5A/3:95–102
Labitzke K (1988) Vulkanismus und Klima. In "Die Erde" (eds Germann K, Warnecke G, Huch M), Springer-Verlag, Berlin Heidelberg New York London Paris Tokyo, 101–114
Labitzke K, Naujokat B (1983) On the variability and on trends of the temperature in the middle stratosphere. Contr Atm Physics 56:495–507
Labitzke K, Brasseur G, Naujokat B, De Rudder A (1986) Long-term temperature trends in the stratosphere: possible influence of anthropogenic gases. Geophys Res Lett 13/1:52–55
Lake JA, Clark MW, Henderson E et al. (1985) Eubacteria, halobacteria, and the origins of photosynthesis: the photocytes. Proc Natl Acad Sci USA 82:3716–3720
Lamb HH (1970) Volcanic dust in the atmosphere; with a chronology and assessments of its meteorological significance. Philos Trans R Met Soc A266:425–533
Lamb HH (1983) Update of the chronology of assessments of the volcanic dust veil index. Clim Monitor 12:79–90
Lamb HH, Johnson AI (1966) Secular Variations of the Atmospheric Circulation Since 1750. Geophys Mem 110, London, Meteor Office, pp 125
Lamb PJ (1982) Persistence of Subsaharan drought. Nature 299:46–48
Lamport DTA (1977) Structure, biosynthesis and significance of cell wall glycoproteins. In "Recent Advances in Phytochemistry" (eds Loewus FA, Runecles VC), 11:79–112
Lande R (1985) Expected time for random genetic drift of a population between stable phenotypic states. Proc Natl Acad Sci USA 82:7641
Landsberg HE, Albert JM (1974) The summer of 1816 and volcanism. Weatherwise 27:63–66
Lasaga AC, Berner RA, Garrels RM (1985) An improved geochemical model of atmospheric CO_2 fluctuations over the past 100 million years. In "The Carbon Cycle and Atmospheric CO_2: Natural Variations, Archean to Present" (eds Sundquist ET, Broecker WS), Am Geophys Union Monogr 32:397–411
Laurmann JA (1986) Future energy use and greenhouse gas-induced climatic warming. Proc of a Symposium on "CO_2 and Other Greenhouse Gases" held at Brussels, Belgium, November 3–5, 1986, by the Commission of European Communities, 39
Laurmann JA, Rotty RM (1983) Exponential growth and atmospheric carbon dioxide. J Geophys Res 88, C2:1295–1299
Lawless JG, Folsome CE, Kvenvolden KA (1972) Organic matter in meteorites. Sci Am 226/6:38–46
Lawrence GM, Barth CA, Argabright V (1977) Excitation of the Venus night airglow. Science 195:573–574
Lawrence JR, Taylor HP Jr (1971) Deuterium and oxygen-correlation: Clay minerals and hydroxides in Quaternary soils compared to meteoric waters. Geochim Cosmochim Acta 35:993–1003
Lawrence JR, Taylor HP Jr (1972) Hydrogen and oxygen isotope systematics in weathering profiles. Geochim Cosmochim Acta 36:1377–1393
Lederberg J (1965) Signs of life. Criterion-system of exobiology. Nature 207:9–13
Leenheer JA (1985) Assessment of organic-solute distributions in water. In "Transport of Carbon and Minerals in Major World Rivers", Pt 3 (eds Degens ET, Kempe S, Herrera R), Mitt Geol-Paläont Inst Univ Hamburg, SCOPE/UNEP Sonderband 58:165–177
Le Fèvre J (1978) Présentation photographique de quelques aspects de l'échouage de l'AMOCO CADIZ. Publ du CNEXO, Série "Actes de Colloques" 6:227–240
Lehninger AL (1966) Supramolecular organization of enzyme and membrane systems. Naturwissenschaften 53:57–63
Lehninger AL (1982) Principles of Biochemistry. New York, Worth Publ Inc, pp 1011
Leighton PA (1961) Photochemistry of Air Pollution. New York, Academic Press
Lennard-Jones Sir J, Pople JA (1951) Molecular association in liquids I: Molecular association due to Jone-pair electrons. Proc R Soc A205:155
Lennon J, McCartney P (1966) Yellow submarine. In "The Beatles Complete", Wise Publ, London New York, 164
Levy A, Henriet H, Bouyssy M (1905) L'ozone atmosphérique. Annales de l'Observatoire Municipal Montsouris 6:18–23
Levy H II (1971) Normal atmosphere: Large radical and formaldehyde concentrations predicted. Science 173:141–143

Lewin R (1984) Alien beings here on earth. Science 223:39
Lewis BTR (1983) The process of formation of ocean crust. Science 220:151–157
Lewis JS, Prinn RG (1984) Planets and their atmospheres. Orlando, Academic Press, Int geoph Ser, pp 470
Li Y-H (1972) Geochemical mass balance among lithosphere, hydrosphere and atmosphere. Am J Sci 272:119–137
Liljas A, Kannan KK, Bergsten P-C et al. (1972) Crystal structure of human carbonic anhydrase C. Nature New Biol 235:131–137
Liou KN (1980) An Introduction of Atmospheric Radiation. New York, Academic Press, Int geoph Ser, pp 392
Lipman CB (1928) The discovery of living micro-organisms in ancient rocks. Science 68:272–273
Lipman PW, Mullineaux DR (eds) (1981) The 1980 Eruptions of Mount St Helens, Washington. Geol Surv Prof Paper 1250, Washington DC, US Governmt Print Off, pp 844
Lohrmann R, Bridson PK, Orgel LE (1980) Efficient metal-ion catalyzed template-directed oligonucleotide synthesis. Science 208:1464–1465
Lohrmann R, Bridson PK, Orgel LE (1981) Condensation of activated diguanylates on a poly(C)template. J Mol Evol 17:303–306
London J (1980) Radiative energy sources and sinks in the stratosphere and mesosphere. In "Proceedings of the NATO Advanced Study Institute on Atmospheric Ozone: Its Variation and Human Influences" (eds Nicolet M, Aikin AC), Washington, DC, US Dept of Transportation, 703–721
Loomis RS (1976) Agricultural systems. Sci Am 235/3:98–105
Lorius C, Jouzel J, Ritz C, Merlivat L, Barkov NI, Korotkevich YS, Kotlyakov VM (1985) A 150,000-year climatic record from Antarctic ice. Nature 316:591–596
Lovelock JE (1979) Gaia. A new look at life on Earth. Oxford Univ Press, Oxford New York Toronto Melbourne, pp 157
Lovelock JE, Margulis L (1973) Atmospheric homoeostasis by and for the biosphere: the gaia hypothesis. Tellus 26:2–10
Lowenstam HA (1973) Biogeochemistry of hard tissues, their depth and possible pressure relationships. In "Barobiology and the Experimental Biology of the Deep Sea" (ed Brauer RW), Chapel Hill, Univ North Carolina, 19–32
Lowenstam HA, Epstein S (1957) On the origin of sedimentary aragonite needles of the Great Bahama Bank. J Geol 65:364–375
Lowenstam HA, Margulis L (1980) Evolutionary prerequisites for early Phanerozoic calcareous skeletons. Biosystems 12:27–41
Luck WAP (1974) Infrared overtone region. In "Structure of Water and Aqueous Solutions" (ed Luck WAP), Verlag Chemie, Weinheim
Luck WAP (1976) Water in biologic systems. In "Inorganic Biochemistry", Top Curr Chem 64, Berlin Heidelberg New York, Springer-Verlag, 113–180
Luck WAP (1980a) Modellbetrachtung von Flüssigkeiten mit Wasserstoffbrücken. Angew Chem 92:29–42
Luck WAP (1980b) Structure of aqueous systems and the influence of electrolytes. Am Chem Soc Symp Ser 127, pp 26
Ludden PW, Burris JE (eds) (1985) Nitrogen Fixation and CO_2 Metabolism. New York, Elsevier
Lupton JE (1983) Terrestrial inert gases: isotope tracer studies and clues to primordial components in the mantle. Annu Rev Earth Planet Sci 11:371–414
Luther FM, Ellingson RG (1985) Carbon dioxide and the radiation budget. In "Projecting the Climatic Effects of Increasing Carbon Dioxide" (eds MacCracken MC, Luther FM), DOE/ER-0237, US Dept of Energy, Washington, DC:25–55
Lutz RA, Rhoads DC (1977) Anaerobiosis and a theory of growth line formation. Science 198:1222–1227
MacCallum AB (1926) The paleochemistry of the body fluids and tissues. Physiol Rev 6:316–357
MacCracken MC (1985) Carbon dioxide and climate change: background and overview. In "Projecting the Climatic Effects of Increasing Carbon Dioxide" (eds MacCracken MC, Luther FM), DOE/ER-0237, US Dept of Energy, Washington, DC, 1–23
MacCracken MC, Luther FM (eds) (1985) Detecting the Climatic Effects of Increasing Carbon Dioxide. DOE/ER-0235, US Dept of Energy, Washington, DC, pp 198
MacLennan DH, Holland PC (1975) Calcium transport in sarcoplasmic reticulum. Annu Rev Biophys Bioeng 4:377
Magaritz M, Holser WT, Kirschvink JL (1986) Carbon- isotope events across the Precambrian/Cambrian boundary on the Siberian Platform. Nature 320:258–259
Maguire WC, Hanel RA, Jennings DE, Kunde VG, Samuelson RE (1981) C_3H_8 and C_3H_4 in Titan's atmosphere. Nature 292:683–686
Maier-Reimer E, Hasselmann K (1987) Transport and storage of CO_2 in the ocean – an inorganic ocean-circulation carbon cycle model. Clim Dynamics 2:63–90
Maisonneuve J (1982) The composition of the Precambrian ocean waters. Sediment Geol 31:1–11
Malone RC, Auer LH, Glatzmaier GA, Wood MC (1986) Three-dimensional simulations including interactive transport, scavenging, and solar heating of smoke. J Geophys Res 91:1039–1053
Manabe S, Bryan K Jr (1985) Co_2-induced change in a coupled ocean-atmosphere model and its paleoclimatic implications. J Geophys Res 90:11, 689–11,707
Manabe S, Wetherald RT (1980) On the distribution of climate change resulting from an increase in CO_2 content of the atmosphere. J Atmosph Sci 37:99–118
Manabe S, Wetherald RT, Stouffer RJ (1981) Summer dryness due to an increase of atmospheric CO_2 concentration. Clim Change 3:347–386
Manheim FT (1976) Interstitial waters of marine sediments. In "Chemical Oceanography" (eds Riley JP, Chester R), vol 6 (2nd ed), Academic Press, London New York San Francisco, 115–186
Manheim FT, Schug DM (1978) Interstitial waters of Black Sea cores. In "Initial Reports of the Deep Sea Drilling Project" (eds Ross DA, Neprochnov YP et al.), 42 Pt 2, Washington, US Govt Printing Office, 637–651
Margoliash E, Fitch WM (1968) Evolutionary variability of cytochrome *c* primary structures. Ann NY Acad Sci 151:359–381
Margulis L (1981) Symbiosis and Cell Evolution. Freeman, San Francisco
Martonosi A, Boland ARD, Boland R et al. (1974) The mechanism of Ca transport and the permeability of sarcoplasmic reticulum membranes. In "Myocardial Biology: Recent Advances in Studies on Cardiac Structure and Metabolism" (ed Dhalla NS), vol 4, Baltimore, Univ Park Press, 473–494
Mason BJ (1977) Zum Verständnis und zur Vorhersage der Klimaschwankungen. Promet 4:1–22
Mason S (1985) Origin of biomolecular chirality. Nature 314:400–401
Mass C, Schneider SH (1977) Influence of sunspots and volcanic dust on long-term temperature records inferred by statistical investigations. J Atmosph Sci 34:1995–2004
Matheja JH (1971) Electron diffraction of membranes. Biophysik 7:163–168

Matheja JH, Degens ET (1971a) Structural Molecular Biology of Phosphates. Gustav Fischer Verlag, Stuttgart, pp 180
Matheja JH, Degens ET (1971b) Function of amino acid side chains. Adv Enzymol 34:1–39
Mather RS, Rizos C, Coleman R (1979) Remote sensing of surface ocean circulation with satellite altimetry. Science 205:11–17
Mathez EA (1984) Influence of degassing on oxidation states of basaltic magmas. Nature 310:371–375
Matson M, Robock A (1984) Circumglobal transport of the El Chichon volcanic dust cloud. Geof Int 23:117–127
Matsui T, Abe Y (1986) Impact-induced atmospheres and oceans on Earth and Venus. Nature 322:526–528
Mattioli GS, Wood BJ (1986) Upper mantle oxygen fugacity recorded by spinel lherzolites. Nature 322:626–628
May RM (1978) The evolution of ecological systems. Sci Am 239/3:118–133
May RM (1986) How many species are there? Nature 324:514–515
Mayr E (1975) Es hat im Laufe der Geschichte ungefähr 600 Millionen Tierarten gegeben – Schätzung! In "Evolution" (ed Scharf J-H), Nova Acta Leopoldina 42:411
Mayr E (1983) The Growth of Biological Thought: Diversity, Evolution, and Inheritance. Harvard Univ Press, pp 974
McCarthy K, Lovenberg W (1968) Optical properties of artificial non-heme iron proteins derived from bovine serum albumin. J Biol Chem 243:6436–6441
McCormick MP, Trepte CR (1987) Polar stratospheric optical depth observed by the SAM II satellite instrument between 1978 and 1985. J Geophys Res 92:4297–4306
McDonald JE (1952) The Coriolis Effect. Sci Am 186/5:72–78
McElroy MB, Yung YL, Nier AO (1976) Isotopic composition of nitrogen: implications for the past history of Mars' atmosphere. Science 194:70–72
McFadden GJ, Preisig HR, Melkonian M (1986) Golgi apparatus activity and membrane flow during scale biogenesis in the green flagellate *Scherffelia dubia* (Prasinophyceae). II. Cell Wall Secretion and Assembly. Protoplasma 131:293–302
McKelvey VE (1980) Seabed minerals and the law of the sea. Science 209:464–472
McKinney ML (1987) Taxonomic selectivity and continuous variation in mass and background extinctions of marine taxa. Nature 325:143–145
McLaren DJ (1970) Time, life and boundaries. J Paleont 44:801–815
McLean DM (1982) Deccan volcanism and the Cretaceous-Tertiary transition scenario: a unifying causal mechanism. Syllogeus 39:143–144
McPeters RD, Heath DF, Bhartia PK (1984) Average ozone profiles for 1979 from the NIMBUS 7 SBUV instrument. J Geophys Res 89D:5199–5214
Menyailov IA, Nikitina LP (1980) Chemistry and metal contents of magmatic genes: the new Tolbachik volcanoes case (Kamchatka). Bull Volcanol 43:197–205
Michaelis W, Mycke B, Vogt J et al. (1982) Organic geochemistry of interstitial waters, Sites 474 and 479, Leg 64. In "Initial Reports of the Deep Sea Drilling Project" (eds Curray JR, Moore DG et al.) 44, Pt 2, Washington, DC, US Govt Printing Office, 933–937
Miller SL (1953) A production of amino acids under possible primitive Earth conditions. Science 117:528–529
Miller SL (1959) Formation of organic compounds on the primitive earth. In "The Origin of Life on the Earth" (eds Oparin AI, Pasynskii AG et al.), Pergamon Press, New York London Paris Los Angeles, 123–135
Miller SL, Horowitz NH (1966) Biology and the exploration of Mars. Natl Acad Sci, Nat Res Counc, Washington, DC, Publication 1296, pp 41
Miller SL, Parris M (1964) Synthesis of pyrophosphate under primitive earth conditions. Nature 204:1248–1250
Milliman JD (1974) Recent Sedimentary Carbonates, Pt 1: Marine Carbonates. Springer-Verlag, Berlin Heidelberg New York
Mishima O, Endo S (1980) Phase relations of ice under pressure. J Chem Phys 73:2454–2456
Mitsui A, Kumazawa S, Takahashi A, Ikemoto H, Cao S, Arai T (1986) Strategy by which nitrogen-fixing unicellular cyanobacteria grow photoautotrophically. Nature 323:720–722
Mix AC, Pisias NG (1988) Oxygen isotope analyses and deep-sea temperature changes: implications for rates of oceanic mixing. Nature 331:249–251
Mizutani H, Mikuni H, Takahashi M, Noda H (1975) Study on the photochemical reaction of HCN and its polymer products relating to primary chemical evolution. Origins of Life 6:513–525
Mobbs SD (1982) Extremal principles for global climate models. QJR Meteorol Soc 108:535–550
Molina MJ, Rowland FS (1974) Stratospheric sink for chlorofluoromethanes: chlorine atom catalyzed destruction of ozone. Nature 249:810–812
Mook WG, Bommerson JC, Staverman WH (1974) Carbon isotope fractionation between dissolved bicarbonate and gaseous carbon dioxide. Earth Planet Sci Lett 22:169–176
Moorbath S, O'Nions RK, Pankhurst RJ (1973) Early Archean age for the Isua iron formation West Greenland. Nature 245:138–139
Moore PB, Asano K (1966) Divalent cation-dependent binding of messenger to ribosomal RNA. J Mol Biol 18:21–26
Morowitz HJ (1967) Biological self-replicating systems. Progr theor Biol 1:35–38
Mosher HS, Morrison JD (1983) Current status of asymmetric synthesis. Science 221:1013–1019
Mostow KE, Friedlander M, Blobel G (1984) The receptor for transepithelial transport of IgA and IgM contains multiple immunoglobulin-like domains. Nature 308:37–43
Mühlenbachs K, Clayton RN (1976) Oxygen isotope composition of the oceanic crust and its bearing on seawater. J Geophys Res 81:4365–4369
Müller G, Irion G, Förstner U (1972) Formation and diagenesis of inorganic Ca-Mg carbonates in the lacustrine environment. Naturwissenschaften 59:158–164
Muenow DW (1973) High temperature mass spectrometric gas-release studies of Hawaiian volcanic glass: Pele's Tears. Geochim Cosmochim Acta 37:1551–1561
Muraoka Y, Petzoldt K, Labitzke K (1987) D-region winter anomaly associated with vertically propagating planetary waves. Ann Geophys 5A:455–461
Murray J, Renard AF (1891) Report on deep sea deposits. Rept Sci Results Voyage HMS Challenger, Deep Sea Deposits, 1–525
Mysen BO, Seifert F, Virgo D (1980) Structure and redox equilibria of iron-bearing silicate melts. Am Mineral 65:867–884
Nagy B (1975) Carbonaceous Meteorites. Amsterdam, New York, Elsevier Scientific Publishing Co, pp 747
Nagy B, Bitz Sister Mary Carol (1963) Long-chain fatty acids from the Orgueil meteorite. Arch Biochem Biophys 101:240–248
Namias J, Cayan DR (1984) El-Niño: implications for forecasting. Oceanus 27:41–47
Narasinga Rao MS (1967) Forms of water in biologic systems. J Sci Ind Res 26:7–10

Narten AH, Thiessen WE, Blum L (1982) Atom pair distribution functions of liquid water at 25° C from neutron diffraction. Science 217:1033–1034

NASA (1984) Ozone Climatology Series Atlas of Total Ozone: April 1970–December 1976. GSFC, Greenbelt, USA

National Academy of Sciences (1975) Petroleum in the marine environment. National Academy of Sciences, Washington, DC

National Academy of Sciences (1984) Global Tropospheric Chemistry: A Plan for Action. Washington, DC, National Academy Press, pp 194

Nature (1978) Climatology Supplement. Nature 276:327–359

Neftel A, Moor E, Oeschger H, Stauffer B (1985) Evidence from polar ice cores for the increase in atmospheric CO_2 in the past two centuries. Nature 315:45–47

Nemethy G, Scheraga HA (1962) The structure of water and hydrophobic bonding in proteins. III. The thermodynamic properties of hydrophobic bonds in proteins. J Phys Chem 66:1773–1789

Neudert MK, Russell RE (1981) Shallow water and hypersaline features from the Middle Proterozoic Mt Isa Sequence. Nature 293:284–286

Neuhaus A (1952a) Orientierte Kristallabscheidung (Epitaxie). Angew Chem 64:158–162

Neuhaus A (1952b) Über Keimbildung und orientierten Stoffabsatz auf artfremden kristallinen Oberflächen. Z Elektrochem 56:453–458

Neuman MW, Neuman WF, Lane K (1970a) On the possible role of crystals in the origins of life. 3. The phosphorylation of adenosine to AMP by apatite. Curr Mod Biol 3:253–259

Neuman MW, Neuman WF, Lane K (1970b) On the possible role of crystals in the origins of life. 4. The phosphorylation of nucleotides. Curr Mod Biol 3:277–283

Newman CM, Cohen JE, Kipnis C (1985) New-Darwinian evolution implies punctuated equilibria. Nature 315:400–401

Newsom HE (1980) Hydrothermal alteration of impact melt sheets with implications for Mars. Icarus 44:207–216

Nicholson SE (1980) Saharan climates in historic times. In "The Sahara and the Nile" (eds Williams MAJ, Faure H), Rotterdam, Balkema, 173–200

Nicholson SE, Flohn H (1980) African environmental and climatic changes and the general atmospheric circulation in the late Pleistocene and Holocene. Clim Change 2:313–348

Nier AO, Hanson WB, Seiff A, McElroy MB, Spencer NW, Duckett RJ, Knight TCD, Cook WS (1976a) Composition and structure of the Martian atmosphere: preliminary results from Viking 1. Science 193:786–788

Nier AO, McElroy MB, Yung YL (1976b) Isotopic composition of the Martian atmosphere. Science 194:68–70

Nieuwolt S (1977) Tropical Climatology: An Introduction to the Climates of the Low Latitudes. New York, Wiley, pp 207

Noller HF, Woese CR (1981) Secondary structure of 16S ribosomal RNA. Science 212:403–411

Northcote DH (1971) The golgi apparatus. Endeavour 30:26–33

Oeschger H, Siegenthaler U, Schotterer U, Gugelmann A (1975) A box diffusion model to study the carbon dioxide exchange in nature. Tellus 27:168–192

Österberg R (1974) Origins of metal ions in biology. Nature 249:382–383

Officer CB, Drake CL (1983) The Cretaceous-Tertiary transition. Science 219:1383–1390

Officer CB, Drake CL (1985) Terminal Cretaceous environmental events. Science 227:1161–1167

Officer CB, Hallam A, Drake CL, Devine JD (1987) Late Cretaceous and paroxysmal Cretaceous/Tertiary extinctions. Nature 326:143–149

Ogura N (1974) Molecular weight fractionation of dissolved organic matter in coastal seawater by ultrafiltration. Mar Biol 24:305–312

Ohno S (1984) Repeats of base oligomers as the primordial coding sequences of the primeval Earth and their vestiges in modern genes. J Mol Evol 20:313–321

Ohno S (1985) Immortal genes. Trends Genet 1:196–200

Ohno S, Epplen J (1983) The primitive code and repeats of base oligomers as the primordial protein-encoding sequence. Proc Natl Acad Sci USA 80:3391–3395

Ohno S, Ohno M (1986) The all pervasive principle of repetitious recurrence governs not only coding sequence construction but also human endeavor in musical composition. Immunogenetics 24:71–78

Okazaki M (1972) Carbonic anhydrase in the calcareous red alga, *Serraticardia maxima.* Bot Mar 15:133–138

Oliver RC (1976) On the response of hemispheric mean temperature to stratospheric dust: an empirical approach. J Appl Meteor 15:933–950

Oliver WA Jr (1980) Relationship of the scleractinian corals to the rugose corals. Paleobiology 6:146–160

Olsen PE (1986) A 40-million-year lake record of early Mesozoic orbital climatic forcing. Science 234:842–848

Oparin AI (1953) The Origin of Life. Dover Publications Inc, New York, pp 270

Oparin AI, Pasynskiĭ AG, Braunshteĭn AE et al. (eds) (1959) The Origin of Life on the Earth. Pergamon Press, New York London Paris Los Angeles

Oró J, Guidry CL (1961) Direct synthesis of polypeptides. I. Polycondensation of glycine in aqueous ammonia. Arch Biochem Biophys 93:166–171

Oskvarek JD, Perry Jr EC (1976) Temperature limits on the early Archean ocean from oxygen isotope variations in the Isua supracrustal sequences, West Greenland. Nature 259:192–194

Ott V, Mack R, Chang S (1981) Noble gas rich separates from the Allende meteorite. Geochim Cosmochim Acta 45:1751–1788

Ourisson G, Albrecht P, Rohmer M (1979) The hopanoids. Paleochemistry and biochemistry of a group of natural products. Pure Appl Chem 51:709–729

Paecht-Horowitz M, Berger J, Katchalsky A (1970) Prebiotic synthesis of polypeptides by heterogeneous polycondensation of amino acid adenylates. Nature 228:636–639

Palca J (1986) Could this be an El Niño? Nature 324:504

Paldridge G (1975) Global dynamics and climate – a system of minimum entropy exchange. QJR Meteorol Soc 101:475–484

Palmer JD (1975) Biological clocks of the tidal zone. Sci Am, Febr 1975, 70–79

Pannella G (1971) Fish otoliths: daily growth layers and periodical patterns. Science 173:1124–1127

Parrish JT (1982) Upwelling and petroleum source beds. Bull Am Assoc Petrol Geol 66:750–774

Pauling L (1930) The structure of the chlorites. Proc Natl Acad Sci US 16:578–582

Pauling L (1948) The Nature of the Chemical Bond and the Structure of Molecules and Crystals. Cornell Univ Press, Ithaca, New York, pp 644

Pauling L (1961) A molecular theory of general anesthesia. Science 134:15–21

Pauling L, Corey RB, Bransen HR (1951) The structure of protein: Two hydrogen-bonded helical configurations of the polypeptide chain. Proc Natl Acad Sci US 37:205–211

Peng TH, Broecker WS, Mathieu GG et al. (1979) Radon evasion rates in the Atlantic and Pacific Oceans as determined during the GEOSECS Program. J Geophys Res 84:2471–2486

Peng TH, Broecker WS, Freyer HD, Trumbore S (1983) A deconvolution of the tree-ring based $\delta^{13}C$ record. J Geophys Res 88, C6:3609–3620

Perner D (1986) Hunting tropospheric OH ions. Nature 322:595

Perry EC Jr (1967) The oxygen isotope chemistry of ancient cherts. Earth Planet Sci Lett 3:62–66

Perry EC Jr, Tan FC (1972) Significance of oxygen and carbon isotope variations in early Precambrian cherts and carbonate rocks of southern Africa. Geol Soc Am Bull 83:647–664

Perry EC Jr, Ahmad SN, Swulius TM (1978) The oxygen isotope composition of 3,800 MY old metamorphosed chert and iron formation from Isukasia, West Greenland. J Geol 86:223–239

Peschel G, Adlfinger KH (1971) Übersichtsreferat: Zur anomalen Struktur des Wassers. Mitt Dt Pharmaz Ges 41/2:29–54

Pettengill GH, Campbell DB, Masursky H (1980a) The surface of Venus. Sci Am 243/2:46–57

Pettengill GH, Eliason E, Ford PG, Loriot GB, Masursky H, McGill GE (1980b) Pioneer Venus radar results: Altimetry and surface properties. J Geophys Res 85:8261–8270

Pettijohn FJ (1975) Sedimentary Rocks (3rd ed). New York, Harper and Row, pp 628

Pflug HD (1978) Yeast-like microfossil detected in oldest sediments of the Earth. Naturwissenschaften 65:611–615

Pflug HD (1982) Early diversification of life in the Archean. Zbl Bakt Hyg, I. Abt Orig C3:53–64

Philander SGH (1983) El Niño Southern Oscillation phenomena. Nature 302:295–301

Phillips RJ, Malin MC (1984) Tectonics of Venus. Annu Rev Earth Planet Sci 12:411–443

Phillips RJ, Kaula WM, McGill GE, Malin MC (1981) Tectonics and evolution of Venus. Science 212:879–887

Piboubes R (1979) AMOCO CADIZ: Conséquences d'une Pollution Accidentelle par les Hydrocarbures. Analyse Bibliographique. Centre Océanologique de Bretagne, Brest, pp 82

Pickett CJ, Talarmin J (1985) Electrosynthesis of ammonia. Nature 317:652–653

Pilbeam D (1984) The descent of hominoids and hominids. Sci Am 250/3:60–69

Pinto JP, Gladstone GR, Yung YL (1980) Photochemical production of formaldehyde in earth's primitive atmosphere. Science 210:183–185

Pirie NW (1937) The meaninglessness of the terms life and living. Perspectives in Biochemistry, Cambridge Univ Press

Pitman WC III (1978) Relationship between eustasy and stratigraphic sequences of passive margins. Geol Soc Am Bull 89:1389–1403

Pitman WC III, Golovchenko X (1983) The effect of sealevel changes on the shelf edge and slope of passive margins. Soc Econ Paleontol Mineral Spec Pub 33:41–58

Pittock AB, Ackerman TP, Crutzen PJ et al. (eds) (1986) Environmental Consequences of Nuclear War. Vol 1: Physical and Atmospheric Effects. SCOPE Report 28. Chichester New York Brisbane Toronto Singapore, John Wiley & Sons, pp 359

Pollack JB (1984) Origin and history of the outer planets: theoretical models and observational constraints. Annu Rev Astron Astrophys 22:389–424

Pollack JB, Black DC (1979) Implications of the gas compositional measurements of Pioneer Venus for the origin of planetary atmospheres. Science 205:56–59

Pollack JB, Black DC (1982) Noble gases in planetary atmospheres: implications for the origin and evolution of atmospheres. Icarus 51:169–198

Pollack JB, Young L (1980) Origin and evolution of planetary atmospheres. Annu Rev Earth Planet Sci 8:425–487

Pollack JB, Toon OB, Sagan OB, Summers C, Baldwin B, Van Camp W (1976) Volcanic explosions and climatic change: a theoretical assessment. J Geophys Res 81:1071–1083

Pollack JB, Toon OB, Ackerman T-P, McKay CP, Turco RP (1983) Environmental effects of an impact-generated dust cloud: implications for the Cretaceous-Tertiary extinctions. Science 219:287–289

Ponnamperuma C, Kirk P (1964) Synthesis of deoxyadenosine under simulated primitive earth conditions. Nature 203:400–401

Pople JA (1951) Molecular association in liquids II. A theory of the structure of water. Proc R Soc Lond A205:163–178

Post WM, Emanuel WR, Zinke PJ, Stangenberger AG (1982) Soil carbon pools and world life zones. Nature 298:156–159

Prelog V (1976) Chirality in chemistry. Science 193:17–24

Preuss H (1962) Grundriß der Quantenchemie. Bibliographisches Institut, Mannheim pp 310

Prigogine I (1979) Vom Sein zum Werden. Zeit und Komplexität in den Naturwissenschaften. München, Piper, pp 261

Prigogine I, Balescu R (1956) Phénomènes cycliques dans la thermodynamique des processus irréversibles. Bull Sci Acad Roy Berg 42:256–265

Prinn RG (1985) The volcanoes and clouds of Venus. Sci Am 252/3:36–43

Prusch RD, Minck DR (1985) Chemical stimulation of phygocytosis in *Amoeba proteus* and the influence of external calcium. Cell Tissue Res 242:557–564

Pullman A, Vasilescu V, Packer L (1985) Water and Ions in Biological Systems. New York London, Plenum Press, pp 823

Pytkowicz RM (1983) Equilibria, Nonequilibria, and Natural Waters. 2 vols. New York Chichester Brisbane Toronto Singapore, John Wiley & Sons, pp 351, pp 353

Raff RA, Mahler HR (1972) The non-symbiotic origin of mitochondria. Science 177:575–582

Railsback LB, Anderson TF (1987) Control of Triassic seawater chemistry and temperature on the evolution of post-Paleozoic aragonite-secreting faunas. Geology 15:1002–1005

Ramage CS (1986) El Niño. Sci Am 254/6:55G–61

Ramanathan V (1981) The role of ocean-atmosphere interactions in the CO_2 climate problem. J Atmos Sci 38:918–930

Ramanathan V, Cicerone RJ, Singh HB, Kiehl JT (1985) Trace gas trends and their potential role in climate change. J Geophys Res 90/D3:5547–5566

Ramanathan V, Callis L, Cess R et al. (1987) Climate-chemical interactions and effects of changing atmospheric trace gases. Rev Geophys 25/7:1441–1482

Rampino MR (1987) Impact cratering and flood basalt volcanism. Nature 327:468

Rampino MR, Self S (1984) The atmospheric effects of El Chichon. Sci Am 250/1:34–43

Rashid MA (1985) Geochemistry of Marine Humic Compounds. New York Berlin Heidelberg Tokyo, Springer-Verlag, pp 300

Rasmusson EM (1984) El-Niño: The ocean/atmosphere connection. Oceanus 27/2:5–13

Rasmusson EM, Wallace JM (1983) Meteorological aspects of the El Niño/Southern Oscillation. Science 222:1195–1202

Rasmussen RA, Khalil MAK (1986) Atmospheric trace gases: trends and distributions over the last decade. Science 222:1623–1624

Raup DM (1976) Species diversity in the Phanerozoic: an interpretation. Paleobiology 2:289–297

Raup DM (1979) Size of the Permo-Triassic bottleneck and its evolutionary implications. Science 206:217–218

Raup DM (1986) The Nemesis Affair: A Story of the Death of Dinosaurs and the Ways of Science. WW Norton and Co, New York London, pp 220

Raup DM, Crick RE (1981) Evolution of single characters in the Jurassic ammonite *Kosmoceras*. Paleobiology 7/2:200–215

Raup DM, Crick RE (1982) *Kosmoceras:* evolutionary jumps and sedimentary breaks. Paleobiology 8/2:90–100

Raup DM, Jablonski D (eds) (1986) Patterns and Processes in the History of Life. Dahlem Konferenzen, Springer-Verlag, Berlin Heidelberg New York Tokyo

Raup DM, Sepkoski JJ Jr (1982) Mass extinctions in the marine fossil record. Science 215:1501–1503

Raup DM, Sepkoski JJ Jr (1984) Periodicity of extinctions in the geologic past. Proc Natl Acad Sci 81:801–805

Raynaud D, Barnola JM (1985) An Antarctic ice core reveals atmospheric CO_2 variations over the past few centuries. Nature 315:309–311

Reid JL, Shulenberger E (1986) Oxygen saturation and carbon uptake near 28°N, 155°W. Deep-Sea Res 33/2:267–271

Reymer A, Schubert G (1984) Phanerozoic addition rates to the continental crust and crustal growth. Tectonics 3:63–78

Rhoads DC, Pannella G (1970) The use of molluscan shell growth patterns in ecology and paleoecology. Lethaia 3:143–161

Riehl H, Meitin J (1979) Discharge of the Nile River: a barometer of short-period climatic variation. Science 206:1178–1179

Riehl H, El-Bakry M, Meitin J (1979) Nile River discharge. Monthly Weather Rev 107:1546–1553

Rieke HH, Chilingar GV, Robertson JO (1964) High-pressure (up to 500,000 psi) compaction studies on various clays. Proc 22nd Internat Geol Congr New Delhi, India

Ringwood AE (1979) Origin of the Earth and Moon. New York Heidelberg Berlin, Springer-Verlag, pp 295

Rittenberg SC, Emery KO, Hülsemann et al. (1963) Biogeochemistry of sediments in experimental Mohole. J Sediment Petr 33:140–172

Robbins ET, Porter KG, Haberyan KA (1985) Pellet microfossils: possible evidence for metazoan life in early Proterozoic time. Proc Natl Acad Sci USA 82:5809–5813

Robinson RS (1964) The duplex origins of petroleum. In "Advances in Organic Chemistry" (eds Colombo U, Hobson GD), New York, The Macmillan Company, 7–10

Robock A (1979) The "Little Ice Age": northern hemisphere average observations and model calculations. Science 206:1402–1404

Robock A (1984) Climate model simulations of the effects of the El Chichon eruption. Geof Int 23:403–414

Robson RL, Eady RR, Richardson TH et al. (1986) The alternative nitrogenase of *Azotobacter chroococcum* is a vanadium enzyme. Nature 322:388–390

Rodhe H, Herrera R (eds) (1988) Acidification in Tropical Countries. SCOPE Rep 36, John Wiley & Sons, Chichester New York Brisbane Toronto Singapore, pp 405

Roeckner E, Schlese U, Biercamp J, Loewe P (1987) Cloud optical depth feedbacks and climate modelling. Nature 329:138–140

Roether W, Münnich KO (1972) Tritium profile at the Atlantic 1970 GEOSECS test cruise station. Earth Planet Sci Lett 16:127–130

Roether W, Münnich KO (1974) The 1971 Transatlantic section of F/S "Meteor" near 40°N. Earth Planet Sci Lett 23:91–99

Roether W, Gieskes JM, Hussels W (1974) Hydrography of a Transatlantic section from Portugal to the Newfoundland Basin. "Meteor" Forsch-Ergebn RA 14:13–32

Ronov AB (1968) Probable changes in the composition of sea water during the course of geological time. Sedimentology 10:25–43

Ronov VI, Khain VE, Balukhovsky AN, Seslavinsky KB (1980) Quantitative analysis of Phanerozoic sedimentation. Sediment Geol 25:311–325

Rosano HL (1967) Mechanisms of water transport through nonaqueous liquid membranes. J Colloid Interface Sci 23:73–79

Rose AH (1984) Physiology of cell aggregation: flocculation by *Saccharomyces cerevisiae* as a model system. In "Microbial Adhesion and Aggregation" (Dahlem Conferences) (ed Marshall KC), Berlin Heidelberg New York Tokyo, Springer-Verlag, 323–335

Rose B, Simpson I, Loewenstein WR (1977) Calcium ion produces graded changes in permeability of membrane channels in cell junction. Nature 267:625–627

Rose MR, Doolittle WF (1983) Molecular biological mechanisms of speciation. Science 220:157–162

Ross DA (1970) Introduction to Oceanography. New York, Meredith Corporation, pp 384

Ross DA, Knauss JA (1982) How the law of the sea treaty will affect US marine science. Science 217:1003–1008

Rosswall T (1979) Nitrogen losses from the terrestrial ecosystems – global, regional and local considerations. In "Proceedings of the Fifth International Conference on Global Impacts of Applied Microbiology" (ed Matang-Kasombut P), GIAM-V Secretariat, Bangkok, 17–26

Rosswall T (ed) (1980) Nitrogen Cycling in West African Ecosystems. Proc workshop arr by SCOPE/UNEP Intern Nitr Unit in coll with MAB (Unesco) and IITA at Intern Inst Trop Agric, Ibadan, Nigeria, 11–15 Dec 1978, pp 450

Rosswall T (1981) The biogeochemical nitrogen cycle. In "Some Perspectives of the Major Biogeochemical Cycles", SCOPE Report 17 (ed Likens G), Chichester New York Brisbane Toronto, John Wiley & Sons, 25–49

Rosswall T (1982) Microbiological regulation of the biogeochemical nitrogen cycle. Plant Soil 67:15–34

Rothman JE (1985) The compartmental organization of the Golgi apparatus. Sci Am 253/3:84–95

Rotty RM (1983) Distribution and changes in industrial carbon dioxide production. J Geophys Res 88:1301–1308

Rotty RM (1986) A look at 1983 CO_2 emissions from fossil fuels. (draft manuscript)

Rubey WW (1951) Geologic history of sea water. An attempt to state the problem. Geol Soc Am Bull 62:1111–1148

Rubinson M, Clayton RN (1969) Carbon-13 fractionation between aragonite and calcite. Geochim Cosmochim Acta 33:997–1002

Ruddiman WF (1985) Climate studies in ocean cores. In "Paleoclimate Analysis and Modeling" (ed Hecht AD), New York, Wiley, 197–257

Rüegg JC (1986) Calcium in Muscle Activation. A Comparative Approach. Zoophysiology, vol 19, Berlin Heidelberg New York, Springer Verlag, pp 300

Russell DA (1979) The enigma of the extinction of the dinosaurs. Annu Rev Earth Planet Sci 7:163–182

Rutgers AJ, Hendrikx Y (1962) Ionic hydration. Trans Faraday Soc 58:2184–2191

Ryaboshapko AG (1983) The atmospheric sulphur cycle. In "The Global Biogeochemical Sulphur Cycle" (eds Ivanov MV, Freney JR), SCOPE Report 19, Chichester New York Brisbane Toronto Singapore, John Wiley & Sons, 203–296

Sackett WM, Moore WS (1966) Isotopic variations of dissolved inorganic carbon. Chem Geol 1:323–328
Sagan C (1985) Nuclear winter: a report from the world scientific community. Environment 27/8:12–15, 38–39
Saimi Y, Kung C (1982) Are ions involved in the gating of calcium channels? Science 218:153–156
Saitou N, Omoto K (1987) Time and place of human origins from mt DNA data. Nature 327:288
Salstein DA, Rosen RD (1984) El-Niño and the Earth's rotation. Oceanus 27:52–57
Saltzman B (1987) Carbon dioxide and the $\delta^{18}O$ record of late-Quaternary climatic change: a global model. Clim Dynamics 1:77–85
Samoilov OY (1965) The Structure of Aqueous Electrolyte Solutions and the Hydration of Ions. Consultants Bureau, New York
Sandberg PA (1983) An oscillating trend in Phanerozoic nonskeletal carbonate mineralogy. Nature 305:19–22
Sano Y, Wakita H, Chin-Wang Huang (1986) Helium flux in a continental land area estimated from $^3He/^4He$ ratio in northern Taiwan. Nature 323:55–57
Sasaki A (1970) Seawater sulfate as a possible determinant for sulfur isotope compositions of some stratabound sulfide ores. Geochem J 4:41–51
Sasaki Y, Hidaka H (1982) Calmodulin and cell poliferation. Biochem Biophys Res Commun 104:451–456
Savin SM (1977) The history of the Earth's surface temperature during the past 100 million years. Annu Rev Earth Planet Sci 5:319–355
Schallreuter R (1984) Framboidal pyrite in deep-sea sediments. In "Initial Reports of the Deep-Sea Drilling Project" (eds Hay WW et al.), vol 75, pt 2, US Government Printing Office, Washington, DC, 875–891
Schidlowski M (1986) $^{13}C/^{12}C$ ratios as indicators of biogenicity. In "Biological Markers in the Sedimentary Record, Methods in Geochemistry and Geophysics" 24 (ed Johns RB), Elsevier Sci Publ BV, 347–361
Schidlowski M (1987) Application of stable carbon isotopes to early biochemical evolution on earth. Annu Rev Earth Planet Sci 15:47–72
Schlanger SO (1986) High frequency sea-level fluctuations in Cretaceous time: an emerging geophysical problem. In "Mesozoic and Cenozoic Oceans", Geodynamics Ser 15:61–74
Schlee S (1973) The Edge of an Unfamiliar World. New York, EP Dutton & Co, Inc, pp 398
Schlögl R (1956) The significance of convection transport processes across porous membranes. Disc Faraday Soc 21:46–52
Schneider SW (1972) Cloudiness as a global climatic feedback mechanism: The effect on the radiation balance and surface temperature of variations in cloudiness. J Atmos Sci 29:1413–1422
Schneider SH (1983) Volcanic dust veils and climate: how clear is the connection? An editorial. Clim Change 5:111–113
Schneider SH (1986) A goddess of the Earth? The debate on the Gaia hypothesis. Clim Change 8:1–4
Schoenbein CF (1840) Recherches sur la nature de l'odeur qui se manifeste dans certaines actions chimiques. CR Acad Sci Paris 10:706–712
Schönwiese CD (1979) Klimaschwankungen. Berlin Heidelberg New York, Springer-Verlag, pp 181
Schönwiese CD (1988) Volcanism-climate relationships within recent centuries. Bull Volcanol (in the press)
Scholtissek C, Naylor E (1988) Fish farming and influenza pandemics. Nature 331:215
Schopf JW (1978) The evolution of the earliest cells. Sci Am 239/3:84–102
Schopf TJM, Hoffman A (1983) Punctuated equilibrium and the fossil record. Science 219:438–439
Schrödinger E (1945) What is Life? Cambridge (England) Univ Press, New York, Macmillan Co, pp 91
Schrön W (1985) Zur Rolle von Fest-Gas-Gleichgewichten in der Geo- und Kosmochemie. Diss, Bergakademie Freiberg, DDR, pp 146
Schubert G, Covey C (1981) The atmosphere of Venus. Sci Am 245/1:44–52
Schufle JA, Huang C-T, Drost-Hansen W (1976) Temperature dependence of surface conductance and a model of vicinal (interfacial) water. J Colloid Interface Sci 54:184–202
Schwartz RM, Dayhoff MO (1978) Origins of prokaryotes, eukaryotes, mitochondria, and chloroplasts. Science 199:395–403
Scott EL (1986) Carl Scheele (1742–1786) and the discovery of oxygen. Nature 322:305
Sears M, Merriman D (eds) (1980) Oceanography: The Past. New York Heidelberg Berlin, Springer-Verlag, pp 812
Seel F, Klos K-P, Recktenwald D, Schuh J (1986) Non-enzymatic formation of condensed phosphates under prebiotic conditions. Z Naturforsch 41b:815–824
Seifert H (1968) Zur Kristallisation von Hochpolymeren an kristallinen Grenzflächen. Kolloid Ztschr u Ztschr Polym 224:97–124
Seiler W (1985) Increase of Atmospheric Methane: Causes and Impact on the Environments. WMO Special Environmental Report No 16
Seiler W, Crutzen P (1980) Estimates of gross and net fluxes of carbon between the biosphere and the atmosphere from biomass burning. Clim Change 2:207–248
Seiler W, Muller F, Oeser H (1978) Vertical distribution of chlorofluoromethanes in the upper troposphere and lower stratosphere. PAGEOPH 116:554–566
Sellers WD (1969) A global climatic model based on the energy balance of the Earth-Atmosphere system. J Appl Meteorol 8:392–400
Selvig KA (1973) Electron microscopy of dental enamel: analysis of crystal lattice images. Z Zellforsch 137:271–280
Shackleton NJ, Hall MA (1984) Carbon isotope data from Leg 74 sediments. In "Initial Reports of the Deep-Sea Drilling Project" 74 (eds Moore TC Jr, Rabinowitz PE et al.), Washington, DC, US Govt Printing Office, 613–619
Shackleton NJ, Kennett JP (1975) Paleotemperature history of the Cenozoic and the initiation of Antarctic glaciation: oxygen and carbon isotope analyses in DSDP sites 277, 279 and 281. In "Initial Reports of the Deep Sea Drilling Project" 29 (eds Kennett JP, Houtz RE et al.), Washington, DC, US Govt Printing Office, 743–755
Shackleton NJ, Opdyke ND (1977) Oxygen isotope and palaeomagnetic evidence for early Northern Hemisphere glaciation. Nature 270:216–219
Shackleton NJ, Pisias NG (1985) Atmospheric carbon dioxide, orbital forcing, and climate. In "The Carbon Cycle and Atmospheric CO_2: Natural Variations Archean to Present" (eds Sundquist ET, Broecker WS), Geophys Monogr 32, AGU, 303–317
Shackleton NJ, Imbrie J, Hall MA (1983) Oxygen and carbon isotope record of East Pacific core V19–30: implications for the formation of deep water in the late Pleistocene North Atlantic. Earth Planet Sci Lett 65:233–244
Shell Briefing Service (1983) Weltenergie – Daten und Fakten. Hamburg, Deutsche Shell AG, pp 12
Shishkina OV (1959) Metamorphization of the chemical composition of muddy waters in the Black Sea. In "Toward Knowledge of Diagenesis of Sediments (Symposium)" (ed

Strakhov NM), Izd Akad Nauk SSSR, Moscow, 29–50 (in Russian)
Short NM, Lowman PD Jr, Freden SC, Finch WA Jr (1976) Mission to Earth: Landsat Views the World. Washington, DC, NASA, pp 459
Siegenthaler U (1983) Uptake of excess CO_2 by an outcrop-diffusion model of the ocean. J Geophys Res 88:3599–3608
Siegenthaler U (1986) Carbon dioxide: its natural cycle and anthropogenic perturbation. In "The Role of Air-Sea Exchange in Geochemical Cycling"(ed Buat-Ménard P), Dordrecht Boston Lancaster Tokyo, D Reidel Publ Co, 209–247
Siegenthaler U, Wenk T (1984) Rapid atmospheric CO_2 variations and ocean circulation. Nature 308:624–626
Sillén LG (1961) The physical chemistry of sea water. In "Oceanography" (ed Sears M), Washington, DC, Am Assoc Adv Sci, 549–581
Sillén LG (1967) Master variables and activity scales. In "Equilibrium Concepts in Natural Water Systems" (ed Gould RF), Adv Chem Ser 67:45–56, Am Chem Soc, Washington
Simkin T, Siebert L, McClelland I, Bridge D, Newhall C, Latter JH (1981) Volcanoes of the World. Smithsonian Institution (Suppl 1985), Stroudsbourg, Hutchinson, pp 233
Simkiss K (1974) Calciumtransport durch Zellen. Endeavour 33:119–123
Simonson BM (1985) Sedimentological constraints on the origins of Precambrian iron-formations. Geol Soc Am Bull 96:244–252
Singer-Sam J, Simmer RL, Keith DH et al. (1983) Isolation of a cDNA clone for human X-linked 3-phosphoglycerate kinase by use of a mixture of synthetic oligodeoxyribonucleotides. Proc Natl Acad Sci USA 80:802–806
Sleep NH (1976) Platform subsidence mechanisms and eustatic sea level changes. Tectonophysics 36:45–56
Sloan RE, Rigby JK Jr, Van Valen LM, Gabriel D (1986) Gradual dinosaur extinction and simultaneous ungulate radiation in the Hell Creek Formation. Science 232:629–633
Smith DC (1984) Coesite in clinopyroxene in the Caledonides and its implications for geodynamics. Nature 310:641–644
Smith MA (1987) Pleistocene occupation in arid Central Australia. Nature 328:710–711
Soderblom LA (1980) The Galilean moons of Jupiter. Sci Am 242/1:68–83
Söderlund R, Svensson BH (1976) The global nitrogen cycle. In "Nitrogen, Phosphorus and Sulphur – Global Cycles", SCOPE Report 7 (eds Svensson BH, Söderlund R), Ecol Bull 22:23–73
Solomon DH, Loft BC (1968) Reactions catalyzed by minerals. Pt 3. The mechanism of spontaneous interlamellar polymerizations in aluminosilicates. J Appl Polymer Sci 12:1253–1262
Spencer CP (1983) Marine biogeochemistry of silicon. In "Silicon Geochemistry and Biogeochemistry" (ed Aston SR), London, Academic Press Inc, 101–141
Spencer DW (1972) GEOSECS II, the 1970 North Atlantic station: hydrographic features, oxygen, and nutrients. Earth Planet Sci Lett 16:91–102
Spitzy A, Degens ET (1985) Modelling stable isotope fluctuations through geologic time. Mitt Geol-Paläont Inst Univ Hamburg 59:155–166
Stacey FD (1980) The cooling Earth: a reappraisal. Phys Earth Planet Inter 22:89–96
Stackelberg M v, Müller HR (1954) Feste Gashydrate II. Z Elektrochem (from 1963 on: Berichte d Bunsengesellschaft) 58:25–39
Stackelberg U v (1986) Entstehung der Manganknollen im äquatorialen Nordpazifik. Fortschr Miner 64/2:151–162
Stackelberg U v, Marching V (1987) Manganese nodules from the Equatorial North Pacific Ocean. Geol Jahrb D87:123–227
Stanley SM (1976) Ideas on the timing of metazoan evolution. Paleobiology 2:209–219
Stanley SM (1979) Macroevolution. San Francisco, WH Freeman & Co, pp 332
Stanley SM (1987) Extinction. New York, Scientific American Books Inc, pp 242
Starr C, Taggart R (1987) Biology: the Unity and Diversity of Life (4th ed). Wadsworth Publ Company, Belmont, California, pp 776
Starr MP, Stolp H, Trüper HG et al. (eds) (1981) The Prokaryotes. A Handbook on Habitats, Isolation, and Identification of Bacteria. Springer-Verlag, Berlin Heidelberg New York, vol 1 and 2, pp 2284
Stephens ER (1987) Smog studies of the 1950s. EOS, Trans Am Geophys Union, Feb 17, 1987, pp 4
Stephens ER, Darley EF, Taylor OC, Scott WE (1961) Photochemical reaction products in air pollution. Int J Air Water Pollut 4:79–86
Stevens CH (1977) Was development of brackish oceans a factor in Permian extinctions? Geol Soc Am Bull 88:133–138
Stevenson DJ, Salpeter EE (1977) The phase diagram and transport properties for hydrogen-helium fluid planets. Astrophys J Suppl 35:239–261
Stierstadt K (1987) Investigations on radioactive fallout (14). Atomkernenergie-Kerntechnik 51/1:52–54
Stillinger FH (1980) Water revisited. Science 209:451–456
Stolarski RS, Cicerone RJ (1974) Stratospheric chlorine: a possible sink for ozone. Can J Chem 52:1610–1615
Stolarski RS, Krueger AJ, Schoeberl MR, McPeters RD, Newman PA, Alpert JC (1986) Nimbus 7 measurements of the springtime Antarctic ozone decrease. Nature 322:808–811
Stommel H, Stommel E (1979) The year without a summer. Sci Am 240/6:134–140
Stommel H, Stommel E (1983) Volcano Weather. Newport RI, Seven Seas Press, pp 177
Stommel H, Swallow JC (1983) Do late grape harvests follow large volcanic eruptions? Bull Am Meteor Soc 64/7:794–795
Strain BR, Cure JD (eds) (1985) Direct Effects of Increasing Carbon Dioxide on Vegetation. DOE/ER-0238, Washington, DC, US Dept of Energy, pp 286
Streyer L (1988) Biochemistry (3rd ed). San Francisco, WH Freeman and Co, pp 1136
Stuermer DH, Harvey GR (1977) The isolation of humic substances and alcohol-soluble organic matter from seawater. Deep-Sea Res 24:303–309
Stuermer DH, Harvey GR (1978) Structural studies on marine humus: a new reduction sequence for carbon skeleton determination. Mar Chem 6:55–70
Stuiver M (1978) Atmospheric carbon dioxide and carbon reservoir changes. Science 199:253–258
Stuiver M, Quay PD, Ostlund HG (1983) Abyssal water carbon-14 distribution and the age of the world oceans. Science 219:849–851
Stumm W, Morgan JJ (1981) Aquatic Chemistry (2nd ed). New York Chichester Brisbane Toronto, John Wiley & Sons, pp 780
Suess H (1906) The Face of the Earth, vol 2. Clarendon Press, Oxford, pp 759
Suess HE (1975) Remarks on the chemical conditions on the surface of the primitive earth and the probability of the evolution of life. Origins of Life 6:9–13
Sugimura Y, Yoshimi S (1988) A high temperature catalytic oxidation method for the determination of non-volatile dis-

solved organic carbon in seawater by direct injection of a liquid sample. Mar Chem 24:105–131

Sunda WG, Huntsman SA, Harvey GR (1983) Photoreduction of manganese oxides in seawater and its geochemical and biological implications. Nature 301:234–236

Sundquist ET, Broecker WS (eds) (1985) The Carbon Cycle and Atmospheric CO_2: Natural Variations, Archean to Present. Geophys Monogr 32, Washington, DC, Amer Geophys Union, pp 627

Supp E (1985) Australiens Aborigines: Ende der Traumzeit? Bouvier Verlag Herbert Grundmann, Bonn, pp 333

Suzuki Y, Tanque E, Sugimura Y (1988) Distribution and chemical nature of DOC and DON in the western North Pacific. Sci Comm Ocean Res (SCORE), JOA Spec Symp "Global Ocean Storage and Fluxes", Acapulco, Mexico, August 1988

Svensson BH, Söderlund R (eds) (1976) Nitrogen, Phosphorus and Sulphur – Global Cycles. SCOPE Report 7, Ecol Bull (Stockholm) 22, pp 192

Sverdrup HU, Johnson MW, Fleming RH (1970) The Oceans, Their Physics, Chemistry, and General Biology (2nd ed). Englewood Cliffs, NJ, Prentice-Hall, Inc, pp 1085

Symons MCR, Blandamer MJ, Fox MF (1967) Is water kinky? New Sci 34:345–346

Szekielda K-H (ed) (1986) Satellite Remote Sensing for Resources Development. London, Graham and Trotman Ltd, pp 221

Szekielda K-H (1987) Investigations with satellites on eutrophication of coastal regions. Pt. VII: Response of the Somali upwelling onto monsoonal changes. In "Transport of Carbon and Minerals in Major World Rivers", Pt 5 (eds Degens ET, Kempe S, Naidu SA), SCOPE/UNEP Sonderband, Mitt Geol-Paläont Inst Univ Hamburg 66:1–30

Szent-Györgyi AG, Szentkiralyi EM, Kendrick-Jones J (1973) The light chains of scallop myosin as regulatory subunits. J Mol Biol 74:179–203

Tabony J, Geyer A de, Braganza LF (1987) Assembly of micelles into higher-order structures with increasing salt concentration. Nature 327:321–324

Takahashi T, Broecker W, Bainbridge A (1981) The alkalinity and total carbon dioxide concentration in the world oceans. In "Carbon Cycle Modeling" (ed Bolin B), SCOPE Report 16, Chichester New York Brisbane Toronto, John Wiley & Sons, 271–286

Takahashi T, Chipman D, Volk T (1983) Geographical, seasonal and secular variations of the partial pressure of CO_2 in surface waters of the North Atlantic Ocean: the results of the North Atlantic TTO Program. In "Proceedings of the Carbon Dioxide Research Conference", Washington, DC, Dept of Energy, Cof.- 820970, II.123–II.145

Tammann G (1900) Über die Grenzen des festen Zustandes. IV. Ann Phys 2:2–31

Taylor FJR (1976) Autogenous theories for the origin of eukaryotes. Taxon 25:377–390

Teal JM, Howarth RW (1984) Oil spill studies: a review of ecological effects. Environ Management 8/1:27–44

The Economist (1987) Volcanoes fight comets for the right to exterminate. The Economist, May 30, 1987:85–86

Theng BKG, Walker GF (1970) Interactions of clay minerals with organic monomers. Israel J Chem 8:417–424

Thilo E (1962) Condensed phosphates and arsenates. Adv Inorg Chem Radiochem 4:1–75

Thode HG, Shima M, Rees CE, Krishnamurty KV (1965) Carbon-13 isotope effects in systems containing carbon dioxide, bicarbonate, carbonate, and metal ions. Can J Chem 43:582

Thomsen L (1980) ^{129}Xe on the outgassing of atmosphere. J Geophys Res 85:4374–4378

Thorpe SA (1985) Small-scale processes in the upper ocean boundary layer. Nature 318:519–522

Tissot BP, Welte DH (1984) Petroleum Formation and Occurrence (2nd ed). Springer-Verlag, Berlin Heidelberg New York Tokyo, pp 699

Tobias PV (ed) (1985) Hominid Evolution: Past, Present and Future. Alan R Liss, pp 499

Toon OB (1984) Sudden changes in atmospheric composition and climate. In "Patterns of Change in Earth Evolution" (eds Holland HD, Trendall AF), Berlin Heidelberg New York Tokyo, Springer Verlag, 41–61

Toon OB, Pollack JB, Ackerman TP, Turco RP, McKay CP, Liu MS (1982) Evolution of an impact-generated dust cloud and its effects on the atmosphere. In "Geological Implications of Impacts of Large Asteroids and Comets on the Earth" (eds Silver LT, Schultz PH), Geol Soc America SP, 187–200

Toon OB, Kasting JF, Turco RP, Liu MS (1987) The sulfur cycle in the marine atmosphere. J Geophys Res 92/C1:943–963

Towe KM (1970) Oxygen-collagen priority and the early Metazoan fossil record. Proc Natl Acad Sci USA 65/4:781–788

Towe KM (1978) Early Precambrian oxygen: a case against photosynthesis. Nature 274:657–661

Towe KM (1981) Environmental conditions surrounding the origin and early Archean evolution of life: a hypothesis. Precambrian Res 16:1–10

Towe KM (1983) Precambrian atmospheric oxygen and banded iron formations: a delayed ocean model. Precambrian Res 20:161–170

Towe KM (1985) Habitability of the early earth: clues from the physiology of nitrogen fixation and photosynthesis. Origins of Life 15:235–250

Trabalka JR (ed) (1985) Atmospheric Carbon Dioxide and the Global Carbon Cycle. DOE/ER-0239, US Dept of Energy, Washington, DC, pp 315

Trabalka JR, Reichle DE (1986) The Changing Carbon Cycle. A Global Analysis. New York Berlin Heidelberg London Paris Tokyo, Springer-Verlag, pp 592

Trafton LM (1981) The atmosphere of the outer planets and satellites. Rev Geophys Space Phys 19:43–89

Treibs A (1936) Chlorophyll and hemin derivatives in organic mineral substances (a review). Angew Chem 49:682–686

Trinkaus E, Howells WW (1979) The Neanderthals. Sci Am 241/6:94–105

Tufty RM, Kretsinger RH (1975) Troponin and parvalbumin calcium binding regions predicted in myosin light chain and T4 lysozyme. Science 187:167–169

Turco RP (1985) The photochemistry of the stratosphere. In "The Photochemistry of Atmosphere" (ed Levine JS), Orlando, Fla, Academic, 77–128

Turekian KK, Clark SP (1975) The non-homogeneous accumulation model for terrestrial planet formation. J Atmos Sci 32:1257–1261

Turing AM (1967) Kann eine Maschine denken? In "Kursbuch 8" (ed Enzensberger HM), Suhrkamp Verlag, Frankfurt, 106–137

Tyndall J (1863) On radiation through the Earth's atmosphere. Philosoph Mag 25:200–206

Unwin N (1986) Is there a common design for cell membrane channels? Nature 323:12–13

Unwin N, Henderson R (1984) The structure of proteins in biological membranes. Sci Am 250/2:56–66

Urey HC (1952) On the early chemical history of the Earth and the origin of life. Proc Natl Acad Sci US 38:315–363

Urist MR (1962) Induced systemic hypersensitivity: Selye's theory. Science 137:120–121

Vail PR et al. (1977) Seismic stratigraphy and global changes of sea level. In "Seismic Stratigraphy – Applications to Hydrocarbon Exploration" (ed Payton CE), Am Assoc Petrol Geol Mem 26, 2 parts, 49–212

Vallee BL (1978) Zinc biochemistry and physiology and their derangements. In "New Trends in Bioinorganic Chemistry" (eds Williams RJP, Da Silva JJRF), New York, Academic Press, 11–58

Vandermeulen JH, Buckley DE, Levy EM et al. (1978) Immediate impact of AMOCO CADIZ environmental oiling: oil behavior and burial, and biological aspects. Publ du CNEXO, Série "Actes de Colloques" 6:159–174

Van Valen LM (1984) Catastrophes, expectations, and the evidence. Paleobiology 10/1:121–137

Veizer J (1983) Geologic evolution of the Archean- Early Proterozoic Earth. In "Origin and Evolution of Earth's Earliest Biosphere: An Interdisciplinary Study" (ed Schopf JW), Chapter 10, Princeton Univ Press, Princeton, NJ, 240–259

Veizer J (1985) Carbonates and ancient oceans: isotopic and chemical record on time scales of 10^7–10^9 years. In "The Carbon Cycle and Atmospheric CO_2: Natural Variations Archean to Present" (eds Sundquist ET, Broecker WS), Geophys Monogr 32, Am Geophys Union, Washington, DC, 595–601

Veizer J, Jansen SL (1979) Basement and sedimentary recycling and continental evolution. J Geol 87/4:341–370

Vidal G, Knoll AH (1982) Radiations and extinctions of plankton in the late Proterozoic and early Cambrian. Nature 297:57–60

Vincent E, Berger WH (1985) Carbon dioxide and Polar cooling in the Miocene: the Monterey hypothesis. In "The Carbon Cycle and Atmospheric CO_2: Natural Variations Archean to Present" (eds Sundquist ET, Broecker WS), Geophys Monogr 32, Am Geophys Union, Washington DC, 455–468

Volchok HL, Chieco N (eds) (1986) A Compendium of the Environmental Measurements Laboratory's Research Projects Related to the Chernobyl Nuclear Accident. Environmental Measurements Laboratory EML-460, Dept of Energy, New York, NY 10014, pp 331

Volkov II, Fomina LS (1974) Influence of organic material and processes of sulfide formation on distribution of some trace elements in deep-water sediments of Black Sea. Am Assoc Petrol Geol Mem 20, Tulsa, Oklahoma, 456–476

Vyas NK, Vyas MN, Quiocho FA (1987) A novel calcium binding site in the galactose-binding protein of bacterial transport and chemotaxis. Nature 327:635–638

Wagener K (1979) The carbonate system of the ocean. In "The Global Carbon Cycle" (eds Bolin B, Degens ET, Kempe S, Ketner P), SCOPE Report 13, Chichester New York Brisbane Toronto, John Wiley & Sons, 251–258

Wald G (1957) The origin of optical activity. Ann New York Acad Sci 69:352–368

Wald G (1962) Life in the second and third periods; or why phosphorus and sulfur for high-energy bonds? In "Horizons in Biochemistry" (eds Kasha M, Pullman B), Albert Szent-Györgyi Dedicatory Volume, Academic Press Inc, New York, 127–142

Walker A, Leakey REF (1978) The hominids of East Turkana. Sci Am 239/2:44–61

Walker A, Leakey RE, Harris JM, Brown FH (1986) 2.5-Myr *Australopithecus boisei* from west of Lake Turkana, Kenya. Nature 322:517–522

Walker GT (1923) Correlation in seasonal variations of weather, VIII: A preliminary study of world weather (World Weather I). Mem India Meteor Dept 24:75–131

Walker GT (1924) Correlation in seasonal variations of weather, IX: A further study of world weather (World Weather II). Mem India Meteor Dept 24:275–332

Walker JCG (1977) The Evolution of the Atmosphere. New York, Macmillan, pp 318

Walker JCG (1982) The earliest atmosphere of the earth. Precambrian Res 17:147–171

Walker JCG (1983) Possible limits on the composition of the Archaean ocean. Nature 302:518–520

Wallace JM, Hobbs PV (1977) Atmospheric Science: An Introductory Survey. New York London, Academic Press, pp 467

Wang F (1985) Middle-Upper Proterozoic and lowest Phanerozoic microfossil assemblages from SW China and contiguous areas. Precambrian Res 29:33–43

Warner DT (1964) Proteins may have hexagonal structure. Rep Art C & En 53–54

Warner Sir F (ed) (1988) The environmental effects of nuclear war: Consensus and uncertainties. Environment 30/6:2–45

Washburn SL (1978) The evolution of man. Sci Am 239/3:146–154

Washington WM, Meehl GA (1983) General circulation model experiments on the climatic effects due to a doubling and quadrupling of carbon dioxide concentration. J Geophys Res 88:6600–6610

Wasserburg GJ, Papanastassiou DA, Tera F, Huneke JC (1977) The accumulation and bulk composition of the moon: Outline of a lunar chronology. Philos Trans R Soc Lond A285:7–22

Wasserman SA, Cozzarelli NR (1986) Biochemical topology: applications to DNA recombination and replication. Science 232:951–960

Watson HD, Walker NPC, Shaw PJ et al. (1982) Sequence and structure of yeast phosphoglycerate kinase. EMBO J 1:1635–1640

Watson SW, Remsen CC (1969) Macromolecular subunits in the walls of marine nitrifying bacteria. Science 163:685–686

Watson SW, Waterbury JB (1969) The sterile hot brines of the Red Sea. In "Hot Brines and Recent Heavy Metal Deposits in the Red Sea" (eds Degens ET, Ross DA), New York, Springer-Verlag New York Inc, 272–281

Watterson DM, Vincenzi FF (eds) (1980) Calmodulin and Cell Function. Ann New York Acad Sci 356:1–443

Weber JN (1965) Evolution of the oceans and the origin of fine-grained dolomites. Nature 207:930–933

Webster F (1984) Studying El-Niño on a global scale. Oceanus 27:58–62

Wedepohl H (1963) Einige Überlegungen zur Geschichte des Meerwassers. Fortschr Geol Rheinland-Westfalen 10:129–150

Weertman J (1976) Milankovitch solar radiations and ice age sheet sizes. Nature 261:17–20

Weidenschilling SJ, Lewis JS (1973) Atmospheric and cloud structures of the Jovian planets. Icarus 20:465–476

Weinstock B (1969) Carbon monoxide: Residence time in the atmosphere. Science 166:224–225

Weiss A (1963) Organische Derivate der glimmerartigen Schichtsilicate. Angew Chem 75:113–122

Weiss AH, La Pierre RB, Shapira J (1970) Homogeneously catalyzed formaldehyde condensation to carbohydrates. J Catalysis 16:332–347

Weiss RF (1981) The temporal and spatial distribution of tropospheric nitrous oxide. J Geophys Res 86:7185–7195

Weiss RF, Lonsdale P, Lupton JE, Bainbridge AE, Craig H (1977) Hydrothermal plumes in the Galapagos Rift. Nature 267:600–603

Wells N (1986) The Atmosphere and Ocean: A Physical Introduction. London Philadelphia, Taylor & Francis, pp 347

Went FW (1960) Organic matter in the atmosphere, and its possible relation to petroleum formation. Proc Natl Acad Sci 46:212–221
Wergedal JE, Baylink DJ (1974) Electron microprobe measurements of bone mineralization in vivo. Am J Physiol 226:345–352
Westbroek P, De Jong EW (eds) (1983) Biomineralization and Biological Metal Accumulation. D Reidel Publ Company, Dordrecht Boston London, pp 533
Westheimer FH (1987) Why nature chose phosphates. Science 235:1173–1178
Weyl WA (1951) A new approach to surface chemistry and to heterogeneous catalysis. Penns State Coll Bull, Miner Ind Experim State Bull 57:3–118
Whitaker AG (1978) Carbon: a new view of its high-temperature behavior. Science 200:763–764
Whitfield M (1974) Accumulation of fossil CO_2 in the atmosphere and the sea. Nature 247:523–525
Wicke E (1966) Strukturbildung und molekulare Beweglichkeit im Wasser und in wässrigen Lösungen. Angew Chem 78:1–19
Wigley TML (1985) Carbon dioxide, trace gases and global warming. Clim Monitor 13:133–148
Wigley TML, Schlesinger ME (1985) Analytical solution for the effect of increasing CO_2 on global mean temperature. Nature 315:649–652
Wigley TML, Jones PD, Kelly PM (1980) Scenarios for a warm, high-CO_2 world. Nature 283:17–21
Wigley TML, Jones PD, Kelly PM (1986) Empirical climate studies. In "The Greenhouse Effect, Climatic Change, and Ecosystems" (eds Bolin B, Döös BR, Jäger J, Warrick RA), SCOPE Report 29, Chichester New York Brisbane Toronto Singapore, John Wiley & Sons, 271–322
Williams DA, Fogarty KE, Tsien RY, Fay FS (1985) Calcium gradients in single smooth muscle cells revealed by the digital imaging microscope using Fura-2. Nature 318:558–561
Williams GP (1979) Planetary circulations. II: The Jovian quasi-geostrophic regime. J Atmos Sci 36:932–968
Williams MAJ, Adamson DA (1980) Late Quaternary depositional history of the Blue and White Nile rivers in central Sudan. In "The Sahara and the Nile" (eds Williams MAJ, Faure H), Rotterdam, Balkema, 281–304
Williams PM, Oeschger H, Kinney P (1969) Natural radiocarbon activity of the dissolved organic carbon in the North-east Pacific Ocean. Nature 224:256–258
Williams RJP (1981) Natural selection of the chemical elements. Proc R Soc Lond B 213:361–397
Williams RS Jr, Carter WD (eds) (1976) ERTS-1, A New Window on Our Planet. Geol Surv Prof Paper 929, Washington DC, US Governmt Print Off, pp 362
Wilson AC (1985) The molecular basis of evolution. Sci Am 235/4:148–157
Wilson JL (1975) Carbonate Facies in Geologic History. Springer Verlag, Berlin Heidelberg New York
Wise SW, de Villliers J (1971) Scanning electron microscopy of molluscan shell ultrastructures: screw dislocations in pelecypod nacre. Trans Am Microsc Soc 90:376–380
Wittfogel KA (1977) Die Orientalische Despotie. Frankfurt, Ullstein, pp 625
WMO (1981) The stratosphere 1981; theory and measurements. WMO Global Ozone Research and Monitoring Project, Report No 11, Geneva, WMO
WMO (1982) Report of the meeting of experts on source of errors in detecting of ozone trends. WMO Global Ozone Research and Monitoring Project, Report No 12, Geneva, WMO
Woese CR, Fox GE (1977) Phylogenetic structure of the prokaryotic domain: the primary kingdoms. Proc Natl Acad Sci USA 74:5088–5090
Wolfe JA (1987) Late Cretaceous-Cenozoic history of deciduousness and the terminal Cretaceous event. Paleobiology 13/2:215–226
Wolfe JA, Upchurch GR (1986) Vegetation, climatic and floral changes at the Cretaceous-Tertiary boundary. Nature 324:148–152
Wollast R, Mackenzie FT (1983) The global cycle of silica. In "Silicon Geochemistry and Biogeochemistry" (ed Aston SR), London, Academic Press Inc, 39–76
Wood B, Martin L, Andrews P (eds) (1986) Major Topics in Primate and Human Evolution. Cambridge Univ Press, pp 364
Woodwell GM (ed) (1984) The Role of Terrestrial Vegetation in the Global Carbon Cycle: Measurement by Remote Sensing, SCOPE Report 23. Chichester New York Brisbane Toronto Singapore, John Wiley & Sons, pp 247
Woodwell GM (1986) Biotic influences on the carbon dioxide content of the atmosphere: the effects of a warming. In "Program of Symposium on CO_2 and Other Greenhouse Gases: Climatic and Associated Impacts", held at Brussels, Belgium, November 3–5, 1986, by the Commission of European Communities, 47
Woodwell GM, Houghton RA, Stone TA et al. (1987) Deforestation in the tropics: new measurements in the Amazon Basin using Landsat and NOAA advanced very high resolution radiometer imagery. J Geophys Res 92, D2:2157–2163
Worthington LV, Wright WR (1970) North Atlantic Ocean Atlas of Potential Temperature and Salinity in the Deep Water including Temperature, Salinity and Oxygen Profiles from the ERIKA DAN Cruise of 1962. Woods Hole Oceanographic Institution Atlas Series, vol 2, Woods Hole, Mass
Wyllie MRJ (1955) Role of clays in well log interpretation. Proc First Nat Conf Clays and Clay Minerals, Div Mines, Calif Bull 169:282–305
Wyllie MRJ (1958) Some electrochemical properties of shales. Science 108:684–685
Wyrtki K (1975) El-Niño – The dynamic response of the equatorial Pacific Ocean to atmospheric forcing. J Phys Oceanogr 5:572–584
Wyrtki K (1982) The Southern Oscillation, ocean-atmosphere interaction and El-Niño. Mar Techn Soc J 16/1:3–10
Yamada T, Sakaguchi K (1980) Nitrogen fixation associated with hotspring green algae. Arch Microbiol 124:161–167
Yariv S, Cross H (1979) Geochemistry of Colloid Systems. Berlin Heidelberg New York, Springer-Verlag, pp 450
Ycas M (1972) *De novo* origin of periodic proteins. J Mol Evol 2:17–27
Yeh, Hsueh-wen, Savin SM (1977) Mechanism of burial metamorphism of argillaceous sediments: 3. O-isotope evidence. Geol Soc Am Bull 88:1321–1330
Yoshino D, Hayatsu R, Anders E (1971) Origin of organic matter in early solar system, III: Amino acids: catalytic synthesis. Geochim Cosmochim Acta 35:927–938
Young JDE (1985) Role of ionic events in the triggering of phagocytosis. J Theor Biol 116:475–478
Zagwijn WH (1975) Chronostratigrafie en biostratigrafie: Indeling van het Kwartair op grond van veranderingen in vegetatie en klimaat. In "Toelichting bij Geologische Overzihtskaarten van Nederland" (eds Zagwijn WH, van Staalduinen CJ), Haarlem, 109–114
Zerefos CS, Ghazi A (eds) (1985) Atmospheric Ozone. Dordrecht Boston Lancaster, D Reidel Publ Co, pp 842

Zimmermann PR, Chatfield RB, Fishmann J, Crutzen PJ, Hanst PL (1978) Estimates of the production of CO and H_2 from the oxidation of hydrocarbon emission from vegetation. Geophys Res Lett 5:679–682

Zoller WH, Parrington IR, Kotra IMP (1983) Iridium enrichment in airborne particles from Kilauea volcano. Science 222:1118–1121

Subject Index

Italized numbers refer to figures

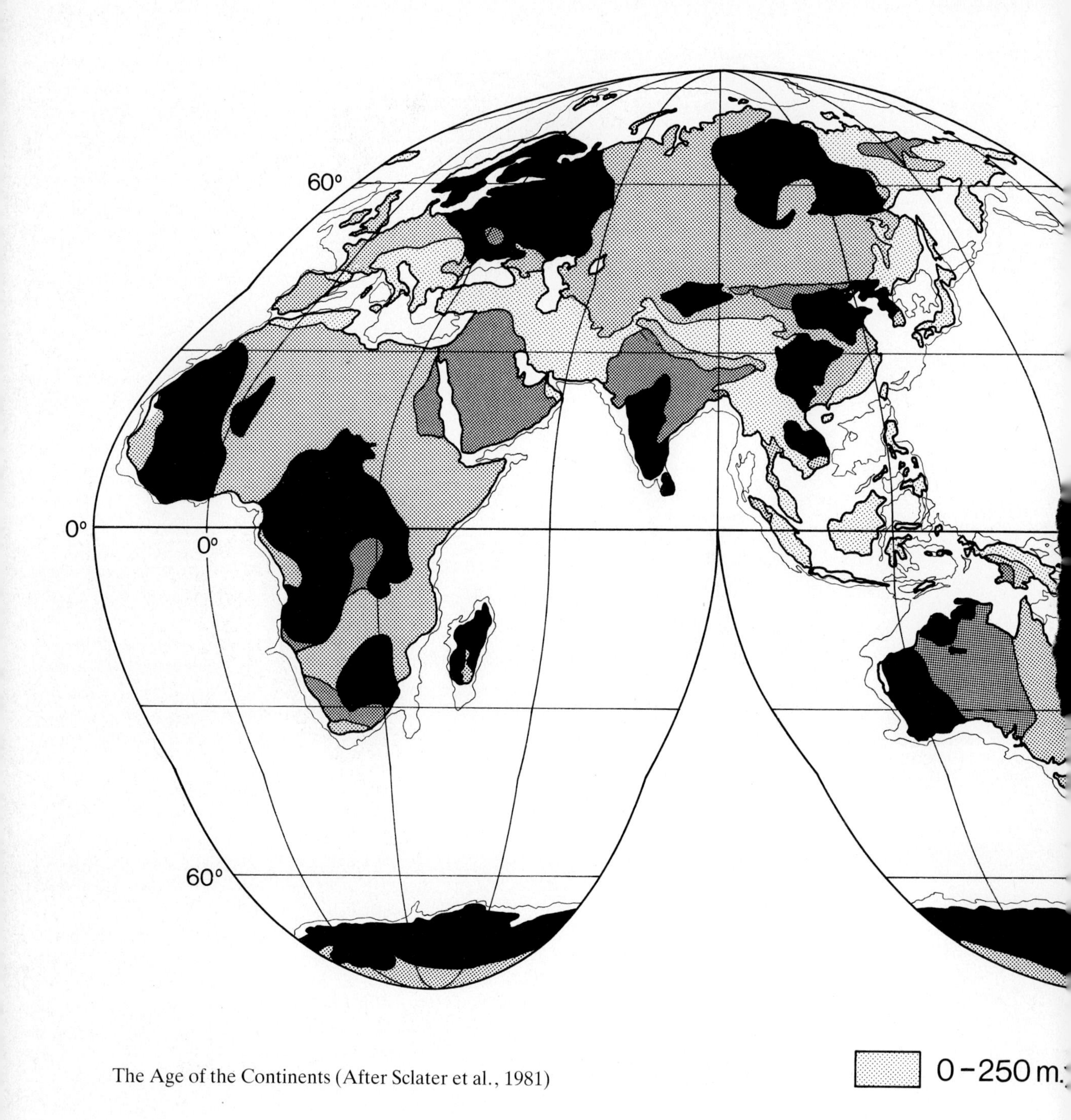

The Age of the Continents (After Sclater et al., 1981)

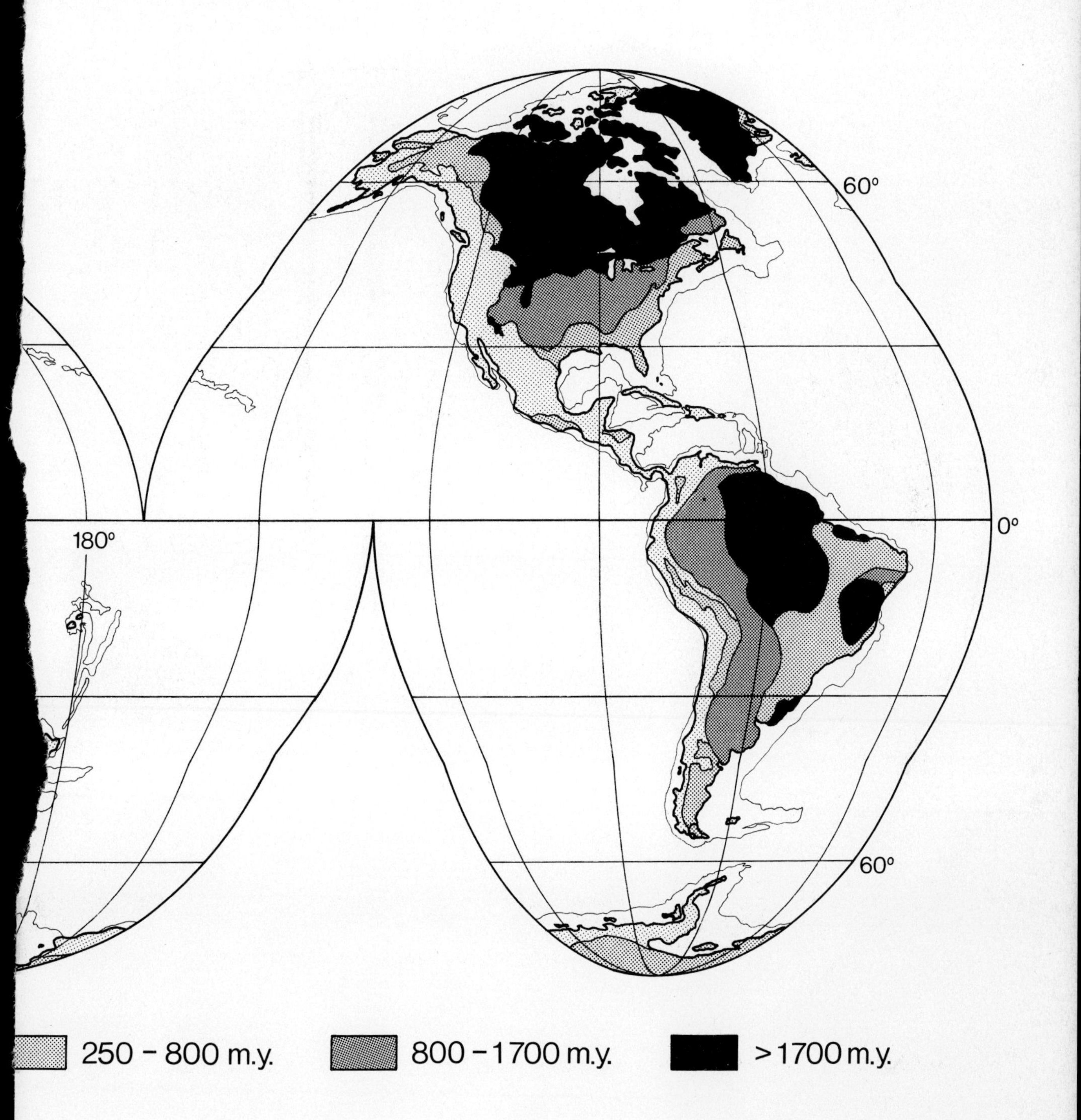
60°
0°
180°
60°
250 – 800 m.y.
800 – 1700 m.y.
>1700 m.y.